# MESOZOIC AND TERTIARY GEOLOGY OF SOUTHERN AFRICA

# MESOZOIC AND TERTIARY GEOLOGY OF SOUTHERN AFRICA

R.V.DINGLE
*University of Cape Town*

W.G.SIESSER
*Vanderbilt University, Nashville*

A.R.NEWTON
*University of Cape Town*

A.A.BALKEMA / ROTTERDAM / 1983

ISBN 90 6191 099 4

Distributed in USA & Canada by: MBS, 99 Main Street, Salem NH 03079, USA
Printed in the Netherlands

# Preface

The study of southern African Mesozoic and Tertiary strata has long had a low priority in both academic and commercial circles, and with the notable exception of diamonds, rocks of this age have not been thought to contain much worth digging up. The former state of affairs was closely related to the latter frame of mind: South African geologists are nothing if not economically pragmatic. A corollary of this legacy has been the generally dismissive approach to palaeontology and Phanerozoic stratigraphy in local training curricula: there is no question that until very recently the 'hard rock brigade' held total sway except in a few outposts such as the University of Durban, Bernard Price Institute for Palaeontological Research (University of Witwatersrand), and at some museums, where the spirit of William Smith was kept alive.

The fact that we are now emerging from this dark age of soft rock studies is due as much to developments in international earth science as it is to local enlightenment. The Deep Sea Drilling Project, preceded and proceeded by geophysical and geological exploration of the continental margins and adjacent offshore areas by overseas institutes, has spawned a small local marine geoscientific community and has brought the processes of continental margin basin formation into focus. The post-Palaeozoic nature of the most-recent phase of continental rifting and drifting gives this period a special significance. The earlier academic marine geophysical surveys undoubtedly had a catalytic effect on the decision to extend the search for oil and gas offshore. At the same time, rocks last studied in earnest in the pioneering days of the Cape and Natal geological surveys were receiving re-newed attention at the hands of a small band of sedimentologists versed in the techniques of the 'new sedimentology' as expounded for example by J.R.L.Allen and R.C.Selley. Particular attention had to be paid to the analysis of fluviatile sediments, and it is no coincidence that this work has gone hand in hand with an upsurge in vertebrate taxonomy and taphonomy. Timely revision of cephalopod taxonomy and biostratigraphy along the SE coast has provided a basis for identifying hiatuses, which placed the regional Cretaceous lithostratigraphic correlation on a firm footing, whilst the logjam in Tertiary studies has been eased somewhat by the application of micropalaeontological techniques to particularly thorny problems of correlation. The state of soft rock geological training and research is much healthier than a few years ago, but it is still a tender plant.

The igneous rocks of post-Palaeozoic age have always had their band of dedicated admirers, and with the facilities provided by the well-organized infra-structure of igneous and metamorphic earth scientists in the subcontinent, have been the subject of a steady stream of internationally significant publications over the years: every geology student in the world knows something about the 'Karoo basalts'. For this reason, and the fact that none of us is an igneous petrologist, we have tended to play down this field of study.

My colleagues and I have written this book because we believe that a review and synthesis of data on the Mesozoic and Tertiary rocks of southern Africa, seen in the context of Gondwana fragmentation, is opportune. Our reference list contains approximately 750 entries and whilst not claiming it to be complete or all the references specifically relating to southern Africa, a breakdown of dates shows that the two earlier comprehensive syntheses of du Toit (1954) and Haughton (1969) were undertaken when 20 and 31 %, respectively, of our citations had appeared. These figures illustrate the increasing tempo of research activity, and hopefully we have been able to cover the more-recent 69 % as successfully as they did the earlier studies. Our presentation is intended as a data and reference source for post-graduate students and research workers who may not have access to some of the more obscure publications cited. We would like to think that it will also be of use to the considerable number of overseas researchers who need data on southern Africa during the course of their Gondwana studies, as well as local undergraduates seeking information on particular pieces of the geological column or areas of study. In a rapidly expanding field, books of this type are soon out of date, but if we can provide a base for the next phase of our understanding of the Mesozoic and Tertiary geological development of the sub-continent we will have been successful.

We gratefully acknowledge fellow workers for supplying us with copies of their publications, and are particularly obliged to Professor I.W.Hälbich (University of Stellenbosch) who placed the results of his research team's activities at our disposal before they appeared in print, and who made constructive criticism of the chapter on the Cape Fold Belt. Dr Q.B.Hendey (South African Museum) kindly made many constructive comments for improvements to the sections on Neogene strata. All three of us are pleased to acknowledge the contribution made by the University of Cape Town in the form of facilities and staff research grants to our past and continuing research programmes (although one of us has since left UCT-WGS). Last but by no means least Mrs Grace Krummeck is thanked for her contribution in undertaking most of the large quantity of typing generated during our compilation.

10 February 1982

R.V.Dingle
Newlands
Cape Town

# Table of contents

# 1

# Southern Africa and its post-Palaeozoic regional setting

## 1.1 INTRODUCTION

Post-Palaeozoic time covers the period during which the modern morphology and outline of southern Africa evolved in a succession of important tectonic events:

i) the boundary between the Palaeozoic and Mesozoic eras roughly coincided with the last orogenic episode to affect the subcontinent;

ii) Triassic-Jurassic time is characterized by intracontinental downwarping and the development of regional tensional structures (rifting);

iii) during the Cretaceous and Tertiary, southern Africa completed separation from its neighbouring parts of Gondwanaland (drifting), and the present interior of the subcontinent was subjected to erosion in which the main direction of sediment transport was to the continental margins and adjacent ocean basins.

Because of these important events, it is a period worthy of special documentation, and as the only substantive section of the local geological column with abundant invertebrate faunas, it is the one part amenable to fine-scale biostratigraphic subdivision and confident correlation with the international stratigraphic time scale.

Although proven commercially-exploitable mineral resources in Mesozoic and Tertiary strata are relatively small in comparison with the overall mineral wealth of the subcontinent, they are not insignificant, and furthermore probably hold greater potential than has been generally conceded (Tab.1). The latter in itself is a good reason to fully document and if possible understand the sedimentary history of the strata in which such minerals are likely to occur.

In terms of the stratigraphic classification previously in use in southern Africa, the rocks under discussion constituted the upper part of the Karoo System, and the Cretaceous and Tertiary systems (e.g. Haughton 1969). Revision of the local stratigraphic code, resulting in the publication of a new scheme of stratigraphic correlation (SACS 1980), has provided the first systematic review of the whole geological column with the object of standardizing stratigraphic practice and nomenclature. As far as the Mesozoic and Tertiary rocks are concerned, this has resulted in few significant deviations from earlier practice, although locally (e.g. in the Upper Triassic) there have been several new formational names introduced. More significantly, what it has helped to achieve is to make the principle of at least attempting to correlate with the international systems and stages an accepted procedure in local stratigraphy. Most of the proposed terminology has been used in this book, but in one or two instances, SACS's justification for a certain terminology or classification is not acceptable and we have deviated from it. Such instances, with reasons, are noted in the text.

Table 2 shows a summary of the stratigraphic classification adopted in this book, and the timing of the main igneous and tectonic events. The right-hand column outlines the temporal extent of the most important lithostratigraphic units which can be correlated with the tectonic events in a manner shown under the heading of phases of crustal movement (annotated A-F). Each phase resulted in a characteristic sediment/igneous sequence of rocks, and these are separated by important non-sequences of one sort or another. The numerous breaks in succession in the epeirogenic phase (F) probably related to eustatic sea level changes, and as such cannot be correlated with local phases of crustal movement.

Of the two important local plate tectonic events, namely the initiation of drift between Africa and South America, and the final separation of the two continental blocks (i.e. the clearing of the Falkland Plateau from the southern tip of the Agulhas Bank), only the former can be approximately correlated with other sedimentary events and crustal movements.

The Kaokoveld lavas north of Walvis Bay, which are clearly related to the Walvis Ridge effusive centre and the opening of the South Atlantic, have been dated between 132 and 108 my, but their climax occurred around 121 my (Siedner & Mitchell 1976). On the Agulhas Bank, the establishment of marine conditions (Lower Sundays River Formation) can be dated as Upper Valanginian (ca.127 my), and both these dates are approximately coeval with that accepted for the initiation of continental drift (i.e. the age of earliest oceanic crust) in the SE Atlantic Ocean and the Natal Valley. This event has been variously dated within the Upper Jurassic-Lower Cretaceous, but the concensus of most workers has been that of 127 my proposed by Larson & Ladd (1973), although a recent review of the oceanic magnetic anomaly data (Martin et al 1982a) indicates a degree of uncertainty as to whether the earliest recognizable anomalies are M10 or M12 (126 or 129 my, respectively, on van Hinte's 1976a time-scale).

In contrast, the final separation of the Falkland Plateau from the southern tip of Africa does not appear to have coincided with any particular crustal movement-related sedimentary episode (in contrast to the deep-sea environmentally-related changes). Dingle & Scrutton (1974) suggested a date of 100 my for this event, whereas Martin et al (1982a) list a variety ranging from 104.3 to 98.3 my, depending on the spreading rate assumed, and the magnetic reversal time scale adopted. Using Van Hinte's (1976a) scale, Martin et al's data (1982a) show a date that lies between 98.3 and

Table 1. Proven and potential mineral resources in the Mesozoic and Tertiary rocks of southern Africa.

| | |
|---|---|
| Tertiary: | diamonds in alluvial and coastal deposits |
| | limestone, construction materials, and road aggregates from duricrusts |
| | phosphate in authigenic, marine Neogene sediments |
| | * potash and uranium in authigenic, marine Neogene sediments |
| | * uranium in duricrusts |
| | * metal-enrichment in late Neogene sediments off the west coast |
| | * methane in late Neogene sediments off the west coast |
| Cretaceous: | * oil and gas in various offshore formations |
| | diamonds in Upper Cretaceous pipes |
| Jurassic: | * mineralization associated with volcanic activity |
| Triassic: | * synsedimentary uranium in the upper Beaufort and Stormberg groups |
| | coal in the Molteno Formation |

* Speculative, or presently commercially not viable

Table 2. Regional stratigraphic and tectono-sedimentary framework of the Mesozoic and Tertiary rocks of southern Africa.

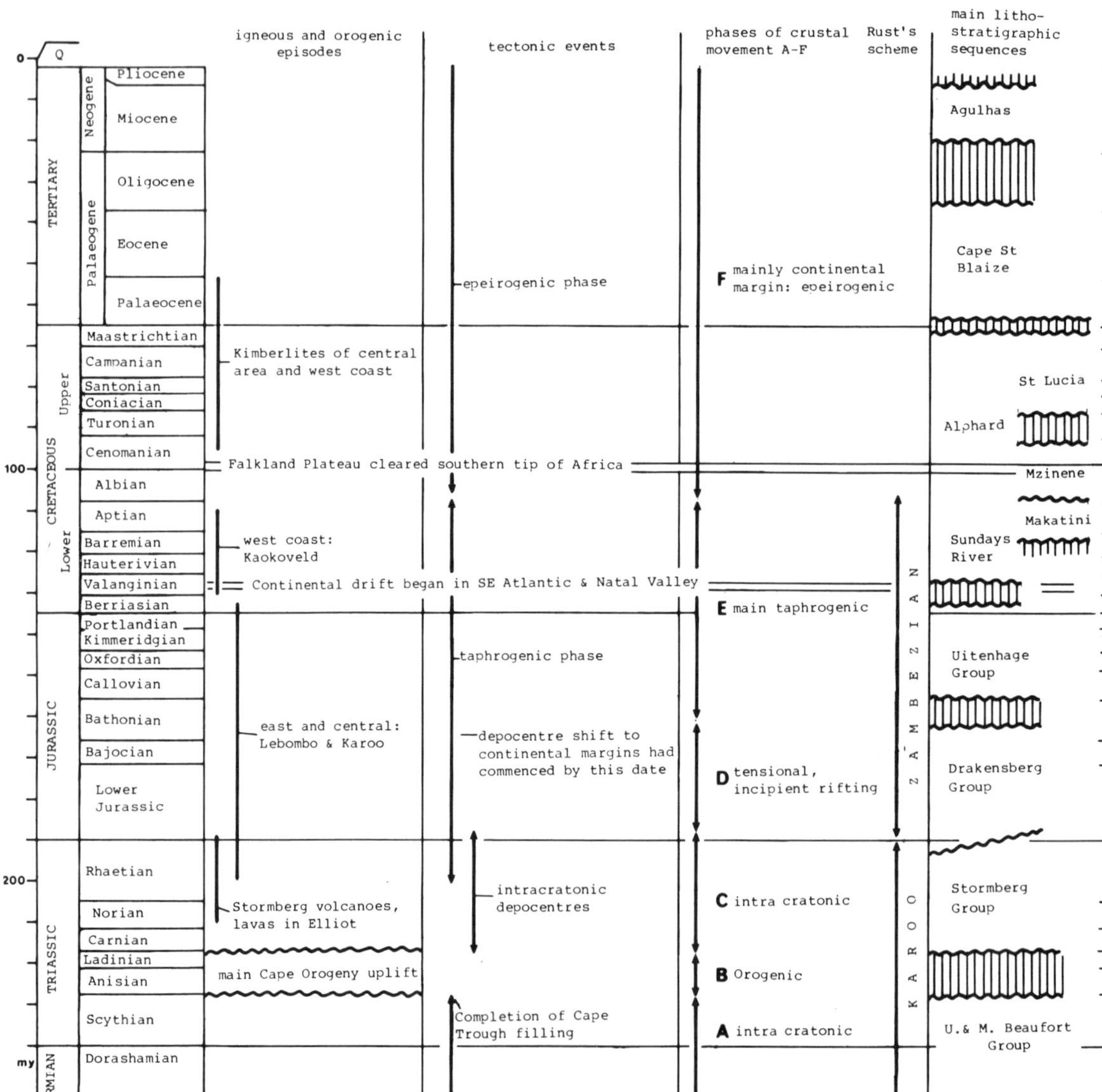

100.4 my (mean of 99.35 my), which is close to the date of 100 my that we have adopted. Whilst this tectonic event could be equated with some of the hiatuses on the western Agulhas Bank, the correlation is tenuous, and certainly was not coeval with the main mid-Cretaceous stratigraphic event in southern Africa, namely the Upper Cenomanian to late Turonian/early Coniacian hiatus that can be recognized on the Agulhas Bank (Horizon 17), in Zululand, in the Orange Basin, and also on the Falkland Plateau (where it spans Upper Cenomanian-Santonian time).

Because the thickest and most extensive Cretaceous and Tertiary sequences in southern Africa lie offshore (under the continental shelf, slope, and rise, and in the adjacent ocean basins), much of the stratigraphic detail presented in this book has been obtained from seismic data (both refraction and continuous reflection), occasionally tied in with borehole information. This situation always presents certain difficulties, such as: correlating important seismic reflectors offshore with structures and stratigraphic events recognized at outcrop onshore; with assessing whether or not seismic horizons are diachronous; and in converting thicknesses on seismic records (which are expressed as soundwave travel time), into thicknesses in metres. In some cases, it is not possible to confidently correlate the seismic records with the onshore succession, and in these we have retained the established acoustic stratigraphy. This is a category not recognized by SACS (1980), although in other parts of the world it is being accorded an equal status in nomenclature with bio- and lithostratigraphic schemes. Consequently, the deep-water stratigraphic schemes for the Cretaceous and Tertiary off the west coast, and in the Natal Valley, are based on the recognition of certain laterally extensive seismic reflecting horizons, which as yet are only poorly dated, and whose significance in purely lithofacies terms can only be speculated upon away from the few areas where boreholes have penetrated them.

Where seismic records can be tied in with numerous boreholes (such as on the Agulhas Bank, and parts of the SW continental margin) it is possible to correlate seismic events with reasonably well-established bio- and lithostratigraphic schemes, and because sufficient interval velocities can be calculated, isopach maps have been compiled. Theoretically, this enables rock units to be characterized by their P-wave velocities, so that the layers identified in, say, refraction or wide angle reflection studies can be tentatively 'dated' and correlated with sequences elsewhere. Whilst this technique has some merit, in that it allows a rough and ready ageing to be achieved, it has to be used with caution because the P-wave velocity of a rock (V) can be expressed as:

$$V = \sqrt{\frac{(K + \frac{4}{3}n)}{\rho}}$$

where K is the bulk modulus, n is the shear modulus, and $\rho$ is the density (McQuillan & Ardus 1977).

Clearly, the velocity depends on the lithology and compaction history of the rock, and not its age, though of course as a general rule, the older rocks are most deeply buried, most compacted, and better lithified, so that they tend to have the higher velocities. Because numerous wide angle reflection and refraction sites have been reported from the Agulhas Bank, and west and east coasts (Ludwig et al 1968a, Leyden et al 1971, Emery et al 1975), we have made use of these data and have followed the scheme produced by Dingle (1979), who compared all the layer models produced from these surveys:

| Velocity range | km/sec | Mean | Age |
|---|---|---|---|
| low | 1.6-2.44 | 2.02 | Cenozoic, and some Upper Cretaceous sediments |
| medium | 2.5-4.24 | 3.37 | Upper Cretaceous, Lower Cretaceous, and locally Jurassic-Lower Cretaceous sediments |
| high | 4.3-4.84 | 4.57 | Jurassic-Lower Cretaceous sediments and lavas, locally Upper Palaeozoic rocks |

Using the mean velocities listed above, it is also possible to produce preliminary isopach maps in areas where no borehole control is available. Anomalies do occur, and emphasize that such schemes are only a first approximation. For instance, 'Karoo-Cape' sediments on the Agulhas Bank have velocities as low as 4.55 km/sec (Spence 1970), whereas in other parts of the same province (SW of Cape St Francis), velocities as high as 4.6 km/sec clearly equate with Jurassic-Lower Cretaceous sediments (Dingle 1979). Recognition of the basement-sediment contact in such areas solely on interval velocity data would obviously be ambiguous.

## 1.2 PLATE TECTONIC SETTING OF SOUTHERN AFRICA

Following Bullard et al's (1965) well-known computer-fitted reconstruction of the South Atlantic along conjugate 500 fm isobaths, numerous refinements have been attempted to locate the precise position of the continent-ocean boundary (COB) off SW Africa, and to date the earliest sea floor spreading magnetic anomalies adjacent to it, hence pinpointing the age of drift onset (Fig.1). Larson & Ladd's (1973) recognition of the Mesozoic (M) series of sea floor spreading magnetic anomalies in the Cape Basin was a significant step in this, and their date of ca.127 my (based on the identification of anomalies M11-M12), has been widely accepted. The situation in the Cape Basin is certainly not straightforward, with serious, and as yet unresolved, complications north of about 32°S caused by anomaly distortions adjacent to the position where the COB

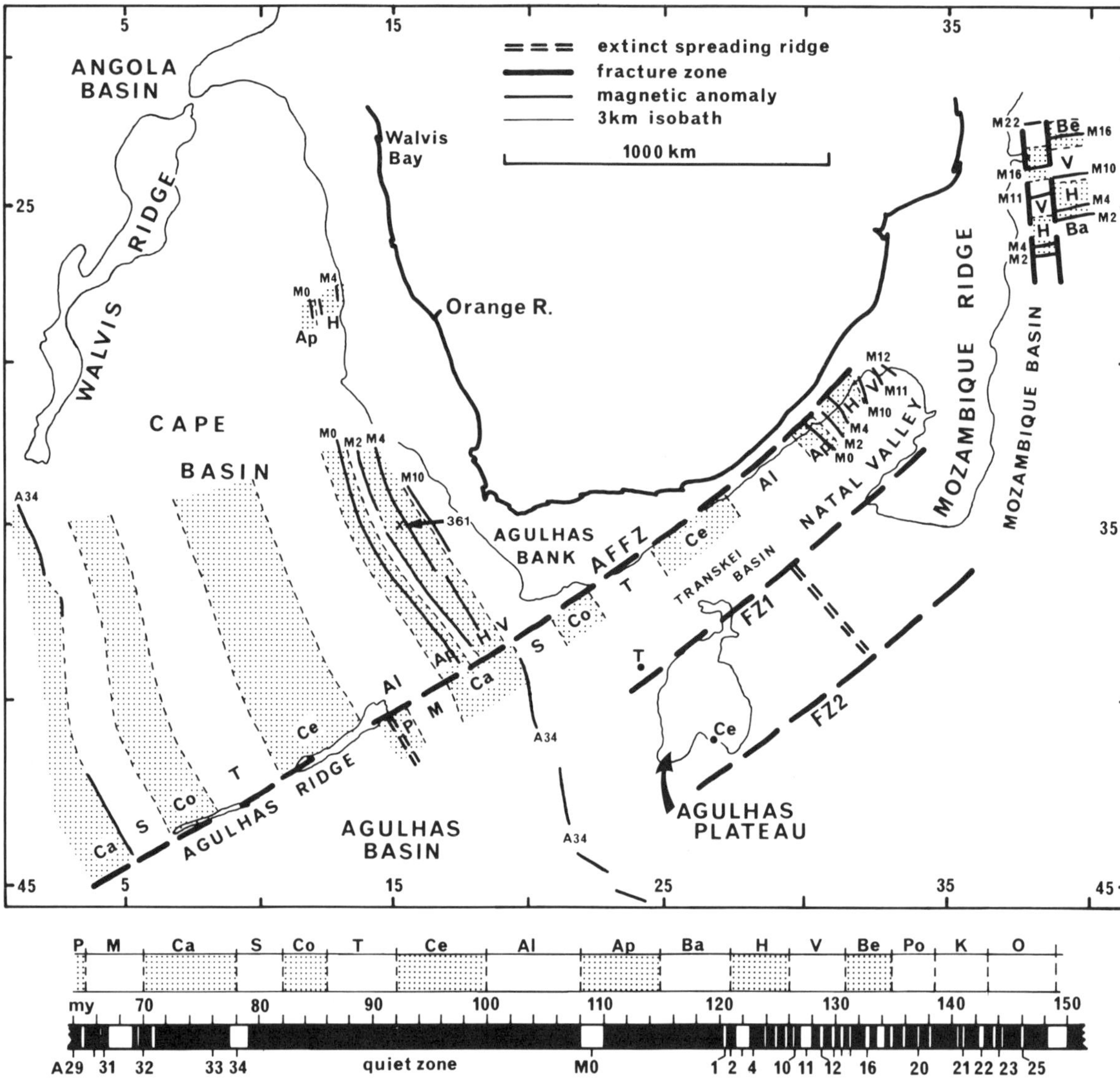


Figure 1. Plate tectonic setting of southern Africa. Based on numerous authors cited in the text, but principally Scrutton (1976a), Simpson *et al.* (1979), and Martin *et al.* (1982a). The time scale is after van Hinte (1976a,b), and the bathymetry (3 km isobath) is taken from Simpson (1974). The extinct spreading ridge south of the AFFZ is based on Barker (1979) and has been dated at 64-61 my, whilst that between the Agulhas Plateau and the Mozambique Ridge is speculative. Magnetic anomalies in the N.Cape Basin have been omitted because of uncertainty in correlation. The two dated samples from the Agulhas Plateau were recorded by Saito *et al.* (1974). Alternated stippling of sea floor is for clarity and has no geological significance. In the Cretaceous quiet zone, see floor has been dated assuming a constant spreading rate. Location 361 refers to DSDP site 361. Abbreviations: P = Palaeocene; M = Maastrichtian; Ca = Campanian; S = Santonian; Co = Coniacian; T = Turonian; Ce = Cenomanian; Al = Albian; Ap = Aptian; Ba = Barremian; H = Hauterivian; V = Valanginian; Be = Berriasian; Po = Portlandian; K = Kimmeridgian; O = Oxfordian; AFFZ = Agulhas/Falkland Fracture Zone; FZ = fracture zone ( 1& 2, see Fig.96).

might be expected. In addition, the anomaly identifications have not been unequivocal, so that, for example, Emery et al (1975) favoured an age within the Jurassic quiet zone (ca.165 my) for the oldest anomalies adjacent to the COB, whilst du Plessis (1979) considered it to be M22 (ca.142 my). All these variations hinged on the recognition of M0 and M1 as the starting point of the M-sequence prior to the late to mid-Cretaceous 'quiet zone', but were further complicated because records north of 32°S were used, as well as data from the apparently more straightforward succession between 32°S and the Agulhas/Falkland Fracture Zone (AFFZ) (see Fig.1).

The unequivocal establishment of continental crust under the eastern Falkland Plateau (Barker et al 1977a) confirmed earlier South America-Africa reconstructions in which the Falkland Plateau was placed adjacent to SE Africa (e.g. Francheteau & Le Pichon 1972, Dingle & Scrutton 1974). The implications of this are that as the South Atlantic drift proceeded, oceanic crust was progressively created from north to south in the Natal Valley, with synchronous ocean crust emplacement in the SE Cape Basin. Furthermore, this geometry requires an initial offset of some 1300 km in the spreading ridge across the AFFZ, whereas only approximately 200 km exist across the present-day spreading ridge, indicating westward ridge jumps of 1100 km since the initiation of drift. Barker (1979) has dated these at 61-64 my (ca.700 km), and 58-60 my (100 km), and has suggested a further jump at about 98 my to account for the remainder. However, Martin et al (1982a) point out that anomaly 34 (ca.78 my on Van Hinte's 1976a scale), as mapped by Bergh & Barrett (1980) is approximately the same distance from the start of the M-sequence on both sides of the AFFZ, implying little or no adjustment before this date. Geophysical evidence for the location of the AFFZ has been compiled mainly on the position of the Cape Slope Magnetic Anomaly adjacent to Africa (e.g. du Plessis & Simpson 1974), and on the offset of anomaly patterns and bathymetric features SW of the continent (e.g. du Plessis 1977, 1979, Martin et al 1982a). A conjugate feature has been recognized adjacent to the Falkland Plateau (e.g. Rabinowitz & LaBrecque 1979).

One factor in the general plate tectonic setting of southern Africa that has caused considerable uncertainty has been the nature and origin of the two relatively shallow sea floor features adjacent to the SE margin, viz. the Agulhas Plateau, and Mozambique Ridge. Although there has been a good deal of speculation on their makeup, mostly based on indirect, geophysical grounds, few substantive data are available. The Agulhas Plateau is probably the better known of the two, and until recently was generally considered to be solely of oceanic origin. Scrutton (1973b) suggested that it was an abandoned oceanic spreading centre, and on the basis of gravity modelling (Scrutton 1976a) showed it to be up to 18 km thick in the north, and ca.15 km thick in the south, with a distinct discontinuity between the two parts, which were separated by what he postulated to be a fracture zone (Figs 1 & 96). Furthermore, a marked difference in the morphology of the upper basement surface between the two parts was commented upon by Emery et al (1975) and Barrett (1977), although the latter, on the basis of seismic refraction studies concluded that all the plateau was oceanic in origin (i.e. basic volcanics). In contrast, a composite origin has recently been proposed by Tucholke et al (1981) on the basis of additional seismic refraction shooting, and dredging. These workers advocate an oceanic origin for the rugged northern part, and a continental origin for the smoother-topped, generally shallower southern part. From the latter area they recovered metamorphic rocks (quartzo-feldspathic gneisses of lower amphibolite to granulite facies, and quartzose metasediments), sediments (arkosic sandstones, foliated argillites, and dense brown cherts), and igneous rocks (?brecciated rhyolites, and intrusive and extrusive basalts).

No such direct data have been obtained from the Mozambique Ridge, but seismic refraction (Hales & Nation 1973), and gravity work (Scrutton 1976a) show that in the south it is about 20 km thick, and that the upper part (ca.8 km) at least, consists of relatively low velocity (<6.0 km/sec) rocks. Like the Agulhas Plateau, the Mozambique Ridge consists of several distinct morphological sections and may also have had a composite origin, although the fact that it continues northward under coastal SE Mozambique, and can be accommodated in palaeogeographical reconstructions without serious overlap suggests that the bulk of the feature is probably underlain by continental crust. The only rock sample obtained from it was weathered basalt in the bottom of DSDP Site 249, situated on the ridge crest (Simpson et al 1974) (see Fig.106).

Finally, the nature of the easternmost part of the area in Figure 1 (Mozambique Basin) has been assumed to be oceanic in all previously-published reconstructions, but confirmation of this, together with the age and magnetic grain of the sea floor have only recently been established (Segoufin 1978, Simpson et al 1979), and shown to be conjugate to sea floor off Antarctica (Bergh 1977).

Figure 1 shows a summary of the sea floor age, structure, and composition around southern Africa, with magnetic anomaly assignments converted to ages using Van Hinte's (1976a) time scale. In the northern Cape Basin, no confident identifications can be plotted because recent work has shown that north of ca.32°S, anomalies adjacent to the margins are distorted (Uchupi & Austin 1981), and previous identifications (e.g. Emery et al 1975, Rabinowitz 1976) are consequently suspect. The M0-M10 sequence can be confidently identified between 32°S and the AFFZ, and using the position of A34, it can be seen that the bulk of the central southern Cape Basin was formed during the Albian-Santonian magnetic quiet period. Note that DSDP Site 361, SW of Cape Town, overlies oceanic crust of probable late Hauterivian age (the borehole did not reach basement, and the oldest sediments sampled were considered Aptian by Bolli et al 1978). The position of the COB is still a point of contention, with, for example, Rabinowitz & LaBrecque (1979) placing it within 70 km of the coast immediately south of the Orange River (corresponding with their magnetic anomaly G), whilst Scrutton (1976b), Dingle (1979), Gerrard & Smith (1980), and Uchupi & Austin (1981) place it much farther westwards (300-350 km from the coast). Farther south, the COB lies closer to shore, and there is less discrepancy in interpretation. The

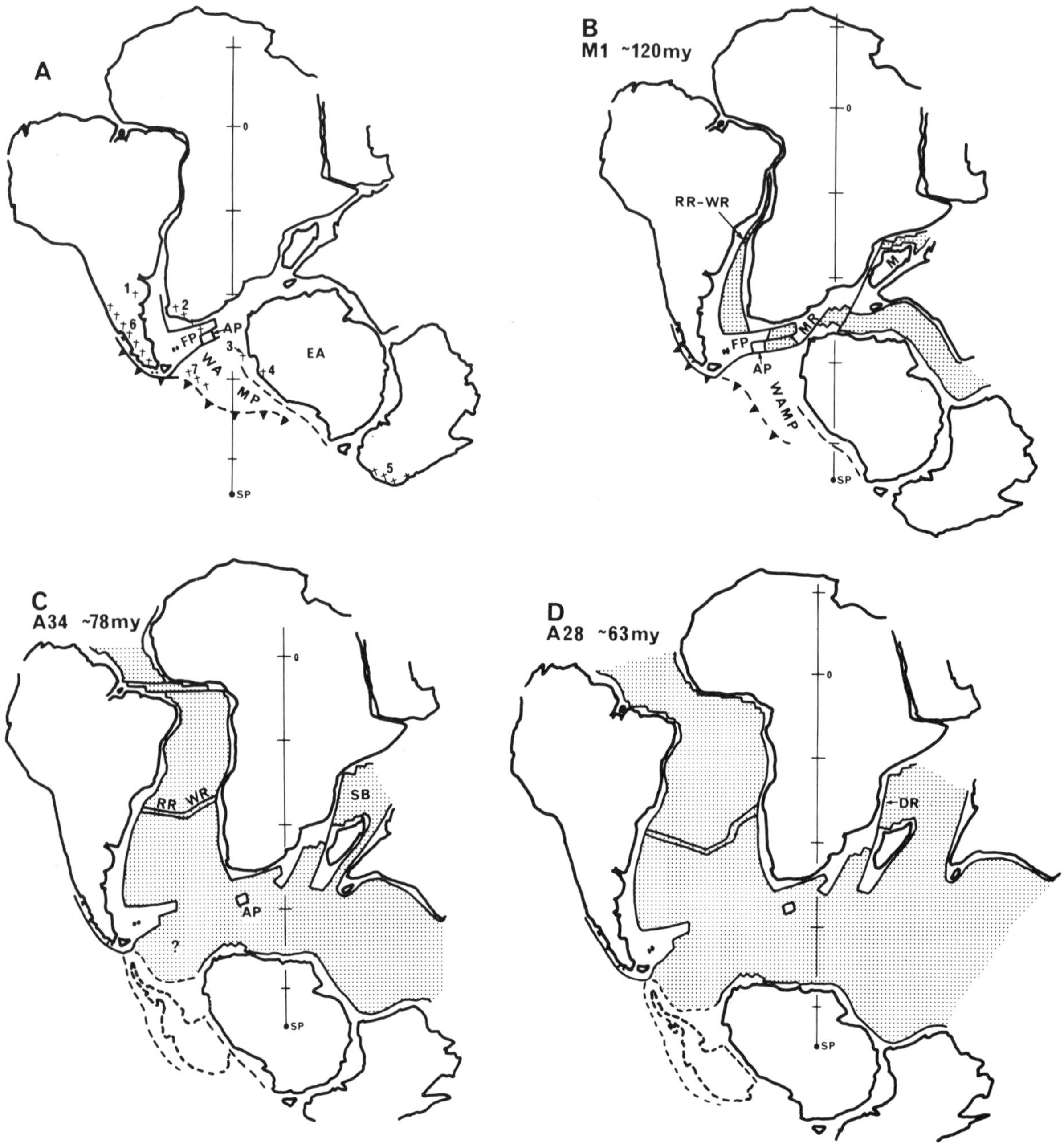


Figure 2. Gondwana palaeogeographies. Based on Norton & Sclater (1979) from the original refits of Elliot (1975a) and de Wit (1977) with modifications from Scrutton (1976a), Simpson *et al.* (1979), Segoufin & Patriat (1980), Miller (1981), Tucholke *et al.* (1981), Watts & Bramall (1981) and Martin *et al.* (1982a). The time scale is that of van Hinte (1976a). Details NE of Madagascar are sketched (i.e. no attempt to refit Seychelles Plateau). Oceanic crust is shaded, Mesozoic orogenic belts shown by crosses, subduction zones shown by solid triangles. Abbreviations and key: FP = Falkland Plateau; AP = Agulhas Plateau; MR = Mozambique Ridge; M = Madagascar; WAMP = West Antarctica microplates (e.g. Ellsworth, Thurston, Marie Byrd Land); EA = East Antarctica; SB = Somali Basin; RR = Rio Grande Rise; WR = Walvis Ridge; SP = South Pole; DR = Davie Ridge. Orogenic features: 1 = Sierra de la Ventana; 2 = Cape Fold Belt; 3 = Ellsworth Mountains; 4 = Pensacola Mountains; 5 = East Australia; 6 = Andes; 7 = Antarctic Peninsula.

A. Pre-drift (early Triassic); B. Anomaly M1, ca.120 my (early Barremian); C. Anomaly A34, ca.78 my (late Santonian/early Campanian); D. Anomaly A28, ca.28 my (early Palaeocene).

Projection is Lambert equal area.

implications of this are that the onset of drift may have been diachronous, and/or that unresolved structural complications mask the true picture to the north of the Orange River (i.e. adjacent to, and north of the Lüderitz Arch, Fig.114).

West of ca.15°E, the AFFZ coincides with the Agulhas Ridge, whose eastern end lines up with the position of the 61-64 my abandoned spreading centre which is postulated to lie south of the fracture zone by du Plessis (1979) and Barker (1979). NE of this point, the whole of the Cretaceous quiet zone and M-Sequence to ?M12 have been mapped by Martin et al (1982a) adjacent to the AFFZ in the northern Agulhas, and Transkei basins, and southern Natal Valley, presumably including the northern section of the Agulhas Plateau. This latter corollary is strengthened by the presence of Turonian sediments close to presumed oceanic basement (Saito et al 1974), in a location adjacent to Coniacian-Turonian oceanic crust immediately south of the AFFZ (Fig.1). The continental (i.e. southern) part of the plateau lies within a separate compartment defined by two further fracture zones (Scrutton 1976a) (FZ1 and FZ2 in figs 1 & 96). Assuming that the southern end of the Mozambique Ridge is continental crust, then a sector of oceanic crust with an abandoned spreading centre must lie between it and the continental part of the Agulhas Plateau. Sediments close to basement in this region of the plateau have been dated as Cenomanian (Saito et al 1974), an age that would be somewhat anomalous if an oceanic origin was ascribed to the feature.

Magnetic anomaly dating in the Mozambique Basin, to the east of the Mozambique Ridge, as well as the strike of both anomalies and fracture zones indicate that creation of oceanic crust in this area was neither synchronous nor structurally related to the spreading that produced the SE Atlantic and Natal Valley. Anomalies possibly as old as 142-143 my (M22), and certainly as old as M16 (ca.133 my, Berriasian) occur in this area, indicating that the onset of drift between SE Africa and Antarctica commenced significantly earlier than the separation of South America and Africa. Recent calculations indicate that drift between the first-mentioned continental couple commenced at the latest by ca.132 my about a rotation pole at 11°N, 41°W (Bergh 1977), and between the latter continents at between 129 and 127 my about a rotation pole at 46.75°N, 32.65°W (Martin et al 1981).

Considering the above plate tectonic structures in a broader, trans-Gondwanaland setting involves the consideration of features which in most instances are beyond the scope of this book. Nevertheless, to understand the significance of much of southern African Mesozoic history, it is important to at least have a broad appreciation of problems of reassembling Gondwanaland, and the progressive disintegration of the super continent, and in particular, southern Africa's relative geographical position at any point in time. Numerous reconstructions have been proposed over the years, of which the earliest and arguably the most celebrated, was that by du Toit (1937). He based the palaeogeographies solely on continental geometry and continuity of regional geological structures and trends, placing particular emphasis on the extent of the so-called Samfrau geosyncline which was the site of late Palaeozoic-early Mesozoic orogenesis in South America (Sierra de la Ventana), South Africa (Cape Fold Belt), Antarctica (Weddell Orogeny sensu Ford & Kistler 1980), and eastern Australia (Brown et al 1968): the Gondwanide Orogeny of du Toit. Other features that placed important constraints were the distribution of Permian glacial deposits, and Permo-Triassic tetrapods and the *Glossopteris* floras.

Du Toit's (1937) reconstruction was a refinement of Wegener's (1924) original proposal, and despite the additional data and constraints provided by plate tectonic modelling and oceanic magnetic anomaly studies, has proved to be essentially correct. Attempts to improve and check du Toit's reconstructions have centred on particular aspects, such as the original position of Madagascar, the relationship of India and Australia, the precise orientation and juxtaposition of Antarctica to Africa, and the whole question of the development and original orientation of the various components that make up West Antarctica. No consensus has yet emerged on many of these points, but certain aspects do appear to have become clearer. For instance, the earliest, and most widely quoted of the 'modern' attempts was made by Smith & Hallam (1970), using computerized continental edge refitting. However, in this, the Antarctic Peninsula of West Antarctica was placed in such a manner as to overlap with the Falkland Plateau. With the evidence currently available, this is inadmissible, and more recent workers have tried to take cognizance of this error, whilst essentially retaining the other relationships established for East Gondwanaland. A good summary has been given by Barron et al (1978) but their own reconstruction has been superseded by data that show unequivocally a northerly location (off East Africa) for Madagascar (e.g. Segoufin & Patriat 1980), constrained further by the anomalies already mentioned in the Mozambique Basin (Segoufin 1978).

Many of the recent advances have stemmed from work in Antarctica and the Scotia Sea area (south of the Falkland Plateau), where geological data are being used to establish the relationship between West Antarctica and its suture with East Antarctica along the Weddell Sea-Transantarctic Mountains-Ross Sea lineament (e.g. Elliott 1975a, 1975b, Ford & Kistler 1980), and the complex sea floor spreading patterns east of the Antarctic Peninsula (e.g. de Wit 1977, Barker 1979, Barker & Jahn 1980). Many of these new data have been incorporated in the reassemblies presented by

Norton & Sclater (1979), and we have used these as a basis for our summaries in Figure 2. This shows four palaeogeographies: A, pre-drift (say earliest Mesozoic); B, early Barremian (ca.M1, 120 my); C, Santonian-Campanian (ca.A34, 78 my); and D, early Palaeocene (ca.A28, 63 my). Points where Figure 2 differs from Norton & Sclater's (1979) reconstruction are relatively minor, but include:

i) the adoption of Elliot (1975b) and de Wit's (1977) fragmentation of West Antarctica;

ii) the incorporation of the Agulhas Plateau as a continental fragment (following Tucholke et al 1981, and the flow lines of Scrutton 1973b);

iii) a slight repositioning of Antarctica and Australia against Africa and India to allow for refitting the southern extension of continental Madagascar and the probable existence of continental crust off SE Mozambique in the Mozambique Channel and Ridge.

In the pre-drift reconstruction (Fig.2A), southern Africa was in an inter-continental situation, in which the southern Cape lay along the northern edge of the late Palaeozoic-early Mesozoic subduction/orogenic zones that ran across the southern edge of Gondwanaland. The relationship of the paired late Permian-early Triassic orogenic belts to each other is not clear: the northern zone, running from eastern Argentina at least as far as the Pensacola Mountains and probably including the Ellsworth Mountains block (Watts & Bramall 1981); and the Andean zone with a possible extension into the late Triassic-early Jurassic belt in the Antarctic Peninsula (Miller 1981). The implications of this reconstruction are that the Cape Fold Belt continued NW under the west coast continental shelf across the southern part of the Orange Basin, and eastwards across the Falkland Plateau and onto the Antarctic plate so that its northern edge passed via the region where the southern Mozambique Ridge and Agulhas Plateau originally abutted.

Evidence of the earliest drift events in Gondwanaland (i.e. the age of the oldest Mesozoic oceanic crust) are equivocal, indeed, large parts of critical areas adjacent to western East Antarctica have not been systematically surveyed, but it is reasonably clear that Antarctica-Madagascar-India-Australia (East Gondwana) separated from West Gondwana (South America and Africa) before the latter subdivided. We have already mentioned that Antarctica had separated from the northern Mozambique Basin by at least M16 (ca.133 my), and possibly M22 (ca.141-142 my, Kimmeridgian), and anomaly data from the Somali Basin north of Madagascar show that the onset of drift between Madagascar and Kenya/Somalia began at a similar time: various ages being quoted at M21 (ca.141 my, Segoufin & Patriat 1980); M22 (ca.142 my, Parson et al 1981); M25 (ca.146 my Rabinowitz et al 1981) (anomaly ages converted using Van Hinte 1976a). Given the palaeogeography in Figure 2A, the two events must have been synchronous, or at least the northern separation cannot have commenced prior to that in the south. This suggests that if Rabinowitz et al's (1981) dating in the Somali Basin is correct, anomalies M23-25 remain to be discovered in the Mozambique Basin.

By the time of the reconstruction in Figure 2B (M1, ca.120 my, Barremian), small oceanic basins existed in the South Atlantic (subdivided by the Walvis Ridge-Rio Grande Rise barrier), and off SE and East Africa. In this reconstruction, we have shown the Agulhas Plateau continental fragment attached to the Falkland Plateau, in contrast to Tucholke et al (1981), who proposed that it remained attached to the Mozambique Ridge in the early spreading episode. Neither of these two alternatives has been fully investigated, but a less-complicated ridge-jump history is necessary to generate our model.

Figures 2C and D show the progressive development of the ocean basins south of the Falkland Plateau and southern Africa. Madagascar probably reached its present position, rotating along the Davie Ridge, by about M0 (108-110 my, late Aptian) times (Segoufin & Patriat 1980), and India is shown separating from it in the Upper Cretaceous (just prior to A34, Santonian) (see Norton & Sclater 1979 for details). This relatively late date for the northward movement of India has also recently been advocated by Colbert (1981) on the grounds of similarities in Gondwanaland dinosaur faunas. In his estimation, a late Cretaceous connection between South America and Antarctica (that was possibly still operative in the early Palaeocene) is also necessary to account for marsupial and early hoofed mammal (condylarths) migrations between the Americas and Australia. Apart from the progressive widening of the surrounding ocean basins, the only significant change between the palaeogeography of southern Africa in the early Tertiary and the present, is the diminution of the role of the Walvis Ridge-Rio Grande Rise barrier in separating deep water circulation between the northern and southern South Atlantic. Other more geographically distant alterations, which, nevertheless have had a profound influence on the climate of the subcontinent (the whole world in fact), have been the separation of West Antarctica from South America, and Australia from East Antarctica. This allowed complete circum-south polar circulation with resultant climatic deterioration and the establishment, by about late Miocene times, of the present climatic regime of the Southern Hemisphere.

## 1.3 LOCAL TECTONIC TRENDS

Having reviewed the plate tectonic setting of southern Africa, it is necessary to outline the main local tectonic controls that affected late Palaeozoic and Mesozoic sedimentation. Rust (1975) has made a perceptive sum-

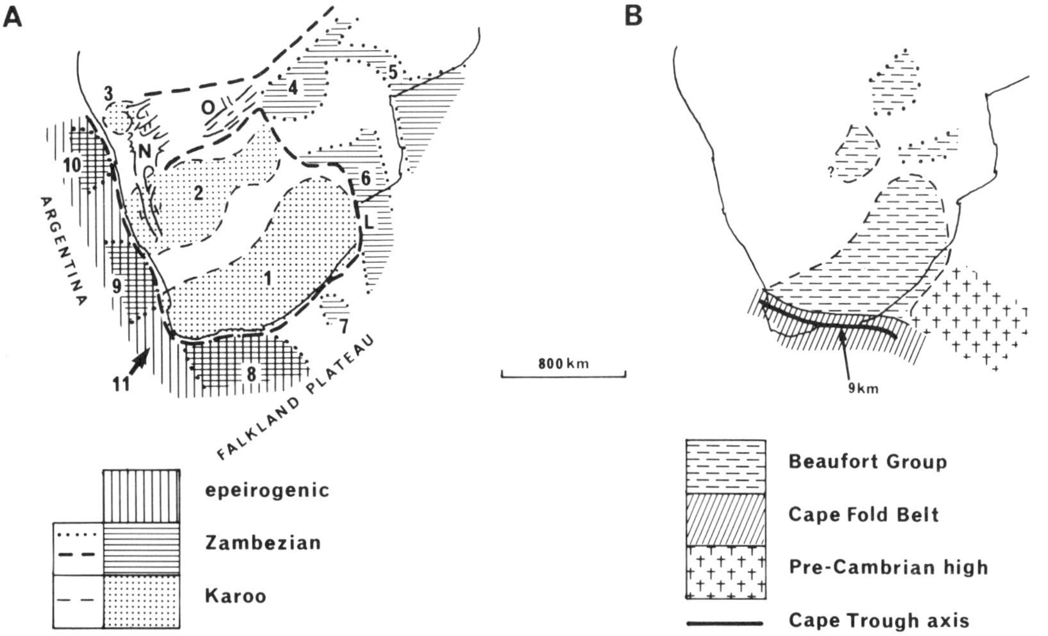


Figure 3. Mesozoic-Tertiary tectono-sedimentary terrains of southern Africa.

A. General setting, based partly on Rust (1975), with additional data from Korn & Martin (1951) and Mallick *et al.* (1981). Karoo-broad downwarping on cratonic blocks with no significant faulting (Permo-Triassic). Zambezian-taphrogenic style (Triassic-Lower Cretaceous). Epeirogenic-regional subsidence, mainly along continental edges, no significant faulting (Upper Cretaceous-Recent). Abbreviations: 1 = main Karoo Basin; 2 = Botswana Basin; 3 = Etjo Basin; 4 = mid-Zambezi Basin; 5 = Zambezi Graben and offshore extension; 6 = N.Transvaal, S.Mozambique, and N.Natal Valley; 7 = St Johns Basin; 8 = Outeniqua Basin; 9 = Orange Basin; 10 = Walvis Basin; 11 = outer continental margin; O = Okavango Graben; N = W.Namibia Karoo and post-Karoo seismic and tectonic zone; L = Lebombo Line.

B. Palaeozoic and early Triassic setting (phases A & B of Table 2). Beaufort basins are late Permian to early Triassic features, whilst the Cape Trough is a late Cambrian to early Permian feature with at least 9 km of fill. The Pre-Cambrian terrain off the east coast probably constituted the eastern flank of the main Karoo Basin, and consisted of the Maurice Ewing Bank and S. Mozambique Ridge.

mary of this topic, and we base our review on his work, although the basic framework had been described by du Toit (1954). Rust recognizes two major late Palaeozoic-early Mesozoic tectono-sedimentary terrains in southern Africa which, with the addition of a third, later style, embrace all the situations that describe the Mesozoic sedimentary history of the subcontinent. Although he did not specifically ascribe a chronology, it is clear from his treatment of the topic, that Rust was fully conscious of the importance of timing in the development of the two terrains that he mapped. Plotted on Table 2, Rust's earliest tectono-sedimentary terrain (Karoo) coincides with our phases A and C (punctuated by the main Cape orogenic uplift, B), whilst the younger style, Zambezian, developed in two stages (phases D and E, Table 2). Subsequent basin development was epeirogenic in nature, and occurred mainly around the continental edges (phase F).

Figure 3A shows the main Meso-Cenozoic tectono-sedimentary terrains in southern Africa. Rust (1975, fig.38.1) restricted his Zambezian terrain to the eastern part of the subcontinent, but geophysical data and satellite imagery (e.g. Reeves 1972, 1978, Mallick et al 1981) have shown that graben continue SW from the Zambezi area of Zimbabwe across Botswana (Okavango) into NE Namibia. In addition, the western edges of Rust's Karoo terrain, at least in the NW, may coincide approximately with the seismic-tectonic lineaments in western Namibia recognized by Korn & Martin (1951). The whole of the offshore area is underlain by large taphrogenic structures, so that the Karoo-style terrains of southern Africa, which are essentially broad, unfaulted cratonic depressions, are ringed by later Zambezian-style depocentres. The youngest phases of regional basin formation (Upper Cretaceous-Recent) affected the continental margin areas, which subsided with little or no faulting or volcanism, and the continental interior in Botswana (Kalahari Basin). What degree of vertical crustal movement has been involved in the latter is not clear, and sediment accumulation may merely be the result of internal drainage patterns caused by continental margin flexing. Zambezian-type terrains are still active in the Okavango Graben, where Neogene dunes have been cut by faulting along lines which had earlier moved to produce Permo-Triassic graben (Mallick et al 1981).

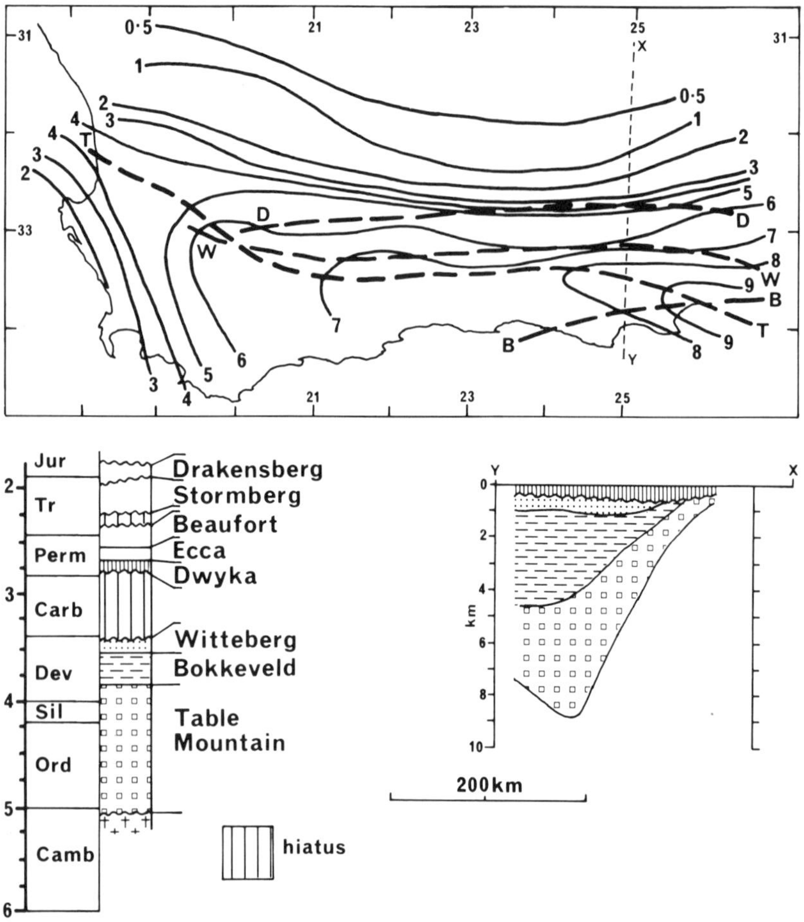


Figure 4. Cambrian to Permian Cape Trough in the S.Cape. Total isopachs in km plotted from data given in Rust (1973), with depocentres (heavy dashed line) for individual groups shown as: T = Table Mountain; B = Bokkeveld; W = Witteberg; D = Dwyka. Horizontal scale is the same for map and section, with time scale (in x 100 my) after SACS (1980) where age of the younger Cape Granites (i.e. maximum age of the Table Mountain Group) is 502 my. The lower three groups constitute the Cape Supergroup, the upper five groups constitute the Karoo Supergroup.

The earliest phase of crustal movement (A) which concerns us here controlled the latest stages in the filling of the Cape Trough and sedimentation on its northern foreland (Fig.3B). Although the sedimentary history of this feature is beyond the scope of this book, we will briefly outline the successive locations of the various depocentres because the main trough axis was the zone that was subsequently folded and uplifted during the late Palaeozoic-early Mesozoic orogeny. Using data published by Rust (1973), total isopachs for the trough infill, and successive axes of maximum sedimentation are shown in Figure 4. At least 9 km of Upper Cambrian to lowermost Permian lie in an E-W to NW-SE zone across the southern Cape, and, with the exception of the Bokkeveld Group axis (?mid-late Devonian), there was a slow northward progression of depocentres onto the northern foreland. Although the southern part of the trough now lies under the continental shelf (Agulhas Bank), the isopachs suggest that in the southern Cape the overall trough axis, for the most part, lay under present-day onshore areas. In the west, it passes seaward north of 33°S, and strikes SE across the continental shelf in the vicinity of Port Elizabeth (25°E). Palaeontological dating of the various lithostratigraphic units is not very precise, but maximum sediment accumulation rates show a fairly steady value of between 37 and 47 m/my for about 150 my of the first 175 my of its existence (the whole sequence is not preserved in the folded area, where mid-Permian to early Triassic rocks are missing), with an apparent short spurt (for about 25 my) during mid-Devonian times, when the rate rose to ca.125 m/my. Some Carboniferous uplift is indicated by the major Carboniferous non-sequence which is locally represented by an unconformity at the base of the early Permian Dwyka Group. Major uplift in the early Triassic produced mountains in the region, where at least some folding had probably already occurred (see Chapter 3), and which now forms the Cape Fold Belt. The main structural trends are reflected in Zambezian-type sediment basins, whose formation began at the latest in mid-Jurassic time, and which probably resulted from subsidence along both new and reactivated major faults.

# 2

# Triassic to early Jurassic rocks of the intracratonic basins

In terms of the crustal development of southern Africa, early Mesozoic time covers the end of intracratonic basinal sedimentation and the regional establishment of intercratonic taphrogenic basins (Tab.2 & Fig.3). Specifically, these events are, in chronological order: the final stages of the infilling of the Cape Trough and subsidence to the north of it (phase A); the uplift of the Cape Trough sediment pile (phase B); continued intracratonic subsidence to the north of the Cape Fold Belt (phase C); massive volcanism, which in the east was associated with the development of taphrogenic lineaments (phase D); and finally, the regional establishment of taphrogenic basins on the west and south coast margins (phase E). Table 2 summarizes these events. In this chapter, phases A, C, and D are covered, whilst phase B, the Cape Orogeny, is dealt with in chapter 3, and phase E in chapter 4.

Strata of Triassic-early Jurassic age crop out in two main areas in southern Africa. The largest and best known is centred on Lesotho in the E Cape, Orange Free State, S Transvaal and E Natal, whilst the other deposits occur as a series of outliers scattered in a broad arc to the north which stretches from Zululand and Swaziland in the SE, through N Transvaal, S Zimbabwe and E Botswana, westwards into NW Namibia (Fig.5). The area around Lesotho contains the youngest strata in the great synform that occupies the central southern part of South Africa. It contains rocks that range in age from Permian to lowermost Jurassic and is known as the Karoo Basin (Fig.3). Whilst the present outlines of this area do not relate directly to the shape of the original depocentre, the term Karoo Basin is a useful and established geographical concept to describe the area which contains the classic Karoo lithostratigraphic sequences of Dwyka, Ecca, Beaufort, Stormberg and Drakensberg groups.

Table 3 and Figure 9 summarize the late Permian to earliest Jurassic succession in the main Karoo Basin and show that Mesozoic sediments related to phase A (final stage of Cape Trough filling) constitute the upper part of the Beaufort Group (Tarkastad Subgroup) and that post-Cape Orogeny sediments (phase C) make up the Stormberg Group. The Mesozoic strata within the Karoo Basin lie towards its eastern side and occupy only about one third of the total outcrop. The succession here is relatively thick and easily differentiated into several well-defined formations which are taken as the type sequences (Tab.3). Farther north, in the archipeligo of Karoo Supergroup outliers, Mesozoic strata are generally much thinner, and with the exception of the uppermost units, the well-differentiated sequence of the main Karoo Basin cannot always be confidently recognized. This relationship was defined by du Toit (e.g. 1954) on lithological grounds (his north and south facies), and has been reiterated more recently by Kitching (1977). Figure 8 shows a section from the main Karoo Basin to the faulted outliers in the northern Transvaal. From the recent work of Kitching (1977) it is apparent that this thinning is due to non-deposition, or slow sedimentation rather than later erosion, and that the Beaufort Group sediments spread onto and across a relatively positive area in the Transvaal where a thin (100-200 m) sequence accumulated on a reduced Ecca succession. It has not been possible to subdivide the Beaufort rocks in these northern outcrops, but it has been suggested (e.g. Ryan 1967 quoted by van Eeden 1973) that the whole Beaufort succession in the northern area is the time equivalent of only the Tarkastad Subgroup of the main Karoo Basin.

During late Palaeozoic/early Mesozoic times, tetrapod faunas worldwide underwent radical evolution, particularly within the reptiles, where mammal-like forms (therapsids), and the earliest dinosaurs first appeared. Details of the phylogenetic changes involved have been studied using the rich vertebrate faunas of the Karoo Supergroup in southern Africa and a very large bibliography exists. The following are a cross-section of contributions: Seeley (1892), Broom (1906), Haughton (1919), Boonstra (1969), Kitching (1970, 1977), Romer (1970), Cox (1973), Keyser & Smith (1979), and Keyser & Cruickshank (1979). Despite their importance, these relatively large terrestrial and semi-aquatic animals are not ideal zone fossils for correlation, either internationally or internally. Coupled with the fact that early vertebrate workers, in their enthusiasm, created a large number of synonyms, often on dubious material, and the inherent difficulties of working in low dip, homogeneous lithologies, it is not surprising that only recently have soundly-based biostratigraphic correlations been

Table 3. Summary of lithostratigraphic and biostratigraphic subdivision of the Triassic and uppermost Permian strata of the main Karoo outcrop and the Lebombo volcanics. Based on numerous authors cited in the text, but specific correlations and definitions follow Lock *et al.* (1974), Cleverly & Bristow (1979), Keyser & Smith (1979), Johnson *et al.* (1976) & Kitching (1977).

| | Group | Lithostratigraphy: Formations and subgroups | | | | | | Biostratigraphy: Vertebrate zonation | | World stages | System | Tectono/ lithostratigraphic phases |
|---|---|---|---|---|---|---|---|---|---|---|---|---|
| | | Western area: Karoo Basin & Lesotho | | | Eastern area: Lobombo Mts. | | | | | | | |
| Karoo Supergroup | Drakensberg Group | Lesotho Fm. | | main flood basalts: thick tholeiitic flows & occasional interbedded pyroclastics | Lebombo Group | Movene Fm. | aphyric basalts | | | Hettangian | Jurassic | D |
| | | | | | | Jozini & Mbuluzi Fms. | rhyolites | | | | | |
| | | Kraai River Fm. | | fine grained tholeiitic basalts & pillow lavas | | Sabie R. Fm. | aphyric basalts | | | | | |
| | | | | | | Letaba Fm. | olivine-rich basalts | | | | | |
| | | Moshesh's Ford Fm. | | lavas, pyroclastics vent fill & interbedded sediments | | Mashikiri Fm. | alkaline basalts | | | to | to | |
| | Stormberg Group | Clarens Fm. | | | yellow & white massive & cross bedded feldspathic sandstones & minor grey-green shales & silts. | | | no formal zonation | occasional light-limbed dinosaurs | | | C |
| | | Elliot Fm. | | | greyish red & purple mudstones with minor yellow & grey sandstones | | | | crocodiles & occ. light-limbed dinosaurs | | | |
| | | | | | | | | | light-limbed Anchisaurid dinosaurs | | | |
| | | | | | | | | | heavy-limbed Melanosaurid dinosaurs | ?Uppermost Ladinian | Triassic | |
| | | Molteno Fm. | | | glittering sandstones, grits & conglomerates with grey & black shale, mudstones & coals | | | | no tetrapods | | | |
| | Beaufort Group | Tarkastad Subgroup (≡ Smithfield Fm. in E.OFS) | Burgersdorp Fm. | | red & bluish-grey mudstones with subordinate sandstone & siltstone | | | *Cynognathus* Zone (≡ Kannemeyeria Zone) | | Scythian | | B |
| | | | Katberg Fm. | | yellowish & pink sandstones & pebbly sandstones with subordinate red siltstone & mudstone. Red mudstones at base | | | *Lystrosaurus* Zone | | | | |
| | | Adelaide Subgroup (partly ≡ Estcourt Fm. in Natal) | W.Karoo Teekloof Fm. | E.Karoo Balfour Fm. | grey, blue & green mudstones with subordinate sandstones | | | *Dicynodon lacerticeps* Zone | | Dorashamian ? | Permian | A |
| | | | | | | | | *Aulacephalodon baini* Zone | | | | |
| | | | | Middleton Fm. | red, bluish-grey & greenish-grey mudstones with subordinate sandstones | | | *Tropidostoma microtrema* Zone | | | | |
| | | | Abrahamskraal Fm. | Koonap Fm. | grey, bluish-grey & greenish-grey mudstones & siltstones with subordinate sandstones | | | *Pristerognathus/Diictodon* Zone | | | | |
| | | | | | | | | Dinocephalian Zone | | | | |
| | Ecca Group | | | | | | | | | | | |

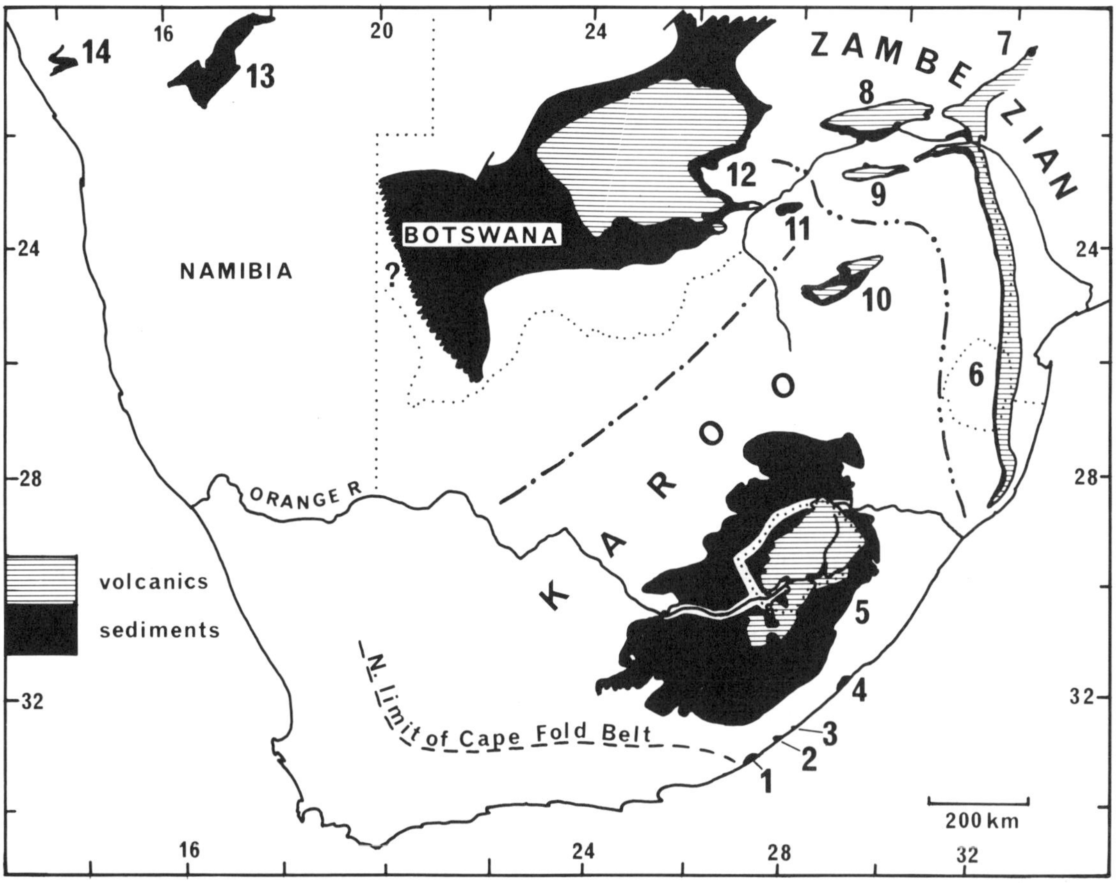


Figure 5. Distribution of Triassic sediments and lavas in southern Africa. After numerous authors cited in the text. Distribution in central and western Botswana is almost exclusively controlled by boreholes. Areas to the north of the main Karoo Basin may contain some Upper Permian strata. Key: 1 = Kidds Beach; 2 = Kwelera; 3 = Kei River; 4 = Mngazana; 5 = main Karoo Basin (centred on Lesotho); 6 = Lebombo Line; 7 = Sabi; 8 = Tuli; 9 = Soutpansberg; 10 = Springbok Flats; 11 = Waterberg; 12 = Eastern Botswana; 13 = Etjo; 14 = Doros. Dot-dash line separates Karoo and Botswana basins; dot-dot-dash lines separates Zambezian and Karoo tectono-sedimentary terrains (see Figure 3).

achieved within the Karoo Basin (e.g. Kitching 1977). Previously, only a crude lithostratigraphic partition into Lower, Middle, and Upper divisions of the Beaufort Group (as for instance expounded by du Toit 1918, 1954) was available, loosely tied in with a reptilian zonation (bedevilled by synonyms). Lithostratigraphic subdivision had been mostly based on coloration differences, which even in the type areas in the eastern Cape provided only a shakey concensus (e.g. Marais & Johnson 1965), whilst farther north and west, the scheme was notoriously unreliable. Recent revision, to a considerable extent spurred on by exploration for uraniferous sediments, has done much to clarify the situation, and regional correlation between litho- and biostratigraphic units is now possible in several areas (e.g. Johnson et al 1976, Kitching 1977, Keyser & Smith 1979, SACS 1980). The rocks most involved in this effort have, however, been the Beaufort Group which means that only the lower Triassic strata have benefited, leaving biostratigraphic schemes no better than Haughton's (1924) preliminary palaeontological subdivisions for the Stormberg Group. The situation at present, therefore, is that whilst there are still areas of disagreement, the Permian-Triassic boundary can be located within the Beaufort Group in rocks that have relatively good palaeontological control. The Triassic-Jurassic boundary on the other hand is a far less tractable problem. The upper Stormberg Group sediments (Clarens Formation) are only sparsely fossiliferous, with poorly known vertebrate faunas, and interfinger with lavas and various pyroclastics of the overlying Drakensberg Group. The underlying Elliot Formation is relatively more fossiliferous and its fauna is Norian/Rhaetian. Radiometric dates on the Drakensberg lavas give ages around 180-190 my (early Jurassic), so presumably on a regional scale the Triassic/Jurassic boundary lies close to the Stormberg/Drakensberg boundary. However, because this is known to be locally

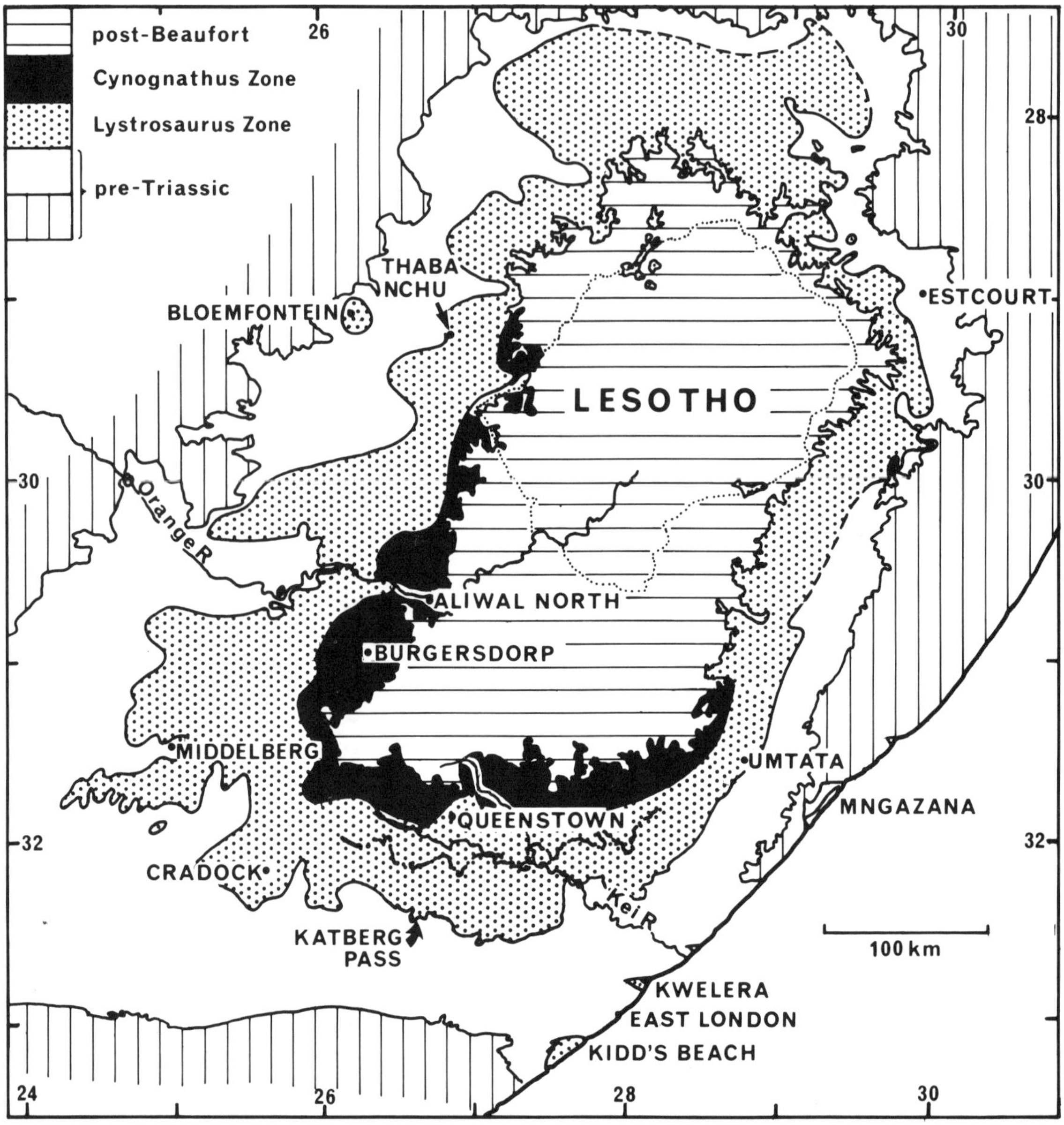


Figure 6. Distribution of strata of *Lystrosaurus* and *Cynognathus* zone ages in the main Karoo Basin. Based mainly on Kitching (1977) with additions from Johnson *et al.* (1976), Keyser & Smith (1979), and Stavrakis (1980). The thick dot-dash line south of Queenstown is the northern limit of the Katberg Formation as mapped by Hiller & Stavrakis (1980).

strongly diachronous, the age of the Clarens Formation can be given no better than Rhaetian-Hettangian and no biostratigraphic zonation is available to even attempt local demarcation.

## 2.1 OUTCROPS OF THE MAIN KAROO BASIN

Figures 5 and 6 show the distribution of the Triassic to early Jurassic strata in the main Karoo Basin, and four small fault-bounded outliers along the coast. These total approximately 159 000 km$^2$ in extent. As previously mentioned, they constitute only the upper part of the Karoo Supergroup sequence and represent an erosional outlier of what was probably a considerably more extensive original development. Some idea of the original distribution can be gauged by considering the basin geometry for the Beaufort Group as a whole and Figure 7 shows the present-day elevation of the base of the Beaufort Group (Upper Permian) contoured from SOEKOR borehole data (Winter & Venter 1970). There are two axes to this depocentre, one forming a

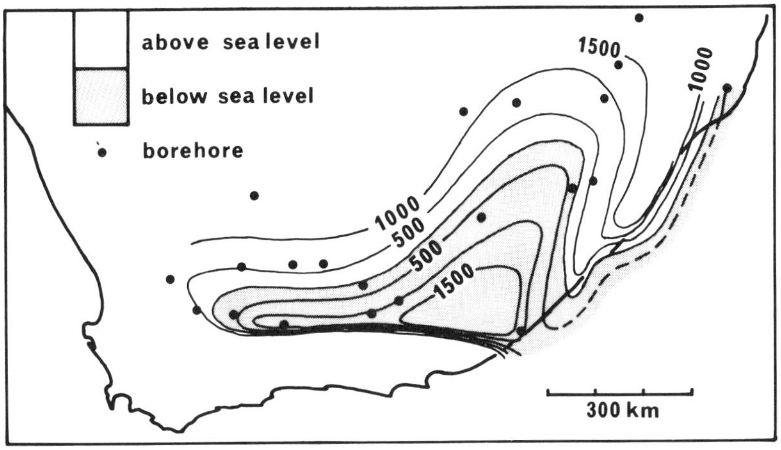


Figure 7. Depth to base of the Beaufort Group (Upper Permian horizon) in metres relative to present sea level. Areas lying below sea level are shaded. Control points are SOEKOR boreholes (data from Winter & Venter 1970).

deep narrow trough along the northern edge of the Cape Fold Belt, and the other a broad basin that trends and shallows NE from the Queenstown area. These two features merge in the Cradock-Queenstown region and continue in a SE direction, crossing the present coastline and continental edge in the vicinity of East London. It would seem that the main depocentre axis coincided with the southernmost trough because isopachs for Permo-Triassic sediments (i.e. excluding the Drakensberg Group) show a similar picture (Fig.8), suggesting that Figure 7 approximates to the original depocentre configuration. A large and unknown thickness of uppermost Permian and all the Triassic has been eroded from the western and southernmost parts of the basin, but Figure 8 shows a cross-section that attempts to reconstruct approximate depths to the Beaufort basin floor using the top of the Stormberg Group as datum. This suggests an original late Permian to Triassic thickness of over 7 km in the vicinity of Grahamstown which thinned gradually westwards to about 6 km about 22°E. Of this at least half was Upper Permian (i.e. Adelaide Subgroup). Northwards from the depocentre axis all three sequences for which we have data (Upper Permian Adelaide Subgroup, Lower Triassic Tarkastad Subgroup, and the Middle and Upper Triassic Stormberg Group), thin rapidly, with the most dramatic diminution being in the lower two. From a maximum estimated thickness of about 7 km at Grahamstown, the trough-fill thinned to around 4.5 km at Queenstown, 1.7 km in the NE Cape (borehole MA1) and about 1 km in eastern Natal. Such an asymmetric sediment wedge indicates that during Upper Permian to Triassic time the area to the north and NE of Queenstown was relatively stable and that the main depocentre lay in an approximately east-west direction in the vicinity of Grahamstown, probably at least until the end of Lower Triassic times. This was the most northerly position attained by the Cape Trough axis, which for most of its history (Cambrian to mid-Permian) was situated within a 90 km wide east-west belt that extended up to 200 km north of the present coast (i.e. under what is now the main Cape Fold Belt) (Fig.4). Rust (1975) has suggested that in the southern Cape Province, the Cape and Karoo Supergroup sediments were deposited in two separate basins. We see no reason to make this distinction and would rather view the Palaeozoic sedimentation in the southern Cape as a slowly evolving miogeosyncline with two distinct stages of infill. During the first stage, the Cape Supergroup was deposited and the main depocentre lay in the south, with sediment accumulation rates peaking at 125 m/my in mid-Devonian times during the deposition of the Bokkeveld shales. The second stage of infilling (Karoo Supergroup) commenced with the accumulation of the early Permian Dwyka glacial/marine sediments in a depocentre whose axis lay very close to the position it occupied in the latter stages of Cape Supergroup sedimentation, so that Witteberg to Ecca strata shared roughly the same trough axis. Sediment accumulation rates rose rapidly through Permian times, peaking (233 m/my) during the deposition of the late Permian Adelaide Subgroup. This increase in sedimentation rate was accompanied by a rapid 35 km northerly shift of the main depocentre. An additional feature of this stage of sedimentation is the far greater extent of the area of accumulation (from the southern Cape to southern Zimbabwe). It culminated in the Cape Orogeny, with uplift of the pre-Permian depocentres in early-mid Triassic times, and it seems reasonable to assume that the rapid northward shift in the depocentre and greater crustal instability during the late Permian were associated with this event.

### 2.1.1 *Lower Triassic (Scythian): Tarkastad Subgroup*

In the Karoo Basin, Lower Triassic (Scythian stage) strata rest conformably upon uppermost Permian (Dorashamian stage) rocks. They consist essentially of two lithostratigraphic units (Katberg and Burgersdorp formations) which represent the upper division of the

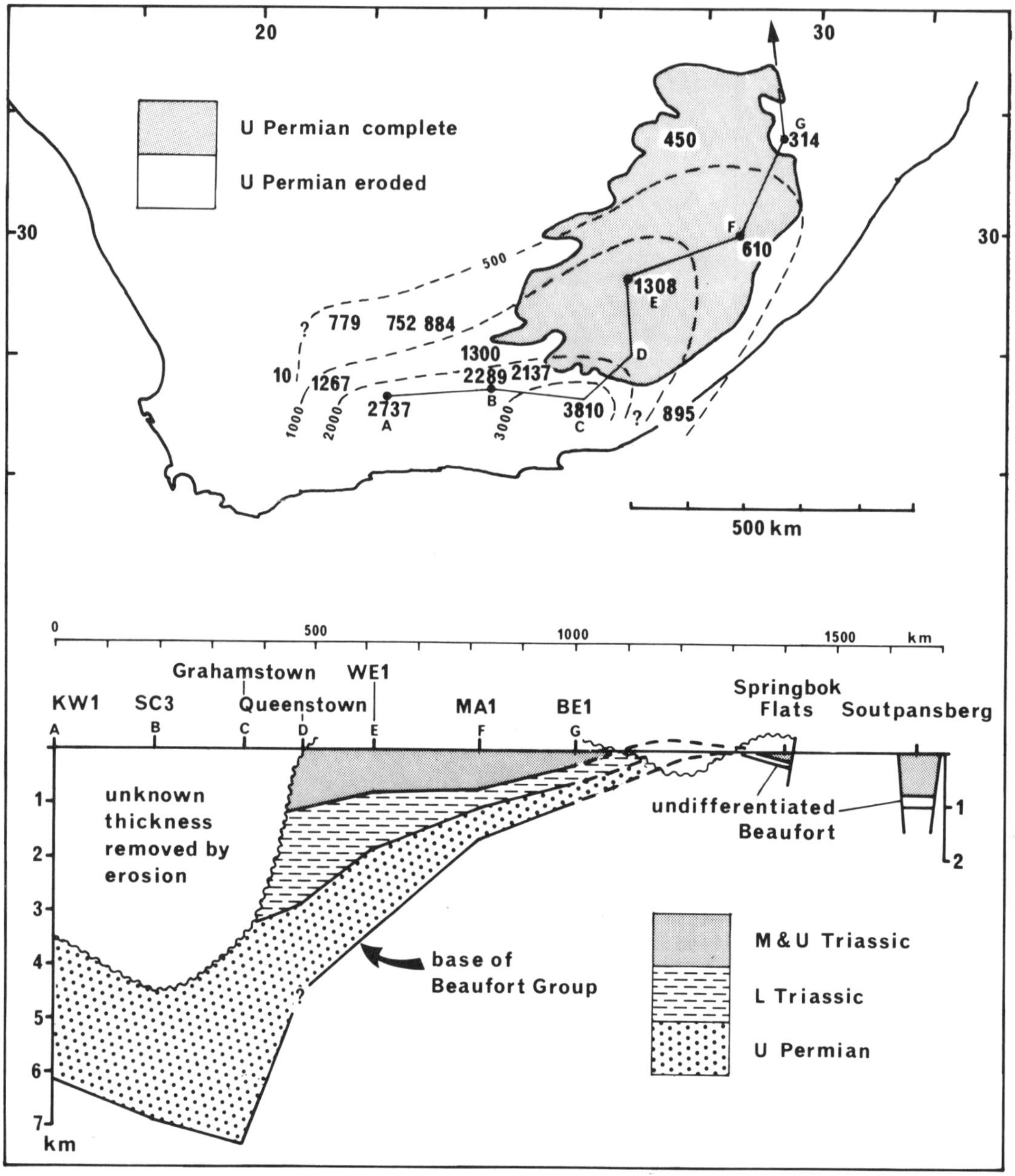


Figure 8. Thickness of Beaufort and Stormberg Groups in main Karoo Basin. Datum is top of Stormberg Group. Map shows isopachs in metres, with control borehole thicknesses indicated. Shaded area is where the Upper Permian succession (Adelaide Subgroup) is complete. Cross-section is along the line A-G. Data mainly from Winter & Venter (1970), Beukes (1970), and Botha & Linstrom (1978).

Beaufort Group as understood, for example by du Toit (1954). This upper division, which has recently been redefined as the Tarkastad Subgroup (Johnson et al 1976, SACS 1980, Tab.3, Fig.9) in the southern areas of Figure 6, contains two biostratigraphic zones (*Lystrosaurus,* and *Cynognathus*). The upper zone (*Cynognathus*) is missing from the northern and eastern areas of the Karoo Basin because of an unconformity and overstep by the Stormberg Group, and correlation of bio- and lithostratigraphic schemes for the Lower Triassic rocks identifies significant diachronism between the Katberg and Burgersdorp formations (Fig.10).

### 2.1.1.1 *Lithostratigraphy*

The Scythian rocks of the southern outcrops of Figure 6 comprise two main lithofacies: a lower arenaceous unit with minor shales and mudstones (Katberg Formation) and an upper argillaceous unit with minor sand-

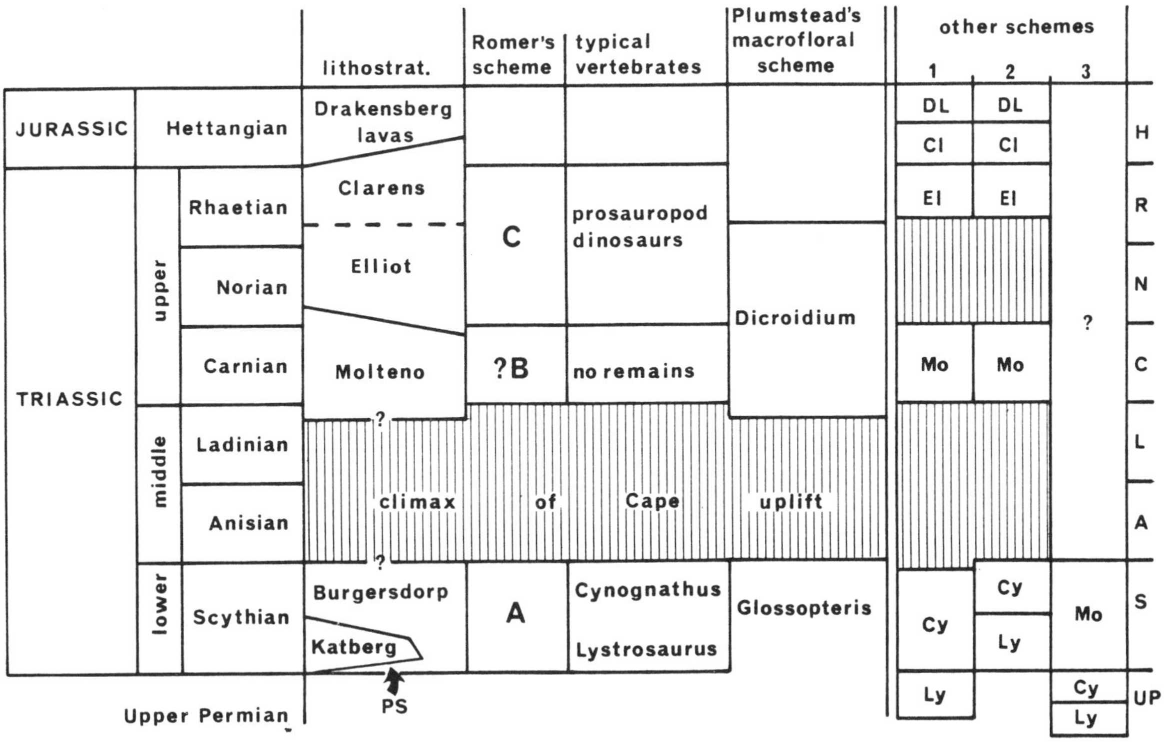


Figure 9. Ages of various Triassic lithostratigraphic units in the main Karoo Basin. Correlation used herein (based on Cox 1973, Kitching 1977) is shown in the left column and is equated with the vertebrate and macrofloral zonation schemes of Romer (1975) and Plumstead (1970), respectively. Some previous classifications are shown in columns 1-3: 1 – Anderson & Anderson (1970); 2 – Anderson & Schwyzer (1977); 3 – Stapleton (1978a). Abbreviations: DL = Drakensberg; Cl = Clarens; El = Elliot; Mo = Molteno; Cy = *Cynognathus* zone; Ly = *Lystrosaurus* zone; PS = Palingkloof Shale. Hiatuses are shaded.

stones and siltstones (Burgersdorp Formation). These rocks overlie a series of grey, blue and green mudstones with minor sandstones (Balfour Formation of the Adelaide Subgroup, see Tab.3) that carry a *Dicynodon* vertebrate assemblage (i.e. are of Upper Permian age).

Figure 10 shows the regional thickness and facies variations in the Lower Triassic rocks between East London and Natal, and clearly brings out the distinction between the unstable southeastern basin and the stable northern platform. In the former, the Tarkastad Subgroup is about 2 000 m thick, but it thins rapidly to about 250 m in the vicinity of Aliwal North and no more than 350 m in Natal. The boundary between these two features must lie just SE of the Aliwal North area because SOEKOR borehole WE1 (about 50 km SE) proved about 850 m of Lower Triassic sediments (Winter & Venter 1970).

Away from the type areas of the eastern Cape, differentiation of the Lower Triassic rocks has not always been satisfactory, and in the eastern Orange Free State Theron (1970, 1975) was unable to recognize the main arenaceous/argillaceous subdivision. He grouped all the rocks between the Stormberg Group and the predominantly grey, green and blue mudstones of the 'middle Beaufort Formation' into one unit; the Smithfield Formation. In Natal, Johnson et al (1976) were not in a position to clarify the relationships of the Estcourt Formation and the overlying red beds, but recent work by Botha & Linström (1978) has demonstrated that the Katberg and Burgersdorp formations of the eastern Cape have lateral facies equivalents in the Belmont and Otterburn formations, respectively, and that the Estcourt Formation is equivalent to the whole of the Adelaide Subgroup farther south. These relationships are shown diagrammatically in Figure 10 and detailed in Table 4.

2.1.1.1.1 *Katberg Formation.* This formation was proposed by Johnson (1966) with its type sections in the Katberg Pass area through the Winterberg Mountains south of Queenstown (Fig.6). In the type area, it consists of pale red and grey, fine-medium feldspathic sandstone with minor maroon mudstones which become more prominent near the top. Johnson (1966) recognized a three-fold subdivision with a lower pale pink and grey speckled sandstone, a middle pale grey sandstone, and an upper pale reddish-purple standstone (Fig.11).

The most comprehensive description of the Katberg Formation lithofacies in its type area has been

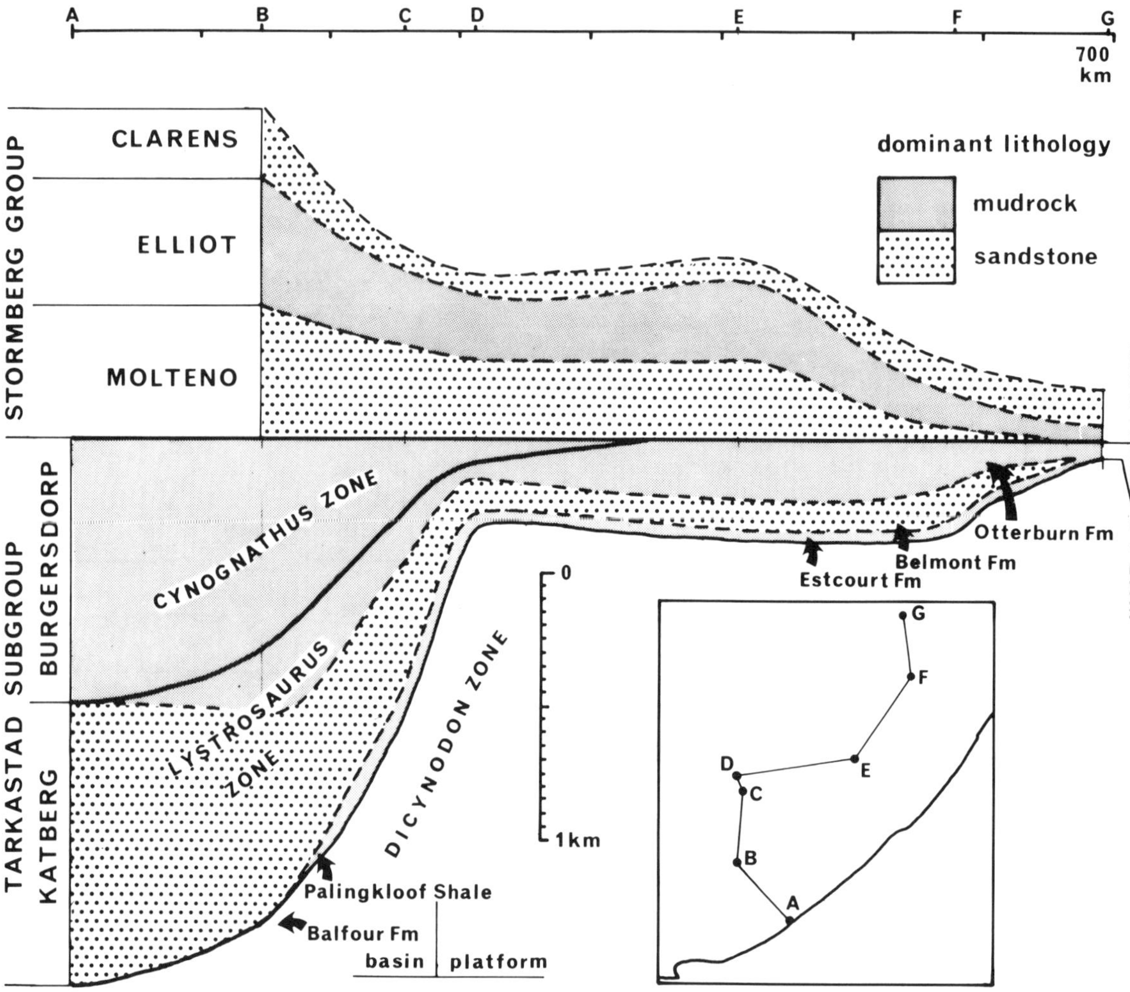


Figure 10. N-S cross-section illustrating diachroneity and thickness variations of main lithofacies in the Triassic rocks of the main Karoo Basin and coastal outliers. Thick solid lines are biostratigraphic boundaries, dashed lines are lithostratigraphic boundaries. Insert shows location of control points. The *Dicynodon* zone marks the top of the Permian. Key to localities: A = East London area; B = Queenstown, Cala, S.crop; C = borehole WE1; D = Aliwal North; E = borehole MA1; F = N.W.Natal; G = N.W.Orange Free State. Data from Figure 12.

given by Stavrakis (1980a, 1980b), who records a wide variety of sandstone and mudrock facies and bedforms. Within the dominant sandstone facies (which occurs in units up to 150 m thick, Johnson 1966), the main component consists of horizontally bedded, fine grained, well-sorted greenish grey sands, which Stavrakis (1980b) interprets as upper flow regime channel fill deposits. They are associated with intraformational conglomerates which consist of reddish-brown to olive-green mud balls and shale flakes (up to 50 cm across), and occasional pebbles and cobbles (up to 65 cm) of quartzite, with minor granite, silicified wood, agate, jasper, and sandstone. According to Johnson (1966), these rudaceous deposits are more common in the upper and lower subdivisions (Fig.11). Transitional and lower flow regime bodies consist of silt-sized material, and occur as horizontally-bedded trough and planar cross-bedded, and relatively rare ripple cross-laminated sets. The last-named type are common in the underlying Balfour and overlying Burgersdorp formations (Stavrakis 1980b). Individual sandstone units can be traced along strike for several kilometres, and often produce conspicuous relief (Johnson 1966).

Within the Katberg Formation, mudrocks are subordinate. They consist of thin reddish mudstones which frequently contain sand wedge polygons, calcareous duricrusts, and mudcracks. Darker, greenish-black mudstones also occur as veneers over thicker sandstone units, and as fill in depressions in the palaeo channels and bars. Stavrakis (1980a, 1980b) notes that nearly all the quartz grains of the Katberg sand-

Table 4. Comparison of lithostratigraphic units of the Beaufort Group (Upper Permian-Lower Triassic) in the Eastern Cape and N.W. Natal (modified slightly after Botha & Lindström 1978).

| Eastern Cape | % | Thickness | N.W.Natal | % | Thickness | |
|---|---|---|---|---|---|---|
| Burgersdorp Formation | | ± 1 000 m | Otterburn Formation | | 130-180 m | |
| grey sandstone | 20-30 | | grey sandstone | 25 | | |
| red mudstone, minor green mudstone | 70-80 | | red mudstone, minor green mudstone | 75 | | |
| Katberg Formation | | 400-1 070 m | Belmont Formation | | 85-165 | |
| brown, yellow, pink, grey sandstone | 30-100 | | | | | Triassic |
| greyish-red mudstone & minor green/grey mudstone | 0-70 | | grey sandstone | 70 | | |
| | | | red & green mudstone | 30 | | |
| Balfour Formation | | 1 220-4 000 m | | | | |
| grey sandstone | 10-60 | | | | | Permian |
| greenish-, bluish-grey, greyish-red & laminated mudstones | 40-90 | | Estcourt Formation | | ± 400 | |
| | | | grey sandstone | 35 | | |
| Middleton Formation | | 1 500-3 480 m | dark grey to black, and laminated mudstones | 65 | | |
| greenish-, bluish-grey & red mudstones | 70-75 | | | | | |
| grey sandstone | 25-30 | | | | | |
| Koonap Formation | | 200-1 450 m | | | | |
| greenish-, bluish-grey mudstones | 70-80 | | | | | |
| greenish-grey sandstone | 20-30 | | | | | |

stone lithofacies have secondary overgrowths, and that some have been enlarged by as much as 50 %. He also records prominent heavy mineral-rich lags up to 50 cm thick and 100 m in length in the upper flow regime flat-bedded sandstone bodies. Typically, these consist of a lower, opaque mineral-rich band, and an upper, coarser, garnet-rich layer. The former is mainly composed of titanomagnetite, with less important goethite/lepidocrite, maghemite, haematite, and ilmenite.

The Katberg sandstones are dominantly feldspathic, although the feldspars are largely altered to kaolin. In the pink and cream sandstones, the feldspars are red-stained, and the rock as a whole contains a high percentage of opaque ore minerals and garnet. Stavrakis (1980b) points out that the opposite is true for the greenish-grey coloured sandstones, which contain low values of opaque minerals and stained feldspars, but frequently have authigenic chlorite.

Overall, Stavrakis (1980b) interprets the Katberg Formation in its type area as representing deposition in swiftly flowing, shallow braided streams in which accumulation took place in flash floods under an arid climate. He draws a close analogy with the Platte River model of Miall (1977), and recognizes the following features in the Katberg Formation as criteria for such a comparison: constant channel switching; transverse bar formation and dissection; non-cohesive banks; deposition of sandy bedload material; scarce ripple cross-lamination; and deposition of mud drapes.

The base and top of the Katberg Formation are gradational, although the former is easily defined, and was taken by Johnson (1966) as the point in the succession in which sandstone predominates over mudrock. The top, as we shall discuss later, represents an interfingering relationship with the Burgersdorp Formation, as more distal parts of the depositional basin are entered. Stavrakis (1980b) has shown that the Katberg Formation in the type area is underlain by 40-80 m of brownish-red shales with numerous siliceous nodules and thin interbeds of light coloured sandstones. He calls this unit the Palingkloof Shale, and notes that both Kitching (1977), and Keyser (1979) have recorded *Lystrosaurus* Zone tetrapods from it. This relationship is diagrammatically shown in Figure 10, and in Stavrakis (1980b, fig.2). The latter author suggests that the Palingkloof Shale is the lowermost representative of the Burgersdorp Formation. As we will mention later, this unit has a lateral equivalent in Natal.

Katberg Formation strata also crop out on the coast, where Mountain (1939, 1962, 1974), Johnson (1966, 1976), and Hiller & Stavrakis (1980) have mapped them in three outliers: Kidd's Beach; Kwelera; and Kei River mouth (Qolora of Hiller & Stavrakis 1980) (Fig.11). The first-named occurs in the core of a broad shallow syncline, whilst the other two are in downfaulted structures. The best exposures are in the vicinity of Chalumna River in the Kidd's Beach outlier (Hiller & Stavrakis 1980), and in the Kwelera outlier (Johnson 1966), where approximately 1 066 m are exposed. Lithologically, both areas are similar, and consist dominantly of pink and grey coloured sandstones that are generally coarser and more feldspathic than in the type area of the Winterberg Mountains (Tab.5).

The formation in the coastal outliers exhibits a crude threefold subdivision, which consists of lower

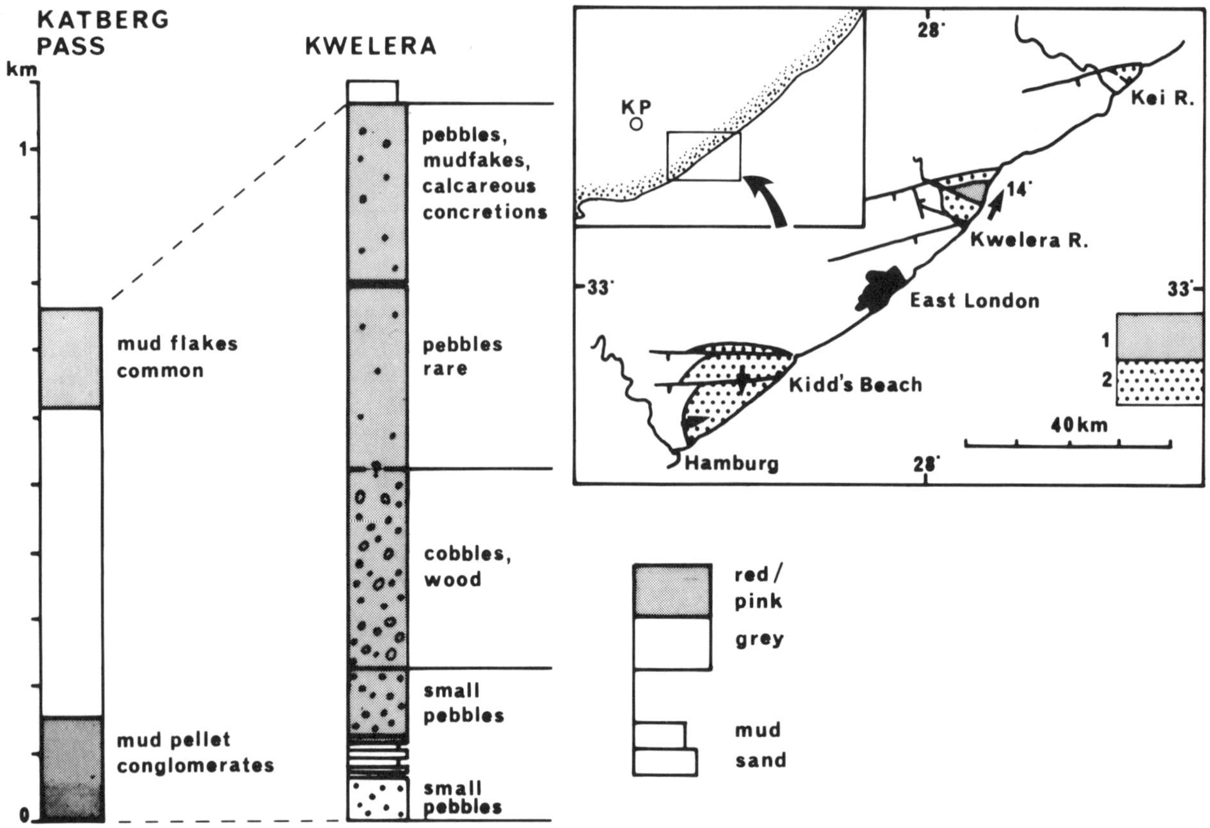


Figure 11. Distribution and thickness of the Katberg Formation in the coastal outliers around East London and a comparison with the Katberg Pass area. In the map, arrows show main dip directions, and 1 = Burgersdorp Fm., 2 = Katberg Fm, KP = Katberg Pass. Data after du Toit (1954), Johnson (1966), Mountain (1974), Hiller & Stavrakis (1980).

pink and grey pebbly sandstones with minor mudstones and sandstones, thick pink and grey sandstones with numerous cobbles and pebble beds, and upper pale purplish-grey sandstones with occasional pebbles, thin mudstones, and calcareous concretions. From their study of the Kidd's Beach outlier, Hiller & Stavrakis (1980) show that the dominant lithology is medium-coarse submature sandstone, which weathers pinkish grey in colour. The sandstones contain local pebble and red shale flake beds which are mostly confined to thin lag conglomerates resting on scoured surfaces, but which are occasionally scattered at random in beds of massive sandstone. The main pebble concentrations occur in the middle unit of Johnson's (1966) subdivision, where Mountain (1962) reported cobbles up to 12.5 cm across.

The dominant lithology is horizontally laminated sandstone which occurs in units 20 cm to 5 m thick, with occasional heavy mineral-rich bands (Hiller & Stavrakis 1980). Trough cross-bedded units are also common, but planar cross-bedding is rare. Massive, upward fining units (40 cm to 3.5 m), with scattered pebble lenses occasionally occur, and the minor mudrock interbeds show mudcracks, and load and flute casts. Primary current lineations indicate a SE source for the detritus. The pebbles in these beds consist of coarse red granite, white granite, pink gneiss, quartz-feldspar porphyry, quartzite, sandstone, fissile fragments of lignite, and silicified wood. No basic igneous rock fragments occur, but Johnson (1966) remarked on the calcic nature of the feldspar grains in the sandstones, and concluded that the source areas were nearby granite/metamorphic terrains. Hiller & Stavrakis (1980) concur with this diagnosis, and add that the presence of wood fragments indicates a Devonian or younger sedimentary component, whilst the heavy mineral (mostly pink garnet, magnetite, epidote, sphene, and zircon), albite-andesine plagioclase, and microcline feldspar suites of the sandstones corroborate a high grade metamorphic or acid igneous source.

A consideration of the lithofacies in the Kidd's Beach outlier, led Hiller & Stavrakis (1980) to interpret the succession as being made up of cycles that begin with lag conglomerates on a scoured surface, overlain by flat-bedded or trough cross-bedded units deposited by fast-flowing shallow braided streams.

Table 5. Components of Katberg Formation sandstones (data from Johnson 1966).

| | Queenstown district | | | Kidd's Beach | | |
|---|---|---|---|---|---|---|
| | 1 | 2 | 3 | 1 | 2 | 3 |
| Quartz % | 40 | 45 | 48 | 37 | 33 | 35 |
| feldspar % | 13 | 9 | 21 | 24 | 9 | 15 |
| rock fragments % | 9 | 7 | 5 | 11 | 7 | 7 |
| matrix (clay) % | 38 | 39 | 26 | 28 | 51 | 43 |

1, 2, 3 refer to the three subdivisions of the formation at each locality, 1 – the base; 3 – the top.

Minor facies, which include fine-grained sandstones, siltstones and mudstones, represent abandoned channel or overbank deposits. The massive sandstones represent particularly strong flood-stage deposits. From studies of the pebble coarseness, and bedforms, Hiller & Stavrakis (1980) concluded that deposition was in the proximal reaches of a river system (probably on the edges of an alluvial fan) in very shallow (1-2 m), narrow (4-30 m) channels. These channels filled with trough cross bedded sands, with periodic sheet flooding giving rise to the flat-bedded sand bodies.

The southeasterly source, and the lithological correlation with the Katberg Formation in the type area led Hiller & Stavrakis (1980) to agree with Johnson (1966, 1976) that the coastal outliers are remnants of a once-extensive, radially symmetric alluvial fan whose proximal part is preserved on the East Cape coast, whose distal part lies in the Katberg Pass area, and which passes northward into a more distal argillaceous lithofacies that constitutes the Burgersdorp Formation. This implies that the Palingkloof Shale was laid down in an arid environment as a temporary low energy facies prior to the injection of coarser material from the Katberg fan. When the latter encroached into the region from the SE, the depositional area for the mudrock lithofacies (Burgersdorp Formation) migrated north- and NW-wards.

A comparison of the texture of the Katberg sandstones from the Queenstown district and those from Kidd's Beach has shown (Johnson 1966) that both are moderately well-sorted and consist mainly of subrounded quartz grains that have a mean grain size of 210 $\mu$. In both areas the pink coloration is caused by haematite staining. A component analysis (Tab.5) shows that while the Queenstown strata have a higher quartz content (40-48 %) than the coastal outliers (33-37 %), the latter have higher rock fragment contents (7-11 % vs 5-9 %). In both areas the clay matrix is greater than 25 %, but whereas it decreases up the succession in the Queenstown district (38 to 26 %) the reverse takes place at Kidd's Beach (28 to 43 %).

Vertebrate fossils have not been recovered from the Katberg Formation of the coastal outliers (Kitching 1977) and consequently the assignment of these strata to the *Lystrosaurus* Zone is tentative.

Narrow outcrops of Katberg Formation have also been mapped farther north, along a 24 km coastal strip in the Mngazana faulted outlier by du Toit (1920). In a recent revision by the Geological Survey (1977) (Figs 6 & 108), Kitching (1977) has recorded *Lystrosaurus* Zone fossils from these outcrops, reinforcing their lithological correlation with the Katberg Formation.

In the Queenstown district vertebrates are rare in the sandstone units of the Katberg Formation, but reasonably common in the mudstone interbeds. Fragmentary skulls and skeletal material tend to be concentrated in the clay pellet conglomerates, and according to Kitching (1977) many skulls show evidence of rolling during water-borne transportation, which in some cases may also be evidence of penecontemporaneous erosion. Other fossil types are limited to silicified logs (up to 30 m long and 1 m in diameter) which are common in the lower part of the formation, and other rare plant fragments.

Away from the type areas of the eastern Cape, the Katberg Formation is thinner and argillaceous horizons are more abundant. This is well illustrated by comparing the successions in the Queenstown/East London areas with those in SOEKOR boreholes farther north (Fig.12). These show a decrease in thickness from 1 066 m near East London to about 220 m near Aliwal North (WE1 borehole). Across the stable platform identified in Figure 10, the Katberg Formation is between 100 and 200 m thick and contains numerous mudstone horizons. Allochthonous pebbles are restricted to a radius of about 100 km from East London (Theron 1975). In NW Natal, the lateral equivalent of the Katberg Formation has been designated the Belmont Formation by Botha & Linström (1978) (Fig.12). It is between 85 and 165 m thick and consists of an arenaceous lithofacies with subordinate purple and green mudstones (Tab.4). The sandstone:mudstone ratio is 2.3:1, and the sandstones are fine to medium, moderately well-sorted quartzose sands with minor feldspar and clay matrix. They are stacked in units that form cliffs up to 40 m high and display trough cross-bedding and scour channels with basal mud pellet conglomerates. Calcareous concretions are common and near the top they reach 3 m across. The bottom and top of the formation is taken at the appearance and disappearance, respectively, of sandstones. According to Botha & Linström (1978), the Belmont Formation is unfossiliferous, but Kitching (1977) has recorded *Lystrosaurus* Zone fossils from beneath this unit, indicating that the Permian/Triassic boundary lies beneath the Belmont Formation, i.e. in the top part of the underlying Estcourt Formation. From primary structures in the sandstones,

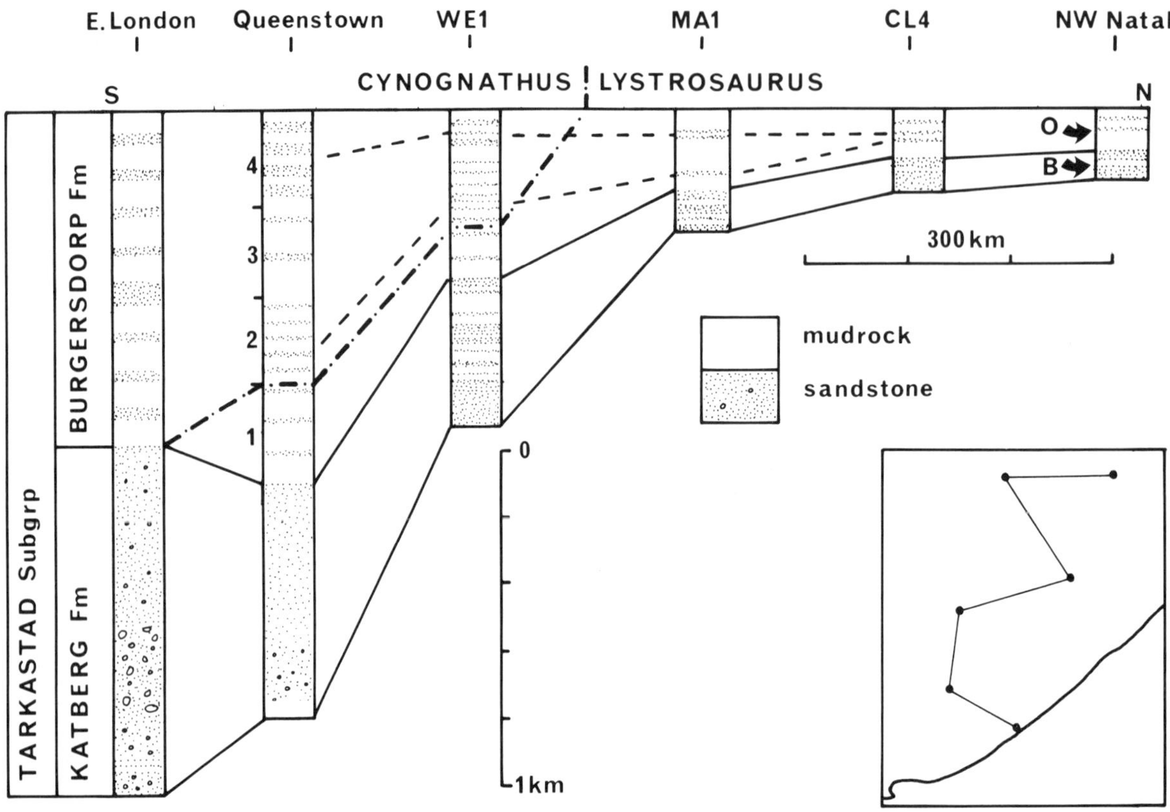


Figure 12. Diachronism in Scythian lithostratigraphic units in an approximate S-N section from East London to N.W.Natal. Solid lines are lithostratigraphic boundaries, thick dot-dash line is the biostratigraphic boundary between the *Lystrosaurus* and *Cynognathus* zones, dashed lines are suggested correlations of minor lithostratigraphic units (1-4) identified in the southern crop by Winter & Venter (1970). East London section is composite from the Kidd's Beach and Kwelera outliers. Other data sources: Marais & Johnson (1965); Johnson (1966); Kitching (1970, 1977); Botha & Linstrom (1978). Abbreviations: B = Belmont Fm.; O = Otterburn Fm.

Botha & Linström (1978) concluded that the Belmont Formation had a SE source. Discussing areas somewhat farther east, around Mooi River (see Fig.13), Theron (1975) identifies a more proximal facies of the Belmont Formation where coarse clastics are again encountered. He interprets this as a second major input point, and has reported clasts up to 18.6 cm across and weighing 4 kg. The rudaceous material is angular to subrounded, and consists of red garnet gneisses (dominant), milky quartz, microcline, red granite, and quartzite whilst the sandstones are generally arkosic.

Detailed descriptions are lacking for the areas on the SW edge of the Katberg Formation's present outcrop, but Kitching (1977), and Keyser & Smith (1979) record that here the formation becomes predominantly argillaceous with bright red, maroon and grey mudstones alternating with greyish-green and light bluish-green shales that frequently contain hard, light grey-green or variegated calcareous nodules. The latter often contain well-preserved vertebrate skulls or almost complete skeletons. Arenaceous beds are less important and consist of light grey-cream/yellow arkosic lenses of medium-coarse, well-sorted and well-rounded grains. They probably represent shoestring sandstone bodies deposited by meandering streams, and the whole sequence must lie close to the lithofacies boundary between the Katberg and Burgersdorp formations. Other fossils consist of amphibia, and rare leaf and stem impressions, which indicate a mixed *Glossopteris/Dicroidium* flora. No fish have been reported, and the only invertebrate fossils are 'millipede-like' organisms at two localities.

2.1.1.1.2 *Burgersdorp Formation.* The term Burgersdorp Formation was first introduced by du Toit (1905) (as the Burghersdorp Beds) for a sequence of red argillaceous rocks beneath the Stormberg Group in the Aliwal North district (Fig.6). At this time, their relationship to other parts of the Beaufort Group was not known, although vertebrate remains had previously been recorded from them (e.g. Seeley 1895), and in subsequent publications it appears that du Toit (1954) regarded the Burgersdorp Formation as equivalent to the whole of the upper division of the Beaufort Group. Johnson et al (1976) and Kitching (1977) have now recognized its true stratigraphic relationship as the highest unit of the Beaufort Group (Tab.3, Figs 10 & 12).

In the type area, the rocks are dominantly argillaceous, and according to Kitching (1977), the upper sections belonging to the *Cynognathus* Zone are more consolidated than those of the underlying *Lystrosaurus* Zone, which they appear to overlie conformably. The formation consists of blue-green, pale greyish-green, and typically red, purple, and maroon mudstones and shales, often variegated and interbedded with subsidiary, widely spaced yellow-buff weathering fine sandstones which are hard and pale bluish grey and grey-brown when fresh. The sandstones seldom produce topographic features. Clay pellet conglomerates are scattered throughout the succession, in association with ferruginous lenticles of coarse gritty sandstone which contain fragments of leaves and stems. Many of the argillaceous units have eroded tops, occasionally with sand-filled hollows, and have mudcracks and rossette shaped inclusions (?fossil desert roses).

Despite the lack of a means to effect a refined internal subdivision, there is good evidence for a broad four-fold lithological sequence; (i) a dominantly argillaceous lower section with thick (6-9 m) mudstones and thin (1.5 m) sandstones, which probably includes most of the *Lystrosaurus* Zone strata; (ii) a section with abundant clay pellet conglomerates with large numbers of complete vertebrate skulls and skeletal elements that usually occur within abraided mudstone clasts up to 45 cm across; (iii) a dominantly argillaceous section; (iv) an upper arenaceous unit. Kitching (1977) gives no thicknesses for these sequences which he recorded from outcrop, but similar sequences can be recognized in some of the SOEKOR borehole data published by Winter & Venter (1970). This is plotted in Figure 12 and shows a significant increase in the frequency and thickness of sandstone bodies in the upper part of the formation, as well as a sandier middle sequence. Percentages of sandstone in the units identified in Figure 12, and which may correspond to Kitching's (1977) subdivisions are: (i) 15 %; (ii) 30 % (the 'conglomeratic' facies); (iii) 15 %; (iv) 50 %.

A thick (ca.1 000 m) sequence of red rocks overlies the Katberg Formation in the Kwelera outlier north of East London. It consists of alternating thick maroon, purple and grey mudstones, and less abundant, but thick, cross-bedded pale maroon or greyish sandstones. Small pebbles occur in some of the sandstones, and those near the top have a 'glittering' aspect that is typical of the Molteno Formation. A particularly thick (70 m) and persistent sandstone unit occurs about 426 m above the base. As Johnson (1966) points out, this sequence typifies the dilemma of the lithostratigraphy of the Beaufort sediments: these beds are probably 'Beaufort', but their upper parts could be included in the Molteno Formation whereas the lower pebbly sandstones are similar to the underlying Katberg facies. No diagnostic vertebrate fossils have been found, but du Toit (1927) has recorded a *Dicroidium* flora from these rocks. On Figure 12 we have shown these beds as equivalent to the Burgersdorp Formation, but this correlation has no palaeontological justification and is based solely on their lithology and relationship to the underlying Katberg Formation (which is itself a correlation based on purely lithological evidence).

North of the type areas of the southern crop (Fig.6), the Burgersdorp Formation thins rapidly to about 250 m in the vicinity of Aliwal North (borehole WE1, Fig.12) and is reduced to 130-180 m in NW Natal. Here, Botha & Linström (1978) describe an argillaceous unit which locally grades laterally into, but regionally overlies the Belmont Formation. They have named this the Otterburn Formation. It consists of red and purple mudstones with minor greyish sandstones (sand:mudstone ratio is 0.33:1). Sandstones are more common in the top part of the sequence and in the north of the area, and consist of fine-medium quartzose sands with crossbedding that indicates both SE and SW sources. Scour channels and mud pellet conglomerates are common in the thicker sandstones which are usually part of upward fining cycles.

Although Botha & Linström (1978) regarded the Otterburn Formation as unfossiliferous, Kitching (1977) has recorded numerous vertebrate sites to the north and south of the Mooi River area. They all indicate *Lystrosaurus* Zone fossils (Fig.10).

A small exposure of Burgersdorp Formation has been recorded at the NE end of the Mngazana faulted outlier by the Geological Survey (1977) (Fig.108). It is underlain by Katberg Formation with *Lystrosaurus* zone fossils, but Kitching (1977) does not record any vertebrates from the red mudrock outcrops so their allocation to the Burgersdorp Formation must remain tentative and in a similar category to the undated red argillaceous beds at Kwelera.

Vertebrate fossils are common and well-preserved in the middle parts of the Burgersdorp Formation along the southern crop of the Karoo Basin, where specimens are most commonly found embedded in the red and purple argillaceous facies, particularly in fine siltstones and hard brown-weathering calcareous concretions in the mudstones. An interesting observation by Kitching (1977) is that the skulls are often sun-cracked, indicating that they had lain exposed to the elements before burial. Vertebrate remains are relatively rare and poorly preserved in the uppermost part of the formation (i.e. in the sandy facies), but fossil leaves have been recorded. Microfloral studies (Anderson & Anderson 1970) indicate that a varied flora existed throughout the rest of the sequence. Fish remains are rare: lung fish (dipnoans) have been reported from Burgersdorp and Rouxville (Gardiner & Jubb 1975) and from the same outcrop at the former locality, Griffith (1978) described a freshwater species of the genus *Saurichthys*. Amphibia are

Table 6. Broom's (1909a) biostratigraphic zonation of the Beaufort Group as amended by Watson (1914).

| Zone | Division | Series |
|---|---|---|
| *Cynognathus* zone | Upper division | Beaufort Series |
| *Procolophon* zone | | |
| *Lystrosaurus* zone | Middle division | |
| *Cistecephalus* zone | Lower division | |
| *Endothiodon* zone | | |
| *Tapinocephalus* zone | | |

relatively abundant. The only invertebrate fossils recovered are worm burrows and the relatively common freshwater lamellibranch *Unio karrooensis.*

The Burgersdorp Formation is unconformably overlain, and progressively overstepped northwards, by Stormberg Group sediments (Fig.10).

#### 2.1.1.2 *Biostratigraphy*

Following the first discovery of vertebrate fossils in the Beaufort Group in 1838 (Bain 1845) it soon became apparent that the Permo-Triassic rocks of the Karoo Basin contain an unrivalled record of the phylogeny of therapsid (mammal-like) and archosaur (including dinosaur) reptiles. Early workers on the sedimentary succession (e.g. Wyley 1859, and Jones 1867 who coined the term Beaufort Beds) also recognized the importance of the vertebrates as aids to stratigraphic studies, and a palaeontological zonation of the Karoo Supergroup was introduced by Seeley (1892). This scheme was modified by Broom (1909a), whose classification (Tab.6), with minor amendments remained in general use until the results of recent taxonomic revision were published by Kitching (1970). Further clarification by Kitching (1977) and Keyser & Smith (1979) has resulted in a reasonably sophisticated biostratigraphic zonation for the whole of the Beaufort Group (Tab.3), and for further details of the historical development of the present scheme the reader is referred to the summaries given in the last two references.

The drastic pruning of synonyms that has been undertaken during the formulation of the present biostratigraphic zonal scheme has also rationalized the time ranges of many species and now allows a good measure of correlation between the bio- and lithostratigraphic classifications of the Beaufort Group. This is shown in Table 3 which is based primarily on Keyser & Smith (1979). The Tarkastad Subgroup contains two biozones (the *Lystrosaurus* and *Cynognathus* zones), but there is evidence that in Natal, at least, the *Lystrosaurus* Zone extends down into the upper part of the Adelaide Subgroup (the Estcourt Formation). Assuming for the moment that the Permian/Triassic boundary coincides with the *Dicynodon/Lystrosaurus* zonal boundary (this is a question we will review later), then the base of the Triassic lies at the base of the Tarkastad Subgroup in the south of the Karoo Basin, and within the upper part of the Adelaide Subgroup in Natal. Diachronism is also evident when the biostratigraphic zonations are applied within the Tarkastad Subgroup, with greater thicknesses of argillaceous facies (Burgersdorp) occurring in the *Lystrosaurus* Zone away from the Queenstown area.

2.1.1.2.1 *Lystrosaurus Zone.* Keyser & Smith (1979) have designated the Lootsberg Pass (about 31°48′S, 24°52′E), near the western extremity of its distribution, as the type area for the *Lystrosaurus* Zone (which is equivalent to the *Lystrosaurus-Thrinaxodon* Assemblage

Table 7. Characteristic vertebrate taxa of the Lower Triassic (Scythian) (after Keyser & Smith 1979).

| | | | A.*Lystrosaurus* zone | B. *Cynognathus* zone |
|---|---|---|---|---|
| Reptilia | | | | |
| Order | Suborder | Infraorder | | |
| Cotylosauria | Diadectomorpha | | *Procolophon trigoniceps* Owen 1876 | |
| Therapsida | Anomodontia | Dicynodontia | *Myosaurus gracilis* Haughton 1917 | *Kannemeyeria simocephalus* (Weithofer 1888) |
| | | | *Lystrosaurus murrayi* (Huxley 1859)<br>*L.declivis* (Owen 1860)<br>*L.oviceps* Haughton 1915<br>*L.platyceps* (Seeley 1898) | *Kombuisia frerensis* Hotton 1974 |
| | Theriodontia | Cynodontia | *Thrinaxodon liorhinus* Seeley 1894 | *Diademodon tetragonus* Seeley 1894<br>*Cynognathus crateronotus* Seeley 1895<br>*Trirachodon berryi* Seeley 1894 |
| | | Bauriamorpha | | *Bauria cynops* Broom 1909 |
| Prolacertilia | Parathecodontia | | *Prolacerta broomi* Parrington 1935 | |
| Thecondontia | Proterosuchia | | *Proterosuchus vanhoepeni* (Haughton 1924) | *Erythrosuchus africanus* Broom 1905 |
| Amphibia; Labyrinthodonta | | | *Lydekkerina huxleyi* (Lydekker 1889) | |

Table 8. Comparison of composition of vertebrate faunas in Upper Permian and Lower Triassic strata of the Karoo Basin. Compiled from faunal lists in Kitching (1977).

| Taxa | | Zone: Dicynodon | | Zone: Lystrosaurus | | Zone: Cynognathus | |
|---|---|---|---|---|---|---|---|
| Order | | No. spp. | % | No. spp. | % | No. spp. | % |
| Cotylosauria | | | | | | | |
| Suborder | Diadectomorpha | 4 | 4 | 3 | 10 | 3 | 14 |
| Millerosauria | | 4 | 4 | – | | – | |
| Therapsida | | | | | | | |
| Suborder | Anomodontia | | | | | | |
| infra-order | Dicynodontia | **30 | 32 | ** 8 | 27 | 2 | 10 |
| Suborder | Theriodontia | | | | | | |
| | Gorgonopsia | 34 | 36 | – | | – | |
| | Scaloposauria | 9 | 9 | 6 | 20 | – | |
| | Therocephalia | 7 | 7 | 1 | 3 | – | |
| | Cynodontia | 3 | 3 | 4 | 13 | ** 5 | 24 |
| | Bauriamorpha | – | | – | | 6 | 29 |
| Thecodontia* | | | | | | | |
| Suborder | Proterosuchia | – | | 2 | 7 | 2 | 10 |
| Eosuchia | | | | | | | |
| Suborder | Younginiformes | 4 | 4 | 3 | 10 | – | |
| Prolacertilia | | | | | | | |
| Suborder | Parathecodontia | – | | 2 | 7 | – | |
| Rhynchocephalia | | | | | | | |
| Suborder | Sphenodontia | – | | – | | 1 | 5 |
| Suborder | Rhynchosauria | – | | – | | 2 | 10 |
| Araeoscelidia (Protosauria) | | – | | 1 | 3 | – | |
| | | 95<br>52 genera<br>40 monospecific | | 30<br>21 genera<br>17 monospecific | | 21<br>20 genera<br>19 monospecific | |

* Early Archosauria
** Zone fossil

Zone of SACS (1980), although the latter authors do not list *Thrinaxodon* amongst the 'characteristic' fossils of the zone). Its lower and upper limits are defined as the horizons at which the genus *Lystrosaurus* appears and disappears, respectively. Remains of the relatively large (skull up to 50 cm long) herbivorous dicynodont are abundant, and Kitching (1977) records seven species of it (see Appendix 1), whilst the only other species of the Infra-order Dicynodontia (*Myosaurus gracilis*) is rare. According to Keyser & Smith (1979), *Procolophon trigoniceps* is also found throughout the *Lystrosaurus* Zone, and is particularly abundant in the upper part. Kitching (1977) records five species of amphibia from this zone (all labyrinthodonts), but no fish remains are known. Fossil tree trunks (*Dadoxylon* sp.) are common, and are particularly abundant in the northern part of the outcrop (north of about 29°S).

A comparison of faunal composition shows that the *Lystrosaurus* Zone vertebrate assemblages differ significantly from those of the underlying *Dicynodon* zone. Basing our statistics on the faunal lists given by Kitching (1977) (Tab 8) a sharp drop in the number of species and genera occurs across (upwards) the zonal boundary (95 spp. and 52 genera to 30 spp. and 21 genera, respectively). Although no reptile species and genera are common to the two zones, both are dominated at the species level by the therapsids (mammal-like reptiles). In the *Dicynodon* Zone, however, species of Dicynodontia and Gorgonopsia dominate (32 and 36 % of the species, respectively) and the genus *Dicynodon* is represented by 21 species (22 % of all species extant). In contrast, within the *Lystrosaurus* Zone, no Gorgonopsia occur and the fauna is dominated at the species level by the Dicynodontia and Scaloposauria (27 and 20 % of the species, respectively). In addition, the dominant dicynodont is *Lystrosaurus* itself with seven species (23 % of all species extant). A further significant difference is the increase in importance of the cynodonts from 3 % in the *Dicynodon* Zone to 13 % in the *Lystrosaurus* Zone. A complete list of species recorded by Kitching (1970, 1977) from the *Lystrosaurus* and *Dicynodon* zones is given in Appendix 1. It should be noted here that Cluver's (1977) demonstration that the name *Daptocephalus leoniceps* is a junior synonym of *Dicynodon lacerticeps* Owen has necessitated changing the name of Kitching's (1970) *Daptocephalus* zone to *Dicynodon* Zone (Keyser & Smith 1979), but that the two zones are identical in concept.

Distribution of *Lystrosaurus* Zone strata are shown in Figure 6. They crop out around the whole of the Mesozoic outlier, but are thickest along the southern crop (Figs 10 & 12). In the SE it seems that the zone coincides with the Katberg Formation and Palingkloof Shale, and if this situation holds for the coastal outliers, the *Lystrosaurus* Zone strata may here be over 1 000 m thick. In the Queenstown district, the upper part of the *Lystrosaurus* Zone is argillaceous, and probably consists of 762 m of Katberg Formation and about 250 m of the overlying Burgersdorp Formation. At present, this correlation has been achieved only by comparing Kitching's (1977) fossil locality map with Johnson's (1966) lithostratigraphic map. Further reductions in thickness take place westwards and northwards. In Natal, *Lystrosaurus* Zone assemblages occur beneath the Belmont Formation (northern facies of the Katberg Formation) in greenish shales and mudstones (Estcourt Formation). These relationships are shown in Figure 10 where the *Lystrosaurus* Zone includes argillaceous rocks above and below the Katberg Formation and illustrates the diachronous relationship of the lithofacies involved.

2.1.1.2.2 *Cynognathus Zone.* This zone was one of those orginally set up by Broom (1909a), and was retained by Kitching (1970, 1977) in his revisions of the Beaufort Group biostratigraphy. Keyser & Smith (1979) have argued for renaming it the *Kannemeyeria* Zone on the grounds that specimens of *Cynognathus* are relatively rare and that the genus still contains many synonyms. They also make the point that it would be appropriate to follow the practice in the other Beaufort Group zones of using a member of the Anomodontia. At present we do not see that any of these are strong enough reasons for deleting a well-established term, especially since the taxonomic validity of *Cynognathus* itself is not in question, and the proposed new zone is an exact time equivalent of the *Cynognathus* Zone (see Keyser & Smith 1979, fig.2.2). The *Cynognathus* Zone as used here is equivalent to the *Kannemeyeria-Diademodon* Assemblage Zone proposed by SACS (1980).

Compared with the underlying *Lystrosaurus* Zone, the *Cynognathus* Zone vertebrate assemblages are relatively sparse at the species level, indicating a further drop in the number of species up the succession (Tab.8): 30 spp. and 21 genera to 21 spp. and 20 genera, respectively. These figures also show that whilst the number of genera in the two zones is not significantly different, there is a slight increase in the number of monospecific genera (17 in the *Lystrosaurus* Zone and 19 in the *Cynognathus* Zone). Several other significant differences are apparent between the two zones, which in some cases indicate the continuation of trends already evident in the older strata. For instance, the importance at the species level of the Dicynodontia continued to decline, and drops from 27 % in the *Lystrosaurus* Zone to 10 % in the *Cynognathus* Zone where only two species occur (*Kannemeyeria simocephalus* and *Kombuisia frerensis*). Also no representatives of the Scaloposauria are known, whilst a new Infraorder, the Bauriamorpha appears and becomes the dominant element in terms of number of species (29 %). The cynodonts become the second most important group in the *Cynognathus* Zone (24 %), where they reach their greatest diversity, having accounted for only 3 % of the number of species in the *Dicynodon* Zone. Overall, the therapsid reptiles maintain their dominance of the vertebrate faunas (63 % in both the *Lystrosaurus* and *Cynognathus* Zones), but other groups which show increases in their relative importance within *Cynognathus* times are the early archosaurs of the Suborder Proterosuchia (*Erythrosuchus africanus* and *Euparkeria capensis*) and the Rhynchocephalia. Only one species has been reported as common to the *Cynognathus* and *Lystrosaurus* zones: *Diademodon tetragonus* Seeley 1894. Kitching (1977) questions the validity of the locality reference for the specimen from the *Lystrosaurus* Zone (Rhenosterberg), so in all probability this species is restricted to the *Cynognathus* Zone.

Kitching (1977) records nine species of amphibia from the *Cynognathus* Zone, and notes that fish remains are rare, the only substantial discoveries being northeast of Aliwal North (Broom 1909b) and at Burgersdorp (e.g. Gardiner & Jubb 1975, Griffith 1978), although Kitching (1977) himself has found isolated dental plates from a number of localities over the whole outcrop.

Strata of *Cynognathus* Zone age occur only in the western and southern parts of the Mesozoic Karoo basin (Fig.6) because they are cut out northwards by the overstep of the Stormberg Group, and in the NE parts of Figure 6 ?Middle and Upper Triassic strata rest upon *Lystrosaurus* Zone sediments (Fig.10). The *Cynognathus* Zone is restricted to the argillaceous Burgersdorp Formation and reaches a probable maximum thickness of about 1 000 m in the SE of the basin. It thins north and westwards to an estimated 800 m east of Queenstown, and a measured 550 m closer to that town. Less than 200 m occur in the vicinity of Aliwal North, and 50 m at outcrop in western Lesotho. Vertebrate fossils are not common in the *Cynognathus* Zone, but tend to be relatively more so in the lower parts of the Burgersdorp Formation.

2.1.1.3 *Permian/Triassic boundary within the Beaufort Group*

There has been general agreement by a wide cross section of workers in other parts of the world, that the zonal sequence for the Beaufort Group vertebrates first applied by Broom (1909a) and recently refined by Kitching (1970 et seq.) represents a valid evolutionary scheme and that the individual zones are isochronous (e.g. Cox 1967, Romer 1970, Rayner 1971, Pomeral 1975, Banks 1978). Application of the zonal concepts in other parts of the world (Antarctica, South America, Madagascar, India, China and Russia) have successfully demonstrated its usefulness for intercontinental correlation of Permo-Triassic terrestrial vertebrate-bearing strata. One area, however, where there is some uncertainty is that of the location of the Permian/Triassic boundary. Because of the complete succession and relative abundance of fossils, the South African Beaufort sequence is frequently used as a reference type, and the local convention of taking the *Dicynodon* (previously *Daptocephalus*)/*Lystrosaurus* zonal boundary as the Permian/Triassic boundary has been widely accepted, although not without some dissention (e.g. Cosgriff 1965, Anderson & Anderson 1970). Macro- and microfloral zonations established by Plumstead (quoted in Falcon 1975) and Hart (1970), respectively add some weight to this assumption, but by and large it is a convention that is not based on unequivocal evidence.

This is borne out when reviewing the stratigraphy of similar-aged rocks in other parts of the world, where one finds a circular argument being employed; viz, lowermost Triassic beds are assigned this age because they contain *Lystrosaurus* Zone fossils. Clearly, this dilemma can only be resolved by relating a *Lystrosaurus* zonal assemblage to a well-dated sequence, which for confident correlation must be one that has been unequivocally tied in with the standard zonal ammonites. Unfortunately, such an ideal association has not yet come to light, and in all localities from where *Lystrosaurus,* the zonal type, has been reported, other faunal and floral taxa do not provide an unequivocal age. The areas in question are: Antarctica (Fremouw Formation), NE India, and in Sinkiang Province, China (Tunghung Shan beds), and of these, the Indian outcrops in the Damodar Valley are the best-studied. Here, in the Raniganj Coalfield, the Raniganj Formation, with undoubted upper Permian micro- and macroflora (at least 46 species of *Glossopteris,* Surange 1975) and fresh-water lamellibranchs (*Anthraconauta simplex,* Ghosh & Mitra 1970) is conformably overlain by the Panchet Formation, which is divided into lower grey/green, and upper red parts. The upper part contains a reptilian fauna with *Lystrosaurus,* whereas the lower part has a mixed *Glossopteris-Dicroidium* flora (Chowdhury et al 1975) and Upper Permian-Lower Triassic estheriids (Tasch et al 1975 take the association of *Schizoneura* and *Glossopteris* to nominally mark the transition between Permian and Triassic strata). In addition, Bharadwaj (1970, fig.2) finds a marked change in microfloral elements across the lower/upper Panchet Formation boundary. In the light of this type of evidence all these workers concluded that the *Lystrosaurus*-bearing upper Panchet beds represent the lowermost Triassic (i.e. Griesbachian subzone), but the *main* evidence that they cite is the fact that the beds in question contain *Lystrosaurus,* and the *Lystrosaurus* Zone of South Africa is Lower Triassic. Neither of the other two areas from where *Lystrosaurus* has been recorded; Antarctica (Colbert 1970) and Sinkiang (Cox 1973, p.81 and Romer 1975, p.470) has produced evidence based on other taxa for a Scythian age for the *Lystrosaurus* assemblage.

We are reduced, therefore, to using indirect methods of correlation. At first sight, an obvious area would seem to be between the Upper Permian to Triassic marine sections in the Salt Range of Pakistan, and the Damodar Valley strata in India. Palynological studies by Balme (1970) (summarized by Anderson & Anderson 1970), and Chowdhury et al (1975) reveal five genera to be common to both areas, but a comparison of the vertical ranges of these genera produces no refinement in the location of the Permian-Triassic boundary in the Raniganj/Panchet sequence, even though Balme (1970) detects a floral discontinuity across the Chhidru Formation (Djulfian, Upper Permian)-Kathwai member (Griesbachian, Scythian) junction in the Salt Range (Waterhouse 1976, tab.2, considers both these rock units to be Upper Permian in age). It is probable that a better correlation will be achieved when possible synonyms have been detected in the floral lists.

One of the few localities from which marine Lower Triassic invertebrates have been described in close association with tetrapods is Scoresby Land in central East Greenland, where the Permian-Triassic boundary is well exposed in the *Martinia* Limestone-Wordie Creek Formation sequence. One level within the lower Wordie Creek Formation, consisting of shallow marine arkoses and conglomerates, contains the following assemblage of stegocephalian amphibia:

*Wetlugasaurus groenlandicus* Säve-Söderbergh 1935
*Stoschiosaurus* (3 spp.)
*Lyrocephalus* (4 spp.)
*Tupilakosaurus heilmani* Nielsen 1954

plus several fish, and the lamellibranch *Claraia,* (Bendix-Almgreen 1976). It also contains a good ammonite fauna which can be confidently assigned to the *Proptychites rosenkrantzi* Zone, placing it unequivocally in the Upper Griesbachian (i.e. the lower part of the Scythian) (Spath 1934, Birkland & Perch-Nielsen 1976). Cox (1973, p.80) claims that *Wetlugasaurus* (or its synonym *Parotosaurus*) occurs in the *Lystrosaurus* Zone of the Karoo, but we can find no record of this (e.g. Haughton 1969, Kitching 1977, du Toit 1954, Anderson & Anderson 1970). It does occur, however, in the *Cynognathus* Zone in South Africa (*Parotosaurus haughtoni* Bolli & Schröder, see Romer 1947) (Anderson & Anderson 1970), as well as in the Andavakaera region of northern Madagascar (*Parotosaurus (Wetlugasaurus)* sp.) (associated with Griesbachian ammonites, although the exact relationship is not clear); the Gosford Subgroup (*Parotosaurus wadei,* whose closest relative is claimed to be *P. nasutus* from the Middle Bunter (mid-Scythian) of Germany) and Hawkesbury sandstone (*Parotosaurus brookvalensis*) of the Sydney Basin in eastern Australia (see Banks 1978); Spitzbergen (*Parotosaurus* sp., Anderson & Anderson 1970); and

Table 9. Alternative positions of Karoo vertebrate zones in relation to Permian-Triassic boundary.
A – placing the Griesbachian substage in the Triassic (e.g. as followed by Grant & Cooper 1973, Birklund & Perch-Nielsen 1976) implies that the *Cynognathus* Zone and probably all of the *Lystrosaurus* Zone belong in the Triassic. This is the scheme we adopt, and nominally place the Permian-Triassic boundary at the base of the *Lystrosaurus* Zone.
B – if the suggestion by Waterhouse (1976) that the Griesbachian substage belongs in the Permian is generally adopted then the Permian-Triassic boundary will fall within the *Cynognathus* Zone.

| | | | | | |
|---|---|---|---|---|---|
| Lower Triassic | Scythian | Dienerian | | *Cynognathus* Zone at least in part Upper Griesbachian in age | |
| | | Griesbachian | Upper | | |
| | | | Lower | *Lystrosaurus* Zone at least in part Lower Griesbachian in age | base of *Lystrosaurus* Zone nominally taken as Permian-Triassic boundary, but there is only circumstantial palaeontological evidence for this |
| Upper Permian A | Dorashamian | | | | |

| | | | | |
|---|---|---|---|---|
| Lower Triassic | Scythian | Dienerian | | *Cynognathus* Zone |
| Upper Permian B | Dorashamian | Griesbachian | Upper | |
| | | | Lower | *Lystrosaurus* Zone |

European Russia (*Parotosaurus* spp.) in Zones V and VII of Olson (1957) and Ochev (1966).

Substantive data for a Triassic (if not specifically Scythian) age for the *Cynognathus* Zone come from the distribution of the fish genus *Saurichthys* Agassiz 1834. Species of *Saurichthys* have a world-wide distribution (Beltan & Tintori 1981) in both fresh and marine Triassic sediments, and have been recorded in the following localities: Lower Triassic – British Columbia, east Greenland, SW Spitzbergen, Germany, France, Nepal, Madagascar, and east Australia; Middle Triassic – Italy, Spain, Turkey, and east Australia; Upper Triassic – Austria, and Italy. The southern hemisphere records are predominantly freshwater (Australia) and paralic (Madagascar), and the record of one jaw in *Cynognathus* Zone strata of the Burgersdorp Formation (Griffith 1978) that have also yielded lung fish, indicates that this South African species was also non-marine.

The weight of this evidence, especially that from Greenland, suggests that the *Cynognathus* Zone is upper Griesbachian in age, i.e. the upper part of the lowermost stage of the Scythian (which is divided into four substages: Griesbachian, Dienerian, Smithian, and Spathian). Because rocks of the *Cynognathus* Zone conformably overlie the *Lystrosaurus* Zone in the Cape, it is reasonable to suppose that the *Lystrosaurus* Zone is all or partly of Lower Griesbachian age, and that the faunal change at the *Dicynodon*/*Lystrosaurus* boundary does indeed mark the Permian/Triassic boundary as accurately as it is likely to be defined in the Karoo in the near future. Two points of local lithostratigraphic interest flow from this conclusion. Firstly, the Permian/Triassic boundary (i.e. the base of the *Lystrosaurus* Zone) in the Cape lies at the base of the Palingkloof Shale, which is about 50 m below the Katberg Formation (SACS 1980). As we have already noted, Stavrakis (1980b) correlated this shale with the Burgersdorp Formation, and not the Balfour Formation (as do SACS 1980, and Keyser 1981). This seems desirable and logical, as it maintains all the Balfour Formation within the uppermost Permian *Dicynodon lacerticeps* Zone. Secondly, the latter zone is presumably Dorashamian in age, and not Djulfian (as proposed by Anderson 1981) because there is no lithological or faunal evidence for a non-sequence between the Balfour and Palingkloof sediments, so that the Permian/Triassic boundary in this area must be considered conformable, and not with a significant hiatus as suggested by Anderson (1981, fig.4).

These relationships are shown in Figure 5 and Tables 3 and 9A, but their nominality must be stressed, and it is important to note that three workers have recently produced data which throw some doubt upon it. Waterhouse (1976) in his survey of worldwide Permian marine classification claims that the Griesbachian in fact belongs to the uppermost Permian. If this re-appraisal gains acceptance then the alternative classification (Tab.9B) would apply, i.e. the Permian/Triassic boundary probably lies within the *Cynognathus* Zone and it might be convenient to take the base of the *Cynognathus* Zone as the nominal base of the Triassic. This is in fact what Cosgriff (1965) has already suggested, although if his scheme is applied and Waterhouse's ideas gain acceptance, then the *Cynognathus* Zone would also have to be assigned to the Permian.

This final position is precisely what Stapleton (1978a) has tentatively advocated. On the basis of palynological studies from the Steynberg-Burgersdorp area he claims that the Permian/Triassic boundary lies somewhere between the upper part of the *Cynognathus* Zone and the lower part of the Molteno Formation (which is generally considered to be Ladinian-Carnian in age). It is difficult to assess Stapleton's data however, because although he collected samples from the Beaufort Group and the Molteno Formation, he considers only the 'total' flora from all three localities. Of the 16 species of spores and pollen that he was able to identify, seven are reckoned to be of Upper Permian age, four of Triassic age and five occur in both periods. It should be noted that the appearance of one species identified by Stapleton (*Densoisporites playfordii*) is considered by Banks (1978) to mark the lowermost Triassic in Queensland. In view of the extremely poor preservation of the material examined by Stapleton, we feel that this evidence is not sufficiently strong to warrant the re-classification suggested by this palynological study (Fig.9, column 3).

2.1.1.4 *Palaeogeography*

Rocks of the *Lystrosaurus* and *Cynognathus* zones were water-lain in a non-marine environment. The arenaceous sediments represent fluvial and deltaic sands, with the high-energy proximal facies lying in the East London and Mooi River areas, a conclusion suggested by palaeocurrent direction studies (Theron 1975) and grain size distribution (du Toit 1954, Johnson 1966, Mountain 1962, Theron 1975). The argillaceous rocks represent lower energy alluvial flat sediments, mostly deposited in large bodies of standing water, that in *Cynognathus* Zone times were prone to desiccation. Periodic flooding, accompanied by short-lived higher energy episodes is indicated by local penecontemporaneous erosion and channelling.

Theron (1975) stressed that the Beaufort Group as a whole represents an essentially complete sedimentary sequence, and that his palaeocurrent studies showed no important alterations throughout Upper Permian to Lower Triassic time. From a consideration of the gross distribution of the sandstone units in time, it is obvious, however, that the energy of the depositional medium changed periodically. We present one palaeogeographical reconstruction (Fig.13) which has been primarily designed to illustrate *Lystrosaurus* Zone times, but which can be modified to apply to *Cynognathus* Zone times.

The main feature is the large alluvial fan/delta emanating from the East London area and giving rise to the Katberg arenaceous facies. Rock types found in the rudaceous component of this facies, as well as the prevalence of garnet in the Tarkastad Subgroup sandstones, which are typically arkosic, indicate that the provenance area lay SE of East London, and consisted of granitic/high grade gneissic terrains. Other heavy minerals in the sandstones include: zircon, rutile, tourmaline, apatite, monzanite and titanite. The sandstones of the Mooi River area are less mature than the Katberg Formation sandstones, and probably indicate a source of pyroxene garnet granulite and garnet biotite schists in the Natal coastal basement complex. Basement rock types on the Falkland Plateau, which lay adjacent to the SE Coast at this time, consisted of high grade gneiss (quartz biotite semipelites), and coarse grained granite pegmatites (DSDP site 330, Bolli et al 1978). As discussed in Chapter 4, the basal Mesozoic sediments at this site contain garnet, tourmaline and zircon, and may therefore have been derived, at a later date, from the same source. The origin of the lignite clasts in the Katberg facies is not known, but must have lain nearby; possibly eroding coal measures or upland peat. It is important to note that there is no evidence at this time that Cape or lower Karoo Supergroup strata in the Cape Trough were being eroded, i.e. the main Cape orogeny uplift had not yet begun on any important scale.

Basin subsidence was strongly asymmetric, with downwarping increasing rapidly south of a line that ran approximately east-west just south of Aliwal North (Figs 10 & 13), whereas farther north, a more stable platform subsided gently by southward tilting. At its maximum extent (during the deposition of the Katberg Formation sandstones), the alluvial fan extended far west, north and northeast into the basin, interfingering with the distal, argillaceous alluvial plain facies between periods of heavy discharge and lower energy input. Van Dijk et al (1978) suggest that this distal region frequently contained large shallow lakes. The continuity in the arenaceous Katberg facies between the eastern Cape and NW Natal indicates that early in *Lystrosaurus* Zone times northward and southwestward movement of fluvial material from the two main input points resulted in the coalescing of the distal parts of the two alluvial fans. This situation did not persist, and well before the end of *Lystrosaurus* Zone times the Mooi River source appears to have become less significant and predominantly argillaceous sediments were accumulating between the two areas.

In the SW part of the main Karoo Basin (i.e. west of 25°E), no Scythian rocks remain, having presumably been stripped off by erosion, but if the situation envisaged by Theron (1975) for Upper Permian times persisted into the Lower Triassic, then here sediment movement was in an easterly to northeasterly direction from an input point in the western Cape. Distal facies from this source would have merged with that at the western end of the Katberg alluvial fan. The situation in the Cape Trough is not clear, but it seems likely that a fresh water basin lay over the southern Cape into which flysch-type sediments were deposited (the lower Beaufort Group rocks have sole markings) (Theron 1975).

During *Lystrosaurus* Zone times, the vertebrate fauna was dominated by numerous species of the medium sized hippopotamus-like herbivore *Lystrosaurus* which, together with amphibia of various sorts in-

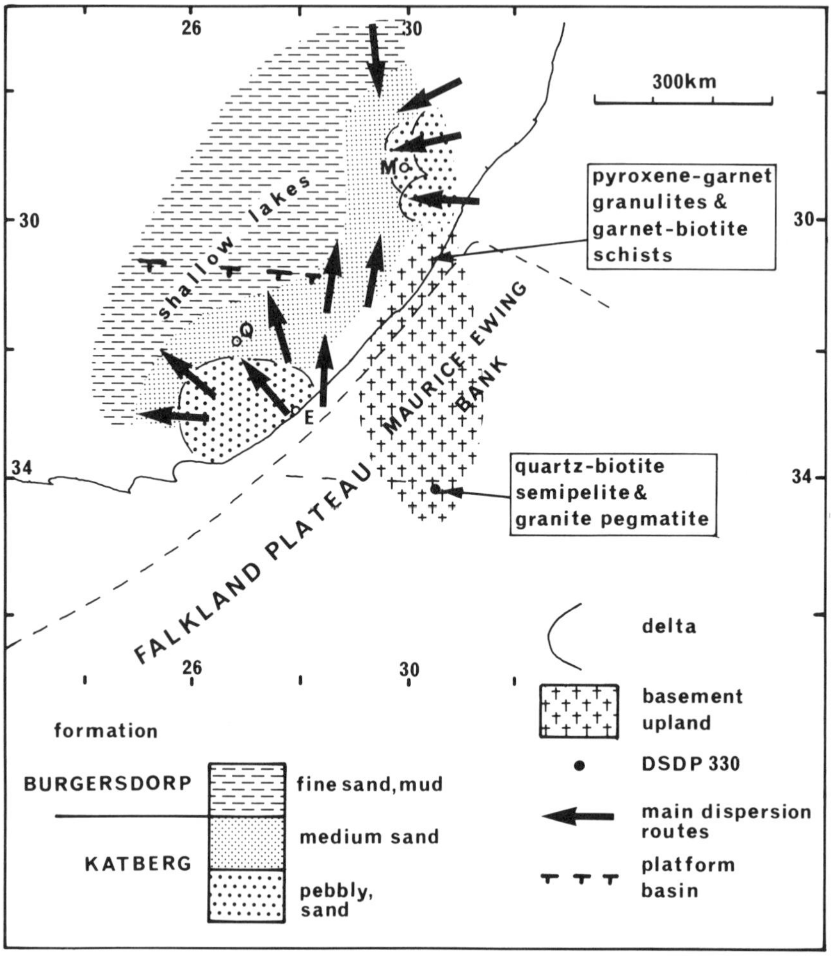


Figure 13. Palaeogeography of main Karoo Basin in Scythian times. See text for discussion. Dispersion routes from data in Theron (1975) and van Dijk *et al.* (1978). Abbreviations: E = East London; M = Mooi River; Q = Queenstown.

dicates an abundance of water in the area. Its skeletons are usually complete in the northern part of the outcrop, whereas in the south they are fragmented, presumably a reflection of different energy conditions. Anderson & Anderson (1970) have made an attempt to assess the environmental significance of the occurrence of the various skeletons, and point to the fact that particular associations commonly suggest a catastrophic overwhelming of the beasts by mud and silt during floods (e.g. the amphibian *Lydekkerinia* occurs at two localities, and in both cases complete skeletons are found in association with complete skeletons of *Lystrosaurus*). An alternative explanation for such death assemblages has been given by Pomeral (1975) who describes the crowding together and expiration of amphibious animals in shrinking muddy pools during bad droughts in East Africa. These two alternatives are diametrically opposed – drought and flood. The fragmental and rolled nature of much of the skeletal material points to transportation after death by running water, and in many instances penecontemporaneous exhumation can be demonstrated (e.g. Kitching 1977, p.27).

An assessment of the relative importance of herbivores to carnivores and aquatic to terrestrial types in the vertebrate populations indicates that aquatic types dominate, and that herbivores were most abundant overall, although it is important to bear in mind that the potential for preservation of the aquatic types would have been considerably greater. These conclusions are based on Benton (1979) and Table IV in Anderson & Anderson (1970). The flora was a mixed *Glossopteris-Dicroidium* assemblage, but it is difficult to assess its abundance because leaves and stems are rarely preserved, although it may be significant that no large terrestrial herbivores are known from the *Lystrosaurus* Zone.

To summarize for *Lystrosaurus* Zone times, we envisage extensive alluvial flats with large bodies of standing water (shallow lakes) and swamp land crossed by meandering streams inhabited by a predominantly aquatic herbivorous fauna. High ground lay to the east and

SE with major rivers entering the area near East London and from eastern and northern Natal. Where these debouched onto the lower ground, large alluvial fans were constructed. Extensive coniferous forests grew on the periphery of the basin.

Projecting our palaeogeography into *Cynognathus* Zone times, the following differences are apparent:

i) an overall lower energy sedimentary environment suggesting a lower runoff from the source area;

ii) evidence of desiccation in the sedimentary record; mudcracks, desert rose crystals, sun-cracked skulls, i.e. a deterioration in the climate;

iii) an abrupt change in the composition of the vertebrate fauna.

The area of sedimentation shrunk in *Cynognathus* Zone times, so that it extended only a short distance onto the stable platform (Fig.13). This may have been due to a slight uplift of the Kaapvaal Craton which foreshadowed the major tectonic events that occurred at the end of *Cynognathus* Zone times with the rise of the Cape Fold Belt.

Sedimentologically, the most important differences between the rocks of the two zones in the more argillaceous nature of the *Cynognathus* strata: the *Lystrosaurus* Zone rocks contain numerous lenticular sandstone bodies; whereas sandstones in the *Cynognathus* Zone are widely spaced. Kitching (1977) notes that the mudrocks of the younger zone are more consolidated. These changes, which are taken to indicate the encroachment of the alluvial flats towards the sediment source areas, and a general lowering of the energy of the sedimentary environment either by climatic or topographic changes or both, were established in the more distal areas before the end of *Lystrosaurus* Zone times. The alluvial flats were frequently desiccated, and this alteration in climate and water supply probably accounts for the differences in the vertebrate fauna which was now dominated by terrestrial types, with a mixture of abundant largish herbivores (e.g. *Diademodon* and *Kannemeyeria*) and carnivores (e.g. *Cynognathus* and *Erythrosuchus*). The abruptness of the change, with no intermediate types, suggests to Kitching (1970) that the new forms migrated into the area. It also implies that terrestrial vegetation was, at least locally, abundant to support the increase in numbers of large browsing and grazing types. A return to somewhat higher energy environments in late *Cynognathus* Zone times (shown by the increase in sandy material) is marked by a decrease in the number and quality of vertebrate remains. Whether this represented an environmental deterioration as far as the vertebrates were concerned, or merely a less-conducive medium for preservation, is not known.

To summarize for *Cynognathus* Zone times, we envisage extensive alluvial flats which were crossed by slow streams associated with bodies of standing water, but which were periodically dry land, either desiccated or covered in soft stemmed vegetation and were inhabited by a varied terrestrial vertebrate fauna.

Sedimentation was temporarily brought to a close by slight uplift, synchronous with the main phase of uplift in the Cape Trough. The effect was naturally greatest in the south, where evidence of erosion and hiatuses between the Beaufort and overlying Stormberg Group strata are well displayed.

2.1.1.5 *Uranium mineralization in the Beaufort Group*

Uranium mineralization was first noted in rocks of the Beaufort Group in 1967 during radiometric logging of SOEKOR boreholes, and active exploration was begun in 1969. Up to 1978, 191 localities had been identified as having uranium occurrences. So far, all the published localities in the Beaufort Group have been in pre-Tarkastad (i.e. Upper Permian) strata, but Turner (1979) noted that continued exploration has led to recent discoveries of mineralization at unspecified localities within the Stormberg Group. These have been in the Elliot Formation of the eastern Orange Free State, and the Clarens Formation and Drakensberg Group lavas in Lesotho. In view of the distinct possibility of further finds within the Lower Triassic Beaufort Group, and the general similarity between the Upper Permian and Elliot Formation red beds, we feel justified in including a short account of the uranium mineralization of the Adelaide Subgroup. Unless otherwise stated, details of sedimentation and mineral occurrences are from Turner (1978b and 1979).

All the reported finds within the Beaufort Group have been west of 25°E (Fig.14) and the most important area lies in a zone in the SW Karoo Basin between Sutherland and Beaufort West. In this area, the Beaufort Group comprises two formations (Abrahamskraal and Teekloof) that make up the Adelaide Subgroup. These rocks are Upper Permian in age and have been divided into five vertebrate zones (Tab.3). The Abrahamskraal Formation has a maximum thickness of 1 800 m and consists predominantly of sandstones with minor mudrocks and chert bands. The sandstones are up to 25 m thick, fine grained, dark grey in colour and contain numerous calcareous nodules. They display horizontal lamination, ripple cross-lamination and cross-bedding and contain abundant plant material, vertebrate bone fragments, and mudstone and calcareous pellet conglomerates. Argillaceous rocks in the Abrahamskraal Formation are massive and unlaminated bluish grey, occasionally maroon and purple mudstones and silty mudstones which are interbedded with greenish grey siltstones and fine sandstones. In the Beaufort West area the upper part of the Abrahamskraal Formation contains a 160 m sequence of regularly interbedded sandstones and mudstones. The sandstones are generally less than 8 m thick and lenticular. They weather a distinctive yellow or golden-brown colour (the so-called zone of golden sands) and contain numerous calcareous

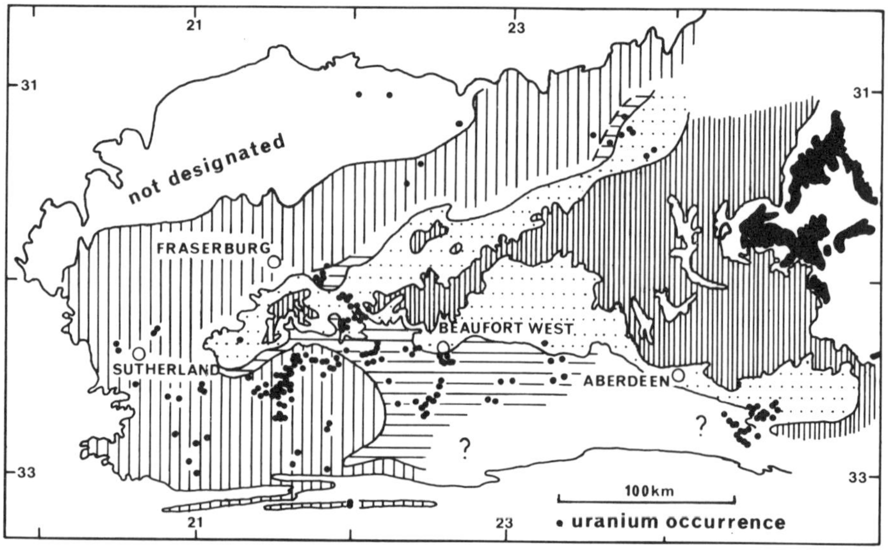


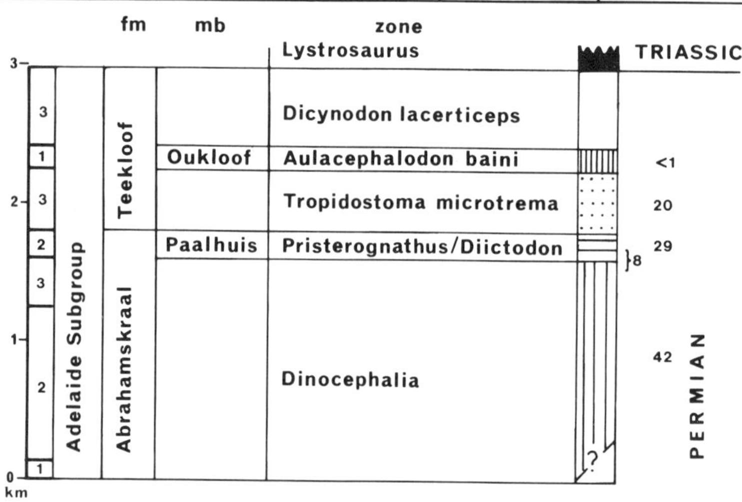


Figure 14. Distribution of published uranium occurrences in the Beaufort Group of the W.Karoo Basin. Lower part of diagram shows thicknesses (for southern crop) and depositional environments of various units according to Turner (1979): 1 = low sinuosity channel facies; 2 = high sinuosity channel facies; 3 = flood basin facies. Figures in right column are percentages of total uranium occurrences in each stratigraphic unit (data from map).

lenses and pods, some of which are mineralized (koffieklip). According to Turner (1979), 90 % of all known uranium occurrences occur within this sequence which has been called the Paalhuis Member and coincides with the *Pristerognathus/Diictodon* Zone of Keyser & Smith (1979). Individual sandstone units at the base of the Paalhuis Member that are known to carry significant mineralization are the Poortjie, Rietkuil, and possibly Rystkuil.

The overlying Teekloof Formation is about 1 000 m thick and consists predominantly of bluish-grey, and maroon to red massive and locally silty mudstones with occasional thin (less than 1 m) interbeds of greenish-grey siltstone and fine sandstones. Calcareous concretions are common throughout the mudstones and they often carry well-preserved vertebrate remains. A 120 m thick arenaceous sequence (Oukloof Member) with sandstone beds up to 28 m thick occurs within the Teekloof Formation (Fig.14).

Turner (1978b) has interpreted this part of the Beaufort Group as a general upward fining succession that was deposited in successively lower energy fluvial regimes: low sinuosity channel, high sinuosity channel, and finally a floodbasin environment (Fig.14). The Paalhuis and Oukloof members represent temporary reversion to higher energy regimes. He links the upward fining succession to a decline in sediment supply coupled with a decrease in channel gradient and a denuded source area. The increase in silt and clay in suspension led to an increase in stream sinuosity. Sediment transport directions were to the north and east.

Although uranium mineralization has been recorded from all three of the facies mentioned above, significant occurrences are mainly in rocks of the high sinuosity facies, where the sandstone to mudrock ratios is about 1:4. These rocks display numerous permeability barriers and the mineralized zones are usually in the thickest parts of the channel sandstone units especially in the lower parts where there are numerous interfingers with mudrocks and discontinuous silt and mudstone beds. Mineralized zones are often separated by wide barren zones, and range from small pods and lenses a few metres long to discontinuous bodies up to 1 km in length. They range in thickness from a few centimetres to 7 m. The most important mineralized zone appears to be in the upper part of the Paalhuis Member.

Mineralized sands are dark grey to black although the rock is sometimes bleached white or yellow-brown. Carbonaceous debris is invariably present, and the host sandstones are arkosic with carbonate cement, sulphides and volcanic debris. They are often interbedded with cherts. The uranium minerals are usually in the insoluble $U^{+4}$ oxidation state and the primary ores are uraninite and coffinite. These are variously associated with secondary hydrated silicates, arsenates, and phosphates of U, Ca and Cu. Recorded secondary uranium minerals are metatorbonite, uranospinite and uranophane.

Based on a study of the state of preservation of plant debris in the Beaufort Group sediments, Stapleton (1977b, 1978b) noted that 80 % of all the known localities of uranium mineralization were within rocks that fell within a 'metamorphic' facies of organic metamorphism (on a facies scale of immature, early mature, late mature, metamorphic, greenschist). This implies that the rocks in question have been heated to about 200°C when the organic debris starts to become opaque and is brown in colour. About two thirds of the Upper Permian outcrop falls within the greenschist organic metamorphic facies where an estimated temperature of 300°C or more has destroyed the organic tissue, leaving only carbon. From this, Stapleton (1977b & 1978b) concluded that the association of uranium mineralization and carbonized debris indicated that mineralization was a post-metamorphic event. Because the metamorphic agents were likely to have been Jurassic dyke and sill intrusion, and in the south the mid-Triassic uplift of the Cape Fold Belt, the mineralization took place long after original sedimentation. The fact that ore bodies are mainly confined to sandstone (i.e. presumably more porous) horizons is thought to strengthen these suggestions.

Turner (1978b & 1979) on the other hand, believes that mineralization was penecontemporaneous with sedimentation, and that it was completed before sediment compaction and lithification. He also rules out diagenetic redistribution, and attributes the mineralization to circulating interpore fluids leaching volcanic debris in the sand bodies (the chert horizons have been attributed to silicification of tuffs by Martini (1974) although Smith (1980) suggests that they may be silicified carbonate hardgrounds). Turner (1978b) appeals to mildly reducing, alkaline ground waters transporting uranium in the form of uranyl carbonate complexes in the porous river channel sands. At permeability barriers (mud interbeds, etc.) stagnation in the presence of a strong reducing agent ($H_2S$ from decaying plant matter) allowed sufficient time for reduction and precipitation of uranium oxides and silicate complexes. The carbonates were precipitated as host rock cement. Contrary to Stapleton's (1978b) assumptions, Turner (1978b) claims that the sandstones after lithification would have been too impervious for later permeating solutions to have produced such peneconcordant ore bodies. Further evidence that some plant debris was also mineralized, presumably before deep burial, is also cited as support for the mineralization episode prior to sediment compaction. Further study is obviously needed to clarify this aspect.

Finally it should be noted that despite considerable prospecting activity and the large area over which mineralization has taken place no economically exploitable discoveries have been reported. Notwithstanding Turner's (1979) claim that 90 % of the uranium occurrences are in the *Pristerognathus/Diictodon* Zone (Paalhuis Member) a comparison of Stapleton's (1978b) locality map and Keyser & Smith's (1979) revised zonal map (Fig.14) gives the following results for the Upper Permian Beaufort Group uranium occurrences:

| Zone | No | % |
|---|---|---|
| *Dicynodon lacerticeps* | none | – |
| *Aulacephalodon baini* | ?1 | 0.5 |
| *Tropidostoma microtrema* | 38 | 20 |
| *Pristerognathus/Diictodon* | 56 | 29 |
| Dinocephalian | 81 | 42 |
| Unresolved *Pristerognathus/Diictodon-Dinocephalian* | 15 | 8 |

### 2.1.2 *Middle Triassic to Lower Jurassic sediments: Stormberg Group*

The sequence of sandstones, mudrocks and volcanics that lie above the Beaufort Group was called the Stormberg series by Wyley (1859). Recently, Lock et al (1974) have placed the upper, volcanic strata into the Drakensberg Group, so that the Stormberg Group as understood here comprises three sedimentary formations – Molteno, Elliot, and Clarens. Following early mapping, Dunn (1878) proposed the following lithological subdivisions that in a slightly modified form remained in use until 1976, when Johnson et al re-classified the sequence in line with more modern criteria:

| Dunn (1878) | Current terminology | |
|---|---|---|
| Volcanic Series | Drakensberg Group | |
| Cave Sandstone | Clarens Formation | Stormberg Group |
| Red Beds | Elliot Formation | Stormberg Group |
| Coal Measures | Molteno Formation | Stormberg Group |

Green (1883) suggested that the Coal Measures be renamed the Molteno Beds, and detailed mapping in the eastern Cape by the infant Cape Geological Commission (Schwarz 1903 and du Toit 1904, 1905) firmly established the lithostratigraphic succession and allowed cor-

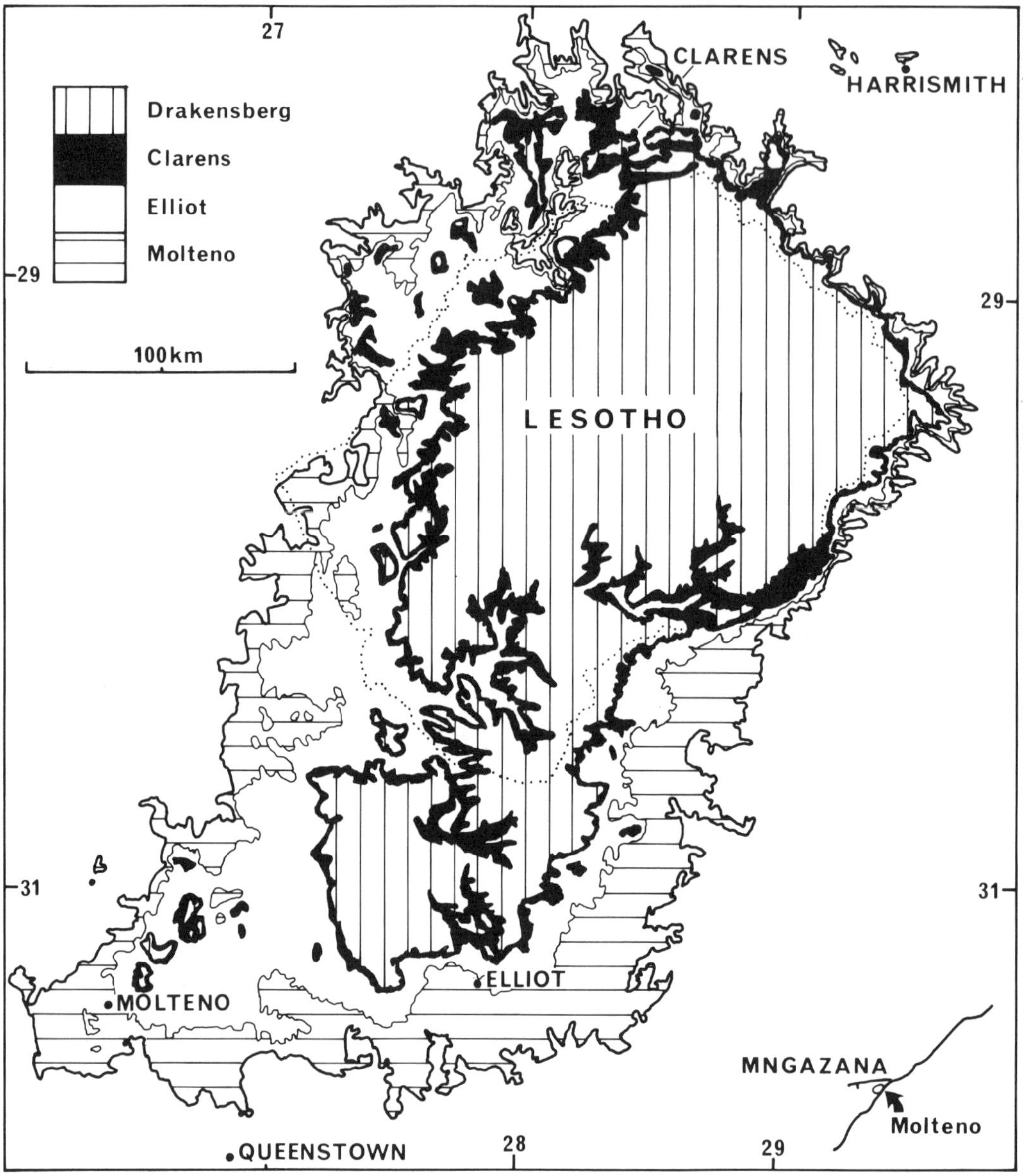


Figure 15. Distribution of the Drakensberg Group and individual formations of the Stormberg Group, in the main Karoo Basin and Mngazana outlier. Based on maps by Beukes (1970), Anderson (1974) and Kitching (1977).

relation to be made around the whole outcrop (Haughton 1924).

Johnson et al (1976) concluded that there is no justification for retention of the term Stormberg, but we feel that it serves a useful purpose in grouping sediments that were deposited to the north, and as a direct result of sediment shedding from, the Cape Fold Belt. They form a distinct tectono-sedimentary sequence between the Beaufort Group, and the overlying Drakensberg volcanic Group.

The Stormberg Group crops out in a large oval centred on Lesotho, in several small erosional outliers in the NE Orange Free State, and at Mngazana on the coast (Figs 15, 18 & 108). The most extensive exposures occur in the eastern Cape between Aliwal North, via Molteno, Dordrecht, Elliot, and Maclear, to Matatielle in Transkei. The country is terraced and hilly with impressive cliff sections formed by the arenaceous units in the Molteno and Clarens formations. Locally, the stratigraphy is difficult to follow because of the large number of dolerite intrusions, but it is here that the type sections of the Molteno and Elliot are found.

Isopachs for the whole group, together with thicknesses for individual formations are shown in Figure 16, and east-west and north-south sections across the main outcrop are shown in Figures 16 and 8. From these it

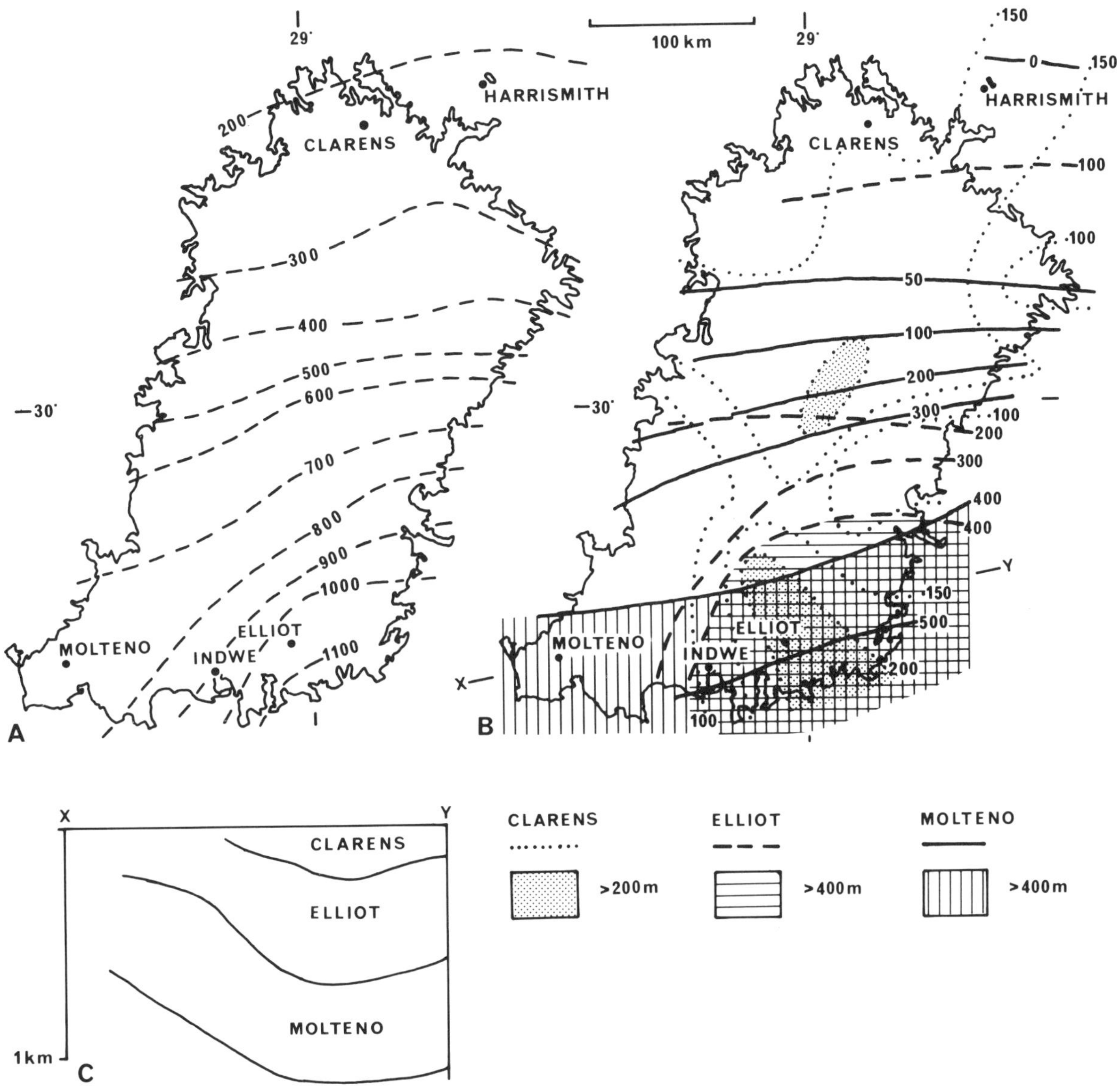


Figure 16. Isopachs in Stormberg Group sediments in the main Karoo Basin. A. Total isopachs; B. Individual formation isopachs; C. Cross-section X-Y which runs NE-SW across southern part of map B. Outlines show base of group outcrop. Date after numerous authors cited in text, but particularly Beukes (1970, from which the Clarens Formation isopachs are redrawn). A N-S section across the same area is shown in Figure 10. All horizontal scales are the same.

is evident that although the southernmost part of the original depocentre has been removed by erosion, the Stormberg Group consists of a SE-thickening wedge. Maximum thicknesses (ca.1 100 m) occur SE of Elliot (in the vicinity of Maclear) and thin rapidly into Lesotho (ca.500 m), after which they thin less dramatically to about 200 m in the Harrismith/Memel district of the northern Orange Free State. This is the trend of the lower two formations (Molteno and Elliot) which combined, dominate the total isopaches (Figure 16B & C). Examination of Figure 16 shows that after the deposition of the Elliot Formation, the basin geometry underwent a radical reorganization and although the thickest accumulation of Clarens Formation does occur in the SE, its main depocentre follows a sinuous northerly trend from the Harrismith/Memel area (120 m thick) to the SE where about 260 m occur. That this is not merely a result of subsequent erosion is shown by the fact that many of the outliers of Clarens Formation on the periphery of the main outcrop are capped by Drakensberg lavas and have their vertical sections preserved intact.

The two lowermost formations (Molteno and Elliot) progressively overstep northwards across *Cynognathus* Zone strata, so that in NE Orange Free State and NW Natal, the Elliot Formation rests on rocks of the *Lystrosaurus* Zone (Fig.10). No vertebrates have been found in the Molteno Formation so that its age is not known, but by the general aspect of its flora, and because the overlying Elliot Formation is definitely of Upper Triassic age, it is generally regarded as being Carnian, locally extending upwards and downwards into the Norian and Ladinian, respectively (i.e. Middle to Upper Triassic). This implies a considerable and progressively northward increasing hiatus between the base of the Stormberg and Beaufort groups, which is marked by a regional erosional discordance between the Molteno Formation and *Cynognathus* and *Lystrosaurus* zone strata. Because of the very low dips, particularly over the stable platform area, this non-sequence appears as a disconformity.

As we discuss in the next section, the provenance area and dispersion patterns for the Molteno Formation are different from those of the Scythian strata, and indicate a change in nature and input direction. This is explained by proposing Anisian times for the main uplift of the Cape Fold Belt which was accompanied by slight regional uplift north of the orogenic zone that was responsible for the hiatus in sedimentation. The Cape Fold Belt is discussed in detail in Chapter 3.

Summaries of lithostratigraphic and biostratigraphic correlations of the Stormberg Group are given in Table 3 and Figure 9. Stapleton (1978a) has recently challenged the accepted dating of the Lower and Middle Triassic rocks in the Burgersdorp area on the grounds that palynological data from the *Cynognathus* Zone and Molteno Formation rocks indicate no significant change across the boundary, and that the microflora are transitional between Permian and Triassic types. He suggests that the Permian/Triassic boundary lies somewhere within the rock sequence included by the *Cynognathus* Zone and Molteno Formation. A corollary of this, if substantiated, is that the climax of the Cape Fold Belt uplift was an Upper Permian-Lower Triassic event. Because many of the taxa in Stapleton's microflora were new, and all were badly preserved, these suggestions have not been adopted here because further research is obviously called for. A revised classification incorporating Stapleton's ideas is shown in Figure 9 (column 3).

#### 2.1.2.1 *Molteno Formation*

Following a series of publications by Turner (1969, 1970, 1971, 1975a, 1975b, 1977, 1978a), the Molteno Formation is the best-studied part of the Stormberg Group. Only its age remains in doubt. The rich flora has been reviewed by Anderson & Anderson (1970), and Anderson (1974), who concluded that it is very similar to that of the Ipswich Coal Measures in Queensland, which have a *Dicroidium* macroflora and an *Alisporites* microflora with ?Middle to Upper Triassic affinities (Ludbrook 1978). No tetrapods have been found in the Molteno Formation and we tentatively assign a ?Ladinian-Carnian-Norian age to it (Fig.9).

The Formation constitutes a coarse sediment wedge, thickest in the SE (500 m, SE of Elliot), which is dominated by thick sandstones and grits with subordinate shales, silts and coals (Figs 16, 17 & 20). Where best-developed (in the Molteno-Indwe-Maclear districts) these are arranged in vertically stacked, upward-fining sequences and in the Molteno-Indwe area, Turner (1970, 1971, 1978a) has recognized six cycles in the succession, the lower three of which terminate in extensively-developed coal measures. These are shown as cycles A to F in Figure 20. The upper cycles are thinner and less well-developed, terminating in shales and carbonaceous shales. A typical cycle (up to 140 m thick) consists of six lithofacies that can be summarized (from base 1 to top 6, Fig.17):

1. A basal conglomerate (facies 1) typically 30-60 cm thick resting on an eroded surface. The conglomerate consists of well-rounded medium to large white quartzite pebbles with a high sphericity, set in a coarse gritty sand matrix. Minor clast types consist of extrabasinal chert and slate pebbles, and intrabasinal logs, shale and coal fragments. There is an overall increase in clast size southwards, and the conglomerates often occur in channels (up to 180 cm deep) cut into the underlying cycle.

The overlying sandstone units which form the dominant rock type of the formation are generally lenticular, ranging in thickness from 20-90 m. They give rise to impressive cliff exposures and can be divided into three distinct lithofacies (2-4).

2. A thick, lower cross-bedded unit (facies 2) consists of coarse, gritty sands with pebble washes of rounded vein quartz, and intraformational shale pebble conglomerates. Granite pebbles have been reported from the sandstones of Transkei (Haughton 1924). Plant fragments, logs and coal lenses occur, particularly near the base. Trough cross-bedding is the dominant sedimentary structure, especially in the coarser sands, although planar cross-bedding does occur. Sets range in thickness from 0.3 to 4.5 m.

3. Facies 3 consists of medium-grained, flat-bedded sandstone, with mica, comminuted carbonaceous matter and primary current lineations on bedding planes.

4. Facies 4 consists of fine-grained, often ripple marked sandstones. Linguoid ripples dominate, but straight crested ripples do occur in the upper parts of the unit.

These sandstones (facies 2-4) are usually pale grey to yellowish grey in colour, but locally (e.g. SW of Aliwal North) are reddish brown. Sorting is generally poor, but becomes better in the finer grain sizes. They are principally composed of quartz, and an average of 27 determinations made by Turner (1972b) in the Aliwal North district give the following composition:

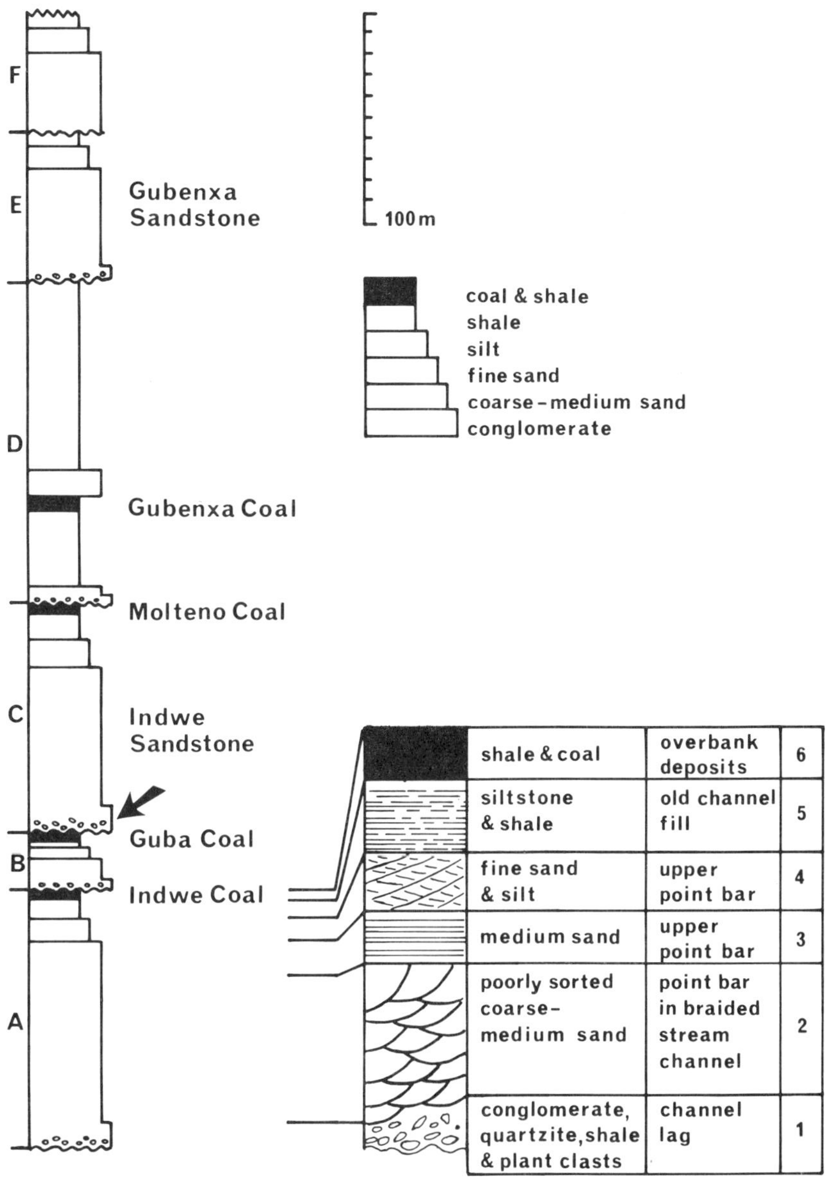


Figure 17. Idealized section of the Molteno Formation in the southern crop (Molteno-Indwe district). Cyclic upward fining sequences are shown A-F, with lithofacies for one cycle shown 1-6. Arrow indicates position of oldest 'glittering sandstone'. Compiled from data in Turner (1970, 1971, 1978a).

quartz 64.7 %, feldspar 2.1 %, matrix (mostly clay) 25.5 %, rock fragments 5.5 %, and mica 2.0 %. The determinations were made on random samples with mean diameters ranging from 0.87 to 2.05 phi. A characteristic of the Molteno sandstones is that they 'glitter'. This phenomenon is caused by secondary overgrowths of silica in optical continuity with the quartz grains, and interstitial cementation by pore space infilling with chalcedony, microcrystalline quartz and megaquartz (authigenic quartz grains $> 20\ \mu$).

Large silicified tree trunks have been reported by Stow (1871) from the thick sandstone units in the vicinity of Dordrecht (about 55 km east of Molteno).

5. The sandstones of facies 4 grade upwards into pale grey to yellowish grey silts, and silty shales with fine sandstone interbeds (facies 5). This unit is typically lenticular and generally less than 6 m thick, and locally, is not present at all. The sediments are flat-bedded and commonly have mica, shale fragments and comminuted plant matter along bedding planes. Cross-bedding and current lineations are lacking in this unit (Turner 1972b).

6. The upper part of the sedimentary cycle consists of a shale/coal composite facies (facies 6). The shales dominate and are grey to black in colour with abundant plant debris (a rich and beautifully preserved flora of leaves, stems and fruits especially of *Dicroidium* and allied types). There are two types of shale; soft, paper thin laminated shale, and hard brittle shale, with a

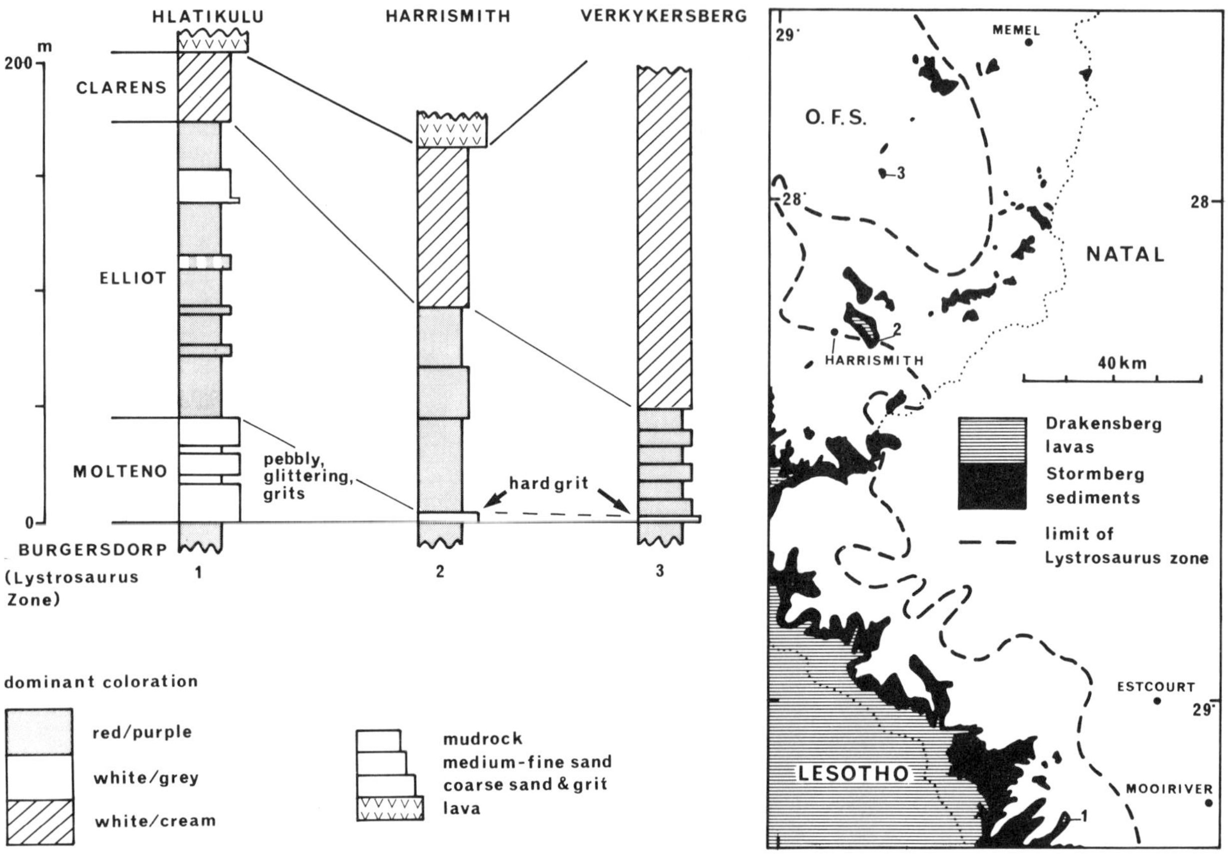


Figure 18. Sections and map of Stormberg Group in N.W.Natal and E.Orange Free State. Shows northward thinning of Molteno Formation, and overstep of Middle and Upper Triassic across the Scythian (*Lystrosaurus* Zone) onto Upper Permian strata. The *Cynognathus* Zone is missing in this area. Data from Haughton (1924), Geological Survey (1970), Turner (1972a) and Kitching (1977).

blocky fracture. Pyrite and ironstone nodules commonly occur within the shales. In the lower half of the succession, they are interbedded with lenticular coals which give rise to three (possibly four) seams. The coals do not have basal underclays and root beds, but the good preservation of plant fossils indicates that the organic debris is essentially autochthonous.

On a regional scale, the Molteno Formation consists of a southward thickening wedge (Figs 16B & 20). The majority of the coarse beds, coals, and well-developed upward fining cycles are only found in the south, but Turner (1977, 1978a) has shown that the sequence as a whole can be divided into three members, each representing a major phase of sediment input from the source area (the still-rising Cape Fold Belt). Members are composed of one or more sedimentary cycles, indicating minor pulses within each sedimentation phase, and on Figure 20 consist of: lower member (cycles A and B); middle member (cycle C, the Indwe Sandstone); upper member (cycles D, E and F). All three members are present in the south, but the lower member (cycles A and B with the Indwe and Guba coals) does not extend more than about 100 km north of the southern outcrops. The middle member (Indwe sandstone and Molteno coal), oversteps the lower member onto the Burgersdorp Formation (*Cynognathus* Zone in the south, *Lystrosaurus* Zone in the north) and can be traced as a prominent unit as far north as the Harrismith district (Figs 18 and 20). It thins progressively northwards from about 100 m in the south to 45 m in Natal, and in the far north is reduced to a metre or so of coarse, hard, pebbly sandstone between the red argillaceous Beaufort and Elliot strata. It probably does not occur at Memel where the Elliot Formation rests directly on the top of the Beaufort Group (Figs 10, 18 and 20). A corresponding reduction in thickness of argillaceous beds within the Indwe Sandstone also occurs northwards, as well as a regional decrease in clast size, although local pockets of particularly coarse material (up to 30 cm) have been reported on the Cape/Natal border at Hlatikulu (Haughton 1924) (Fig.18). The Indwe Sandstone is the stratigraphically lowest member of the Molteno Formetion to exhibit the 'glittering' phenomenon, and represents the most extensive injection of detritus during the deposition of the formation.

The upper member of the Molteno Formation comprises several sediment cycles (D, E and F on Fig.20). Turner (1970, 1978a) is vague on the exact number that can be recognized above the Gubenxa/Molteno coal horizons. We have shown three on Figures 17 and 20, but there may be four. The uncertainty arises because in the upper member, the cyclic nature of deposition is less well-developed; no coals are known, and the conglomeratic facies is locally absent. Turner (1978a) suggests that the upper member can be traced about 300 km north of the main southern outcrops.

The exact stratigraphic relationship of the Molteno Formation with the overlying Elliot Formation is not well understood. Some workers suggest that there is a large non-sequence between the two (e.g. Anderson & Anderson 1970) but field evidence indicates that the Molteno and Elliot formations interfinger, implying a diachronous junction. For this reason we have shown a younger age for the top of the Molteno Formation in the south (lowermost Norian) than in the north (Carnian) (Figs 9 and 28).

A small outlier of the Molteno Formation occurs on the downthrow side of the Mngazana Fault on the Transkei coast (Figs 15 & 108). Here, du Toit (1920a) estimated a thickness of about 450 m of glittering, coarse, gritty, porous and pebbly sandstones, overlying *Lystrosaurus* Zone Beaufort rocks. They dip steeply towards the fault plain (20-30°NE) and are intruded by several dolerite dykes and sills. The pebbly horizons are composed of white and grey quartzites set in a coarse feldspathic quartz sand, whilst the other sandstones are fine grained and grey in colour. No shaley rocks occur and no plant remains have been reported. Correlation is, therefore, purely on lithological similarities with exposures about 90 km away to the west.

No fossil tetrapods have so far been recorded from the Molteno Formation, creating a blank spot in the evolutionary picture in southern Africa between the end of the Scythian to latest Carnian or early Norian. Previous records have been disproved by Turner (1972a) who referred all the vertebrate remains at the Tsoaning River section in Lesotho to the underlying Burgersdorp Formation (Fig.19). Other faunal remains are also rare, and of the 42 fossil plant sites recorded by Anderson (1974) only nine have produced animals: 1 species of crustacean, 32 species of insects, and 1 species of fish (*Semionotus capensis* Jubb 1973). The insect faunas, which have been described by Riek (1974) consist mostly of Blattodea (cockroaches, 45 spp), Coleoptera (beetles, 65spp.) and Homoptera (cicadids etc. 65 spp.) and show a close relationship to Australian Triassic faunas of the Mt Crosby Formation in the Clarence-Moreton Basin of northern New South Wales, which Banks (1978) dates as Carnian.

Fossil plants are abundant and beautifully preserved, especially in the coal and shale parts of the sedimentary cycles. Du Toit (1927) gave a comprehensive summary of the earlier records, and Anderson (1974) has given a preliminary account of recent intensive collecting. She records the following floral elements:

| | |
|---|---|
| Leaves | – 32 genera, 104 spp. |
| Fertile structures (fruits, seeds, etc.) | – 31 genera, 42 spp. |
| Spores, pollen grains and acritarchs | – 30 genera, 53 spp. |

but points out that the record of 31 genera for fertile structures probably represents only half that number of plant genera, and eight species of acritarchs cannot be included in a listing of macroflora. A grand total of 33 genera and 106 species is reckoned to be a good estimate of the Molteno macroflora (Anderson 1974, p.5). The vegetation includes trees (e.g. conifers and ginkgos), woody shrubs (e.g. cycads and larger *Dicroidium*), ground cover (e.g. ferns and smaller *Dicroidium*) and swamp growth (e.g. Sphenophytes and possibly *Phoenicopsis*). The distribution of vegetation was apparently fairly uniform, and the most diversified and abundant genus was *Dicroidium* (21 spp.) with locally abundant and diverse *Neocalamites* (4 spp.), *Phoenicopsis* (1 spp.) and *Baiera* (4 spp.). Anderson (1974) postulates that this vegetation was supported in a well-watered environment that extended over an area of 1 500 by 750 km. Several of the taxa that occur in the Molteno Formation have also been recorded in the Carnian Ipswich coal measures of Queensland (Flint & Gould 1975).

2.1.2.1.1 *Stormberg Coalfield.* Dunn (1878) applied the term Coal Measures to the whole of the Molteno Formation, but it was soon realised that this was an inappropriate description. Nevertheless, coal mining took place on a small scale in various parts of the Aliwal North-Maclear area after 1864, in what has become known as the Stormberg Coalfield (Fig.21). Larger-scale commercial mining started in 1882 with the formation of the Stormberg Coal Co., and reached a peak in the period 1900-1904 when an average of about 194 000 short tons p.a. were produced. Production declined thereafter and ceased in 1948, mainly due to the availability of cheaper coal from the Transvaal and Orange Free State.

Coals and shaley coals occur at the tops of the lower three sedimentary cycles (A, B & C on Figs 17 & 20) and possibly within cycle D, and four seams have been mined: Indwe, Guba, Molteno and Gubenxa, but it is not clear whether the last two are separate, or developments of the same seam. Figure 21 shows the distribution and representative sections through each of the main coal seams. The thickest and most extensively developed is the lowermost, the Indwe, which occurs over the whole area. It is a composite seam up to 4.3 m thick, composed of alternating bands of shale and coal

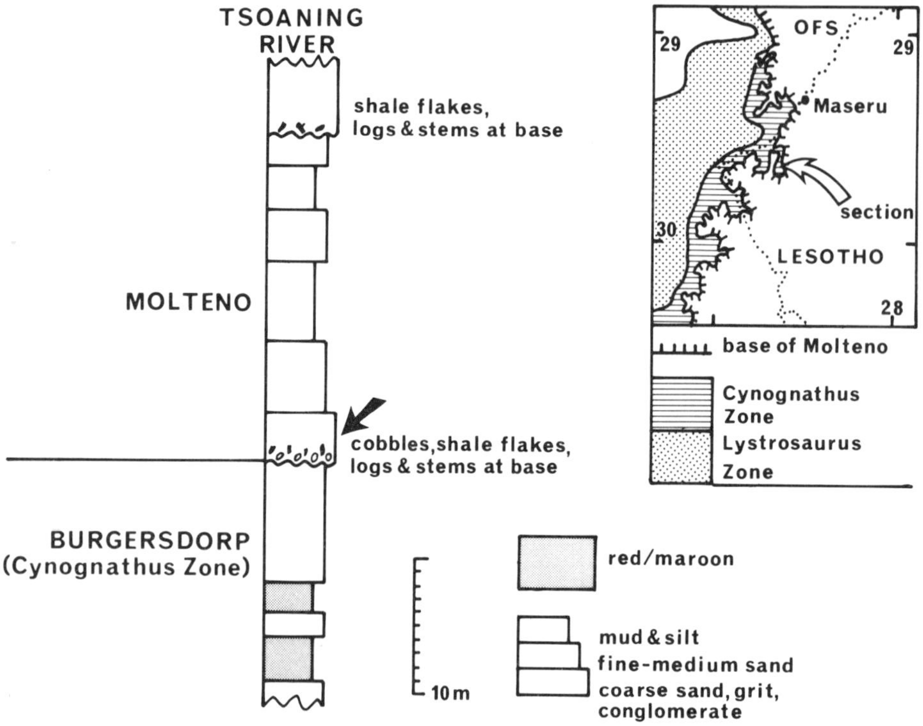


Figure 19. Beaufort (*Lystrosaurus* Zone, Burgersdorp Fm.)/Molteno contact at the Tsoaning River section in Lesotho. Although the top Burgersdorp sandstones are fawn-khaki in colour, the coarse, feldspathic, glittering basal Molteno beds rest on an erosion surface. The latter are Indwe Sandstone equivalents. The Burgersdorp strata contain abundant reptile remains, whereas the Molteno is barren of vertebrates, but contains abundant plant fossils. This section probably shows Molteno cycles C and D. Data from Turner (1972a).

with fireclays, which exhibit frequent, abrupt changes in thickness and lithology. This is partly due to penecontemporaneous erosion, which has locally removed much of the upper part, but also to differential compaction and the original basin geometry (Fig.22). In the Indwe district, erosion has locally removed all the top part of the coal-bearing layers, down to and including the Twelve Inch seams and this affected the commercial viability of the mining operations because the best coals in the Indwe seam are generally near the top. Nevertheless, the most productive mining areas were in the Indwe district, which seems to have been the main centre of coal formation. To the west and south, the shale content increases.

The Guba coal is best-developed in the Guba Valley SW of Indwe, but extends from Molteno to the vicinity of Maclear (Fig.21). It lies between 24 and 30 m above the Indwe Coal (Figs.17 & 20), but the two seams seldom both carry good coals in the same area. The best Guba coals tend to occur towards the bottom of the seam (Fig.21) so that unlike the Indwe Coal, erosion of the top did not significantly affect its commercial prospects. In addition, because the lower part carries only a few shale partings, the Guba was commercially more viable than the Indwe coal.

Two coal seams occur above the Indwe Sandstone; the Molteno and Gubenxa, but because of the complexity of intervening dolerite sills it has not been possible to determine the precise relationships between them. Turner (1971) treats them as two separate seams. The Molteno seam is generally of poor quality; the coals are shaley and impersistently developed, and locally, only represented by carbonaceous shale and vitrain stringers. It has been traced from just east of Molteno to the SW of Maclear, but was never of large-scale commercial importance. The best coals occur in the east where about 30 cm was the maximum thickness of workable coal.

The Gubenxa coal possibly lies at a slightly higher level than the Molteno seam, and is confined to the SE part of the coalfield, where, in the Engcobo-Maclear area, it was commercially the most important coal and was worked on a small-scale by the Tembuland Coal Syndicate. The best coals occur in the upper part of the seam and range from 60-120 cm thick.

Structurally, the Stormberg Coalfield is simple. There is a gentle regional dip of between 1 and 5° to the SE, onto which has been superimposed two sets of broad, open anticlines and synclines (N-S, and NE-SW). Faulting is also on a small-scale and in two directions: N-S and E-W. It appears to have been normal, and throws are generally small and variable, although locally

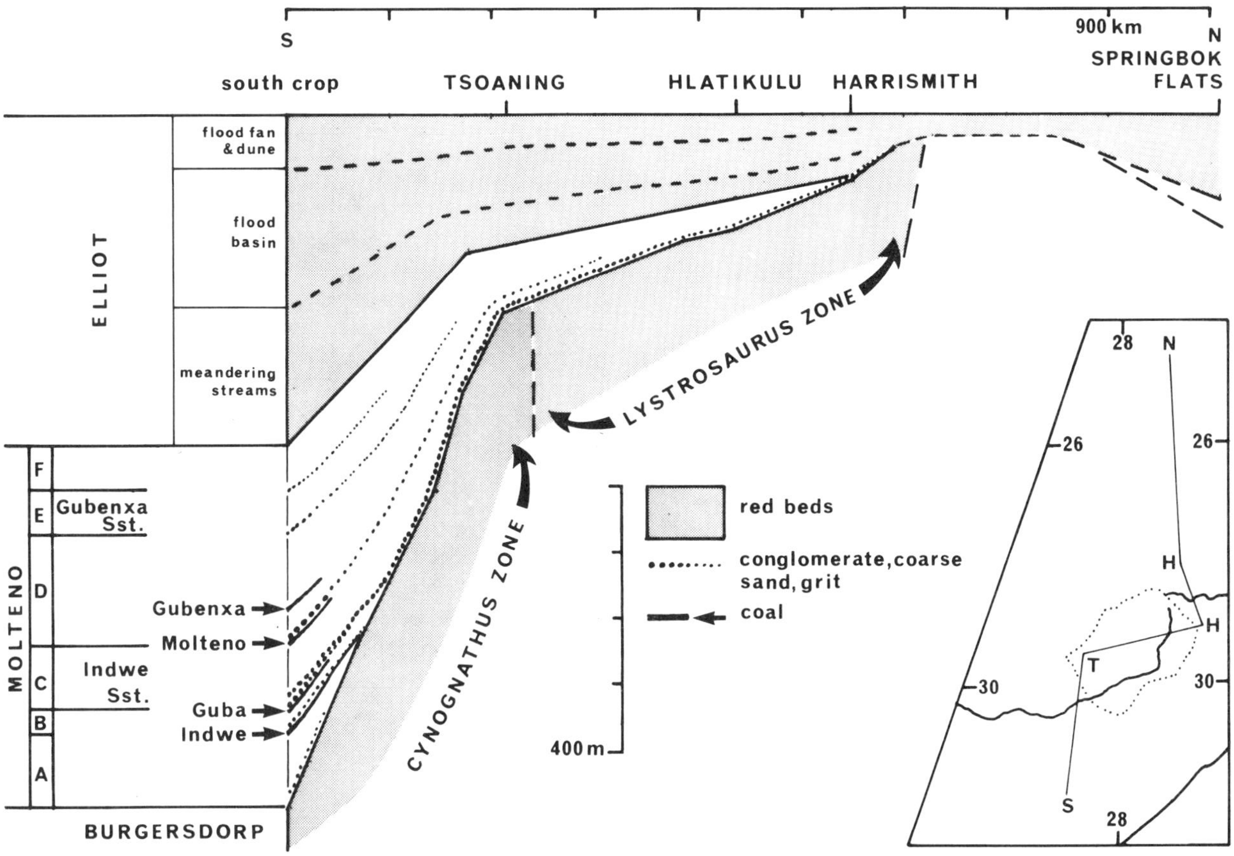


Figure 20. N-S thickness and lithofacies changes in Molteno and Elliot formations across the main Karoo Basin and Springbok Flats outlier. Molteno cycles A-F lose their identity northwards, where the whole of the formation has been referred to cycle C (Indwe Sandstone) by Turner (1978a). Elliot lithofacies and depositional environments are based on an analysis of the sandstones by Visser & Botha (1980). Note overstep by Molteno Formation across the *Cynognathus/Lystrosaurus* boundary.

(e.g. on the Witkop farm near Jamestown, and immediately south of Indwe) throws of 300 m have been reported (Turner 1971). The area has been invaded by many dolerite dykes and sills, the former along fault lines, so that correlation of sedimentary horizons is often hampered. Locally, the igneous intrusions have altered the quality of the coal by partial distillation of the volatile content, producing pseudoanthracite. The change in chemical composition of the coal is accompanied by a deterioration in its physical properties, so that close to intrusions the coal is fractured, deformed (by plastic flow) and contains numerous cavities. Loss of volatiles is accompanied by an increase in ash content caused by the deposition of mineral salts in the fractures and cavities. Farther away from the intrusion, mild thermal metamorphism is confined to an expulsion of volatiles and an increase in the rank of the coal.

Generally speaking, the quality of the coals from the Stormberg Coalfield is low (Turner 1971). This is mainly due to their high ash content (on average 25 % of the Indwe and Guba seams) which makes them unsatisfactory for the production of coke, even though their volatile content is theoretically high enough. In addition, the Stormberg coals have a low hydrogen content (2.85 % average of eight determinations quoted by Turner 1971) making them unsuitable for the production of tar and light oils. For all other commercial activities (electric power generation, domestic heating, manufacture of gas, locomotive stoking) the high ash content, coupled with a relatively low calorific value make the Stormberg Coals economically unattractive.

Turner (1971) has made a revised assessment of the reserves of the Stormberg Coalfield based on new drilling surveys. Assuming a 65 % recovery rate, he estimated that the coalfield originally held total reserves of 216 million tons, of which 5.4 million tons have already been extracted. However, from a commercial point of view, only seams of 1 m or more in thickness are now considered viable, and on this estimate the original coal reserves amounted to 43.7 million tons of which 3 million tons have already been extracted. Turner (1971) estimates, therefore, that the Stormberg Coalfield contains about 40.7 million tons of workable coal (mostly in the Guba and Indwe seams), although under present

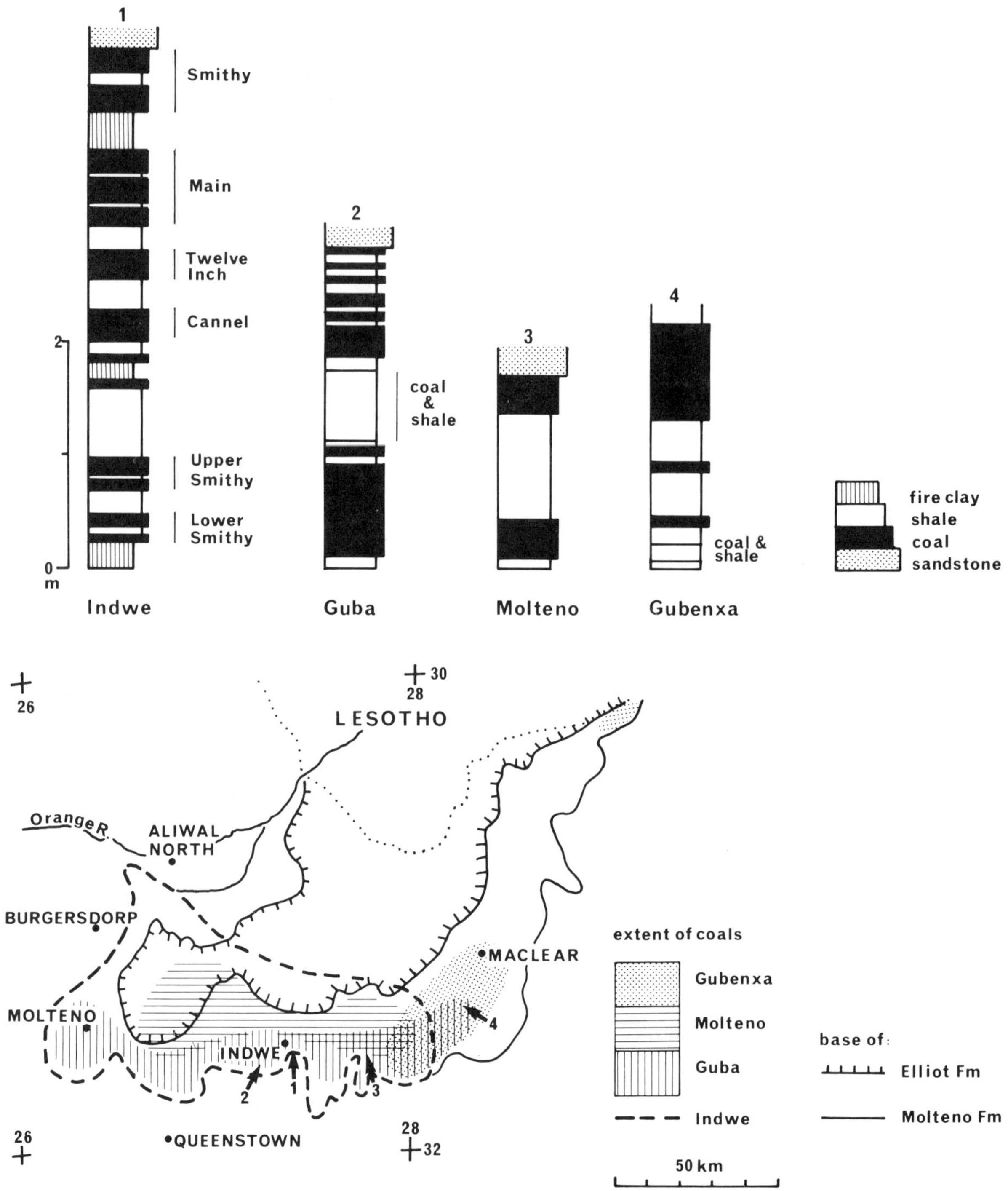


Figure 21. Typical sections and distribution of the coal seams in the Stormberg Coalfield. Section locations arrowed 1-4 on map. Based on data in Turner (1971).

market conditions its extraction is not economically attractive.

2.1.2.2 *Elliot Formation*

Formerly known as the Red Beds because of their colour, the Elliot Formation forms a continuous outcrop around Lesotho and in the extreme north rests directly on *Lystrosaurus* Zone rocks (Burgersdorp and Otterburn formations), with which they share a similar lithology (Figs 10, 15 & 20). The Elliot Formation

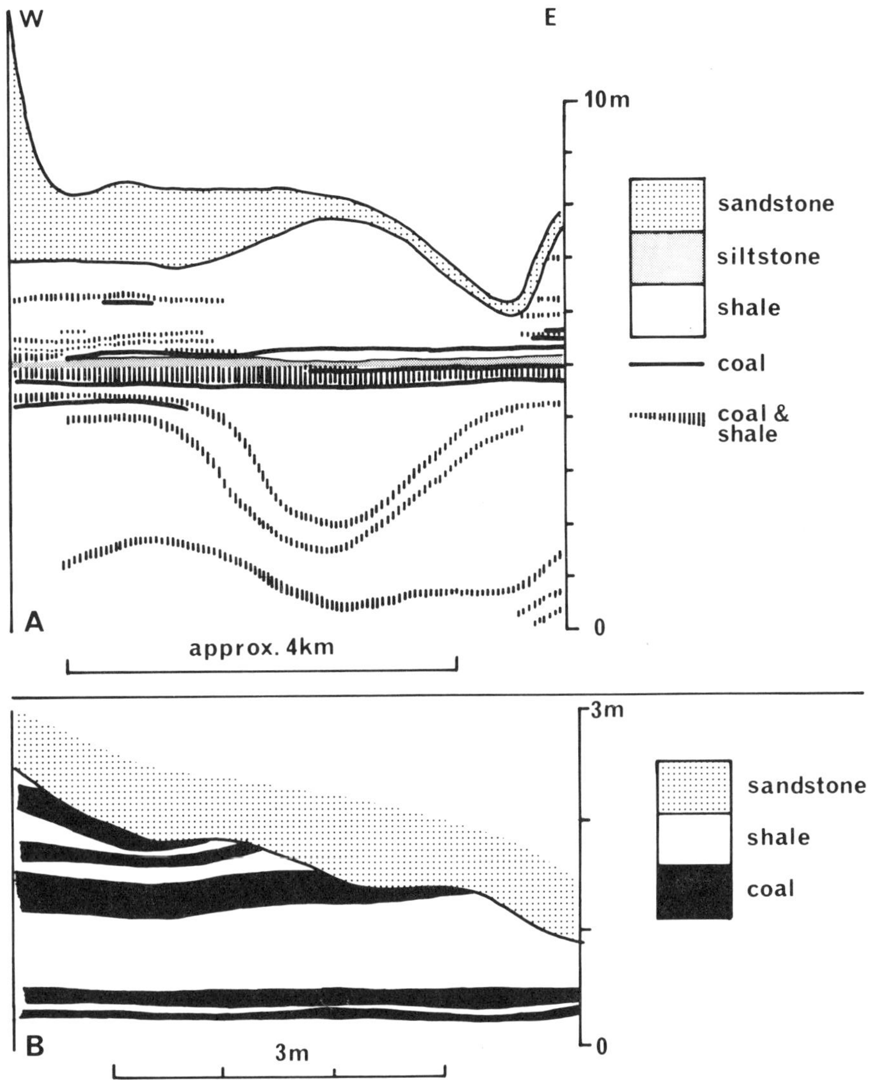


Figure 22. A. Lateral facies and thickness variations in the Indwe Coal around Kramberg (Aliwal North).
B. Contemporaneous erosion in Indwe coal measures, Indwe district.
Both sections drawn from borehole data quoted in Turner (1971).

reaches a maximum thickness of about 490 m along the south crop and thins northwards into Lesotho (200 m), thereafter reducing gradually to the Harrismith/Memel area where it is about 48 m thick (Figs 10, 16, 18 & 20). It is locally exposed in good sections around the southern crop, and the more resistant horizons give rise to large cliffs, but generally the Elliot Formation forms the more subdued slopes beneath the impressive cliffs of the overlying Clarens Formation sandstones.

Relatively little work has been published on the sedimentology and stratigraphy of the Elliot Formation following early descriptions from the Aliwal North and Elliot districts by du Toit (1904, 1905) and the good general summary by Haughton (1924). The only recent accounts are by Botha (1968), Le Roux (1974) and Visser & Botha (1980) on the type section in the Barkly Pass near Elliot, and some general comments by Johnson (1966). No comprehensive regional lithostratigraphic correlation has been attempted and because there is no recognized biostratigraphic zonal scheme, it is not possible to assess the degree of diachronism exhibited by the formation in relationship to the Clarens and Molteno formations. Tetrapod remains are relatively abundant in the Elliot Formation, and the reptiles faunas are characterized by the presence of archosaurid types, both pseudosuchians and early dinosaurs (Cox 1973). This fauna is typical of Romer's Fauna C (1975) which has been found in Keuper rocks (Norian/Rhaetian) of NW Europe. Because a similar assemblage also occurs in the overlying Clarens Formation, we have allocated the bulk of the Elliot/Clarens sediments to the Norian/Rhaetian and suspect continuous sedimentation from Carnian (possibly uppermost Ladinian) through to Rhaetian and locally Hettangian times for the Stormberg sediments (Fig.9).

The base of the Elliot Formation has been defined

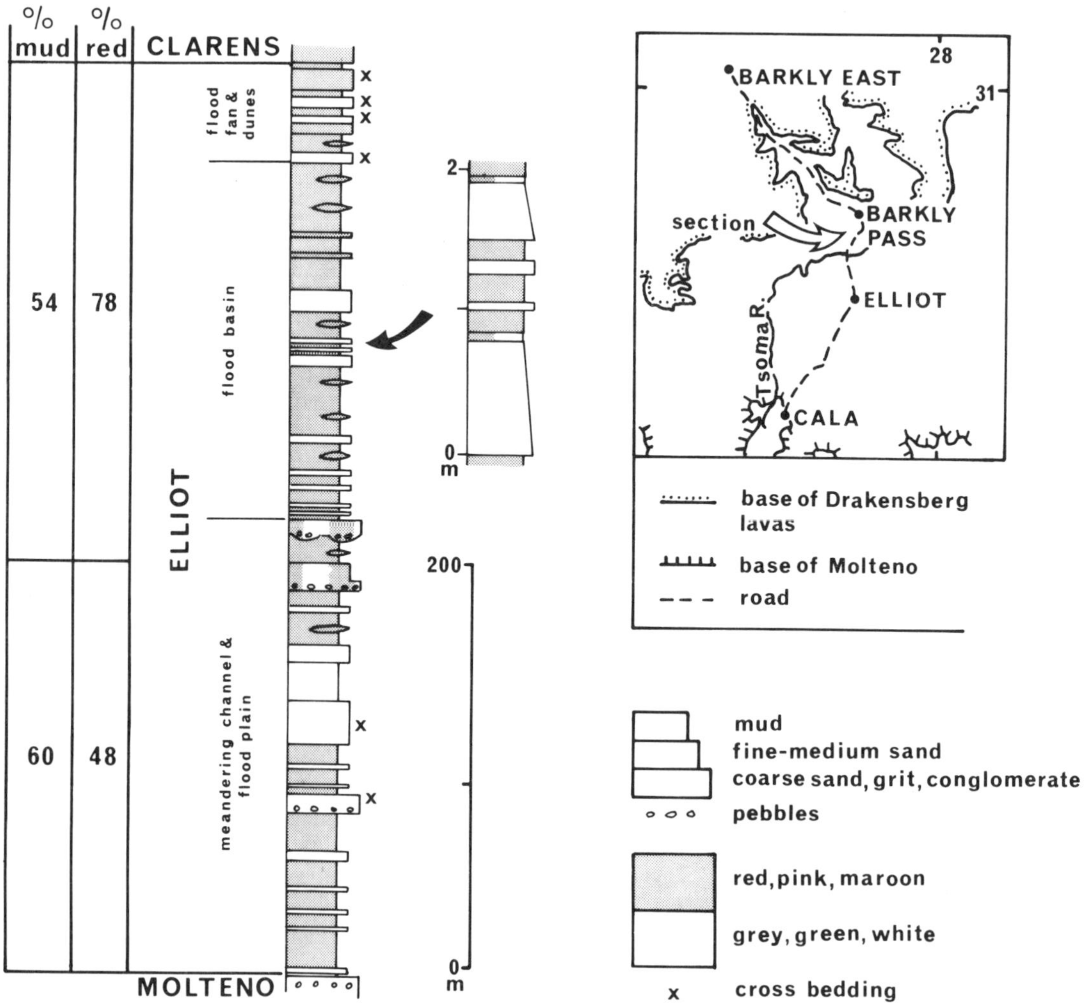


Figure 23. Type section of the Elliot Formation in the Barkly Pass. Data from Botha (1968) and Visser & Botha (1980).

in the type section on the Barkly Pass north of Elliot, at the top of the uppermost 'glittering' pebbly sandstone/small pebble conglomerate (Botha 1968) (Fig.23). Here, the basal beds consist of thin grey shales and yellow sandstone (4 m), which are overlain by 55 m of purple and greyish red mudstones with thin interbeds of grey, red and yellowish sandstones and siltstones. Whilst this has provided a clear cut boundary at the type locality, it is plain, from past experience, that lateral correlation based purely on lithological changes of this nature are likely to prove difficult because reddish coloured 'glittering' sandstones are known to occur well above the first appearance of red shales (e.g. in the Struman district near Elliot, and near Dordrecht (du Toit 1904), and in the Wodehouse area (Haughton 1924)). In addition, in some areas (e.g. Wodehouse) the Elliot Formation has been bleached and is similar in appearance to the underlying Molteno Formation.

Examples of the difficulties involved can be cited from the Indwe district, where du Toit (1904) quotes an abnormally low thickness of Molteno Formation (300 m compared with the expected 450 m), and a change in thickness of 260 m within 22 km in the Elliot Formation. Johnson (1966) re-examined the area, and decided that these unusual thickness variations were due entirely to the application of different criteria for locating the Molteno/Elliot boundary. From his studies in Lesotho, Turner (1972a) concluded that it is best to use a 'package' of criteria for defining the basal beds of the Elliot Formation; abundance of reptile remains; presence of persistent (laterally and vertically) red mudstones; absence of carbonaceous shales, and a low sandstone/shale ratio.

At the type section, the Elliot Formation consists predominantly of thick (20-65 m) greyish red and purple mudstone units, with minor yellow and grey sandstone bodies, and rare alternating greyish yellow and red sandstone/mudstone units (ca.15 m thick) (Fig.23). Argillaceous beds predominate throughout the sequence, but are slightly more abundant in the lower part: mud/sand ratio 1:1.5 in the lower half, and 1:1.2 in the upper half. They are relatively soft and crumble readily on weathering, and towards the top of the sequence, take on brilliant red hues, and are more silty. The mudstones are frequently bioturbated, and some horizons have a thin (up to 20 cm) iron-rich palaeosol with scattered calcareous nodules beneath overlying sandstone units. The arenaceous units are usually fine to medium-grained, but coarse, gritty, pebbly sands occur at three horizons in the middle and lower part of the succession, and a quartz pebble 7.5 cm across was reported by Botha (1968) from the middle section.

Visser & Botha (1980) have recently made a study of the lithofacies and bedforms of the arenaceous units >0.3 m thick in the Elliot Formation type section in the Barkly Pass, and they recognize four main subdivisions, that can be interpreted in terms of depositional environments:

A. Channel sandstones. These occur as thick (>3 m) lenses with irregular basal contacts (scour and fill), that fill channels with thickness:width ratios up to 0.2. The sands are generally fine grained, poorly-sorted, with mud clasts near the base, and are usually structureless except for abundant sole markings and trough cross-lamination in the scour-fill depressions. Visser & Botha (1980) interpret these as sediments laid down under relatively high velocity currents typical of an upper flow regime.

B. Point-bar sandstones. These occur as large lenses (up to 3.5 m thick) that extend laterally for up to 100 m (i.e. they have minimum thickness:width ratios of 0.035). The lenses are often stacked to form units 25 m thick, in which individual basal contacts are sharp and smooth. The sands are coarse-medium grain sized, with occasional thin gritty bases grading upwards into fine sands, and at the top silt and mud. Trough cross-bedding is the main structure, whilst the tops of lenses are bioturbated.

C. Crevasse splay sandstones. Three varieties were recognized. Type C1 forms continuous beds up to 1.5 m thick, which have sharp, but uneven basal contacts. They are fine-grained, and generally upward fining, intensely bioturbated, and show cylindrical dewatering structures. Type C2 is finer grained than C1, and does not exhibit upward fining, whilst type C3 shows upward coarsening, with gradational lower, and sharp upper contacts, mainly fine grain sizes, and frequent slump structures.

D. Dune sandstones. These form fairly continuous beds, up to 5 m thick consisting of stacked bodies. Each bed is typically divided into a lower part with low angle planar, or horizontal bedding, and an upper part with cross-bedding (dips up to 25°) and slumps. The two parts are separated by a desiccation surface with dewatering structures.

These lithofacies are vertically distributed in an uneven way throughout the succession, with types A and B being more important in the lower 200 m, type C dominating the middle 200 m, and type D to the exclusion of other types in the top 40-70 m. On the basis of this distribution, Visser & Botha (1980) interpret depositional environments as shown in Figure 23: a lower zone representing a meandering channel and flood plain facies; a middle flood basin facies; and an upper flood fan and dune facies. This succession indicates a progressive decrease in depositional energy through time, with the establishment near the top of predominantly aeolian conditions. In the lower sequence (meandering channel, flood plain facies), type A sandstones are more abundant in the lower half (further emphasizing an upward decrease in environmental energy), with a strong tendency for the cycle A/B/C to occur, i.e. crevasse channel-point-bar-crevasse splay. This is interpreted as indicating deposition by laterally migrating, high sinuosity streams. The abundance of crevasse splays and overbank muds suggests a high-volume, low gradient river system flowing from the south and SE. In the upper part of this lower sequence, the lower incidence of type A sandstones suggests that the meander belt was abandoned, and that sedimentation took place in a flood plain environment that was subjected to periodic large-scale flooding with mud settling and injection of fine sand via crevasse splays.

This latter depositional environment passed upward into one which was even more remote from the active channel system (Visser & Botha's flood basin facies). It was inundated only occasionally during large floods, and was of great lateral extent. The presence of numerous desiccation cracks and calcareous duricrusts testifies to aridity, the progressive intensification of which is indicated by the onset of the upper facies (flood fan and dune facies) which is represented by type D sandstones and bright red silty muds. Visser & Botha (1980) consider the lower part of the D sandstone beds as representing aeolian sands reworked by ephemeral sheet floods, and the upper parts of the type D beds as aeolian/playa lake deposits in which slumping of the damp sand gave rise to distorted layers. Sediments were fed into the depositional basin via wadis and associated sheet flood processes, which is why Visser & Botha use Reineck & Singh's (1973) term 'flood fan' to identify the extensive areas of sheet sands which were the source for the windborne material. The provenance of the silty, bright red mudrocks are not discussed by Visser & Botha (1980), but presumably are considered to represent either wind-blown dust, and/or settling and reworking of fine material associated with playa lakes established after flash floods.

This youngest depositional environment is clearly

transitional between the fluviatile main part of the Elliot Formation, and the predominantly aeolian main body of the Clarens Formation, although it is similar to that proposed by van Dijk et al (1978) for the basal part of the Clarens Formation in Natal.

The top of the Elliot Formation in the Barkly Pass section is taken by Botha (1968) as a greyish-red mudstone (which grades laterally into a pale red massive sandstone) that is conformably overlain by pale red or greyish yellow massive fine grained sandstones. Botha 1968) describes the formal transition from one formation to the other as being both somewhat arbitrary and difficult to define. While in gross terms, therefore, the glittering, coarse and pale Molteno sandstones, the red Elliot mudrocks, and the pale, massive or cross bedded Clarens sandstones are very distinctive units, transition from one to the other is gradual, and probably regionally diachronous.

No similarly-detailed sections have been published from other parts of the Elliot Formation outcrop, although du Toit (1904, 1905) and Haughton (1924) have provided general accounts. The formation is thickest in the area around Elliot but thins rapidly across Lesotho into Natal (Figs 16 and 20). Its isopachs follow the same general trend as those of the Molteno Formation and it was deposited as a similar wedge-shaped body, although its main depocentre is distinctly more restricted (compare the isopachs on Figure 16B). In the northern part of the outcrop, the Elliot Formation consists of thick red and purple mudrock sequences with massive fine and medium grained sandstones and thinner grits which Haughton (1924) mentions as being soft and feldspathic on the Verkykersberg mid-way between Harrismith and Memel. There is a general increase in the mud/sand ratio northwards from 1.3:1 (Elliot area) to 2.8:1 (Hlatikulu) and 2.6:1 (Harrismith), with a sudden increase in coarse material in the northern-most section at Verkykersberg (1.3:1). Overall, the arenaceous rocks tend to fine up the succession, where the mudrocks often become a more brilliant red hue. Red coloration is caused by finely divided haematite. Carbonate nodules, occasionally abundant enough to form limestones, occur at Herschal and Maclear, where they weather out to give a pocked appearance to outcrops.

Tetrapod fossils are relatively abundant in the Elliot Formation, whilst plant remains are rare. Silicified logs are occasionally found, and in the Aliwal North and Herschel districts some are more than 3 m long, but leaves and stems of *Thinnfeldia* and *Schizoneura* have been reported from only two unspecified localities by Haughton (1924).

Haughton (1924) first drew attention to signs of increasing aridity up the Elliot Formation on the basis of its vertebrate fossil content. Additional sedimentological features, such as suncracks, and tetrapod and worm tracks in the mudstones, along with the discovery of a dreikante pebble and a fossil silcrete by Haughton (1924) also indicate periods of desiccation, which contrast with the apparently well-watered environments in which at least the southern facies of the Molteno Formation were laid down. So far, however, no gypsum or halite have been found. The tetrapod remains in the Elliot Formation are, with few exceptions, found in the red mudrocks and fine sandstones. Skeletons are usually not complete, but individual bones do not show signs of severe wear and are often articulated. This evidence points to relatively low energy, short-distance water transport. Locally, vertebrate bones are very common, and Turner (1972a) records that in the Tsoaning River area of Lesotho 'fragments of prosauropod (*now palaeopod*) can be virtually picked up at will'.

The scheme devised by Haughton (1924) for subdivision of the Elliot (and Clarens) formations has not yet been developed into a formal biostratigraphic zonation, neither has it been applied even in an informal sense to a regional mapping of the Stormberg Group. Nevertheless it does highlight marked changes in the reptile faunas within these rocks, and is potentially suitable to effect a regional biostratigraphic correlation. Haughton's (1924) original criteria and taxonomic grouping have been somewhat revised in a recent review by Anderson & Anderson (1970), and it is their updated version, including taxonomic amendments to the Order Saurischia by Galton & Cluver (1976) and van Heeden (1979) that we discuss here. Table 10 lists the known genera and species from the Elliot and Clarens formations.

Approximately the lower half of the Elliot Formation contains a reptile fauna that consists almost exclusively of plateosaurid dinosaurs. These were heavy-limbed, broadfooted quadrupedal types up to 12 m in length that probably inhabited lowland marshes. Following the revision by van Heeden (1979) only one valid species has been recognized in this group (*Euskelesaurus browni*), with several indeterminate species that may belong to the same genus. Anderson & Anderson (1970) also refer to very rare occurrences of anchisaurid reptiles in this zone. Haughton (1924) suggested that this lower part of the Elliot Formation was deposited under conditions of heavy seasonal rainfall, corresponding to Ellenberger's (1970) Stormberg climatic phase A (warm and humid).

Most of the upper part of the Elliot Formation has a reptile fauna that indicates more arid climatic conditions (Ellenberger's Stormberg climatic phase B – warm but steppe-like and arid) and is dominated by anchisaurid dinosaurs. These were semi-bipedal, light-limbed, slender-footed types, delicately built with hollow bones, and were up to 6 m in length. Following a revision of this group by Galton & Cluver (1976), two genera are now recognized in the Elliot Formation: *Massospondylus* (4 spp.) and *Thecodontosaurus* (2 spp.). Other smaller reptiles that occur in this 'zone' are small (up to 22 cm long) therapsid (mammal-like) herbivorous

Table 10. Summary of tetrapod species recorded from the Clarens and Elliot Formations (after Anderson & Anderson 1970, Galton & Cluver 1976, and van Heerden 1979).

A Clarens Formation

- Reptilia
  - Archosauria
    - Crocodylomorpha
      - *Notochampsa istedana* Broom 1904
      - *Pedeticosaurus leviseuri* van Hoepen 1915
    - Saurischia
      - Palaeopoda
        - Anchisauridae
          - *Anchisaurus capensis* Broom 1906
          - *Thecodontosaurus skirtopodus* Seeley 1894
          - *Massospondylus dubius* (Haughton 1924)
          - *M.palustris* (Broom 1911)
          - *M.africanus* (Broom 1911)
          - *M.taylori* (Haughton 1924)
    - Ornithischia
      - *Geranosaurus atavus* Broom 1911
  - Synapsida
    - Therapsida
      - Tritylodontia
        - *Tritylodontoides maximus* Fourie 1962
      - Ictidosauria
        - *Diarthrognathus broomi* Crompton 1958

B Elliot Formation

- Reptilia
  - Archosauria
    - Crocodylomorpha
      - *Erythrochampsa longipes* Broom 1904
      - *Orthosuchus stormbergi* Nash 1968
      - *Sphenosuchus acutus* Haughton 1915
    - Saurichia
      - Palaeopoda
        - Anchisauridae
          - *Massospondylus gracilis* (van Hoepen 1920)
          - *M.carinatus* Owen 1854
          - *M.harriesi* Broom 1911
          - *M.schwarzi* Seeley 1895
          - *Thecodontosaurus browni* Seeley 1895
          - *T.minor* Haughton 1918
        - Plateosauridae
          - *Euskelesaurus browni* (Huxley 1866)
          - cf. *Euskelesaurus* 4 spp. indet.
    - Ornithischia
      - *Fabrosaurus australis* Ginsburg 1964
      - *Heterodontosaurus tucki* Crompton & Charig 1962
      - *Lycorhinus angustidens* Haughton 1924
  - Synapsida
    - Therapsida
      - Tritylodontia
        - *Tritylodon longaevus* Owen 1884
        - *Likhoelia ellenbergi* Ginsburg 1962
      - Ictidosauria
        - *Trithelodon riconoi* Broom 1912
        - *Pachygenelus monus* Watson 1913
- Mammalia
  - Eotheria
    - Docodonta
      - Morganucodontidae
        - *Erythrotherium parringtoni* Crompton 1964
        - *Megazostrodon rudnerae* Crompton & Jenkins 1968

tritylodontids (2 spp.) and insectivorous ictidosaurs (2 spp.), as well as herbivorous ornithischidian archosaurs (about 1 m long). Two very small (up to 13 cm long) insectivorous mammals have also been recorded from these upper vertebrate assemblages (Tab.10).

In the upper part of the Elliot Formation, where interfingers with the Clarens Formation occur, the anchisaurid assemblage contains faunal elements that were probably restricted to isolated ponds in a relatively arid setting. They consist of three species of crocodylomorpha (small, early crocodiles up to 70 cm long), together with fish, crustaceans, and insects.

Good illustrations of nearly all the Elliot Formation tetrapods are given by Cluver (1978). Benton (1979) places both the Elliot and Clarens formations' assemblages in his lowland category.

#### 2.1.2.3 *Clarens Formation*

Formerly known as the Cave Sandstone because of its characteristic, deeply-pitted weathering surfaces, the Clarens Formation represents the youngest sediment accumulation in the Karoo Basin area. It was brought to a close by the extensive Upper Triassic to Lower Jurassic Drakensberg volcanism, and by the time this had ceased, the drainage geometry of the subcontinent had been radically altered by taphrogenic basin formation around its edges. Thereafter, sediment dispersion in South Africa has been mainly in directions opposite to those that operated during Triassic times, and the Karoo Supergroup has been subjected to massive denudation with substantial diminution of the Stormberg Group outcrops.

The Clarens Formation forms a relatively narrow outcrop around Lesotho (Fig.15), where it frequently forms impressive and weirdly eroded cliffs and buttresses on top of the more rounded hills formed by Elliot Formation strata. These cliffs, which are often topped by resistant cappings of Drakensberg lavas stand out as prominent landmarks with attractive scenery, such as in the Golden Gate National Park. The formation reaches its maximum thickness of about 243 m in the eastern Cape, but its main depocentre does not follow the southward thickening wedge of Molteno and Elliot times (Fig.16) and in contrast, was deposited in a trough which strikes NE-SW from the Harrismith area into Lesotho, where it swings sharply SE in the Stormberg Mountains of the eastern Cape. The establishment of this trough is probably related to local tectonism that predates and is partly contemporaneous with Drakensberg volcanism. It exhibits rapid local lateral thickness variations, particularly in the eastern Cape on the flanks of the main depocentre, where it is occasionally missing. On the other hand, in contrast to the underlying sequences, the Clarens Formation maintains a relatively constant regional thickness (ca.150 m) to the northern edge of the Karoo outcrop and, as we

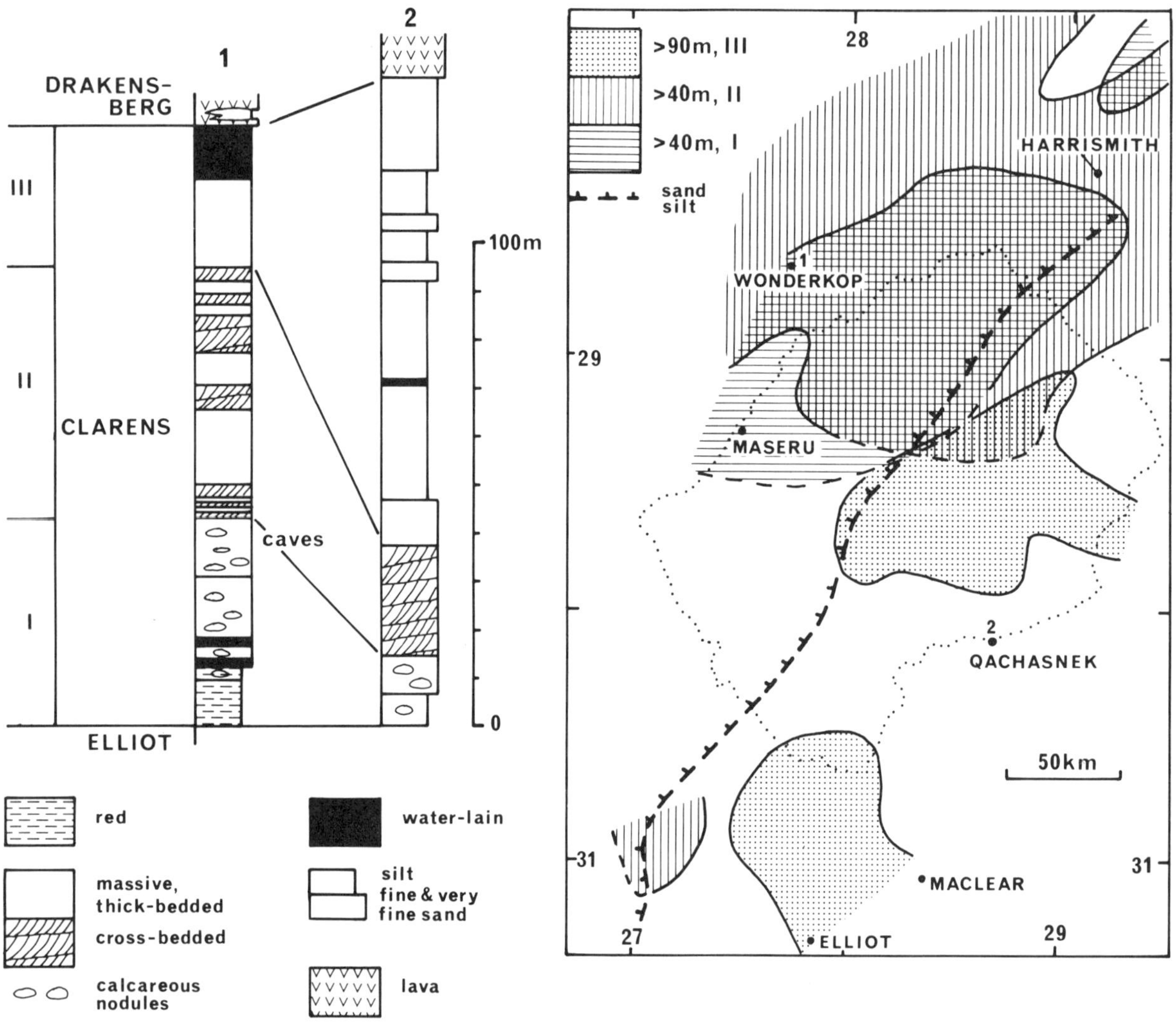


Figure 24. Main depocentres for Clarens Formation members I-III in the main Karoo Basin. Typical sections are for west (1-Wonderkop) and east (2-Qachasnek) localities. The sand/silt boundary on the map is for member III and denotes western limit of silty horizons. Data from Beukes (1970).

shall discuss later is the most widely developed of all the Stormberg Group sediments in the extra-Karoo areas. It has not been extensively studied, and no detailed regional mapping has been carried out. This is because of its relatively homogeneous lithology, lack of well-developed marker horizons, relatively sparse fossil content, and absence of known mineral deposits. Vertebrate bones and tracks have been found in the Orange Free State, Lesotho and NW Natal, and the reptilian faunas are closely allied to those found in the Elliot Formation. They have the characteristics of Romer's C fauna (1975) and probably indicate a Rhaetian to Hettangian age (Ellenberger 1970, Cox 1973) (Fig.9).

Beukes (1970), who coined the name (reported in Johnson et al 1976), has undertaken the only recent regional work on the formation, following the pioneering studies of du Toit (1904, 1905) and Schwarz (1903). Haughton (1924) added a few details in his review of Karoo rocks, and detailed studies of outcrops in Natal have been reported by van Dijk et al (1978), and Eriksson (1979).

The base of the formation has been taken by Beukes (1970) and earlier workers at the point where red and purple mudstones and cross-bedded sandstones become subordinate to lighter and 'massive' sandstones and grey/green shales. In practice, this junction is usually fairly obvious (within a few metres) at any locality, but the gradational nature of this facies change, both laterally and vertically, means that it is almost certainly an irregularly diachronous junction on a regional scale. Potentially confusing variations on this definition can be introduced as, for example, that applied by Eriksson (1979) in the Kamberg area of Natal (see Fig.25), where the boundary between the Elliot and Clarens formations is taken at the top of the last continuous red bed. A similar, but more complex stratigraphic problem is found at the top of the Clarens Formation, where the sediments are overlain by, and interfinger with, the lavas of the Drakensberg Group. This situation has been complicated by local large-scale erosion prior to extrusion, synchronous extrusion of lavas and sediment deposition, and tectonism in the country sediments adjacent to individual volcanic centres.

The Clarens Formation typically consists of yellowish- and white-weathering massive (i.e. apparently unbedded), and cross-bedded fine, feldspathic sandstones with minor grey and green shales and silts. Beukes (1970) has subdivided the formation into three members, by recognizing a middle division, in which cross-bedding is common, over- and underlain by thick massive units with minor waterlain sands and mudrocks (Fig.24). The Basal member (I) is thickest in the east Orange Free State and NE Lesotho (ca.43 m) and is absent or less than 10 m over most of the eastern Cape (Fig.24). It consists of thick-bedded (> 3 m) massive light brown and pale reddish, very fine sandstones and silty sandstones, which become more yellowish and grey towards the top of the member. Calcareous concretions (few mm up to 1 cm) are scattered throughout the sandstones, and in one broad horizon, the calcareous content of the matrix is particularly high and its susceptibility to erosion gives rise to the cave-like weathering. Beukes (1970) reports local conglomerates composed of penecontemporaneously eroded calcareous concretions, indicating that they are synsedimentary features. Reddish rocks are common near the base of the formation, and emphasize the gradual alteration in sedimentary environments across the Elliot/Clarens boundary. They are locally accompanied by thin, lenticular muds, shales, and fine sandstones, which exhibit ripple marks, suncracks, and clay pellet conglomerates, all indicating deposition under ephemeral shallow water conditions. These deposits, which probably represent playa lakes, have proved fossiliferous in the eastern Cape (Siberia), at Ficksberg and Harrismith in the Orange Free State, and in NW Natal, and contain a fauna of *Estheria,* insects, ostracods, fish and dinosaur tracks.

In NW Natal (Kamberg Nature Reserve), Eriksson (1979) has described the succession immediately above the top of the Elliot Formation as consisting of (from the base up) 50 m of pale orange-pink to white, medium-coarse sandstone with planar cross-bedding and channel fill (with basal mud clast lags), which is overlain by 15 m of alternating fine to very fine-grained sandstones and siltstones, and 55 m of thin-bedded, silty, fine to very fine-grained sandstones and siltstones with minor mud partings. The upper sequence contains ripple and tool marks, mudcracks, tetrapod footprints and coprolites. He interprets this succession as representing:

i) an early period of fluvial transport caused by floods in a dry desert terrain that redistributed aeolian sands in a bajada (arid alluvial plain with no alluvial fans) by braided wadi channels;

ii) superseded by deposition in temporary playa lakes which were created during infrequent, but heavy rainfall. No original aeolian sediments were recorded by Eriksson (1979) in the 120 m section, but the presence of fine, well-sorted aeolian sand grains clearly relates to a nearby dune field. Such a source would have occurred at Giant's Castle (see Fig.25 for relative locations) from where van Dijk et al (1978) have described a similar situation to that recorded by Eriksson (1979), although unfortunately they do not state the precise stratigraphic level within the Clarens Formation of the outcrop with which they were dealing. The lower part of the exposure is a fluvial cross-bedded sandstone which is overlain by bioturbated mudstones (ca.80 cm) suggesting the establishment of a shallow pond. These are overlain by 3 m of fine sandstone and siltstone with shallow water sedimentary structures (rill marks, etc.), suncracks, thin gypsum layers and reptile footprints that indicate a playa lake-type environment. Ephemeral streams filled in this lake with coarser sands, and the whole sequence occurs adjacent to aeolian sands. This situation suggests that a small playa lake formed where streams were dammed by shifting desert dunes (van Dijk et al 1978).

Cross-bedded sandstones are restricted to the middle member (II) of the Clarens Formation. This member is thickest along the western part of the outcrop, and reaches its maximum development in the zone between northern Lesotho and the Harrismith area (ca.60 m), whilst a smaller area about 40 m thick occurs between Aliwal North and Barkly East (Fig.24). In the type area in the Orange Free State, Member II consists of alternating cross-bedded and massive sandstone units (Fig.24). The latter are very fine-grained, whilst the cross-bedded sands are fine-grained, and better-sorted (Fig.25B). There is a progressive decrease in grain size ESE in the cross-bedded sands from about 170 $\mu$ in the Orange Free State, to 80 $\mu$ in the eastern Cape, which suggests a westerly source (Fig.25A). Beukes (1970) recognizes two main cross-bedding types: Omikron (units 1-10 m thick), and pi (units 2-3 m thick) which he equates with straight-crested transverse, and bachan dune deposits, respectively. Lenticular, laminated argillaceous facies with shallow water and desiccation structures are rare in Member II, although in the SW parts of the outcrops (between Dordrecht and Barkly East) this facies is well-developed.

Clarens Formation Member III consists of thick, massive units of very fine sandstone in the west Orange Free State, and western Lesotho which grade into thick, massive, silty sands and silts in the east (Fig.24). They reach a maximum thickness of over 240 m between Maclear and Elliot, where a narrow trough-like depocentre strikes NE-SW. This feature was first described by du Toit (1905), and may be associated with tectonism which accompanied the early phase of Dra-

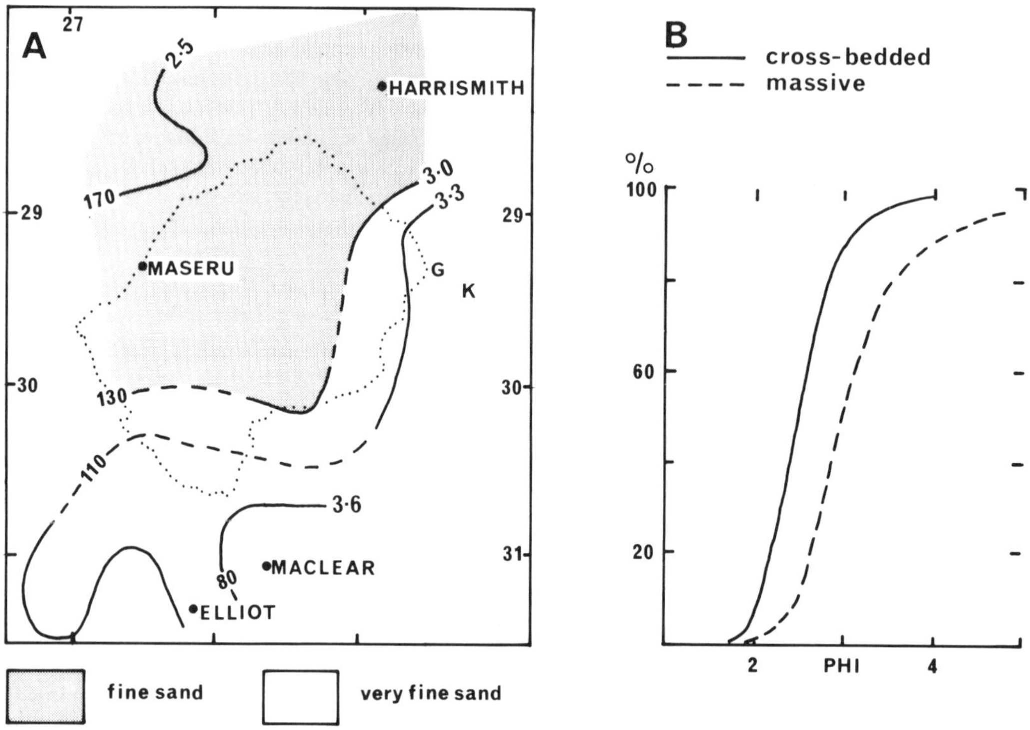


Figure 25. A. Regional variation in mean grain size of cross-bedded units of the Clarens Formation. Large letters indicate phi values, small letters show micron equivalents. G = Giants Castle locality of van Dijk *et al.* (1978), K = Kamberg locality of Eriksson (1979).
B. Cumulative frequency curves for cross-bedded and massive sandstones in the Clarens Formation. They show better sorting in the cross-bedded units in member II.
Data from Beukes (1970).

kensberg volcanic activity. A somewhat thinner, smaller basin occurs under central Lesotho (ca.90 m) and in the west, Member III is less than 30 m thick. These thickness variations are accompanied by a progressive eastward fining in grain size (similar to that shown in the cross bedded units of Member II) and again suggest a westerly source for the detritus. Thin, lenticular, shallow water deposits are abundant in Member III, particularly near the top, in association with the earliest lavas.

Mineralogically and texturally, the arenaceous sediments of the Clarens Formation are very homogeneous. The sand grains are composed predominantly of clear quartz, although some have what Beukes (1970) calls a 'metamorphic' texture, and the bulk of them are rounded and 90 % are frosted and pitted. Feldspar (orthoclase, microcline and plagioclase) invariably occurs, but published accounts give no values. Heavy minerals are minor constituents and include zircon, garnet, epidote, apatite, andalusite, titanite, and tourmaline. Sorting varies from well- to very well-sorted in the fine sandstones, and from medium- to well-sorted in the very fine sandstones, whilst the matrix content of the siltier rocks reaches 60 % and consists of zeolite, kaolinite, montmorillonite and chlorite.

It has been usual practice to take the base of the earliest ash, lava or agglomerate band as the local top of the Clarens Formation, where with only limited fossil evidence and few radiometric dates there is no way of identifying an isochronous datum. This introduces considerable diachroneity into stratigraphic mapping because ashes and lavas are frequently found interbedded with sediments, particularly in the southern areas, and regional isopach variations (especially in Member III) must owe a certain amount to this factor. The higher incidence of lava/sediment interbeds, together with the local occurrence of lavas, as low down as Elliot Formation rocks, indicates that volcanism commenced earlier in the south, a conclusion reached by Haughton (1924).

In some areas, large-scale erosion preceded the start of the extrusive phase of volcanism, whereas in others, lavas and ashes are interbedded with or overlie sediments apparently conformably. In the former areas, an hiatus is implied, and good examples have been described by Beukes (1970) and Lock et al (1974). Those shown in Figure 30 suggest erosion by running water and in two such localities identified by Beukes (1970, p.328) water flow was estimated to have been from west to east. Tectonic activity prior to large-scale effusion, presumably by earthmovement associated with explosive eruption, also occurred in many areas where small-scale folding and faulting affected the main Clarens Formation sandstones and underlying Elliot strata, but not the overlying basalts. Du Toit (1904, 1905) described examples of faulting followed by erosion prior to lava extrusion from several localities (e.g. Elliot, Dordrecht) and Beukes (1970) cites an 18 m fault near Fouriesberg in the eastern Orange Free State. The top of the Clarens Formation, therefore, varies in nature and age from place to place: locally an hiatus occurs between the cessation of sedimentation and lava extrusion; and in others the base of the Drakensberg Group is placed low in the succession and encompasses a good deal of sedimentary material.

Fossils are not common in the Clarens Formation. Silicified logs occur intermittently throughout the sandstones, and although Ellenberger (1970) records conifer, cycad, equisetes and taxodineae, plant remains are rare. Animal fossils are locally found in the water-lain mudrock interbeds that represent temporary lakes in the otherwise arid terrain. Apart from the small invertebrates already mentioned, the most important fossils are reptiles. Table 10 shows the revised list of known species given by Anderson & Anderson (1970), and a strong resemblance to that of the upper part of the Elliot Formation is apparent. Anderson & Anderson (1970) divide the succession into three parts (following Beukes 1970) and recognize faunas that reflect variations in aridity. The lower zone (equivalent to Beukes's Member I) has a fauna adapted to semi-arid conditions in which the aeolian structure of the original dune bodies has been destroyed by intermittent, but torrential rain, plant roots and sand flows. The reptile fauna in the lenticular bodies of water-lain sediments in this zone are dominated by three genera of anchisaurid (light-limbed) reptiles: *Anchisaurus* (1 spp.), *Thecodontosaurus* (1 spp.) and *Massospondylus* (possibly 4 spp.) with rare crocodylomorpha and single finds of a medium sized tritylodont herbivore (25 cm skull) and a small ictidosaurian insectivore (5 cm skull). This fauna is very similar to that in the upper part of the Elliot Formation with two anchisaurid genera common to both although the overall tetrapod fauna is apparently less diverse. No plant or animal fossils were known to Anderson & Anderson (1970) from the middle zone (Member II) which was apparently the most arid, but it seems that the outcrop described by van Dijk et al (1978) was from this level and they recorded reptile tracks. The upper zone (Member III) indicates a return to semi-arid conditions, but no vertebrate remains other than tracks have been reported so far (Kingsley 1964, quoted in Beukes 1970).

It is appropriate to mention here that fossils have been found in some sediment interbeds within the Drakensberg volcanics. Anderson & Anderson (1970) mention charred tree trunks, and Ellenberger (1970) has listed equisetes, conifers (4 spp.), cycads, salisburiea and ?medulloseae. Vertebrates are represented by footprints including possible mammalian types. Other fossils in the interlava sediments include various arthropods (including ostracods), insects and numerous unidentified invertebrate tracks. Ellenberger (1970) records various fossils at 50, 300, 500-560, 650-700 and 1 000-1 700 m above the base of the volcanic sequence, although the highest recognizable vertebrate tracks were from the 500-560 m levels.

#### 2.1.2.4 *Palaeogeography*

All available evidence leads us to the conclusion that the main uplift of the Cape Fold Belt took place in post-*Cynognathus* Zone – pre-Molteno Formation times (i.e. Anisian-early Ladinian). The Molteno Formation strata were thus deposited to the north of the newly-formed and still intermittently rising mountain range. Subsequent Elliot and Clarens formation sediments accumulated in a similar geographical setting although under somewhat different climatic and depositional-energy regimes, the latter being a function mainly of continued topographic discontinuities between the source and accumulation areas. From Figures 8, 10, 16 and 20 it can be seen that the main Molteno and Elliot depocentre lay adjacent to the Cape Fold Belt (the southern part of the basin having since been eroded away) in a relatively narrow intracratonic trough. The relatively rapid subsidence of the crust in this trough (a minimum of about 50 m/my) was probably a rheologically induced response to the orogenic uplift. This response had waned by Upper Triassic times, so that the Clarens Formation was deposited in an approximately N-S trough, and the main asymmetry in this younger downwarp was probably induced by local violent volcanic activity and attendant crustal deformation. Compared to the Molteno and Elliot formations, the Clarens Formation maintains a relatively constant thickness in a north-south direction, indicating that the Cape Fold Belt was no longer the main sediment source and that the zone immediately to the north of it had ceased to subside rapidly. In this respect, it is possible to consider the Molteno and Elliot sedimentary successions as being related, with the coarse lithologies representing proximal facies and the finer sediments more distal sedimentary environments.

Theron (1962) has suggested that when first formed, the Cape mountains were up to 4 200 m high, and it is probable that, locally at least, they were snow and ice covered for parts of the year. This elevated source area was the main contributor of sediment to the Karoo Basin during Molteno times, when high energy agents were capable of transporting coarse clasts (average size 16.5 cm) up to 180 km northwards. Turner (1978a) has used these figures to estimate that a maximum palaeoslope of about 23 m/km was generated during Molteno times. The overall fineness of the overlying Elliot Formation indicates that by Upper Triassic times

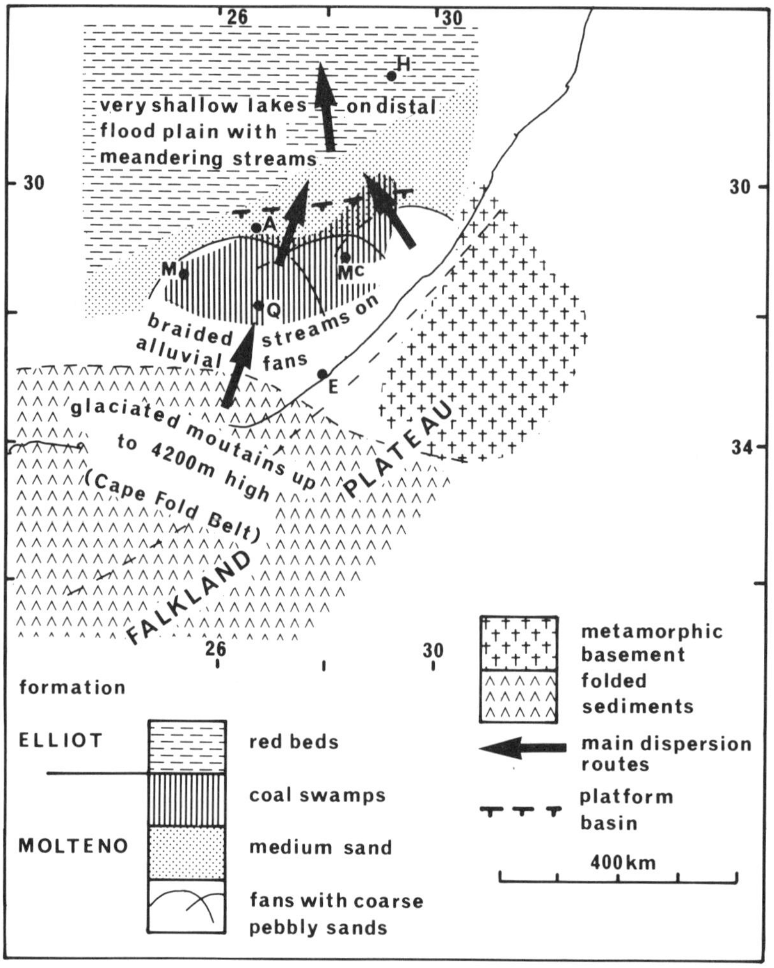


Figure 26. Palaeogeography of main Karoo Basin in Carnian times. See text for discussion. Transportation directions from Turner (1977, 1978a). Abbreviations: M = Molteno; Q = Queenstown; E = East London; H = Harrismith; A = Aliwal North; Mc = Maclear.

this slope was very much more gentle, as a result of considerable denudation of the mountain range.

The three main phases of Molteno sedimentation indicate periodic rejuvenation and progressive degradation of the source area relief, with a particularly large upward tectonic pulse and resultant widespread distribution of coarse detritus during the second phase (cycle C, Indwe Sandstone Member). Minor pulses (cycles A-F), indicating smaller adjustments in sediment supply, were superimposed on the major phases. All the textural and component evidence amassed by Turner (1969 et seq.) points to the Molteno sediments being water-lain, and from his studies on the cross-bedding in the sandstone units, Turner considers them to have been deposited in a fluvial environment (e.g. 1975b, 1978a). Cross-bedding is predominantly of the trough type probably representing asymmetric megaripple bedforms migrating down channel under a high energy, lower flow regime, and planar cross-beds, which are less common, all point to deposition as megaripples and bars in a braided river environment. Meandering is indicated by only one set of epsilon ripples. The Molteno Formation sediment wedge, as seen today, was probably formed as an alluvial plain on the distal part of an alluvial fan complex by low sinuosity, braided rivers. Current direction studies (Turner 1977) indicate two main transport directions (from the SE & SSW), which coalesced in the vicinity of central Lesotho. The more proximal parts of the alluvial fan are no longer preserved. As sedimentation in any one cycle proceeded, progressive reduction in the energy of transportation resulted in finer detritus reaching any point and a southward migration of the more distal parts of the alluvial plain.

During the deposition of the Molteno Formation we envisage palaeogeographies typified by Figure 26. In the early stages of each sedimentary cycle, coarse sandstones with large well-rounded and well-sorted clasts were swept down from the mountainous region in the south and SE, where lack of vegetation and preponderance of mechanical weathering sustained a high supply rate, possibly seasonally augmented by ice and snow melting, and glacial debris. As the cycle proceeded and sediment supply became less coarse, the lower energy distal environments encroached southwards, culminat-

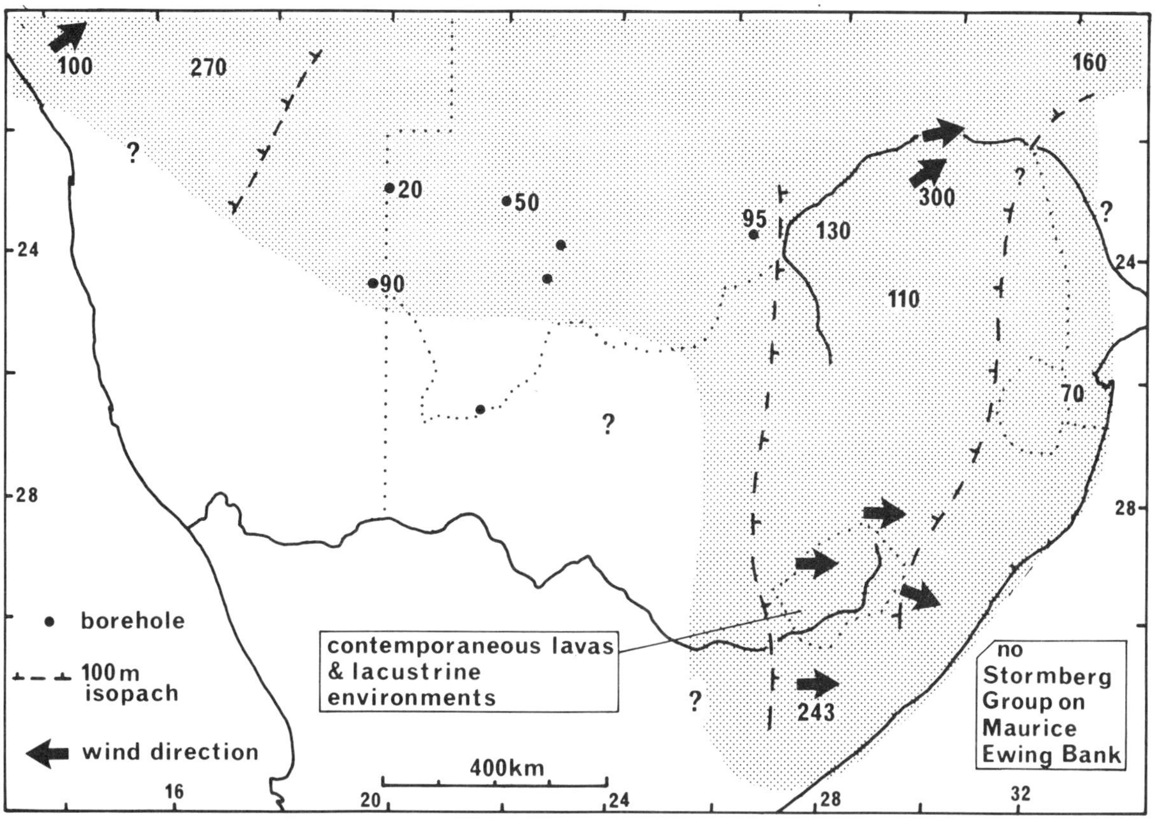


Figure 27. Extent of the Clarens Formation desert in southern Africa. It persisted from Rhaetian to ?Hettangian in the SE, ?Carnian to Lower Jurassic in the Etjo area, and Rhaetian to Berriasian in the extreme N.W. (Doros area). Wind directions were consistently westerly. Locally, aeolian conditions gave way to lacustrine environments, and over large areas the upper sediments are interbedded with lavas. Maximum accumulation was in NW Namibia, N.Transvaal, and in the SE. Spot thickness values are shown on the map. The figure of 600 m quoted by van Eeden *et al.* (1955) in the Soutpansberg is probably not accurate (see their original account) and a value of 300 m has been assumed. Data are from various sources, but principally Haughton (1969), Green (1969), Hodgson (1970), Beukes (1970), Bigarella (1970).

ing in the establishment of swampy and lacustrine, low energy conditions between the main river channels. Individual lakes were probably surrounded by, and sustained within their flood zone limits, a thick vegetation which accumulated in situ to produce autochthonous peat deposits (see Falcon 1975, fig.8). The preponderance of leaves (compared with logs and bark) accounts for the high ash content of the resultant coals. Periodic, possibly seasonal, floods terminated plant growth and deposited fine terrigenous debris in the coal swamps, locally causing erosion of the peat. Lack of evidence of evaporites, desiccation cracks, and the fresh state of feldspars indicates a cool, damp climate on the lower part of the alluvial fan, and the proximal parts of the alluvial plain. Conditions for the formation of coals were never as favourable after the early stage of cycle D, when only carbonaceous shales were deposited, possibly reflecting a slightly drier climate (and less luxuriant plant growth). The bulk of the coal measure plants belong to the lowland *Dicroidium* flora, but elements of upland floras (*Glossopteris* and *Ginkgo*) do occur. Curiously, no tetrapod remains have been found despite the apparently favourable environmental conditions. It may have been too cool and damp.

It was probably not coincidental that the onset of cooler, damper conditions took place after the establishment of a major physiographic barrier over the southern Cape. The mountain range clearly affected local climates immediately to the north of it, and the degree of aridity may have been inversely proportional to their height. A general increase in aridity accompanied by a rapid southward encroachment of red-beds marks the termination of the Molteno and the establishment of the Elliot lithofacies. In all probability, the deposition of red beds of this type was continuous in the northern part of the basin from mid-Triassic to Upper Triassic times (Figs 26 & 28). Numerous examples of ripple marks, cross bedding, local erosion surfaces, and clay pellet conglomerates indicate that the Elliot Formation was deposited in very shallow, non-marine conditions, probably on the distal parts of a flood plain, typified by large shallow lakes and high sinuosity mean-

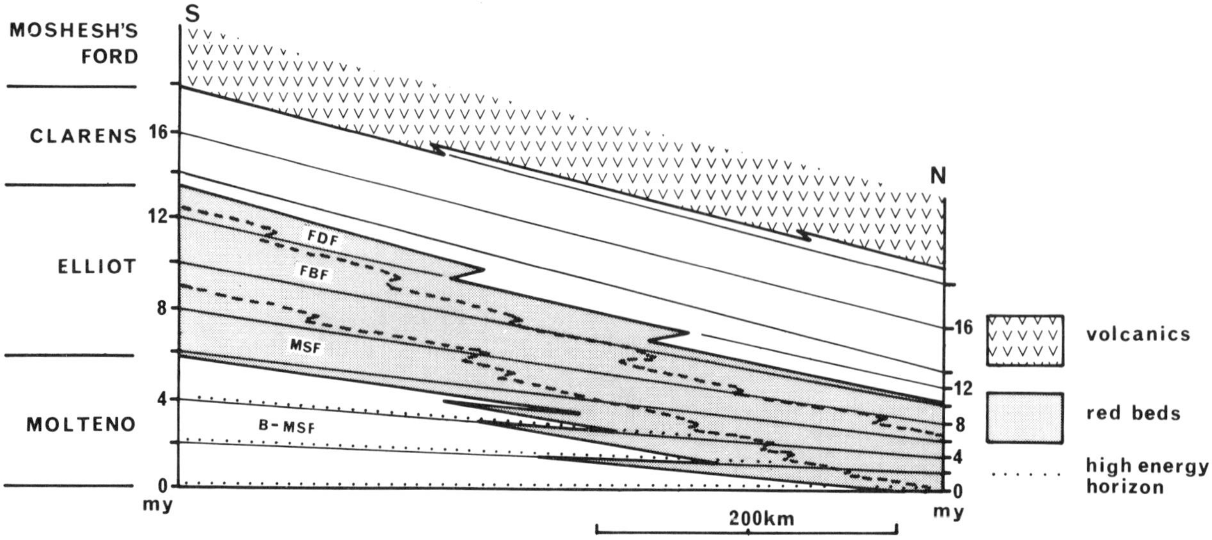


Figure 28. Schematic distribution of Stormberg and Drakensberg lithofacies in time and space. Diagram in the form of a N-S section across main Karoo Basin. Vertical scales are in my above datum which is taken at the base of the Molteno Formation at approximately 220 my BP (top Ladinian). Dates are speculative, but give relative values that can be related to accumulation rates. High accumulation rates persisted in the south for the whole period, but were low in the north for the first 14 my. Note the strongly diachronous nature of the major lithofacies boundaries which indicate an earlier onset of red bed deposition (arid/semi-arid conditions) in the north. Abbreviations: FDF = flood fan/dune facies; FBF = flood basin facies; MSF = meandering stream facies; B/MSF = braided and meandering stream facies. Data on Elliot depositional environments are from Visser & Botha (1980).

dering streams. Desiccation of the lakes between flood periods is indicated by the presence of sun cracks and tetrapod footprints, whilst periodic increased flow by the streams resulted in the deposition of coarser detritus in the form of crevasse splay sands. The red coloration of the sediments is due to fine haematite and is more prominent in the upper parts of the formation when deposition took place in areas more remote from the stream channels. If the degree of redness is taken as a measure of aridity, then the appearance of aeolian sediments near the top of the formation confirms the suggestion of a progressively deteriorating climate. The green shales were probably deposited in low ph, standing bodies of water.

During Elliot Formation times therefore, we envisage southward encroachment of the distal flood plain/ lacustrine environments into the eastern Cape, and the general establishment of relatively dry conditions. Nevertheless, the sediments are waterlain, so dry periods followed by heavy (?seasonal) rains may be indicated. Towards the top of the Elliot Formation a further deterioration in climate is indicated by the appearance of aeolian sands interfingering with water-lain red rocks. The drier upper Elliot Formation conditions were accompanied by a change in tetrapod faunas with the progressive incoming and final dominance of light-limbed types, in contrast to the numerous large browsers that inhabited the area at the start of Elliot deposition. Vegetation on the lower flood plains was relatively sparse, compared to the more proximal and locally densely-vegetated area around the Molteno coal swamps, as fossils are restricted to occasional silicified tree trunks and rare leaves and stems. Volcanism locally occurred in the eastern Cape in Elliot times, with the extrusion of lava flows and deposition of pyroclastic debris.

The onset of aeolian conditions was probably diachronous, but with its regional establishment (Clarens Formation) radical changes in climate and tectonic conditions are indicated. Firstly, the southern intracratonic trough ceased large scale subsidence and maximum sedimentation during Clarens times was roughly along a north-south axis. Secondly, the provenance and mode of sedimentation shows strong contrast with the Molteno/Elliot situation. Unequivocal waterlain sediments are relatively rare and are restricted to thin, lenticular bodies with fresh water fossils, tetrapods and shallow-water sedimentary structures, while the bulk of the Clarens Formation is 'massive' and cross bedded fine sandstones and silty rocks. Texture and sedimentary structures indicate that the cross-bedding is of aeolian origin, whereas the 'massive' units are thought to have been deposited initially under aeolian conditions, but that the primary structures were penecontemporaneously destroyed by violent rain storms which produced sand flows. Temporary playa lakes produced by particularly heavy rains were responsible for the lacustrine waterlain units. Directional studies on the aeolian cross-bedding indicate a consistent westerly wind regime, and the finer silty units have been likened to modern loess.

During Clarens Formation times therefore, we envisage a generally dry, windy climate with occasional heavy storms and the development of playa lakes with short term establishment of vegetation, and vertebrate and invertebrate faunas (Fig.27). Aeolian cross-bedding is most consistently preserved (i.e. fewer storms occurred to rework the aeolian structures) in the middle member (II) of the Clarens Formation, so that the steady deterioration in climate that commenced at the end of Molteno times (early Norian) reached its climax at this level (?mid-Rhaetian). A slight amelioration, with a return to a higher frequency of heavy storms and playa lake formation is evident in the upper member (III) of the formation, (?late Rhaetian-early Hettangian) where large water-eroded channels locally occur at the top of the Clarens sandstones. These, along with other erosional landforms, were eventually filled in and smoothed out by lava flows.

Explosive volcanism was widespread in the eastern Cape during much of Clarens Formation times, and the pillow lavas in the lower parts of the Drakensberg Group attest to the presence of the water bodies. Most of the sandstones in the eastern Cape contain quantities of pyroclastic debris (Fig.35) and many of the vents were choked and filled in with windblown and water reworked detritus and pyroclastic sands and silts. Local collapse structures and regional depressions caused by volcanic activity can be recognized by rapid thickness variations in the Clarens Formation.

With the onset of large-scale fissure eruptions, sedimentation was brought to a close. Weathering surfaces between the lavas are rare, indicating that sedimentation did not have a chance to re-establish itself between successive eruptions of the highly mobile flows.

Figure 28 schematically summarizes the diachronous relationships within the Stormberg Group that we have outlined above. It illustrates that relatively slow sedimentation rates prevailed in the north of the Karoo Basin until low in the Clarens Formation (shown by convergence of the time lines northwards), after which there was more or less uniform accumulation over the whole basin. However, large-scale volcanism commenced earlier in the south, whilst sedimentation proceeded for longer in the north. The diagram also expresses our interpretation of the strongly diachronous nature of the Molteno/Elliot and Elliot/Clarens lithofacies showing that predominantly red-beds were being deposited in the north for the whole period during which the Molteno Formation was accumulating in the south, though the former accumulated at a relatively slower rate. Increased aridity and aeolian deposition probably commenced earlier in the north, not spreading to the south until several million years later.

### 2.1.3 *Early Triassic to early Jurassic volcanic rocks*

The accumulation of Karoo Supergroup rocks was completed by the extrusion of thick lava sequences during early Jurassic times. The original extent of this cover is not known, but possibly it stretched from Namibia, across Botswana, Lesotho, and the Lebombo Line into SW Mozambique (Fig.5). Extensive denudation has reduced it to large, isolated outliers, the most impressive of which is centred on Lesotho and the eastern Cape, where stacked flows form rugged high ground in the Stormberg and Drakensberg ranges (Fig.15).

In this section we will deal only with the stratigraphic aspects of the so-called 'Karoo volcanics', leaving remarks on the petrography and geochemistry to Chapter 7. Three main phases of volcanic activity can be recognized in the late Palaeozoic to early Mesozoic rocks of South Africa, and although these can all be loosely referred to as the 'Karoo volcanic episode', only the later two phases can be mapped as the Drakensberg Volcanic Group. They are:

i) an early explosive phase for which no local volcanic centres are known;

ii) a middle phase with development of discrete volcanic centres which produced extrusive, intrusive and ejected material;

iii) a late phase of widespread lava extrusion associated with dyke and sill intrusion.

At present, a formal lithostratigraphic classification for all these rocks has not been erected, so we have adopted a scheme based upon the suggestions of Lock

Table 11. Classification of Karoo Supergroup volcanic rocks in the main Karoo Basin and along Lebombo Line (modified from Lock *et al.* 1974, Johnson *et al.* 1976, Cleverly & Bristow 1979). Phases shown in Figure 194.

| Group | KAROO Formation | Lithology | Phase |
|---|---|---|---|
| Drakensberg | Lesotho<br>Kraai River | flood basalts | III |
| | Moshesh's Ford | lavas, pyroclastics, sediments, vent fill | II |
| Stormberg | Clarens<br>Elliot<br>Molteno | sediments containing finely disseminated volcanic debris | I |
| Beaufort | Burgersdorp | | |

| Group | LEBOMBO Formation | Lithology |
|---|---|---|
| Lebombo | Movene | basalt and sediment interbeds |
| | Jozini | rhyolites |
| | Sabi River | basalts |
| | Letaba | |
| | Mashikiri | nephelinite |

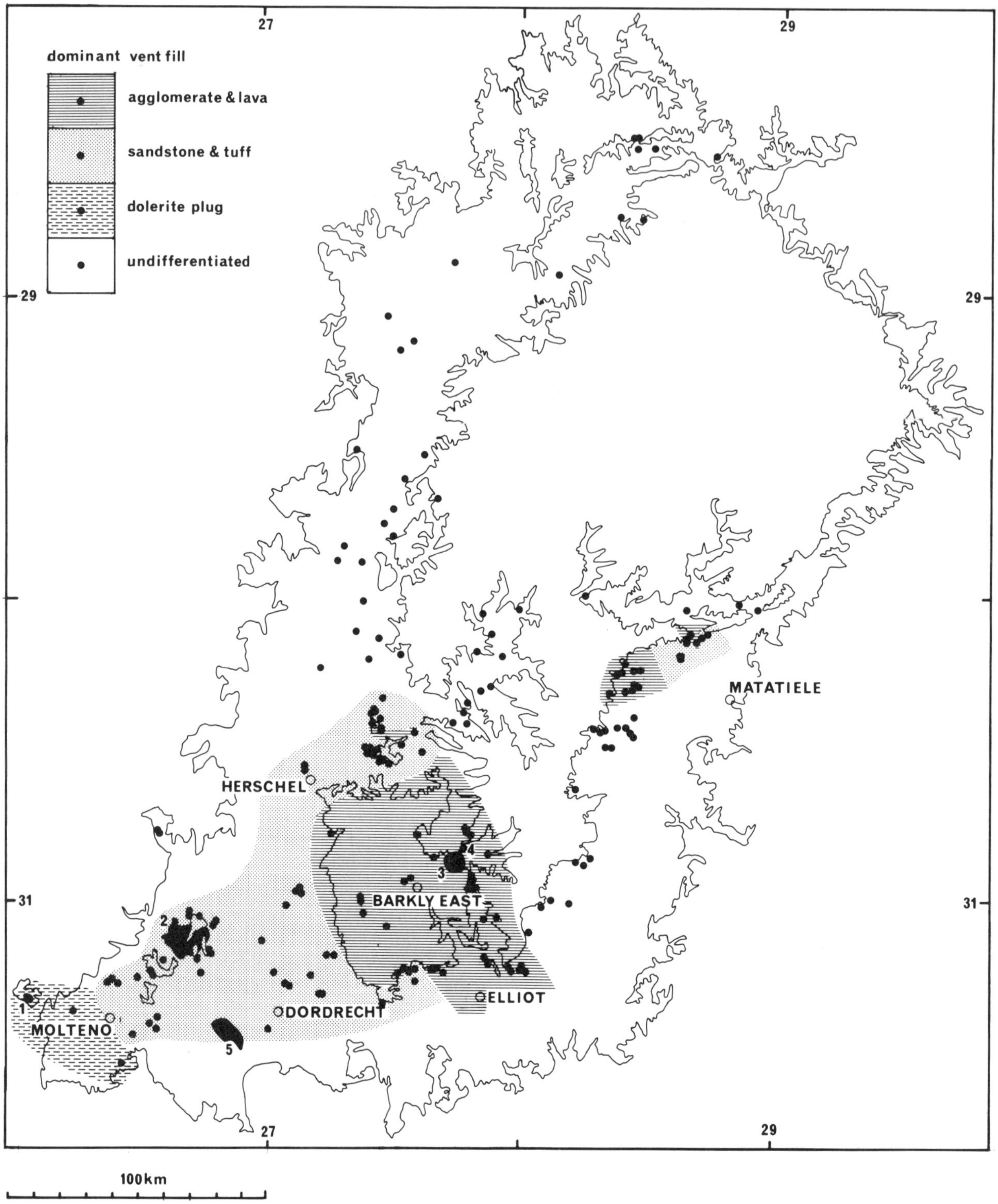


Figure 29. Upper Triassic-Lower Jurassic volcanic vents, plugs, and diatremes in the main Karoo Basin. Outlines show bases of Molteno and Lesotho/Kraai River formations. Mainly compiled from data in du Toit (1904, 1905), Gevers (1929), Geological Survey (1970), Lock *et al.* (1974). Localities: 1 = Gladkopjes; 2 = Roodepoort/Zwartfontein; 3 = Kelvin Grove; 4 = Belmore; 5 = Birds River.

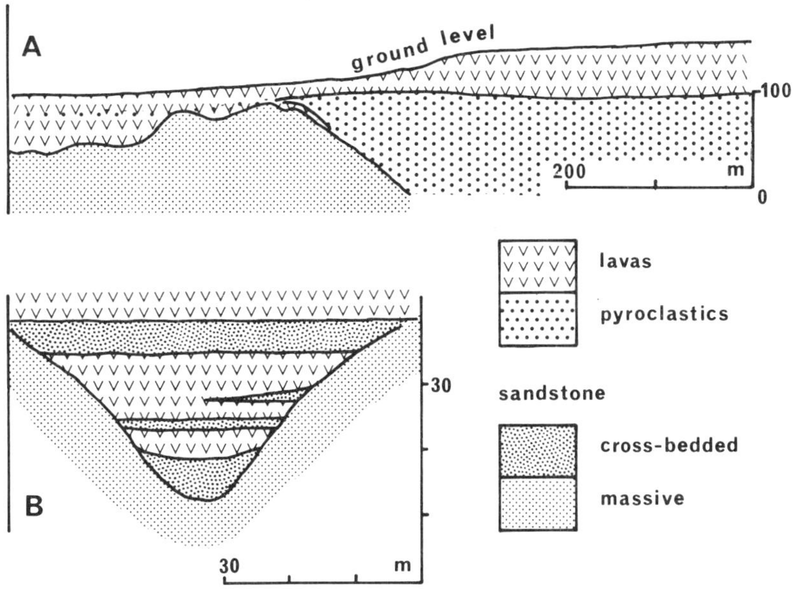


Figure 30. Sketches of exposures showing evidence of erosion of Clarens Formation before lava eruption, and penecontemporaneous lava eruption and sedimentation. A. Valley section in Rebels Glen, 8 km N of Barkly East (from a photograph and interpretation by Lock *et al.* 1974). B. Unlocated section in member III (after a sketch by Beukes 1970).

et al (1974), Johnson et al (1976) and SACS (1980), which is summarized in Table 11. For comparison, we have included the classification of the Lebombo volcanics, but because it is uncertain whether or not the two sequences were ever physically connected, correlation between the rocks of the Karoo Basin and the Lebombo Line has not been attempted. The Lebombo Group is dealt with separately under the section Extra-Karoo outcrops (2.2.5.1).

2.1.3.1 *Early phase of volcanism (?early to late Triassic)*
Volcanic debris is widespread in late Palaeozoic/early Mesozoic rocks of the southern and eastern Cape, and appears to be concentrated at two horizons: Lower to Middle Permian; and Triassic.

In the Herschel district, fine wind-borne volcanic matter (represented by the zeolite laumonite, which is thought to be an alteration product of volcanic glass) has been reported by Fuller (1970) in all the sediments of *Cynognathus* Zone and younger age, up to and including Clarens facies interbeds within the Drakensberg lavas. The laumonite occurs in red and grey/green beds of both arenaceous and argillaceous lithofacies, as 0.5-1.0 cm spots in quantities varying from a trace to nearly 50 % of the sediment.

Similar occurrences (along with unaltered glass shards) have been described by Elliot & Watts (1974) and Lock & Wilson (1975) from the Lower and Middle Permian (Dwyka and Ecca) of the Cape Trough, and the Permo/Triassic of Antarctica. Turner (1978b) also records acid volcanic fragments (along with minor, originally tuffaceous cherts) from the Upper Permian Abrahamskraal Formation of the western Karoo Basin. Elliot & Watts (1974) have suggested that all these occurrences (including those reported by Fuller 1970) have a common source, namely, volcanism associated with the Cape orogeny and its lateral extension in Antarctica. We feel that this is unlikely, at least for the Upper Beaufort and Stormberg strata because there is no evidence that tectonism in the Cape Trough was still active as late as Upper Triassic times. Fuller (1970) suggested that the volcanism in the eastern Cape and Lesotho commenced in Scythian times with widespread explosive activity, and that it continued unabated until Upper Triassic times when extensive effusion took over. Although no local volcanic centres definitely older than Clarens Formation (Rhaetian) are known, this would seem to be the most likely explanation as undoubtedly a large number of vents are buried beneath the main lava fields of Lesotho and eastern Cape, and in addition, some of the larger vents could have been active intermittently since Lower Triassic times. A less-likely source, but one which seems possible in view of the westerly palaeowind directions in Upper Triassic times is volcanism in early taphrogenic zones along the west coast continental margins (i.e. eastern Argentina). We will review 'Karoo volcanism' in terms of Gondwanaland reconstructions in Chapter 7.

2.1.3.2 *Drakensberg Group*
2.1.3.2.1 *Phase of shield volcanoes and explosive vents (Upper Triassic): Moshesh's Ford Formation.* Large-scale volcanic activity in the Karoo Basin area commenced in Upper Triassic times with the formation of shield volcanoes, diatremes and subsidence caldera. About 200 individual centres have been mapped in the eastern Cape, Lesotho and northern Orange Free State, (an area of over 150 000 km$^2$), but because they do not penetrate the youngest basalts (Lesotho Formation), more must lie hidden (Fig.29). Although many of the

pipes seem to have been formed solely by the escape of high pressure gases, some of the larger volcanoes acted as local centres for the dissemination of lava and ejecta on a large scale. Each of these centres will, therefore, have produced a local stratigraphic sequence, which in many cases consists of interbedded lavas, ashes and sediments, and obviously the establishment of a regionally-applicable lithostratigraphic scheme will require a good deal of careful mapping. This is only in its early stages.

The volcanic activity of this phase proceeded synchronously with the accumulation of the Clarens Formation, so that in the south, thick interbeds of typical Clarens lithofacies occur within the volcanic sequences. Farther north, in the northern Orange Free State only a thin sequence of this mixed lithofacies intervenes between the Lesotho Formation lavas and the main body of the Clarens Formation and in many areas, thick-bedded flood basalts rest directly on the Stormberg Group sediments with no intermediate phase.

The volcanic and associated strata of the eastern Cape have been described in detail by du Toit (1904, 1905, 1911, 1912a) and Gevers (1929), but only recently has systematic mapping commenced (e.g. Watters 1967, Eales & Booth 1974, and particularly Lock et al 1974). A wide variety of rock types is found associated with this phase of volcanism which have been grouped by Lock et al (1974) into the Moshesh's Ford Formation, which has its type sections NE of Barkly East (Figs 29, 32 & 34). Two main associations can be recognized: vent fill; and extra-vent effusive and ejected material (Figs 33 & 34).

The vents vary in size from a few metres to several kilometres (17 km at Roodepoort/Zwartfontein), and have been classified by du Toit (1904, 1905) and Gevers (1929) into two types: those which contain no appreciable quantities of magmatic material and were probably formed by the forceful blast of gases (diatremes); and those which do contain magmatic debris, flows or plugs. Gevers (1929) believed that this classification is artificial in as much as the distribution of the two types closely follows the stratigraphic level to which erosion had taken place. This certainly seems the case, because nearly all the vents which are seen cutting the Clarens Formation contain appreciable quantities of magmatic material, whereas those at the level of the Elliot Formation mostly do not. At still lower levels (i.e. in the Molteno Formation) the vents are mostly plugged with dolerite, whilst no vents are known from the Beaufort Group (Fig.29). Overall, therefore, the distribution appears to be stratigraphically controlled with the vents developed only at levels above the Beaufort/Molteno boundary, which before erosion must have lain between 1 000 and 2 000 m beneath the surface. These relationships are summarized in Figure 36, whilst Table 12 gives lithological details of the various fills.

In those cases where little or no coarse magmatic debris is involved in vent filling (i.e. mostly those seen cutting the Elliot Formation), the diatremes are plugged with white siliceous sandstones, tuffs, sedimentary agglomerates and sandstones. The fine material is a mixture of pulverized sediment (fragments of Elliot and Clarens strata predominate), later sedimentary infill, and some ash, whilst agglomerates contain a jumbled mass of large and small sedimentary fragments set in a fine ground mass. The resemblance of this whitish-yellow plug infill to 'normal' Clarens lithofacies, and the fact that typically the country rock shows no sign of tectonic disturbance sometimes led early workers to consider these plugs to be outliers of the Clarens Formation. Gevers (1929) believed that this association represents the earliest phase of activity, and that in some areas the vents did not develop into true volcanoes.

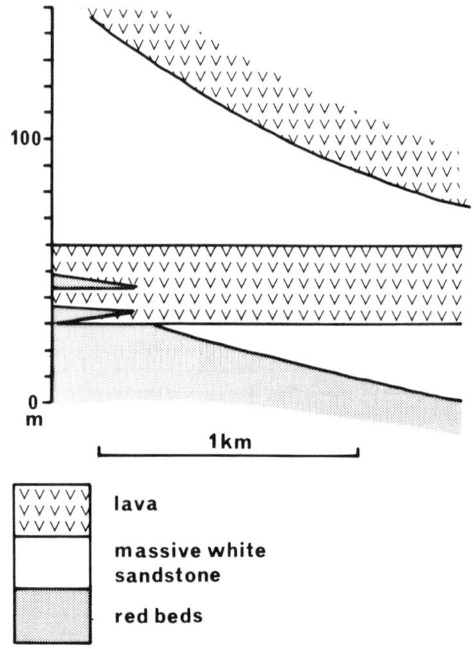


Figure 31. Diagrammatic relationship of lavas, and Clarens (white sandstone) and Elliot (red beds) formations west of Raasay (location on Fig.32). After the description by du Toit (1904, p.200). Scales approximate.

Almost all the vents seen cutting the Clarens Formation contain abundant magmatic material, interbedded with lavas and sediments (Fig.29). The bulk of the volcanic rocks are basaltic in composition, but andesites have been reported from the Barkly East area by du Toit (1905). The relationship of these volcanoes to the country rocks is often complex, the latter having sometimes suffered considerable tectonism (faulting and tilting), whilst whole areas seem to have undergone rapid subsidence. In this respect, it may be significant that the main depocentre for the upper part of the Clarens Formation (Member III) lies along the zone where the major volcanic centres are thought to have been situated, and it is in this region that large lateral thickness variations have been recorded. The various rock types listed in Table 12 are often intimately interbedded, and the agglomerates and tuffs can be further subdivided on the grounds of whether they show bedding or merely consist of jumbled and chaotic masses. Solid, basic, plug-like bodies are typical of the vents seen cutting the lower stratigraphic levels.

Extra-vent rocks consist of interbedded lavas, pyroclastics (tuffs, tuffites, and agglomerates) and sandstones. Each large volcanic centre produced its own suite, which interfingered with the products of adjacent volcanoes (Fig.34), whilst the earliest volcanic rocks at any one locality rest on the country rock, often burying locally eroded features (Figs 30 & 34). Widespread volcanic activity seems to have begun after an considerable thickness of the Clarens Formation had been deposited, but in a few areas lavas and tuffs rest directly on (e.g. at Maclear, and Raasay, Fig.31) or occur within (near Jamestown, Botha & Theron 1966) the Elliot Formation. This indicates that some centres were active well before the main phase, and lends support to Fuller's (1970) contention that some were in existence as early as Lower Triassic times. Activity was intermittent because sedimentation was locally re-established with the deposition of sandstones with varying quantities of both fine and coarse volcanic debris. Major centres for lava and pyroclastic dispersion have been identified at

Table 12. Lithology of vent fill in the Stormberg volcanoes of the Drakensberg Group.

| | |
|---|---|
| Magmatic vent fill | |
| 1. Tuffaceous lavas | highly vesicular and pumaceous basalt with boulders, bombs and lapilli of lava and agglomerate |
| 2. Tuffs and agglomerates of volcanic matter | poorly sorted (fine to 1 t boulders) basaltic debris and glassy matter, with embedded bombs and lapilli of amygdaloidal basalt and agglomerate |
| 3. Tuff and agglomerate of sedimentary rock set in volcanic matter | sedimentary fragments set in a matrix of pulverized sediment and highly pumaceous volcanic matter |
| 4. Tuffaceous sands | sandstones with lapilli and fragments of volcanic origin |
| 5. Plugs | coarse basaltic plugs |
| Essentially non-magmatic vent fill | |
| 6. Sandstone tuff | fine pulverised sedimentary material |
| 7. Tuffs and agglomerate of sedimentary rock set in sedimentary material | fine to coarse fragments and blocks of country rock set in pulverized sedimentary matrix |
| 8. Sedimentary fill | sedimentary fill, either by collapse of coutry rock or later sedimentation |

Moshesh's Ford (the Kelvin Grove and Belmore vents) (Figs 29, 33 & 34), the Broadford vent (Fig.34), the huge Roodeport/Zwartfontein centre (Fig.29), in the Donnybrook area (Fig.34), as well as composite centres in the Katberg/Majuba Nek, Barkly Pass/Tembe Pass, Birds River, and Matatiele areas (Fig.29).

Working in the Moshesh's Ford region, Lock et al (1974) have established a stratigraphic sequence which they have been able to map for a radius of over 35 km around the large vents at Kelvin Grove and Belmore. Lavas (Drumbo Member), pyroclastics (Jennerville and Bell Kop members), tuffite (Dingle Member) and vent fill (un-named) which make up the Moshesh's Ford Formation (Figs 33 & 34) are overlain unconformably by flood basalts of the Kraai River and Lesotho formations. In the type area, the basal lavas reach a thickness of at least 90 m where they consist of tholeiitic pahoehoe and rarer pillow types. Individual flows are up to 5 m thick and usually have pipe amygdales at their bases. Westwards, similar lavas (Donnybrook Member, interbedded with sandstones) from the Donnybrook volcano occur slightly higher in the sequence and reach a maximum thickness of at least 70 m (Figs 32 & 34). The overlying pyroclastics are typically chaotic agglomerates, consisting of basaltic glass and ash (<20 %), and boulders and fragments of sedimentary rock (mostly shales), dolerite and rare meta-quartzites. This material is thought by Lock et al (1974) to have been emplaced as lahas (ignimbrites). The upper part of the pyroclastic pile consists of bedded material (possibly water-lain reworked equivalents of the underlying chaotic strata), with occasional bombs, and an upper 'tuffite' (mixture of pyroclastic and detrital material, i.e. a tuffaceous sandstone) which forms a thin (up to 6 m) widely-developed and easily recognized light coloured top to the Moshesh's Ford Formation. In the type area, a maximum thickness of about 170 m has been recorded for the lavas and pyroclastics of the Moshesh's Ford Formation, but at least 240 m have been measured by du Toit (1905) west of Barkly East.

Similar sequences have been described from the Roodeport/Zwartfontein centre (Gevers 1929), and the Majuba Nek area in the Herschel district (Watters 1967) (Fig.29). No detailed correlation is yet possible between the former and the Moshesh's Ford area, but Watter's work (1967) allows an extension of Lock et al's (1974) succession to the western edge of the volcanic province (Fig.32). Here, the basal basalts are up to 73 m thick and are overlain by sandstones, tuffaceous sandstones and fine agglomerate with locally interbedded thin basalts. The clastics occasionally show signs of water-lain bedding and Watters (1967) assumes that some of the lavas were subaqueously extruded. No radiometric dates are available for the Moshesh's Ford Formation.

An interesting variant amongst the Moshesh's Ford Formation rocks is a large cauldron subsidence complex

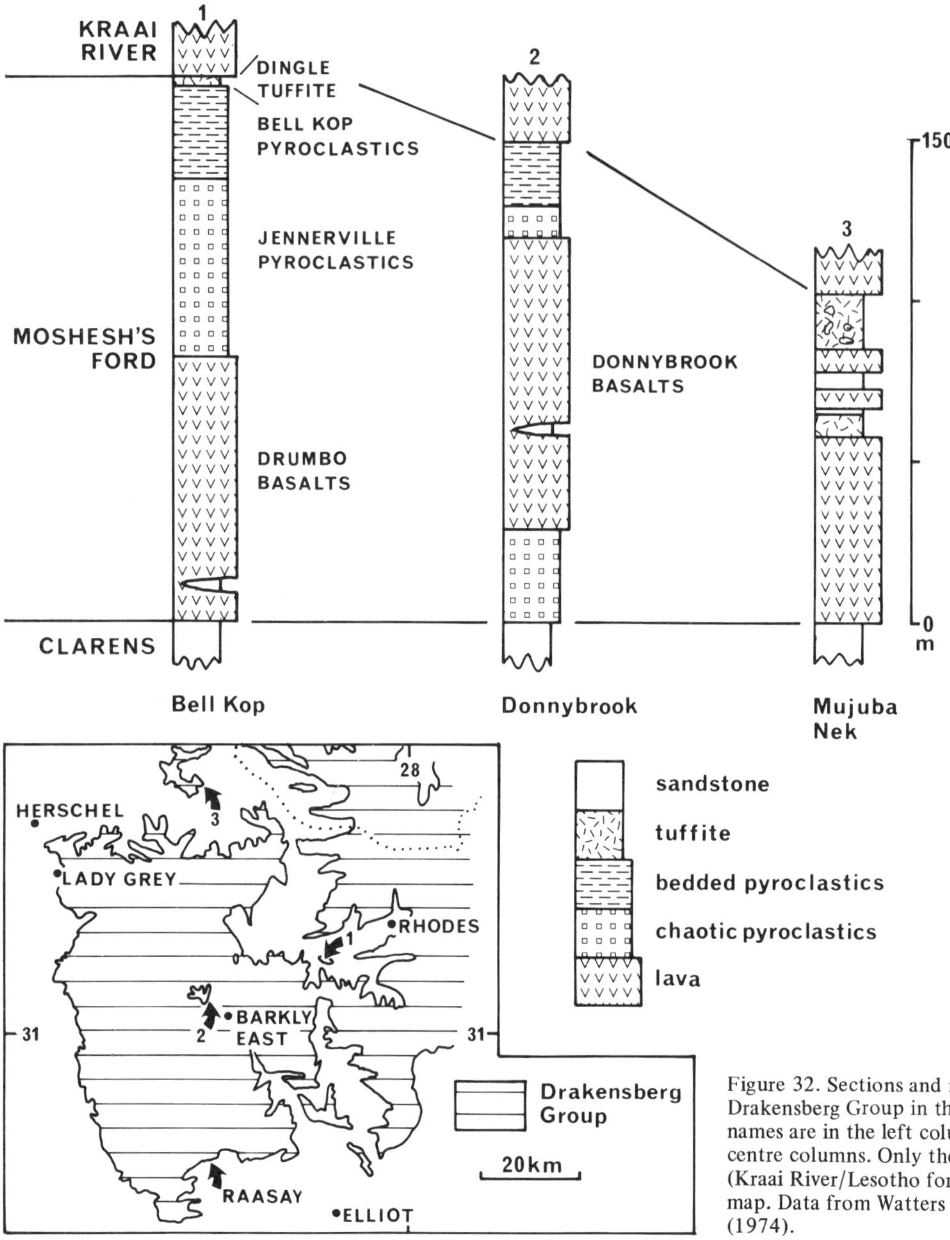


Figure 32. Sections and map of lower part of Drakensberg Group in the E.Cape. Formation names are in the left column, member names in centre columns. Only the extent of the upper lavas (Kraai River/Lesotho formations) is shown on the map. Data from Watters (1967) and Lock *et al.* (1974).

(13.5 km in length) formed mainly of gabbro, which occurs SW of Dordrecht near Birds River station (locality 5 on Fig.29). It cuts across the Molteno-Elliot formation contact and has huge xenoliths (up to 4 km across, and as much as 700 m above their usual stratigraphic level) of these rocks, as well as Clarens Formation, on its upper surface that represent a collapsed roof structure. Reference was first made to this feature by du Toit (1905, 1911), who later described it as a 'bell jar'-shaped intrusion, but a complete account of its structure and igneous rocks has only recently been given by Eales & Booth (1974). Apart from the fact that its gabbroic composition is unusual amongst the Stormberg volcanic centres, the Birds River complex is important because a detailed geochemical and petrographic analysis of the various metamorphic and differentiation phenomena have allowed Eales & Booth (1974) to propose a petrologic and intrusive model for it which they believe is generally applicable to other large intrusive Stormberg structures, such as Mount Arthur. Their model can be summarized as follows:

i) emplacement of a sheet-like body of basic magma beneath updomed sediment cover. In the case of the Bird's River complex, this probably occurred during

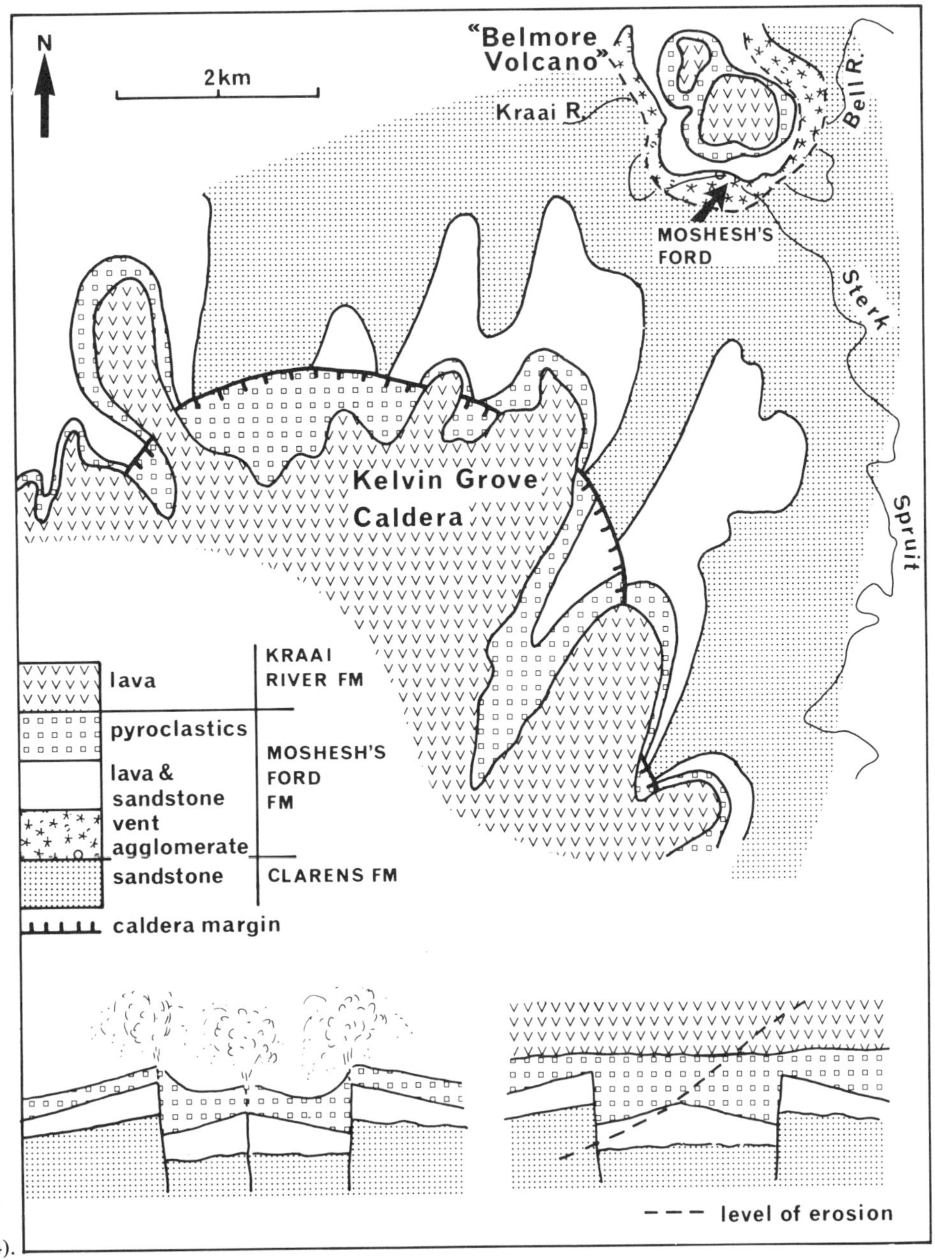


Figure 33. Distribution and speculative development of volcanic centres in the Moshesh's Ford area. The lower sketches relate specifically to the Kelvin Grove caldera. Locations on Figure 29. Data from Lock *et al.* (1974).

the deposition of the Elliot or Clarens formations. Explosive and minor extrusive activity accompanied this event, with the production of tuffs and breccias;

ii) partial crystallization of the magma, leading to iron and alkali enrichment of the melt, but only limited $SiO_2$ enrichment;

iii) development of ring fractures in the roof, which allowed collapse of large blocks of overlying sediment into the partially crystallized hypabyssal body;

iv) iron and alkali-enriched magmas intruded into the ring fractures to crystallize as ferrogabbro and ferrotholeiite. By traversing highly oxidized sediments such as the Elliot Formation, even more oxidized conditions were achieved and with increased precipitation of iron oxides, the magmas suffered silica-alkali enrichment. In addition, loss of water to the permeable country rocks resulted in extensive metasomatism of the sediments, with the production of metasomatic granophyres, and the reduction of ferric iron in sediments to magnetite;

v) renewed hypabyssal injection of basic magma, in the form of olivine gabbro, filled the 'bell jar', and may possibly have been accompanied by extrusion of flood basalts. In contrast to the earlier emplacement of tholeiites (which resulted in cauldron collapse), this later

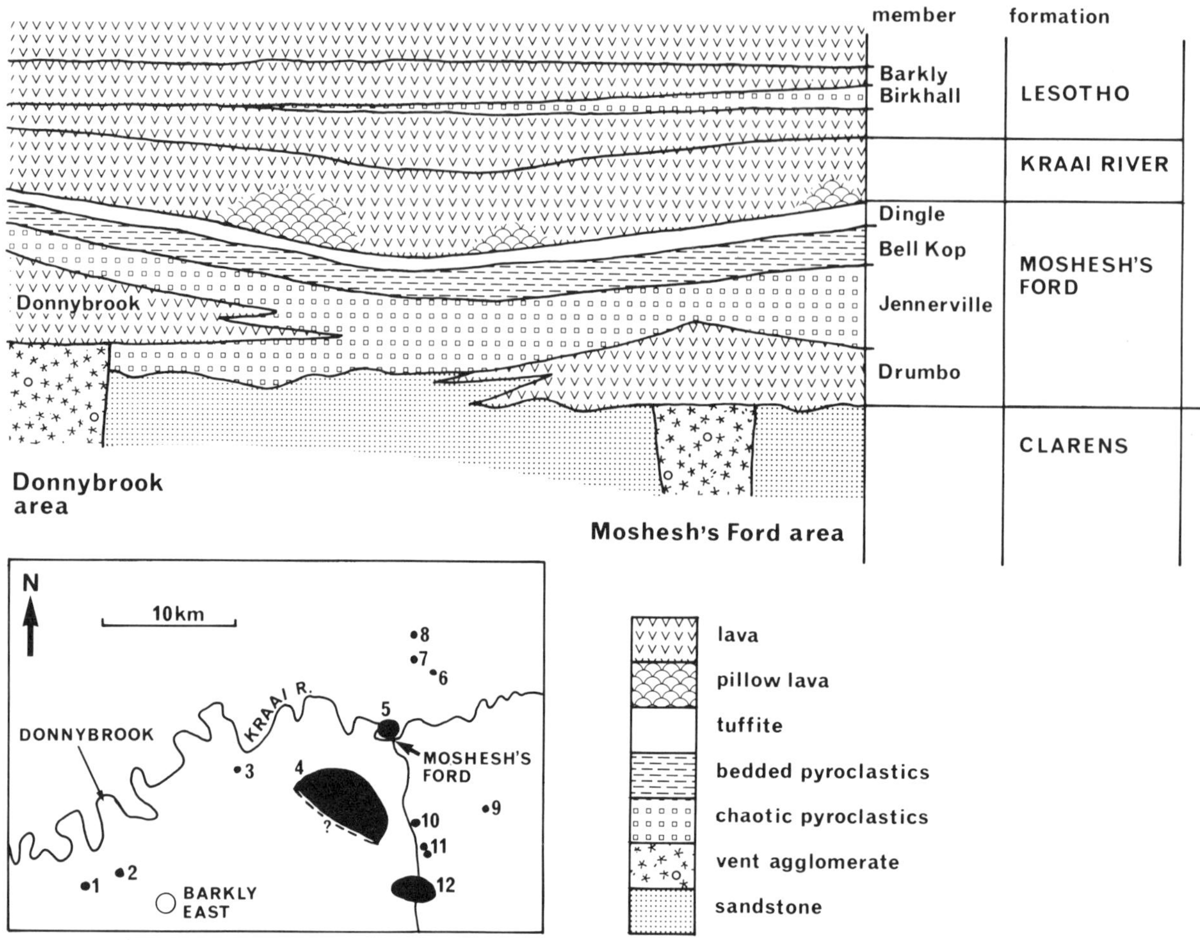


Figure 34. Idealized stratigraphic relationships of lower Drakensberg Group in the Barkly East area (based on data in Lock *et al.* 1974). Location map shows vents, etc. recorded by du Toit (1905) and Lock *et al.* (1974): 1 = Wittekrans Spruit; 2 = Careleon; 3 = Lilliput; 4 = Kelvin Grove; 5 = Belmore; 6 = Holderness; 7 = Glenfillan; 8 = Fetcani Glen; 9 = Clontare; 10 = Branksome; 11 = Caerlave rock; 12 = Broadford.

phase displaced the roof structure upwards. Movement was along the old ring fractures, but some lateral emplacement also occurred. Some of the large roof fragments became engulfed by the magma with local fusion and tridymite dyke emplacement;

vi) gravitational settling of olivine probably took place, and picrites have been observed in some areas (e.g. Elephant's Head, though not at Bird's River);

vii) late stage fractionation and concentration of $H_2O$-enriched magmas locally resulted in the production of pegmatites.

Eales & Booth (1974) point to the fact that no fragments of the Drakensberg flood basalts have been found in the roof structure and breccias along with the Stormberg Group sediments, as support for their contention that the Bird's River gabbro complex was emplaced prior to the extrusion of the main flood basalts of the Lesotho Formation.

2.1.3.2.2 *Flood basalts: Kraai River and Lesotho formations.* Although much has been written on the geochemistry, petrology and general tectonic setting of the main Drakensberg flood basalts, little or no attempt has been made to map them, nor to establish a stratigraphy for them. In the eastern Cape, they overlie the Moshesh's Ford Formation unconformably (Fig.33), frequently cutting out the uppermost parts and resting directly on the basal lavas or vent fills. The latter rarely penetrate into the lower parts of the flood basalt flows, indicating that the discrete volcanic centres represented by the vents ceased to function shortly after the establishment of large-scale effusive activity, when highly mobile lavas quietly welled up from numerous fissures, fed by the dykes that cut the Stormberg sediments and Moshesh's Ford Formation volcanics.

The upper part of the lava pile has suffered severe erosion, and nowhere is the top of the sequence seen.

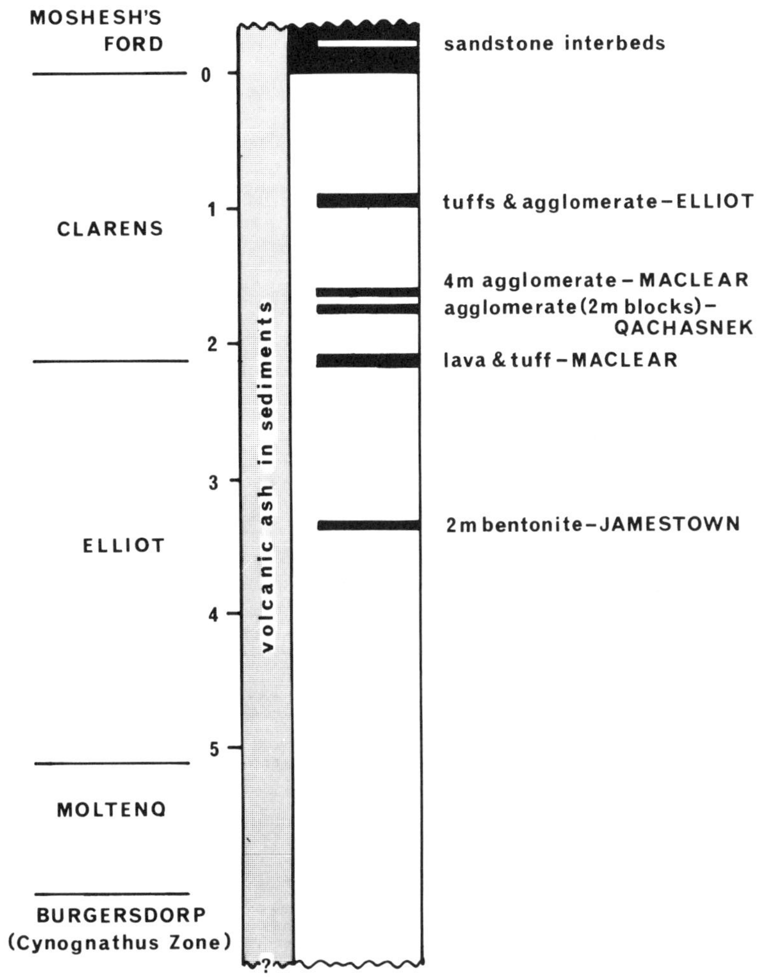


Figure 35. Relative stratigraphic positions of volcanic rocks in the Triassic-Lower Jurassic sediments of the main Karoo Basin. Thicknesses are x 100 m and volcanic material is shown stippled and black. Data from du Toit (1954), Botha & Theron (1966) and Fuller (1970).

A maximum of 1 400 m occur in the Mount aux Source region of NW Natal, but probably no more than 1 000 m are preserved in the eastern Cape, where Lock et al (1974) have recognized a twofold division: a lower sequence of fine-grained basalts with numerous pillow lavas – Kraai River Formation; and a thick upper sequence of basalts and pyroclastics – Lesotho Formation (Fig.34).

The Kraai River Formation has so far been recognized only in the eastern Cape, where it rests on various members of the Moshesh's Ford Formation. It is up to 78 m thick and consists of fine grained tholeiitic basalts, typically with vertical columnar jointing. Individual flows are up to 15 m thick and can be traced laterally for many kilometres. Pillow lavas often occur at the base of individual flows (Botha 1971, Lock et al 1974), and similar structures have been recorded near the base of the flood basalts at Cathedral Peak in NW Natal (McCarthy 1970). Together with the pillow lavas noted by Watters (1967) at Majuba Nek, these records indicate the presence of bodies of standing water at a time when Clarens Formation sedimentation had all but ceased.

McCarthy (1970) also notes that in NW Natal, the lower flood basalts of the Drakensberg Group (equivalents of either the Kraai River or Lesotho formations) rest directly on the Clarens Formation, indicating that no representatives of the Moshesh's Ford Formation were laid down in this area. Significantly, Figure 29 shows no volcanic vents in the NE part of the Karoo Basin, which suggests that south and westwards from NW Natal, the upper part of the Clarens Formation is progressively replaced by lavas and debris of the middle phase of volcanism set out in Table 11. If the regional upwelling of the flood basalts was a relatively synchronous event, which seems likely, then it implies that while large scale volcanism of the middle phase was taking place in the eastern Cape (i.e. the Moshesh's Ford Formation was being laid down) Clarens Formation sediments were still being deposited in the northern and eastern areas of the Karoo Basin.

The main flood basalts have been grouped together as the Lesotho Formation by Lock et al (1974). They

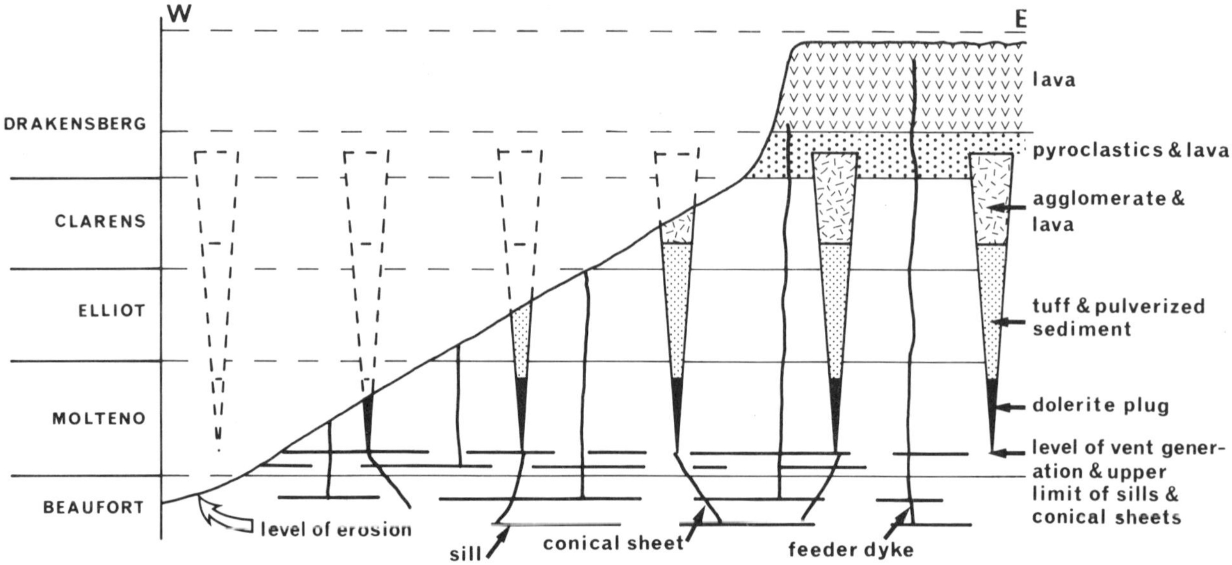


Figure 36. Idealized E-W section showing stratigraphic relationships of Drakensberg intrusive and extrusive volcanic rocks in the main Karoo Basin. See text for explanation.

are composed of thick, widespread tholeiitic flows with pipe amygdales at their bases, and are occasionally interbedded with pyroclastics which are up to 37 m thick and contain abundant sedimentary rock fragments. Great mobility of lavas in the upper part of the Drakensberg Group is indicated by the large aerial extent of individual flows (e.g. Lock et al 1974 have traced one 20 m thick flow over an area of at least 300 km$^2$ in the Barkly East area).

Most of the outcrop of the Karoo Supergroup is intruded by innumerable, extensive, thick sills (both concordant and discordant) and dykes of fine grained dolerite. Their form and nature have been described in numerous publications (see du Toit 1954 for a summary) and it is their lateral persistence and resistance to weathering that produces the typical mesa and butte (kopje) topography of large tracts of the Karoo landscape. One fascinating and unresolved aspect of this massive injection of basic material, apart from the fundamental question of crustal rheology, is the relationship of the intrusive bodies to the vent volcanism and flood basalts described above.

It has generally been accepted that the flood basalts were generated by fissure eruptions (e.g. du Toit 1954, Haughton 1969, Cox 1972, Lock et al 1974, etc.), and evidence in favour of this hypothesis hinges around the following points:

i) the vents do not penetrate the main flood lavas;

ii) explosive volcanic products are scarce in the flood lavas;

iii) wedge-shaped lava accumulations, as would be expected from shield volcanoes, are lacking.

Feeders for the flood basalts are assumed to be the numerous large dykes which cut the Karoo sediments and which penetrate through the superimposed lavas. In this respect it is significant that the number of dykes decreases up the stratigraphic column, presumably indicating waning volcanic activity. Sills, which are the dominant intrusive body in the Beaufort and lower Stormberg sediments, are relatively rare higher up, their upper limit approximately coinciding with the lower levels at which the explosive vents first appear. Taken together, this evidence suggests the following sequence of events and overall disposition of volcanic rocks (Figs 35 & 36):

i) earliest upward movement of magma in Lower Triassic times, resulting in the explosive activity which produced the fine ejecta and local lavas found in the sediments of the *Cynognathus* Zone and Elliot Formation. Location of vents unknown, but probably within the eastern Cape/Lesotho area;

ii) increased scale of upward movement of magma and emplacement of large sill-like magma chambers at 1-2 000 m beneath the surface in Elliot/Clarens times (Norian-Rhaetian). Degassing of magma and contact with ground water produced explosive action and creation of diatremes. Many of these vents erupted considerable amounts of magmatic material in the form of lava and ejecta and are thought to have tapped the magma chambers via inclined conical sheetlike structures (Gevers 1929, Eales & Booth 1980). Crude stratification within the vents was controlled by temperature/pressure relationships as well as mechanical properties of the country rock, so that similar features were formed at approximately the same levels. This results in the same vent-fill types clustering within particular

Stormberg formations in present-day outcrops (Figs 29 & 36);

iii) final phase of igneous activity culminating in dyke injection from the level of, and above, the upper level of sill intrusion. The sills presumably still acted as the main magma chambers. Massive, quiet effusion of lavas proceeded from these dykes.

2.1.3.2.3 *Age of the Drakensberg Group.* No systematic study has been published on radiometric dates for volcanic rocks of the Karoo Basin to determine the chronology of the various phases, although numerous isolated determinations are available. The earliest known lavas occur in the Elliot Formation, which places them within the Norian (ca.210 my), and the main vent phase (which cuts the Clarens Formation) is probably upper Rhaetian or Lower Hettangian (around 190 my). McDougall (1963) determined K-Ar ages of 190-154 my (Hettangian–Callovian) for Karoo dolerite dykes, and K-Ar dates of 189-160 my (Hettangian-Bathonian) on lavas (probably Lesotho Formation), and 187-137 my (Hettangian-Portlandian) on dykes and sills have been published by Fitch & Miller (1971). A range of 182-158 my (Sinemurian-Bathonian) has also been obtained by the K-Ar method on dykes and sills by Sauvan et al (1975). Fitch (1972) further remarked that the main dykes and flood basalts of the Karoo Basin were probably confined to a 20 my period and quoted ~172 my (Aalenian) for the dolerites. From these results it seems that the flood basalts and their feeder dykes range in age from about 190 to 172 or 154 my (i.e. Lower to Middle Jurassic).

## 2.2 EXTRA-KAROO TRIASSIC TO LOWER JURASSIC ROCKS

Karoo Supergroup rocks occur in several large outliers in Namibia, Botswana, northern Transvaal/southern Zimbabwe, and along the western edge of the Lebombo mountains (Figs 5 & 38). The relatively undisturbed strata in the western areas belong to the Karoo tectono-sedimentary terrains (Doros, Etjo, Botswana, Springbok Flats and main Karoo), whereas in the eastern outcrops (Soutpansberg, Tuli, Nuanetsi, Lebombo), strong tectonism (mostly block faulting and tilting) has affected the Triassic rocks within the Zambezi tectono-sedimentary terrain (Figs 3, 38 & 39).

Bio- and lithostratigraphic correlation between these outliers is hampered by lack of palaeontological control, regional lateral facies variations, and poor exposure. Figure 37 shows representative sections from each area, and approximates to an E-W traverse along the line shown in Figure 38. Except in the west (where it rests on metamorphic basement) the local base of the Triassic cannot be located so that any diachronism between individual formations in these outcrops and the standard sequences farther south cannot be assessed. Only one rock unit, the so-called Bushveld Sandstone (and its western equivalent the Plateau Formation in the Etjo area), can be confidently correlated across the whole area. It is a fine grained, homogeneous, white, yellow or pinkish sandstone, with both massive and cross-bedded units, and in the east it locally contains numerous calcareous nodules. On the basis of its stratigraphic position within the sequence, its lithofacies, and the fact that in Namibia Keyser (1973) has dated it as Upper Triassic (Carnian-Norian), we have no hesitation in equating it with the Clarens Formation of the main Karoo outcrops, and take its base as our datum in Figure 40.

Immediately beneath the Clarens Formation, the rock sequence is rather variable in thickness and lithofacies. It consists of 100-300 m of red, green, and yellowish shales and sandstones overlying a thick succession of carbonaceous shales, coals and feldspathic sandstones which is conventionally equated with the Ecca Group. Accounts by previous workers (e.g. du Toit 1954, Boocock & van Straten 1962, Green 1969, van Eeden 1973, Cooper 1974a) indicate a degree of uncertainty as to whether the upper, essentially non-carbonaceous rocks that locally overlie the Ecca Group are lateral facies equivalents of younger Ecca horizons, or whether they are local representatives of the Beaufort and Stormberg groups. On purely lithostratigraphic grounds, the latter interpretation must be accepted and is strongly supported by the record of *?Euskelesaurus* sp. from the Springbok Flats outlier (Haughton 1924). This plateosaurid reptile is typical of the lower part of the Elliot Formation of the main Karoo outcrops. Such a correlation, as set out in Figures 37, 40 and 42, implies that the rocks between the Clarens Formation (≡Bushveld Sandstone) and the coal-bearing Ecca Group are equivalent to the whole of the Beaufort Group and the Molteno and Elliot formations farther south. Whether or not they represent the same time interval (Upper Permian to Upper Triassic) is not known, but presumably the Permian/Triassic boundary lies somewhere within this sequence, unless of course the Ecca facies persisted longer in the north than it did farther south. This suggestion has in fact been proposed by Ryan (1967), who implies that the boundary, which occurs in the top part of the Beaufort Group in the south, lies low down in the Beaufort Group farther north. At the moment there is no way of testing this.

In addition, unpublished palaeontological evidence of van Heeden (quoted in Visser & Botha 1980) suggests that the Molteno-Elliot-Clarens equivalents in the northern Transvaal may be older than their analogues in the main Karoo Basin. These possible relationships have been only partly included in Figure 42, mainly because they cannot be quantified at present, but imply a shorter hiatus (or no hiatus) during Anisian-Ladinian times in the north.

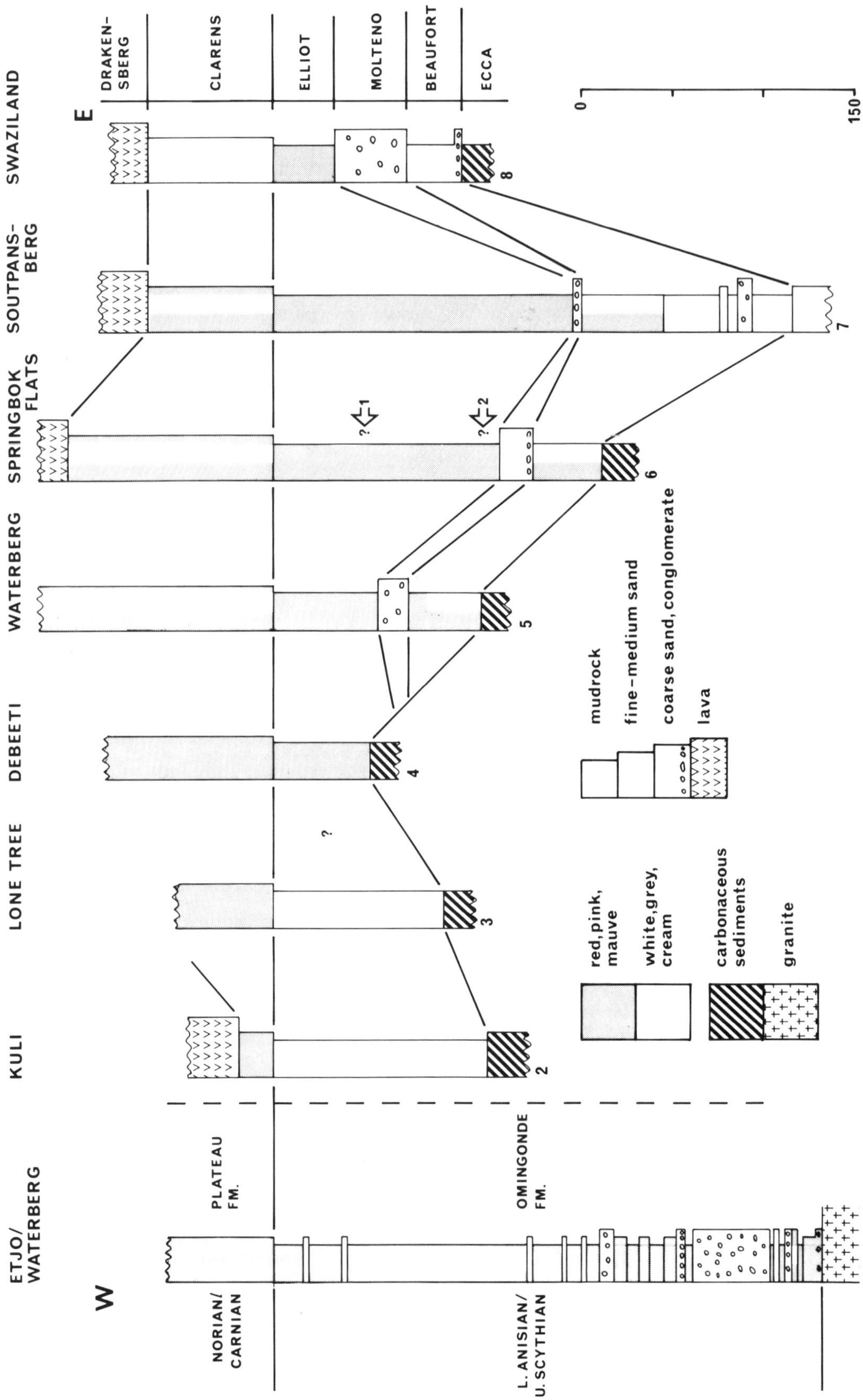


Figure 37. Correlation of Triassic rocks in Swaziland, N.Transvaal, Botswana and Namibia. Datum taken at base of Clarens Formation, and localities shown in Figure 38. No subdivision of the Beaufort Group is yet available, and the Permo-Triassic boundary probably falls within it. Correlation of the Namibian sequences with other areas is uncertain. In the Springbok Flats succession 1 and 2 refer to horizons from which a dinosaur femur *Gigantoscelus molegraafi* van Hoepen, and *Euskelesaurus* sp., respectively were reported by du Toit (1954). Data from numerous sources, but particularly Gevers (1936), van Eeden *et al.* (1955), Visser & van der Merwe (1959), Scogings & Lenz (1961), Green (1969), van Eeden (1970), and Keyser (1973). See text for explanation and details of correlations.

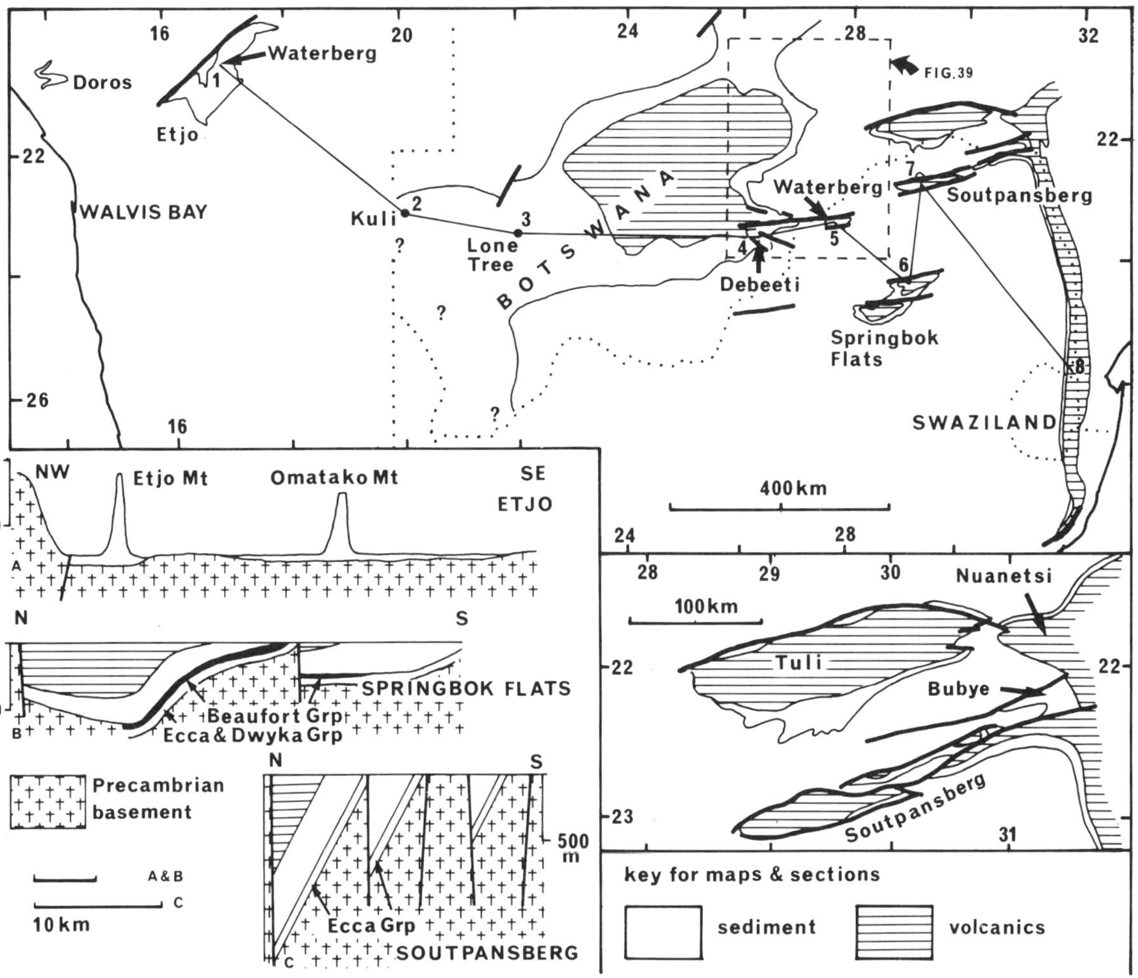


Figure 38. Distribution of Triassic strata in Swaziland, N.Transvaal, Botswana and Namibia. Localities and section line for Figure 37 are shown in upper diagram. Beaufort Group is shown undifferentiated, and in W.Botswana 'Karoo' strata may include rocks of Permian age (Dwyka and Ecca equivalents). Small map at lower right shows detail of outcrops in N.Transvaal. Sketch cross-sections are to illustrate the nature of the Etjo, Springbok Flats and Soutpansberg outliers and are not meant to be comprehensive. Data from: Gevers (1936); van Eeden *et al.* (1955); Visser & van der Merwe (1959); Vail (1964); Haughton (1969); Green (1969); Crockett & Jones (1975); Jansen (1975). The location of Figure 39 is shown in the main map.

### 2.2.1 *Eastern Botswana, northern Transvaal, southern Zimbabwe and the Lebombo Line*

This area is illustrated by localities 4 to 8 in Figures 37, 38, 39 and 40, and although exposures are locally available, the succession can only be satisfactorily established with the aid of borehole data (e.g. van Eeden et al 1955, Visser & van der Merwe 1959, Green 1969, and van Eeden 1973). Outcrops and subcrops of Karoo Supergroup strata in this region form a series of erosional and/or faulted outliers strung across the boundary zone between Rust's (1975) two tectono-sedimentary terrains, with the Springbok Flats outlier in the Karoo terrain, and the Soutpansberg, Tuli, and Nuanetsi/Lebombo in the Zambezian. The occurrences in eastern Botswana (and the extension into the Transvaal in the Waterberg outlier) also lie within the latter terrain, but are an extension of the main body of Karoo Supergroup rocks underlying the Kalahari region (Green et al 1980) (Fig.5). The latter is classified by Rust (1975) as a Karoo-type terrain (see Fig.3), so that the structural relationship between the eastern Botswana outcrops, and their westerly counterparts is obscure at present.

The uppermost sediments over the whole of this area consist of pale yellow, pink, white or red massive and cross-bedded sandstones which are locally silicified, or weather into curious shapes. Various names have been applied to these sandstones (Bushveld Sandstone, Ssake Sandstone, upper Samkoto Series, with the first most widely used), but their obvious correlation with the Clarens Formation of the Karoo Basin obviates the need for further use of these terms. Typically, the se-

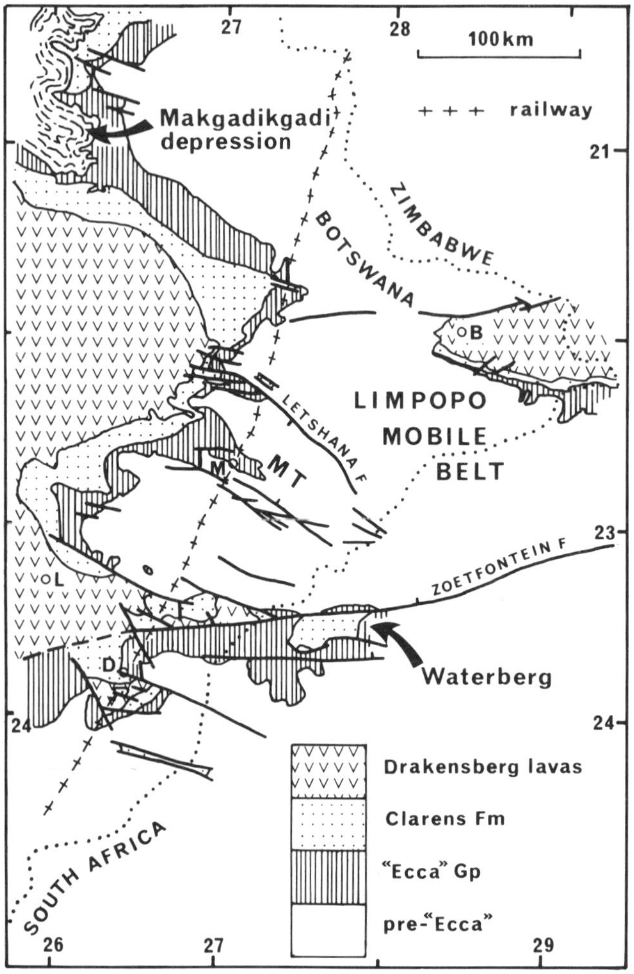


Figure 39. Permian-early Jurassic sediments and lavas in S.E.Botswana and their relationship to re-activated lineaments in and adjacent to the Limpopo Mobile Belt. Abbreviations: D = Debeeti; M = Mokoro; MT = Mokoro Trough; L = Lephepe; B = Bobonong. State boundaries are dotted. The top of the 'Ecca' Group is probably a lateral facies equivalent of the Elliot Formation. Simplified after Green *et al.* (1980).

quence lying between the Clarens and Ecca lithofacies consists of various green and red shales and sandstones, the latter locally being pebbly and coarse. It varies in thickness from about 50 m in eastern Botswana to over 300 m in the Soutpansberg, and about 120 m along the central Lebombo Line in Swaziland.

In all these areas, the upper beds (i.e. immediately beneath the Clarens Formation) consist of a homogeneous sequence of red, brown and purple shales, mudstones and siltstones with local calcareous nodules. Lithologically, and in relation to the Clarens Formation, a correlation with the Elliot Formation of the main Karoo Basin is strongly suspected and this contention is supported by the presence of *?Euskelesaurus* reworked into grey shales in the Springbok Flats outlier (quoted by Haughton 1924). In eastern Botswana, the Elliot Formation (50 m) rests directly on the carbonaceous-rich Ecca Group, but eastwards it thickens, and is separated from the Ecca by a progressively thickening series of green and grey shales and sandstones (Fig.40). The Elliot Formation reaches its maximum thickness in the Soutpansberg outlier (about 160 m), before thinning eastwards towards the Lebombo Line (40 m).

Except in the Debeeti area, the red Elliot Formation is underlain by a thin, but persistent white to brown, fine to coarse, gritty sandstone. It locally contains pebbles, calcareous nodules and a basal conglomerate, and in Swaziland has a 'glittering' aspect. Falcon (1975) reports coal seams from this horizon in the Springbok Flats (and in the Sabi Basin farther NE) which is presumably the same occurrence as the 8 m seams noted by du Toit (1954, p.313) from SW of Warmbaths. These lithological characteristics, together with its stratigraphic position beneath extensive red beds, suggests a correlation with the Molteno Formation of the main Karoo Basin (Figs.37 & 40). Its non-appearance in the west is probably a result of lateral facies variation, where it passes into a red mudrock facies, similar to that at the northern end of the Karoo Basin. The Molteno Formation varies in thickness from a maximum of 40 m in Swaziland to about 5 m in the Soutpansberg.

Sediments below the Molteno Formation are less

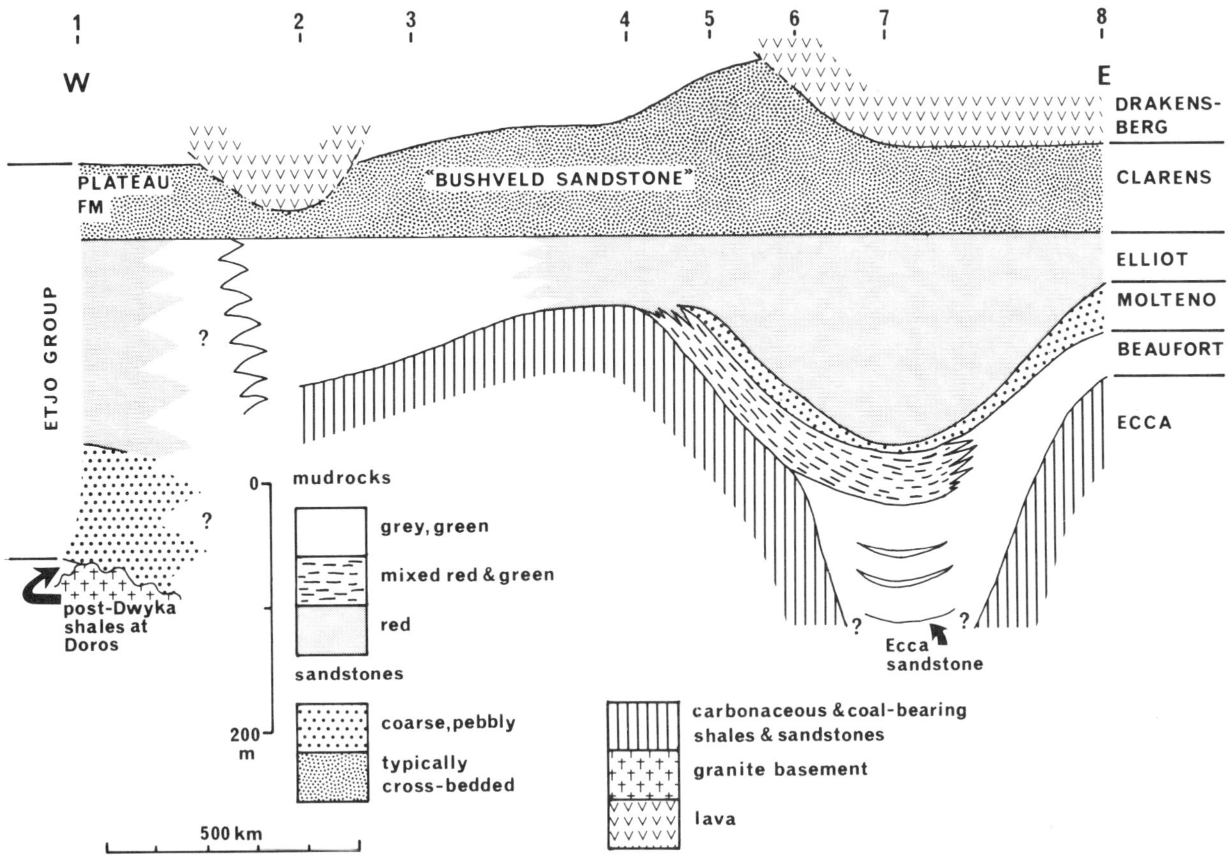


Figure 40. Lithostratigraphic correlation of Triassic strata as a W-E section from Namibia to Swaziland. Location of control points is on Figure 38. Degree of diachronism is illustrated by reference to Figure 42. Reference to post-Dwyka shales at Doros serves to illustrate the unusual stratigraphic setting of the Etjo outlier. Data from Figure 37.

easy to allocate to units established at type sections in the south. In the Waterberg, Springbok Flats and Soutpansberg areas, a 40-45 m sequence of red, brown and green variegated mudrock occurs immediately below the Molteno Formation, but this unit is missing in Swaziland. Furthermore, in the Waterberg and Springbok Flats these variegated rocks rest directly on the Ecca Group, whereas farther east an additional sequence of green and grey mudrocks with fine to medium sandstones, and grits intervenes (Fig.40). In Swaziland, this lower unit constitutes the whole of the succession between the Molteno and Ecca units. Lithologically these rocks can best be accommodated within the Beaufort Group, but so far no correlatable lithostratigraphic subdivisions have been achieved and no fossil vertebrates have been reported with which to date the strata. Figure 40 shows our interpretation of the lateral and vertical lithological relationships in these 'Beaufort' rocks. A further complication arises in the Soutpansberg where boreholes have not detected the boundary between these rocks and a carbonaceous-rich Ecca lithofacies. The Beaufort Group in this area varies in thickness from about 40 m in the Waterberg outlier to perhaps 120 m in the Soutpansberg and about 30 m in Swaziland.

Apart from *?Euskelesaurus,* the only other positive vertebrate identification is of *Gigantoscelus molengraafi* van Hoepen, (quoted by du Toit 1954). It is not clear from du Toit's account whether this occurrence is from the Clarens Formation or from a sandstone unit within the Elliot Formation. We have nominally placed it in the latter in Figure 37. As mentioned previously, van Heeden (quoted in Visser & Botha 1980) is currently working on the reptile faunas of this region, and clarification of the stratigraphical problems outlined above can be expected in the future.

### 2.2.2 *West and central Botswana (Kalahari region)*

Karoo Supergroup rocks in central Botswana are generally horizontal, or have very low dips, and are commonly obscured by regolith, calcrete, or aeolian sands. Even in areas not covered by such deposits, exposures are very poor, and most information has come from boreholes (e.g. Heath 1966, Green 1969, van Eeden 1973, Boocock & van Straten 1962). The rocks have been previously grouped into the Bushveld Sandstone (above) and the Bushveld Mudstone (below). As shown in Figures 37 and 40, we equate the arenaceous section with the Clarens Formation, and accepting the equation of the lower carbonaceous-rich rocks with the Ecca Group, it follows that the so-called Bushveld Mudstone is lithostratigraphically equivalent to all or parts of the Beaufort /Molteno/Elliot succession in the eastern Botswana/ northern Transvaal area. The sequence here is a monotonous succession of grey and green mudstones and shales up to 120 m thick from which no fossils have so far been reported. Presumably, it passes eastwards into the red rocks of the Debeeti district of eastern Botswana, but the true stratigraphical relationships are not known, in particular whether they are the lateral equivalents of the Elliot Formation only. From the boreholes of Heath (1966) quoted by van Eeden (1973) it appears that this sequence has not yet been recorded from the eastern part of Namibia.

Despite the poor exposure and extensive superficial cover in the Kalahari region of Botswana, it is sometimes possible to photogeologically identify areas underlain by Karoo Supergroup, and Mallick et al (1981) have noted that a distinction can be drawn between the Clarens Formation, and older, waterlain sediments. Particularly in the more eastern parts of the area, the Clarens Formation weathers to produce a thick regolith sand which in places has been blown up into migrating dunes. These contrast with the more earthy regoliths on the older Karoo strata. Because the para-autochthonous weathering products of the Clarens Formation are similar in colour to the allochthonous Kalahari Sands, photogeological differentiation between the two is difficult, although occasionally patches of slightly redder sand (indicated by greener colours on false-colour Landsat images) in southern Botswana possibly represent areas of thinner superficial cover of residual, rather than allochthonous material (Mallick et al 1981).

### 2.2.3 *NW Namibia (Doros and Etjo regions)*

Triassic sediments occur at two localities in NW Namibia (Figs 5, 38 & 41) and although they are relatively close together, correlation between them is not well-established.

The largest of these outliers lies in north central Namibia along the southern side of the large, reversed Waterberg Fault (Fig.38) where a series of impressive inselbergs (e.g. Waterberg, Etjo Mountain, Omatako Mountain) rest on Pre-Cambrian granitic basement. The outlier as a whole has been given several different names, but we will refer to it as the Etjo outlier. The best geological descriptions of the area have been given by Gevers (1936) (although a geological map of the region was previously prepared by Reuning 1923) and more recently Keyser (1973) has identified a vertebrate fauna which allows a reasonably good age range to be determined for the rocks of the outlier. Keyser (1973) divides the succession into two formations: Omingonde (lower) and Plateau (upper) (Fig.37). The Plateau Formation is a lateral equivalent of the Clarens Formation and consists of fine feldspathic sandstones with numerous aeolian cross-bedded units (according to Gevers 1936). It contains footprints of saurischian reptiles which suggest a Carnian or younger age.

The Omingonde Formation (which Gevers 1936 called the Lower Etjo Beds) consists of red mudstones and sandy shales with calcareous nodules interbedded with lenticular reddish feldspathic muddy sandstones, brownish quartzites and reddish and white arkoses. Arenaceous units are more common towards the base, where they are locally thick (up to 17 m), coarse and conglomeratic with subrounded granitic pebbles and boulders. Gevers (1936) considers these coarse basal layers to have been laid partly by running water and partly as scree.

Keyser (1973) has found a good vertebrate fauna in the Omingonde Formation, mostly in the argillaceous facies, both as skeletal remains and tracks. It contains five genera found in the type *Cynognathus* Zone in the Karoo Basin, with two cospecific forms: *Kannemeyeria simocephalus* and *Diademodon tetragonus* but also has elements which Keyser (1973, p.11) considers to be slightly more advanced than those found in the type beds viz. a cynodont (*Titanogomphodon*) and several dicynodonts. He suggests, therefore, that these strata are possibly slightly younger than the type *Cynognathus* Zone sediments in the main Karoo outcrops, i.e. possibly late Scythian or early Anisian. A list of the species reported by Keyser (1973) is given in Table 13.

Because of the unevenness of the basement surface, the thickness of the Etjo outlier strata (Etjo Group) varies between 600 m (Etjo Mountain) and 400 m (Great Waterberg), whilst the whole sequence dips in an easterly or ESE direction under the Kalahari Sands. As

Table 13. Vertebrate fauna of the Omingonde Formation (?Upper Scythian-?Lower Anisian), Etjo Outlier (after Keyser 1973).

Amphibia
- ? *Micropholis* sp.

Reptilia
- Anomodontia
  - Dicynodontia
    - *Kannemeyeria simocephalus* (Weithofer 1888)
    - *Rhopalorhinus etionensis* Keyser 1973
    - *Dolichuranus primaevus* Keyser 1973
  - Bauriamorpha
    - ? *Sesamodon browni*
  - Cynodontia
    - ? *Cynognathus* sp.
    - *Diademodon tetragonus* Seeley 1894
    - *Trirachodon* sp.
    - *Titanogomphodon* sp.

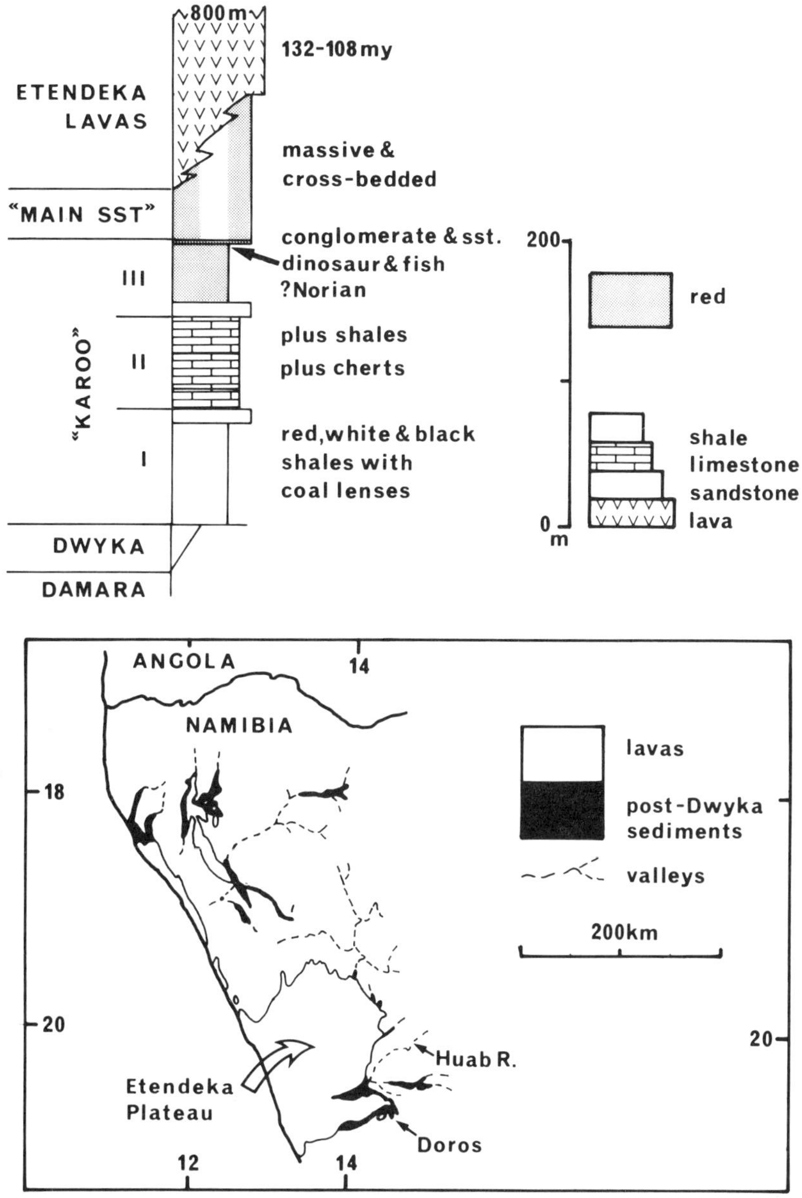


Figure 41. Distribution of Permo-Triassic sediments in coastal N.W.Namibia. Composite section for the Huab/Doros area shows that the 'Main Sandstone' (previously referred to as Etjo Sandstone) may be younger than the Plateau Sandstone in the Etjo area because the underlying red shales have been dated as Upper Triassic (?Norian), whereas the equivalent red mudrocks in the Etjo area are Scythian-Anisian. The Plateau Sandstone is Carnian-Norian. The classification 'Karoo I-III' follows Hodgson (1970). Map shows all post-Dwyka (Lower Permian) – pre-Etendeka lava (132-108 my) sediments lying in deeply-incised pre-Dwyka valleys. Data from: Siedner & Miller (1968); Hodgson (1970); Keyser (1973); Gidskehaug *et al.* (1975); Martin (1976).

seen in Figure 40 this sequence is relatively thick compared with other outliers in Botswana, northern Transvaal, etc., and differs further in resting directly on basement, i.e. no representatives of the Beaufort or lower groups of the Karoo Supergroup are present.

In this respect, the Etjo outlier also differs significantly from the coastal zone of Namibia where several small, irregular outliers of post-Dwyka (Lower Permian) strata are preserved in a series of deeply incised pre-Dwyka valleys (Martin 1976). They are partly covered by Lower Cretaceous lavas, and are best known from the Doros/Huab River area in the extreme SE (Fig.41). Du Toit (1954) previously described these rocks as lying in narrow graben, but this was probably a mistaken reference to the valleys in which the outliers are preserved. The geological succession in the Doros region consists of thin isolated outcrops of tillite on a Pre-Cambrian basement surface, which are overlain by 2-300 m of shales, sandstones and limestones, and a further 800 m of lavas. Hodgson (1970) subdivided the lower part of the upper sedimentary succession into three units which he called Karoo I, II & III (Fig.41) (Johnson et al 1976 report that the name Twyfelfontein Formation has been suggested for these beds). On the basis of its lithology he equated the lower tillite horizon with the Dwyka tillite (Lower Permian) of the Karoo Basin, but because of reported finds of *Mesosaurus* in the overlying 'Karoo I to III' considered this

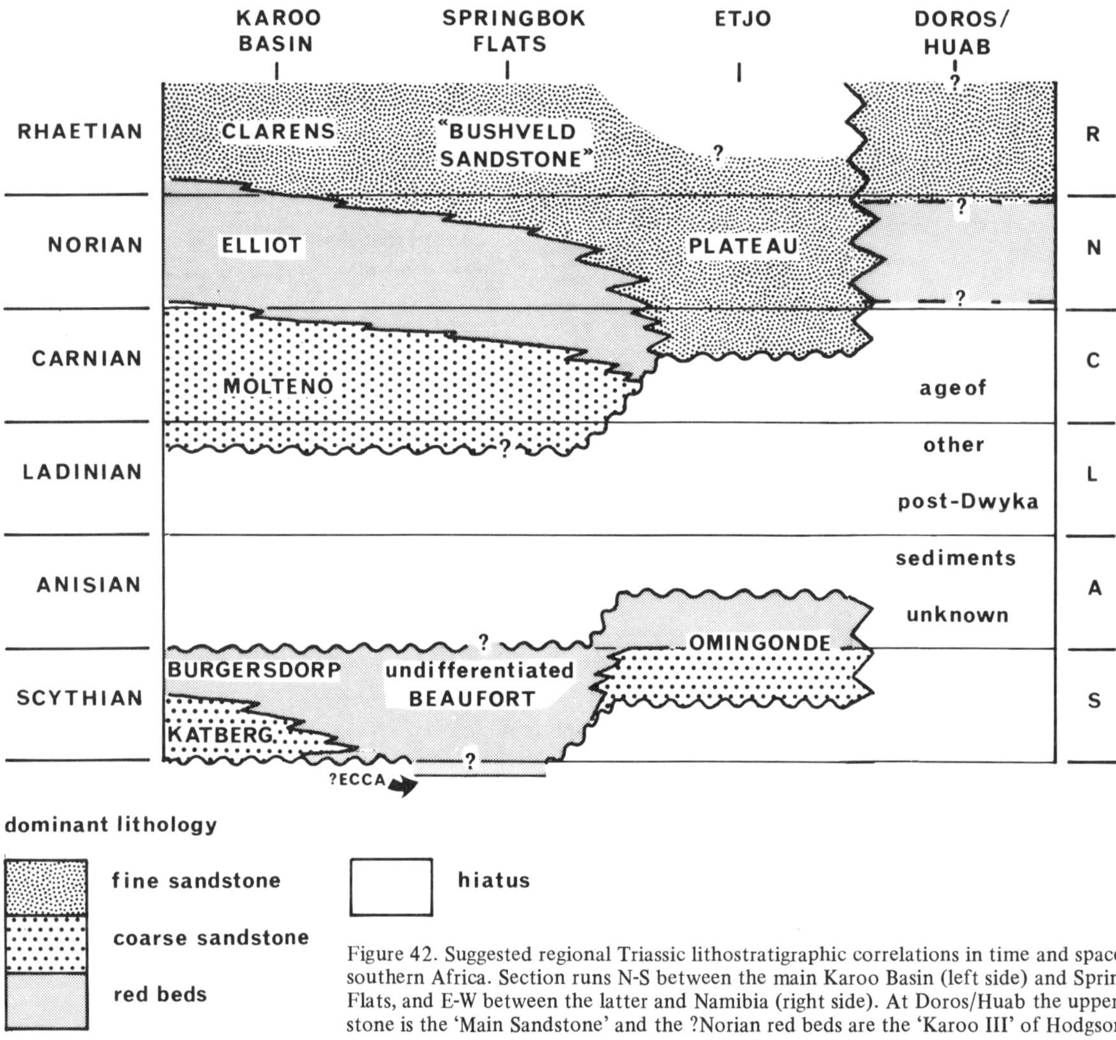


Figure 42. Suggested regional Triassic lithostratigraphic correlations in time and space in southern Africa. Section runs N-S between the main Karoo Basin (left side) and Springbok Flats, and E-W between the latter and Namibia (right side). At Doros/Huab the upper sandstone is the 'Main Sandstone' and the ?Norian red beds are the 'Karoo III' of Hodgson (1970).

sequence to also belong to the Dwyka Group. Martin (1970) in a discussion of Hodgson's (1970) paper, cast some doubt on the accuracy of these earlier records (e.g. Reuning & van Huene 1925), and Keyser (1973) after his visit to the area reported that he could not relocate the *Mesosaurus* locality. He did, however, report dinosaur remains from the red mudrocks that lie beneath the main sandstone (i.e. Karoo III of Hodgson 1970, Keyser 1973, p.13) which indicate a ?Norian age for this unit. Whether Hodgson's Karoo I & II are Permian or Lower Triassic in age remains to be satisfactorily established. Hodgson (1970) called the upper, main sandstone the Etjo Sandstone. This terminology can no longer be applied because Keyser (1973) has shown that the Doros sandstones are significantly younger than those at Etjo, and the name Etjo Sandstone is no longer used for a unit in the Etjo outlier. We have colloquially called them the 'main sandstones' in Figure 41. The rocks in question reach a maximum thickness of 100 m and consist of yellowish, massive, upward coarsening sandstones and interbedded laminated sandstones in the west, and red well-sorted fine-grained, aeolian cross-bedded sandstones in the east. They are overlain by lavas (Etendeka lavas) which in the west interfinger with, and locally metamorphose the sandstones. Siedner & Miller (1968) and Gidskehaug et al (1975) have dated these lavas between 114 and 128 my (Lower Cretaceous) and the point of interest here is that these dates suggest that the 'main sandstone' in the Doros/Huab coastal outliers ranges in age from Upper Triassic to Lower Cretaceous. Sediment accumulation was presumably intermittent, but this is the only record from southern Africa of a sedimentary sequence that spans the whole of the Jurassic time interval.

### 2.2.4 *Regional correlation of Triassic sediments*

Figure 42 summarizes the bio- and lithostratigraphic

correlation of Triassic rocks in the Karoo and extra-Karoo outcrops of southern Africa. There is remarkable similarity in the sedimentary succession throughout the subcontinent, with all major lithofacies widely developed.

Lower Triassic red or red/grey variegated mudrocks and sands and pebbly sands are found in all areas, with the probable exception of Doros/Huab, and they generally constitute an upward fining sequence. In the Etjo area this lithofacies is possibly slightly younger than equivalents farther south, but may reflect less severe mid-Triassic erosion. In all areas, again with the possible exception of Doros/Huab, there is a major mid-Triassic (Anisian-Ladinian) non-sequence during which the Cape Fold Belt was uplifted. Sedimentation commenced again earliest in southern areas with the coarse Carnian Molteno Formation wedge, but northwards this facies is thin and most Carnian rocks are probably of red-bed facies. Elliot Formation red mudrocks of probable Norian age are found in all areas except Etjo, where post-mid-Triassic sedimentation possibly commenced in upper Norian times with an arenaceous facies (the Plateau Formation), that was an early lithofacies equivalent of the Clarens Formation which developed into the most widespread of all the Triassic sedimentary rocks. Figure 27 shows that throughout large tracts of southern Africa it represents an arid, mostly aeolian lithofacies under the influence of prevailing westerly winds.

In the Karoo Basin and the outcrops to the north and east, the accumulation of Clarens Formation was brought to a close by Lower Jurassic volcanic activity, but evidence from Namibia suggests that in the extreme west, where no Jurassic lavas occur, aeolian sedimentation continued (as the 'main sandstone' at Doros/Huab) until Lower Cretaceous times when it was terminated by eruption of the Etendeka lavas.

### 2.2.5 *Igneous rocks*

Thick lava flows overlie the Triassic sediments in most of the northern outliers, although their age and petrology show considerable variation compared to the relatively homogeneous basalts in the Karoo Basin.

On the NW Namibian coast (Kaokoland) intrusive and extrusive (Etendeka lavas) are late Jurassic to early Cretaceous in age and the Lower Jurassic Karoo Basin volcanics have no age equivalents in this area. The only igneous rocks in Namibia of comparable age to those in the main Karoo Basin occur in the south central part of the country. Immediately north of Mariental up to 360 m of basaltic lavas, petrologically comparable to those of the Lesotho Formation, crop out over 10 000 km$^2$ in the Hoachanas Basin. They overlie Permian Dwyka and Ecca sediments, and have been dated as Lower to Middle Jurassic (173-161 my) by Gidskehaug et al (1975). 200 km south of Mariental, dolerite sills intrude Dwyka sediments in the vicinity of Keetmanshoop. These have been dated as 199-178 my (Upper Triassic to Lower Jurassic) by Gidskehaug et al (1975).

Du Toit (1954) suggested that Drakensberg lavas

Table 14. Classification of volcanic rocks of the Lebombo Group (after Cleverly & Bristow 1979).

| Division | Formation | Lithology | Maximum thickness |
|---|---|---|---|
| Upper Basalts | Movene Basalt | dark, fine-grained aphyric basalts. Olivene and plagioclase-phyric varieties occur. Interbedded rhyolite flows (600 m), pumice, pitchstone, acid breccia and sandy tuff (Sica Beds). | 2 000 m |
| Rhyolites | Jozini Rhyolite and Mbuluzi Rhyolite | reddish plagioclase-phyric rhyolites. A local development in Swazliland has quartz phenocryst and is a grey colour (Mbuluzi Rhyolite). Other local developments are: Oribi Beds (very abundant quartz phenocrysts); Bumbeni Complex (rhyolitic lavas, tuffs, pumice and basalt with a thin basal layer of trachybasalt and trachyandesite – Mpilo Member) | 5 000 m |
| Lower Basalts | Sabi River Basalt | main basalt of lower division – dark finegrained aphyric basalts with olivine and plagioclase-phyric varieties, contains local developments of thin plagioclase-phyric rhyolite flows (Mkutshane Beds); thin rhyolites of similar composition to Jozini Rhyolites (Twin Ridge Beds) | 5 000 m |
| | Letaba Basalt | confined to northern half of range. Olivine rich basalts with olivine-poor flows and interbedded alkali-basalts and shoshonites | 5 500 m |
| | Mashikiri Nephelinite | confined to northern part of range. Alkaline (nepheline) lavas (limburgites) | 170 m |

Major intrusive bodies include:
1. Rooirand Dyke Swarm: major dolerite dyke swarm in the lower Sabie River Basalt in south of range
2. Komatipoort Complex: elongate gabbro intrusion near Komatipoort

Granophyres: Numerous granophyres are intruded close to the boundary between Jozini Rhyolites and Sabie River Basalts in Swaziland.

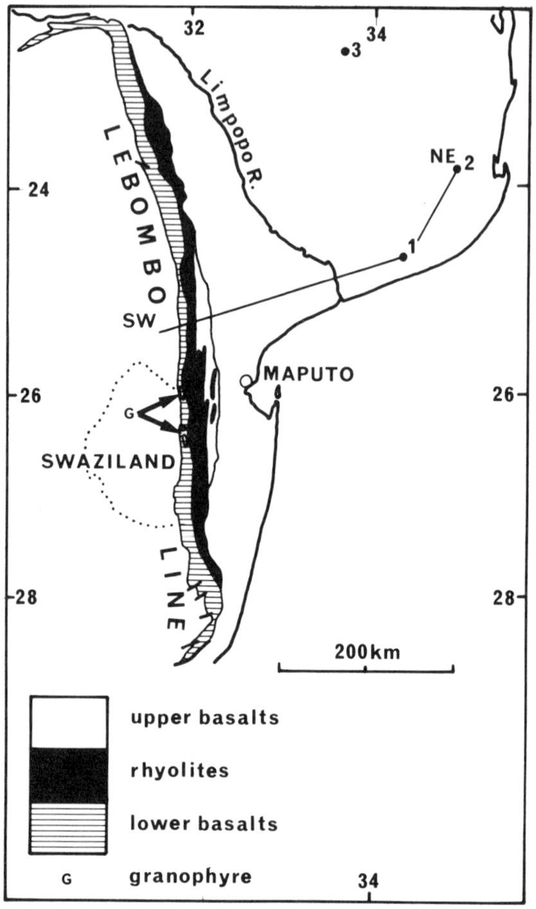


Figure 43. Distribution of Triassic-Lower Cretaceous igneous rocks along the Lebombo Line. NE-SW section across the S. Mozambique coastal plain is shown in Figure 45.

overlie the Clarens Formation under cover of Kalahari sands over much of central Botswana, and borehole data presented by Green (1969) indicate that they extend to about 23°E (Figs 5 & 38). Numerous small outcrops occur near the southern edge of the Karoo outcrop in Botswana (near Lephepe), and along the southern side of the Makgadikgadi Depression, whilst further outcrops occur on the SW margin of the Okavango Graben. Mallick et al (1981) report other occurrences in boreholes on the floor of the Okavango Graben. Lavas crop out sporadically in eastern Botswana and in the Waterberg outlier, but occur extensively in the Springbok Flats and Soutpansberg areas. In the former, up to 365 m of alternating amygdaloidal and massive olivine basalts overlie the Clarens Formation (Visser & van der Merwe 1959). A fine sandstone interbed (up to 6 m thick) locally occurs near the base, and sporadic tuffs and agglomerates are developed.

Up to 1 277 m of massive and amygdaloidal olivine basalt and limburgite lava flows (3-15 m thick) overlie the uneven upper surface of the Clarens Formation in the Soutpansberg (van Eeden et al 1955). Tuffs have been reported from one locality, and lenses of red shale have been noted 1 220 m above the base. Igneous sequences in the vicinity of the Soutpansberg contain alkaline elements (nepheline basalts), trachytic lavas, ignimbrites, and tuffs, and possibly carbonatites (Jacobsen & McCarthy 1975), which are indicative of the different tectonic setting of this eastern region. The volcanic suites in the Soutpansberg, Tuli and Nuanetsi are tectonically associated with the Lebombo Line volcanics and Fitch (1972) quotes radiometric ages 190-195 my for the lower alkaline and tholeiitic basalts of this area.

2.2.5.1 *Lebombo Group*

The thickest and most spectacular Triassic-Jurassic igneous rocks occur along the Lebombo Line, where up to 13 km of lavas and rare interbedded sandstones overlie the Clarens Formation, and crop out in a zone 20-50 km wide over a distance of about 660 km. They give rise to the narrow, impressive Lebombo Mountain range (Fig.43). The extrusive rocks have an overall tripartite stratigraphy, and are intruded by numerous dykes and sills. The whole sequence is draped over a major crustal discontinuity and dips eastward under the Cretaceous and Cenozoic sediments of the Zululand/Mozambique coastal plain at between 5 and 40°, in a monoclinal arrangement (Fig.45). Good general descriptions of the whole structure, and the relationships between the various rock types have been given by du Toit (1929), and Wachendorf (1973), amongst others, whilst detailed geochemical studies (Bristow 1980) and revised stratigraphic correlations (Cleverly 1979, Cleverly & Bristow 1979) have been made in recent years as part of a regional geochemical/structural synthesis of 'Karoo volcanism' as a contribution to the International Geodynamics Project, sponsored by the Council for Scientific and Industrial Research.

Earlier workers (e.g. du Toit 1929) considered the succession of mainly extrusive igneous rocks along the Lebombo Line in terms of three main rock divisions: Lower Basalts, Rhyolites, and Upper Basalts, and notwithstanding the revised regional stratigraphic correlations proposed by Cleverly & Bristow (1979), which we show in Figure 44, it is convenient (often for cartographic reasons) to use the tripartite scheme to illustrate the gross structure of the Lebombo volcanic pile. Cleverly & Bristow (1979) show that the Lower Basalt division can be divided into three formations, which are found successively overstepping from the north onto the Clarens Formation of the eastern Transvaal and Swaziland: Mashikiri nephelinites; Letaba basalts, and the most extensive unit, which crops out along almost the whole length of the range, the Sabie River basalts.

The Mashikiri Nephelinite Formation crops out in the northern part of the Kruger park, and consists of alkaline lavas, which carry stellate glomeroporphyritic aggregates of clinopyroxene, and rare nepheline crystals

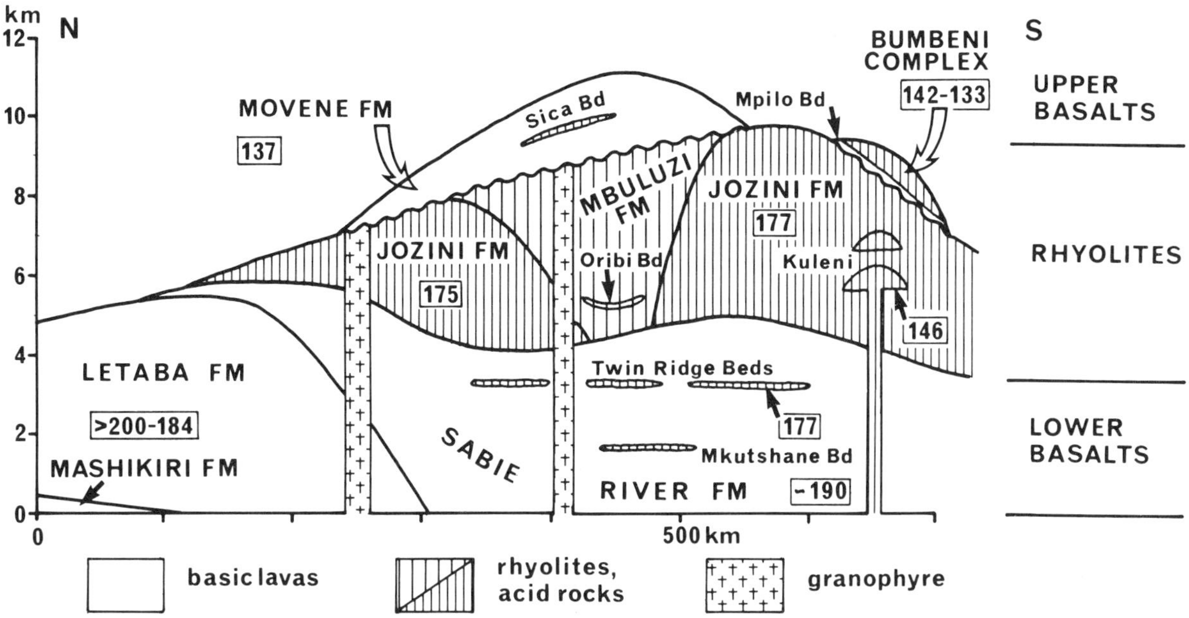


Figure 44. Schematic N-S section along the Lebombo Line showing stratigraphic relationships and thicknesses of main igneous formations. Radiometric dates are in rectangles. Nomenclature and thicknesses are after Cleverly & Bristow (1979), dates are from Burger & Coertze (1973), Flores (1973), and Bristow (1980). Right column shows the previous informal classification.

and biotite flakes. These rocks reach a maximum thickness of ca.170 m. More extensively-developed (ca. 5.5 km thick), but still confined to the northern Lebombo area, are the Letaba basalts, which are mainly olivine-rich, but occasionally contain flows of olivine-poor, and alkaline basalts, and shoshonites, South of about 25°S, the Sabie River Basalt is, to all intents and purposes, equivalent to the Lower Basalt division of earlier workers. It reaches a maximum thickness of about 5 km along the south Swaziland border, and extends to the southern end of exposures in the Lebombo range. The rocks are dark, fine-grained, generally aphyric basalts in which two minor interbedded units occur: Mkutshane Beds, and Twin Ridge Beds. The former are thin rhyolite flows in the lower part of the sequence, which occur only in the central part of Swaziland (Cleverly 1979), whilst the latter are a further, more extensively developed set of thin rhyolite flows in the upper part of the main basalts. The Twin Ridge rhyolites are geochemically similar to the overlying main (Jozini) rhyolites, whereas the Mkutshane rhyolites differ significantly on trace-element and isotopic grounds (Cleverly & Bristow 1979).

The Middle, Rhyolite division has been divided into two formations by Cleverly & Bristow (1979). The most extensive, running almost the whole length of the range, and forming most of the high ground, is the Jozini Formation (ca.5 km thick), whilst a quartz-phenocryst-rich variety, the Mbuluzi Formation (ca. 5 km thick) occurs only (and appears to replace the Jozini Formation) in the vicinity of Swaziland. The exact stratigraphic relationship between these two is not clarified by Cleverly & Bristow (1979), but reference to Cleverly's (1979) map of the Swaziland sector of the Lebombo suggests that the Mbuluzi rhyolites are a local variant that grades laterally into the Jozini type, although Cleverly (1979) does specifically mention that the former overlies the latter. A particularly quartz phenocryst-rich variety within the Mbuluzi Formation, the Oribi Beds, crop out in Swaziland and locally reach a thickness of 250 m. In Swaziland, Cleverly (1979) has recognized over 30 individual rhyolite flows, each averaging over 200 m in thickness and extending along strike for distances up to 60 km. The tops of the flows are autobrecciated, with glassy laminated bases. He concludes that they were emplaced by pyroclastic flow at relatively high temperatures (similar to ignimbrites), but that on cooling, flows collapsed to form fluid masses. A further localized development of rhyolitic lavas, ash, and tuff beds occurs above the Jozini Formation at the southern end of the range: the Bumbeni Complex. According to Cleverly & Bristow (1979) and Bristow (1980), this complex is separated from the underlying rocks by an erosion surface, and its stratigraphic relationship to the rest of the Lebombo sequence is not precisely known. The Bumbeni Complex has a thin (50 m) basal sequence of trachybasalts and trachyandesites (the Mpilo Member), and is intruded by small syenitic bodies.

The Upper Basalt division has been re-named the Movene Formation by Cleverly & Bristow (1979). It crops out only in Mozambique, and is restricted in

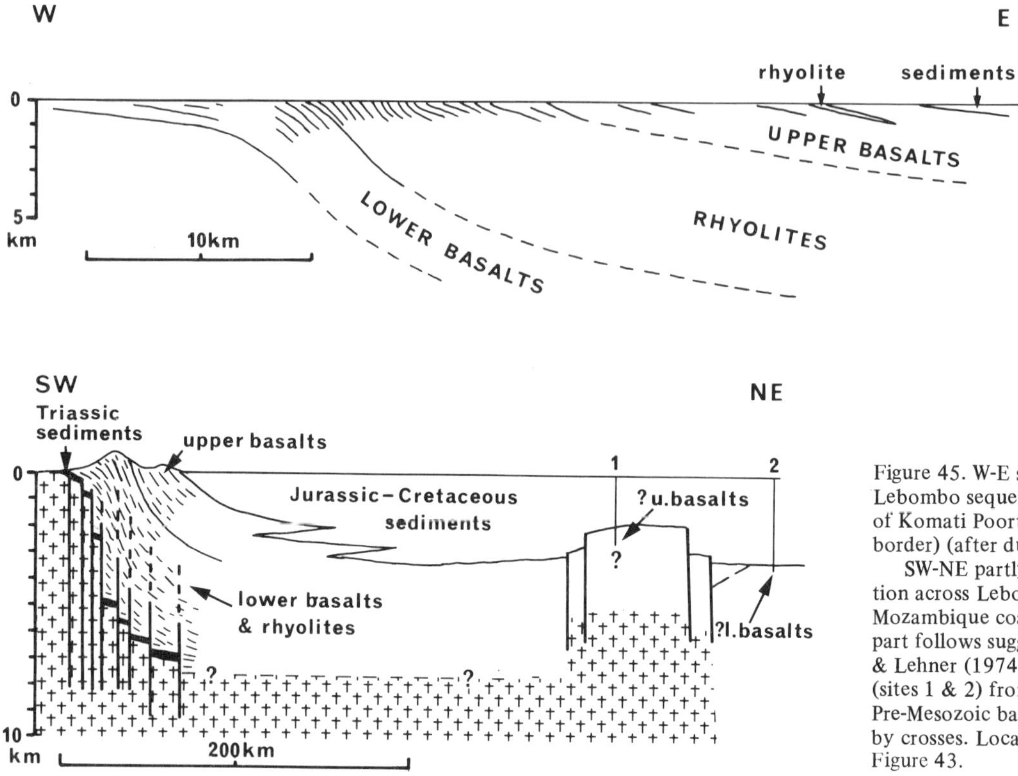


Figure 45. W-E section across the Lebombo sequence in the vicinity of Komati Poort (N.Swaziland border) (after du Toit 1929).

SW-NE partly speculative section across Lebombo Line and S. Mozambique coastal plain. The NE part follows suggestions by Beck & Lehner (1974). Borehole data (sites 1 & 2) from Flores (1973). Pre-Mesozoic basement indicated by crosses. Location shown in Figure 43.

lateral extent to the central part of the Lebombo Line (Fig.44). Petrologically, it appears to be similar to the Sabie River Formation basalts, although nepheline-normative varieties are common. A local interbedded subdivision is the Sica Beds, a group of rhyolitic flows.

The whole sequence of extrusive Lebombo rocks is intruded by a large number of dykes (mostly subparallel to the regional tectonic trend) and occasional larger bodies. The former are thought to have acted as feeders for their respective lava types, so that the Lower Basalts are cut by one set of dolerite dykes (with occasional gabbros), the lower two divisions are cut by felsite feeders for the acid Jozini and Mbuluzi formations, whilst the whole succession is cut by a second set of dolerite and felsite feeders for the Upper Basalts and rhyolites (Sica Beds). Along the western edge of the mountain range, these successive intrusions have produced a myriad of intrusions into the Sabie River Formation and underlying sediments (e.g. the Rooirand dyke swarm). Other intrusive bodies include granophyres in the northern and central parts (which probably post-date the Jozini rhyolites according to Cleverly 1979), and rhyolitic domes and dykes (Kuleni structures) which interfinger and intrude the Jozini rhyolites. Maximum dips (30-40° to the east, Cleverly 1979) occur in the Rhyolite division, and it is here that the tuffs and other evidence of explosive activity are most common. Such evidence suggested to du Toit (1929) that maximum crustal downwarping to the east took place during this phase of extrusion, which radiometric dates on the rhyolites place around 175-177 my (Cleverly 1979, Bristow 1980).

Radiometric age determinations (either quoted or produced by Bristow 1980, and shown on Figure 44) show that the lower, basic division was extruded over a period that extended from slightly > 200 my to 184 my (in the north) and 177 my (in the south). These rocks were very closely followed by the 175-177 my Middle, Rhyolite division, during which phase, maximum crustal downwarping and eastward tilting occurred. A significant break (about 40 my long) separates the main rhyolites from the Movene basalts (Upper Basalts) (137 my) and the Bumbeni Complex (133-142 my). During this hiatus, it is probable that the Kuleni rhyolites (146 my) and the granophyres were intruded. Hence, the early phase of Lebombo volcanic activity (Lower Basalts, Rhyolites) was approximately synchronous with the widespread igneous activity in the main Karoo Basin, whilst the later phase (Upper Basalts, Bumbeni Complex) was synchronous, or slightly earlier than early Cretaceous activity in the Zambezi Graben (Flores 1973).

Rocks of the earlier episode (Lower Basalts and Rhyolites) are more widespread, and in view of the greater downwarping associated with them, indicate that the main phase of crustal subsidence along the

Table 15. Igneous rocks encountered in boreholes under the Mozambique coastal plain (after Flores 1973) (see Figures 43 and 45 for locations).

| Borehole | Depth below ground level | Rock types | Thickness m |
|---|---|---|---|
| 1 | 1 832 | amygdaloidal dolerite with weathered basalt horizons and two acid flows. ?Upper Basalt division | >860 |
| 2 | 3 130 | intercalated olivine basalt and altered tuffs. ?Lower Basalts | > 56 |
| 3 | 3 008 | limburgitic basalt and claystone. ?Lower Basalts | >109 |

eastern edge of the Kaapvaal Craton took place during late Triassic-early Jurassic times. It may have continued intermittently, with sporadic volcanism, but the second phase of volcanic activity (Upper Basalts) probably indicates, increased crustal subsidence that heralded the establishment of large sediment basins in southern Mozambique and the northern Natal/Zululand area.

Basement descends rapidly eastwards across the Lebombo Line discontinuity (Fig.45), but depths to it under the Mozambique plain are not known. However, an extension of the Lebombo Group under the plain is proved by borehole data from Mozambique (Flores 1973), where the rock types listed in Table 15 have been encountered (Figs 43 & 45). None of these rocks were dated, but on petrological evidence Flores (1973) assumes they are the same age as the Upper Triassic-Lower Cretaceous Lebombo sequence, rather than the mid-Cretaceous age (115 ± 10 my) of the younger intrusions associated with the Zambezi Graben and its southerly extension.

## 2.3 SOME STRUCTURAL CONTRASTS IN THE SEDIMENT BASINS

Throughout Rust's (1975) Karoo tectono-sedimentary terrain, there is no evidence of penecontemporaneous faulting, and all earthmovements appear to have been of a regional epeirogenic nature, with later gentle flexing and normal faulting largely responsible for producing the disposition of present erosional outliers. In the Springbok Flats, this has resulted in a bipartite saucer-shaped depression, whereas the Etjo outlier, which is faulted along its NW boundary and dips gently eastwards, probably represents the eroded western remnant of a large synclinal outlier under eastern Namibia (Fig.38). In both these areas the faulting post-dates the Drakensberg Group igneous activity, and although we do not have an accurate upper age limit for these earthmovements, a mid-Jurassic age is suspected for their commencement as there may be a link between them and the taphrogenic movements in marginal areas of the continent, and other parts of the Zambezian terrains.

In strong contrast, Beaufort and Stormberg sedimentation in the Soutpansberg took place in a linear zone in which crustal subsidence and intermittent faulting had had a long history. Jansen (1975) suggests that a total thickness of about 7 km of pre-Permian strata occur in the Soutpansberg basin and that penecontemporaneous faulting took place during the deposition of the Permian-Triassic sediments and Jurassic lavas. Post-lava faulting is merely reactivation of older fault lines. He suggests that the basin is the western extension of a taphrogenic belt superimposed on the Limpopo Mobile Belt, and although the sedimentary sequence here compares favourably with that in the Waterberg area, it was deposited under very different tectonic circumstances. As described by Flores (1973), and discussed in Chapter 4, later developments of the easterly extension of this tectono-sedimentary environment are seen in the mid-Jurassic to early Cretaceous sequences of the Zambezi Graben, suggesting that the history of faulting and vertical crustal movements in the Soutpansberg lineament spans the period from about 2 000-160 my BP.

Further evidence for late Palaeozoic-early Mesozoic reactivation of old fault lines within, and adjacent to the Limpopo Mobile Belt has been found by Green et al (1980) in eastern Botswana (Fig.39). Here, Karoo Supergroup strata, including Stormberg and Drakensberg groups were deposited in NW-SE and ENE-WSW graben and half graben. Distribution of older rocks in the area confirms Jansen's (1975) observations that vertical crustal movement in the boundary zone between the Rhodesian and Kaapvaal cratons has taken place over very long periods of time and that pre-, syn- and post-Karoo Supergroup faulting can be recognized in association with Permian to Jurassic sediment and lava accumulations. These movements, which appear to be reactivations of older fault lines, produced a variety of sediment basin types, which Green et al (1980) recognized, with specific examples in brackets:

a) post-Karoo faulting that has merely resulted in erosional isolation of basins (e.g. across the Letshana Fault);

b) pre-Karoo faulting that provided depressions for subsequent sediment accumulation (e.g. the NW-SE Makoro Trough, in which the sediment fill, whose youngest representatives are the 'Upper Ecca' (=Beaufort Group), are not affected by the faults);

c) synsedimentary faulting that ceased after deposition of the Clarens Formation, synchronous with the onset of Drakensberg lava extrusion (e.g. to the north of the Zoetfontein Fault).

On the northern side of the Limpopo Mobile Belt (around Bobonong in the western end of the Tuli structure), the fault movement was probably pre-Karoo in age, and Green et al (1980) consider that the structure may be analagous to the Makoro Trough.

Permo-Triassic lithofacies in these various isolated basins was strongly affected by their geometry, particularly where synsedimentary faults were active. Although Green et al (1980) quote no examples from the presumed-Mesozoic strata, they cite examples of arenaceous horizons in argillites as evidence of local topographic discontinuities. As in the Soutpansberg, this relatively young phase of sedimentation has been along similar lineaments and in similar shaped receptacles to earlier phases in the same area (i.e. Waterberg Group), indicating remarkable continuity in the style of crustal movement over a long period of time (at least as old as the Ventersdorp Supergroup, and possibly to 3 000 my).

The Lebombo Line is thought to represent a major crustal discontinuity between a stable craton to the west and the marginal zone of vertical tectonics to the east, and we concur with Stratten's (1970) speculation that the latter is a southerly extension of a major taphrogenic belt that runs down the length of the east African continental margin from Ethiopia to northern Natal. Part of Rust's (1975) Zambezian tectono-sedimentary terrain is coincidental with this, and features such as the Soutpansberg graben are intracontinental extensions. The earliest evidence for differential movement along the Lebombo Line dates from the late Triassic-early Jurassic, but it may have commenced earlier (it is at least Permian in age in East Africa), and continued into Cenozoic times.

# 3

# The Cape Fold Belt

## 3.1 INTRODUCTION

The Cape Fold Belt extends from Cape Town northwards as far as Clanwilliam, and eastwards as far as East London (Fig.46). Deformation involved a trough which had been filled with sediment since about late Cambrian time, and contained rocks of the Cape Supergroup (consisting of the Table Mountain, Bokkeveld, and Witteberg Groups) and the Karoo Supergroup (Dwyka, Ecca, and Beaufort Groups). On the basis of dates obtained by the argon-release method, Hälbich (personal communication) deduces deformation episodes at 273-279 my and 256-259 my, i.e. entirely Permian, but there are no large stratigraphic breaks in the sequence at these times and it must be inferred that deformation at depth was contemporaneous with continuing deposition at the surface. The earliest evidence of uplift and erosion, possibly due to orogenic activity, is found in the Katberg Formation of the Beaufort Group which is of Scythian age, and might correspond to the third episode deduced by Hälbich, dated at 225-234 my. However, the overlying rocks (Burgersdorp Formation), reflect a diminution of uplift in the source area, and the next sedimentary sequence that can be related to a major uplift event is the Molteno Formation, i.e. approximately Middle to Upper Triassic. It seems possible that we may have to distinguish between the ages of deformational events in the Cape Fold Belt, and the ages of major uplift and erosion, and that the climax of the latter may be later than the former.

The two branches of the Cape Fold Belt result from two distinct axial trends almost at right angles, which converge near Cape Town, and continue in a SW direction to the coast as a single trend. In places, a northward-convex arcuation is superimposed on the overall east-west trend, e.g. near Cape Agulhas and in the Uniondale area of the eastern Cape. Major east-west faults are also a prominent feature of the structure, and most of these downthrow to the south: the two most important are the Cango and Worcester faults. Thrusting occurs only in the southern branch and is of relatively minor importance, but increases eastwards with intensity of folding. It is of minor importance around Laingsburg, but quite distinct major thrusts occur east of Meiringspoort and into the Baviaanskloof.

Fold styles vary considerably from place to place, and also vertically in the stratigraphic sequence. The NW or Cedarberg branch consists of generally upright open folds, although locally, flank dips may be close to vertical, and the Cape Supergroup disappears eastward beneath Karoo cover along a major north-south monocline. Horst and graben structures accompany the folding near Clanwilliam. In the east-west branch there is an overall tendency for fold intensity to increase from south to north, from open upright folds to near-recumbent isoclines, but in detail the pattern is more complex. Along a central N-S traverse, fold style and intensity are zonally developed. Up to three east-west cleavages develop, and locally even a fourth, N-S crenulation cleavage has formed.

Fold intensity seems markedly higher just north of the major faults, and while only folds of large amplitude and wavelength are found in the lower Table Mountain, lower Dwyka, and lower Ecca groups, additional smaller scale folds are common in the upper Table Mountain, Bokkeveld, Witteberg, Upper Dwyka and upper Ecca groups. Axial planes, when not vertical, are usually southward dipping, but some areas of Witteberg rocks show northward-dipping axial planes. Metamorphism is of very low grade, and ranges from unmetamorphosed (ca.150°C, 1-2 kb) along the northern margin of the Fold Belt, through anchi-metamorphism to the border of anchi-epi-metamorphism (lower border of the greenschist facies (ca.300°C, 2-3 kb) in the Swartberg range. From there southward it increases very slowly to reach lower greenschist facies (biotite) in the Outeniqua range at 350°C or more (Hälbich & Cornell 1983).

The only evidence of contemporaneous igneous activity consists of volcanic fragments in greywackes, and some partly volcaniclastic cherts and crystal tuffs.

The first attempt at a synthesis of the Mesozoic deformations was by du Toit (1926) who regarded the 'Natal foldings' (Lebombo monocline) as the beginning of a fold belt most of which was concealed under the ocean, but which probably continued southwards to meet the E-W folds in a syntaxis corresponding to that where the E-W folds and the Cedarberg trends meet.

The 'Cedarberg Foldings' he regarded as having been initiated prior to Dwyka deposition, but with continued or renewed movement in the Permian and later, so that

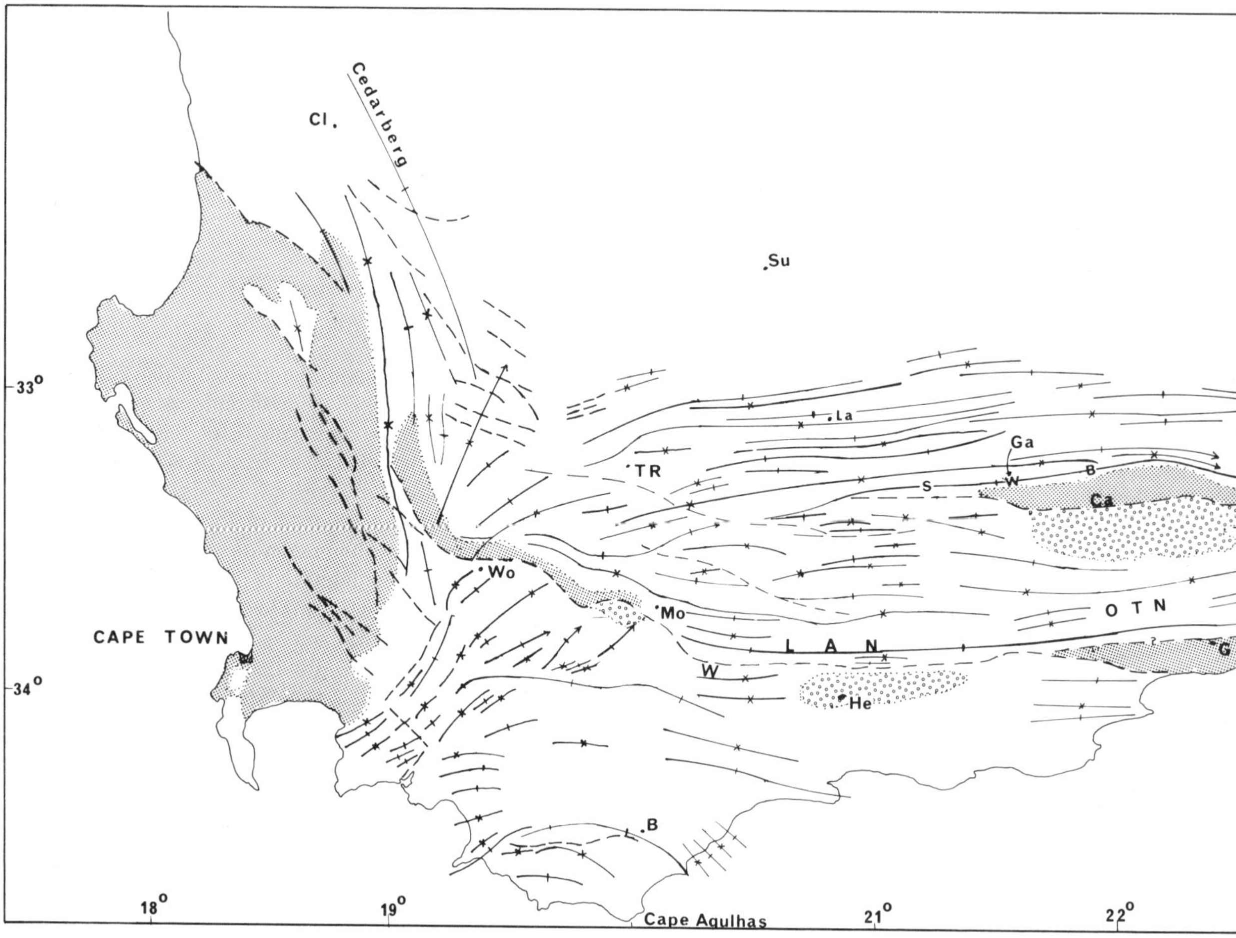


Figure 46. Major folds and faults of the Cape Fold Belt, and localities mentioned in the text. Localities: B = Bredasdorp; BK = Baviaans Kloof; Ca = Cango; Cl = Clanwilliam; EL = East London; G = George; Ga = Gamkapoort; He = Heidelberg; La = Laingsburg; Mo = Montagu; MP = Meirings Poort; OTN = Outeniqua Mountains; PE = Port Elizabeth; SK = Skoorsteenberg; SWB = Swartberg Mountains; Su = Suurberg; TR = Touws River; Un = Uniondale; W = Worcester Fault; Wo = Worcester; Zu = Zuurberg Fault.

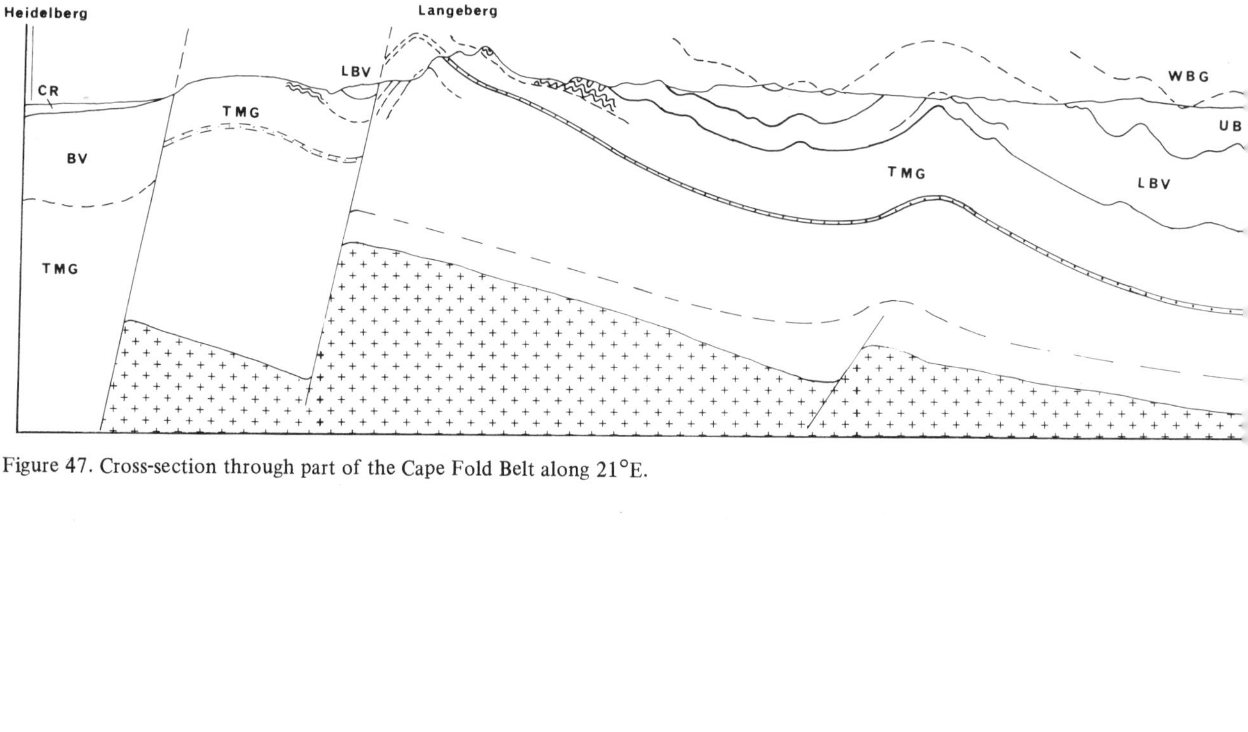


Figure 47. Cross-section through part of the Cape Fold Belt along 21°E.

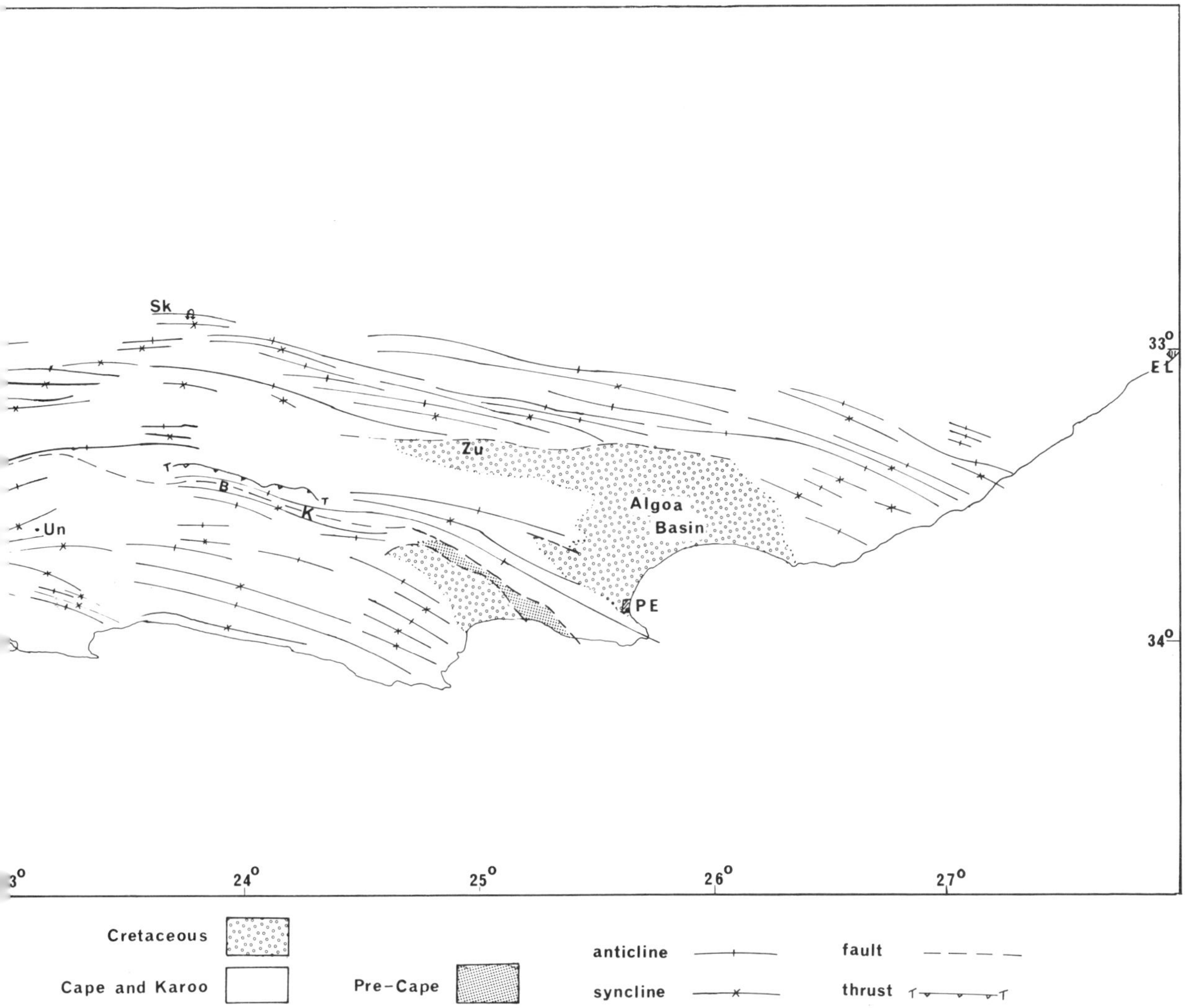
Sk
33°
EL
Zu
B
K
Un
Algoa
Basin
PE
34°
3°
24°
25°
26°
27°
Cretaceous
Cape and Karoo
Pre-Cape
anticline
syncline
fault
thrust

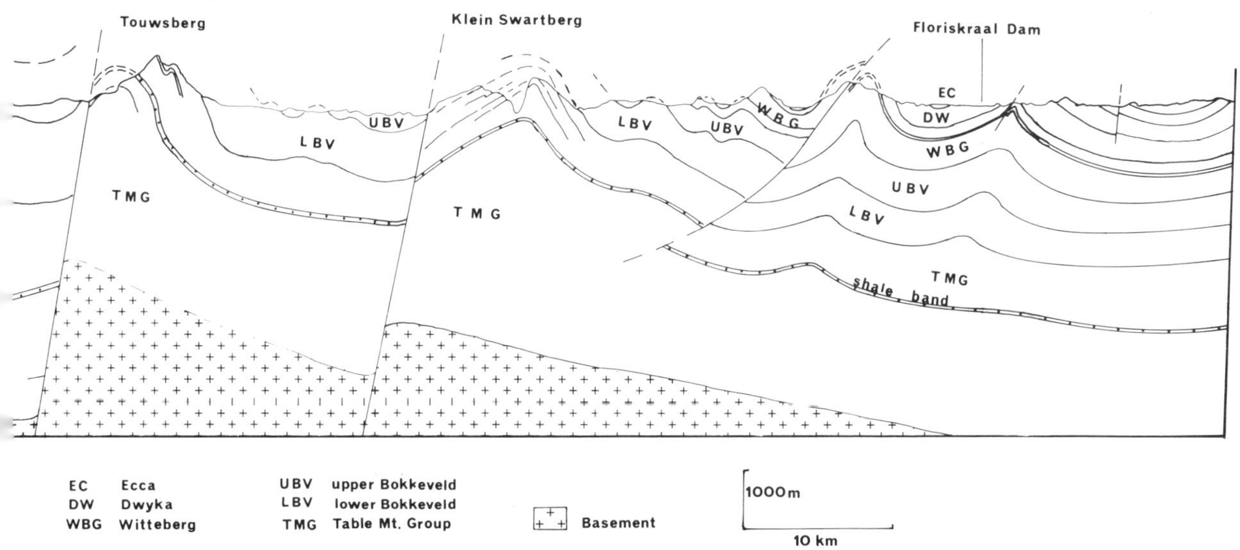
Touwsberg
Klein Swartberg
Floriskraal Dam
EC
DW
WBG
UBV
LBV
TMG
shale band
EC Ecca
DW Dwyka
WBG Witteberg
UBV upper Bokkeveld
LBV lower Bokkeveld
TMG Table Mt. Group
Basement
1000m
10 km

both Dwyka and Ecca rocks were involved. He attributed the Cedarberg trend to a 'previously established zone of compression' which had deformed the pre-Cape rocks of the area along the same lines. The E-W 'Cape Foldings' he stated to form a series of roughly concentric arcs concave southward with overturning in places suggesting pressure exerted from the south, and involving formations up to Lower Beaufort in age.

Du Toit believed, therefore, that deformation occurred in three episodes – pre-Dwyka in the Cedarberg, post-Lower Beaufort in the main E-W belt, and late Jurassic-early Cretaceous in the Lebombo, but that in all cases renewal of movement took place at a later date, the main event being the faulting of the otherwise almost undeformed Cretaceous deposits, which he took to be fracturing associated with the breakup of Gondwanaland. He did not speculate as to the cause of the fold-generating stresses, but in a contribution to the review by de Villiers (1944), du Toit mentions the possibility that the overall northerly drift of Africa during the Permo-Triassic was temporarily reversed during the Triassic, and that this reversal was responsible for the compression of the Cape trough. De Villiers himself closely follows du Toit's reconstruction of the sequence of events, but invokes a different mechanism, namely one proposed by Lawson (1927) for the Canadian Rockies. This involves a transfer of lower crustal material outwards in response to downward buckling of a trough under sediment load, resulting in crustal thinning beneath the trough. When this thinning reached a critical point, the crust failed and there was a sudden horizontal collapse in response to an assumed 'primary, pervasive, tangentially compressive stress', and the resulting shortening was responsible for the folding and thrusting.

De Villiers (1944) extended this model to the Cape Fold Belt, and suggested that since the rock flow was away from the trough, it would combine with the primary stress to form a vertical couple which was responsible for the observed asymmetry and overfolding. Later, de Villiers (1956) modified this concept, suggesting that the couple acted in a plane having a NE strike and steep SE dip, resulting in arcuation of the fold axes and giving them their plunging *en echelon* arrangement. He also noted that the syntaxis shows no signs of folds crossing, but is in fact formed by a gradual divergence from a NE direction into easterly and NNW directions respectively. A NE-SW stress direction was suggested for the Cedarberg folds, the swing into a SW trend in the south being the result of compression from the south against a rigid buttress whose SE edge ran approximately from Gordon's Bay to Worcester.

No new proposals were made concerning the mechanism of the Cape folding until Newton (1973) put forward the idea of gravity gliding. This is discussed in more detail below, but the basic concept is that the southern slope of the trough became steepened by uplift along major strike faults until eventually the overlying sedimentary pile began to move northwards, with intense folding occurring on the northern sides of the faults, and the overall intensity of folding increasing until the axis of the trough was reached, where reversal of the slope brought all movement to a halt. The Cedarberg folding would be due to an eastward movement down a gentler slope, and the syntaxis would be the result of interference between the two sliding masses. Rifting associated with the breakup of Gondwanaland would result in a reversal of movement on the major faults, so that today most of them have a southerly downthrow.

Almost immediately after, de Swardt et al (1974) published a completely contradictory viewpoint: they regard the basement and cover rocks as having deformed as a single unit, by slip on cleavage planes which represent a regional plane of transport. This feature, coupled with the northward vergence, variations in intensity and style of folding, and the dying out of the deformation against a relatively undisturbed foreland, suggests to them an orogeny of Alpine type 'which is generally ascribed to gravity sliding away from a median zone of excessive uplift, with movement of the cover over a more rigid basement; this is accompanied by folding and thrusting, with vergence away from the median zone. The Cape orogeny would thus be a typical 'doppel-orogen', whose component with opposite vergence is perhaps concealed under the ocean or preserved on another continent. The best explanation advanced so far for orogenies of Alpine type is that of the 'collision of crustal plates' (de Swart et al 1974, p.56).

This hypothesis is also further discussed below. More recently, the suggestion that the Fold Belt can be related to a subduction model has been made by several workers (Rhodes 1974, Martini 1974, de Beer et al 1974, Lock 1980) and this point is also discussed below.

## 3.2 CHARACTERISTICS OF THE CAPE FOLD BELT

The structures within the Belt vary considerably in style and orientation and an attempt is made here to describe a sufficient number of examples to illustrate the range of types involved and give an overall impression of the way in which the rocks have been deformed.

### 3.2.1 *Folds*

In the E-W part of the Fold Belt, many examples of upright open folds and asymmetric folds with south dipping axial planes can be found. Figure 47 shows some typical examples along a cross-section at 21°E, and it is apparent that there is a gradation from upright to inclined structures in a northward direction, with dips on the steeper limbs becoming overturned in the extreme north, near Laingsburg. Small near-bedding plane thrusts

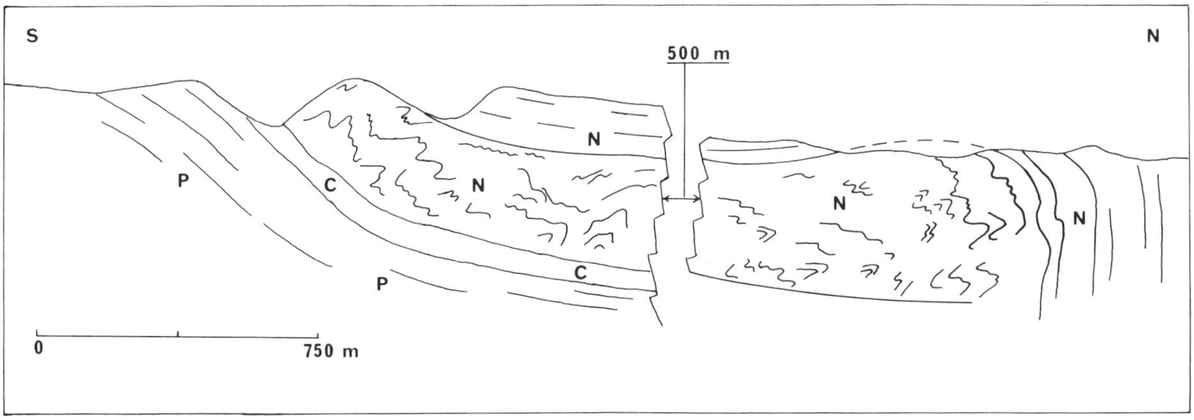


Figure 48. Structures in the northern part of a cross-section through the Swartberg Pass (22°E). Note the regularity of dips in the more competent Peninsula and upper Nardouw units, and the disharmonic lower order disturbances in the lower Nardouw sequence just above the major decollement horizon in the Cedarberg Formation. Abbreviations: P = Peninsula Formation; N = Nardouw Formation; C = Cedarberg Formation. After Halbich (1983c, fig.11-3).

occur on the south dipping flanks of some anticlines, as shown by a repetition of distinctive quartzites, and are clearly a part of the overall finite strain pattern. Cases where the asymmetry is in the opposite sense, i.e. axial planes dipping north, are fairly rare, but have been observed, particularly in the Witteberg rocks near Touws River. In Witteberg rocks, the variation in lithology results in complex folding by a variety of mechanisms (Hälbich 1977, fig.9), including flexural slip and tangential longitudinal strain in the quartzites, development of hinge fractures and thrusts, and largely passive behaviour of shales and siltstones. No penetrative cleavage is developed, and disharmony on a small scale is common.

Data on the detailed profile shapes of some megafolds and the amount of shortening in some parts of the Fold Belt was provided by Hälbich (1977, 1978). He used photogrammetric techniques combined with field measurements to construct three profiles across major anticlines, two of which were composed of Table Mountain Group and one of Witteberg Group rocks. In one of these, the Tradouw profile, sufficient marker beds were traceable to allow classification of individual units according to the method of Ramsay (1967). Considerable variation, both in individual units and between units, was apparent, and Hälbich states that folding probably developed by a combination of flexural slip, tangential longitudinal strain, and flattening, with flexural slip probably predominating in the massive quartz arenite sequence (Hälbich 1978, p.257). It is claimed that the profiles show a sudden increase of thickness of many units on the north limbs of the major folds, and this is interpreted as an original depositional feature which controlled the location of the major fold. Measurement of Hälbich's diagrams, however, shows that the overall thickness increase is $<10\%$ and that as some units thicken, others thin, thus throwing doubt on his interpretation. The presence of minor folding on the northern (and steeper) limbs, which Hälbich (1978) attributes to local gravity sliding, should also be allowed for in inferring thickness changes. What the profiles demonstrate beyond doubt is that the disharmonic folding (Figs 48 & 49) is accommodated in the shaly Cedarberg Formation.

Calculations of shortening by Hälbich (1977) are based on (1) measurements along folded competent beds, and (2) deformed fossils in Bokkeveld Group rocks, the average figure for the latter being 23-27 %, depending on the amount of volume loss, and the former giving closely comparable figures. He favours tangential compressive shortening as the cause of the folding, and believes that the basement rocks are probably involved in the deformation, but does not speculate on the causes of such shortening.

Disharmonic folding can be seen in many places, being a natural consequence of the generally concentric style of folding determined by the more competent members of the sequence, and the alternation of these with more incompetent horizons. Examples are:

i) chert and siltstone bands in the Dwyka shales near Laingsburg, which are intensely buckle-folded on a small scale, whereas neither the White Band above nor the tillite below are affected;

ii) the uppermost Table Mountain Group north of Heidelberg (Fig.50) where the folding dies out at the contact with the overlying Bokkeveld;

iii) between the first and second Bokkeveld sandstones east of Worcester (Fig.51);

iv) in the Witteberg rocks east of Touws River;

v) in the Gamkapoort anticline (Fig.49).

Shallow-dipping, near-isoclinal folds are fairly rare, but are present in the Outeniqua mountains north of George (Fig.52, and see de Swardt et al 1974) and at Schoor-

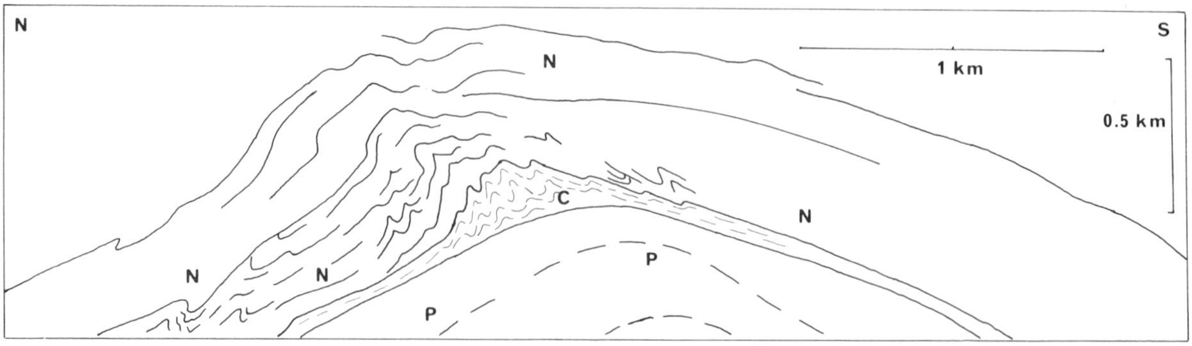


Figure 49. Structural profile through Gamka Poort (21°40′E), demonstrating decollement of Nardouw Formation along the Cedarberg Formation on the southern limb of an originally symmetric mega buckle-fold in the Table Mountain Group. Asymmetry, disharmony, and decoupling occur within the Nardouw Formation. After Halbich (1977 fig.6 & 1982c fig.11-1).

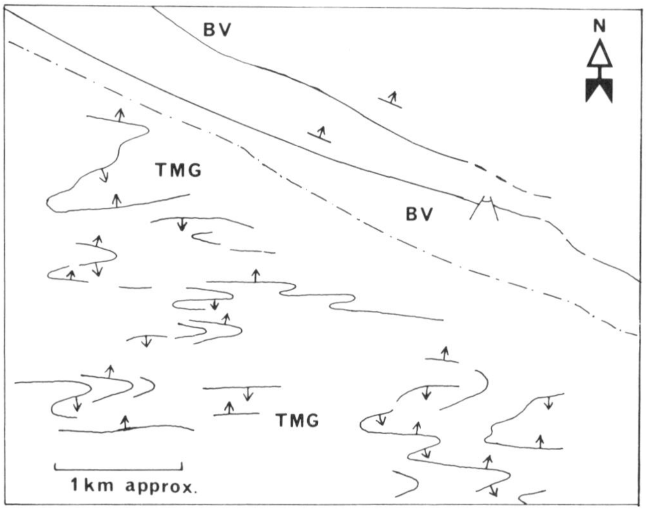


Figure 50. Small-scale disharmonic folding in upper Table Mountain Group north of Heidelberg. Abbreviations: TMG = Table Mountain Group; BV = Bokkeveld Group. Traced from aerial photographs.

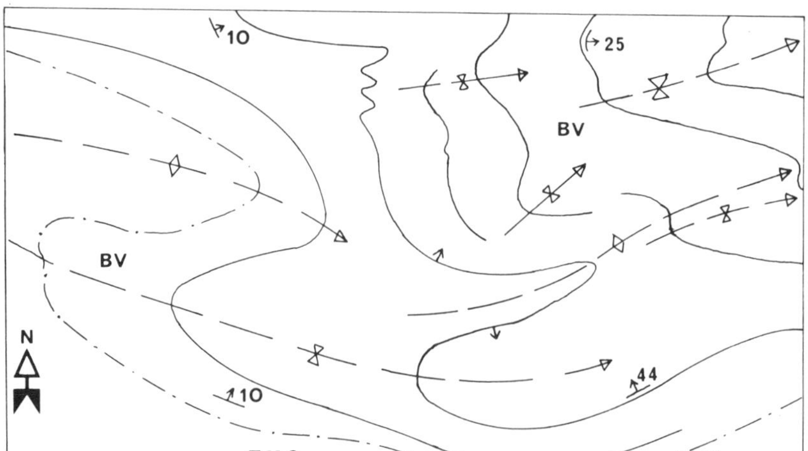


Figure 51. Disharmonic folding within the Bokkeveld Group NE of Worcester. Abbreviations as for Figure 50.

Figure 52. Flat-lying, near-isoclinal fold in the Table Mountain Group, Outeniqua Mountains.

steenberg on the north edge of the fold belt (Haughton et al 1953). The latter is an interesting case, in that the structure at the surface is an apparently simple open anticline, but drilling revealed that this was actually the crest of a near recumbent, near-isoclinal fold, with the upper Witteberg shales completely squeezed out of the limbs (Fig.53). It is possible that many more such structures are present but have not been revealed by drilling, since similar structures occur in Bokkeveld sandstones north of Meiringspoort (Coetzee 1979 & 1983).

Other folds with near-horizontal axial planes, but with a more angular, chevron-like profile, are common in some parts of the Fold Belt, e.g. near Montagu (Fig. 54) and in Meiringspoort (Fig.55). These are classic examples of cascade folds (Harrison & Falcon 1934), which conform to the criterion of de Sitter (1956, p.289) that the inverted limbs of gravity folds are not thinned, whereas in compression structures they show intense shearing and thinning. Because of their geometry they are disharmonic, dying out rapidly up and down the sequence. Their enveloping surfaces are parallel and dip steeply north, and they are usually found as second-order structures on the steep or overturned limbs of mega-folds north of major faults. It is possible that they are gravity-induced collapse structures and result from slip along the bedding down the steeply north-dipping limb of first order folds. Maximum stress in quartz grains is oriented north-south and parallel to bedding on both limbs of the chevrons (Coetzee 1979, 1983). Their possible connection with faulting is discussed below. Other examples occur in the Baviaanskloof Range, where they were described by Theron (1969) as 'horizontal chevron folds'. An incipient stage in the formation of such folds may be represented in the same area by the stepfolds described by de Villiers (1944), which are a series of monoclines whose axial planes dip about 45°S, and which are almost closed at the base, but open out into more symmetrical and rounded synclines and anticlines higher up.

In the NW or Cedarberg branch of the Fold Belt,

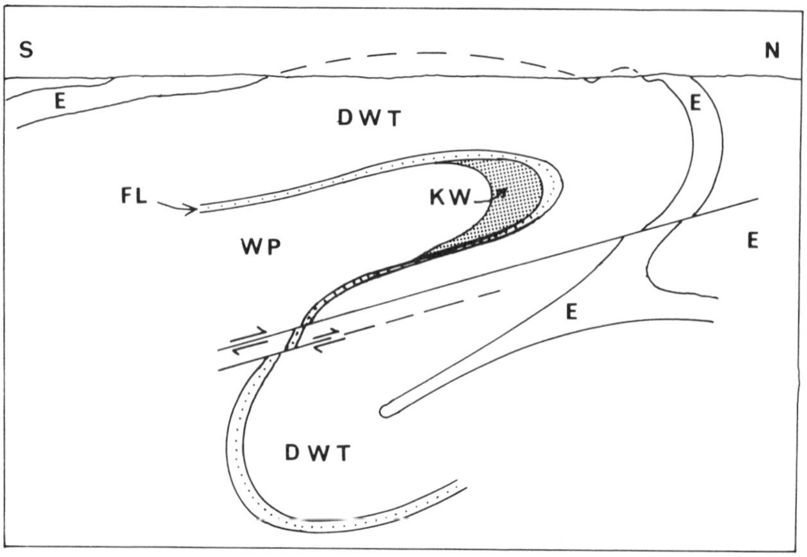


Figure 53. Isoclinal folding near Skoorsteenberg, as revealed by the Alandale borehole (Haughton *et al.* 1953). Abbreviations: WP = Witpoort Sandstone; KW = Kweekvlei Shale; FL = Floriskraal Quartzite; DWT = Dwyka Tillite; E = Ecca.

the folds are generally open and dip moderately, but in a few places dips are very steep and locally slight overturning occurs. Fold axial planes are upright, and it may be that in this area basement faulting plays a part in the formation of folds. In the southern part of the range, the Dwyka tillite is conformable with the Witteberg and is completely involved in the folding, but as the folding dies out northward a disconformity appears at the base of the Dwyka which becomes an unconformity as the tillite over-steps on to pre-Cape rocks. This has been interpreted as demonstrating a pre-Dwyka phase of folding, but nowhere does the Dwyka lie on previously folded Cape rocks.

Study of a number of sample areas in the Fold Belt (Newton 1975) showed that there is in many places a consistent relationship between fold wavelength and lithology. The major competent members of the sequence are the Peninsula Formation of the Table Mountain Group and the Dwyka tillite, and these form large folds with wavelengths of several kilometres, sometimes combining to form major anticlinoria with a wavelength of up to 20 km. The Bokkeveld rocks, while of course conforming to the larger structures, often show in addition many folds of smaller wavelength, on average about 1 km. The Witteberg, consisting of rapid alternations of quartzites and shales, show large numbers of folds of the order of half a kilometre in wavelength, and in the Nardouw Formation (upper Table Mountain Group), folds may be even smaller and more numerous. The same applies in the case of the Dwyka shales mentioned above. Generally, amplitude decreases with wavelength, so that fold geometry does not vary greatly, although there is a tendency for the smaller folds to be tighter than the larger ones. There is a clear relationship, then, between wavelength and competent member thickness, as was demonstrated from examples covering a wide range of magnitudes by Currie et al (1962).

In plan, the folds show a number of interesting features, apart from the overall convergent pattern seen in Figure 46. First, the fold axes, although mainly horizontal, plunge at each end, and in general length is roughly proportional to wavelength. Second, where a fold dies out it is often replaced by an adjacent one, giving an en echelon pattern. Third, a rather rare type of structure is seen east of Worcester, in which an apparently simple major fold is in fact made up of a number of smaller folds oblique to its main trend (Fig. 56), and these minor folds are parallel to the regional trend, whereas the major structure is somewhat oblique to it. Folds with very similar features have been investigated both in the field and in model studies by Dubey & Cobbold (1977), who show that they can form by geometric bending and flexural slip. Fold propagation develops from an initial irregularity along the hinge direction, with adjacent new folds developing serially. This leads eventually to interference with folds developing from other centres, the important controlling factor being relative phase, out-of-phase folds link obliquely, and bifurcations may appear. Other folds are prevented from developing, and terminate at a steep plunge. All these features can be observed at different places in the Cape Fold Belt.

Profound N-S variations in fold style occur (Hälbich 1983a, 1983b). Higher intensities and cascade folding are found in the more competent quartzitic units north of the major strike faults, while the less intensely folded strata south of these faults are intensely cleaved. Strain marker analyses reveal similar amounts of crustal shortening across both zones (Hälbich 1983b). Le Roux (1977, 1983) found that the distribution of second order fold intensities is linked to stratigraphy and posi-

Figure 54. Cascade folds in Table Mountain Group, north of the Worcester Fault in Kogman's Kloof near Montagu (photo by A.O. Fuller).

tion relative to the limbs of mega-folds. The northern limit of folding is fairly sharp, and it seems significant that it coincides with the axis of maximum thickness of the Cape-Karoo trough (Ryan 1968). North of this line there are only a few monoclines with axial planes dipping north.

### 3.2.2 *Faults*

Faulting in the Fold Belt consists of a few very long faults with major displacement, and a large number of smaller faults. The largest is the Worcester fault, which can be traced almost continuously for about 300 km. Malmesbury (pre-Cape Supergroup) rocks are exposed in places on its northern (upthrow) side, and several Jurassic-Cretaceous basins formed on its southern side. The throw is at a maximum near Worcester itself, where stratigraphic considerations suggest a figure approaching 6 000 m and progressive unroofing is evident from the stratigraphic sequence of the pebbles in the Enon conglomerates. The other major faults are the Cango and Zuurberg, which show similar characteristics to the Worcester fault. Large-scale northward-downthrowing faults are rare, but there is one just south of Touws

Figure 55. Cascade folds in Meirings Poort, Table Mountain Group.

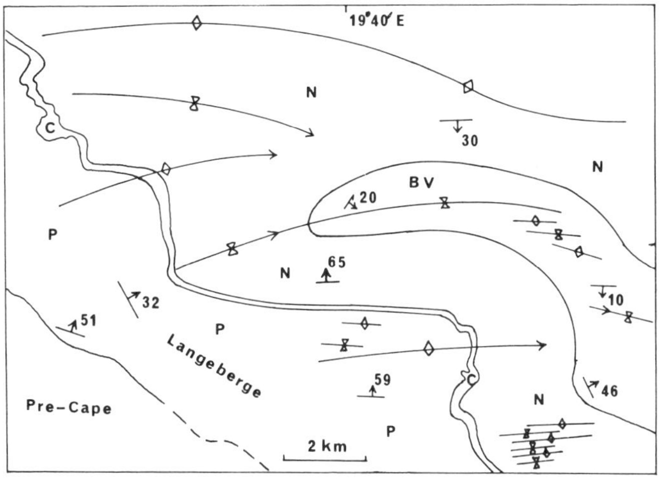


Figure 56. Complex en echelon buckle folding east of Worcester.

River, and there is evidence for another north of George (Newton 1975).

The majority of faults are approximately along strike or somewhat oblique, so that in general they follow the fold trends. However, where the main and Cedarberg fold trends meet in a syntaxis, the faults merely swing from E-W to approximately NW or NNW. It may not be coincidental that this is the predominant direction of the pre-Cape faults also. South of Worcester, the strike faults trend SW, and the faulting here is very complex because of interference between the SW and NW trending sets (Fig.57). Elements of the NW trend seem to extend almost as far as Cape Agulhas. North of the Swartberg range only minor thrusting is seen in the west around Laingsburg (Newton 1966). Further east, around Prince Albert, major thrusts are initiated as the fold intensity increases. These thrusts seem to be closely linked with fore-limb and back-limb attenuations and decollement respectively (Hälbich 1983c) along certain stratigraphic horizons. They may extend for several tens of kilometres along strike east of Meiringspoort, and northward displacements of up to 3 km can be measured. Along a profile through Willowmore (23½° longitude) the fold belt reaches its maximum width, highest intensity of deformation and most pronounced thrusting (Theron 1969).

## 3.3 PROBLEMS AND DISCUSSION

It will be clear from the preceding sections that our knowledge of the Cape Fold Belt is still very incomplete, and in this section some of the major problems are discussed in the light of available evidence, this discussion forming the basis for the models proposed in the next section.

The main problems encountered in any attempt to explain the evolution of the Cape Fold Belt are listed below, and are treated in the same order in the discussion which follows:

i) the two trends of the Fold Belt which are almost at right angles, and which meet in the Cape syntaxis;
ii) the lack of significant regional metamorphism and igneous activity prior to, and during, the main deformation;
iii) the extent of involvement of the pre-Cape basement in the folding;
iv) the age, history and significance of the major faults;
v) the deep structure of the Fold Belt;
vi) the mode of deformation;
vii) the relationship of the Fold Belt to Gondwanaland and the possible role of plate tectonic mechanisms.

### 3.3.1 *The two trends*

The major part of the Cape Fold Belt has an average E-W trend and extends from East London in the east to Ceres in the west. There it converges with the NNW-trending Cedarberg ranges forming a syntaxis on the north side of the Worcester fault (Fig.46). Detailed mapping has shown no interference effects or refolding relationships between folds of the two trends, and they must therefore be of the same age. The overstep of the base of the Dwyka rocks in the north of the Cedarberg branch has been invoked as evidence of an earlier age for the NNW folds (du Toit 1926), but it can equally be regarded as due to a continuation of the tilting and subsidence which characterized the early development of the Cape trough (Newton 1973). Also evident from Figure 46 is the overall arcuation of the main part of the Fold Belt, within which there are smaller arcs, as at Cape Agulhas and in the Longkloof area, all of which are convex northward. South of the Worcester fault near Worcester the dominant trend is SW, and is the result of the convergence of the NNW and E-W trends.

### 3.3.2 *Metamorphism and igneous activity*

The low degree of igneous and metamorphic activity in the Fold Belt is well known and needs no further emphasis. Elliott & Watts (1974), and Martini (1974), following earlier work by Fuller (1970) in the Beaufort and younger strata, have shown that some sandstones and mudstones of Dwyka, Ecca, and Beaufort age contain some volcanic detritus, and that some cherts could be regarded as reworked tuffs. However, the total proportion of volcanic material is small, and it is mainly fine-grained, so that Martini (1974) is of the opinion that the source area was several hundred kilometres away, and suggests the granitic basement of Patagonia as a source area. The possibility of contemporaneous igneous activity led to the dating of some granitic pebbles from the Katberg sandstone (Elliott & Johnson 1972). The ages obtained ranged from 390 to 210 my, but argon loss cannot be excluded and the results must therefore be interpreted with caution.

With regard to the metamorphic grade in the Fold Belt, de Swardt & Rowsell (1974) found evidence in the form of pyrophyllite, chlorite porphyroblasts, authigenic sphene and epidote, etc., to suggest that the rocks belonged to the lower border of, or perhaps within, the lower greenschist facies. It is interesting that most of their specimens are from unfolded areas, suggesting that the mineralogical changes were due to the effects of depth of burial rather than to deformation, and thus that there was no further metamorphism associated with the folding. Illite crystallinity measurements conducted along a geotraverse through the Cape Fold Belt between 22 and 23° longitude (Hälbich & Cornell 1983) revealed that in the extreme north the Karoo rocks are unmetamorphosed (T ca.150°C and P = 1-2 kb), but approach the border of the upper anchimetamorphic grade (T ca.300°C, P = 2-3 kb) in the Swartberg range, and in the coastal range reach lowermost greenschist facies (T ca.350°C). The pressures were derived from estimated geothermal gradients and extrapolated stratigraphic thicknesses. The first two temperatures were measured on fluid inclusions from syntectonic quartz veins, and the third is estimated from the recrystallized nature of the quartz in the Outeniqua mountains. In the Bokkeveld Group, north and south of the Swartberg range there is a cleavage mimetic to bedding and a later tectonic cleavage transecting this: mica and chlorite occur parallel to both directions, and the illite crystallinity therefore reflects both events (Hälbich & Cornell 1983).

Throughout the profile it is clear that the sediments were first dehydrated along a solution cleavage (Gray 1977a, 1977b, 1978), followed by crystallisation of micas along the new cleavage $S_1$. Finally, slip along and rotation of $S_1$ occurred as $S_2$ and $S_3$ developed as crenulation cleavages. No recrystallisation occurred on the last two cleavages, but evidence of additional dewatering can be seen (Hälbich & Cornell 1983).

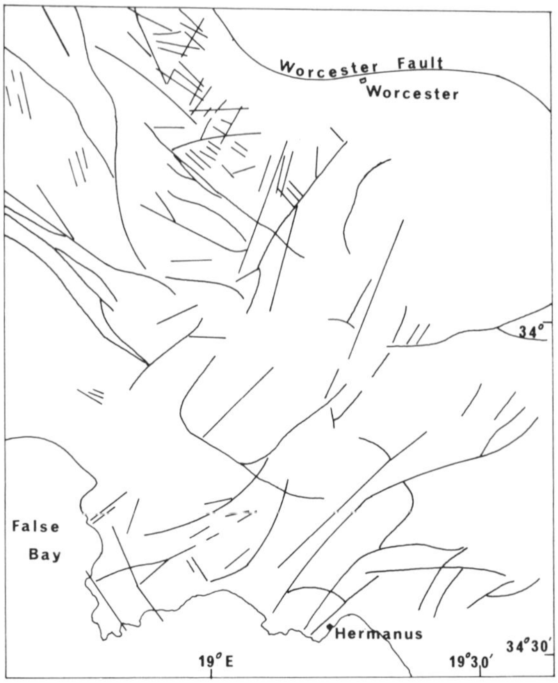


Figure 57. Fault and fracture pattern in the syntaxial area south of the Worcester Fault.

Chloritoid appears in places along the Cape/pre-Cape contact, and its significance is discussed in the next section.

### 3.3.3 *Basement involvement*

That the pre-Cape rocks are involved in the deformation with their cover is implicit in most accounts of the structure of the Cape Fold Belt. However, a review of earlier work by Newton (1975) showed that the evidence was rather sparse, and Hartnady (1969) stressed the difficulty of disentangling pre-Cape and post-Cape structures in the Malmesbury Group, but considered that only the latest structures could possibly be attributed to the Cape Folding, and these are both rare and weakly developed. This and other lines of evidence therefore led Newton (1973) to postulate a largely uninvolved basement and disharmonic gravity-induced folding of the cover rocks. In the George area, however, the strong parallelism between the main cleavages in the pre-Cape and the cover rocks led de Swardt et al (1974) to conclude that this cleavage was due to the post-Cape deformation, and was the major plane of tectonic transport. They acknowledge, nevertheless, that these structures could have evolved under gravity sliding conditions such as can be found in Alpinotype mountain belts (de Swart et al 1974, p.56).

Three phases of deformation are traced in the pre-Cape rocks south of the coastal range by Gresse (1976,

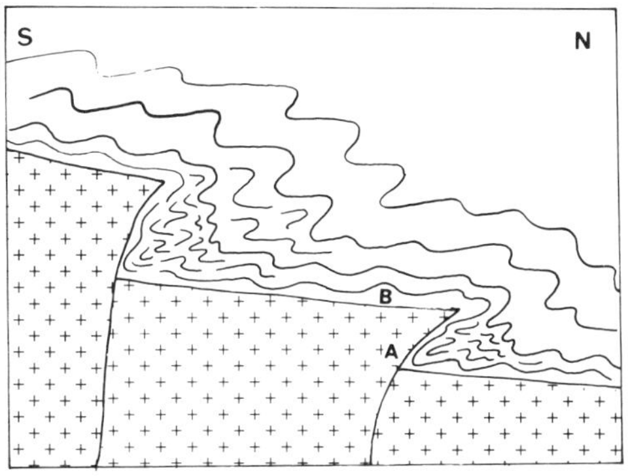


Figure 58. Suggested mechanism for the formation of cascade folds and overturned basement contacts (after Newton 1973).

1983). The first two are associated with syntectonic granite intrusion and regional amphibolite facies and contact metamorphism (Krynau 1983) dated at 500-600 my and therefore definitely of pre-Cape age. The third phase of deformation produces east-west folds with final rupture and/or shearing of granites and metasediments alike. This happened at rather low temperatures with retrograde metamorphism to greenschist facies. In some places these youngest structures can be traced across the unconformity into the Cape Supergroup cover rocks. According to le Roux (1977, 1983), very similar structural relationships hold for the basement north of the coastal range, where the basement has a different composition, granite intrusions are absent, basic sheets occur and greenschist facies metamorphism was never exceeded. It seems that both types of basement were heavily involved in the post-Cape folding.

Another interesting aspect of the Cape/pre-Cape contact is the development of chloritoid (ottrelite) and the possible extent of shearing on and parallel to the contact. Originally recorded by Schwarz (1905), the distribution of ottrelite in the Swellendam area was described by Scholtz (1946) where the ottrelite was found in shaly basal members of the Table Mountain Group, and in the basement (Malmesbury Group). It grades downward into a chlorite zone. The same distribution was found near Worcester by Hartnady (1966), and Potgieter (1950) also mentions the presence of this mineral in a broad zone beneath the unconformity in the George area. The stability field of ottrelite is not yet well-established. From experimental work, Hoschek (1969) suggests that a typical genesis is from reaction between chlorite and kaolinite or pyrophyllite at between 400 and 425°C. Whether such temperatures were attained in the vicinity of the Cape/pre-Cape contact may be doubted, but it does seem inescapable that large quantities of heat must have been generated there, and sustained for a long enough time to enable the ottrelite porphyroblasts to form. Although Hälbich (personal communication) suggests that the temperature resulting from depth of burial may have been just sufficient to generate ottrelite, its unique association with the unconformity, and the absence of any higher grade minerals below the ottrelite zone, argue against such an origin. It seems, therefore, that this heat can only have come from friction, and that large amounts of slip must have occurred on this contact. While this does not in itself form proof of folding by gravitational gliding, it is not easily compatible with slip folding of the type envisaged by de Swardt et al (1974).

### 3.3.4 *The major faults*

These have been traditionally regarded as mid-Mesozoic structures (de Villiers 1956), i.e. initiated only after the folding, and associated with the uplift of the Cape Ranges and formation of valleys in which the Jurassic-Cretaceous deposits accumulated. Evidence of activity at this time comes from the occurrence of Enon conglomerates banked up along fault planes, and the inversion of stratigraphy as represented by the vertical sequence of pebble types, demonstrating continuous uplift and erosion to successively deeper levels. Displacement of the Enon Formation rocks themselves suggest that some movement continued in post-Enon time. However, Newton (1973) suggested the possibility of an earlier phase of activity for these faults, and this aspect is dealt with here.

With regard to a maximum age limit, there seems to be no evidence that faulting had any effect on facies development during the subsidence of the Cape trough. However, many of the faults are parallel to faults and shear zones in the basement, and it is possible that at least some of the younger faults are reactivated lines of weakness. For instance, along the western edge of the Fold Belt, major faults in the Malmesbury rocks (pre-Cape) produce only small displacements in the unconformity (Hartnady et al 1974). The evidence for the faults having been active during the folding is the close relationship between the intensity of folding and thrusting and the distribution of major faults, which frequently form the southern boundary of zones of more intense folding (e.g. in the Langeberg and Swartberg mountains). The major thrust at Baviaanskloof (Theron 1969) occurs in a similar position with respect to major faulting. These factors imply a close relationship between folding, faulting and thrusting (see also Söhnge 1934) and it was suggested by Newton (1973) that a mechanism involving gravitational gliding tectonics could best explain such an association. The slope on which gliding would have taken place was the southern flank of the Cape Trough, but instead of a smooth northward slope it was suggested that differential uplift along faults created a series of steps (Fig.58). The weight of the raised blocks and their superincumbent load would

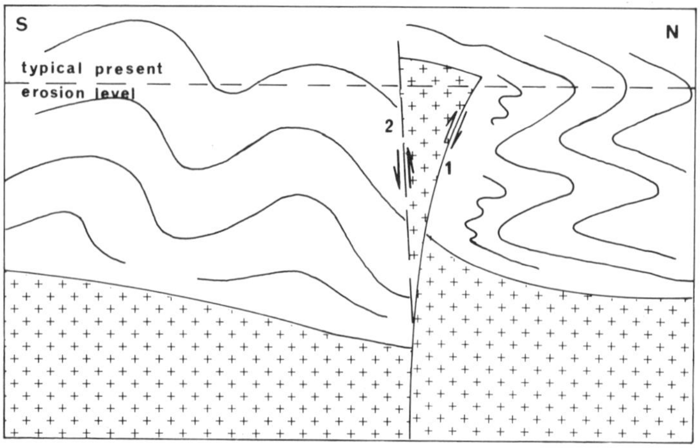


Figure 59. Schematic representation of proposed relationships across the Worcester (2) and Cango (1) faults.

result in a lateral spreading which caused the faults to become inclined southward, while the steep slopes on the step risers gave intense folding. This concept thus explains:

i) the overturning in many places of the Cape/pre-Cape contact;

ii) the localization of intense folding on the northern side of major faults;

iii) the presence of cascade folds with approximately horizontal axial plane;

iv) the development of ottrelite due to intense differential slip between basement and cover rocks.

The occurrence of thrusting can be attributed to more brittle deformation of the competent members in the later stages. A certain amount of basement involvement in the folding would not vitiate the model, since it would depend on the degree of diagenesis of the sediments and their rheological contrast with the basement, which would be primarily depth-dependent.

Important implications of this model are:

i) The 'unconformity' at the base of the Cape sequence may in places actually be a fault surface (region A in Figure 58) or a surface of decollement (region B in Figure 58);

ii) the outcrops of pre-Cape rocks on the northern sides of faults are in fact remnant slices which were on the southern sides of the original faults (Fig.59);

iii) the present displacements on the major faults are in the reverse sense to the original movements, and involve the initiation of new segments (section (2) in Figure 59). This would take place during a subsequent phase of subsidence and tension prior to and accompanying the breakup of this part of Gondwanaland. It is possible that many additional faults were initiated at this stage, and not all of these need have been southward downthrowing. As described above, there are numerous small faults in the Cape Fold Belt, and the directions of throw are often alternate, giving minor horst and graben tectonics: the largest of the northward downthrowing faults are the Touws River fault, and the Zuurberg and Commando Kraal faults in the Algoa Basin.

An alternative model, put forward by Hälbich (personal communication) is as follows: along the weakest lines or zones in the basement (which are present in each mobile belt where basement-cover relationships are involved) the earliest mega-folds are initiated. Where Newton (1973) invokes a single fault to lift the basement and thus to let it (implicitly) take part in the deformation, it is perhaps more natural to let this happen along countless joints and faults along an inherited weak zone. Fieldwork by D.M.le Roux (1983) and Gresse (1976, 1983) has shown that the two types of basement did take part in the Cape orogeny not only by faulting but also by penetrative folding across the contact and deeply into the older rocks. The association mentioned above between faulting, overturned Cape/pre-Cape contacts and ottrelite growth fits the more conventional model equally well. The cascade folds are of second order and represent collapse structures on steep or overturned limbs of mega-folds. Thrusting and decollement are confined to and partly even triggered by the same first-order structures (Hälbich 1983c, 1983d), which need not have been the result of reversed normal faulting. At T ca.300°C and P = 2-3 kb (Hälbich & Cornell 1982) the quartzites of the Swartberg were only beginning to recrystallize along narrow mm-wide zones. Excessive strain hardening held sway and dynamic equilibrium was never reached (Coetzee 1979, 1983, Hälbich 1983a, 1983b, 1983c). It is doubtful therefore whether the basement beneath the Swartberg could really have yielded in the manner envisaged by Newton (1973).

In the Outeniqua range there is no stratigraphic necessity for faulting along or near the Cape/pre-Cape contact (Krynau 1977), but the quartzites have attained dynamic equilibrium by total recrystallization with development of a new a-lineation fabric (Hälbich & Cor-

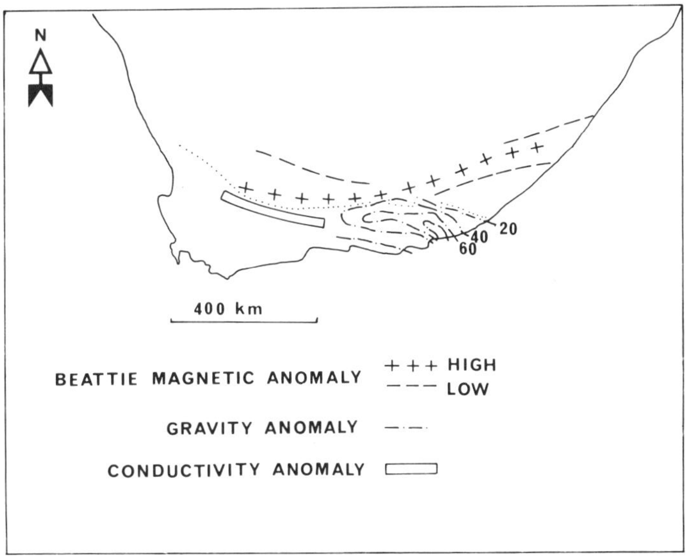


Figure 60. Geophysical anomalies associated with the Cape Fold Belt (dotted line).

nell 1983). This is most readily explained as having formed during a phase of near-horizontal compression which developed a north-verging classical shear couple with northward and upward extension of the tectonite in the direction of least resistance. This fabric also penetrates deeply into the basement.

### 3.3.5 *The deep structure*

Knowledge of the deep structure of the Fold Belt is very incomplete. Hales & Gough (1960) reported a –80 mgal gravity anomaly beneath the Algoa Basin (Fig.60) and suggested that it was caused by a relict root, or thickening of the crust, which has provided Airy-type compensation for a mountain chain since eroded away. Subsequently, a magnetometer array revealed the existence of an electromagnetic induction anomaly which Gough (1973) interpreted as a ridge of highly conductive mantle material under the west central part of the fold belt, at a depth of not more than 50 km. It could represent the line of ascending mantle material, and the gravity anomaly could mark the eastward extension of this line. The absence of the gravity anomaly in the west would be interpreted as the result of uplift (isostatic compensation?) having already occurred there. De Beer et al (1974) discuss the same data, and also the Beattie magnetic anomaly (Fig.60) which consists of a linear maximum flanked over most of its length by linear minima. The authors state that unpublished geo-electric soundings show that the body causing the anomaly is in the basement beneath the Karoo sediments, most likely in the middle or lower crust, and that it cannot be explained in terms of the mantle upwelling model of Gough (1973). They propose an explanation for the anomaly in terms of plate tectonics, which was disputed by Truswell (1975). Subsequently, as a result of further magnetometer array studies, de Beer (1978) suggested that a conductivity anomaly coincides with the magnetic (Beattie) anomaly and that both relate to the composition of the pre-Cape rocks rather than the Cape-Karoo succession. The latest interpretation of the magnetic and conductivity data (de Beer 1983) is summarised in Figure 61. The geophysically 'more competent' crust in the south is correlated with metamorphosed and recrystallized sediments (amphibolite grade) of the Kaaimans Group intruded by syntectonic stocks and sheets of 500-600 my granite, granitoids, and pegmatites. North of this, the 'less competent' crust up to the Beattie anomaly is probably formed of Kango Group basement lacking granite intrusion and never subjected to more than greenschist facies metamorphism. During the Cape orogeny this basement was certainly less competent than the Kaaimans type to shearing and folding. The southern margin of the southern Cape conductive belt (de Beer 1982) very nearly underlies a rather wide zone of basement faulting (formerly regarded as a single fracture, the Cango fault) east of 22° longitude.

The Beattie anomaly separates normal magnetic crust in the north from abnormal (more strongly) magnetic crust in the south. According to de Beer (1982), this may mean basaltic, possibly oceanic crust. In the cover rocks this feature more or less marks the change from visible folding in the south to a zone of minor listric thrusting and almost imperceptible foldings (dips less than 5°) to the north. The southern margin of the Karoo dolerites coincides approximately with the northern margin of the fold belt and with a line which, at least east of 22° longitude, coincides with the sub-Karoo pinchout of the Cape Supergroup, since

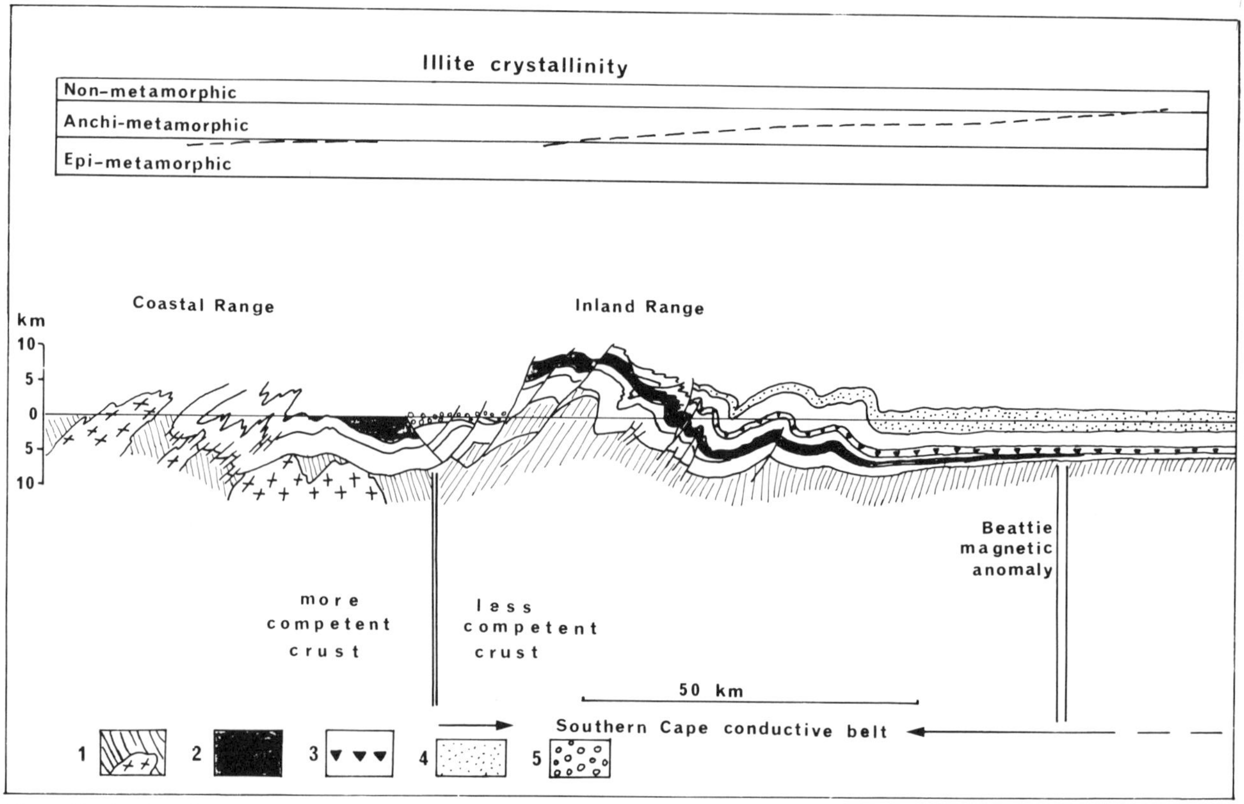


Figure 61. Stratigraphic and structural profile through the Cape Fold Belt between 22 and 23°E. 1 = pre-Cape sequences with granitoids, 2 = Bokkeveld Group, 3 = Dwyka Group, 4 = Beaufort Group, 5 = Jurassic-Cretaceous (after Halbich 1981, fig.2).

1 000 my basement was found underlying the Karoo in several boreholes by SOEKOR.

### 3.3.6 *Mode of deformation*

The only point of agreement amongst various writers on this topic is that the folding is due to lateral compression, but views on the exact way in which the rocks reacted to this stress and the ultimate origin of the stress cover a wide spectrum. For de Villiers (1944) the mechanism was a horizontal collapse due to primary compressive stress, for du Toit (in de Villiers 1944) it was due to a temporary southward drift of Africa. For Newton (1973), the stress was due to piling up of material gliding down a tectonically steepened slope, while for de Swardt et al (1974) the deformation took place by slip along cleavage planes in an Alpino-type orogen, in which gravitational gliding could have been the generative force.

The main lines of evidence for the more recent viewpoints are set out below.

Gravity gliding is suggested by the following features.

i) Folding is entirely confined to the southern half of the Karoo depositional trough, i.e. the folding dies out rapidly along the axis of maximum subsidence.

ii) Folding intensity is at a maximum on the northern side of the major faults, which are therefore contemporaneous with the folding episode. Overall, fold intensity increases northward (i.e. downslope), and the only known major recumbent isoclines occur in the north just before folding dies out.

iii) The existence of two fold trends almost at right angles which were formed simultaneously and combine to form a syntaxis. The pre-Dwyka fold phase postulated previously is unsupported by evidence, the sub-Dwyka break being disconformable and not unconformable.

iv) Fold style and vergence. Fold vergence is overwhelmingly northward, and where the folding is most intense the dip of the axial planes becomes very shallow. In profile, many of these folds are of the type called cascade folds (Harrison & Falcon 1934), which are typical of gravity folding, and which conform to the criterion of de Sitter (1956, p.289) that the inverted limbs of gravity folds are not thinned (Figs 52 & 55). Disharmonic folding is common, as is the case in most examples of gravity tectonics.

v) Isopach maps for the Cape-Karoo trough (Rust 1973) show that after Bokkeveld (Witteberg?) time a northward slope existed on the south flank of the trough, and an increasing amount of detritus was derived from this direction. Clearly, uplift was taking place to the south, and the axis of maximum downwarp migrated northward with time. A simple continuation of uplift to the south would be sufficient to eventually induce northward creep down the slope so formed.

vi) The ottrelite zone at the base of the cover sequence indicates considerable differential movement of basement and cover.

vii) The northward arcuation seen in several parts of the fold belt is compatible with a gliding mechanism.

For de Swardt et al (1974) the most important lines of evidence are:

i) The pervasive cleavage which affects both Cape and pre-Cape rocks. It is axial planar, and although variable in dip it has a constant E-W strike.

ii) The folds show apparent thickening in the hinges indicating typical shear folds, and the cleavage therefore represents a regional plane of transport.

iii) Anticlines are tighter than synclines, and have moved farther, resulting in overriding of Bokkeveld by Table Mountain group, and Table Mountain group by Cango Formation (pre-Cape).

iv) Folds in the pre-Cape show a variety of plunges, attributed to the effect of superimposing a later phase of folding, due to transport along the regional cleavage, on a pre-existing set of folds.

v) Gneissic foliation in the George granites has the same orientation as that in the Cape rocks.

As part of the South African contribution to the International Geodynamics Programme, Prof. I.W.Hälbich and his co-workers investigated a N-S strip through the Cape Fold Belt between 22 and 23°E longitude, and their results are presented in the following publications: Coetzee (1979, 1983); Gresse (1976, 1983); Hälbich (1977, 1978, 1983a, 1983b, 1983c, 1983d), Hälbich & Cornell (1983); Hälbich et al (1983); Krynau (1983); Krynau & Gresse (1980); D.M.le Roux (1977, 1983); J.P.le Roux (1983).

Their main conclusions can be summarized as follows:

i) Fold style is zonally developed, and no deformation is found outside the confines of the original Cape Supergroup basin, of which only the northern part is preserved on the continent. The Karoo strata are deformed only in the area south of the axis of maximum subsidence.

ii) Cape structures penetrate the basement everywhere in the geotraverse area. A conventional tectonism involving basement and cover is envisaged with early mega-folds triggered at the weakest spots in the basement (and/or cover), and the largest fold amplification occurred there.

iii) Pre-Cape trends commonly run parallel to the Cape direction, or occasionally very nearly at right angles to it.

iv) Basement composition and configuration seem to control the style and intensity of deformation of the Cape-Karoo rocks to some extent.

v) Profile reconstruction using deformation style studies reveals that, at the time of folding, 20 km of crustal rocks was involved in places.

vi) Dynamic equilibrium, i.e. a near-steady-state balance between deformation and metamorphism, was only reached in the southern part of the fold belt where temperature, pressure, and strain rate conditions were favourable.

vii) The deformation is polyphase, with up to three east-west cleavages developing in succession. Paroxysms as determined by the Ar/Ar step-wise heating method occurred at 273-279 my, 256-259 my and 225-234 my. Most of the deformation therefore took place in the Permian and not in the Triassic as was hitherto thought.

viii) The 'Little Karoo' region between the two main east-west mountain ranges experienced lower intensity folding possibly because it was underlain by a more competent basement than the strata north of the Cango fault zone.

ix) The sudden increase in deformation intensity along the coastal range is accompanied by a change in style that fits the grade of metamorphism of the tectonites. It seems as if critical conditions of pressure, temperature and strain rate were reached here (see also (vi) above). Higher grades of metamorphism can be expected on the continental shelf.

x) Cretaceous faulting occurred along old weak zones where early megafolds were initiated which spread northwards as fold trains.

### 3.3.7 *Summary*

Although our knowledge of the Cape Fold Belt is still largely lacking in detail, sufficient new data have emerged as a result of the National Geodynamics Programme, through the workers at Stellenbosch University, to enable us to sketch in at least the outlines of the story. During the main depositional phase, which lasted perhaps for nearly 300 million years, there evolved a deep east-west trough whose main axis of subsidence moved progressively northward in time, from Bokkeveld to upper Beaufort. Correspondingly, the southern shoreline of the basin migrated northward also, so that there was an increasing contribution of detritus from the southern source area. Dating of micas by the stepwise heating method implies that cleavage formation began in the early Permian, with a second phase in mid-Permian and a final phase in the earliest Triassic, which means that deformation in the depths of the trough was contemporaneous with the uninterrupted accumulation above of the Ecca and lower Beaufort groups. In the upper Beaufort and Molteno groups there is evidence of sedimentation controlled by pulsatory uplift, and it may be that we now have to recognize that the orogenesis *sensu stricto*, i.e. mountain-forming up-

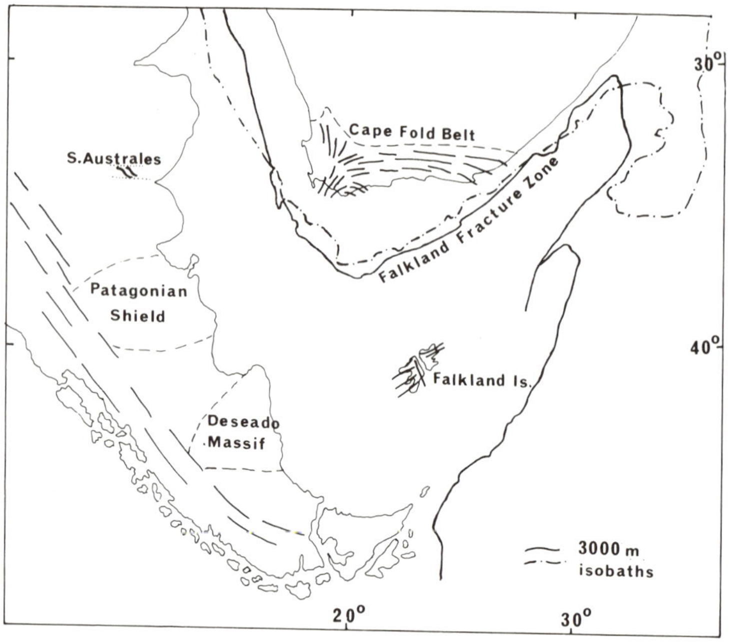


Figure 62. Spatial relationship of the Cape Fold Belt, Falkland Islands, and Sierras Australes before breakup of Gondwanaland using the refit of Martin *et al.* (1982a).

lift, was of lower to middle Triassic age, and largely post-dated the folding and cleavage formation.

There can be little doubt now that at least in the central part of the east-west fold belt, the basement was involved in the deformation with the cover rocks, or that clearly differentiated zones of differing style of deformation can be recognized, with corresponding increase in the metamorphic grade from north to south, although still of very low grade. In other parts of the fold belt the evidence for basement involvement is not so clear, and differential movement is implied by the presence of chloritoid confined to the rocks adjacent to the unconformity.

All authors agree that gravity played a role in the deformation, but there is much difference of emphasis, which may, at least in part, be due to the study of different areas. It should be mentioned that gravity can play many different roles, from soft-sediment sliding on a depositional slope, through local slippage on a steep fold limb (Hälbich 1977), to the formation of large scale, basement-involving nappe structures, as in the Alps, and it seems possible that a significant part of this spectrum may be present in the Cape Fold Belt. The exact age and part played by faulting and fault zones still needs closer investigation, but the importance of at least some pre-Cape features, and their activity during deformation, can be recognized.

It is probable that we shall never understand the Cape Fold Belt fully from studying only that part which is at present exposed in South Africa or largely hidden on the continental shelf, and we must also take into account the contemporaneous belts of both east and west Gondwanaland. The next section attempts to outline the correlatives of west Gondwanaland.

## 3.4 THE CAPE FOLD BELT AND GONDWANALAND

It has long been recognised that the Cape Fold Belt is part of a much more extensive zone of deformation, possibly extending as far as the sub-Andean ranges to the west and the Hunter-Bowen orogeny in Australia to the east. This was called the Gondwanide orogeny by du Toit (1937). A good review of the multiple orogens making up this belt is by Cradock (1974), who distinguishes early Palaeozoic orogens (Tasman in Australia, Borchgrevink in Antarctica, and an 'Acadian' orogeny in South America), and Permo-Triassic orogens (Hunter-Bowen in Australia, Ellsworth in Antarctica, the Cape Fold Belt, and the Sierra de la Ventana and sub-Andean zone in South America). Attention will be confined here to the Permo-Triassic components of this belt, and to 'West Gondwanaland', since the Africa-Antarctica fit is still somewhat ambiguous.

In the reconstruction previously discussed in Chapter 1, with the Falkland Plateau fitting into the Natal Valley between the Natal continental margin and the southern Mozambique Ridge (Figs 2 & 62), it will be noted that the Falkland Islands and the Sierra de la Ventana are approximately equidistant from the Cape Fold Belt as exposed on land, i.e. about 700 km apart. It is remarkable that in spite of this there is a very close similarity in the lithostratigraphic successions (du Toit 1927, Halle 1912), implying that the sequence of

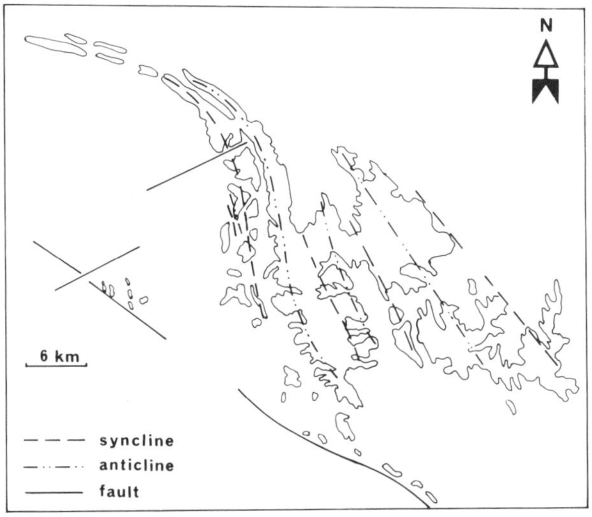


Figure 63. Major folds and faults of the Sierras Australes, Buenos Aires Province, Argentina (after Kilmurray 1975).

depositional environments must have been nearly identical: all have a thick basal arenaceous sequence, followed by impure sandstones and shales; a diamictite of glacial affinity; and above this a sequence of sandstones and shales with evidence of increasing aridity with time. Fossils of similar type, of lower Devonian and Permian age, are present in all sequences.

Structurally, there is somewhat less similarity. The Southern Hills of Buenos Aires Province (of which the Sierra de la Ventana form a part) have an overall sigmoid shape (Fig.63) with fold axes curved in plan and plunging shallowly NW and SE. The stratigraphic units either thin drastically or disappear under younger cover to NW and SE, and appear in ascending stratigraphic order from west to east. Folding in the lower part of the sequence consists of tight, asymmetric, inclined to overturned folds with axial planes dipping SW and only gentle plunges. Folds over a large range of scales have almost identical geometry. These folds die out rapidly in the lower part of the diamictite, however, and the upper and more massive part shows only gentle monoclinal dips. Further up in the sequence, gentle, open folding appears, with axial planes vertical or inclined steeply SW. Cleavage is strongly developed in the tightly folded rocks, and is parallel to cleavage seen in the few exposures of basement, so the situation resembles that in the George area of South Africa as described by de Swardt et al (1974).

The main structural elements of the Falkland Islands (Greenway 1972) are:

i) a major NE-trending monocline following Falkland Sound and increasing rapidly in amplitude southwards, which is responsible for the presence of Upper Lafonian rocks on East Falkland;

ii) E-W to WNW folds which dominate in the north, and have a marked tendency to converge on the north end of the monocline. Folding is tighter on the east side of the monocline;

iii) in the south of West Falkland, folds swing from WNW to N-S, converging on the south end of the monocline, but folding is very gentle;

iv) dolerite dykes are confined to West Falkland, and consist of two sets, one (NNE) parallel to the monocline, the other WNW-trending;

v) on East Falkland a strong fracture direction is parallel to the monocline.

The Falkland Sound monocline thus seems to be a fundamental structure – other folds, dykes and fractures are all related to it, and it seems likely that it is a drape structure over a hinge-fault in the basement. In fact, Greenway (1972) suggests that basement faulting may be the fundamental control for both folding and dyke emplacement, and this seems very likely.

The three areas – Cape Fold Belt, Sierra de la Ventana and Falkland Islands – thus show rather different structural features which were probably formed in different ways. In the Cape Fold Belt, the main shortening was N-S, with subsidiary E-W shortening in the west. In the Sierra de la Ventana shortening was NE-SW and folds were subsequently bent into sigmoid form. In the Falkland Islands, there was N-S shortening on both sides of a major N-S fault, possibly controlled by basement faulting. The folding is gentler on the west side than in the east. On both sides folds die out rapidly southwards.

None of the three areas show features typical of a subduction-generated orogeny such as major overthrusting, large scale igneous activity or high grade metamorphism, any more than they show trench-type sediments, ophiolites, or melanges. They could be regarded as the distal elements of such an orogeny, and this raises the question of where such an orogenic zone would be sited. There is evidence for intrusives and extrusives of Triassic age in Argentina, south of Sierra de la Ventana, and mainly in the northern half of the Patagonian shield, whilst a 227-240 my granite occurs (Lopez Lecube) about 30 km west of Sierre de la Ventana. In the Pampean ranges, too, age determinations of 200-450 my have been obtained, and it is possible to postulate a zone of Palaeozoic igneous activity running SSE across Argentina. The geology of these areas is not sufficiently well-known to confirm that this activity is a result of subduction zone processes, and in fact age determinations show that intrusion and extrusion have gone on throughout Mesozoic and Tertiary times in the Patagonian Shield. The eastward continuation of this postulated subduction zone raises further problems because the close stratigraphic similarity of the Falkland Island and South African sequences implies that they have always been on the same plate, and that any subduction zone must lie farther south. This implies that the Cape Fold Belt lay at least 1 500 km from such a plate margin (Fig.64). Lock (1980) has suggested that the way round this problem is for the sub-

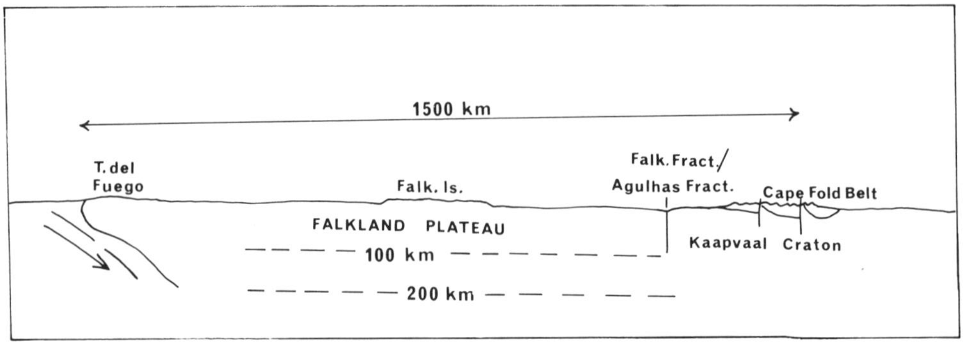


Figure 64. Schematic pre-drift cross-section from Tierra del Fuego to the Kaapvaal craton.

duction zone to become nearly horizontal beneath the Falkland Plateau, finally turning down again beneath the Cape Fold Belt. In the horizontal section the two plates would be coupled together, and crustal shortening would be concentrated at the northern edge, where the subduction zone resumed its dip. Because of the frictional coupling, this phase would be fairly short-lived, and a new subduction zone would form along the Andean continental margin. Difficulties arise with this theory in explaining the initial causes for the 'flat-plate subduction', and in the small degree of crustal shortening recorded for the Cape Fold Belt – some 25 %, which would represent only 1 cm/yr for 5 my.

The alternative to the subduction model is that the Cape Fold Belt and the Southern Hills of Argentina should be regarded as intraplate orogens, as suggested by Dingle (1973b) and Newton (1975, 1980). In spite of numerous hypotheses, little is really known about the controls of vertical movements within plates, and no detailed models can be reliably put forward at this stage, but the probable variety of such controls in an inhomogeneous lithosphere would fit in better with the differing structural styles noted above, and could better explain the varying trends and intensities of folding and different roles played by folding and fracturing in different areas.

# 4

# Middle Jurassic to Lower Cretaceous

## 4.1 INTRODUCTION TO JURASSIC-CRETACEOUS STRATIGRAPHY

By mid-Jurassic times, large complex depocentres had begun to develop along what were later to become the continental margins of southern Africa, and today, mid-Jurassic-Cretaceous sediments are known only from the coastal fringes and under the continental shelf and slope (Fig.65). This radical shift in depocentres from a previous mid-continental location, was accomplished by horst and graben development (taphrogenic tectonics) around the edges of the old cratonic blocks, and ultimately led to the breakup of this part of Gondwanaland. Farther north (in Somalia, Kenya, Tanzania and Madagascar), taphrogenic structures had started to develop in late Permian time, and in Tanzania and Madagascar, up to 9 km of shallow marine Permo-Triassic rocks occur in narrow coastal basins (Kent 1974). Figures 65 and 67 show the distribution of Jurassic-Cretaceous strata around southern Africa.

In southern Africa, the earliest dates so far available for the commencement of this tectonic phase are 162 my (Bathonian) in the northern Algoa Basin (Hill 1972), 166 my (Bajocian) in the Zambezi Graben (Flores 1973), mid-late Jurassic in the Orange Basin, and late Triassic-early Jurassic along the Lebombo Line. In all these areas, extrusive volcanism accompanied or excluded sedimentation. The large, localized downward crustal movements that occurred at these times produced a rugged terrain in which sedimentation was locally very rapid, in tectonically unstable environments (taphrogenic phase of sedimentation). These movements ceased in late Aptian-early Albian times in all areas, with the possible exception of southern Mozambique, and graben infill sediments are overlain, often with a marked discordance, by thick Upper Cretaceous and Cenozoic strata which dip gently seawards and locally prograde over the continental margin into the deep ocean basins (epeirogenic phase of sedimentation). It is tempting to equate these two tectonic/sedimentary episodes with the 'rifting' and 'drifting' phases of continental separation, and in many places around the world, where similar sequences occur, it is fashionable to do so (e.g. Falvey 1974). However, Kent (1977) has produced evidence, on a worldwide basis, to suggest that this is not necessarily the case, although the two phenomena may be related. We will return to this question later, but suffice to say at this point that the commencement of the drifting stage of continental separation between South America and southern Africa, has been dated, using sea floor spreading anomaly patterns, and depending upon points of interpretation, at between about 150 and 127 my (see Chapter 1), a minimum of 18 my before the end of the taphrogenic phase.

The separation of South America and southern Africa was accomplished by a clean-break tensional rift along the west coast, and shearing along the south and east coasts, with the effect that new oceanic basins were immediately created adjacent to the continental margin to the north of the Agulhas/Falkland Fracture Zone, and progressively created from north to south along the east coast as the Falkland Plateau slid past the truncated St Johns and Outeniqua basins (Dingle & Scrutton 1974).

Sedimentation has probably been more or less continuous since at least Upper Jurassic times in the basins that underlie the continental margins, but inland of the present-day coast has been sporadic in time and space (Figs 65 & 67). On the west coast, only one tiny outlier of marine Mesozoic sediment (Cenomanian at Bogenfels) is known between Moçamedes and Cape Town, whereas in the intermontaine coastal basins of the southern Cape, sedimentation was locally rapid (up to 4 km) for a short period (Middle Jurassic to ?Hauterivian), then apparently ceased altogether. Between East London and Durban, intermittent transgressions are recorded in small, isolated outliers, and between the Tugela River and southern Mozambique, a relatively thin (generally $<1$ km) veneer of Barremian-Maastrichtian sediments represents the western feather edge of a thicker offshore succession which laps against the eastern side of the Lebombo mountains. Northwards the sediment basin under the Zululand coastal plain expands rapidly into southern Mozambique and forms the only extensive onshore Mesozoic basin with marine sediments. Total thicknesses are not known on a regional scale, but locally over 2 km are suspected.

Offshore, thick Mesozoic sediment basins have been proved by drilling and geophysical exploration (Fig.65). These basins were originally clearly defined by base-

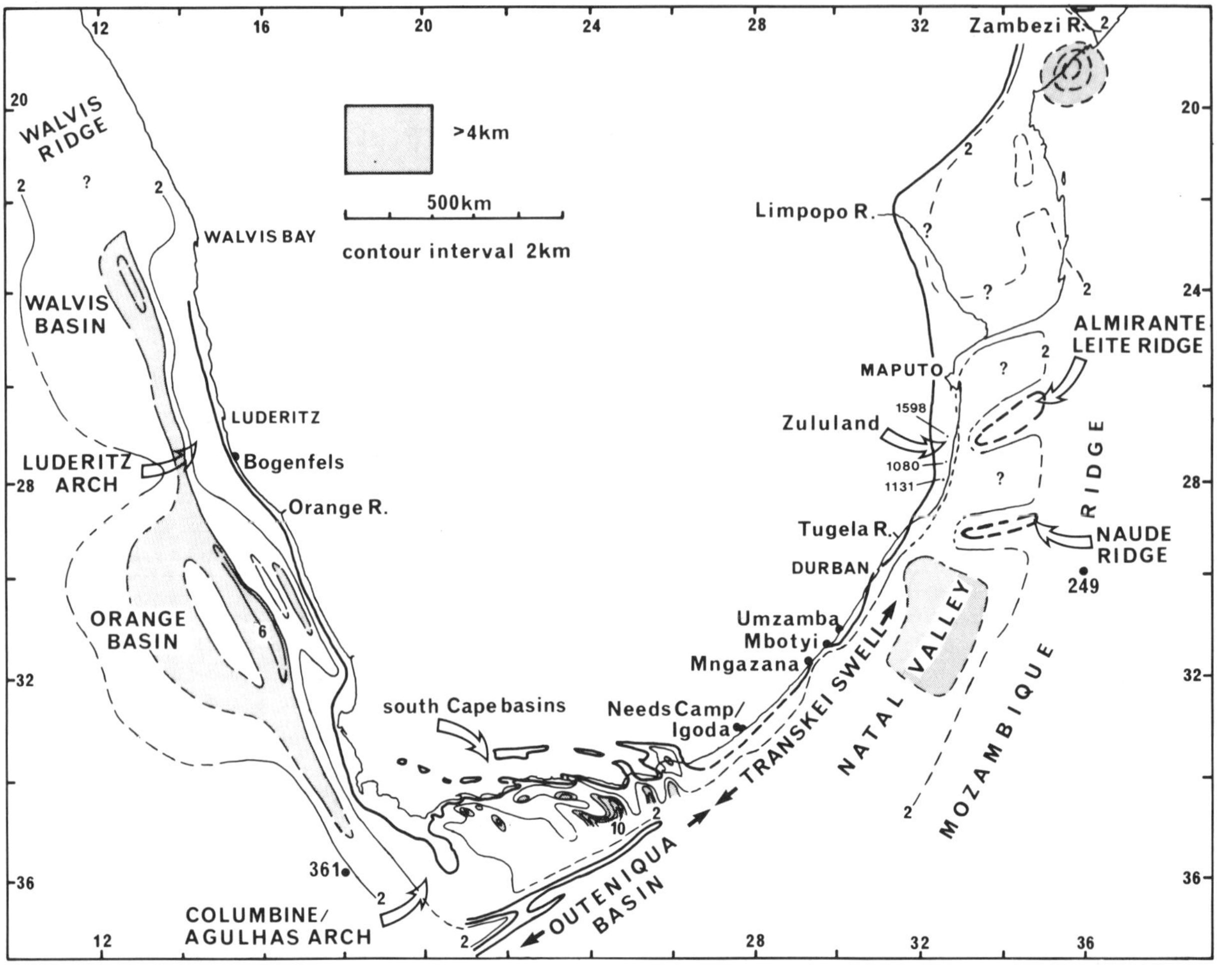


Figure 65. Isopachs (km) of Jurassic-Cretaceous strata on the continental margins of southern Africa. Contoured at 2 km intervals, with 0 km shown thickened. On the west coast the isopachs probably include thick Jurassic lavas, but age equivalents in S.Mozambique (i.e. extensions of the Lebombo Group) have been omitted. Data quality and type varies greatly (see text for details), with numerous sources cited in the text. Offshore locations 249 and 361 are Deep Sea Drilling Project sites.

ment structural control (e.g. Dingle & Scrutton 1974), but by Cenozoic times they had lost much of their identity through the cumulative effects of broad epeirogenic downwarping, and the oceanward shift of the depocentres. On the west coast, two large basins (Walvis, and Orange, both >6 km thick) are separated by a relatively buoyant basement block west and SW of Lüderitz. The continental margin between Cape Columbine and the southern tip of the Agulhas Bank (Columbine/Agulhas Arch) has remained a buoyant feature since Palaeozoic times and forms the angle of the continental fracture line between the rifted west coast and the sheared south and east coasts. To the east of this arch lies the large, complex Outeniqua Basin (locally up to 10 km thick), which is subdivided by several elongate tilted basement blocks which are the extensions of the Cape Fold Belt anticlines. The Outeniqua Basin narrows eastwards and is bounded in the east by the Port Alfred Arch at the southern end of the Transkei Swell which probably has a thin cover of undifferentiated Karoo Supergroup strata and possibly isolated outcrops of Cretaceous sediments. Only one small sediment basin is known to contain Jurassic-Cretaceous sediments offshore the Transkei – St Johns Basin (ca.2 km), and this is separated by another basement block (Port Shepstone Arch) from the southern end of the Natal-Zululand-Mozambique Basin which stretches to just south of Durban on the continental shelf. As the coastal plain widens into Zululand, the feather edge of this basin encroaches onshore, although maximum Mesozoic sediment accumulation occurred under the Tugela Cone (>4 km) and in the Natal Valley. The latter is a large, graben-like basin bounded in the west by the Lebombo Line, and in the east by the Mozambique Ridge and its buried onshore extension, and is subdivided laterally by at least three large basement ridges (Naude, Almi-

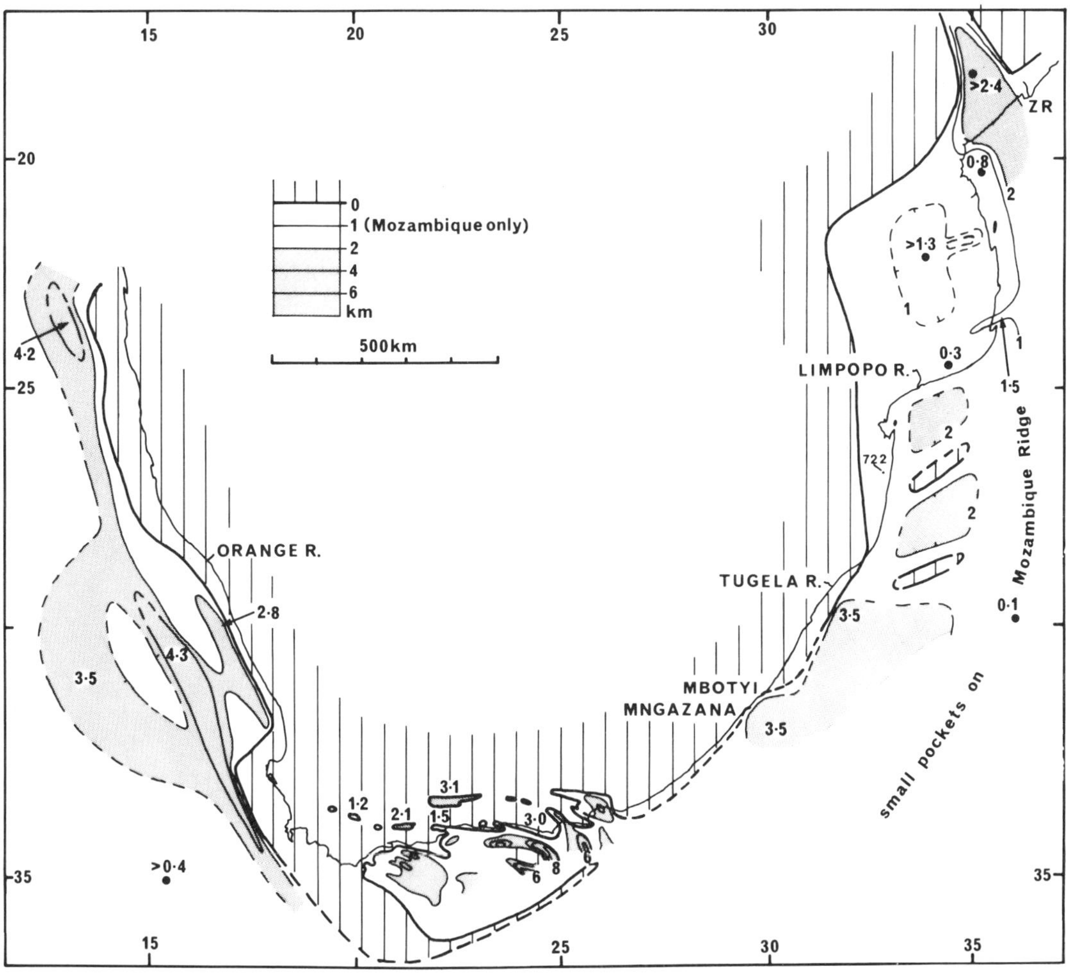


Figure 66. Isopachs (km) of Jurassic to Lower Cretaceous strata on the continental margins of southern Africa. Contour interval is 2 km except for addition of the 1 km isopach. Over most of the map the strata are mid-Jurassic to Albian, but in S.Mozambique Jurassic lavas have been omitted and sediments as young as Turonian are included (Domo Formation). West coast high velocity strata (i.e. between reflectors T and R on Figures 120 and 121) have been included. See text for further details of data selection and author citation.

rante Leite, and an unnamed ridge in the coastal zone to the east of the Limpopo River estuary).

## 4.2 REGIONAL DISTRIBUTION OF MIDDLE JURASSIC-LOWER CRETACEOUS ROCKS

Distribution of the earliest Mesozoic sediments in the marginal basins around southern Africa is shown in Figures 66 and 67. Outcrops are limited to small outliers in the southern Cape, a narrow offshore strip close to the coast on the Agulhas Bank, two tiny outliers on the Transkei coast, and a poorly-exposed zone along the foot of the Lebombo mountains. Offshore, several extensive basins are concealed beneath younger cover, although identification of these rocks using geophysical data, particularly differentiation between the lower part of the sequence and 'basement', is difficult and not always unequivocal.

Because these rocks, for the most part, represent the earliest sedimentary infill in newly-formed, and actively developing taphrogenic basins, their distribution and lithofacies are strongly controlled by basement tectonics. This early fault control of the basin geometries and resulting rapid thickness variations can be contrasted with later epeirogenic control by comparing the isopachs in Figures 66 and 124. On the west coast margin the two major sediment basins are probably bounded by coast-normal marginal offsets that define the limits of basement highs, and the Orange Basin at least appears to be internally subdivided by ridges (horsts) parallel to the coast. Maximum thicknesses of 4.2 km and 4.3 km have been estimated from seismic data for the Walvis and Orange Basins, respectively (Fig.66). The former is poorly known from published information, but the

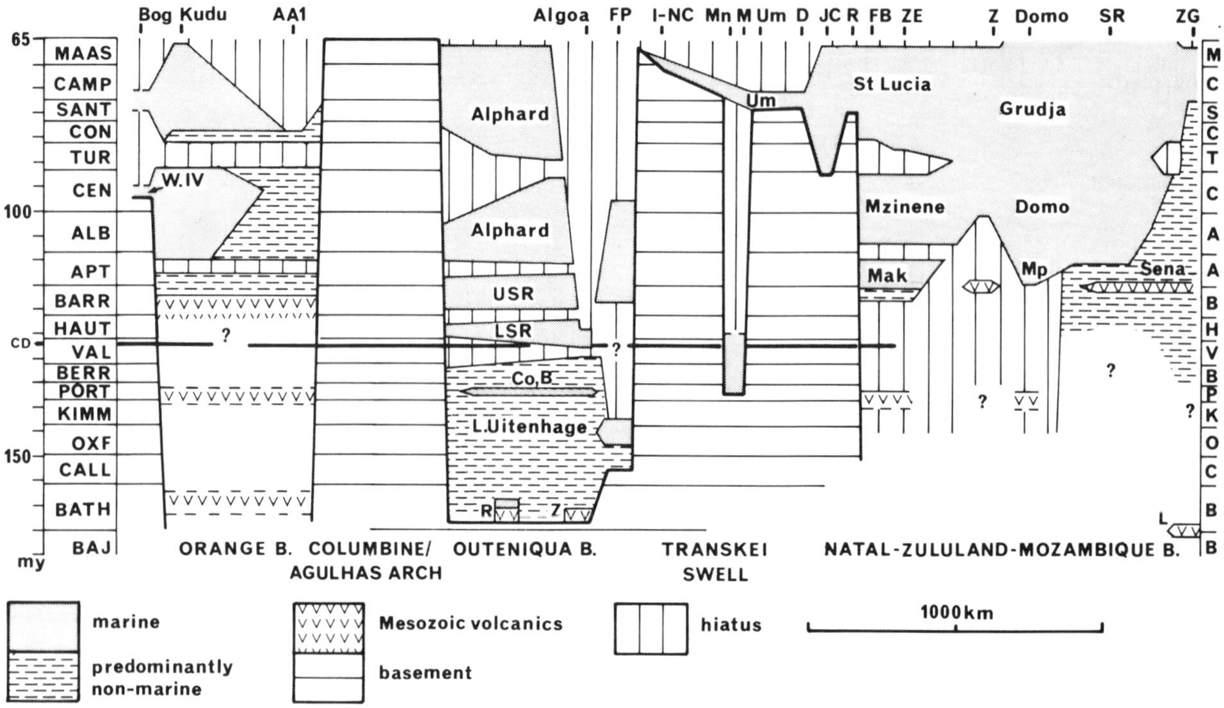


Figure 67. Summary regional correlation of Middle Jurassic to Cretaceous strata around the continental margins of southern Africa: Namibia on the left to the Zambezi Graben on the right. The Jurassic-Lower Cretaceous succession on the Falkland Plateau has been included in its pre-drift location. Horizontal lines in the basement correspond to stage boundaries to aid correlation. Only the main lithofacies are shown. Time scale is that of van Hinte (1976a). The diagram is after Dingle (1982a), but is based on numerous authors cited in the text.
Abbreviations:
Localities: Bog = Bogenfels area; Kudu = Kudu 9A-1 borehole; AA1 = A-A1, Ba-A1,A2 boreholes in S.E.Orange Basin; Algoa = Algoa Basin; FP = Falkland Plateau; I-NC = Igoda/Needs Camp; Mn = Mngazana; M = Mbotyi; Um = Umzamba; D = Durban; JC = J(C)-1 borehole; R = Richards Bay (particularly borehole BH9); FB = False Bay; ZE = borehole ZE, N.Zululand; Z = Zandamela borehole; Domo = Domo borehole; SR = Save River mouth; ZG = Zambezi Graben.
Stratigraphy: W.IV = Wanderfeld IV beds; Alphard = Alphard Formation; USR = Upper Sundays River Formation; LSR = Lower Sundays River Formation; Co = Colchester Formation; B = Brenton Member; L.Uitenhage = Lower Uitenhage Group; R = Robberg Member; Z = Zuurberg Member; Um = Umzamba Formation; St Lucia, Mzinene, Grudja, Domo, Sena = all named formations; Mak = Makatini Formation; Mp = Maputo Formation; L = Lupata Group.

Jurassic-Lower Cretaceous Orange Basin is clearly a complex structure in which thick (>2 km) fill extends up to 420 km offshore (i.e. it extends across the COB). Along the southern coast, between the Columbine/Agulhas and the Port Alfred arches, extensional tectonics operated as a shearing couple that cut obliquely across the line of the Cape Fold Belt. This had the effect of truncating the mid-Jurassic taphrogenic basins that had developed along the structural grain of the Cape Fold Belt, and judging from the variations in thickness displayed in the Upper Jurassic-Lower Cretaceous sediments, also had the effect of enhancing the differential vertical movements between the basement blocks (i.e. the lateral thickness variations are more pronounced in the upper part of this sedimentary episode). Sediment thicknesses in this depocentre (Outeniqua Basin) range from less than 2 to >8 km.

Along the east coast, the Agulhas Fracture Zone shear truncates the St Johns Basin, but locally 3.5 km are estimated from seismic data at the base of the present continental slope, and clearly the basin extends across the COB to merge with fill in the Natal Valley basin, where a further 3.5 km of Jurassic-Lower Cretaceous sediments lies under the Tugela Cone. The relationship between the infill of the northern Natal Valley, and the large-scale vertical tectonics in Mozambique is obscure, but the earliest marine Mesozoic sediments in Zululand lap onto surfaces of deeply weathered lavas which were extruded over an active crustal fracture during Jurassic time. Lava extrusion continued well into Cretaceous times in southern Mozambique, so here again the mid-Jurassic to Lower Cretaceous depocentre locations were strongly influenced by active fault lines.

Large-scale taphrogenesis had ceased by the end of Aptian times around southern Africa, and Albian sediments spread over and beyond the previous boundary faults as a new, epeirogenic phase of continental sub-

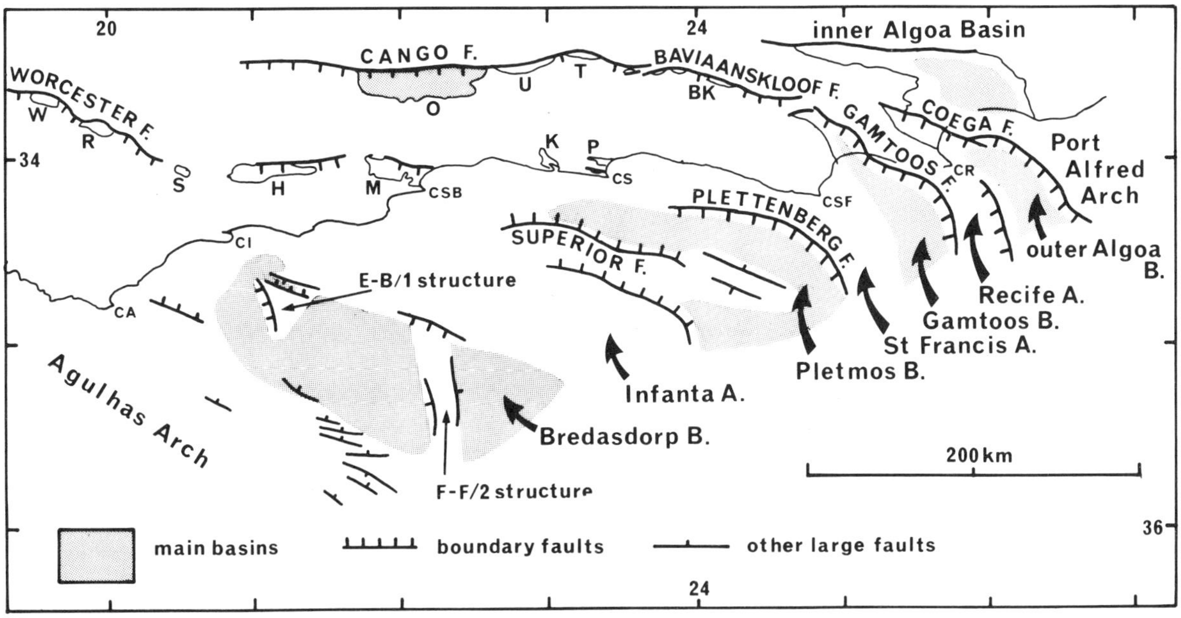


Figure 68. Main structural features that controlled sedimentation during Jurassic to Lower Cretaceous times in the southern Cape. Only the major faults are shown. Abbreviations: F = fault; B = basin; A = arch; W = Worcester; R = Robertson; S = Swellendam; H = Heidelberg; M = Mossel Bay; K = Knysna; P = Plettenberg & Pisang; O = Oudtshoorn; U = Uniondale; T = Toverwater Poort; BK = Baviaans Kloof; CA, CI, CSB, CS, CSF, CR = capes Agulhas, Infanta, St Blaize, Seal, St Francis, and Recife, respectively.

sidence began. Depending upon the local lithostratigraphy, some of the Albian sequences will be dealt with in this section (e.g. in Zululand and Mozambique) whereas in other areas (e.g. most of the Agulhas Bank, and the west coast) their description is more conveniently handled in the Upper Cretaceous section.

A summary of the distribution of mid-Jurassic to Lower Cretaceous strata in time and space is shown in Figure 67. The following accounts will be given on a geographical basis, starting with the southern Cape area because it is the best known both onshore and offshore.

## 4.3 OUTENIQUA BASIN AND ADJACENT AREAS IN THE SOUTHERN CAPE (UITENHAGE GROUP)

In the southern Cape, Middle Jurassic to Lower Cretaceous sediments occur as a series of small elongate outliers in north tilting half graben within the Cape Fold Belt, that stretches from Worcester in the west to Algoa Bay in the east, a distance of about 600 km (Fig.68). These structures continue offshoıe under the Agulhas Bank, where they form graben and half graben subbasins within the Outeniqua Basin, the name given to the regional depocentre that lies between the buoyant Columbine/Agulhas Arch in the west and the Port Alfred Arch in the east and is bounded to the SE by the Agulhas Marginal Fracture Ridge (see Fig.112). Algoa Basin, the largest of the onshore outliers, has a thick succession of fossiliferous marine and freshwater sediments, which have long attracted the attention of palaeontologists (the earliest reference is by Hausmann 1837), and it is here that the majority of stratotypes for the southern Cape area have been established (Figs 67 & 72). The other onshore basins are less well-known because their lithofacies (predominantly continental and fluviatile) are sparsely fossiliferous and are difficult to confidently correlate on a regional basis. McLachlan & McMillan (1976) have given an excellent summary of fossil records from the Uitenhage Group of the southern Cape area.

### 4.3.1 *Structural and basement features*

In the southern Cape, the elongate asymmetric anticlines and synclines of the Cape Fold Belt trend approximately E-W between about 20 and 24°E, before swinging sharply SE in the vicinity of Port Elizabeth. The major anticlines are typically bounded on their southern sides by large, southward throwing normal faults, and it is within the resultant, asymmetric, northward-tilting half grabens that the onshore mid-Jurassic to Lower Cretaceous taphrogenic basins developed. Although individual faults are not usually continuous for more than about 200 km, two major fault lines controlled the location of the Worcester-Mossel Bay, and Oudtshoorn-Gamtoos series of outliers (Fig.68). Both fracture zones can be traced under the Agulhas Bank (as the Plettenberg and Gamtoos faults, respectively), where

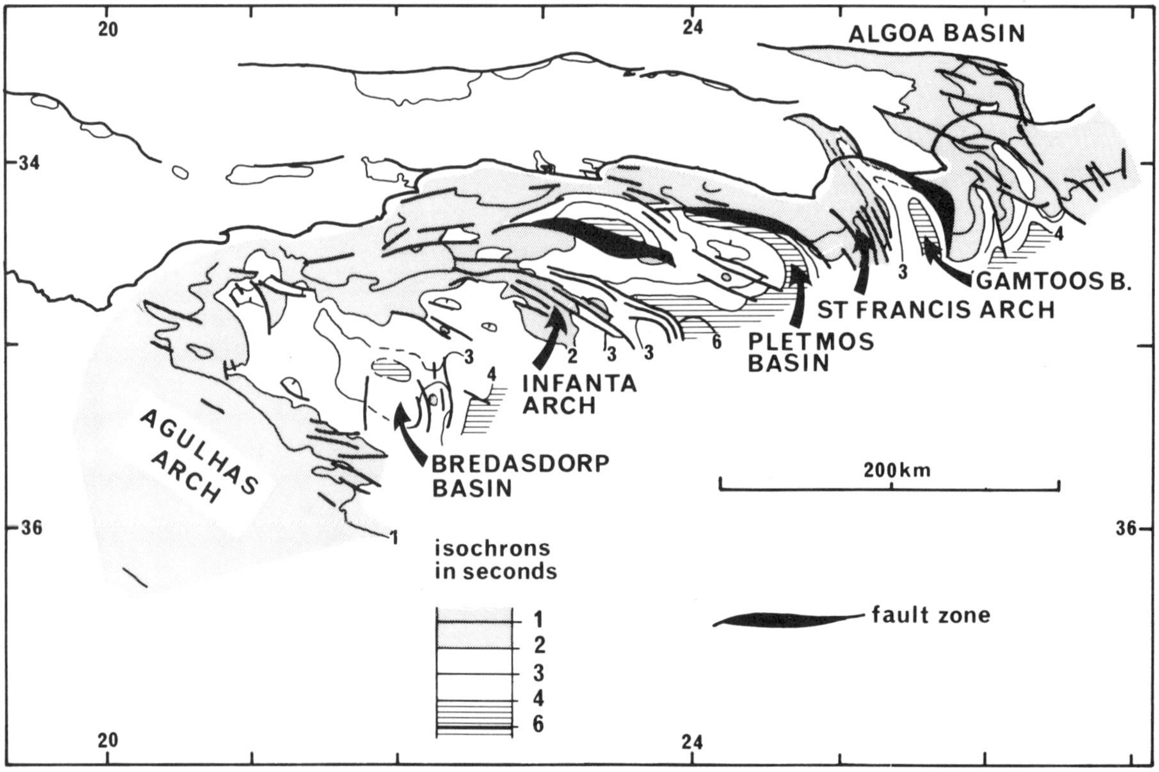


Figure 69. Basement structure and topography of the Outeniqua Basin. Depths to basement are in seconds of P-wave travel time from sea level datum. Contour interval 1 sec except the 5 sec isochron has been omitted for clarity. Approximate conversions to metres are: 1 sec = 1.25 km; 2 sec = 2.5 km; 3 sec = 4.75 km; 4 sec = 7 km; 6 sec = 10.5 km. Data mainly from SOEKOR (1976).

they form the northern boundaries of two major offshore subbasins: Pletmos; and Gamtoos. The latter is a direct continuation of the onshore outlier of the same name.

In addition to the two main series of outliers, other basins occur in the Knysna-Plettenberg area, and north of Algoa Bay. The former are partially fault-bounded on their southern sides, and rest on top of the main basement high which separates the Oudtshoorn-Gamtoos and Worcester-Pletmos basin lineaments. This high feature continues under the Agulhas Bank as the St Francis Arch. The Algoa Basin is probably partially fault-bounded along its northern edge, and is subdivided longitudinally by several other large fractures (Coega, Addo and Colchester faults). Although it is the largest (and probably deepest) of the onshore basins, there is no corresponding series of outliers to the west of it.

Under the Agulhas Bank, a further large taphrogenic depocentre lies to the south and SW of the Worcester-Pletmos lineament, from which it is separated by the Infanta Arch. This basin (Bredasdorp Basin) has no onshore equivalent, although it does lie between two anticlinally-dominated basement highs at Cape Infanta and Cape Agulhas. Present-day elevations onshore are generally produced by resistant Table Mountain Group quartzites in anticline cores which cross the coast at prominent headlands (viz. capes Infanta, St Blaize, Seal, St Francis and Recife). Their offshore extensions, where they have been drilled, also consist of hard quartzites which can be correlated, on purely lithological grounds, with Table Mountain Group rocks. Similarly, both onshore and offshore basins are floored with less-resistant Bokkeveld-like shales. This evidence suggests that the taphrogenic basins developed along the lines of synclines originally formed during the Cape Orogeny, particularly in view of the fact that shale clasts are subordinate in the basin fill, i.e. the high ground adjacent to the Middle Jurassic-Lower Cretaceous basins was already stripped bare of Bokkeveld cover.

Before discussing structures in detail it is necessary to comment on the parameters we have used to calculate thickness of Middle Jurassic-Lower Cretaceous sediment infill, and to differentiate basal sediments from 'basement' in the offshore areas. Interpretation of seismic reflection data by SOEKOR (1976) have revealed a strong, regionally developed reflecting horizon (D) which correlates well with the Mesozoic/Palaeozoic junction in boreholes. It is from this datum that figures such as 69, 70, 71, 73, 74, etc. are compiled. Average P-wave velocities calculated for the localities where

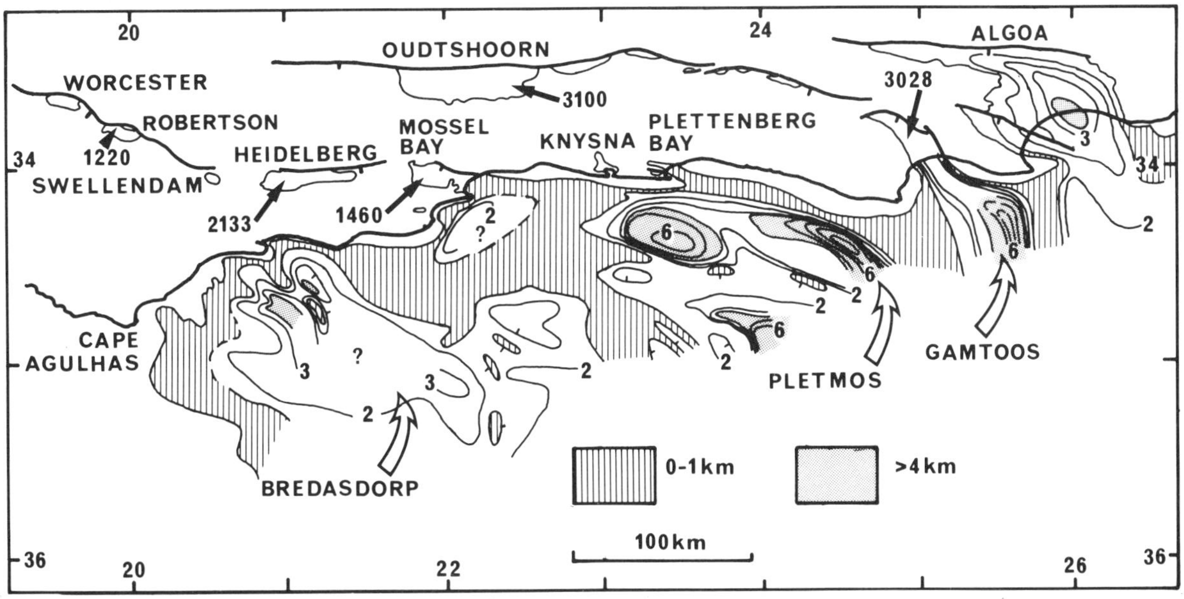


Figure 70. Isopachs of Jurassic-Lower Cretaceous strata in the Outeniqua Basin and southern Cape. Maximum known or estimated thicknesses (m) in the latter are indicated. Data reduced from SOEKOR (1976) using the following average P-wave interval velocities; pre-Sundays River Formation – 3.670 km/sec; Sundays River Formation – 3.196 km/sec. Contour interval is 1 km.

borehole and reflecting seismic data are available give the following average values:

i) Sundays River Formation: 3 196 m/sec (range = 2 848-3 732 m/sec);

ii) pre-Sundays River Formation: 3 670 m/sec (range = 4 000-3 548 m/sec);

and we have used these to calculate isopachs. Other regionally-developed reflecting horizons which correlate with important lithostratigraphic markers used by SOEKOR (1976) are: A = top Sundays River Formation, C = mid-Sundays River Formation, B = base Sundays River Formation. There have been some recent revisions to the acoustic stratigraphic correlation used by SOEKOR (1976) and du Toit (1976) between the east and west sectors of the Outeniqua Basin (McLachlan & McMillan 1979), but we have used their basic scheme and will comment upon amendments in the appropriate sections. At present, we have neither seismic nor borehole data from the region between the shelf break and the Agulhas Marginal Fracture Ridge which relate to Jurassic-Lower Cretaceous rocks.

In the accounts of individual structures that follow, depths to basement are in seconds of travel time for sound waves used during seismic reflection profiling. Datum is taken at sea level, and the times quoted are those for sound waves to travel from the source to reflectors and back to the surface i.e. two way or total time. To convert to thicknesses in metres these times, therefore, need to be halved. Seconds have been used as a scale because insufficient interval velocities are known for a conversion of the total sediment sequence to thicknesses in metres.

4.3.1.1 *Bredasdorp Basin*

This is a wide basement depression which lies between the Infanta and Agulhas Arches (Figs 68 & 69). Although its cross-section is asymmetric (Fig.73B), it is best described as a graben, in contrast to most of the other major taphrogenic basins in the southern Cape, which are half graben. The southern edge of the Bredasdorp Basin is formed by the faulted and gently northward tilting flank of the Agulhas Arch, and can be traced as a prominent subsurface feature for over 150 km SE of Cape Agulhas. The upper part of the arch is bare of post-Palaeozoic rocks, except for a thin, impersistent cover of Cenozoic sediments, and is composed of Table Mountain Group quartzite and Bokkeveld shales over a granite core. Locally, faults along this margin rapidly drop the basement to the floor of the Bredasdorp Basin, which typically lies between 2 and 4 sec depth (i.e. between about 2.5 and 7 km). The floor is relatively flat, except in the western part (Fig.73A), and dips generally to the NW so that the deepest parts of the Bredasdorp Basin (between 4 and 5 sec) lie close to its northern margin (Figs 69 & 73B).

Basement descends rapidly via numerous small boundary faults from the Infanta Arch into the northern Bredasdorp Basin. The linearity of this northern margin is broken by several smaller, deep graben which form embayments onto the Infanta Arch, and at least three horsts which project SE and southwards into the basin. These horsts stand between 2 and 4 km above the general level of the basin floor and are bounded by short curved faults. They have been targets for oil exploration drilling and the largest of them (the so-called

F-F/2 structure, Fig.68) forms a barrier across the SE end of the Bredasdorp Basin and during Middle Jurassic to Lower Cretaceous times probably isolated the main basin from areas to the east. Limited seismic data suggest that on the eastern side of the F-F/2 structure basement descends rapidly to at least 7 km.

The western end of the Bredasdorp Basin does not extend onshore. It breaks up into a series of small, narrow, NW-SE horsts and graben which extend to within 30 km of the coast between Cape Agulhas and Cape Infanta (Fig.69).

Two Tertiary igneous centres lie within the Bredasdorp Basin, both on subsidiary basement highs, and both consisting of numerous plug-like features:

i) Alphard Banks, which lie in the vicinity of borehole D-A/1 on a ridge that subdivides the western part of the main basin (Fig.73A).

ii) on the F-F/2 structure at the SE end of the main basin.

4.3.1.2 *Infanta Arch*

The Worcester-Mossel Bay-Pletmos basin line lies to the north of several anticlinal ridges which cross the coast in a series of capes and prominent headlands between capes Infanta and St Blaize. This broad basement high continues out to sea as the Infanta Arch. It has an east-west trend until about 23°E, where it turns to the SE (Figs 68 & 69). Close to the coast, the Infanta Arch lies at relatively shallow depths (<1 sec or <1.5 km) but farther east, it gradually descends to between 3 and 4 sec (Fig.69) and is covered with progressively thicker sediments (compare Figs 73A,B & 74C,D). In addition, its eastern part is cut by numerous, long faults which break its upper surface into several very narrow horsts and graben, that contain thicker pockets of sediment. As we will discuss later, the Infanta Arch was an important barrier or positive axis between the Bredasdorp and Pletmos basins for the whole of mid-Jurassic to Lower Cretaceous time.

4.3.1.3 *Worcester-Pletmos basin line*

This forms a continuous east-west line of half graben for over 400 km and is bounded on its northern side by a series of massive, southward throwing normal faults, one of which (Worcester Fault) is still active (Fig.68). The estimated throw on the Worcester Fault is about 6 km, and the measured throw across the Plettenberg fault of the order of 9 km (5 sec) (Fig.74D). The onshore basins are probably remnants of originally more extensive depocentres which may have formed a continuous deposit. The Pletmos Basin offshore is a large, complex structure which curves sharply to the SSE at its eastern end, and is subdivided longitudinally by a relatively shallow (2-2.5 sec) tilted basement fault block that continues to the SE as a narrow horst (Fig.74C,D). The boundary fault to this block (Superior Fault) throws to the north at least 3.5 sec, with the result that the structure between the Superior and Plettenberg faults is a strongly asymmetric graben whose floor tilts to the SW in the west, and NE in the east (Fig.69).

The northern edge of the Infanta Arch lies about 20 km south of the Superior Fault, and is marked by a series of northward throwing faults. These drop the basement to progressively greater depths in a NE direction, resulting in another half graben along the southern edge of the Pletmos Basin (Fig.74D). Depths to basement in all three major depressions in the Pletmos Basin locally exceed 4 sec (?7 km), and reach maxima of just over 6 sec.

4.3.1.4 *St Francis Arch*

This asymmetric basement high separates the Pletmos and Gamtoos basins, and has a flat top with small graben filled with young sediments. The crest, with < 200 m of Jurassic-Lower Cretaceous cover, lies near its western margin which is marked by the curved (in vertical and horizontal planes) Plettenberg Fault. Its NE flank descends relatively gently via numerous faults into the Gamtoos half graben. At its extreme SE end, the St Francis Arch trends SSE, the depth to its crest has increased to 1.6 sec (2 km), and its width has narrowed to about 15 km (compared to about 45 km close to Cape St Francis).

4.3.1.5 *Oudtshoorn-Gamtoos basin line*

This line of half graben can be traced for nearly 350 km and is bounded on its northern side by three long southward throwing faults (Fig.68). In the west, the large Oudtshoorn outlier, with an estimated maximum depth to basement exceeding 3 km, is separated by a small transverse basement high from the smaller Uniondale and Toverwater Poort basins. These outliers are linked to the major Gamtoos Basin via a string of small Mesozoic outliers in the Baviaans Kloof. These are somewhat unusual because they lie at present elevations up to 900 m above sea level, which is much higher than the other basins (generally below 300 m), and the floors of the three largest appear to be predominantly composed of Table Mountain quartzites (Schwarz 1904). The Baviaans Kloof outliers are separated from the Gamtoos Basin by an E-W cross fold and the respective boundary faults of the two areas are both deflected by it. The Gamtoos Basin widens and deepens in a SE direction as its axis strikes progressively more southerly adjacent to the strongly curved northern boundary fault. The maximum known throw on this fault is about 3 km onshore and in excess of 6 sec in the vicinity of the present day shelf break (Fig.69). Because of reflection interference from the steep fault plane on seismic records, the maximum depths to basement in the deepest parts of the Gamtoos basin are not accurately known.

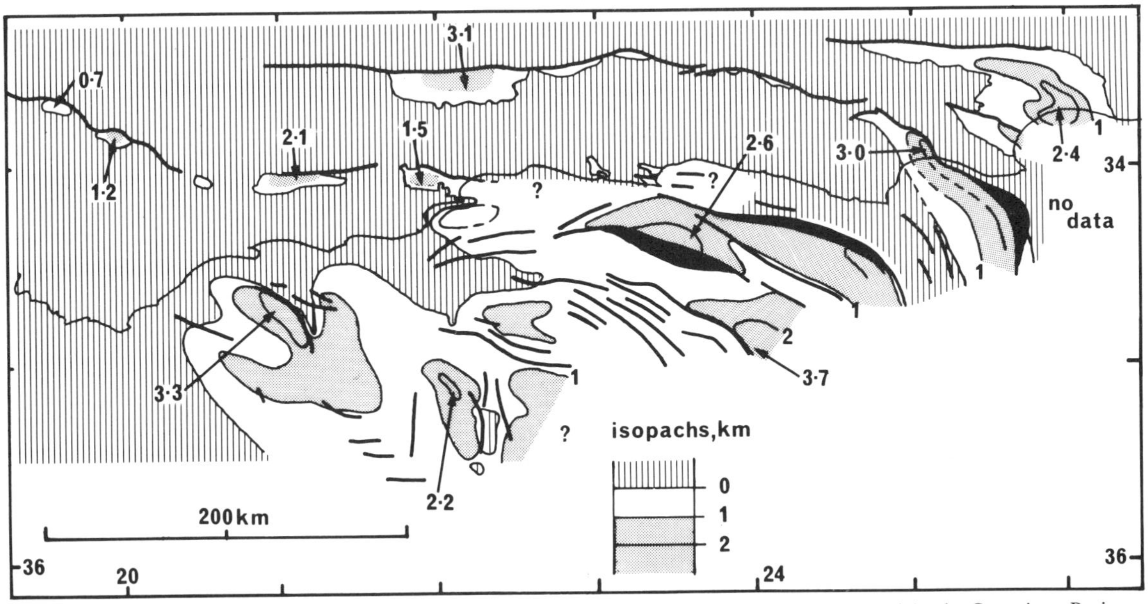


Figure 71. Distribution and thickness of Lower Uitenhage Group (?Middle Jurassic to Valanginian) in the Outeniqua Basin and southern Cape. Contour interval 1 km, maximum thicknesses shown as spot values. Data converted from SOEKOR (1976) isochron maps using an average P-wave interval velocity of 3.67 km/sec.

4.3.1.6 *Algoa Basin and Recife Arch*

The Algoa Basin is a composite structure, but is composed essentially of three curved half graben, two of which (Uitenhage and Sundays River troughs) have onshore extensions (Figs 68, 69 & 75). The largest (Sundays River Trough) lies in the north, and stretches from the NW panhandle to just offshore in the SE part of Algoa Bay. Its northern boundary is locally formed by a southward throwing fault, but there is some dispute as to whether the whole of this boundary is faulted, as shown on most older maps and the Geological Survey 1:1 000 000 sheets (1970). For instance, Lock et al (1975) claim that a large northern boundary fault would be in conflict with their suggested Mesozoic sediment dispersion patterns, and consider that the structure of the northern part of the Algoa Basin is a southward tilted half graben with a major northward throwing boundary fault along the Addo Fault (called by them the Commando Kraal Fault) (Fig.75), and that only a series of small southward throwing fractures coincides with the present north edge of the Mesozoic outcrop. This is not the solution favoured by SOEKOR (1976) who interpret the structure as a small northward throwing fault 15 km south of the edge of the Mesozoic outcrop, to the south of which lie several large south throwing faults that drop basement about 2.5 km into the Sundays River Trough (Fig.75E). The picture is complicated by the eastward encroachment of the Addo Ridge basement high which subdivides off a small fault trough in the SW (Coega Embayment, of Winter 1973). The maximum depth to basement in this sector is 4 100 m near Colchester.

The second major half graben lies on the southern side of the southward throwing Coega Fault (Fig.75) which can be traced for about 120 km to the SE edge of the Port Alfred Arch, where it forms a boundary fault (Figs 68 & 69). Maximum depth to basement in the Coega half graben exceeds 4 600 m. It is closed at its southern end, and separated from a third half graben by the crest of a narrow, curved, tilted basement block, which has two large curved faults on its southern side. These faults throw to the south and they can be traced to within about 10 km of Cape Recife. The floor of this southerly basin (at south end of Fig.75E) dips off the eastern side of the Recife Arch towards the boundary faults, and the basin opens and deepens southwards. The rather complex basement topography created by the fault patterns and tilted blocks in the Algoa Basin suggests that during Jurassic-Lower Cretaceous times, the three main half graben may have been more or less isolated from each other, at least during the early stages of sedimentation, and that only the southern basin had a deeper water connection to depocentres on the Falkland Plateau. For convenience of description we have called the basins north of the Coega Fault the Inner Algoa Basin, and the complex region to the south the Outer Algoa Basin.

The Recife Arch, which separates the Algoa and Gamtoos depocentres is the narrowest and shallowest of the offshore basement highs. It curves strongly to the south a short distance east of Cape Recife, and at the edge of the present day continental shelf (about 45 km offshore) lies about 1 sec beneath sea level (<1 km of sediment cover). It was apparently also an

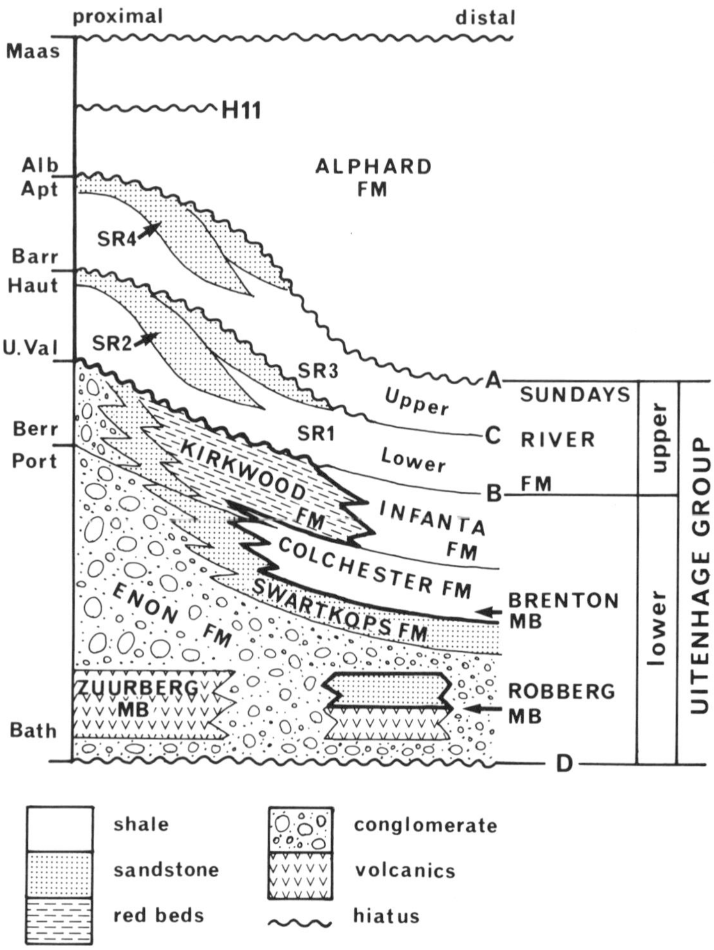


Figure 72. Schematic relationships of Jurassic to Cretaceous lithofacies in the E.Outeniqua and southern Cape basins. The diagram refers specifically to the Pletmos and Algoa basins. See Figure 92 for correlations of these areas to the Bredasdorp Basin in the west. A-D are prominent seismic reflecting horizons. Thickened line marks the marine (above)/non-marine (below) boundary. Based on Dingle (1973a, 1978, 1980a), du Toit (1976), SOEKOR (1976), and McLachlan & McMillan (1976, 1979).

impressive barrier in Jurassic-Lower Cretaceous times because so far it has not been possible to confidently correlate borehole and seismic horizons across it (SOEKOR 1976, McLachlan & McMillan 1979).

### 4.3.2 *Stratigraphy*

Renewed interest in the stratigraphy and facies of the sediments of the southern Cape taphrogenic basins has been generated by commercial and academic research on the Agulhas Bank. Although complete consensus does not exist, good agreement has been reached on the main points of lithostratigraphic correlation in the area, and Figure 72 is based on the various classifications proposed by Hill (1972), Dingle (1973a, 1978), Winter (1973, 1979), du Toit (1976) and McLachlan & McMillan (1976, 1979), and is close to the scheme formally submitted for consideration to the South African Committee for Stratigraphy (du Toit et al 1975, SACS 1980).

Figure 67 summarizes the distribution of Middle Jurassic to Lower Cretaceous strata in the Outeniqua Basin in relation to main depocentres and basement highs in adjacent areas. The whole of the sequence comprises the Uitenhage Group, which is subdivided into seven formations and several members and beds. With the exception of the Upper Sundays River and Infanta formations, all the rock units are represented and crop out in the Inner Algoa Basin, and have been well-described in the literature since the latter part of the 19th century.

Atherstone (1857) first proposed the classic subdivision of the Algoa Basin:

Sundays River Beds (*top*)
Variegated Marls and Wood Beds
Enon Conglomerate (*bottom*)

and it was assumed that there was a certain amount of lateral and vertical interfingering between the three units, and that they represent a straightforward gradation from continental (Enon Conglomerate), through fluviatile/estuarine (Variegated Marls and Wood Beds) to shallow marine (Sundays River Beds). This view was not seriously challenged until borehole data became available, and in the most modern summaries based solely on field relationships (e.g. du Toit 1954, Haughton 1969), the Sundays River Beds are considered to grade laterally into the non-marine sequences. Although

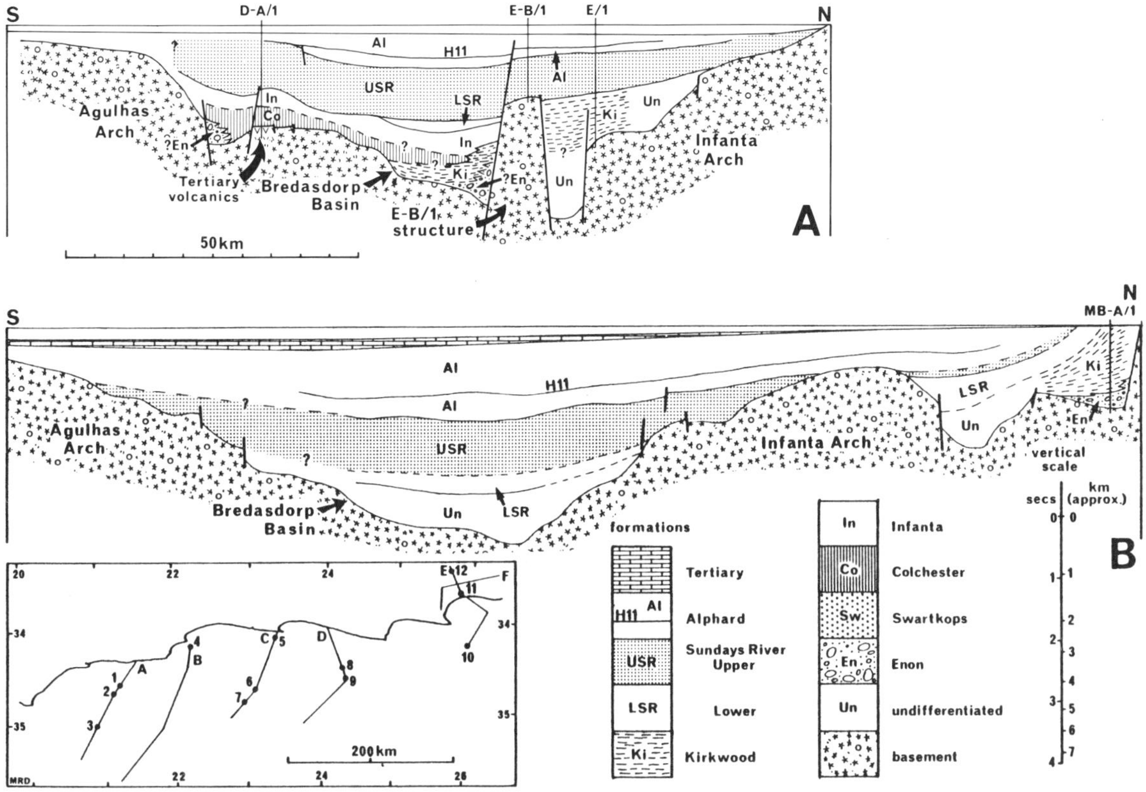


Figure 73. Sections across Bredasdorp Basin. A – west, B – east. Datum is sea level. Constructed from isochron maps in SOEKOR (1976), but is partly speculative in deeper areas. Subdivision of the Sundays River Formation is uncertain, but is on the assumption that the base of USR = reflector C in the Pletmos Basin and reflector B in the Bredasdorp Basin (as used by du Toit 1976), following the revision of nomenclature by McLachlan & McMillan (1979). Insert shows location of sections in Figures 73, 74 and 75. Borehole location: 1 = E/1; 2 = E-B/1; 3 = D-A/1; 4 = MB-A/1; 5 = PB-A/1; 6 = G(a)-C/1; 7 = F-D/1; 8 = G(b)-Gemsbok/1; 9 = G(b)-Springbok/1; 10 = H(b)-Hartebeest/1; 11 = AL1/69; 12 = PA1/68.

the evidence is still equivocal, it seems more likely that an hiatus separates the Sundays River Formation from the lower units (e.g. McLachlan & McMillan 1977), and that a regionally significant change in the sedimentary environment occurs across this boundary. It is, therefore, fruitful to consider the Sundays River and pre-Sundays River strata as representing fundamentally different stages in the taphrogenic basin infill. In addition to the three main rock units, early workers recognized several less extensively developed horizons from outcrops and boreholes, which they attempted to fit into the accepted scheme. The most significant of these were the volcanoclastic and lava suite in the north of the Algoa Basin (Zuurberg Member), and the 'lower marine beds' (Colchester Formation), which appeared to lie at the base of the Variegated Marls and Wood Beds. To some extent, the relationships of these units still present problems, particularly because they have such an important bearing on two significant events in the Algoa Basin, and by implication, over the whole of the Outeniqua Basin. These are, respectively:

i) dating the earliest taphrogenic movements and infill sequences in the southern Cape and assessing the degree of volcanic activity associated with the early basin development;

ii) the age of the earliest marine incursion into the Outeniqua Basin area.

Both these questions will be referred to on numerous occasions in the following sections, and as far as the development of a satisfactory tectono-sedimentary model for the southern Cape area is concerned, their importance cannot be overstressed. Taken with the evidence from Deep Sea Drilling Project sites 327 and 330 on the Falkland Plateau, they are the oldest potentially well-datable units in the area.

Tate (1867) introduced the term Uitenhage Beds to group the units described by Atherstone (1857), whilst Rigassi & Dixon (1972) and Winter (1973) proposed the first revisions to the original lithostratigraphic nomenclature of the Algoa Basin succession that have led to the classification shown in Figure 72. A further important step was the incorporation of the offshore drilling and seismic data by du Toit (1976), which allowed the full extent of the lateral facies variations

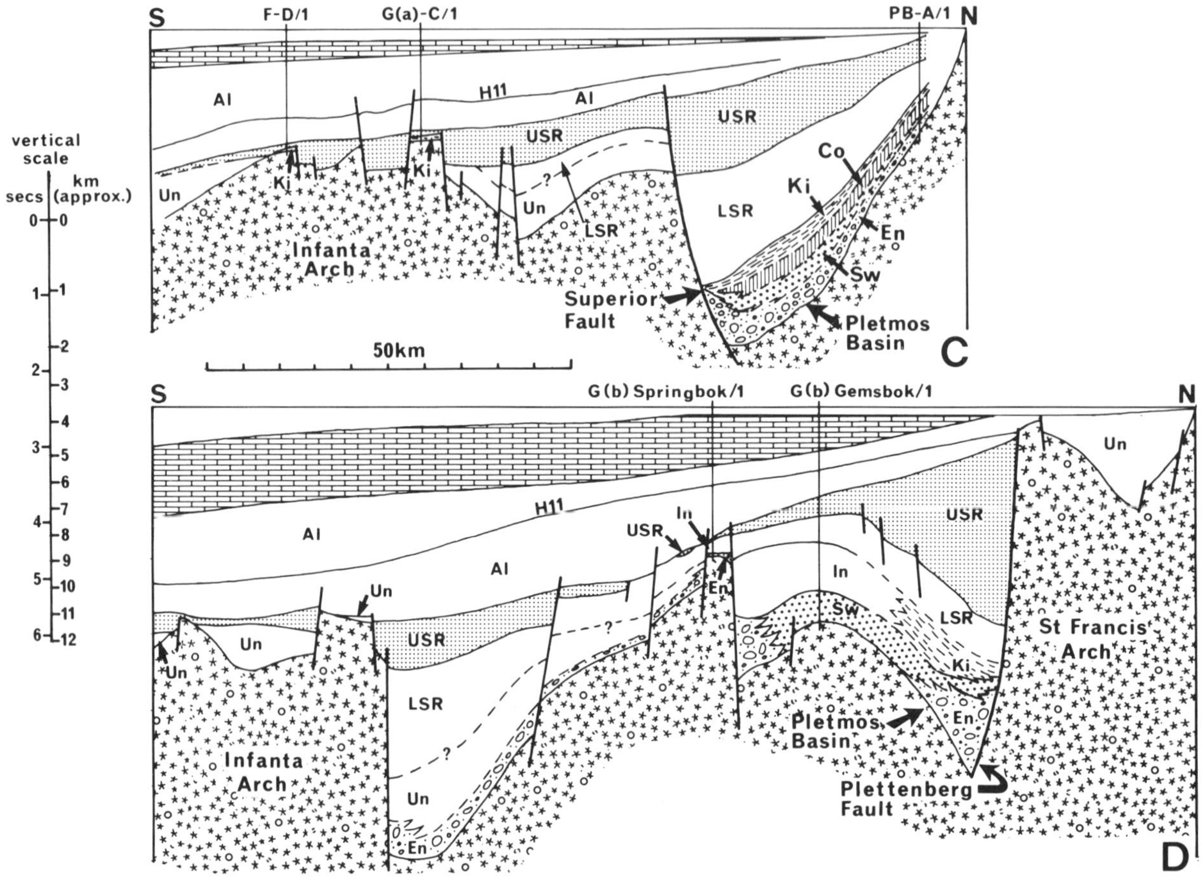


Figure 74. Sections across central Outeniqua Basin. C – W.Pletmos Basin. D – E.Pletmos Basin. See Figure 73 for location, key, and further explanation.

of the Kirkwood Formation (Variegated Marls and Wood Beds in the previous terminology) to be appreciated, and the recognition of the cyclic nature of the Sundays River Formation, only the lower cycle of which is represented onshore. Where seen, the base of the Middle Jurassic-Lower Cretaceous sediments rests with a strong discordance on the Palaeozoic Cape Supergroup and probably locally upon pre-Cape metamorphic rocks. On the other hand, the upper limit of the Uitenhage Group is not exposed onshore, but has been recorded in several boreholes in the Outeniqua Basin, where it is represented by a non-sequence, and possible unconformity, of varying duration (du Toit 1976).

In terms of the lithostratigraphic classification shown in Figure 72, we identify two distinct stages in the infill of the taphrogenic basins of the southern Cape, which are represented by the Lower and Upper Uitenhage Group. The lower stage is probably mid-Jurassic to ?Berriasian (or possibly lower Valanginian) in age and involved continental, paralic, and restricted marine sedimentation, with local volcanic activity. It was followed by a Valanginian to Albian (locally ?Cenomanian) paralic and shallow marine sequence (Upper Uitenhage Group), during the deposition of which there was an increase in the differential subsidence rates in the block-faulted basins. A subsidiary tectonic event (SOEKOR seismic reflector C, Figs 72 & 92) is represented by local hiatuses and possibly erosion horizons within the Upper Uitenhage Group, and was accompanied by a marked southward shift and contraction in area of the depocentres, as well as a further increase in differential subsidence rates. This is reflected in increased sedimentation rates in the Upper Sundays River Formation depocentres, which were restricted to the Outeniqua Basin. The thickness and distribution of the Uitenhage Group (i.e. the Middle Jurassic-Lower Cretaceous rocks) in the Outeniqua Basin and adjacent areas is shown in Figure 70. Maximum thicknesses (>6 km) occur in the Pletmos Basin adjacent to the Plettenberg and Superior faults (Fig.74C,D) and in its southern sector, and in the Gamtoos Basin. Large areas over the basement highs identified in Figures 68 and 69 have cover of <1 km whilst thicknesses in the onshore basins vary from over 3 km (Oudtshoorn, Gamtoos and Algoa) to 1 220 m (Robertson). The isolation of the inner Algoa Basin from its offshore (outer)

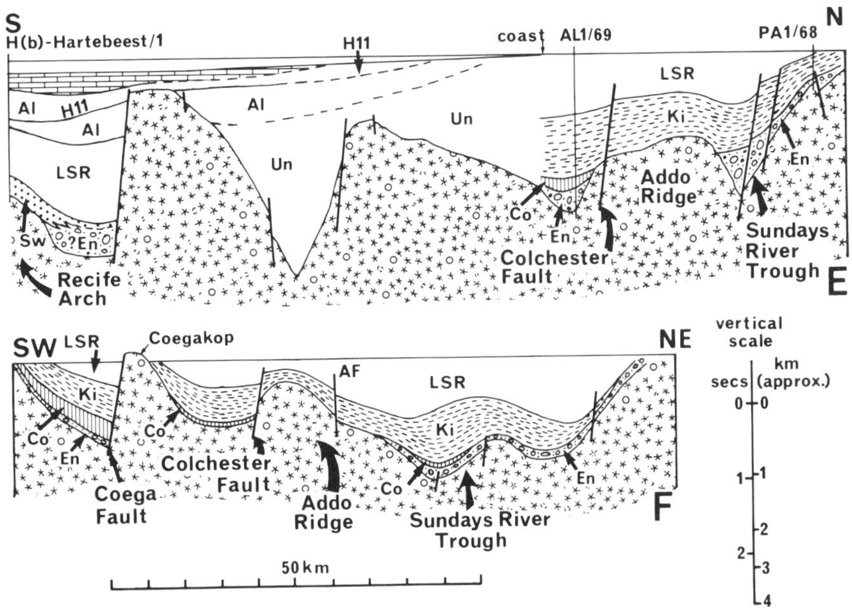


Figure 75. Sections across Algoa Basin. E – outer basin, F – inner basin. Note that no satisfactory correlation has been published for strata beneath the Alphard Formation in the outer basin. Terminology for inner basin follows Winter (1973). AF = Addo Fault. See Figure 73 for location, key and further explanation.

parts, and the asymmetry of the Pletmos Basin is well-illustrated in Figure 70.

4.3.2.1 *Lower Uitenhage Group (Middle Jurassic to Berriasian or ?Lower Valanginian)*

These strata represent the earliest stage of fill in the taphrogenic basins, and occur in all the known Mesozoic depocentres in the southern Cape. Their distribution and thickness are shown in Figure 71. It is clear that all the southern Cape taphrogenic basins began their development during the Middle Jurassic to Lower Cretaceous period, and that maximum infill, at least in the larger basins was approximately the same (i.e. between 2 and 3 km). This suggests that local crustal collapse along lines of pre-existing weakness in the Cape Fold Belt was more or less synchronous and regionally of the same magnitude.

In the Algoa, Oudtshoorn-Gamtoos, and Worcester-Pletmos lineaments, the basins are strongly asymmetric, and sedimentation was controlled by differential subsidence across major boundary faults, so that maximum thicknesses invariably lie adjacent to these main fractures. The intervening basement highs are either bare or covered by a thin veneer (< 500 m) of Lower Uitenhage Group sediments, and obviously had the effect of isolating the main depocentres. In the Bredasdorp Basin, the thickest sediment accumulations lie adjacent to a small subsidiary horst in the NW part of the main depression (at least 3.3 km), and in a somewhat thinner pocket (ca.2.2 km) west of the F-F/2 structure, which has a crest bald of Lower Uitenhage Group strata. In general, sediment accumulation appears to have been less in the Bredasdorp Basin than in the other major basins with large areas having less than 2 km. In addition, crustal subsidence took place in a more symmetric fashion. Although there are no borehole, refraction seismic, or deep reflection seismic data published from the area between the present shelf break (ca.200 m water depth) and the Agulhas Marginal Fracture Ridge (the COB), it is possible that several of the sub-basins widen and deepen to the south and SE, so that even thicker sequences might be expected to lie under the continental slope of the SE Agulhas Bank.

The Lower Uitenhage Group comprises a complex of continental, fluviatile, estuarine and shallow marine lithofacies which interfinger laterally and vertically in response to basin geometry. Because the individual sediment basins are relatively narrow, elongate, and were to all intents and purposes, separate depocentres, the various lithofacies boundaries are probably diachronous from basin to basin, depending upon elevation discontinuities across particular boundary faults. As a general rule, the coarse continental-finer fluviatile sediments (Enon and Kirkwood formations) represent a proximal lithofacies pair, and they grade distally into lower energy environment sediment with an increasing marine influence in the younger rocks (Colchester and Infanta formations).

4.3.2.1.1 *Enon Formation.* This is the most widely distributed rock unit in the southern Cape basins, although its distribution is extremely patchy and it displays rapid thickness variations (Fig.78A). It consists of poorly-sorted conglomerates with interbedded and subordinate sandstones, silts, mudstones, lavas, tuffs and bentonites.

The type area is in the panhandle of the northern Algoa Basin, where all the main rock types are well-exposed. They were first formally described by Ather-

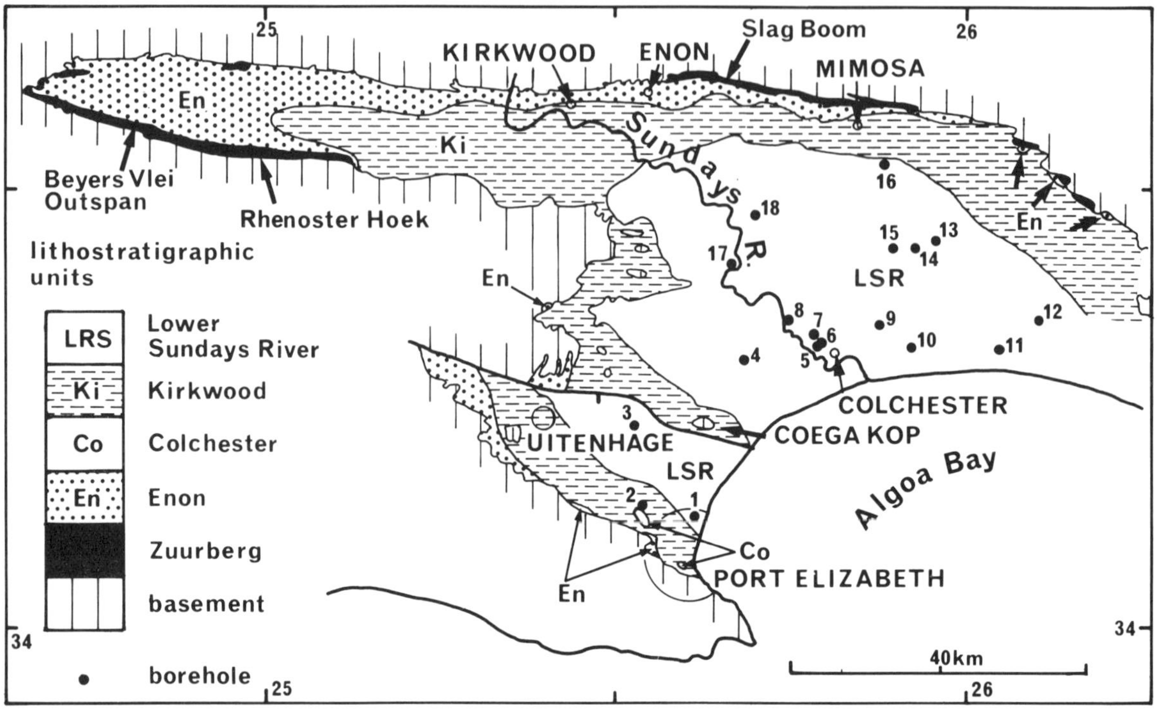


Figure 76. Distribution of Middle Jurassic-Valanginian strata in the Inner Algoa Basin. Modified from Winter (1973), and McLachlan & McMillan (1976). Boreholes: 1 = SW1/08; 2 = BT1/74; 3 = ST1/71; 4 = BR1/71; 5 = CO1/67; 6 = CO3/71; 7 = CO2/70; 8 = SH1/74; 9 = KE1/71; 10 = AL1/69; 11 = SM1/76; 12 = SV1/71; 13 = NA1/69; 14 = NA2/70; 15 = NA3/70; 16 = PA1/68; 17 = AD1/68; 18 = CK1/68.

stone (1857), and modern descriptions have been given by Rigassi & Dixon (1972), Hill (1972), Lock et al (1975), and Shone (1976), and Figures 76 and 77 show details of outcrops and sections in the Algoa panhandle. The Enon Formation is intermittently well-exposed along the northern boundary of the Algoa Basin, where the type sections occur near the Enon Mission station. Here it is approximately 490 m thick, and consists of a thin (20 m) basal angular conglomerate, a middle 170 m of volcanic rocks (tuffs and basalts) and an upper sequence of about 300 m of conglomerates and sandstones. Field relationships along the southern crop of the panhandle indicate that the volcanogenic strata constitute a laterally impersistent lens within the main body of the Enon Formation, and we will refer to this, and its equivalents elsewhere as the Zuurberg Member, whilst retaining the classification of Hill (1972) for its constituent parts (Fig.77).

In the Enon Mission area, the basal part of the formation consists of angular, poorly-sorted cobbles and pebbles of light grey orthoquartzite and tuff set in a quartz sand matrix. The clasts are up to 25 cm in diameter, but usually between 0.5 and 5 cm across. Local lenses of more rounded pebbles occur within the angular conglomerate. Hill referred to these rudaceous beds as the Slagboom Beds.

The lower part of the Zuurberg volcanic Member (Coerney Beds) consists of pale orange, pink and white tuffs. Near the mission, they are about 70 m thick, and are predominantly fine-grained, with occasional lapilli layers (1-5 m), and small quantities of angular quartz and chalcedony clasts. Bedding is poorly developed, but rare layers of reworked, waterlain tuffaceous sands and small pebble conglomerates have been recorded by Hill (1972). These beds have proved to be fossiliferous, with both dinosaur and plant fragments occurring in the Slagboom section (Haughton & Rogers 1924, Hill 1972). The tuffs are cut by dolerite dykes. As interesting observation by Hill (1972) is the occurrence of the Coerney tuff resting directly on Palaeozoic basement at a high level in the Klein Winterhoek Mountains indicating the existence of rugged relief prior to its ejection.

The upper part of the Zuurberg Member (Mimosa Beds) consists of lavas which range in composition from tholeiitic basalts to ferrolatites with subsidiary thin interbedded tuffs. The basalts are compositionally indistinguishable from basalts in the Lesotho Formation of the Drakensberg Group (Marsh et al 1979). At Slagboom, these rocks are about 80 m thick, and appear to conformably overlie the Coerney tuffs. Individual flows are difficult to distinguish, but according to Hill (1972) are probably several metres thick. Vesicular and amygdaloidal varieties occur, and the vesicules are usually filled with zeolites, chalcedony and calcite.

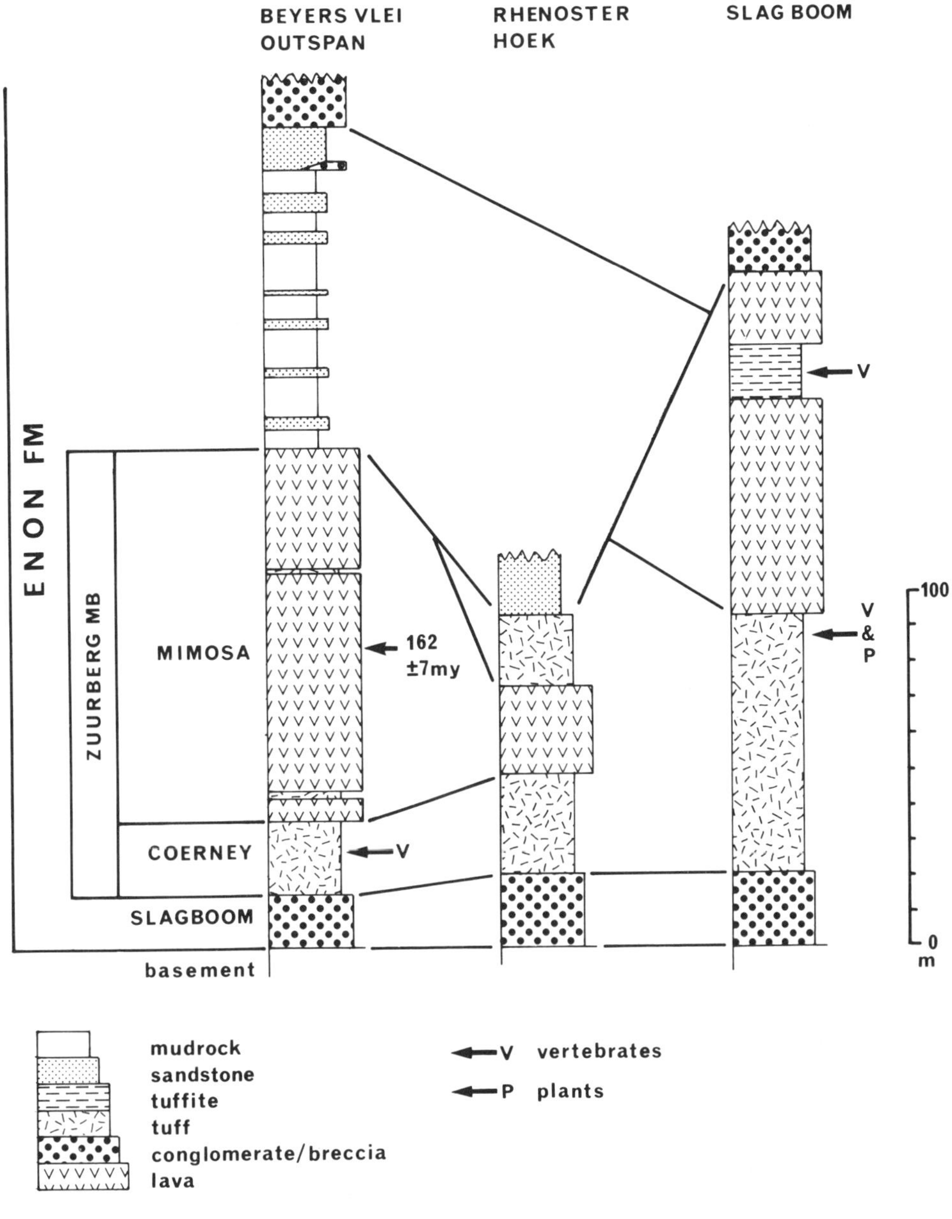


Figure 77. Sections in the Zuurberg Member of the Enon Formation, N.W.Algoa Basin. See Figure 76 for locations. Data from Haughton & Rogers (1924), Hill (1972), McLachlan & McMillan (1976).

Several thin, pale red tuff bands occur near the base of the Mimosa Beds, whilst near the Slagboom section lenses of greenish-pale orange and pink tuffs between 5 and 15 m thick occur near the top of the sequence. Hill (1972, p.17) has also reported possible vertebrate remains in the lower red tuffs near Slagboom. Between Beyers Vlei Outspan and Rhenoster Hoek along the southern crop, the Zuurberg Member thins from 125 m to about 70 m. This is accompanied by a sharp reduction in thickness of the Mimosa Beds and the appearance of a second tuff layer about them. North-south correlation across the Algoa panhandle compared with the Slagboom section farther east, suggests that there was considerable interfingering between the main basalt and tuff horizons (Fig.77).

In the type section, the Zuurberg basalts are overlain by about 300 m of clast-supported, poorly-sorted and generally unbedded conglomerates with a sandy or argillaceous matrix. The clasts, which mostly show a preferred imbrication direction, usually consist of well-rounded orthoquartzites, sandstones, shale, and vein quartz. Rare pebbles of amygdaloidal basalt, agate and tuff also occur and on petrographic grounds these are thought to have been derived from the underlying Zuurberg volcanics. Fragments in the conglomerate are pebble and cobble sizes, but pieces range up to 50 cm

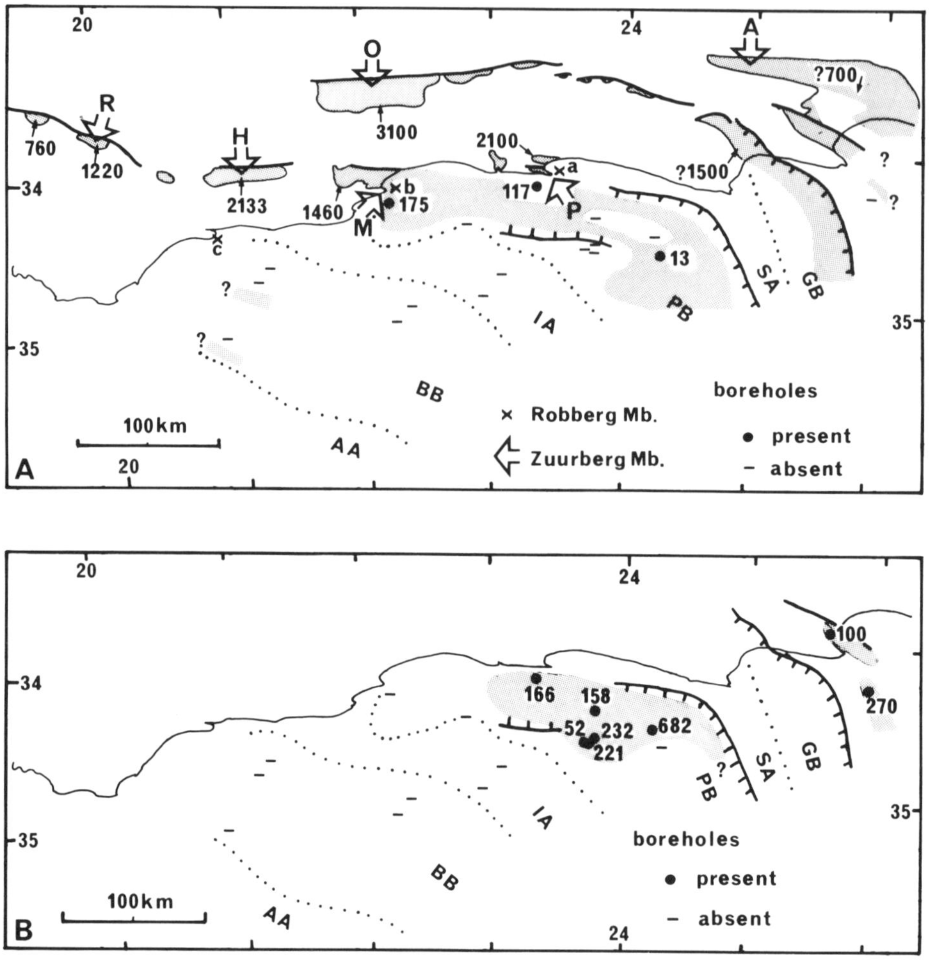


Figure 78. A. Distribution and thickness of Enon Formation in Outeniqua Basin and southern Cape. Speculative occurrences marked with query. Borehole and spot thicknesses shown in metres. Data mainly from du Toit (1976). Zuurberg Member occurs: R = Robertson – tuff; H = Heidelberg – tuff; M = Mossel Bay – tuff; O = Oudtshoorn – 4 m tuffite; P = Robberg Peninsula – 1 m bentonite; A ≐ Algoa Basin – tuffs and lavas. Robberg Member occurs: a = Cape Seal (Robberg); b = Cape St Blaize; c = Cape Infanta.

B. Distribution and thickness of Swartkops Formation in Pletmos and Algoa basins. Data mainly from du Toit (1976). Abbreviations: AA = Agulhas Arch; BB = Bredasdorp Basin; IA = Infanta Arch; PB ∓ Pletmos Basin; SA = St Francis Arch; GB = Gamtoos Basin.

in diameter, and large blocks exceeding 4 m have been recorded close to faults. Percussion marks are common on the larger clasts (Shone 1976). Dips along the northern crop vary from over 50°N, west of the Enon Mission, to averages $>$20°S farther east.

This upper part of the formation is the classic Enon Conglomerate of the early workers, and displays typically bright red and yellow coloration, locally giving rise to impressive pitted cliff faces. The conglomerate typically occurs in massive lenticular units with interbedded red and grey planar and cross-bedded sandstones. Local basement rock types are well-represented, although of limited extent beyond the present outcrop: Table Mountain quartzites, Witteberg quartzites, and Bokkeveld shales (now seen as a deeply weathered red matrix set with dark shale clasts). Interestingly, Shone (1976) reported a total absence of Karoo Supergroup lithologies even though Dwyka Group outcrops occur adjacent to some sections.

In addition to previously mentioned thickness variations in the Mimosa basalts, other points of difference emerge between the north and south crops of the Algoa panhandle. In the south there is a generally more rounded aspect, as well as greater variety of clast type in the basal conglomerate, and to the west, thick (up to 180 m) essentially non-conglomeratic trough-bedded sandstones and greenish mudstones intervene between the Mimosa basalts and the main Enon conglomerates

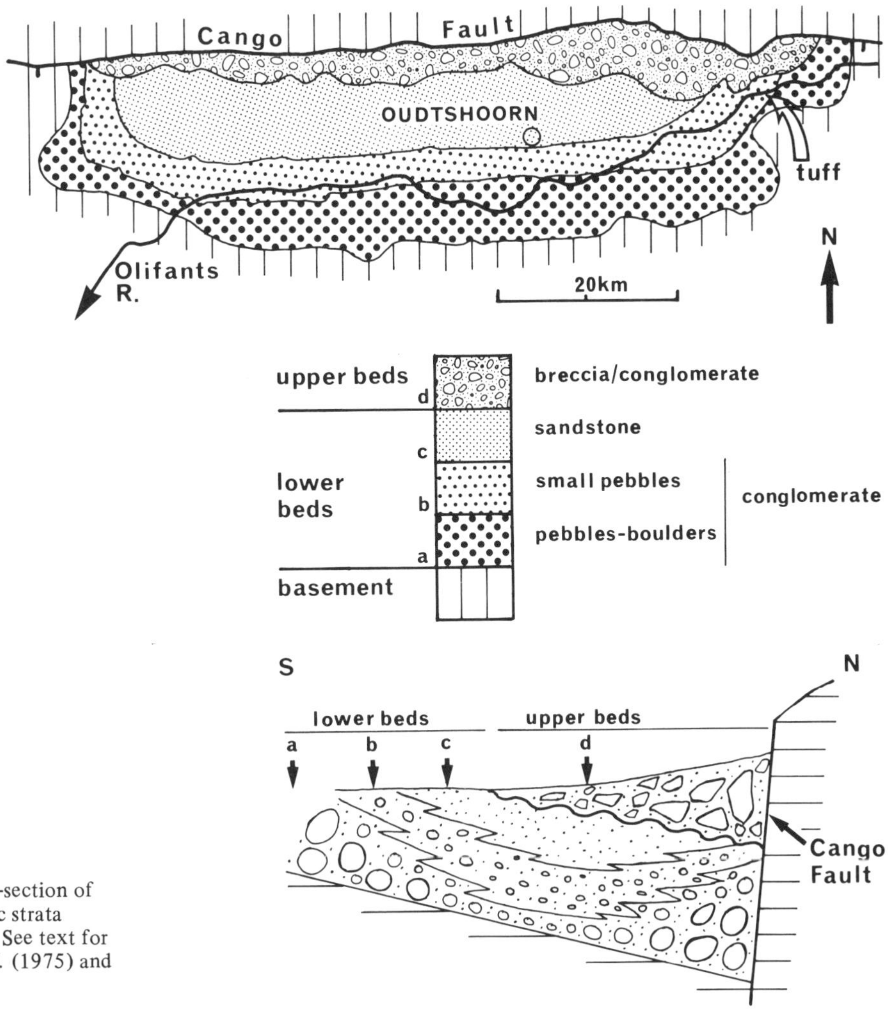


Figure 79. Sketch map and cross-section of Oudtshoorn Basin. Post-Mesozoic strata removed. Section is not to scale. See text for explanation. Based on Lock *et al.* (1975) and Lock (1978).

(Hill 1972). In the extreme west, both the Mimosa basalts and Slagboom rudaceous member fail, and Coerney tuff rests directly on basement. Dips along the southern crop vary from 15 to 40°N in the west, to about 10°N farther east.

It is from the western end of this southern crop that a date of 162 ± 7 my (whole rock, K-Ar) was obtained from the Mimosa basalt (McLachlan & McMillan 1977).

The Enon Formation crops out at several other places around the periphery of the Algoa Basin (Fig.76), and has been intersected in numerous boreholes (Winter 1973). Its maximum known thickness at outcrop is 490 m in the type area and in boreholes 197 m (AL 1/69) at the SE end of the Sundays River Trough (Winter 1973), but a consideration of basement topography leads us to suspect considerably greater thicknesses (? up to 700 m) in the trough axis (Fig.75E). It forms the basal element of the Uitenhage Group everywhere in the Algoa Basin except over the basement highs of Addo Ridge and the Coega Kop fault block. No records of the volcanic member have been made from boreholes or in outcrops from the SE or SW parts of the basin.

The regional distribution, and known maximum thicknesses of the Enon Formation are shown in Figure 78A. Thick, widespread accumulations occur in the Oudtshoorn-Gamtoos and Worcester-Pletmos basins, but so far no occurrences have been reported from the Bredasdorp Basin. In the Gamtoos Basin, the Enon Formation interfingers with variegated clays of the Kirkwood Formation to the extent that Haughton et al (1937) could not satisfactorily map the boundary between the two formations. The Enon Formation conglomerates predominate, and over 1 128 m were proved by SOEKOR in borehole LO 1/69 (McLachlan & McMillan 1977). They comprised the normal reddish-coloured massive units with interbedded yellowish and red cross-bedded sands. The Lower Uitenhage Group is at least 3 100 m in the offshore extension of the Gamtoos Basin, but how much of this belongs to the Enon Formation is not known. In Baviaans Kloof, north-dipping outliers of reddish and yellowish-white conglomerates occur in narrow basins adjacent to impressive Table Mountain quartzite fault scarps. Locally, the Enon rocks have been folded and faulted, presumably by later movements along the faults, although Hill (1972) suggests that the brecciation is due to explosive volcanism.

In the Olifants River valley, three separate basins contain extensive deposits of Enon Formation, the largest and best known of which is the Oudtshoorn Basin. This has been described in detail by du Preez (1944) and Lock et al (1975) who recognized a fourfold subdivision of the succession (units a-d in Fig.79). A lower reddish conglomeratic unit (a) crops out around the southern edge of the outlier, and contains well-rounded cobbles and boulders of Table Mountain quartzites (up to 1 m across) and Bokkeveld shales, the latter decreasing rapidly in number upwards. It is overlain, and grades laterally into yellowish sands interbedded with small pebble conglomerates (b). Pebbly horizons become less abundant towards the north and east where they are replaced by yellowish sands and green and red mudstones (c). The total thickness of units b and c is at least 200 m. Lithologically, unit c is reminiscent of parts of the Kirkwood Formation in the Algoa Basin, and it is a moot point as to whether it ought to be classified as a finer interval in the Enon Formation, or an interbed of Kirkwood Formation.

The three lower lithofacies (a-c) are unconformably overlain by a northward thickening wedge of coarse red conglomerates and breccias interbedded with sandstone lenses (d). Clast sizes increase towards the Cango fault plane, where many boulders are > 1 m across. The estimated 3 000 m of Enon Formation present in the Oudtshoorn Basin (Rigassi & Dixon 1972) dips northwards (i.e. towards the boundary fault) at between 10 and 20° (locally 30°). It is interesting to note that no pre-Cape Supergroup rock clasts are known from the conglomerates, despite the extensive exposures immediately north of the Cango Fault.

Enon Formation occurs in all the basins of the Pletmos-Worcester lineament, and in two places in subsidiary depressions on the general high feature that lies to the north in the Knysna and Plettenberg Bay areas. Offshore, boreholes have proved the formation at the base of the taphrogenic fill in three places in the Pletmos basin, with a maximum of 175 m near Mossel Bay (Fig.78). All the sites were on relatively high features, so it is probable that considerably thicker deposits occur in the deeper parts of the main Pletmos Basin south and SE of Plettenberg Bay, although the thicknesses shown in Figures 74C and D are speculative and have not been substantiated. No details have been published of the lithology of the offshore sequences.

At Knysna, and two areas near Plettenberg Bay (Piesang and Bitou), reddish conglomerates consisting predominantly of rounded Table Mountain quartzite clasts set in a sandy matrix laterally interfinger with finer sandstones and mudstones reminiscent of the Kirkwood Formation. The relationship of the Enon Formation to marine clays and sandstones on the southern side of the Knysna Outlier (Brenton) is not clear. Farther west, varying thicknesses of Enon Formation occur in the Mossel Bay, Heidelberg, Swellendam, Robertson, and Worcester outliers (Fig.78). All contain interbedded conglomerates with subordinate sands, pebbly sand and argillaceous rocks and local clast types vary from granitic-gneisses at Mossel Bay adjacent to the pre-Cape metamorphic complex southwest of George, to Bokkeveld at Swellendam, and Ecca, Dwyka and Witteberg in the Worcester area. It is interesting to note that in the latter, no Table Mountain quartzites or Malmesbury clasts have been recorded, despite extensive present-day outcrops. The proportion of conglomerates to other lithofacies varies from basin to basin, but is usually higher in the lower parts of any individual sequence, although in the Heidelberg outlier the rudaceous component is subordinate, except in the basal beds.

No Enon conglomerates have been found in the five boreholes so far reported from the Bredasdorp Basin (du Toit 1976). All these were on relatively high features, but in comparison to the occurrences in the Pletmos Basin, it might have been expected that in boreholes DA/1 and E/1 (Fig.73A), at least, this facies would have been found. Certainly, the Bredasdorp Basin is generally shallower, wider and less asymmetric than the other offshore basins, and except locally, the boundary fault scarps appear to be less steep. It is possible that the Enon conglomeratic facies was not developed in the Bredasdorp Basin, or that it is restricted to small pockets in the NW and SE adjacent to particularly steep boundary faults (see Fig.87).

Two members of the Enon Formation are widely, but sparsely developed in the depocentres west of Algoa Basin. The Zuurberg volcanic member is represented by thin bands of tuff in the Robertson, Heidelberg, Mossel Bay, Oudtshoorn (at least 4 m), and Plettenberg Bay (1 m bentonite) basins (e.g. McLachlan & McMillan 1976, Lock et al 1975), as well as an unconfirmed report from the Worcester Basin (Marsh et al 1979). Assuming that these deposits are synchronous with the Coerney tuffs in the Algoa Basin, the implication is that sedimentation had begun in taphrogenic basins over most of the southern Cape by Middle Jurassic times. It could be argued that the tuffs represent ash bands of different ages, but given the long distances from source (wherever it happened to be) it is more difficult to explain the occurrences of bands of different ages within different basins, than to assume a single volcanic episode whose eruptive products were widely dispersed. Until more positive dating is available we prefer the simplest explanation.

At three places on the south coast (capes Seal, St Blaize and Infanta), a tough silicified variant of the local basal Enon Formation contains a limited marine fauna (Fig.78). This has been called the Robberg Formation by Rigassi & Dixon (1972), but we agree with McLachlan & McMillan (1976) that it is more easily explained as a facies variant of the Enon Formation, which locally debouched into the sea, and call it the Robberg Member. At the type section on Cape Seal it consists of conglomerates and relatively well-sorted, locally ripple-marked sandstones, at two localities crowded with marine lamellibranchs. The whole sequence it silicified and near the base Rigassi & Dixon (1972) recorded a tuffaceous siltstone. At Cape St Blaize, Rigassi & Dixon (1972) measured 70 m of the Robberg Member, which has reduced to a few metres at Cape Infanta. They found trace fossils and plant remains at each locality. There is no firm palaeontological evidence to correlate the Robberg Member marine horizon with the Colchester Formation as McLachlan & McMillan (1976) have done, but clearly it represents an early marine incursion into the Pletmos and Bredasdorp basins. Its intimate association with volcanic debris on Cape Seal suggests that it may be older than Portlandian, and on Figure 72, we have placed it at the same stratigraphic level as the Zuurberg Member (Bathonian/Callovian), recognizing, however, that it could represent a marine tongue anywhere in the Upper Jurassic. The only palaeontological evidence we have of the age of the Robberg Member is Cooper's (1974) identification of *Megatrigonia conocardiiformis* (Krauss) from Cape Seal which he compares to species recorded for the Upper Kimmeridgian-Portlandian of northern Mozambique.

As might be expected in a continental rudaceous lithofacies, fossils are rare and fragmentary in the Enon Formation. Dinosaur bone fragments have been reported from tuffs in the Coerney and Mimosa beds of the Zuurberg Member (Haughton & Rogers 1924, Hill 1972) as well as fragmentary silicified wood and plant material (Haughton & Rogers 1924), but no positive identification appears to have been possible on this material. Within the conglomeratic facies of the Algoa Basin, only rare wood fragments have so far been reported (Rogers & du Toit 1909, Haughton & Rogers 1924), but none have been recorded from the Gamtoos Basin.

Farther west, within interbedded sandstones and mudrocks, fossils are abundant although it is difficult to decide whether the rocks in question should be considered part of the Enon Formation or assigned to the Kirkwood facies. In the Knysna area, Schwarz (1906) found indistinct plant remains in grey shales, but a more abundant flora has been identified by Seward (1903, 1907) from the Mossel Bay outlier (Tab.16). Fresh-water fossils have been found in the Heidelberg outlier (Rogers & Schwarz 1900, 1902a, Jones 1901, Rogers & du Toit 1909): *Estheria anomala* Jones, fish scales, ostracods, wood and plant stems, beetle wing cases and *Unio uitenhagensis.* Further plant fragments are known from the Oudtshoorn outlier (lignite pockets, fragment of *Cladophlebis* sp. as well as a few dinosaur teeth) and Worcester (silicified wood). None of these can be considered age diagnostic although the plant fragments have a similar aspect to those from the Kirkwood Formation in the Algoa Basin.

At Robberg, the silicified conglomerate has an abundant lamellibranch fauna. In addition to *Megatrigonia conocardiiformis* already mentioned, Schwarz (1900) identified *Trigonia* and *Gervillia.* McLachlan & McMillan (1976) did not consider the identifications made so far from this assemblage to be as age diagnostic as believed by Cooper (1974). Nevertheless they do suggest an Upper Jurassic age for part of the Enon Formation, which together with the K-Ar date for the Mimosa basalts, and the local interfingering relationship with the Kirkwood Formation indicate to us that the Enon Formation represents a highly diachronous lithofacies that ranges in age from Middle Jurassic to Berriasian. This assessment compares with Bathonian to Portlandian (Winter 1979); Bathonian-Callovian (volcanic facies) and Berriasian-Valanginian (conglomeratic facies) (McLachlan & McMillan 1976); and Bathonian (volcanic facies and Robberg Member) and Portlandian-Hauterivian (conglomeratic facies) (McLachlan & McMillan 1979).

4.3.2.1.2 *Swartkops Formation.* Distal representatives of the Enon Formation are medium to fine-grained, poorly-sorted, angular to sub-angular quartzitic sandstones, interbedded with subsidiary brownish and dark grey shales. Du Toit (1976) equated these with the Swartkops Formation, first recognized by Atherstone (1857) in the vicinity of Uitenhage, and named by Winter (1973) for a thin sandstone unit (100 m) which was intersected above 20 m of Enon conglomerate in the Swartkops (SW 1/08) borehole NE of Port Elizabeth (Fig.76). The sands are poorly developed onshore, and are confined to the Pletmos and Outer Algoa basins offshore. In the former, they locally reach 682 m in the G(b)-Gemsbok/1 borehole, which du Toit (1976) designates as the offshore stratotype. Its vertical distribution in boreholes, where it either overlies Enon Formation or basement, and its lateral extent within the Pletmos Basin (compare Figures 78A and B) indicate that it is a distal (low energy) vertical and lateral facies equivalent of the Enon Formation. In the Pletmos Basin, the Swartkops Formation is probably thickest in the axial parts, generally decreasing towards the fault planes, but is typically best-developed on the medium-sized structural highs where the Enon Formation is not present (du Toit 1976). No fossils have been recorded from the Swartkops Formation, but textural characters and lithostratigraphic relationships suggested to McLachlan & McMillan (1976, 1979) that it is of fluvial or estuarine origin.

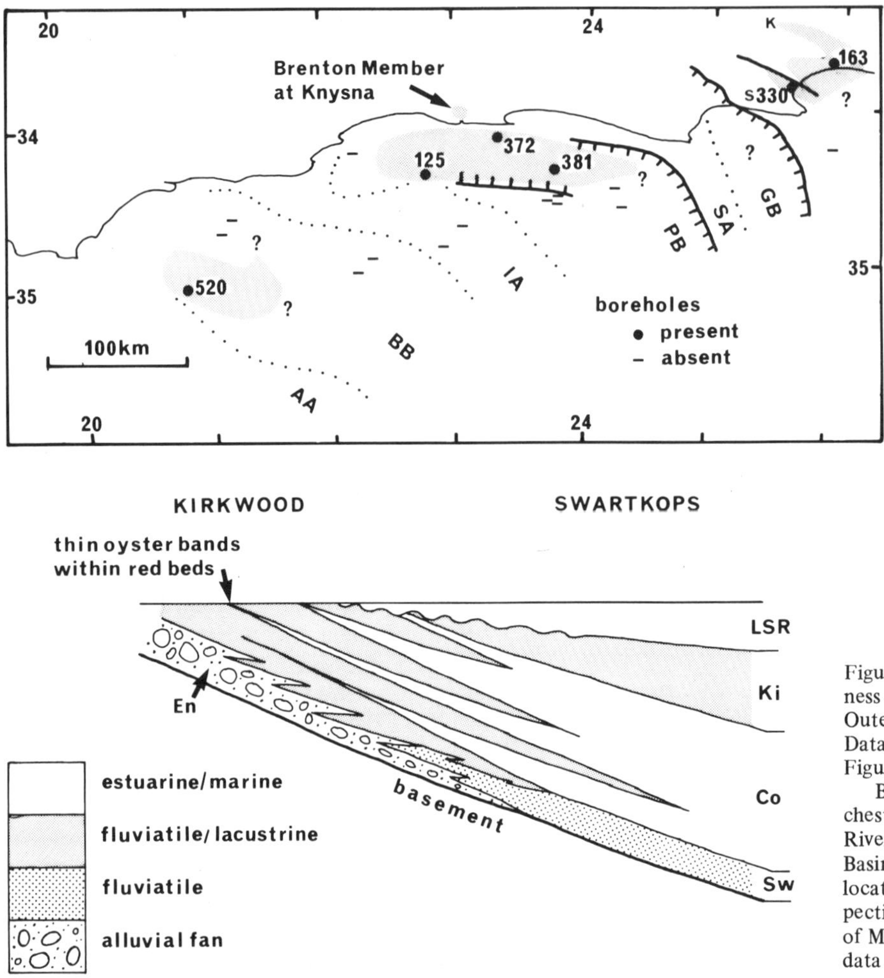


Figure 80. A. Distribution and thickness of Colchester Formation in the Outeniqua Basin and southern Cape. Data mainly from du Toit (1976). See Figure 78 explanation for abbreviations.

B. Schematic relationship of Colchester, Kirkwood and Lower Sundays River formations in the Inner Algoa Basin. Section runs (left to right) from locations K to S on the map (A), respectively. Based on the interpretations of McLachlan & McMillan (1976), and data in Haughton (1928).

4.3.2.1.3 *Kirkwood Formation.* The term Kirkwood Formation was proposed by Winter (1973) to replace the long-used name 'Variegated Marls and Wood Beds'. This was a welcome move because the old term was a mouthful, and something of a misnomer, especially with regard to the 'marls'. The type area for the Kirkwood Formation is near Kirkwood on the northern edge of the Algoa Basin (Fig.76) and Winter (1973) proposed that the sections measured by Haughton (1928) should be the reference stratotype, although because neither the top nor bottom of the formation is seen in these sections, he also proposed that the boreholes CO 1/67 and CO 2/70 in the south central Algoa Basin should be designated the subsurfaces stratotypes. Figure 81 gives details of these sections, including the composite CO 1/67 - CO 2/67 borehole log (Winter 1979).

At present, there is still some uncertainty as to the nature of the upper and lower limits of the Kirkwood Formation. It had been generally accepted in earlier works (e.g. du Toit 1954) that the Enon Formation passed laterally and vertically into the Kirkwood Formation, which in turn passed laterally and vertically into the Sundays River Formation. In the reappraisals carried out by SOEKOR geologists (Winter 1973, 1979, du Toit 1976, McLachlan & McMillan 1976, 1979) the first relationship was considered to be valid, but extensive seismic, borehole sample and geophysical log data suggested that the boundary between the Kirkwood and Sundays River formations was either disconformable or in proximal areas even unconformable. Field work in the Algoa Basin on the other hand, has led Shone (1976, 1978) to the conclusions that whilst the Enon and Kirkwood do not interfinger, there is a gradation between the Kirkwood and Sundays River formations. Shone (1978) counters suggestions that a sharp change in compaction values (as measured by specific gravity) across the boundary in boreholes (2.53 Sundays River to 2.60 Kirkwood, Stewart 1973) indicates a significant disconformity, by appealing to differential water loss caused by variation in permeability in the sandstone units in the two formations. The problem is certainly not unequivocally solved by the data

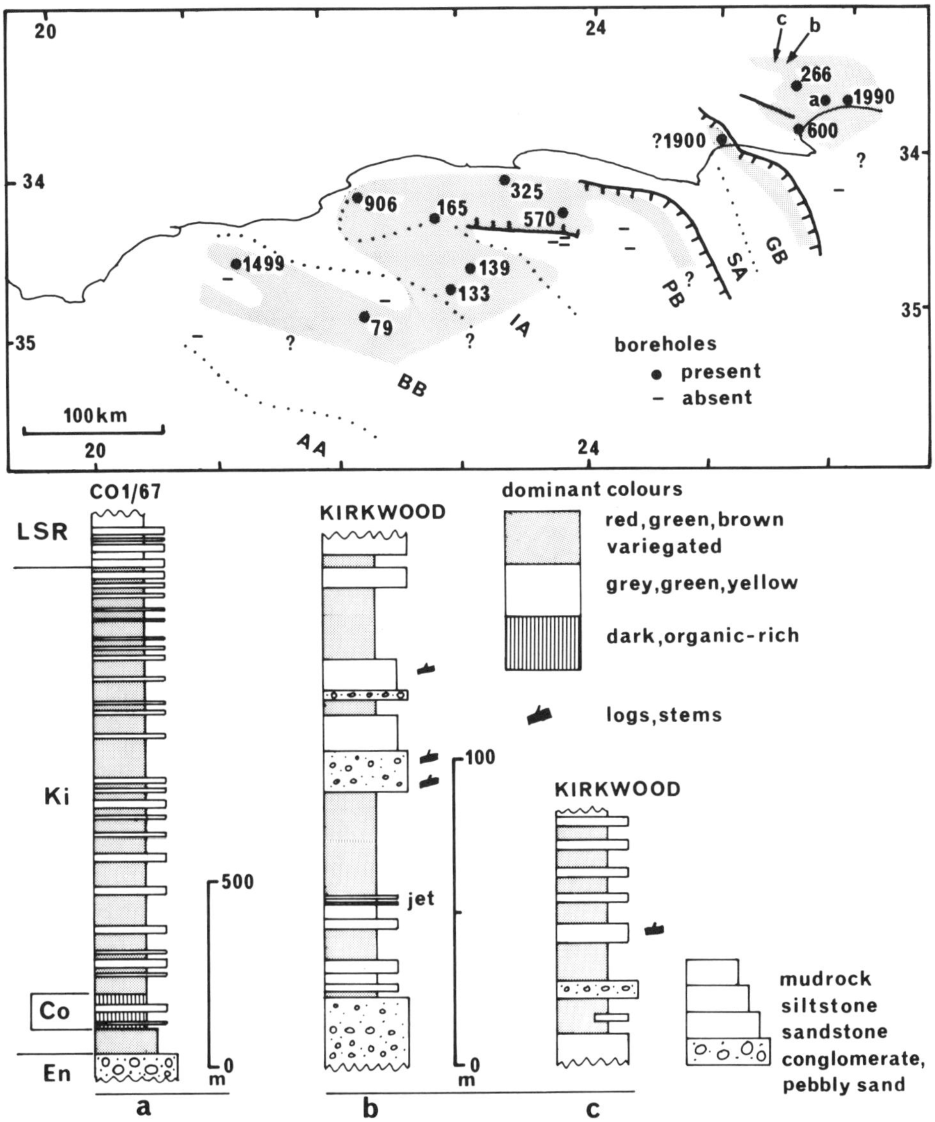


Figure 81. Distribution and thickness of Kirkwood Formation in the Outeniqua Basin and southern Cape. Data from Winter (1973), du Toit (1976), and McLachlan & McMillan (1976). Sections a-c are located in the Algoa Basin (see map). Borehole CO1/67 (the bottom 150 m are taken from borehole CO3/71) is the sub-surface stratotype for the formation selected by Winter (1973). Sections b and c are those measured by Haughton (1928) and taken as the formation types by Winter (1973). Section b is the canal cutting east of Kirkwood (Haughton 1928 p.20-21) (locality 7 of McLachlan & McMillan (1976), and section c is in the Sundays River at the Kirkwood Bridge (Haughton 1928 p.19-20) (locality 6 of McLachlan & McMillan 1976). Neither the top nor the bottom of the formation is seen in sections b and c. Winter (1973) indicates a possible unconformity at the top of the Kirkwood Formation in borehole CO1/67. See Figures 73 & 78 for abbreviations.

to hand, but we believe that the balance of evidence, particularly taken in a regional context, where the Sundays River Formation is invariably found overlying, and not grading laterally into the Kirkwood Formation, is in favour of a scheme such as Figures 72 and 80, which follow the suggestions of du Toit (1976), McLachlan & McMillan (1976) and Winter (1979). The Kirkwood Formation grades laterally into and interfingers with shallow marine sediments: the Colchester; and Infanta formations. The latter, which is not known from any present-day onshore localities is overlain by the Sundays River Formation.

The Kirkwood Formation occurs extensively in the taphrogenic basins of the southern Cape (Fig.81) and comprises two main facies: a dominantly arenaceous, pebbly facies (formerly Wood Beds), and a predominantly argillaceous facies (formerly Variegated Marls). They were deposited contemporaneously and inter-

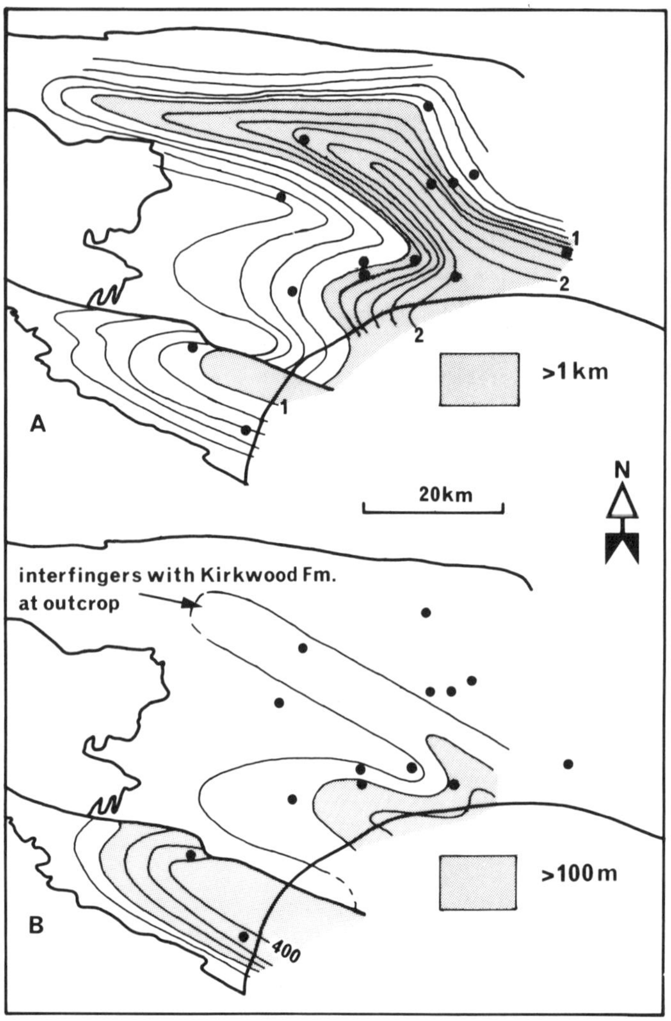


Figure 82. Thickness of Kirkwood (A) and Colchester (B) formations in the Inner Algoa Basin. Contour interval 200 m (A) and 100 (B). Modified from Winter (1973). Dots are control boreholes (see Figure 76 for identification).

finger, but because of variations in the local basin geometry, and fluctuations in sediment supply, are not developed at consistent levels from place to place, although within the Algoa Basin there is a general increase in the number of sandstone bodies towards the axial parts of the depocentres (Fig.83). At the same time, however, the ratio of thickness of argillaceous to arenaceous sediments also increases away from the original basement highs, so that it is clear that whilst the sandstone units are more numerous away from the source areas, they are thicker in the proximal parts of the basins.

In the type area, the Kirkwood Formation consists of interbedded yellowish, yellowish-brown, white, and pale grey current-bedded sandstones, and pebbly sandstones, and subordinate red, green and grey mudstones (up to 60 m thick), alternating with units of red and greenish grey silty mudstones with subordinate sandstone (up to about 30 m thick) (Fig.81). In the arenaceous facies, the sandstone bodies are often coarse, massive or planar and cross-bedded, upward fining lenticular units (up to 2 m thick) with erosive bases and pebbly, bone and wood fragment lags on their floors. These units are commonly stacked in sequences up to 20 m thick. The sandstones at outcrop are typically soft and gritty, and locally the wood fragments are large (tree stumps and trunks). Conglomerates and reddish grey mudstones with roots and plant fragments are subsidiary interbeds. The conglomerates contain quartzite, shale, tillite, amygdaloidal basalt and pink tuffs.

In the argillaceous facies, mudstones are typically red, but frequently variegated red-greenish grey in colour, and they contain subsidiary sandstones up to 2 m thick. The mudstone units frequently contain roots and are bioturbated, and both mudstones and sandstones contain plant and fresh water lamellibranch fossils (Haughton 1928).

Over 2 100 m of red and green mudstones and minor sandstones occur in the central part of the Algoa Basin (Sundays River Trough) (Fig.82), and in borehole CO

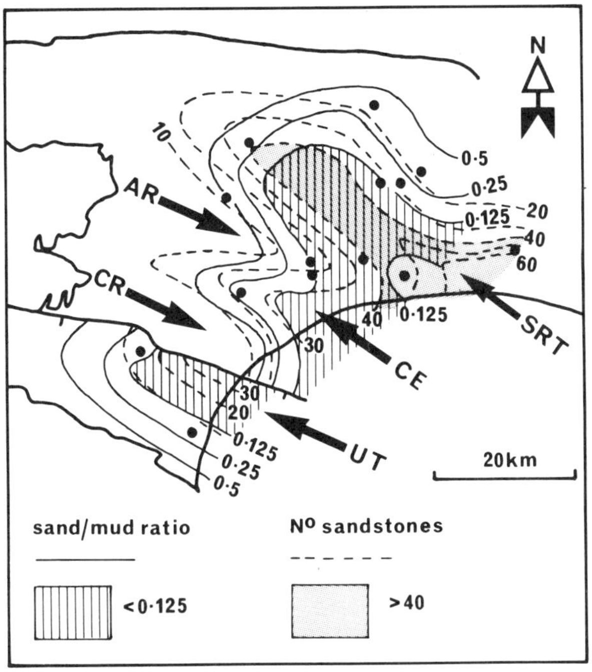


Figure 83. Sandstone/mudrock variation in the Kirkwood Formation of the inner Algoa Basin. The ratio is based on gamma ray log measurements, and the number of sandstones is for units >3 m thick. Data from Winter (1973). Dots are control boreholes (see Figure 76 for identification). Abbreviations: UT = Uitenhage Trough; CR = Coega Ridge; CE = Coega Embayment; AR = Addo Ridge; SRT = Sundays River Trough.

1/67, which is situated on the SE nose of the Addo Ridge, 1 264 m of alternating red, brown and green mudstones and green and grey fine sandstones are recorded by Stewart (1973). The 1 000 m isopach penetrates into the Algoa panhandle, outlining a narrow, southerly opening trough, but both the Addo and Coega ridges (Fig.83) have variable and generally thin cover of Kirkwood Formation (Fig.75). A comparison of the petrography of the shales and sandstones between the basinal and ridge sites reveals little difference (Stewart 1973). In both environments, the sandstones are medium to fine-grained with abundant rock fragments (ca.20 %), and the only significant variation is the frequently high calcite matrix values at the CO 1/67 (basinal) site (at least 20 %). In both environments, the argillaceous rocks are also very similar, although in CO 1/67 the percentage of montmorillonite in the clays is consistently <20 %, compared to average values of >30 % on the Addo Ridge (borehole AD 1/68). At both sites carbonaceous material is relatively rare and is restricted to the sandstones. Faultward thickening of the Kirkwood Formation against the Colchester and Coega faults (Fig.75E,F) indicates significant boundary fault movement during deposition. Distribution of the Kirkwood Formation in the offshore parts of the Algoa Basin is not known, but it does not occur in the Hartebeest borehole at the extreme southern end of the basin, in an area which was probably isolated from the depocentres to the north.

Lithofacies similar to the sandstones and variegated mudrocks of the Algoa Basin Kirkwood Formation are known from all the major, and most of the minor basins along the southern Cape coast, but in most cases they are associated with typical Enon Formation conglomerates and cannot be mapped as a distinct lithostratigraphic unit. This emphasizes the interfingering nature of the two lithofacies and poses a problem of classification. For convenience of thickness estimates, etc., we have included these finer grained variants within the Enon Formation on Figures 78A and 79, whilst appreciating the reality of the rudaceous to argillaceous gradation shown in Figure 72. Only at the SE end of the Gamtoos Basin have thick units been encountered consistently enough to allow an estimate of Kirkwood Formation thicknesses. Boreholes LO 1/69 and MK 1/70 (McLachlan & McMillan 1976) proved 1 128 m of Enon and 3 028 m of Enon and Kirkwood, respectively, suggesting a thickness of about 1 900 m in the latter. Surface outcrops in the Gamtoos Basin have been studied by Haughton et al (1937), but no satisfactory distinction could be made on a regional scale between the Enon and Kirkwood lithofacies. The latter consists of cross-bedded, light coloured sandstones with red mudstones and grey sandy shales, and near Lourie, lignitic clays with amber, ferruginous concretions, and plant-bearing sandstones have been unsuccessfully prospected for coal (Rogers 1906).

Offshore, the Kirkwood Formation has been intersected in several boreholes (du Toit 1976) (Fig.81). The thickest development is in the Bredasdorp Basin (E-1) (Fig.73A), where 1 499 m of fine sediments (no colour reported but presumably includes red mudrock) rest directly on basement, although there is a suggestion that the contact may be faulted. No Kirkwood Formation occurs in D-A/1, which is unexpected, considering its basement elevation in relation to that in borehole E-1; possibly the Kirkwood Formation laps against the northern side of the D-A/1 basement ridge. The second occurrence (79 m) is on the crest of a small horst on the northern margin of the Bredasdorp Basin which suggests that thicker developments of the Kirkwood Formation lie farther south in the axial part of the main basin.

A thin (133-165 m) capping of Kirkwood Formation rests directly on basement over the faulted crest of the Infanta Arch (F-D/1, G(a)-C/1, and F-B/1) (Fig. 74) where reddish brown and brick red shales and mudstones interfinger with pebbly, lignitic light-medium grey sandstones (du Toit 1976). This is the only published record of lignite from the Kirkwood Formation of the Outeniqua Basin, and significantly it is from the top of a structural high which probably lay in an analogous position to the more proximal areas in the

Algoa Basin where the arenaceous, wood-rich facies is typically developed.

Three boreholes in the Pletmos Basin prove an extensive, thick development of Kirkwood Formation (du Toit 1976). All consist of red or brown shales, which in borehole MB-A/1 (906 m) overlie and interfinger with Enon Formation, in PB-A/1 (325 m) overlie Colchester Formation, and in G(a)-B/1 (570 m) interfinger with grey shales of the Infanta Formation and overlie Colchester Formation. The Kirkwood Formation is missing from the Superior Ridge and the G(b)-Springbok/1 – G(b)-Gemsbok/1 high where it is replaced by the Infanta Formation, but we speculate that a thick development occurs in more proximal settings against the Plettenberg Fault (Fig.74D).

McLachlan & McMillan (1976) have given a good account of the fossils recorded from the Kirkwood Formation. Vertebrate remains appear to be widespread although the quality and quantity is not high: abraided fragments are relatively common in the sandstones and conglomerate, whilst more complete specimens occur in the grey siltstones. No complete skulls or skeletons have been reported. The only positive identifications made so far are: an *Iguanodon*-like dinosaur (by Atherstone 1857), *Palaeoscincus africanus* by Broom (1910), *Algoasaurus bauri* by Broom (1904); and cf. *Acanthopholis* by Tate (1867). There has been no modern reappraisal of these taxa, nor apparently, any description of numerous other vertebrate fossils in various collections. Invertebrates are of fresh water or estuarine taxa and include eight species of lamellibranchs (including oysters), five species of gastropods, and the fresh water ostracods *Cypridea, Bisulcocypris* and *Darwinula* (Shone 1976).

Plant remains are common and often very well-preserved, particularly at the fossil leaf localities around Dunbrodie (SE of Kirkwood on the Sundays River, Fig.76). The most comprehensive account of the flora is by Seward (1903) who recorded six species of ferns (sometimes unusually well preserved), eight species of cycads, and four species of conifers (Tab.16). Brown & Gow (1976) have commented on the remarkable preservation of some of this material and concluded that it could only have been achieved by rapid and complete burial with little or no post-depositional compaction. They also considered the environment to have contained a relatively diverse flora (including large coniferous trees on the margins of water bodies), which was locally and very rapidly overwhelmed by floods carrying a high suspended sediment load resulting in accumulation at rates of between 0.5 and 0.75 m per flood cycle. The proximity to estuarine fossil localities showed that the vegetation was growing at no great distance from the head of local estuary systems.

Thirty-eight species of miospores have been recorded in the Kirkwood Formation by Scott (1976) from boreholes in the south of the Inner Algoa Basin. The assemblages, which are dominated by *Classopollis* spp. (57-89 %), indicate a coastal vegetation of Upper Jurassic-Barremian age. Other assessments of the age of the Kirkwood Formation are also vague, and have been reviewed by McLachlan & McMillan (1976). The plants, which are the best preserved element have been dated as 'Wealden' by Seward (1903) (i.e. Portlandian-Aptian), and comparisons with the flora of the Rajmahal Series in India (which is probably Neocomian) also suggest a Lower Cretaceous age (e.g. Tate 1867). In agreement with McLachlan & McMillan (1976) and Winter (1979), we have tentatively placed the Kirkwood Formation in the uppermost Jurassic-lowermost Cretaceous (i.e. Portlandian-Berriasian or possibly lower Valanginian) (Fig.72).

Table 16. Plant fossils from the Lower Uitenhage Group.

| | |
|---|---|
| A. Mossel Bay outlier (Seward 1903, 1907). ?Enon Formation | |
| | *Cladophlebis denticulata* (Brongniart) forma *atherstoni* |
| | *Taeniopteris* sp. cf. *T.arctica* (Heer) |
| | *Taxites* sp. |
| | *Osmundites kolbei* |
| B. Algoa Basin (Seward 1903), Kirkwood Formation | |
| Ferns | *Onychiopsis mantelli* (Brongniart) |
| | *Cladophlebis browniana* (Dunker) |
| | *C.denticulata* (Brongniart) forma *atherstoni* |
| | *Sphenopteris fittoni* Seward |
| | *S.* sp. |
| | *Taeniopteris* sp. |
| Cycads | *Zamites recta* (Tate) |
| | *Z.morrisii* (Tate) |
| | *Z.africana* (Tate) |
| | *Z.rubidgei* (Tate) |
| | *Nilssonia tatei* Seward |
| | *Cycadolepsis jenkinsiana* (Tate) |
| | *Benstedtia* sp. (cf. *Coniferocaulon colymbeaeforme* Fliche) |
| | *Carpolithes* |
| Conifers | *Araucarites rogersi* Seward |
| | *Brachyphyllum* sp. |
| | *Conites* sp. A |
| | *C.* sp. B |

4.3.2.1.4 *Colchester Formation.* Early workers (e.g. Atherstone 1857, Rogers & Schwarz 1901, Haughton 1928) mapping in the vicinity of the confluence of the Sundays and Bezuidenhouts rivers (Dunbrodie area SE of Kirkwood) in the Algoa Basin, described numerous sections with greyish sandstones and clays, and sandy limestones which locally contain a rich estuarine and shallow marine invertebrate fauna, as well as vertebrate bones and plant fragments. These beds interfinger with the red and green shales and yellowish sandstones of the Kirkwood Formation. Further outcrops of grey clays and sandstones with estuarine and shallow marine faunas overlain by, and interfingering with, red and green mudstones occur near Port Elizabeth at Bethelsdorp Salt Pan (site 2 on Fig.76) (Atherstone 1857,

Table 17. List of fossils from the Colchester Formation in the vicinity of Port Elizabeth (Bethelsdorp Salt Pan & North End Lake) (mainly after Tate 1867 and Stow 1871 as summarized by McLachlan & McMillan 1976).

| |
|---|
| Lamellibranchs |
| *Ostrea (Exogyra) jonesiana* |
| *Ostrea* sp. |
| *Astarte* sp. |
| *Ptychomya* sp. |
| *Pleuromya* sp. |
| *Cyrena?* sp. |
| *Tellina* sp. |
| *Gervillia* sp. |
| *Trigonia goldfussi* |
| **Melania* sp. |
| **Modiola* sp. |
| Gastropods |
| *Turritella rubidgeana* |
| *Cerithium* sp. |
| *Natica* sp. or *Phasianella* sp. |
| **Limnaea?* sp. |
| **Viviparus (Paludina)?* sp. |
| Echinoids |
| *Cidaris pustulifera* |
| Indet. spines |
| Brachiopods |
| *Orbicula?* sp. |
| Crustacea |
| *Cyzicus* sp. |
| Cycad |
| *Zamia* sp. |

* Swartkops borehole (Rogers 1910)

Stow 1871), and North End Lake (Stow 1871, Haughton 1928). All these beds were correlated with the 320 m of grey and blue clays, the so-called 'Lower Marine Beds', that occur between the Swartkops sandstone and Kirkwood Formation red and green shales in the Swartkops (SW 1/08 borehole), which until recently (e.g. Haughton 1928, du Toit 1954, Rigassi & Dixon 1972), had been considered to represent the lower part of the Sundays River Formation interfingering with the Kirkwood Formation. SOEKOR borehole data have now clarified the situation, and Winter (1973) formally proposed the name Colchester Formation (which had first been introduced by Rigassi & Dixon 1972) for the dark grey waxy shales (1 m) and overlying grey shales (100 m) with an estuarine or shallow marine microfauna that occur near the base of the boreholes CO-3/71 and CO-1/67 (Winter 1979, Fig.81). In these boreholes, the Colchester Formation is separated from the Enon Formation by about 70 m of red-brown mudrocks, and is overlain by red shales, both of the Kirkwood Formation. Exposures of Colchester Formation at Dunbrodie, near Port Elizabeth, as well as the interbeds at the base of the Kirkwood Formation in the Colchester and Swartkops boreholes, thus represent a temporary marine incursion into the more proximal parts of the depocentre (Figs 72 & 80). McLachlan & McMillan (1976) give a good summary of the numerous older records which relate to the status of this formation in the Algoa Basin.

Within the Inner Algoa Basin, the Colchester Formation reaches a maximum thickness of about 200 m at the SE end, where it was deposited in two sub-basins separated by the Addo Ridge (Fig.82B). It appears to be underlain by Kirkwood Formation only around the periphery of the basin and elsewhere rests directly upon either the Enon or Swartkops formations. Evidence of syn-sedimentary tectonism is seen in the Coega sub-basin, where the Colchester Formation thickens towards, and reaches at least 400 m in the vicinity of the Coega boundary fault (Figs 75F, 80 & 82B).

Rigassi & Dixon (1972) recorded the Colchester Formation from two other coastal basins in the southern Cape. Their report of the 'Lower Marine Beds' in boreholes in the Gamtoos Basin has not been substantiated by SOEKOR geologists (McLachlan & McMillan 1976), but outcrops of marine beds in the Knysna outlier have been known since 1906 when Schwarz recorded trigonias from Brenton. Because of the uncertainty that has surrounded the age and lithostratigraphic correlation of the Colchester Formation at Brenton, we will describe it in some detail at the end of this section.

Offshore, the Colchester Formation has been encountered in the Bredasdorp and Pletmos basins (du Toit 1976) (Figs 73A, 74C & 80). In the former, 520 m of thinly-bedded, dark grey shales with ammonite impressions are intruded by Tertiary volcanic rocks, and their base was not reached. They are overlain by light grey shales of the Infanta Formation, and the boundary between the two has been selected on the basis of downhole geophysical logs. The Colchester Formation does not occur on the crest of the Infanta Arch, although 125 m of light grey shales with subsidiary sandstones intersected in borehole F-B/1 on the faulted northern flank of the arch have been referred to this formation bý du Toit (1976). In the Pletmos Basin, the Colchester Formation occurs in the PB-A/1 (372 m) and G(a)-B/1 (381 m) boreholes, and at both locations consists of grey shales (lignitic towards the base of the latter), which conformably overlie the Swartkops Formation, and are conformably overlain by red and brown mudstones of the Kirkwood Formation. Organic-rich (anoxic) shales, potentially of oil source grade, are widespread in the Colchester Formation in the Outeniqua Basin (McLachlan & McMillan 1979), with organic carbon values ranging from less than 1 to 4 %. Slight enrichment in molybdenum is also often present, suggesting deposition in reducing conditions favourable for the preservation of organic material, and small quantities of oil were encountered in the Colchester Formation in borehole CO-1/67 in the Inner Algoa Basin (Leith & Rowsell 1979).

Everywhere that it has been recorded onshore and offshore, with the sole exception of the D-A/1 borehole, the Colchester Formation is overlain by the Kirkwood

Formation. It is absent from the crests of all the major basement highs, and is probably thickest and most extensively developed in the deepest parts of the main depocentres.

Within the Algoa Basin, fossils have been recovered from Colchester Formation outcrops in the vicinity of Port Elizabeth, the Swartkops borehole, and from thin interbeds in the Kirkwood Formation near Dunbrodie. At the former sites, nine species of lamellibranch, and three species of gastropod make up the bulk of the fauna along with echinoids, brachipods, crustacea and plant fragments (Tab.17). As pointed out by McLachlan & McMillan (1976) this is a mixed marine/estuarine assemblage, which corroborates identifications quoted by Rogers (1910) of fresh water and estuarine ostracods and molluscs in the Swartkops borehole. At present, little data are available on the fossil content of the Colchester Formation in boreholes from the Outeniqua Basin. Du Toit (1976) recorded ammonite impressions from shales in the Bredasdorp Basin although no identifications of these have been published, and McLachlan & McMillan (1979) mention the presence of belemites and echinoids from offshore boreholes. The most complete descriptions, however, are of the microfaunal content of the formation in borehole PB-A/1 south of Cape Seal (McLachlan & McMillan 1976) (Tab.18). Nine species of ostracoda and five species of benthic foraminifera have been identified, with the fauna as a whole showing strong similarities to those from the Sundays River Formation (six ostracod and two foraminifera common) and the Brenton beds of the Knysna Outlier (five ostracod and four foraminifera common). Presence of the ostracod genus *Cypridea* and a record of the calcareous algae *Chara* are the only definite indicators of fresh or brackish water influence, although a somewhat restricted marine environment is suggested by the relative abundance of agglutinated benthic foraminifera. No microfossils have so far been described from the Colchester Formation outcrops in the Algoa Basin.

For an assessment of the age of the Colchester Formation, the palaeontological evidence summarized above is equivocal, but on the basis of its general stratigraphical relationships, Dingle (1973b) concluded that the Colchester Formation was Upper Jurassic in age. Recent micropalaeontological data (as yet unpublished) have confirmed this assignment (Portlandian) (McLachlan & McMillan 1979, Winter 1979) with important implications for dating other units of the Lower Uitenhage Group. It provides further evidence that locally the Enon and Swartkops formations are Upper Jurassic or older, and implies that the Kirkwood Formation is likely to be locally Upper Jurassic, and generally lowermost Cretaceous (?Berriasian) in age.

Table 18. Microfossils from the Colchester Formation of borehole PB-A/1 (data from McLachlan *et al.* (1976a).

| | Ostracoda | | Foraminifera |
|---|---|---|---|
| * + | *Rostrocytheridea chapmani* | * + | *Epistomena caracolla* |
| * + | *Oligocythereis? dubia* | * + | *Vaginulinopsis gr. prima* |
| + | *Cytherella algoaensis* | | *Haplophragmoides* sp. A |
| * + | *C.inaequivalva* | * | *Astacolus* sp. A |
| * + | *C.schwarzi* | | *Dorothia subtrochus* |
| * + | *Amicytheridea brentonensis* | | |
| | *P.* sp. A | | |
| | *Cypridea* sp. | | |
| | *Cytherelloidea* sp. A | | |

Also found in: + Sundays River Formation; * Brenton Member

4.3.2.1.4.1 *Brenton Member.* Small outcrops of marine sediments occur along the south side of Knysna Lagoon in the Knysna Outlier (Fig.84). Because these beds are the only extensive onshore marine deposits outside the Algoa Basin, and because of their proximity to outcrops of Enon Formation on the northern side of the outlier, the Brenton beds have attracted considerable attention over the last few years. For reasons which will be discussed below we consider the Brenton beds to be a member of the Colchester Formation, a correlation first proposed by Rigassi & Dixon (1972).

The outcrops were mentioned by Schwarz (1906), but the first detailed descriptions of the stratigraphy and palaeontology were given by Rogers (1909), and these have been added to by Schwarz (1915), du Toit (1954), Dingle & Klinger (1972), and Miller (1975). The most comprehensive account is that by McLachlan et al (1976a), who formally designated measured sections at five localities along the lagoon shore as the stratotype for their Brenton Formation (Fig.84). At the base, unfossiliferous reddish shales (unit VI) grade rapidly upwards into fine-grained greenish sandstones which are ripple marked, cross-bedded, and bioturbated in the lower part (unit V). These are overlain by at least 6 m of soft grey shales (unit IV) which contain abundant glauconite, shell fragments, and carbonaceous debris. Angular and subangular quartz and large white mica flakes are common. The invertebrate fauna is well-preserved and tends to be concentrated in lenticular units, with occasional hard calcareous nodules. It is from this unit that the bulk of the micro- and macrofossils described in the literature have been collected. These fossiliferous shales are overlain by about 15 m of grey sandy shales with subordinate interbeds of coarse sand and small-pebble conglomerates (unit III). This unit is poorly exposed and apparently unfossiliferous. It is overlain by a prominent 0.5 m thick rudaceous unit (II), the lower part of which is typically composed of well-rounded quartzite cobbles (up to 10 cm) set in a grey or white quartz and glauconite sand matrix. This is overlain by a hard shell conglomerate with abundant oyster, trigonid and echinoid clasts (Cooper 1979). The record of conglomerate at the base of the sequence at

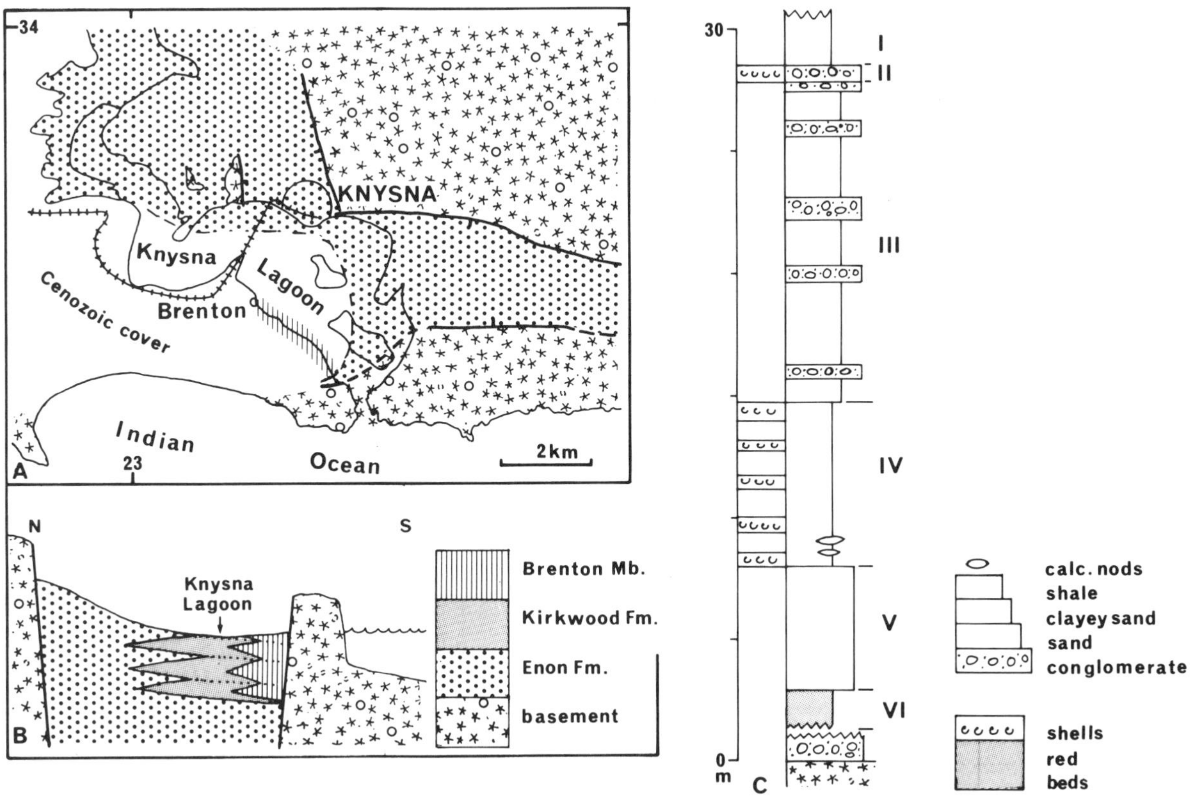


Figure 84. Details of the Brenton Member in the Knysna Outlier. A. Sketch map after Miller (1975) and McLachlan *et al.* (1976a). B. Speculative N-S cross-section after Dingle & Klinger (1972). C. Composite section on S side of the lagoon compiled from profiles in McLachlan *et al.* (1976a) and retaining their VI-I notation.

the NW end of the section (Rogers 1909) is considered by McLachlan et al (1976a) to represent fallen blocks of unit II partially buried on the lagoon shore. The upper part of the Brenton beds (unit I) is unfossiliferous grey silty shale (seen to 2 m).

A further section to the SE, near the entrance to Knysna Lagoon, exposes a conglomerate with angular to sub-rounded pebbles with a red clay or sand matrix. Lithologically, this is similar to typical Enon Formation, but its relationship to the marine beds is uncertain. From its position it would seem to overlie the Brenton beds, although McLachlan et al (1976a) suggest that a small fault may intercede between the two areas, or that the rocks represent a fault breccia or post-Cretaceous talus.

The relationship of the rocks on the south side of the lagoon to the thick development of Enon Formation on the north side of the lagoon is not known. The presence of a southern boundary fault to the outlier has been suggested by several authors (e.g. Miller 1963, Dingle & Klinger 1972, Miller 1975, and McLachlan et al 1976a), but there is no evidence for a fault across the middle of the lagoon separating the marine from the non-marine (Enon and Kirkwood) strata. The presence of Kirkwood Formation east of Knysna, and the red and green shales and sandstones below unit IV, together with conglomeratic and coarse sand horizons in units III and II suggests that a rapid facies change from high energy fluviatile, to low energy fluviatile and shallow marine sediments may exist. In this case, the conglomeratic interbeds would represent temporary injections of coarse material directly into the sea and would be analogous with (though not necessarily the same age as) the Robberg Member of the Enon Formation. An alternative interpretation, favoured by McLachlan et al (1976a) is that the Brenton beds overlie the Enon Formation, although in our opinion the NE dips of the marine sediments, with their conglomeratic interbeds, do not support this suggestion. A speculative section across the outlier is given in Figure 84.

Faunal evidence from unit IV undoubtedly indicates deposition in normal marine salinities, though the possibility of a restricted environment with slow sedimentation rates is suggested by the presence of authigenic glauconite and abundant comminuted plant material. A low energy environment is also suggested by the preservation of the lamellibranchs (often intact pairs), whilst the occurrence of belemnites, nautiloids and ammonites indicates at least periodic connections to the open ocean. The other fossiliferous horizon (II) indicates a beach setting. The available evidence suggests that whilst the shape of the original depocentre may not have been exactly as seen today, the Brenton beds were deposited during a temporary incursion of the sea into a small basin along the northern edge of the main Pletmos Basin.

The age of this marine incursion is in dispute and hinges upon an interpretation of the palaeontological evidence, and particularly the marine invertebrates. Until recently (e.g. du Toit 1954, Haughton 1969), the Brenton beds were unequivocally correlated with the Lower Sundays River Formation in the Algoa Basin because the marine macrofauna of the former was considered identical with that of the latter. However, this was based on a misconception of the original identifications, even though Rogers (1909) noted that 'the fossils . . . obtained show that the fauna of these beds is not identical with that of the Sundays River Beds . . . but very like it,' and the difference was further highlighted by the work of Schwarz (1915). Dingle & Klinger (1972) pointed out this discrepancy, and a recent revision of the trigonias (Cooper 1979) and the accompanying revised macrofaunal list (Tab.19A) shows that none of the macrofossils identified to species level have yet been recorded in the Sundays River Formation in the Algoa Basin. In particular, the ammonite originally found by Rogers (1909), and redescribed by Klinger et al (1972) as *Hybonoticeras* aff *H.hildebrandti* (Beyrich) points firmly to an Upper Jurassic age for the Brenton beds. Suggestions by McLachlan et al (1976a), and Cooper (1979) that this is a misidentification are unjustified.

Dingle & Klinger's (1972) original suggestion of an Upper Jurassic age for the Brenton beds was made on a comparison of its ostracod fauna with assemblages from Madagascar. McLachlan et al (1976a) have re-examined the Brenton fauna and have made additional identifications, comparing the assemblage with that of the Lower Sundays River Formation from the Algoa Basin and borehole PB-A/1 southeast of Cape Seal Table 19 lists the ostracoda and foraminifera recorded from the Brenton beds, Colchester, Kirkwood and Lower Sundays River formations from the southern

Table 19. Fauna of the Brenton Member.

A. macrofossils

| Faunal list after revision of Tigoniidae by Cooper (1979) | Faunal list after Schwarz's (1915) revision |
|---|---|
| *Iotrigonia* cf. *I.vau* (Sharpe) | |
| *Pterotrigonia (P.) etheridgei kitchini* (Schwarz) | *Trigonia kitchini* Schwarz |
| *P.(P.) savagei* Cooper | |
| *Steinmanella (S.) kensleyi* Cooper | *T.holubi?* Kitchin |
| *Isognomon theseni* (Schwarz) | *Perna theseni* Schwarz |
| *Isognomon brentonensis* (Schwarz) | *P.brentonensis* Schwarz |
| *Hybonoticeras* aff. *H.hildebrandti* (Beyrich) | *Acanthodiscus* sp. |
| *Nautilus?* sp. | *Nautilus* sp. |
| *Hibolites* sp. | *Belemnites* sp. |
| echinoid spines | *Cidaris* spines |
| ? no revision | *Pentacrinus* |

B. microfossils

| 1 | 2 | 3 | 4 | Ostracoda |
|---|---|---|---|---|
| X | X | X | X | *Cytherella inaequivalva* Dingle |
| | X | X | X | *C.knysnaensis* Dingle |
| X | X | X | X | *C.schwarzi* Dingle |
| | X | X | X | *Cytherelloidea zoetgeneugdensis* Dingle |
| | X | X | | *C.brentonensis* Dingle |
| | | X | X | *Sondagella theloides* Dingle |
| X | X | X | | *?Oligocythereis dubia* Dingle |
| X | X | X | X | *Rostrocytheridea chapmani* Dingle |
| | | | | *R.* sp. |
| | | X | X | *Majungaella nematis* Grekoff |
| | X | X | | *Progonocythere reticulata* Dingle |
| X | X | X | | *Amicytheridea brentonensis* (Dingle) |
| X | X | X | | Genus A. |

| 1 | 2 | 3 | Foraminifera |
|---|---|---|---|
| X | | | *Astacolus* sp. |
| | X | X | *A.beerae* |
| | X | X | *A.gibber* |
| | X | X | *Citharina cristellarioides* |
| X | | X | *Epistomina caracolla* |
| | | X | *Globulina prisca* |
| | | X | *Haplophragmoides* sp. |
| | X | X | *Lagena apiculata neocomiana* |
| | X | X | *L.hauteriviana hauteriviana* |
| | X | X | *Lenticulina nodosa* |
| | X | | *L.loryi* |
| | | X | *?Neoflabellina* sp. |
| | | X | *N.* cf. *malakialinensis* |
| | | X | *Planularia* gr. *madagascariensis* |
| | | X | *Reinholdella valendiscensis* |
| | | | *Tristix acutangulus* |
| | | | *T.insignis* |
| X | X | X | *Vaginulinopsis prima* |

1 – Colchester Fm PB-A/1
2 – Kirkwood Fm PB-A/1
3 – Lower Sundays River Fm PB-A/1
4 – Lower Sundays River Fm, Algoa Basin (U.Val.)

Table 20. Similarity of Brenton Member ostracod and foraminifera faunas at species level with those from other Uitenhage Group formations (data from McLachlan *et al.* 1976a).

| | Total | Common | % similarity |
|---|---|---|---|
| Ostracoda | | | |
| [1] Colchester Fm | 16 | 6 | 38 |
| [1] Kirkwood Fm | 18 | 10 | 56 |
| L.S.River Fm | | | |
| PB-A/1 | 24 | 12 | 50 |
| Algoa | 25 | 7 | 28 |
| Foraminifera | | | |
| [1] Colchester Fm | 20 | 3 | 15 |
| [1] Kirkwood Fm | 20 | 8 | 40 |
| [1] L.S.River Fm | 39 | 14 | 36 |

1 – Borehole PB-A/1

Cape, and it shows that despite claims to the contrary by McLachlan et al (1976a), correlation between the Brenton beds and the Lower Sundays River Formation is not particularly strong (in the case of the Algoa Basin the links are relatively weak)(Tab.20). In both microfaunal groups, the similarity at the species level is greatest with the Kirkwood Formation faunas of borehole PB-A/1 (56 % for ostracods and 40 % for foraminifera). For a comparison, ostracod species similarity between the Upper Valanginian assemblages of the borehole PB-A/1 and the Algoa Basin is 48 % (27 species, 13 common). Of particular importance for age determination are the progonocytherid ostracods which Dingle (1982a) has shown to have regional significance for correlation in southern Gondwanaland during Callovian to Cenomanian times, and which are represented by three species at Brenton. We consider it significant that of the five species recorded from Brenton and the Algoa Basin, only one is common to both areas (*Majungaella nematis* Grekoff), and that this ranges geographically from Argentina to India and temporally from Portlandian to Cenomanian. Within this group we suspect that *Amicytheridea brentonensis* (Dingle) is the most useful species, with a wide distribution (southern Africa to India) and a Middle Jurassic to Portlandian or Berriasian range. It has been recorded from Brenton, and the Colchester, Kirkwood and Lower Sundays River formations of borehole PB-A/1, but not from the Algoa Basin (Brenner & Oertli 1976), and we suggest that this might indicate a slightly older age for the Lower Sundays River Formation in borehole PB-A/1 than the supposed time equivalent rocks in the Algoa Basin.

Stapleton & Beer's (1977) examination of the calcareous nannofossils and palynomorphs does not clarify the uncertainty, and does not provide convincing evidence for a Valanginian age for the 48 species of palynomorph recorded: 52 % have been recorded from Jurassic rocks; 98 % from Berriasian to Barremian rocks; and 92 % from Aptian/Albian rocks, whilst the calcareous nannofossil assemblage was described as rare and badly preserved.

In summary, whilst not being conclusive, we believe that the weight of evidence, particularly from the *Hybonoticeras* and progonocytherid ostracods, points to an Upper Jurassic (?Portlandian) age for the Brenton beds and indicates a correlation with the Colchester Formation, as originally suggested by Rigassi & Dixon (1972). Final clarification of the stratigraphic position of what we will regard as the Brenton Member is only likely to be achieved by the discovery of a more extensive ammonite, calcareous nannofossil, or a planktonic foraminiferal assemblage.

4.3.2.1.5 *Infanta Formation.* Du Toit (1976) introduced this term to describe a thick sequence of light grey to medium grey (occasionally dark grey) siltstones, clays and shales which are known only from boreholes in the Outeniqua Basin (Fig.85). Because well-developed sequences of both Kirkwood and Infanta formations had not been encountered in the same borehole, and yet they occupy similar positions in the succession with respect to the underlying arenaceous-rudaceous Swartkops-Enon pair, du Toit (1976) concluded that the Infanta Formation is a distal facies equivalent of the Kirkwood Formation. Further evidence to support this assumption has been found in borehole G(a)-B/1 on a basement high in the Pletmos Basin, where thin grey Infanta-type shales interfinger with Kirkwood Formation red mudrocks (Fig.85). These relationships (summarized in Fig.72) suggest that the Colchester Formation represents an early, more restricted facies of the Infanta Formation, and in at least one borehole (D-A/1, Fig.73), du Toit (1976) records the latter overlying the Colchester Formation.

Du Toit (1976) designated a section of borehole G(b)-Gemsbok/1 as the stratotype for the Infanta Formation (Fig.85), where 1 132 m of grey silty, locally calcareous shales are overlain (probably unconformably) by the Lower Sundays River Formation, and rest conformably on Swartkops Formation sandstones. The upper and lower boundaries are well-marked by downhole geophysical parameters, especially the gamma log (porosity). In the type borehole, the Infanta Formation is homogeneous, poorly fossiliferous and well bedded (especially in the lower parts).

Elsewhere in the Outeniqua Basin, thicknesses vary from 82 m (northern edge of the Bredasdorp Basin) to 609 m (Pletmos Basin). The formation is characteristically best-developed on basement rises, away from the axial parts of the main depocentres (e.g. Fig.74D). Because the depocentres were originally strongly asymmetric, proximal facies coincides with the present axial portions (i.e. adjacent to boundary faults) and accounts for the distal equivalent of the Kirkwood Formation (i.e. the Infanta Formation) occurring away from the thickest parts of depocentres. The Infanta Formation maintains an homogeneous lithology over more than 300 km in the Outeniqua Basin (grey silty shales), although minor sandstones occur in the west

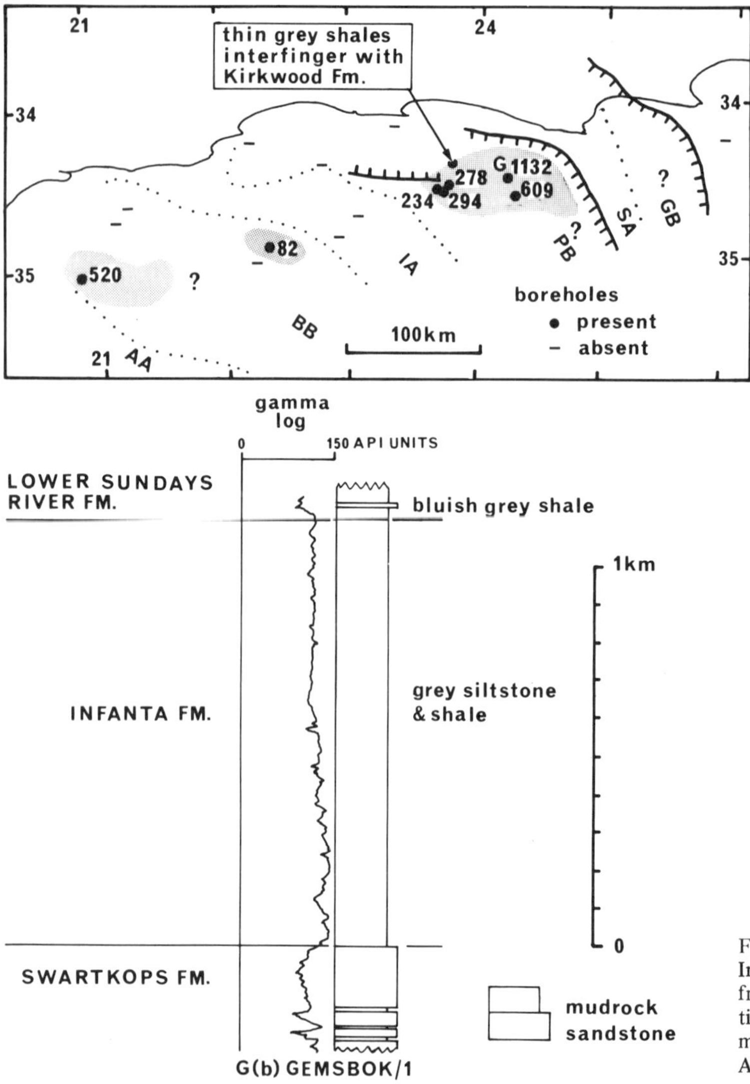


Figure 85. Distribution and thickness of the Infanta Formation in the Outeniqua Basin. Data from du Toit (1976). Log shows stratotype section selected by du Toit (1976), marked G on map. See Figure 78 for abbreviations. API = American Petroleum Institute.

(Bredasdorp Basin) and numerous hard, calcareous sandstones occur on the crest of the Superior Ridge. It is absent from the Infanta, St Francis and Recife arches, and was not recorded in the Gamtoos and Algoa basins by du Toit (1976).

4.3.2.1.6 *Palaeogeography.* Sedimentation of the Lower Uitenhage Group was initiated, and its lithofacies controlled by taphrogenesis along the line of the Cape Fold Belt, which underwent crustal collapse. Fracture zones, which were probably lines of overthrusting during the Cape folding, reversed their roles and acted as major south throwing boundary faults along the northern sides of developing graben and half graben.

Large-scale, localized, vertical movements in the southern Cape were probably initiated by Middle Jurassic times and were accompanied, locally, by volcanism. They resulted in the formation of elongate, asymmetric graben and half graben, and generated considerable local relief on the upthrow sides of the boundary faults, which continued to move intermittently until the end of Aptian times. However, it was only during Lower Uitenhage Group times (Middle Jurassic to Berriasian) that large elevation differences occurred across the boundary faults, and during this period rejuvenation of the local relief occurred on several occasicns. Response to these movements is reflected in readjustments of the sedimentary processes, which resulted in interfingering and lateral variations in the extent of the major lithofacies. These are set out in Table 21 and designated (i)-(iv). Proximal environments are represented by two lithofacies (high and low energy-related), and as the elevation discontinuities were degraded by weathering, lithofacies (iii) encroached both upvalley and laterally towards the sediment source, displacing

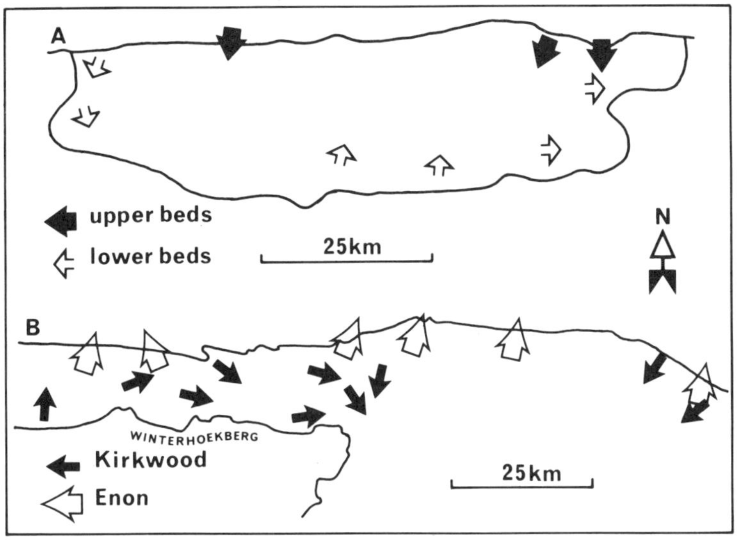


Figure 86. Palaeocurrent directions in Lower Uitenhage Group.
A. Oudtshoorn Basin Enon Formation. Lower beds refer to 'a, b, c', and upper beds refer to 'd' of Figure 79, respectively.
B. Inner Algoa Basin Enon and Kirkwood formations.
Data from Lock *et al.* (1975).

Table 21. Sedimentary environments of the main lithofacies of Lower Uitenhage Group.

| | Proximal | Distal |
|---|---|---|
| High energy | (i) rudaceous & arenaceous deposits which accumulated at the feet of the boundary fault scarps, and represent transverse sediment shedding across fault lines into valleys: piedmont & alluvial fans: Enon Formation | (ii) medium-fine fluviatile sands: Swartkops Formation |
| Low energy | (iii) arenaceous & argillaceous deposits accumulating in the valley floors by longitudinal fluviatile, and lacustrine processes: a) arenaceous braided stream facies b) argillaceous alluvial plain facies: Kirkwood Formation | (iv) estuarine/shallow marine clays and sands, best developed in the SE parts of the Outeniqua sub-basins but temporarily transgressing northwards into Algoa, northern Pletmos and western Bredasdorp basins: Colchester & Infanta formations |

lithofacies (i) as basin fill proceeded. Minor fault movements resulted in interfingering of facies (i) and (iii), as higher energy input temporarily increased. Shallow marine environments were established by Oxfordian times over the Falkland Plateau (Barker et al 1977a), and possibly about the same time encroached onto the outer Agulhas Bank area, with local transgressions into the lower reaches of the intermontaine basins, resulting in the interfingering of the fluviatile lithofacies (iii) with estuarine-shallow marine sediments of lithofacies (iv).

Because the tectonic environment was one of continued fault movement, individual lithofacies are likely to be diachronous and not physically continuous from basin to basin. In addition, rejuvenation of topography led, especially in the proximal areas, to successive re-establishment of lithofacies(i) at the expense of lithofacies (ii) and (iii). Good examples of this are the superposition of the upper coarse conglomeratic unit over the sands and mudstones in the Oudtshoorn Basin (Fig.79), and wedges of Enon conglomerate within the Kirkwood Formation in borehole MB-A/1. On the other hand, regional crustal subsidence probably superimposed a crude synchroneity on major periods of differential movement across boundary faults so that, for instance, the establishment of the Colchester or Infanta formations was probably accomplished within a period of less than a few (?1-2) million years along the whole of the northern parts of the Bredasdorp-Pletmos-Algoa basins. This will be in contrast to the Enon Formation whose intermittent accumulation may have spanned as much as 28 my.

Sedimentological studies on the Lower Uitenhage Group have been made by Lock et al (1975), Shone (1976) and Stewart (1973), and although these deal only with the two largest onshore basins, it is reasonable to suppose that their findings are of general application. In the Oudtshoorn Basin, Lock et al (1975) investigated the sedimentary processes and dispersion patterns of the proximal lithofacies (i) and (iii). Early sedimentation in the basin is represented by the lower conglomerates (Fig.79 beds a & b), and imbrication studies of the rounded pebbles indicate a transverse dispersion pattern towards the axial part of the basin (Fig.86A). The two resultant conglomeratic wedges graded laterally into the axial part of the basin, where lower energy, fluvial processes transported small pebbles,

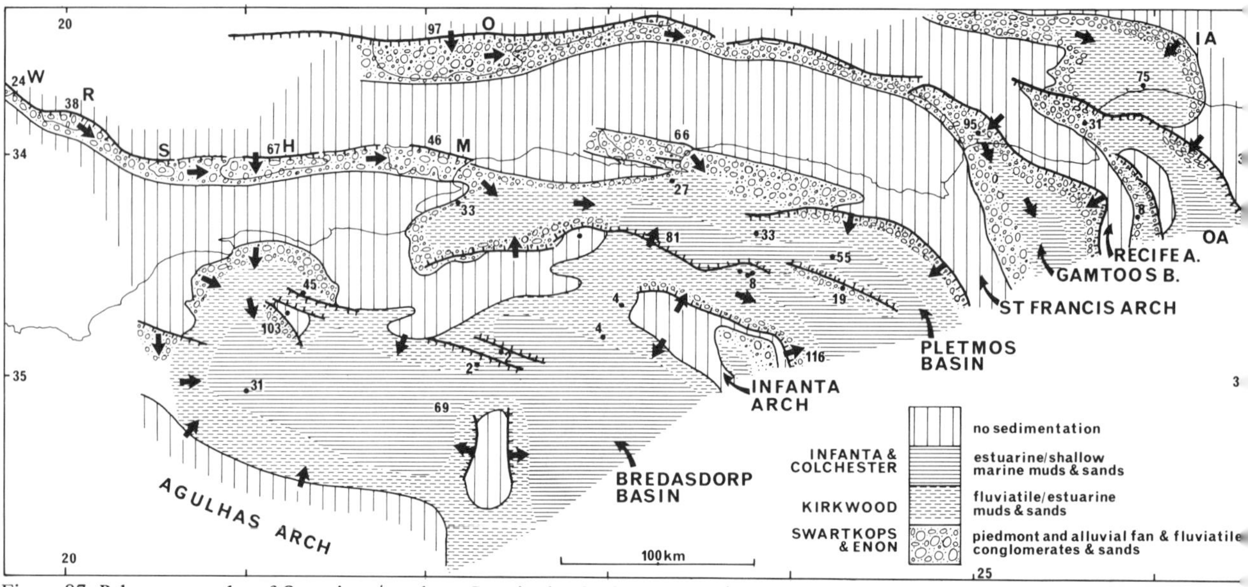


Figure 87. Palaeogeography of Outeniqua/southern Cape basins for late Jurassic/early Cretaceous time. Control points are boreholes (dots), figures are average sedimentation rates for the period 162 my (mid-Jurassic) to 130 my (Lower Valanginian). Sediment dispersion routes shown by arrows. Heavy barred lines are major synsedimentary faults. Based on Dingle (1973b) with numerous additional data, particularly from du Toit (1976). Abbreviations: W = Worcester; R = Robertson; S = Swellendam; H = Heidelberg; O = Oudtshoorn; M = Mossel Bay; IA = Inner Algoa; OA = Outer Algoa.

sands and silts in an easterly direction parallel to the depocentre axis. A rejuvenation of the boundary fault resulted in southward shedding of a further wedge of angular conglomerates and breccias (Fig.79, beds b,d). Imbrication is absent or poorly developed in the rudaceous components, but suggests a southward dispersion of a rapidly southward fining wedge (Fig.86) which Lock et al (1975) interpret as large coalescing fanglomerates (piedmond fans). These studies suggest that initially, the Oudtshoorn Basin may have been more symmetric than at present, and that the dominance of the northern boundary scarp as a sediment source only dates from the initiation of the second phase of conglomerate deposition. Continued movement along the Cango Fault is indicated by considerable shearing and crushing of the Enon Formation strata. Exposure of the Cango Group basement rocks to the north of the fault scarp must date from a later (?Upper Cretaceous or even Tertiary) period, because none of the Pre-Cambrian lithologies are found in the Lower Uitenhage Group sediments.

Similar studies in the northern Algoa Basin led Lock et al (1975) to conclude that during the early stages of basin infill, the northern boundary fault had not generated significant relief on its north side, and that sediment dispersion was by northward flowing, high energy, unbraided streams (Fig.86B). The sediment source must have lain in the Winterhoekberg and its easterly extension, the Addo Ridge, with the main depocentre in the southern part of the panhandle. A reversal of the dispersion pattern is evident from lithofacies (iii) (Kirkwood Formation), where south and SE current directions are indicated from the sandstone cross-bedding (Fig.86B). This lithofacies represents a lower energy, fluviatile environment within the intermontaine basin during a more mature stage of the infill, and was deposited in the axial parts of the valley, where sediment dispersion was by braided rivers, passing into lower-lying alluvial plains. The arenaceous facies (iiia) represents point bar and channel fill units, and the argillaceous facies (iiib) flood plain muds. Large accumulations of bone and wood fragments probably represent deposition in standing bodies of water such as small lakes or meander cut offs.

Stewart (1973), from his study of the argillaceous lithofacies (iiib) of the Kirkwood Formation came to similar conclusions, attributing the red colour of the muds and their frequent gypsum seams to desiccation in inter-bank areas. He also detected a change in provenance from the Enon to the Kirkwood formations. The heavy mineral suite of the former is richer in zircon than it is in garnet, whilst the reverse is true of the Kirkwood lithofacies (iiib). This is consistent with a Cape Supergroup source for the Enon Formation, and Dwyka, Ecca, and possibly lower Beaufort group sources for the argillaceous Kirkwood Formation where the abundance of clay minerals, and the dominance of illite and chlorite is also seen as supporting evidence (following observations by Hofmeyer 1971).

The typical red and green mudstones of the Kirkwood Formation represent a non-marine environment, but local interfingering of grey shale, sandstones and limestones of the proximal Colchester Formation (lithofacies (iv) of Tab.21) indicates that on at least one occasion a marine transgression penetrated well into the Algoa Basin. Here, oyster beds are prominent, suggesting a brackish-water (estuarine) environment, but the occurrence of echinoid fragments, and lamellibranchs such as *Trigonia, Tellina, Gervillia* indicate that normal marine conditions obtained temporarily. Evidence from the deep boreholes in the Algoa Basin (e.g. CO-1/67) also suggests that during the early part of this marine incursion, the sedimentary environment was locally anoxic resulting in the accumulation of dark, waxy carbonaceous shales.

Applying similar lithofacies models to the whole of the southern Cape area it is possible to present a schematic palaeogeographical picture for a hypothetical point in time which we will call late Jurassic/early Cretaceous. It must be remembered that the various lithofacies are locally diachronous, and in many areas vertically stacked, so that in Figure 87 it is not possible to represent the constantly changing character of the scene and there are insufficient data to accurately construct a palaeogeography for a specified point in time. Figure 87 presents, therefore, many valid generalizations.

One of the most important points seen in the reconstruction is the virtual isolation of the individual basin lineaments: only the Infanta Arch was covered with a relatively thin sediment veneer by the end of Lower Uitenhage Group times. This being so, the presence low down in the sequence of marine incursions that reached to the northern limits of all the major basins (except the Gamtoos), indicates that they must have all opened at their SE ends to the sea that lay over the Falkland Plateau in Middle to Upper Jurassic times. A further feature is the regional control of lithofacies by the large boundary faults. Across and against topographic discontinuities, piedmont and alluvial fans were generated by transverse dispersion (Enon Formation), whereas longitudinal movement down the valley axes produced the lower energy fluviatile Kirkwood sands and muds. Lock et al (1975) argue in favour of the original continuity of the Oudtshoorn-Gamtoos and Worcester-Mossel Bay basins, citing as evidence the fact that the cross folds separating them today are known in several cases to postdate the sedimentation. We also find this suggestion more appealing than trying to account for sedimentation in isolated basins with internal drainage.

The abundance of wood fragments in the Kirkwood Formation attests to well-vegetated valley floors, and the frequent preservation of delicate fern impressions indicates rapid sedimentation in bodies of quiet or standing water. Dark shales of the Colchester Formation (relatively organic rich, indicating somewhat restricted environments) represent the earliest major marine incursion into each of the basins, whilst the lighter coloured Infanta shales suggest more widespread, better oxygenated conditions. Although the latter facies penetrated into the Bredasdorp and Pletmos basins, it seems not to have reached as far north and east as the 'Colchester incursion'. Obviously, prior to the general establishment of marine conditions in the distal parts of the basins, the proximal lithofacies (i) and (iii), and higher energy distal lithofacies (ii) would have been more widespread.

Overall, the picture is of a rugged hinterland (the present-day southern Cape) with steep intermontaine basins (locally with high and fault-bounded hill scarps), adjacent to a highly indented coastline with elongate, rocky peninsulas and islands that stretched SE and south out to sea in a classic ria setting. The lower reaches of the mud-filled valleys with meandering streams supported a luxuriant vegetation of conifers, cycads and ferns, and passed into wide sediment-choked estuaries. Small islands, and shoals representing the tops of isolated horsts and tilted fault blocks remained local sediment sources and facies barriers for the whole of the period, which was probably about 32 my in duration. Highest sediment accumulation rates are generally close to the main boundary faults and range from 95 m/my (Gamtoos Basin) to 116 m/my in the southern Pletmos Basin. Lower rates, in the range 27-75 m/my are found in the more distal and median parts of the basins, whilst over the tilted horsts net sedimentation was nil or very low (4 m/my on the Infanta Arch).

The unconformably overlying Lower Sundays River Formation, which has been dated as Upper Valanginian, provides a minimum age for the top of the Lower Uitenhage Group. On seismic records this unconformity (or disconformity) has been identified as the regionally-developed reflector B (SOEKOR 1976, du Toit 1976).

#### 4.3.2.2 *Upper Uitenhage Group (Upper Valanginian to Lower Aptian, locally Cenomanian)*

This is a sequence of shallow marine sediments which was deposited in the last stage of taphrogenic basin infilling. At present, the distribution of these rocks is restricted to the Outeniqua and Algoa Basins, and there is no evidence to suggest that they were deposited in any of the other basins of the southern Cape (Fig.88). With the exception of small bald zones on the Recife and western Infanta arches, and the crest of the tilted fault block near the H(b)-Hartebeest/1 borehole (Fig. 75F), Upper Uitenhage Group rocks form a continuous cover from the southeastern end of the Agulhas Arch, to the central part of the Algoa Basin (Fig.88). They effectively filled in and smoothed out the complex basin geometry inherited from the earlier stage of basin infill (Lower Uitenhage Group), and at the close of Upper Uitenhage Group times (Aptian), few of the large boundary faults which had controlled depocentre development at the beginning of the period, remained active. Nevertheless, synsedimentary movement along the main boundary faults was the major cause of thick-

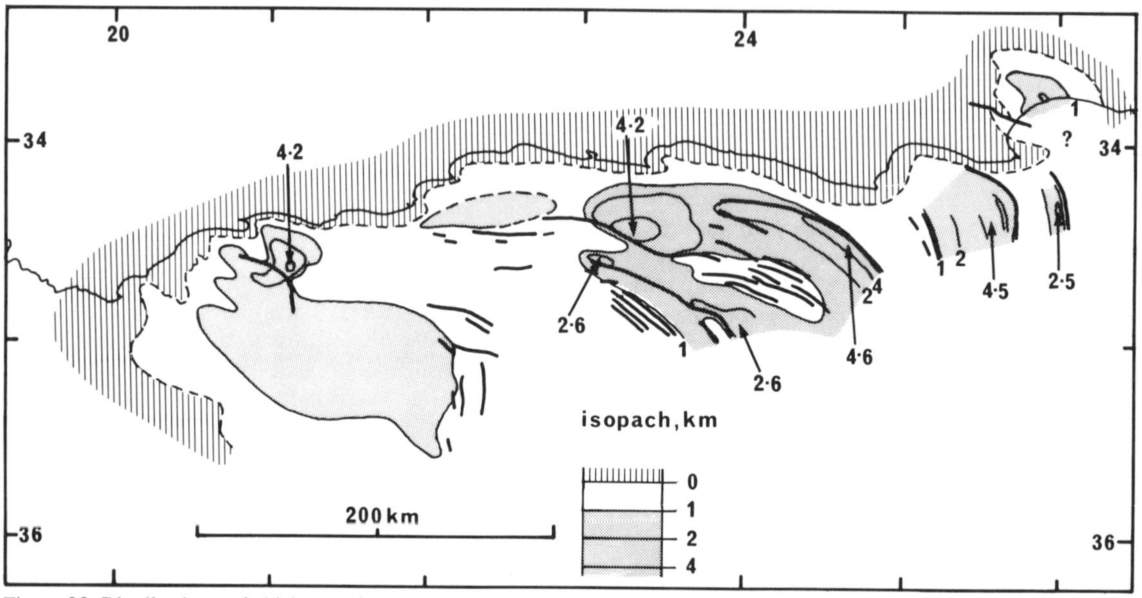


Figure 88. Distribution and thickness of Upper Uitenhage Group (Sundays River Formation) in the Outeniqua and southern Cape basins. Compiled from isochron data in SOEKOR (1976) using an average P-wave interval velocity of 3.196 km/sec.

ness variations during the deposition of the Upper Uitenhage Group. A reduction in the number of large fault lines that were active during the successive stages of basin infill represented by the Lower and Upper Uitenhage Group sediments can be gauged by comparing Figures 71 and 88.

Maximum sediment thicknesses occur in the Pletmos Basin, where over 4 km lie adjacent to the Superior (4 283 m) and Plettenberg (4 602 m) faults (Fig.88). These two asymmetric wedges (Fig.74C,D) are separated by a NNW-SSE faulted cross-high with less than 2 km cover. Similarly thick, but less extensive deposits occur in the Gamtoos Basin (4 522 m against the western flank of the Recife Arch), and in the NW extremity of the Bredasdorp Basin (4 187 m). The former probably expands and thickens in a southerly direction, but the pocket in the Bredasdorp Basin lies in a curious circular depression on the northern side of a small horst (close to D-B/1 on Fig.73A). In contrast to the thick, extensive, asymmetric wedges in the other basins, the main Bredasdorp Basin has a symmetrical, elongate fill somewhat under 2 km thick (Fig.73B), which shows signs of strong fault control only in the NW and SE, although several smaller faults occur on the SE edge of the Agulhas Arch, where the group progressively oversteps onto the basement (Fig.73B). In the Algoa Basin, the former geometry seems to have been retained during the deposition of the Upper Uitenhage Group (only the lower part occurs here), which reaches a thickness of about 2 km in the vicinity of the present coast near the mouth of the Sundays River.

Although mostly buried by the end of Aptian times, the original large basement highs clearly played an important role in isolating the major depocentres. Thicknesses of sediment over the highs vary: Agulhas Arch – nil; Infanta Arch – generally $<500$ m, with occasional very small patches bare, and local pockets $>2$ km; St Francis and Recife arches – $<500$ m; whilst the F-F/2 structure at the SE end of the Bredasdorp basin has several areas with $<500$ m. Obviously, not all these values faithfully reflect original sediment cover because there is locally a strong unconformity at the base of the overlying Alphard Formation (for example, on the crest of the Superior Ridge, and the eastern parts of the Infanta Arch), but the regional sediment distribution in Figure 88 is probably a fair reflection of relative thicknesses.

The Upper Uitenhage Group comprises the Sundays River Formation, which is a complex of upward coarsening sedimentary cycles subdivided by a regionally developed hiatus that shows as the prominent reflector C on seismic records (SOEKOR 1976) (Fig.72). The sequence was originally interpreted as two simple upward coarsening cycles (e.g. SOEKOR 1976, du Toit 1976), but a revision of the acoustic stratigraphy of the Bredasdorp Basin (McLachlan & McMillan 1979) has shown that the picture is more complicated, with two cycles above Horizon C in the Bredasdorp Basin, the upper one of which is apparently missing in the Pletmos Basin (Fig.92). We have designated these cycles SR1 to SR6, following as closely as possible du Toit's (1976) original scheme, but taking into account the revision in stratigraphic correlation. Sediments below the hiatus represented by Horizon C (SR1 and SR2) have been called the Lower Sundays River Formation, and sediments above Horizon C (SR3 to SR6)

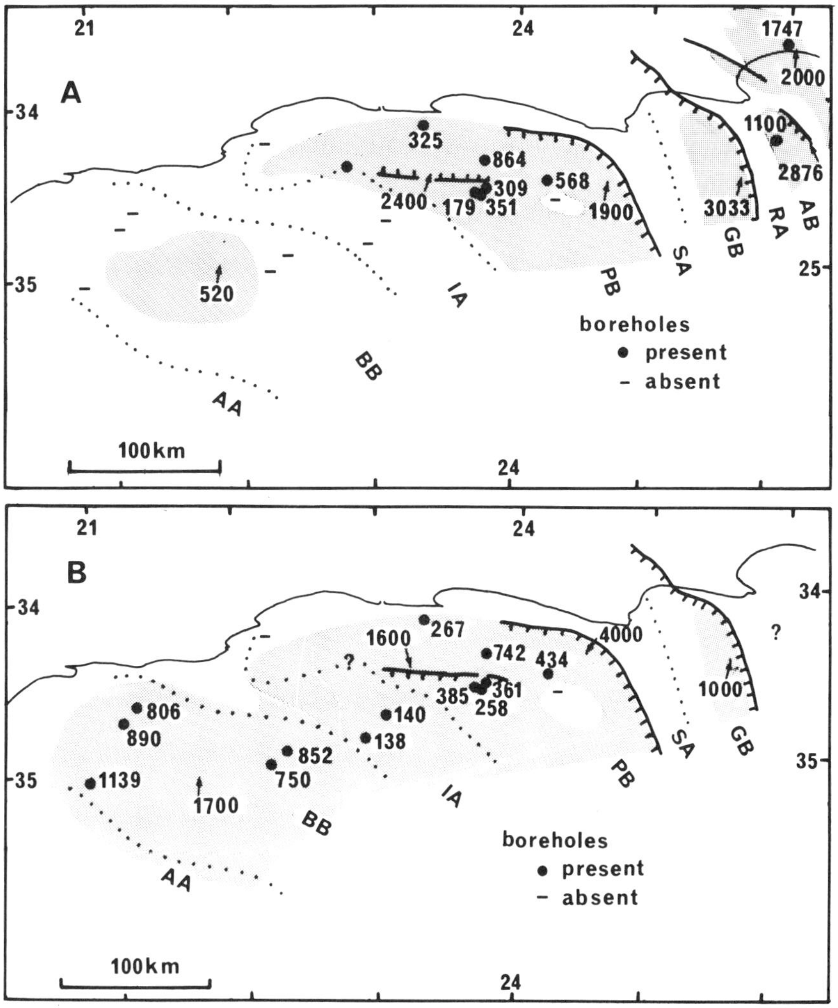


Figure 89. Distribution and thickness in the Outeniqua and southern Cape basins: A. Lower Sundays River Formation; B. Upper Sundays River Formation. Data from du Toit (1976), compiled using the seismic reflector assignments of McLachlan & McMillan (1979) (see Fig.92). See Figure 78 for abbreviations.

have been called the Upper Sundays River Formation, again following as closely as possible du Toit's (1976) nomenclature.

The Upper Sundays River Formation is the more extensively-developed of the two because it generally laps higher up the flanks of structural highs, where it oversteps onto Lower Uitenhage Group and basement. Both sequences locally thicken sharply towards major boundary faults, producing strongly asymmetric sediment wedges, except in the Bredasdorp Basin, where a regionally symmetric axial downwarping took place. The Upper Sundays River Formation is known only from the offshore areas, so that the well-known type sections of the Sundays River Formation in the onshore Algoa Basin are restricted to the lower part of the formation.

The base of the Upper Uitenhage Group is generally unconformable on the Lower Uitenhage Group.

4.3.2.2.1 *Lower Sundays River Formation (Upper Valanginian to ?Lower Hauterivian).* These beds are well-exposed at a number of localities in the Algoa Basin along the Sundays River (Fig.76), from where sections have been described by numerous authors, for example: Atherstone (1857), Stow (1871), Rogers (1906), Haughton (1928), Dingle (1969) and Shone (1976). The name Sundays River was first used, more or less in its modern sense, by Atherstone (1857), and type sections were formally proposed by Winter (1973) for which he selected the long cliff sequence in the Zoetgeneugd area (Fig.90) and a section in borehole CO-1/67 (Fig.91). Brenner & Oertli (1976) have zoned the Algoa Basin Sundays River Formation using ostracods from numerous SOEKOR boreholes and recognized a five-fold subdivision. The Zoetgeneugd Cliff probably occupies the lower two of these zones, and in Figure 90 we have included the Colchester cliff section, which probably contains the upper three. All five zones (A-E) are present in the CO-1/67 borehole (Fig.91).

Winter (1973), and McLachlan & McMillan (1976) have shown that the Lower Sundays River Formation rests unconformably upon the Kirkwood Formation,

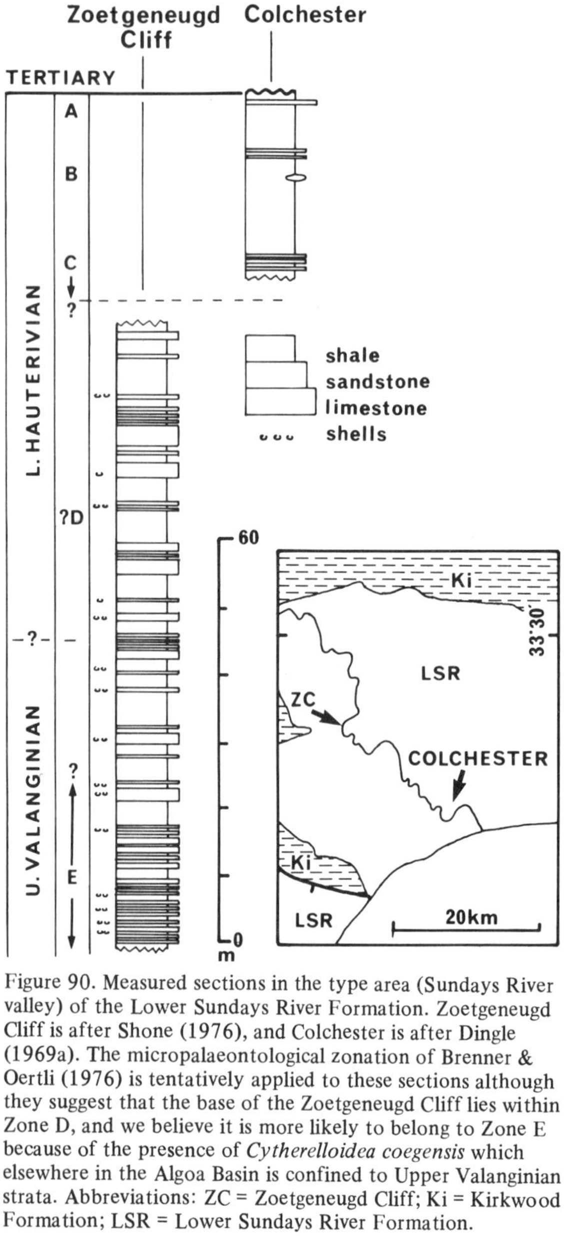


Figure 90. Measured sections in the type area (Sundays River valley) of the Lower Sundays River Formation. Zoetgeneugd Cliff is after Shone (1976), and Colchester is after Dingle (1969a). The micropalaeontological zonation of Brenner & Oertli (1976) is tentatively applied to these sections although they suggest that the base of the Zoetgeneugd Cliff lies within Zone D, and we believe it is more likely to belong to Zone E because of the presence of *Cytherelloidea coegensis* which elsewhere in the Algoa Basin is confined to Upper Valanginian strata. Abbreviations: ZC = Zoetgeneugd Cliff; Ki = Kirkwood Formation; LSR = Lower Sundays River Formation.

and is itself overlain unconformably by the Upper Sundays River Formation in the Outeniqua Basin. This upper contact does not occur in the present-day onshore Algoa Basin, where a maximum thickness for the sequence of just over 2 000 m occurs near the coast to the east of Colchester (Fig.93A). The main depocentre coincides with the Sundays River Trough, with the 1 km isopach extending inland towards the NW panhandle (Fig.75 & 93) and is separated from the subsidiary Coega Trough (?500 m) by the bare basement ridge of Coega Kop (Fig.75F).

Contact with the underlying red Kirkwood Formation shales is thought to lie about 4.8 km SE of Dun-

Table 22. Modal analyses of two Lower Sundays River Formation sandstones from CO 1/67 (after Stewart 1973).

| 408 m | | 835 m | |
|---|---|---|---|
| Quartz | 31.0 % | Quartz | 39.0 % |
| Feldspar | 7.6 % | Feldspar | 7.4 % |
| Rock fragments | 15.0 % | Rock fragments | 13.0 % |
| Matrix | 34.4 % | Matrix | 10.4 % |
| Opaques | 12.0 % | Opaques | 1.8 % |
| | | Calcite | 28.4 % |

brodie station (see McLachlan & McMillan 1976), and the lowest beds exposed consist of grey sandy shales and sandstones with marine fossils. Because of the intermittent nature of exposures in the Algoa Basin, it has not yet been possible to produce a detailed map of either litho- or biostratigraphical subdivisions, but application of Brenner & Oertli's (1976) ostracod zonation indicates a general younging towards the SE (Fig.90). Ammonite faunas indicate that the whole of the Lower Sundays River Formation in the Algoa Basin falls within the Upper Valanginian (Spath 1930, Cooper 1973), although no subdivision has yet proved possible.

In the Algoa Basin, the Lower Sundays River Formation consists of compact grey clays, silts, and sandy clays, with subordinate interbedded and nodular limestones and sandy limestones, and grey sandstones. Thin gypsum stringers occur locally. The argillaceous sequences are usually less than 60 m thick, but occasionally much thicker, particularly towards the basin centres. They are predominantly light to medium grey, laminated silty mudstones with frequent very thin lenses and laminae of fine to very fine sand. Lignite and comminuted plant material is locally abundant in the darker laminae, and there is a moderately abundant shelly marine fauna throughout. Occasional thin shaly, greenish grey, medium to fine sandstones also occur. The sequence is characterized by being finely laminated and displaying small-scale slumping, differential compaction structures, minor channeling, numerous minor erosion surfaces, and frequent bioturbation.

The arenaceous sequences are up to 75 m thick, but are generally much less (ca.10 m), and consist of numerous sandstone units (up to 2 m thick) separated by thin shale interbeds. Individual sandstones are typically greenish grey, fine to medium grained, quartzose sands with abundant rock fragments. They frequently have a high calcareous content in the matrix, which is normally composed of clay material probably derived from original detrital fragments. The sandstones are usually finely-laminated, with clay and carbonaceous particles, and display minor channelling and differential compaction structures. On a macro-scale they are planar and cross-bedded, with ripple laminations and occasional channel-fill sequences with basal lags that contain wood, shell and mud pellet clasts. Lignite and comminuted plant matter is common in the darker laminae. Table

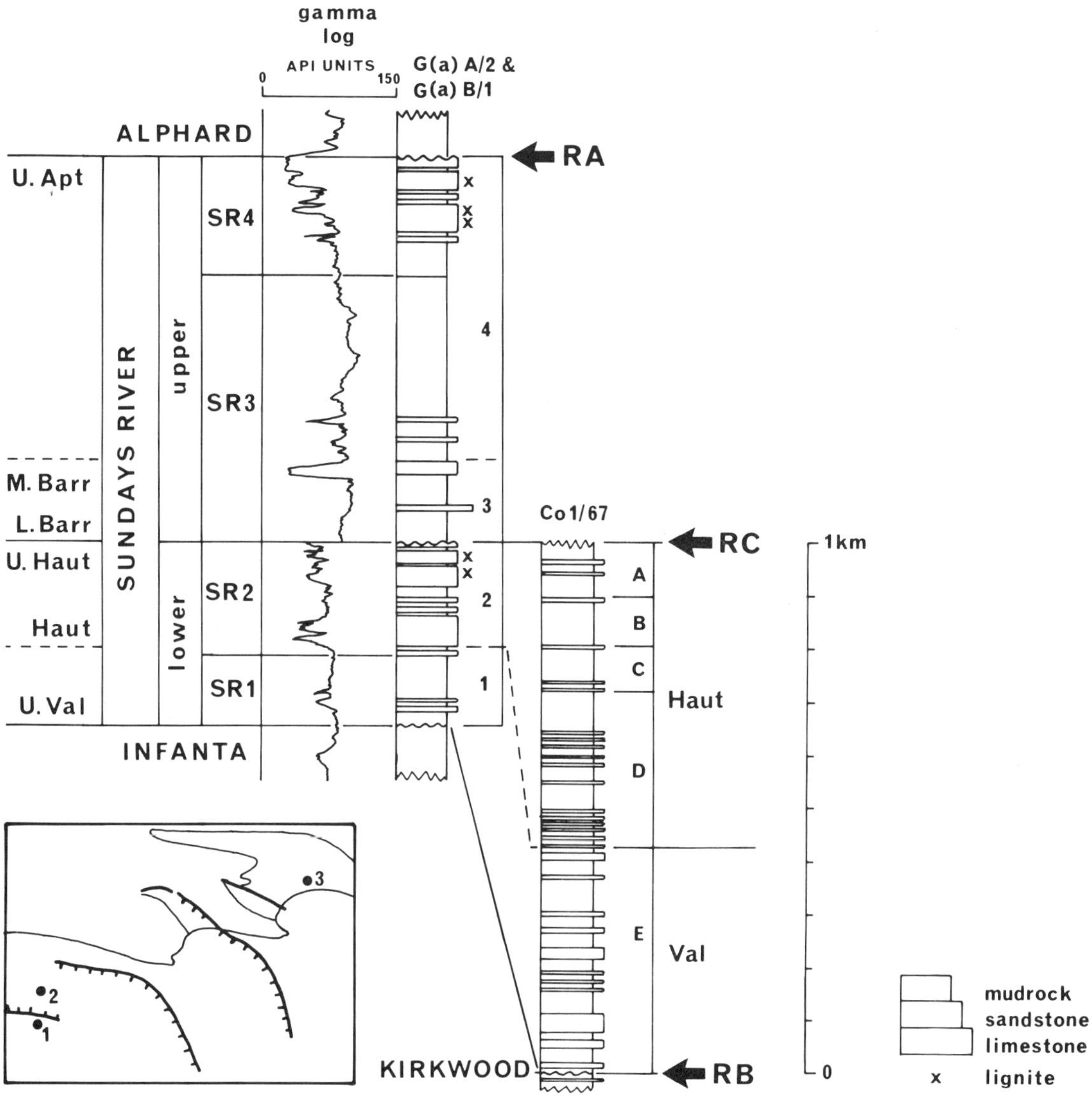


Figure 91. Borehole sections in the Sundays River Formation in the Outeniqua and Inner Algoa basins. Boreholes G(a)-A/2 and G(a)-B/1 (locations 1 & 2 on insert, respectively) comprise the composite sub-surface stratotype described by du Toit (1976), and CO1/67 (No.3 on insert) is the sub-surface stratotype described by Winter (1973). In the offshore sections, SR1 and SR2 are from G(a)-A/2, and SR3 and SR4 are from G(a)-B/1. Arrows show horizons correlated with seismic reflectors B, C and A of SOEKOR (1976), and du Toit (1976).

Palaeontological zonation in CO1/67 is based on the ostracod scheme (A-E) of Brenner & Oertli (1976). No zonation has been published for the Outeniqua Basin Sundays River Formation, but assemblages and associations 1-4 are from McLachlan & McMillan (1979) (see Table 25 for details).

22 gives modal analyses of two typical sandstones from CO-1/67.

Towards the main depocentres in the Algoa Basin there is a sharp decrease in the number of sandstone units, and in the sand/mud ratio, i.e. the sandstones are more numerous and thicker around the edges of the Sundays River and Coega troughs (Fig.93B). This contrasts with the sand distribution in the Kirkwood Formation, where there is an increase in the number of sandstone units >3 m thick towards the basin centres, emphasizing that the main dispersion media (i.e. the rivers) operated axially down the valleys during Kirkwood times, whereas sediment input was laterally down the basin margins into a standing body of water (shallow sea) in Lower Sundays River times.

Thin tuff bands have been reported from the Lower Sundays River Formation of the Algoa Basin by Winter (1973), and Rigassi & Dixon (1972).

On a regional scale, the Lower Sundays River Formation is known to occur extensively in the Pletmos Basin (from borehole evidence, du Toit 1976), and in the Bredasdorp and Gamtoos basins on seismic evidence

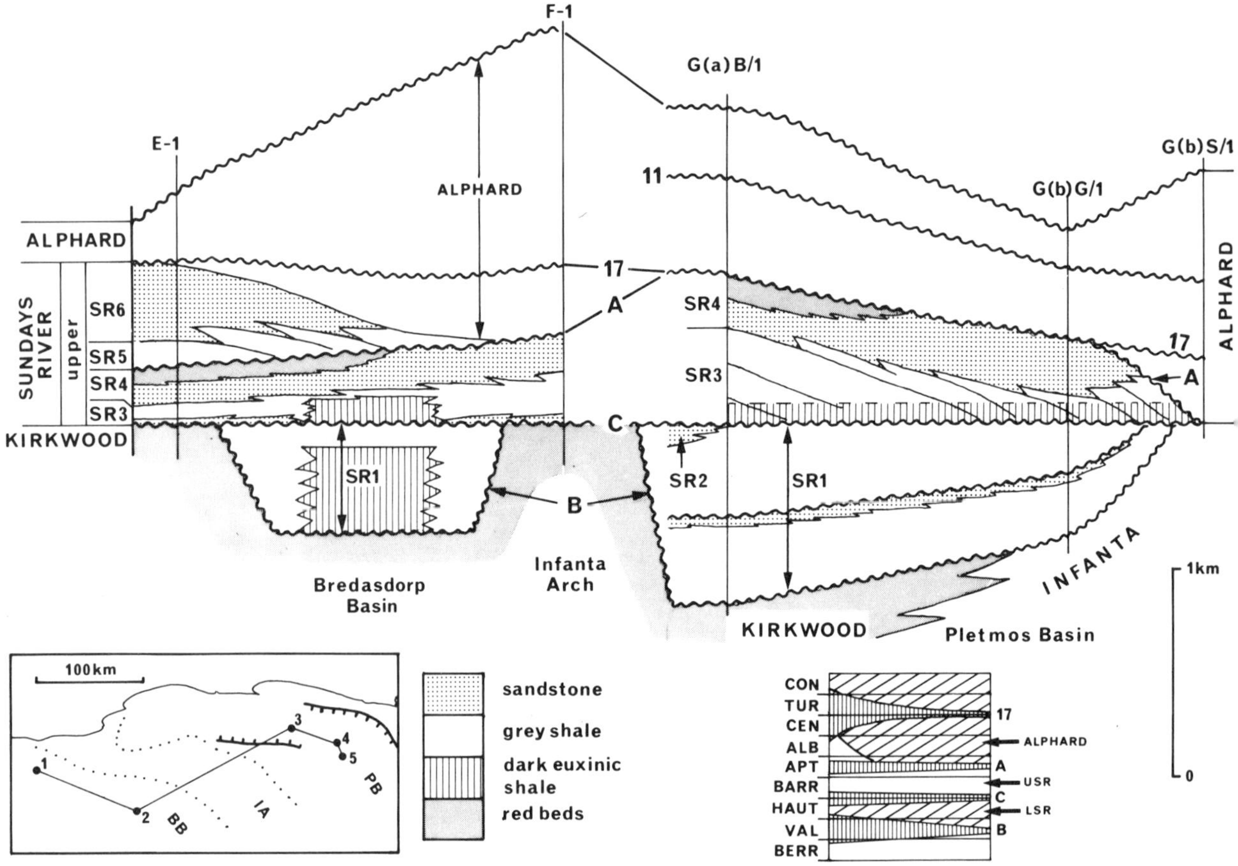


Figure 92. Correlation of Sundays River Formation between Bredasdorp and Pletmos basins. Borehole data are from du Toit (1976) and seismic reflector correlation follows McLachlan & McMillan (1979). The Upper/Lower classification follows du Toit (1976) (i.e. above C = Upper, below C = Lower), but SR5 and SR6 are additional notations added by us for completeness. Right side insert shows diachronous relationships of reflectors and lithofacies (hiatuses are vertical shading, and Alphard and Lower Sundays River formations have diagonal shading to facilitate identification). Left side insert shows localities: 1 = E/1; 2 = F/1; 3 = G(a)-B/1; 4 = G(b)-Gemsbok/1; 5 = G(b)-Springbok/1, where barred lines are the Superior and Plettenberg faults. Note variable horizontal scale of section.

(SOEKOR 1976) (Figs 89A & 92). Its distribution in the Outer Algoa Basin is not known, although it probably forms the upper part of the so-called Hartebeest Formation (du Toit 1976, see below). Du Toit (1976) selected borehole G(a)-A/2 in the Pletmos Basin as the subsurface reference section for the Lower Sundays River Formation in the Outeniqua Basin and on lithological and seismic evidence subdivided it into two units; a lower argillaceous member (SR1) and an upper arenaceous member (SR2) (Figs 91 & 92). The lower member (SR1) is a medium to dark, bluish grey shale with subordinate, thin fine-grained medium to light grey sandstones near the base. It is typically lignitic, and its upper part is relatively more fossiliferous. Above, and apparently prograding over the lower shales, the arenaceous (SR2) member consists of medium grey, medium to fine grained, well-sorted angular to subangular lignitic sandstones with subordinate lignitic light grey shale interbeds. On a regional scale, the arenaceous members are thickest in the proximal parts of the basins, whilst the argillaceous members are thicker in the more distal areas. Also, seismic evidence suggests that the lower member (SR1) contains a minor hiatus or unconformity above a median arenaceous unit (Fig.92). Borehole records cited by du Toit (1976), and values calculated from SOEKOR (1976) seismic data, indicate that within the Pletmos Basin, the Lower Sundays River Formation varies in thickness from less than 200 m over the Superior Ridge and nil in borehole G(b)-Springbok/1 to about 2.4 km adjacent to the Superior Fault and about 1.9 km against the Plettenberg Fault (Figs 74 & 89). Although some of these variations may be due to later differential erosion across the basement highs, it is mostly the result of contemporaneous differential subsidence across the major boundary faults i.e. continuing taphrogenic earth movements.

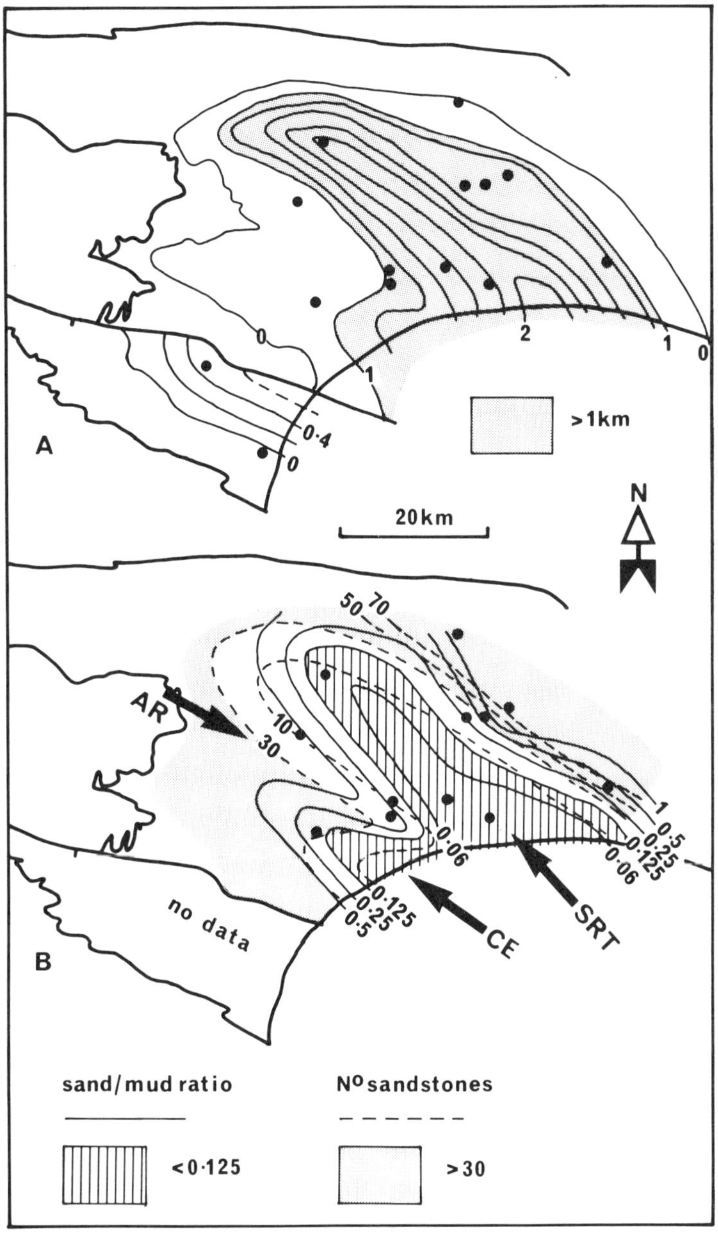


Figure 93. A. Thickness of Lower Sundays River Formation in Inner Algoa Basin.
B. Sandstone/mudrock variation in Lower Sundays River Formation. The sand/mud ratio is based on gamma log measurements, and the number of sandstones is for units >3 m thick. Abbreviations: SRT = Sundays River Trough; CE = Coega Embayment; AR = Addo Ridge. Data from Winter (1973).

Although du Toit (1976) recognized extensive deposits of the Lower Sundays River Formation in boreholes in the Bredasdorp Basin, recent revision of the seismic stratigraphic correlation (McLachlan & McMillan 1979) indicates that this is not the case, and the whole of the succession is represented by a grey argillaceous member (which we equate with SR1 in the Pletmos Basin) with no representative of the arenaceous SR2 member. These shales in the Bredasdorp Basin have a thick, dark euxinic section and seismic data suggest that they may reach 520 m in the centre of the basin (Figs 73A,B, 89A & 92).

The Lower Sundays River Formation has not been recorded from boreholes or on seismic records over the crests of the Infanta, St Francis or Recife arches, and its upper boundary is formed by a regionally developed hiatus (seismic reflector C) which cuts across the crests of the basement highs. McLachlan & McMillan (1979) estimate that this non-sequence, which is, locally at least, an erosion surface spans the Hauterivian/Barremian boundary with a slightly longer time gap in the west (Fig.92). Wherever du Toit (1976) quoted borehole data, the Lower Sundays River Formation rests on either the Kirkwood or Infanta formations, from which it is separated by seismic reflector B, which over basement highs merges with reflector C. The time gap represented by B spans the late Berriasian to early Hauterivian in the west, and mid Valanginian in the east (McLachlan & McMillan 1979). The implications of these data are that SR1 is of Hauterivian age in the Bredasdorp Basin, whilst SR1 and SR2 are late Valanginian to Hauterivian in the Pletmos and Algoa basins (Fig.92).

The Sundays River Formation in the Algoa Basin contains an abundant fauna, particularly of invertebrates, which has been the subject of numerous publications (e.g. Hausmann 1837, Tate 1867, Newton 1896, Kitchin 1908 and Spath 1930). Unfortunately, few modern revisions have been carried out so that the faunal lists presented by Engelbrecht et al (1962) are based mainly on the early work, and it is difficult to assess the age and ecological relevance of many of the taxa in terms of current nomenclature. McLachlan & McMillan (1976) have provided a good summary of current knowledge.

Vertebrates are rare and there has been no general discussion or systematic revision of the material available: dinosaur vertebrae (Engelbrecht et al 1962); a lacertilian reptile (Tate 1867); and a marine reptile *Plesiosaurus capensis* Andrew (1910). There is, on the other hand, a large and varied invertebrate fauna, which is dominated by the molluscs: nine species of gastropods with the most diverse genera *Pleurotomaria* (2 spp.), *Natica* (2 spp.) and *Turbo* (2 spp.); and 62 species of lamellibranchs with the most diverse taxa trigoniidae (9 spp.), *Astarte* (4 spp.), and *Pecten* s.l. (3 spp.). Oysters have been recorded, but no freshwater molluscs are known from the formation as understood by McLachlan & McMillan (1976). The cephalopods, whilst not being particularly common, have proved the most useful group within the invertebrates for age determinations, although McLachlan & McMillan (1976, p.205) point out that ammonites have not been recorded from the lowest beds in the Sundays River Formation. The most diverse genus is *Olcostephanus* (14 spp.), in addition to which Cooper (1981) records seven other genera (10 species) (Tab.23). A coral, worm tracks, an echinoid and a crustacea have also been recorded (Engelbrecht et al 1962). Plant remains are very rare and limited to wood fragments and pieces of amber.

The most intensive modern studies on the fauna have been on microfossils, and although the results of the work on foraminifera still await completion that on ostracods has been published (Dingle 1969, Brenner & Oertli 1976). Brenner & Oertli's (1976) study resulted in the establishment of a zonal scheme (assemblages A-E), and the recognition of an important faunal dichotomy within the sequence. Typical taxa of these two assemblages are listed in Table 25, whilst overall the ostracod fauna comprises 35 species and 20 genera (Tab.24). The Cytherellidae constitute the most important group both numerically (37 % fauna) and in diversity (8 spp.), but it is the presence and relative importance of *Majungaella, Rostrocytheridea* and *Sondagella* that is particularly significant (Dingle 1982a). These genera are endemic to south Gondwanaland, and in the Algoa Basin they constitute 56 % by number of individuals, and 14 % by number of species. As we will discuss later, these taxa form important links in establishing the position of the Algoa Basin in early Cretaceous palaeogeographies of south Gondwanaland.

Table 23. Cephalopoda from the Lower Sundays River Formation (after Engelbrecht *et al.* 1962, Cooper 1973, following the revision of Cooper 1981).

| |
|---|
| Ammonoidea |
| *Partschiceras rogersi* (Kitchin) |
| *Bochianites glaber* Kitchin |
| *B.africanus* (Tate) |
| *Olcostephanus (Olcostephanus) atherstoni* (Sharpe) |
| *O. (O.) densicostatus* (Wegner) |
| *O. (O.) rogersi* (Kitchin) |
| *O. (O.) victoris* Spath |
| *O. (O.) fascigerus* Spath |
| *O. (O.)* aff *durangensis* (Cantu Chapa) |
| *O. (O.) baini baini* (Sharpe) |
| *O. (O.) baini* var. *sphaeroidalis* (Spath) |
| *O. (O.) ventricosus* (Von Koenen) |
| *O. (O.) uitenhagensis* (Kitchin) |
| *O. (O.) riccardii* Cooper |
| *O. (O.) astieriformis* (Bose) |
| *O. (O.) coahuilensis* Imlay |
| *O. (O.)* cf. *perinflatus* (Mathcron) |
| *Neohoploceras subanceps* (Tate) |
| *Distoloceras spinosissimum* (Hausmann) |
| *D.* cf. *irregulare* Imlay |
| *Eodesmoceras haughtoni* Spath |
| Coleoidea |
| *Belemnopsis africana* (Tate) |
| *B.gladiator* Willey |
| Nautiloidea |
| *Eutrephoceras uitenhagense* |

An assessment of the age of the Sundays River Formation in the Algoa Basin has been the subject of discussion for nearly 150 years, since Hausmann (1837) proposed a Lower Cretaceous age. Although several workers have proposed a Jurassic age (e.g. Bain 1856, and Tate 1867) there has been a general concensus that Hausmann's diagnosis was correct. This date was refined by Kitchin (1908), who suggested a Hauterivian age, followed by Spath (1930), who concluded that the whole formation was Upper Valanginian. Cooper (1973, 1974, 1981) has corroborated Spath's age determination and prefers an uppermost Valanginian age. Micropalaeontological work has indicated a compromise between these later mollusc-based studies with McLachlan & McMillan (1979) identifying the planktonic foraminifera *Hedbergella ?hoterivica* from the Lower Sundays River Formation in the Bredasdorp Basin. This is a Hauterivian zonal species, which, taken together with the ammonite data suggest an Upper Valanginian-Hauterivian age for the Lower Sundays River Formation in the Outeniqua Basin. This record does not of course prove that Hauterivian strata occur in the Algoa Basin, although Brenner & Oertli (1976) assume (for no sound palaeontological reasons) that their upper ostracod assemblage (i.e. zones A-D in Figs 90A, 91) is Hauterivian in age, with the lower assemblage (E) Upper Valanginian. Comparing Figure 90 with the fossil locality map of McLachlan & McMillan (1976, fig. 8), this interpretation certainly is at variance with the ammonite evidence of Spath (1930) and Cooper (1973), because ammonite-localities definitely occur within rocks of Brenner & Oertli's (1976) zones A-D, and Cooper (1974) specifically rejected a Hauterivian age for the youngest beds because none of the typically fine-ribbed Lower Hauterivian species of *Olcostephanus* have been recorded. Evidence from the Algoa Basin itself is, therefore, slightly equivocal, although taken in a regional context there can be no doubt that the Lower Sundays River Formation ranges in age from Upper Valanginian to Hauterivian.

4.3.2.2.2 *Upper Sundays River Formation (Barremian to Cenomanian).* The Upper Sundays River Formation does not occur onshore in the Algoa Basin, and is known only from boreholes on the Agulhas Bank (du Toit 1976) (Fig.89B). It lies above a regionally developed unconformity (seismic reflector C) and locally below reflectors A, 17, or a conformable boundary into the Alphard Formation (McLachlan & McMillan 1979) (Fig. 92). The stratotype is in borehole G(a)-B/1 (du Toit 1976), where the 742 m section can be divided into lower argillaceous (SR3) and upper arenaceous (SR4) members (Figs 91 & 92). The lower member (SR3) consists of medium to dark grey lignitic shales and clays, with subordinate inter-bedded sandstones, and thin hard dolomitic layers in the lower part. This member grades upwards via thick (121 m) siltstones into medium to coarse, porous, well-sorted, angular to rounded, medium to light grey sandstones, interbedded with thin clays and shales (SR4). The argillaceous component in SR4 decreases in importance up the succession, whilst lignite is common throughout, and some of the shales are radio-active. Lignite is a common component throughout the basin, and McLachlan & McMillan (1979) mention extensive bands over 4 m thick in both the Pletmos and Bredasdorp basins.

Seismic studies cited in du Toit (1976) suggest that this upward coarsening sequence represents a basinward prograding deltaic succession, and additional information given by McLachlan & McMillan (1979) indicates that the lowermost shales in the SR3 member (i.e. prodelta clays) are a dark, organic-rich euxinic facies, whilst the top part of the SR4 member consists of a thin red bed lithofacies.

The simple succession described above is found in the Pletmos Basin, where thicknesses of the Upper Sundays River Formation vary between about 250 and 4 000 m (Figs 74 & 89). Rapid thickening towards the major boundary faults indicates substantial differential movement during accumulation of these rocks, but it is significant that displacement of the upper boundary of the sequence across these faults is small, and in some cases negligible, i.e. movement had all but ceased by the time sedimentation of the Upper Sundays River Formation in the Pletmos Basin was completed.

About 140 m of Upper Sundays River Formation occurs on the crest of the Infanta Arch (Fig.89B), and the succession is entirely sandy (boreholes F-1 & F-E/1, du Toit 1976). It expands into the Bredasdorp Basin, where between 750 and 1 139 m have been reported in boreholes by du Toit (1976), and a maximum thickness of about 1 700 m is anticipated from seismic evidence (SOEKOR 1976) (Fig.73). The succession here is more complex than in the Pletmos Basin and has resulted in some uncertainty in correlation between the two areas (see McLachlan & McMillan 1979 for a statement of the problem). The most recent seismic correlations suggest a development as shown in Figure 92, with two upward coarsening sequences SR3-SR4 and SR5-SR6. Published borehole data are available for only the proximal area in the NE of the basin (D-A/1, E-B/1 and E-1) and du Toit (1976) emphasizes the sandy, lignitic nature of the succession, which in E-B/1 rests directly on basement (?Bokkeveld Group) on a horst crest. Further lithological information in McLachlan & McMillan (1976) indicates an anoxic shale sequence in the basin centre, and red bed development at the top of the SR4 member. This strengthens our correlation of the two members across the Infanta Arch (Fig.92). Seismic data cited in du Toit (1976) suggests that both these upward coarsening sequences represent basinward prograding deltas, although the intervention of an erosion surface (seismic reflector A) has locally removed the top of SR4. McLachlan & McMillan's (1979) latest seismic correlations indicate therefore, that the erosion surface which forms the top of the Lower Sundays River Formation in the Pletmos Basin (reflector A) is locally overlain by members SR5 and SR6 in the Bredasdorp Basin. An assessment of the time intervals represented by the various non-sequences in Figure 92 (shown in the insert) indicates that whilst the bulk of the Upper Sundays River Formation is Barremian to Lower Aptian in age, some sections in the west of the Bredasdorp Basin are locally as young as Upper Albian or even Cenomanian.

No detailed lists have been published of the fauna or flora of the Upper Sundays River Formation, although large amounts of micropalaeontological work have been undertaken by SOEKOR and its consultants. Table 25 shows the marker microfossils (mostly benthic foraminifera) that provide SOEKOR with its zonation scheme for the Sundays River Formation (upper and lower divisions) in the Pletmos Basin (McLachlan & McMillan 1979), and as already mentioned, *Hedbergella ? hoterivica* occurs as an addition in assemblage 2 in the Bredasdorp Basin. This scheme, and its extension in the

Table 24. Upper Jurassic – Cenomanian ostracods in South Africa (after Dingle 1982a). Data from numerous sources mentioned in the text.
Col. = Colchester Fm; Kirk. = Kirkwood Fm; LSR = Lower Sundays River Fm; USR = Upper Sundays River Fm; ?P-B = ?Portlandian-Berriasian; ?B-V = ?Berriasian-Valanginian.● = presence; O = cf; Neoc. = Neocomian; Port. = Portlandian; Val. = Valanginian; Haut. = Hauterivian; Barr. = Barremian; Apt. = Aptian; Alb. = Albian; Cen. = Cenomanian.

| | Algoa fauna | | Mngazana | Brenton | Outeniqua Basin Offshore, PB-A/1 | | | |
|---|---|---|---|---|---|---|---|---|
| | | | | | Col. | Kirk. | LSR. | USR. |
| | No. | % | ?Neoc. | ?Port. | ?P-B | ?B-V | U.Val. | Haut. |
| *Polycope* spp. | 2 | <1 | | | | | | |
| *Cytherella inaequivalva* | 40 | <1 | ● | ● | ● | ● | ● | ● |
| *C.knysnaensis* | 500 | 3 | | ● | | ● | ● | ● |
| *C.hausmanni* | 2500 | 16 | | | | ● | ● | ● |
| *C.algoaensis* | 2000 | 13 | ● | | ● | ● | ● | ● |
| *C.schwarzi* | 180 | 1 | ● | ● | ● | ● | ● | ● |
| *C.* sp. C | | | ● | | | | | |
| *Cytherelloidea zoetgeneugdensis* | 600 | 4 | | ● | | ● | ● | ● |
| *C.coegensis* | 15 | <1 | | | | | ● | |
| *C.agyroides* | 17 | <1 | | | | | ● | |
| *C.brentonensis* | | | | ● | | ● | ● | |
| *C.agulhasensis* | | | | | | | | |
| *C.* spp. | | | | | | | | |
| *Bythocypris* sp. (≡ *Bairdiacypris* sp.) | | | ● | | | | | |
| *B.stroggylae* | 1200 | 8 | | | | | ● | ● |
| *B.nodosa* | 700 | 4 | | | | | | |
| *Pontocyprella* sp. | 1 | <1 | | | | | | |
| *Argilloecia?* sp. | ?10 | <1 | ● | | | | | ● |
| *Saxellacythere* sp. | ?10 | <1 | | | | ● | ● | ● |
| *Acrocythere costata* | 200 | 1 | | | | | ● | ● |
| *A.alexandria* | 10 | <1 | | | | | | |
| *A.* sp. | | | ● | | | | | |
| *Sondagella theloides* | 600 | 4 | | ● | | | ● | ● |
| *Oligocythereis? dubia* | 14 | <1 | | ● | ● | ● | ● | ● |
| *Mandelstamia* cf. 962 Grekoff. | 18 | <1 | | | | | | ● |
| *Rostrocytheridea chapmani* | 2500 | 16 | | ● | ● | ● | ● | ● |
| *R.ornata* | 32 | <1 | | | | | | |
| *R.* sp. | | | | ● | | | | |
| *Majungaella nematis* | 1100 | 7 | | ● | | | ● | ● |
| *M.bifurcata* | 3000 | 19 | | | ● | ● | ● | |
| *M.hemigymnae* | 28 | <1 | | | | ● | ● | |
| *Progonocythere plettenbergensis* | | | | | | | | ● |
| *P.reticulata* | | | | ● | | ● | ● | |
| *P/Amicytheridea brentonensis* | | | | ● | ● | ● | ● | |
| *P?* sp. A. | | | | | ● | | | |
| *Arculicythere tumida* | | | | | | | | |
| *Fastigatocythere? grekoffi* | 45 | <1 | | | | | | |
| *Annosacythere?* sp. A | 1 | <1 | | | | | | ● |
| *Procytherura maculata* | 150 | 1 | ○ | | | | ● | ● |
| *P.beerae* | 35 | <1 | | | | | | |
| *P.dinglei* | 115 | 1 | | | | | | |
| *Metacytheropteron* sp. A | 1 | <1 | | | | | | |
| *M.* sp. | | | ● | | | | | |
| *Cytheropteron (Infracytheropteron) persica* | 100 | 1 | | | | | ● | ● |
| *Cytheropteron?* sp. | | | | | | | | |
| *Macrocypris?* sp. | | | ● | | | | | |
| *Paracypris* sp. | | | | | | | | |
| *Paranotacythere?* sp. | | | ● | | | | | |
| *Cythereis agulhasensis* | | | | | | | | |
| *'Cythereis'* spp. | | | | | | | | |
| *Isocythereis sealensis* | | | | | | | | |
| *Pirileberis* spp. | | | | | | | | |
| *Eucytherura (Vesticytherura) tuberculata* | 7 | <1 | aff | | | | | |
| Genus A | ?30 | <1 | | ● | ● | ● | ● | |
| *Cypridea* sp. | | | | | ● | ● | | |
| *Ilyocypris* sp. | | | | | | ● | | |

| Algoa Basin LSR. U.Val. | L.Haut. | Offshore outcrops U.Apt.-Alb. | Zululand Apt. | Alb. | Cen. |
|---|---|---|---|---|---|
| ● | ● | | | | |
| ● | ● | | | | |
| ● | ● | | | | |
| ● | ● | | | | |
| ● | | | | | |
| ● | | | | | |
| ● | ● | | | | |
| ● | | | | | |
| ● | | | | | |
| | | ● | | | |
| | | | ● | ● | ● |
| | ● | | | | |
| | ● | | | | |
| ● | | | | | |
| | ● | | | | |
| ● | ● | | | | |
| | ● | | | | |
| | ● | | | | |
| ● | ● | ● | | ● | |
| | ● | | | | |
| | ● | | | | |
| ● | ● | | | | |
| ● | ● | | | | |
| | ● | ○ | ● | ● | ● |
| ● | ● | | | | |
| | ● | | | | |
| | | ● | | | |
| ● | | | | | |
| ● | | | | | |
| | ● | | | | |
| | ● | | | | |
| | ● | | | | |
| | ● | | | | |
| ● | ● | | | | |
| | | ● | | | |
| | | | ● | | |
| | | ● | | | |
| | | | ● | ● | ● |
| | | ● | | | |
| | | | ● | ● | |
| | ● | | | | |
| | ● | | | | |

Bredasdorp Basin, have been used by McLachlan & McMillan (1979) to date the hiatuses represented by reflectors B, C, A and 17 in Figure 92.

4.3.2.2.3 *H(b)-Hartebeest/1 borehole.* Du Toit (1976) was unable to correlate much of the succession in the H(b)-Hartebeest/1 borehole with sequences in the Algoa and Pletmos basins. The borehole lies on the eastern flank of the Recife Arch, in a steep half graben developed to the south of the Coega boundary fault (Fig. 75E). The basal sediments, which he correlated with the Swartkops Formation, rest on metamorphosed slates, but the overlying succession, below the base of the Alphard Formation, was said to have affinities with both the Sundays River, Infanta and Colchester formations, all of which are characterized by the abundance of grey argillaceous rocks. To avoid confusion, du Toit (1976) placed these rocks in a temporary category (Hartebeest Formation), but in subsequent publications SOEKOR geologists (1976) assign them to the Sundays River Formation, and imply that they belong to the Lower division of this formation. No palaeontological evidence has been published to substantiate this claim and the seismic data are equivocal.

The lower part of the sequence consists predominantly of well-bedded grey shales, whilst the upper part is coarser with more abundant sandstones. Lignite occurs throughout, and some of the sandstones have thin limestones above them. The shales are said to contain macrofossils, including marine forms (belemnites), but du Toit (1976) gave no details of taxa or their vertical distribution.

Du Toit (1976) considers that the sequence can be divided into 'a number' of cyclothems representing alternating deep and shallow water environments, and each cyclothem displays the following sequence (i) base to (iv) top:

iv) thin, unfossiliferous microcrystalline limestone (?fresh water);

iii) sandstones with brownish grey shale and lignite. Upper boundary is sharp;

ii) lignitic grey shale which grades into (iii);

i) glauconitic, fossiliferous grey shale grading into (ii).

From this description it seems as if this is an upward coarsening sequence starting from medium water depth marine shales and ending in shallow marine or nonmarine rocks. It could correlate with any of the deltaic cycles in the Sundays River Formation, but because onshore evidence suggests that large-scale movement along the Coega Fault ceased at the end of Lower Sundays River Formation times the correlation shown in Figure 75 seems reasonable.

4.3.2.2.4 *Palaeogeography.* During deposition of the Upper Uitenhage Group (Upper Valanginian to Aptian, locally Cenomanian), the taphrogenic basins of the

Table 25. Marker microfossils from the Sundays River Formation in the Pletmos Basin (after McLachlan & McMillan 1979, fig.3).

| Assemblage[1] | Member[2] | | |
|---|---|---|---|
| 4 | SR4 | *Ammobaculites aequale* group<br>*Lenticulina nodosa*<br>*Uvigerinammina manitobensis* | Aptian |
| 3 | SR3 | *Verneuilina howchini*<br>*Trochamminoides coronus* | Barremian |
| 2 | SR2 | *Epistomina caracolla*<br>*Astacolus gibber* | Hauterivian |
| 1 | SR1 | *Cytherella algoaensis* | Valanginian |

1 – Assemblages 1-4 as used on Figure 91
2 – Members SR1-SR4 as used on Figures 91 and 92

southern Cape silted up and the highly irregular block-faulted topography initiated during Lower Uitenhage times was effectively smoothed out. This was brought about by a general cessation of large-scale fault movement and differential subsidence between the original horsts and graben. Although a few of the major boundary faults remained active until the close of (and locally, beyond) the Upper Uitenhage sedimentary episode, their importance as sediment thickness and facies controls had become minimal by this time.

At the close of the Lower Uitenhage Group sedimentary episode (Berriasian) a withdrawal of the shallow Infanta Formation sea, accompanied by earth movements and erosion, gave rise to a regional hiatus that is represented by seismic reflector B. A further marine transgression occurred in the Upper Valanginian and led to the establishment of widespread marine conditions and the start of the Sundays River Formation cycles of sedimentation. The basal (argillaceous) member of the Lower Sundays River Formation (particularly the calcareous sandstones and sandy limestones) has a well-preserved normal marine fauna (including ammonites), but the fauna has a low diversity, and with the admixture of abundant comminuted plant material indicates a shallow marine environment of deposition in the proximity of large river systems. Although normal marine conditions prevailed, direct access to the open ocean was probably limited, as witnessed by the lack of planktonic foraminifera, notwithstanding the fact that Lower Cretaceous assemblages are not as ubiquitous as their Upper Cretaceous counterparts. Planktonic foraminifera first appear in numbers in the Upper Sundays River Formation (Aptian). Earlier accounts of the Lower Sundays River Formation have (in the Algoa Basin), included hyposaline taxa in the faunal lists (e.g. Rogers 1906, Haughton 1928), but SOEKOR geologists have shown from the field relationships in the Dunbrodie area; that the beds from which these assemblages

Table 26. Foraminifera from the Mngazana Beds (after Mclachlan *et al.* 1976b).

| | |
|---|---|
| *'Allomorphina'* | *L.* sp. |
| *Ammobaculites* sp. | *Marginulinopsis gracilissima* |
| *Astacolus calliopsis* | *Neoflabellina* sp. |
| *A.gibber* | *Nodosaria sceptrum* |
| *A.mutilatus* | *Nubecularia lucifuga* |
| *A.perobliqua* | *Patellina subcretacea* |
| *A.* spp. | *Planularia* cf. *madagascariensis* |
| *Citharina sparsicostata* | *Planularia* spp. |
| *Dentalina communis* gr. | *Pseudoglandulina humilis* gr. |
| *D. linearis* | *Pseudopolymorphina* sp. |
| *D.* spp. | *Pyrulina cylindroides* |
| *Dorothia kummi* | *Quinqueloculina minima* |
| *Eoguttulina anglica* | *Reinholdella hofkeri* |
| *E.* cf. *inovroclaviensis* | *R.valendisensis* |
| *E.* cf. *liassica* | *R.* sp. |
| *Epistomina caracolla* | *Saracenaria bronni* |
| *Globulina prisca* | *S.* cf. *compacta* |
| *Lagena apiculata neocomiana* | *Spirillina minima* |
| *L.hauteriviana hauteriviana* | *S.tenuissima* |
| *L.* cf. *sulcata* | *Tristix acutangulus* |
| *Lenticulina bifurcilla* | *T.insignis* |
| *L.cultrata* gr. | *Trocholina infragranulata* |
| *L.nodosa* | *Vaginulina* spp. |
| *L.subtilis* | *Webbinella subhemisphaerica* |
| *L.* spp. | |
| *Lingulina bettenstaedti* | |

Table 27. Typical ostracods from the Lower Sundays River Formation, Algoa Basin (after Brenner & Oertli 1976).

A Upper assemblage (A-D) – Hauterivian
*Majungaella hemigymnae* Brenner & Oertli
*M.nematis* Grekoff
*?Bythocypris nodosa* Brenner & Oertli
*?B.stroggylae* Brenner & Oertli
*Acrocythere costata* Brenner & Oertli
*A.alexandria* Brenner & Oertli
*Procytherura maculata* Brenner & Oertli
*P.beerae* Brenner & Oertli
*P.dinglei* Brenner & Oertli
*Eucytherura (V.) tuberculata* Brenner & Oertli
*Mandelstamia* cf. 962 Grekoff
Genus A Brenner & Oertli

B Lower assemblage (E) – Upper Valanginian
*Cytherella algoaensis* Brenner & Oertli
*C.schwarzi* Dingle
*Cytherelloidea agyroides* Dingle
*C.coegensis* Dingle
*?Fastigatocythere grekoffi* Brenner & Oertli
*?Annosacythere* sp.A Brenner & Oertli
*Cytheropteron (I.) persica* Brenner & Oertli

were obtained are referable to the Colchester Formation, and that none of the estuarine assemblages can be attributed to the Sundays River Formation (McLachlan & McMillan 1976).

A regressive phase involving delta front advancement into the depocentres is indicated by the upward coarsening, and outward prograding of the upper arenaceous member (SR2) and the thickening of sediment sequences

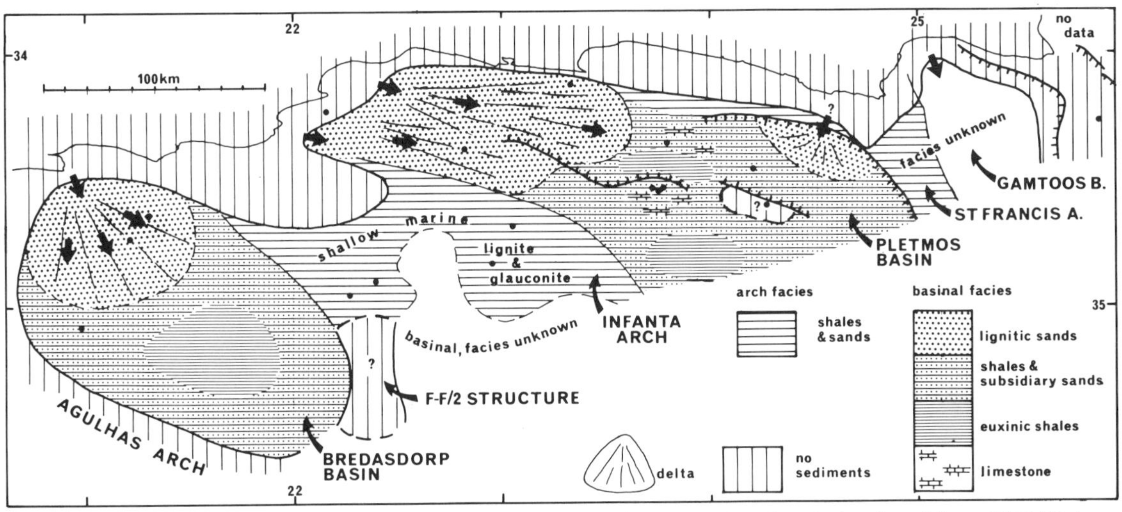


Figure 94. Palaeogeography of Outeniqua Basin in Upper Barremian/Lower Aptian time during deposition of SR3/SR4 members of the Upper Sundays River Formation. Control boreholes are dots. Euxinic shales in basin centres are speculative features. Bare areas over the F-F/2 and G(b)-Springbok/1 structures may be due to either non-deposition or later erosion. Upper Sundays River Formation occurs in the Gamtoos Basin (SOEKOR 1976), but no lithofacies details have been published.

into boundary fault-controlled depocentres suggests that this was triggered by uplift in the source areas.

A similar cycle of marine retreat, erosion, re-establishment of shallow marine conditions, followed by a regressive phase of delta outbuilding and ?fluviatile red bed deposition is repeated during the deposition of the Upper Sundays River Formation and, at least in the Pletmos Basin, there is evidence of a substantial increase in differential movement across major boundary faults (Superior and Plettenberg faults). Boundary fault movements then appear to have progressively decreased with time, so that there is very little displacement (perhaps 100 m) across the top of the Upper Sundays River Formation over the crest of the Superior Ridge.

Figure 94 shows a schematic palaeogeographical reconstruction of the Outeniqua Basin for Upper Barremian/Lower Aptian times (SR3/SR4). The coastline lay close to its modern counterpart in several places, but elongate, low peninsulas projected seaward over the main basement highs, isolating the sedimentary basins except for local shallow water (probably intermittent) connections over their crests. Wide delta fronts dispersing sediments longitudinally (SE) at the proximal ends of the basins and possibly laterally off the peninsulas, advanced into the deeper parts of the basins, progressively decreasing the area in which deeper water marine shales were deposited (SR4). Thin sand sheets were periodically injected by slides and turbidity flow in advance of the delta fronts, giving rise to the lower argillaceous member (SR3) with interbedded sand stones and sandy limestones, whilst the deepest parts of the basins were poorly oxygenated, resulting in euxinic shales. In the proximal parts of the basins, the complete cycle is sandy (e.g. boreholes E-1, E-B/1 in the NW Bredasdorp Basin). The whole area was a relatively low energy environment, with few large topographic discontinuities across the major boundary faults, implying that sediment accumulation kept pace with crustal subsidence in the Pletmos and Bredasdorp basins. Assuming a sedimentation period of Barremian to Upper Aptian (ca.10 my) in the Pletmos Basin suggests high accumulation rates adjacent to the Plettenberg (400 m/my) and Superior (160 m/my) faults. Maximum rates were considerably lower in the Bredasdorp Basin (perhaps 80 m/my), although precise dating in the western part precludes a satisfactory comparison. It is clear, however, that despite the large quantity of terrigenous debris being supplied to the basins, it was insufficient to fill them and extend the river flood plains and estuaries as far east and SE as had been the case during Lower Uitenhage times. Subaerial exposure of the delta tops at the end of the SR4 cycle is indicated by the development of red beds in the proximal facies. The abundance of comminuted plant material and lignite in both the delta top and prodelta clays attests to the well-vegetated nature of the catchment areas of the rivers.

By the end of the SR4 cycle, we envisage a coastline of wide, delta-filled embayments separated by extensive low lying peninsulas whose southern and SE extensions were marked by occasional low islands and wide submarine shoals with shifting sand bars and high currents which prevented extensive argillaceous sedimentation and encouraged authigenic mineral (glauconite) production.

A very similar general palaeogeography is envisaged for Lower Sundays River Formation sedimentary environments, the major differences being: a less extensive area of sedimentation in the west, in particular the isolation of the Bredasdorp and Pletmos basins; more extensive sedimentary areas in the east, where marine conditions extended into the Inner Algoa Basin.

A consideration of sedimentation on the Agulhas Bank in mid-Cretaceous times, in the context of the separation of SE Africa and the Falkland Plateau will be dealt with in section 4.4.3.

## 4.4 EAST COAST AND ADJACENT OFFSHORE AREAS

### 4.4.1 *Structure and basement features*

Figures 66, 67 and 95 show the distribution of Middle Jurassic to Lower Cretaceous strata in the eastern part of southern Africa, whilst Figures 1, 95, 96 and 104 show the main structural features in the region. It is immediately clear from a comparison of these that sediment basin development has been strictly controlled by major crustal lineaments and that there are two strongly contrasting styles. North of about 30°S, sediment basins form a series of large compartmented graben which trend approximately N-S and which extend from the vicinity of Cape St Lucia into southern Mozambique (Fig.96A). These depocentres are separated from a further basement low in the Mozambique Basin/Zambezi Graben areas by the northerly extension of the Mozambique Ridge basement high. South of 30°S, major depocentres are restricted to areas immediately adjacent to and beyond the various COB's that delimit the southern Natal Valley/Transkei Basin, which has for its eastern edge the Mozambique Ridge (Figs 1, 96).

These two contrasting regions have fundamentally different crustal structures, which in turn have given them different sedimentary histories (Figs 2 & 96). The northern region is underlain by thinned continental crust which, at least in the Zambezi Graben, had begun taphrogenic basin development by Middle Jurassic times. Early graben fill consists of coarse continental clastics and volcanics, which are followed by shallow marine sediments in Barremian to Cenomanian times. In lithofacies and timing therefore the sediment basins of Zululand and southern Mozambique have a similar history to those in the southern Cape. Because of the relative sparcity of borehole and seismic data, particularly offshore, it is not possible to give confident estimates of maximum thicknesses of Middle Jurassic to Lower Cretaceous in these basins, but with the probable exception of the coastal plain of southern Mozambique they all exceed 2 km, with generally less than 500 m on the basement highs (Fig.66).

The southern region constitutes the southern Natal Valley. It is underlain by oceanic crust and flanked by continent in SE Africa, and probably thinned continental crust in the Mozambique Ridge. Sedimentation in the Natal Valley did not commence, therefore, before the separation of Africa and South America, and the earliest infill in this basin cannot be older than about Valanginian in the north and Aptian/Albian in the south (Fig.2). The continental flanks to this oceanic basin have a history that is parallel to that of the eastern Falkland Plateau, and apart from the small St Johns Basin with its onshore extensions at Mngazana and Mbotyi, the SE African side has remained a buoyant feature since Lower Triassic times, (the Upper Cretaceous cover over this feature was the result of high sea stands and not large-scale basement subsidence). Maximum sediment thicknesses in this region are about 3.5 km close to the COB, although in the north this is a result of progradation from the Tugela cone which actually lies on continental crust at the southern end of the taphrogenic region.

The main structural elements that controlled Middle Jurassic to Lower Cretaceous sedimentation in the east coast areas of southern Africa can be summarized as:

i) long, approximately N-S, and NNW-SSE normal fractures bounding wide horsts and graben on continental crust north of 30°S;

ii) relatively short NE-SW fractures and basement highs which offset and compartmentize the northern Natal Valley-Limpopo grabens;

iii) long, approximately NE-SW shear fractures bounding the Natal Valley south of 30°S (similar features along the western edge of the Mozambique Basin trend approximately N-S);

iv) various arcuate fractures, and E-W to NW-SE lineaments in pre-Jurassic rocks on the continental edge between 29 and 33°S.

#### 4.4.1.1 *Taphrogenic province of southern Mozambique, Zululand and the northern Natal Valley*

This area represents the southernmost extension of the taphrogenic zone that stretches for 7 000 km from the Dead Sea/Red Sea area, through Ethiopia and down the eastern side of Africa, including the western side of Madagascar: East African Rift System. A good deal of controversy surrounds the nature and underlying causes of this line of crustal weakness, which at its simplest, can be regarded as a manifestation of continental separation (or attempted separation) (for instance Dietz & Holden 1970). However, this explanation does not satisfactorily take into account the long history of tectonism and crustal subsidence that has characterized it (e.g. Kent 1974, McConnel 1980). In addition, the various branches of the system result in a complex pattern of crustal fracturing which is not representative of zones along which known examples of continental separation have proceeded to completion. There is general agreement that the lineament is a zone of tension. A discussion of the whole question of vertical tectonics in continental areas is beyond the scope of this book

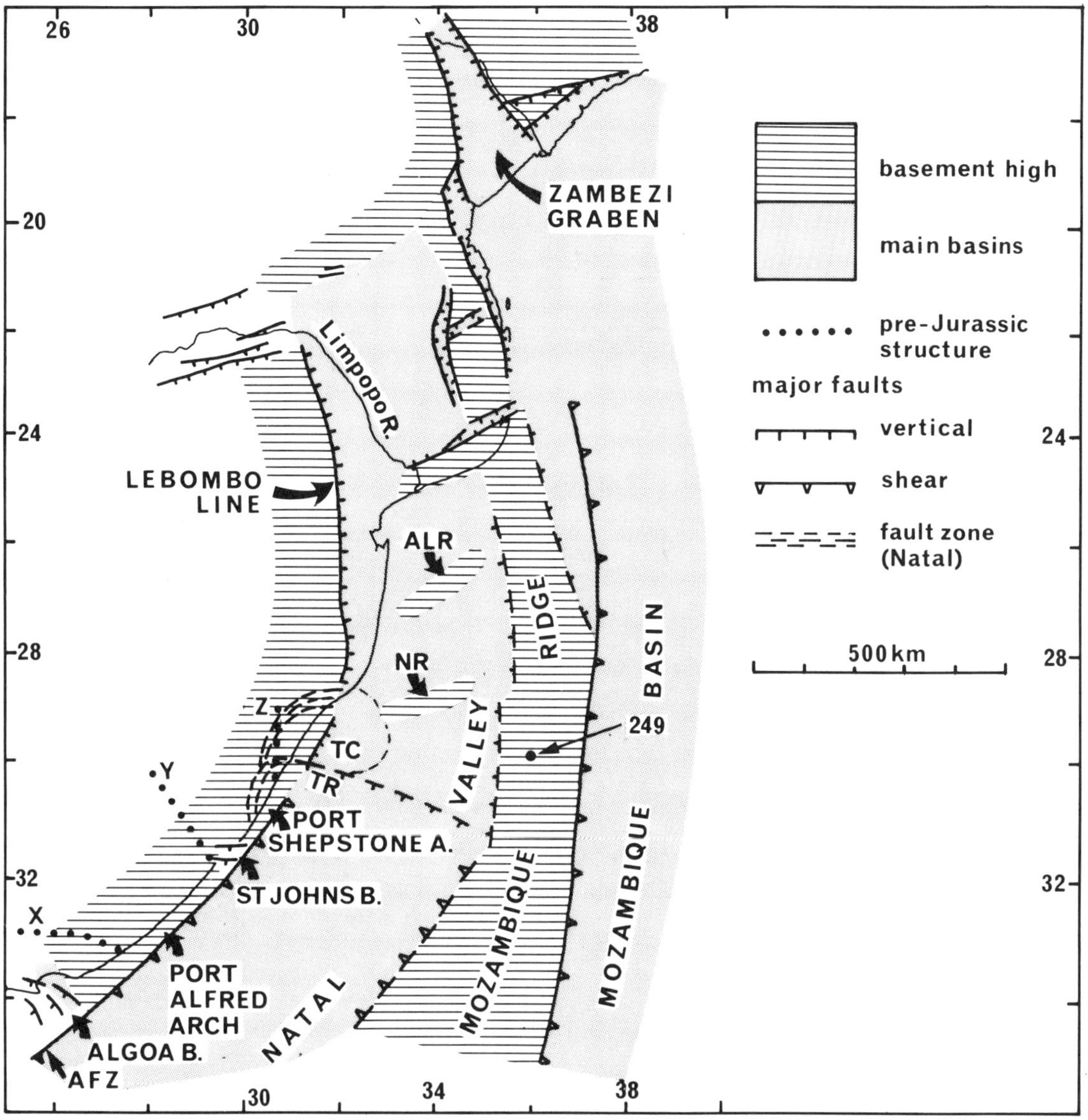


Figure 95. Regional tectonic setting of Jurassic-Lower Cretaceous sediment basins on the continental margin of S.E.Africa and adjacent offshore areas. Based on numerous sources, but particularly Flores (1973) and Dingle (1980a). The pre-Jurassic axes (X-Z) are further illustrated in Figure 97 and are after Ryan (in Haughton 1969). Abbreviations: AFZ = Agulhas Fracture Zone; NR = Naude Ridge; ALR = Almirante Leite Ridge; TC = Tugela Cone; TR = South Tugela Ridge; B = basin; A = arch; X, Y, Z = pre-Jurassic structures; 249 = DSDP site.

and we will restrict ourselves to observation on the style and crustal morphology in SE Africa.

Figure 95 shows that the area consists of two wide graben-like depressions separated by a horst line. The western graben is flanked by the Lebombo Line which marks the eastern edge of the buoyant Kaapvaal Craton. This is a major N-S crustal discontinuity which was the site of large-scale volcanism from latest Triassic to late Jurassic times (ca.200-137 my). At its southern end, the Lebombo Line is cut by an arcuate fault zone in the vicinity of Empangeni and Richards Bay (ca.29°S) where Pre-Cambrian basement is offset to the east and the coastline has a sharp deflection of about 55 km (e.g. Figs 133 & 142). Farther south, the structural edge of the Natal Valley continues with a NE-SW trend under the present day Tugela Cone, and separates relatively shallow basement on the continental shelf (<1.5 km) from deep basement in the Natal Valley (3.5-4 km) (e.g. Fig.106B). At about 30°20′S, a second arcuate fracture zone in Natal crosses the coast and continues east as the Tugela Ridge (Martin et al 1981) and appears to terminate the NE-SW structures that underlie the Tugela Cone. Data on the trend of this discontinuity farther east are very sparse, but it coincides

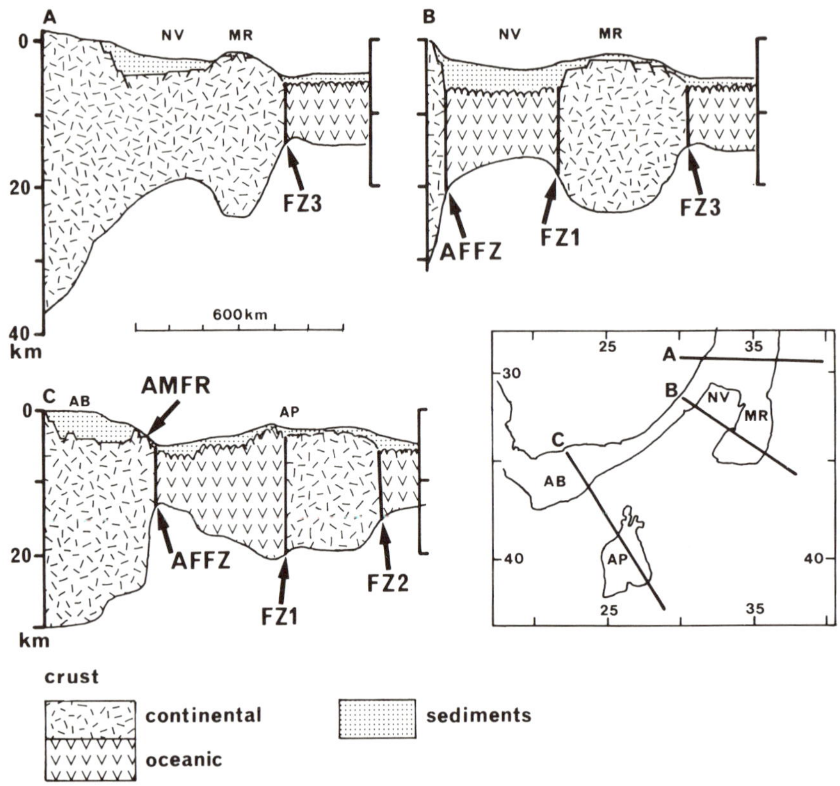


Figure 96. Crustal structure off S.E.Africa. Modified slightly from the gravity/seismic refraction models of Scrutton (1976a), with data of Tucholke *et al.* (1981). See Figures 2 and 3B for fracture zone notations. Abbreviations: FZ = fracture zone; AF = Agulhas/Falkland; AB = Agulhas Bank; AP = Agulhas Plateau; NV = Natal Valley; MR = Mozambique Ridge; AMFR = Agulhas marginal fracture ridge.

with a fundamental change in crustal structure (see also Fig.1). To the north of it, basement lies at between 3 and 4 km below sea level and crustal thicknesses are of the order of 15 km, whilst to the south basement lies at depths greater than 5 km and the crust is about 9 km thick (Figs 96 & 106). We take this discontinuity to mark the junction between thinned continental crust to the north and normal oceanic crust to the south.

Northwards, the Natal Valley is crossed by three NE-SW lineaments which divide it into four compartments. In the offshore areas, these basement highs constitute the Almirante Leite and the Naudé Ridges (Dingle et al 1978), and the latter lies approximately in line with the Empangeni-Richards Bay fault zone at the southern end of the Lebombo Line. Data on the age and structure of these two ridges are sparse, but seismic records indicate that they were emplaced before Cenomanian times and the Almirante Leite Ridge is at least partly volcanic in origin (tuffs and agglomerate have been dredged from it). The nothern end of the Natal Valley (about 25°S) coincides with a major basement high across which the coastline is deflected by about 240 km and whose eastern extension into the Mozambique Basin is seismically active (Flores 1973). The Natal Valley graben continues northwards under the south Mozambique coastal plain and is terminated at its northern end by basement highs and fault lines along a 300 km wide front. These faults extend SW into northern Transvaal and SE Botswana where they delimit the Limpopo Mobile Belt, and have had a history of vertical tectonism since at least 2 000, and possibly, 3 000 my BP.

A second, parallel graben extends down the Zambezi Valley into the northern Mozambique Basin. Its southern end probably lies about 24°S and its western edge can be traced for nearly 700 km down the Zambezi Graben and along the eastern side of the large horst which is the extension of the offshore Mozambique Ridge. Between 20 and 24°S, this structural discontinuity coincides with the present coastline.

Data on the deep structure and composition of the Mozambique Ridge are scarce and equivocal, but most workers agree that it is a continental fragment (e.g. Dingle & Scrutton 1974, Norton & Sclater 1979), a viewpoint that is supported by gravity and seismic refraction work which indicate crustal thicknesses of over 20 km (Scrutton 1976) (Fig.96) (see Chapter 1).

#### 4.4.1.2 *Transkei Swell, Agulhas Fracture Zone, and Southern Natal Valley*

The continental edge off SE South Africa is narrow, precipitous and straight, and follows the trace of one of the longest (ca.1 300 km) continental margin offsets in the world: the Agulhas Fracture Zone. This lies along a small circle about the early opening pole for the South America/Africa continental separation

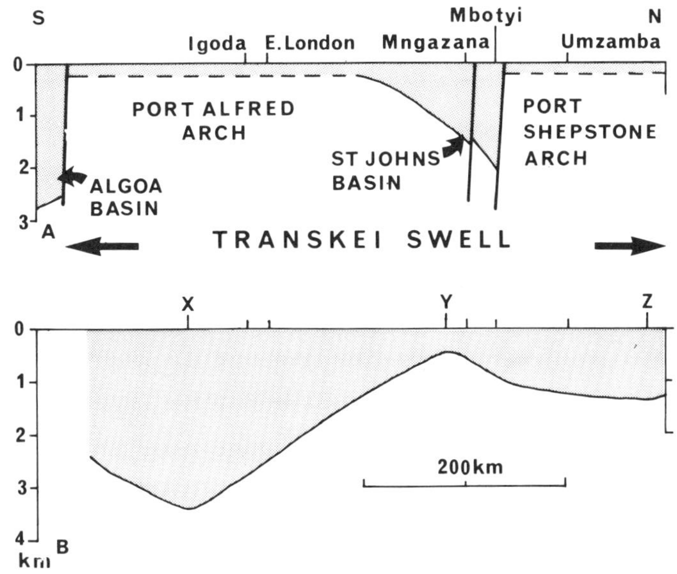


Figure 97. N-S sections across the Transkei Swell.
A. Mesozoic (probably all Jurassic-Lower Cretaceous) sediments on the continental shelf (<10 km offshore). The Port Alfred and Port Shepstone arches may be bald.
B. Lower-Middle Permian sediment basins plotted using the same control points (ticked) as in section A. Features X, Y and Z are located on Figure 95. Data from Ludwig *et al.* (1968), Ryan (in Haughton 1969), Simpson & Dingle (1973).

(46.75°N, 32.65°W, Martin et al 1981) and truncates all pre-drift structures in SE Africa, and along the northern edge of the Falkland Plateau, where the conjugate shear zone is called the Falkland Fracture Zone (Dingle & Scrutton 1974). SW of about 34°S, the Agulhas Fracture Zone is marked by a prominent basement ridge (Agulhas Marginal Fracture Ridge) (Scrutton & du Plessis 1973) which forms an outer barrier along the truncated SE side of the Outeniqua Basin (Figs 96 & 112). North of 34°S the position of the COB lies at the foot of the continental slope and is marked only by the slope gravity anomaly (free air edge effect up to 130 mgal peak to peak) and a strong positive magnetic anomaly (Scrutton 1976a). The fracture zone here truncates a wide buoyant feature (Transkei Swell) which underwent an interesting reversal of crustal stability, certainly after, and possibly prior to continental separation. Du Toit (1937) was the first person to comment on the possible significance of the truncation of the Cape Fold Belt to the NE of the Algoa Basin, and also the fact that the eastern end of the Karoo Basin is missing – presumably it is now on the Falkland Plateau. Figure 97B shows late Palaeozoic/early Mesozoic basin morphology along a coastal section between East London and 30°30′S. During Ecca and Beaufort times (Permo-Triassic) the main depocentre (with about 3.5 km of Lower-Middle Permian sediments) lay approximately E-W, and crossed the present continental margin slightly to the SW of East London (Fig.7, and X on Figs 95 & 97). A large basement high (running NE-SW) lay on the northern side of this trough and its axis (Y) crossed the coast to the south of Port St Johns. A further shallow basin at the northern end of the section lay between Port St Johns and 30°30′S (Fig.7 and Z on Figs 95 & 97). This whole sector (X to Z) has remained as one buoyant feature since Triassic times, with the exception of the crest of the old high (Y), which appears to have collapsed along two major fault lines (Mngazana and Egoso) to produce a half graben basin containing about 2 km of medium velocity (3.1-4.1 km/sec) sediments (St Johns Basin). These two faults cross the coast in the vicinity of Port St Johns, where two small Lower Cretaceous outliers are preserved. The northernmost fault (Egoso) forms the southern edge of the Port Shepstone Arch, and deflects the coastline and the continental shelf about 3 km to the east (Figs 108 & 109). The southern post-Triassic basement high (Port Alfred Arch) has its crest in the vicinity of East London (apparently coinciding with the Upper Palaeozoic/Lower Triassic trough axis), and is truncated on its southern end by a SE extension of the Coega Fault on the eastern side of the Algoa Basin. The buoyant northern sector of the Transkei Swell is the Port Shepstone Arch. Depth to basement at the outer edge of the Port St Johns Basin is over 6 km (Chetty & Green 1977), and the presence of a similar small deep basin on the northern edge of the Falkland Plateau which aligns with the Port St Johns basin in our palaeogeographic refit (Fig.104) suggests that the initial collapse of the basement high Y began prior to continental separation.

Onshore basement structures in the northern part of the Transkei Swell are complex, but consist essentially of two large arcuate fault zones with their long, curved faults throwing downwards on their convex (i.e. western) sides (Fig.95). These two zones enclose cores of granitic Pre-Cambrian, and granitic and Lower Palaeozoic basement in the north and south respectively, and are bare of Jurassic-Lower Cretaceous sediments. The southern fault zone lies adjacent to the

Port Shepstone Arch basement high on the narrow continental shelf (depths to basement < 500 m beneath sea level), and extends to the vicinity of Durban (30°S), where it swings sharply eastwards to cut the coast at right angles. This point coincides with a sudden widening of the adjacent continental shelf and the steep southern flank of the Tugela Cone and the buried Tugela Ridge. It is also in this vicinity that the gravity anomaly marking the Agulhas Fracture Zone ceases and we take it to mark the COB relating to the palaeoposition of the Falkland Plateau (Fig.104).

The northern arcuate fault zone stretches from south of Durban to the southern end of the Lebombo Line, where it swings sharply eastwards to cut the coast south of Richards Bay. This coincides with a further alteration in continental shelf morphology (it narrows rapidly to the north), and with a NE deflection of the coastline. Several of the faults on the outer, northern part of the arcuate zone curve NE, and the Eteza Fault terminates the Mesozoic lava outcrops where it crosses the Lebombo Line (Figs 95, 133 & 142). There is no direct evidence for the age of these faults, but Maud (1961) considered them to be Jurassic, and this is supported by the fact that they cut the Jurassic Lebombo lavas, and yet are overlain by undisturbed Upper Cretaceous sediments. Presumably they were formed in response to the major tectonic events that affected this area in Jurassic to Lower Cretaceous times, but whether this was the widespread taphrogenic faulting, or the SW movement of the Falkland Plateau, or a combination thereof is not known.

The southern Natal Valley lies between the Agulhas Fracture Zone and the Mozambique Ridge, and differs morphologically from the northern sector: it is wider (390 vs 280 km), has greater water depths (3-4.5 vs 0.5-2.5 km), and generally has a much thinner sediment infill, particularly in the south. The differences are attributed to its relative youth and different crustal structure.

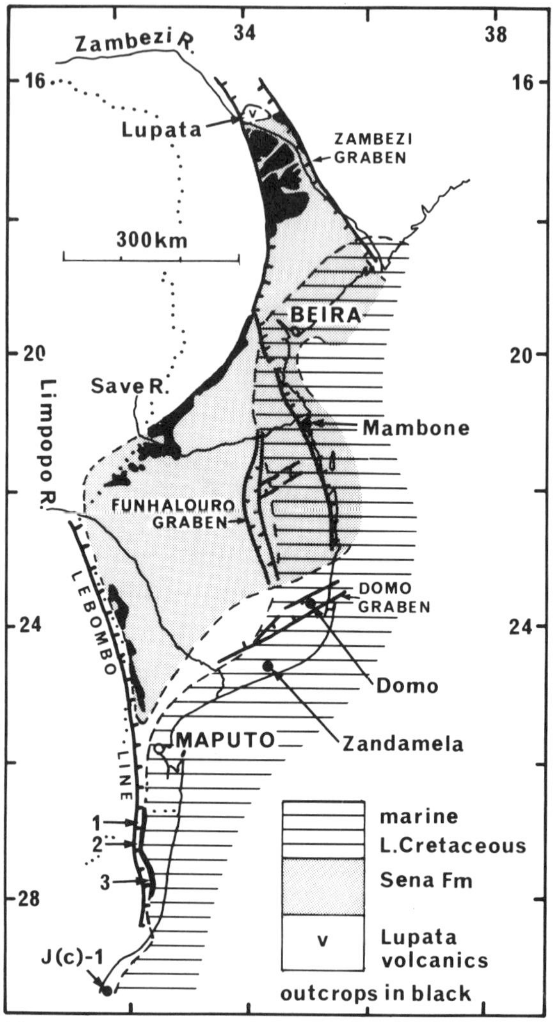


Figure 98. Distribution of Middle Jurassic-Lower Cretaceous rocks in Zululand and S.Mozambique. In Mozambique, Barremian to Turonian strata have been included, whereas in Zululand the upper limit has been drawn at the top of the Albian. Important boreholes are indicated by dots. Data mainly from Flores (1973), but Zululand localities 1-3 are after Kennedy & Klinger (1975) (their locality numbers are given in brackets): 1 = Mlambongwenya (170); 2 = Mfongozi (168-169); 3 = Mzinene (51).

### 4.4.2 *Stratigraphy*

The strong asymmetry between Jurassic-Lower Cretaceous sediment accumulation on the Transkei Swell and in the coastal basins farther north is diagrammatically shown in Figure 67. South of the Richards Bay area, rocks of this age are confined to the isolated Mngazana and Mbotyi outliers, whilst to the north in Zululand, sediments as old as Barremian occur, and locally in Mozambique a thick Middle Jurassic to Turonian sequence is developed. Numerous boreholes and exposures have allowed a satisfactory litho- and biostratigraphic zonation to be developed and it is in this area that the standard palaeontological zonal sequences for the mid-Cretaceous of southern Africa has been established. Offshore areas locally contain thick successions which are thought to be at least as old as early Cretaceous (offshore Zambezi Graben, Tugela Cone, Central Terrace area of the Natal Valley), but no boreholes have been drilled, so correlation is entirely an acoustic stratigraphy projected from the J(c)-1, Zandamela and DSDP 249 boreholes on the peripheries of the main basins (Figs 95 & 98). Under these circumstances only generalizations can be made on possible age and lithofacies of the basin fill.

For the area south of 30°S the situation is more complex. Prior to continental separation (~Valanginian) the southern Natal Valley area was occupied by the Falkland Plateau, and from this we have excellent stratigraphic control in the form of DSDP boreholes 327,

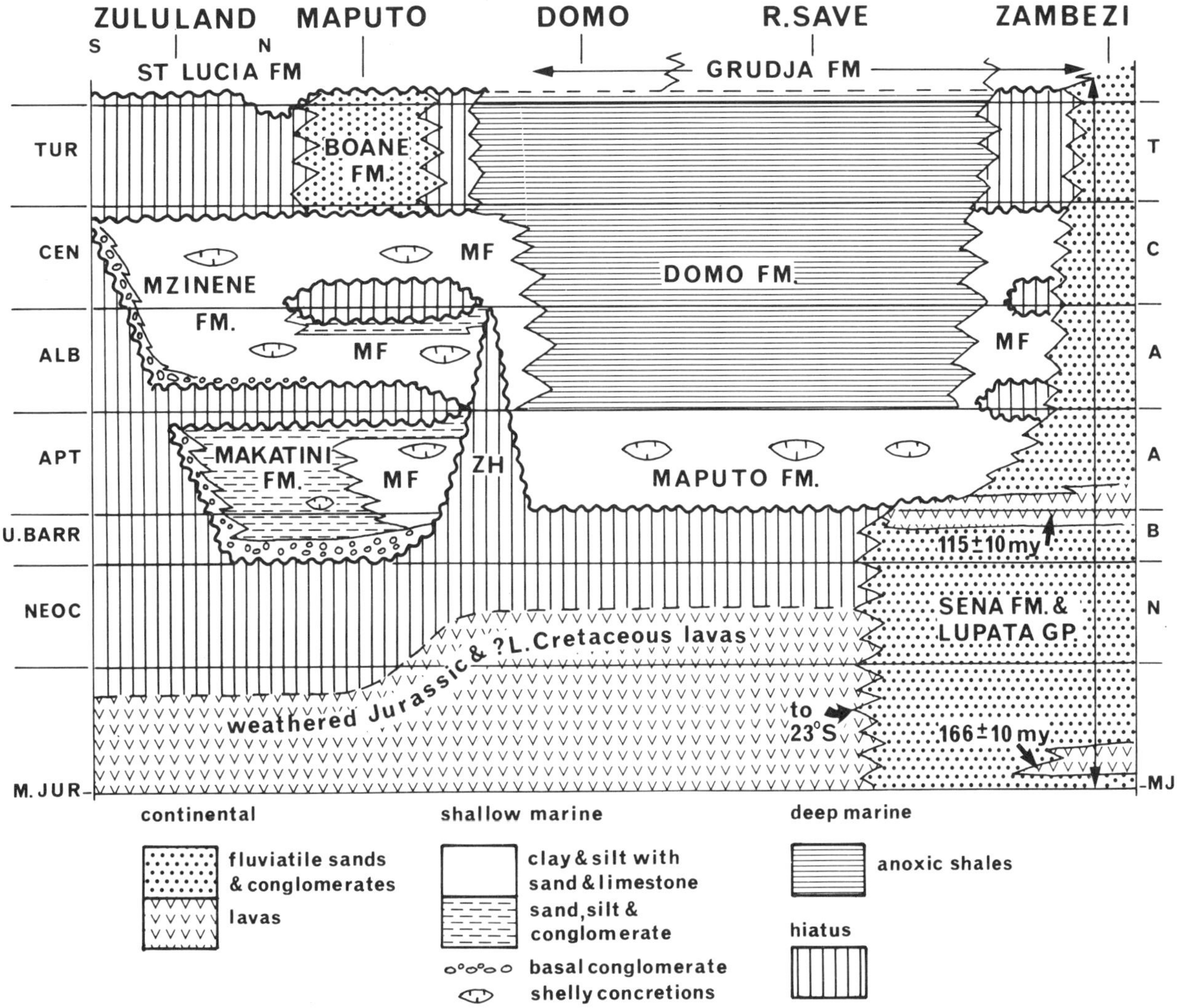


Figure 99. Schematic lithostratigraphic correlation of Jurassic to mid-Cretaceous strata between Zululand and the Zambezi Graben. Approximate length of section is 1 200 km. Based mainly on the work of Flores (1973), Kennedy & Klinger (1975), Forster (1975a, 1975b), McLachlan & McMillan (1979). Neocomian includes all unresolved Lower Cretaceous older than Upper Barremian. Abbreviations: MF = Maputo Formation; ZH = Zandamela High. Not to scale.

329 and 330. In addition, there are the small outliers at Mngazana and Mbotyi on the Transkei Swell which lay adjacent to the Falkland Plateau at the time of their deposition. Of the sediments deposited in the ocean basin that was progressively created as the Falkland Plateau moved SW we know practically nothing, although we can make some general predictions based upon palaeogeography and regional lithofacies trends.

4.4.2.1 *Natal, Zululand and southern Mozambique*
The distribution of Middle Jurassic-Lower Cretaceous strata in coastal basins of Natal, Zululand and southern Mozambique is shown in Figures 98 and 100, and a litho- and biostratigraphic correlation chart for the area is given in Figure 99. The oldest strata (Middle and Upper Jurassic) are found in the Zambezi Graben and over much of the coastal plain of Mozambique, where the succession consists of continental clastics and lavas (Sena Formation and Lupata Group). In the areas to the east of the Lebombo Line, these overlie and may interfinger with older ('Karoo') lavas which themselves vary in age from late Triassic to late Jurassic. As previously discussed (section 2.2.5.1), the extent of lava cover in southern Mozambique is not well-known, but boreholes described by Flores (1973) indicate that it must originally have been extensive (Fig.45), and some authors (significantly, oil company geologists such as Kent 1974, Beck & Lehner 1974) consider that the whole of the coastal plain of Mozambique is underlain by a more or less complete cover of lavas, and that they

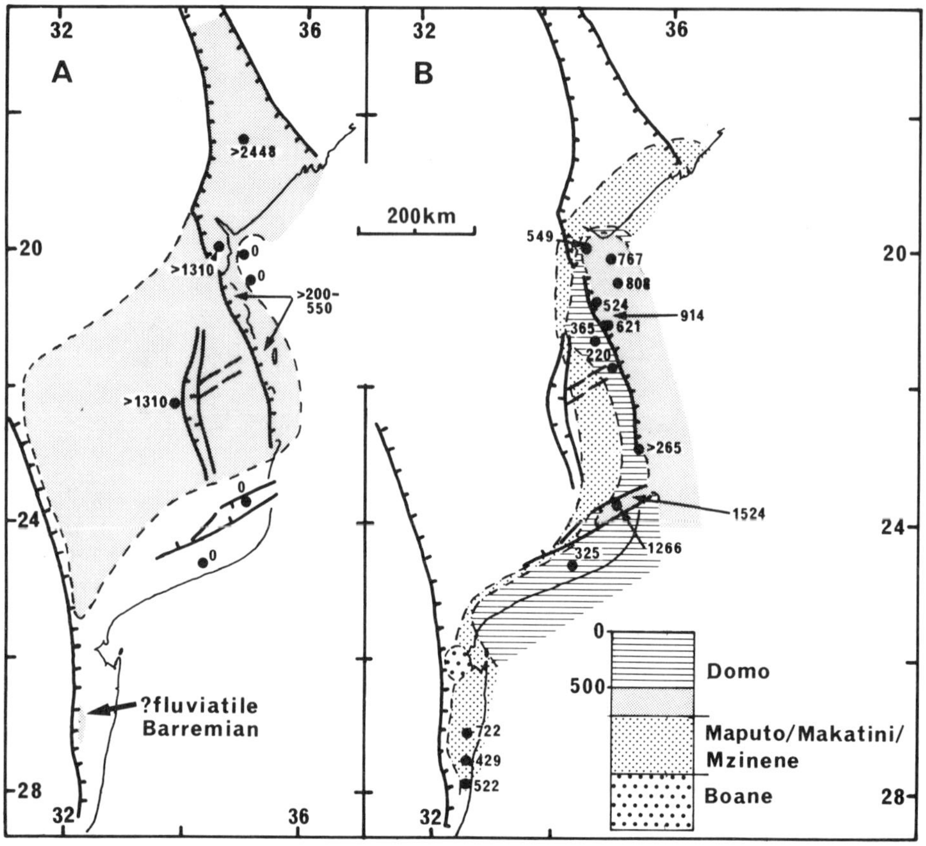


Figure 100. Distribution and thickness of Upper Jurassic-Lower Cretaceous sediments in Zululand and S.Mozambique.
A. Continental sands, conglomerates and interbedded lavas of the Sena Formation. Basal unfossiliferous conglomerates in Zululand may be fluviatile and laterally equivalent. Sena thicknesses are mostly minimum values quoted by Flores (1973).
B. Upper Barremian to Turonian marine sediments and Turonian fluviatile sands and conglomerates near Maputo (Boane Formation). Proximal facies is the Maputo, Makatini, and Mzinene formations. Distal (anoxic) facies is the Domo Formation for which the 500 m isopach is shown. Data from Flores (1973) and Forster (1975a), with borehole values in Zululand calculated from McLachlan & McMillan (1979).

extend eastwards under the Mozambique Channel. Volcanic rock types encountered in the deep boreholes include: olivine basalt, tuffs, rhyolites, and amygdaloidal dolerite (Tab.15). Red clays representing subaerial weathering products were reported in the Zandamela borehole, and Lower Cretaceous alkaline lavas in the Mambone borehole. No radiometric dates have been obtained from these rocks, and all the 'ages' quoted are from petrological comparisons with dated outcrops inland (Flores 1973). If it is accepted that they were extruded onto a relatively even surface, then a consideration of the present depths to the tops of these lavas (Fig.45) indicates that a considerable amount of differential movement must have taken place since extrusion. Numerous boreholes in Zululand (McLachlan & McMillan 1979) (sites on Fig.100B) have bottomed in weathered lavas, indicating that the relatively narrow coastal plain in the south is also undertain by eastward extensions of Lebombo volcanics.

Earliest marine sediments in the area occur in Zululand (Upper Barremian) and these are followed quickly in time (Aptian) by the establishment of widespread marine conditions along the whole of the present Mozambique coastal zone. Sedimentation was intermittent in the region south of Maputo, and ceased altogether from late Cenomanian to Coniacian (locally Turonian) times, but farther NE it appears to have locally been continuous until Eocene times (Fig.99). Marine environments were not established across the central coastal plain of Mozambique until Upper Cretaceous times, so that over a large part of the area under consideration, Lower Cretaceous sediments are of continental lithofacies (Sena Formation). Because of the relatively poor and isolated exposures in country often covered by dense bush, early geological work tended to concentrate on descriptions of the rich molluscan assemblages, especially the ammonites: for example: Neumayr (1885), Spath (1925) and Newton (1924) in Mozambique; and van Hoepen (1926), and Spath (1921) in Zululand. Despite this, no satisfactory regional litho- or biostratigraphic correlation

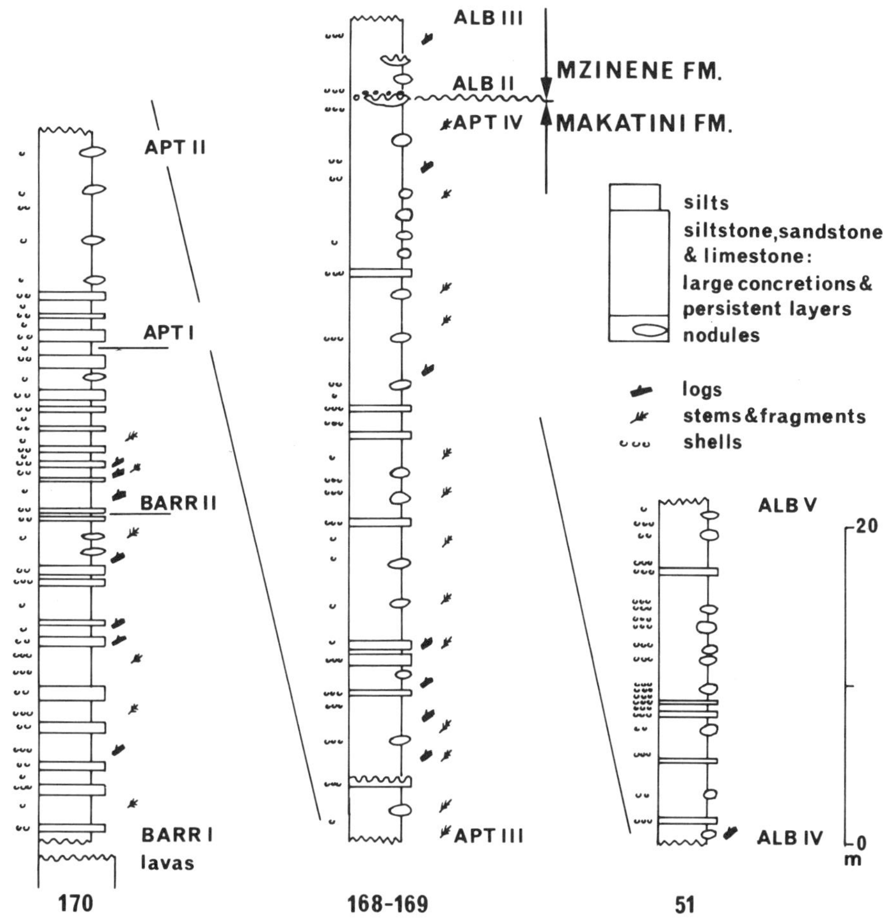


Figure 101. Sections in the Makatini and Mzinene formations, Zululand. Plotted from data in Kennedy & Klinger (1975) whose locality numbers have been retained. See Figure 98 for locations: 170 = Mlambongwenya Spruit; 168-169 = Mfongozi Spruit; 51 = Mzinene River near Hluhluwe. Locality 51 is the type section of the Mzinene beds as described by van Hoepen (e.g. 1926).

scheme was available until recently, when Flores (1973) described the geological framework of southern Mozambique from oil company data and Cooper (1974) summarized previous biostratigraphic work. Kennedy & Klinger (1975) and Förster (1975a, 1975b) produced regional litho- and biostratigraphic classifications applicable to various parts of Zululand and southern Mozambique. The work of Kennedy & Klinger (1975 et seq.) in particular, has made the biostratigraphy of Zululand the best known Mesozoic succession in southern Africa, and we will describe this area first.

4.4.2.1.1 *Makatini and Mzinene formations in Zululand (Upper Barremian to Upper Cenomanian).* In Zululand, the Lower Cretaceous rocks of the coastal plain thicken eastwards and northwards, reaching a calculated maximum of 722 m in boreholes on the Makatini Flats (McLachlan & McMillan 1979) (Fig.100B). They rest with a strong unconformity on the weathered surface of the Lebombo Group, which at Namahacha near the Mozambique-Swaziland border has interbeds of red sandstones bearing plant fragments (*Otozamites*) and freshwater lamellibranchs (*Unio*) (du Toit 1954) between lavas dated at 165 my (Flores 1973). In the northern part of Zululand, there is a general overstep to the south so that the oldest beds (Barremian) are found only in the north, whilst farther south, Albian marine sediments rest directly on the lavas. In the extreme south of the outcrop the lower unit (Makatini Formation) appears again, resting on lavas (Fig.133).

Kennedy & Klinger (1975) subdivided the Lower Cretaceous of Zululand into two formations (Fig.99): Makatini (below) and Mzinene (above), which are separated by a non-sequence across the Aptian/Albian boundary. The Makatini Formation is a series of shallow marine clays and sands which interfinger with basal fluviatile coarse sandstones and conglomerates (Figs.99 & 101). The coarsest lithofacies occurs in the west (Fig.100A) and consists of matrix and clast-supported, poorly sorted conglomerates with volcanic, quartzite and agate clasts, interbedded with cross-bedded, medium to coarse-grained sandstones. Large (5-6 m long, 50-60 cm diameter) logs, often bored, frequently occur crowded along particular horizons. Sandy silts and sandy shelly limestones and concretions with an abundant marine fauna interfinger with, and overlie

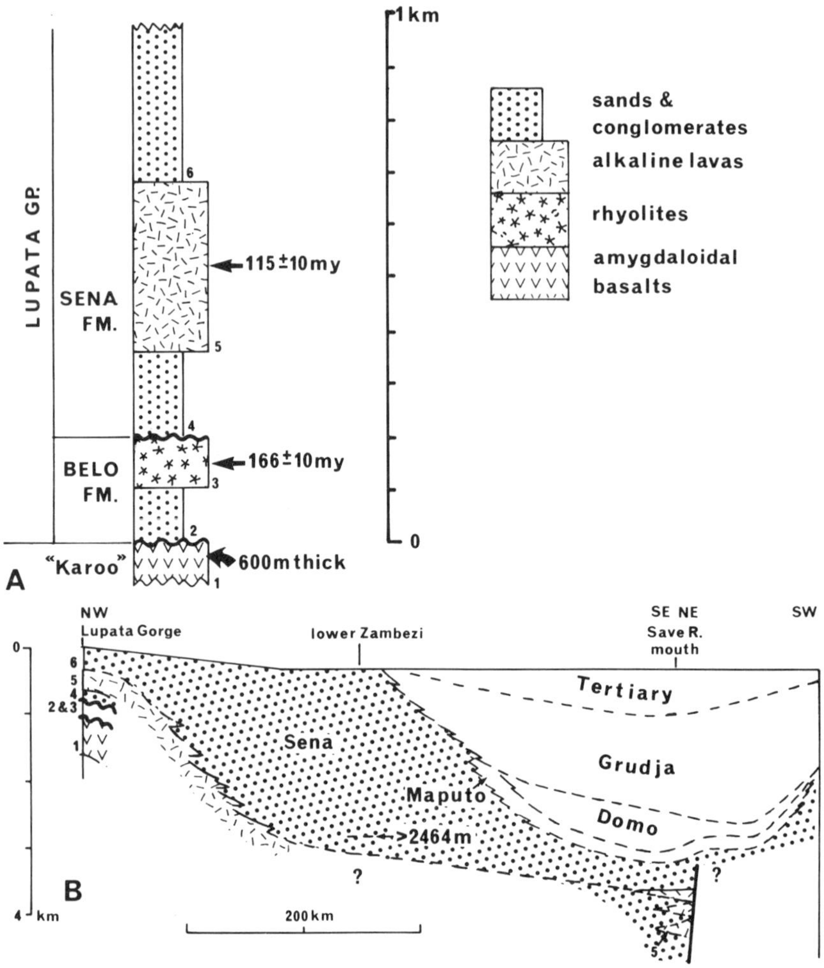


Figure 102. A. Representative section in the Lupata Group, Lupata Gorge, Zambezi Graben. The lowest basalts are undated and have been equated informally with the Lebombo Group. Data from Flores (1964).

B. Schematic cross-section down the W side of the Zambezi Graben from Lupata Gorge to a position SW of Save River mouth. Constructed from data in Flores (1973). Small numbers (1-5) refer to members of the Lupata Group in upper diagram.

the coarse non-marine sediments. The silts are often cross-bedded (with occasional pebble lags), intensely bioturbated, and contain abundant logs and plant fragments. Shell banks are common.

The marine facies of the Makatini Facies can be traced northwards into Mozambique as the Maputo Formation (Fig.99), and extends as far south as Hluhluwe, but the non-marine facies extends for only about 30 km into Mozambique in the Catuane area. The best exposures of this latter facies, along with the lowest marine horizons (Barremian), are in NW Zululand along the foothills of the Lebombo range in the western Makatini Flats.

The Makatini Formation is overlain by the Mzinene Formation, from which it is separated by a non-sequence which can be traced from Ndumu on the Zululand/Mozambique border to Mtubatuba in the south (170 km). The base of the Mzinene Formation locally rests on Lebombo lavas, and the junction with the Makatini Formation is marked by a hard ground of eroded concretions bored by lithodomous bivalves and encrusted with oysters, serpulids and other epizoans (Figs 99 & 101).

Lithologically, the Mzinene Formation represents a somewhat lower energy lithofacies than the underlying sediments, and consists of shallow marine silts, and sands (often glauconitic) with subordinate shelly and sandy concretions, bedded limestones and local, pebbly horizons. In addition to the main non-sequences at the top and bottom of the formation, many of the upper surfaces of the nodule and limestone horizons are bored, and suggest periodic increases in current velocity producing minor non- or condensed sequences and minor erosion surfaces. These surfaces are usually overlain by a current aligned basal shell lag (10-20 cm thick), and up to 3.5 m of bioturbated buff-coloured silts. Clearly, these minor cycles represent waxing and waning water current strengths. Similar features occur in the overlying St Lucia Formation, and the Maputo Formation (Fig.135). A major, regionally-developed non-sequence occurs at the top of the Mzinene Formation (Upper Cenomanian) and is marked by a thick hard ground.

4.4.2.1.2 *Lupata Group of southern Mozambique (Middle Jurassic to ?Santonian).* This group has been

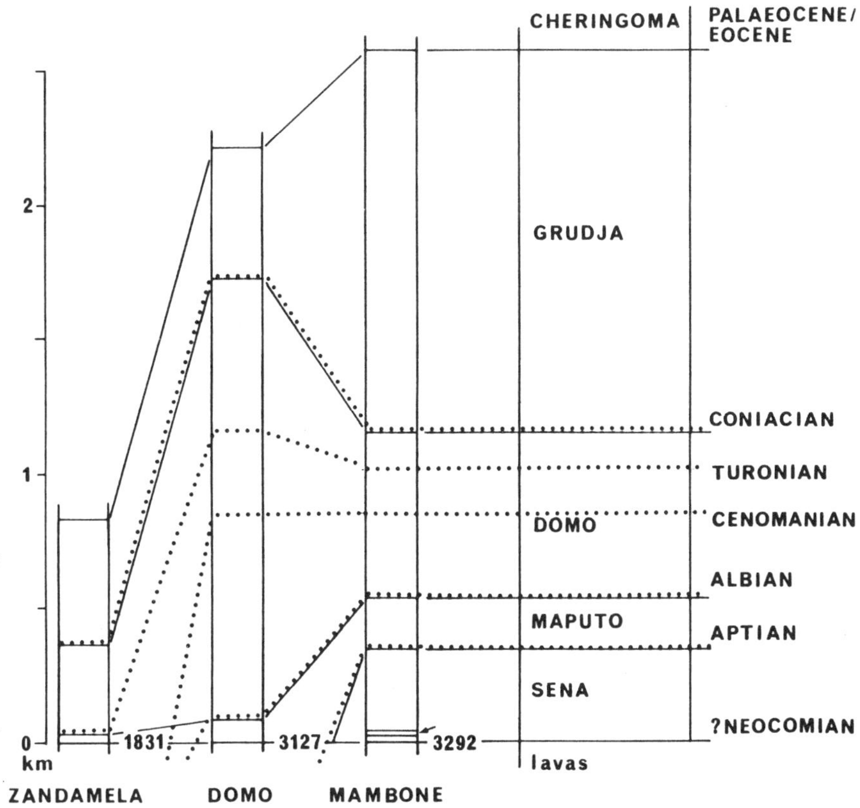


Figure 103. Thicknesses of Lower Cretaceous-Palaeocene sediments in three important boreholes in coastal S.Mozambique. Locations shown in Figure 98. Figures (m) at lower right side of each column are depths from surface to volcanic basement. Arrow near base of Mambone indicates an alkaline lava flow. Age assignments vary in accuracy, but approximately synchronous horizons shown by dotted lines (see text for discussion). Formation boundaries shown as solid lines. Compiled from data given in Flores (1973).

described by Flores (1964, 1973) from outcrops and boreholes in the Zambezi Graben and the southern coastal plains. It has been divided into two formations (Belo and Sena), but only an arenaceous/rudaceous lithofacies of the latter is widely developed (Figs 98, 99, 100A & 102), with the Belo Formation apparently restricted to the Zambezi Graben area. In the type section of the Lupata Gorge (about 300 km upstream from the Zambezi River mouth), Flores (1964) recognized two sequences of fluviatile rudites/arenites and lavas, resting on an eroded surface of 'Karoo' lavas (Fig.102A). These lower lavas are not dated, but on petrological comparisons and stratigraphic position, Flores (1964) assigns them a 'Stormberg' age (presumably this implies correlation with the Drakensberg Group and a late Triassic-early Jurassic age).

The basal sequence of the Lupata Group (Belo Formation) consists of about 100 m of reddish brown to light green coarse-medium-grained arkosic sandstones, with small fragments of 'Karoo' lavas and a calcareous matrix cement, overlain by about 80 m of columnar rhyolite and thin tuffs. The rhyolites have been radiometrically dated (K/Ar, whole rock) at 160 + 10 my (Flores 1964).

The Belo Formation is unconformably overlain by a thick sediment sequence (at least 400 m) with a thick (at least 300 m) intercalated lava unit: Sena Formation. The lower sediment member rests on eroded Belo rhyolites, and locally oversteps onto the Belo sandstone member. It is about 100 m thick and consists of a grey or yellow-grey, poorly-sorted conglomeratic sandstone with angular and sub-angular quartz and rhyolite clasts, and a calcareous cement. This is overlain by at least 300 m of alkaline basalts, and phonolites, which have been radiometrically dated (K/Ar, whole rock) at 115 ± 10 my (Flores 1964). These lavas are overlain, apparently conformably, by about 300 m of yellowish grey to light red and reddish brown, poorly-sorted conglomeratic, angular to sub-angular sandstones, with interbedded breccias and tuffs.

The Sena Formation thickens SE down the Zambezi Graben, and reaches a maximum development to the north of Beira (at least 2 448 m) (Figs 100 & 102B). It crops out along the western side of the Mozambique coastal plain, where it rests directly on Lebombo lavas, and has been intersected by boreholes in the Save River plain, and along the coast. It does not occur in the coastal regions south of about 23°S, nor in offshore boreholes on the south side of the Zambezi Cone, but in the vicinity of the Save River mouth (Mambone borehole) has been found interbedded with alkaline lavas which Flores (1973) correlates with the lava sequence in the Lupata Gorge. Palynological data indicate an Upper Jurassic to Coniacian age for the Sena sandstones,

Table 28. Ammonite and planktonic foraminifera biostratigraphy of Zululand and Outeniqua Cretaceous (after Dingle 1982a).

| | Zululand | | | | Outeniqua Basin | | | |
|---|---|---|---|---|---|---|---|---|
| | Ammonites (Kennedy & Klinger 1975) | Correlation with international planktonic foraminifera zones (Van Hinte 1976a) | Fms. | Planktonic foraminifera recognized by McLachlan & McMillan (1979) | Planktonic foraminifera recognized by McLachlan & McMillan (1979) | International zonal equivalents (Van Hinte 1976a, Postuma 1971) | Formations/groups | |
| Maastrichtian | – no ammonites; *Inoceramus*<br>– Pachydiscids<br>abundant *Eubaculites* | *Globotruncana contusa* | St Lucia | *Globotruncana contusa* | *Rugoglobigerina rugosa*/*Globotruncana arca* | *Globotruncana gansseri* | | M |
| Campanian | abundant *Submortoniceras* spp. | | | | | | | C |
| Santonian | abundant *Texanites* s.s. | upper *Globotruncana renzi*/*G.sigali* | | | *Globotruncana primitiva*/*G.renzi* | | Alphard Fm | S |
| Coniacian | *Kossmaticeras theobaldianum* | | | | | *Globotruncana renzi*/*G.sigali* | | C |
| Turonian | | | | *Praeglobotruncana helvetica* | *P.helvetica* | *Praeglobotruncana helvetica* | | T |
| Cenomanian | *Calycoceras* ex gr. *naviculare*<br>abundant *Sharpeiceras* & *Mariella* | upper *Rotalipora cushmani* | Mnzinene | *Rotalipora cushmani*<br>*Rotalipora apenninica* | *Rotalipora reichi*<br>*Hedbergella washitensis* | *R.reichi*<br>*Planomalina buxtorfi*/*Rotalipora apenninica* | Alphard Fm<br>? | C |
| Albian | *Douvilleiceras* cf. *mammillatum* | upper *Ticinella bejaouaensis*/*Globigerinelloides gyroidinaeformis* | | no planktonic foraminifera | | | ? | A |
| Aptian | *Diadochoceras nodostocatum*<br>*Prochelinoceras* spp. | *Globigerinelloides ferreodensis*/*Ticinella bejaouaensis* | Makatini | | no plantonic foraminifera | | Upper Sundays R. Fm. | A |
| Barremian | *Colchidites* spp. | upper *Hedbergella sigali* | | | | | ? | B |
| Hauterivian | | | | | *Hedbergella ?hoterivica* | *H.hoterivica* | Lower Sundays R. Fm. | H |

although marine interbeds suggest that they may range into the Santonian (Flores 1973). It clearly represents a strongly diachronous proximal graben fill facies, and the rapid lateral thickness variation reported from boreholes (Flores 1973)(Fig.100A) suggests that considerable differential crustal subsidence occurred during its deposition. This is particularly well seen south of Beira, where thin cover occurs on top of the coastal horst, adjacent to thick sequences in the lower Zambezi Graben. Eastward thinning is a result of lithofacies changes to marine deposits. Förster (1975a) reports a continental/estuarine facies in the southern part of the Funhalouro Graben. The Sena lithofacies peters out in the vicinity of Maputo, where only a narrow thin outcrop, equivalent to the basal fluviatile Makatini Formation (Barremian) occurs (Figs 99 & 100).

4.4.2.1.3 *Maputo and Domo formations of southern Mozambique (Aptian to Cenomanian and Albian to Turonian, respectively).* In the coastal area of Mozambique, the Sena Formation sandstones pass eastwards into marine sediments, which had transgressed to about 34°E by the end of Lower Cretaceous times (Figs 98, 100B). Two distinct lithofacies can be recognized within the marine sediments, a proximal facies (Maputo Formation) which is contiguous with the Makatini and Mzinene formations of Zululand and occurs along the whole length of the Mozambique coastal region, and a distal facies (Domo Formation) which is known only from boreholes and is restricted to coastal and offshore areas between about 20 and 26°S. Figure 99 summarizes the lateral and vertical relationships between the marine rocks of the Zululand-Mozambique region, and the non-marine Sena Formation which is restricted to western and northern areas.

Earliest records of the Maputo Formation are slightly younger than the equivalent strata in Zululand (i.e. Aptian) and over the basement high east of the Limpopo River mouth (i.e. in the Zandamela borehole) sedimentation did not commence until latest Aptian or early Cenomanian times. The Maputo Formation crops out west of Maputo, but elsewhere is known only from boreholes. It is locally up to 600 m thick and consists of fine grained glauconitic sandstones and sandy marls with occasional calcareous nodule horizons and sandy limestones. As in Zululand, the calcareous nodules frequently have bored upper surfaces and are overlain by upward fining shell lags indicative of cyclic increase and decrease in bottom current strengths and the production of minor non-sequences along hard grounds. The formation has a basal rudaceous member that resembles that of the Makatini Formation, and a late Aptian-early Albian non-sequence (as in Zululand) that is marked in the proximal areas by a pebbly conglomerate and accumulations of thick-shelled bivalves. A further hiatus (signifying a regression according to Förster 1975a) is marked by pebble conglomerates in the late Albian-early Cenomanian. The top of the formation coincides with a regional late Cenomanian regression in the proximal marine facies. The Maputo Formation has a rich invertebrate marine fauna with well-preserved and abundant ammonites and bivalves. Its biostratigraphy has recently been studied by Förster (1975b) and it has a maximum local age range of Aptian to Cenomanian (Flores 1973, Förster 1975a) (Fig.103).

A somewhat more distal marine facies occurs in boreholes farther east (Fig.100B), where the Maputo Formation is overlain by, and grades laterally into dark grey, pyritic, thinly bedded marly shales with minor sand horizons: Domo Formation. It is known only from boreholes, and appears to lie in a narrow, deep trough along the SW extension of the Zambezi Graben which is bounded to the west by the northerly extension of the Mozambique Ridge (Fig.95). The Domo Formation is thickest immediately offshore (914 m) and in a small subsidiary graben (Domo Graben) at about 24°S (1 524 m), and thins rapidly across the main coastal horst (<400 m) west of which it passes into the Maputo Formation. South of 24°S dark Domo shales are thin over the Zandamela High and are not known as far west as Maputo (Figs 99 & 100B).

Although the Domo Formation contains open marine elements (planktonic foraminifera), its euxinic nature suggested to Flores (1973) that it represented a restricted environment which is replaced eastwards by a more open water facies under the Zambezi Cone. The Domo Formation has been dated as Albian to Turonian and it grades upwards, with no apparent hiatus, into the overlying Grudja Formation. West of Maputo the Maputo Formation (Upper Cenomanian) is separated from the overlying Grudja Formation by a marked non-sequence that spans Upper Cenomanian to Upper Turonian, and locally Campanian time. Clearly, the important mid-Cretaceous hiatus described by Kennedy and Klinger (1971) from Zululand was effective only in southernmost Mozambique.

4.4.2.1.4 *Palaeontology of the Makatini, Mzinene and Maputo formations.* The macrofauna of Zululand and south Mozambique is dominated by an abundant and diverse fauna of lamellibranchs, gastropods and cephalopods, but only the last have been systematically studied and revised in line with modern taxonomy. Other groups are poorly represented (usually <10 %, Tab.32) and in Zululand, Kennedy & Klinger (1975) reported the following distribution: belemnites and brachiopods occur in numbers only in the Aptian, echinoids are rare except for a few horizons in the Albian and Cenomanian; corals are common only in the Cenomanian, and arthropods are rare throughout. Worm casts and tracks are common.

Vertebrate remains are not common, and Kennedy & Klinger (1975) noted large reptile remains only in the Albian – otherwise only fish teeth were recorded.

Table 29. Ammonite zonation of Zululand Cretaceous rocks (amended slightly after Kennedy & Klinger 1975).

MAASTRICHTIAN

Maastrichtian III: No ammonites present. Inoceramid debris is abundant.

Maastrichtian II: Coarsely ornamented baculitids of the *Eubaculites ootacodensis* type are abundant. Pachydiscids are also present.

Maastrichtian I: Feebly ornamented to smooth *Eubaculites* are common. Other ammonites include *Saghalinites* sp., *Pachydiscus (Neodesmoceras), Menuites, 'Epiphylloceras'* and *Hoploscaphites.*

CAMPANIAN

Campanian V: Giant *Bostrychoceras* are abundant, with scarcer *Saghalinites* and compressed pachydiscids.

Campanian IV: *Saghalinites cala* and *Pachydiscus (P.)* sp. are common. Other ammonites include *Gunnarites antarcticus, Nostoceras?* sp., *Pachydiscus (Neodesmoceras)* sp.

Campanian III: Faunas are sparse but distinctive. A feebly nodose *Baculites* is abundant, and giant (1 m) pachydiscids (probably *Eupachydiscus*) are very common.

Campanian II: *Menabites (Australiella)* is abundant in the lower part of this division but species including *M.(A.)australis* and *M.(A.)besairei,* together with *Bevahites* spp. range throughout. *Baculites sulcatus* is abundant throughout, whilst pachydiscids become common in the higher parts: e.g. *Anapachydiscus subdulmensis, A.wittekindi, A.arialoorensis* and *Pachydiscus manambolensis.* Other ammonites include *Hoplitoplacenticeras plasticum plasticum, Maorites* sp., *Neogaudryceras* sp., *Gaudryceras* sp. and *Bostrychoceras* sp.

Campanian I: *Submortoniceras woodsi* (Spath) and related forms are common; other ammonites include *Bevahites* spp. and *Menabites* spp., *Hauericeras gardeni. Pseudoschloenbachia, Bostrychoceras* spp., *Vendegiesiella* sp. cf. *spinosa, V.trituberculata, Karapadites* and diplomoceratids.

The local base is drawn below the level of abundant *Submortoniceras.*

SANTONIAN

Santonian III: *Hauericeras gardeni* is abundant. The remainder of the fauna is relatively scarce and is made up of: *Plesiotexanites stangeri* and varieties, *Texanites soutoni, T.* spp., *Pseudoschloenbachia, ?Eupachydiscus, Hyphantoceras, Reginaites, Submortoniceras, Bevahites* and diplomoceratids.

Santonian II: *Plesiotexanites stangeri* and varieties are abundant. The remainder of the fauna consists of: *Texanites soutoni, T.* spp., *Hauericeras, Pseudoschloenbachia, ?Eupachydiscus, Hyphantoceras* and diplomoceratids.

Santonian I: *Texanites oliveti, Plesiotexanites stangeri densicosta* and *sparsicosta, Eutexanites, Paratexanites. Hauericeras gardeni, Pseudoschloenbachia, Pseudophyllites indra, ?Karapadites, ?Eupachydiscus, Gaudryceras, Hyphantoceras* and diplomoceratids.

The local base is drawn at the level of the appearance of *Texanites* s.s. in numbers.

CONIACIAN

Coniacian V: Abundant baculitids ornamented only by growth striae. Also forms resembling *Pseudoschloenbachia primitiva* Collignon. *Scaphites, Tetragonites, Protexanites, Texanites* and *Paratexanites* occur.

Coniacian IV: *Baculites* gr. *capensis* are abundant, and compressed, finely ornamented peroniceratids, *Zuluites,* and robustly ornamented *?Gauthiericeras* (e.g. 'Falsebayites', 'Fluminites', 'Hluhluweoceras' and 'Andersonites' of van Hoepen) are locally common. *Tetragonites, Protexanites* and *Paratexanites* also occur.

Coniancian III: *Placenticeras,* coarsely ornamented peroniceratids *(Zuluiceras), Protexanites, Miotexanites, Paratexanites, Baculites bailyi, Kossmaticeras* and *?Praemuniericeras* are common.

Coniacian II: *Proplacenticeras kaffrarium* is abundant. *Peroniceras* gr. *tridorsatum,* and *Forresteria* are common. Other types include: 'Eedenoceras' *multicostatum, Basseoceras krameri, Kossmaticeras sparsicosta, K.sakondryense, Puzosia* spp., *Pachydesmoceras, Lewesiceras australe, Yabeiceras* spp., *Pseudoxybeloceras matsumotoi, Hyphantoceras reussianum, Allocrioceras* spp., *Baculites bailyi, Scaphites meslei* and *Protexanites.*

Coniacian I: *Proplacenticeras kaffrarium* is abundant. Other types include *Kossmaticeras theobaldianum, Bostrychoceras indicum, Pachydesmoceras denisonianum* and *P.* sp.

The local base is drawn at the level of appearance of *Kossmaticeras theobaldianum.*

TURONIAN

No Turonian rocks known from outcrop in Zululand. They occur at depth in boreholes in northern Zululand.

CENOMANIAN

Cenomanian IV: Sparsely fossiliferous. Fauna includes: *Calycoceras* gr. *choffati, C. nitidum, C.* gr. *naviculare* and *Eucalycoceras.* Highest parts of the Cenomanian are missing.

Cenomanian III: *Turrilites acutus* is abundant throughout, whilst *Acanthoceras* spp. is abundant in the lower part and *Calycoceras* gr. *choffati* abundant in the upper part. Other forms include: *Turrilites costatus, T.scheuchzerianus, Acanthoceras cornigerum, Forbesiceras largilliertianum, F. sculptum, Calycoceras gentoni paucinodatum, Desmoceras, Hypophylloceras, Borissiakoceras, Anisoceras, Stomohamites, Sciponoceras, Scaphites, Puzosia, Bhimaites.*

Cenomanian II: *Neostlingoceras rorayensis* is common, with the remainder of the fauna consisting of: *Hypoturrilites carcitanensis, H.gravesianus, H.tuberculatus, H.nodiferus, Mariella* spp., *Sciponoceras roto, Scaphites, Desmoceras latidorsatum, Tetragonites subtimotheanus, Forbesiceras largilliertianum, Sharpeiceras laticlavium, Mantelliceras* spp. and a number of desmoceratids.

Cenomanian I: *Sharpeiceras florencae, S.falloti* and *Mariella oehlerti* are abundant. Other forms include: *Desmoceras latidorsatum, Sciponoceras roto, Scaphites* cf. *simplex, Mariella, Ostlingoceras, Hypoturrilites, Mantelliceras* and *Flickia.*

The local base is drawn at the incoming of abundant *Sharpeiceras* and *Mariella oehlerti.*

ALBIAN

Albian VI: Characterized by the appearance of *Durnovarites* spp. and *Stoliczkaia* spp. *Idiohamites, Hamites* and *Anisoceras* are abundant, whilst *Lechites, Mariella, Hypengonoceras, Tetragonites* and puzosiids are scarcer.

Albian V: Characterized by the abundance of mortoniceratids. Genera present include: *Hysteroceras, Oxytropidoceras, Tarfayites, Dipoloceras, Diplasioceras, Mortoniceras, Deiradoceras, Erioliceras, Arestoceras, Cainoceras, Puzosia, Bhimaites, Desmoceras, Hypophylloceras, Anagaudryceras, Gaudryceras, Tetragonites, Hamites, Anisoceras, Labeceras, Myloceras, Jouberticeras* and *Protetragonites.*

Albian IV: *Oxytropidoceras, Manuaniceras* and *Androiavites* are common. Other genera present include: *Pseudhelicoceras, Mojsisovicsia, Hypophylloceras velledae, Jouberticeras, Argonauticeras* and desmoceratids.

Albian III: *Douvilleiceras, Anagaudryceras sacya, Eubrancoceras* aff. *aegoceratoides* and *Oxytropidoceras* are abundant, *Carinophylloceras* is common, and *Lyelliceras* is frequent. Other forms include: *Umsinenoceras, Hypophylloceras, 'Beaudanticeras', 'Cleoniceras', 'Sonneratia', Rossalites, Ammonoceratites, Alopecoceras, Argonauticeras* and *Pictetia.*

Albian II: *Douvilleiceras,* including *D.orbignyi* and *D.mammillatum* is abundant. Other forms are scarce, but include *Ammonoceratites, Pictetia* and democeratids and lytoceratids.

Table 29 (continued).

Albian I: Absent.
Local base drawn at appearance of *Douvilleiceras*.

APTIAN
Aptian IV: Giant, fine-ribbed forms of *Tropaeum* are abundant, and *Lytoceras* is common. Other forms include: *Tonohamites, Acanthoplites, Diadochoceras nodosocostatum, Australiceras, Sinzovia, Toxoceratoides, Helicancyloceras* and *Nonyaniceras*.
Aptian III: Characterized by an abundance of *Acanthoplites, ?Diadochoceras, Valdedorsella, Phylloceras, Ancyloceras, Protanisoceras, Tonohamites* and *Lytoceras*. Other forms include: *Tropaeum, Australiceras, Toxoceratoides, Helicancyloceras* and *Nonyaniceras*.
Aptian II: *Cheloniceras s.s.* becomes frequent, together with *Valdedorsella* or *Pseudohaploceras, Ancyloceras, Tropaeum* and *Australiceras*. Other forms include: *Lytoceras, Adouliceras* and *Toxoceratoides*. A non-sequence may separate Aptian II and III.
Aptian I: *?Procheloniceras* is abundant. Other forms include: *Tropaeum, Ancyloceras, Theganeceras, Lytoceras, Adouliceras* and *Australiceras*.
Local base drawn at appearance of cheloniceratids.

BARREMIAN
Barremian II: *Colchidites* occur in very large numbers. Other forms include: *Sanmartinoceras, Phylloceras, Lytoceras* and *Ancyloceras*.
Barremian I: Characterized by an abundance of crioceratitids, including 'Emericiceras', 'Acrioceras', hemihoplitids, *Heteroceras*, aconeceratids, *?Sanmartinoceras, Phylloceras serum, Eulytoceras phestum*. Occasional *Colchidites, Lytoceras* and *Ancyloceras* also occur.
Local base drawn at appearance of ammonite faunas, and the presence of *Colchidites* indicates an Upper Barremian age.

Plant remains, on the other hand, are extremely abundant, with logs up to several metres long common on many levels and log beds in the Makatini Formation. Lignite chips are a significant sediment component throughout the succession.

Kennedy & Klinger in a series of publications (1975 et seq.) have detailed the ammonite faunas of the Zululand Cretaceous, and have erected a biostratigraphic zonation scheme which includes the Lower Cretaceous Makatini and Mzinene formations (Tab.29). No complete species list has yet been published, but Table 30 shows the genera recorded during their revisions, whilst Table 31 shows trends at generic level within the various Cretaceous stages. In the Lower Cretaceous, greatest diversity occurs in the Albian (43 genera), with the most restricted faunas in the Barremian (9). Similarity between faunas of adjacent stages is relatively low within the same formations: 12 % in the Makatini Formation, and 13 % in the Mzinene Formation, but across the Aptian/Albian non-sequence that separates the two formations, similarity is even lower – 2 %, indicating a high turnover rate at the generic level during the short time gap involved (Upper Aptian IV to Lower Albian II). Within the Makatini Formation, the Barremian strata are typified by *Colchidites*, which suggests that they correspond to the Upper Barremian of Europe (Kennedy & Klinger 1975), and the Aptian is marked by incoming of the early cheloniceratids (*?Procheloniceras* in Aptian I, and *Cheloniceras* in Aptian II). The presence in Aptian I-IV of various *Lytoceras, Ancyloceras*, and *Phylloceras* provide a link between the lower (Barremian) and upper (Aptian) parts of the Makatini Formation. The very low similarity across the Makatini/Mzinene boundary may have been influenced by the somewhat deeper water environments that were established following the Aptian/Albian non-sequence, and only *Rossalites* is common to both formations. The lowermost Mzinene Formation is Albian II in age and is characterized by the appearance of *Douvilleiceras* which is abundant in Aptian II-III, but is replaced in Aptian IV-V by abundant mortoniceratids. Genera such as *Mariella, Puzosia, Desmosceras* and *Tetragonites* appear in the Upper Aptian and form a relatively strong similarity link with the overlying Cenomanian part of the Mzinene Formation, which is characterized by the appearance of abundant *Sharpeiceras* and *Mariella*. Towards the top of the Mzinene Formation (Cenomanian IV), ammonite faunas become sparse as water depths shallowed and the energy of sedimentary environments increased. The top of the Cenomanian is missing, and although the similarity between the Cenomanian and overlying Coniacian ammonite faunas are relatively high (6 %) for such a long non-sequence, it is significant that this is only in relatively long-ranging genera: *Puzosia, Scaphites* and *Tetragonites*. Endemism in the Lower Cretaceous ammonite genera of Zululand reaches a maximum of 27 % in the Albian, after a sharp increase across the Makatini/Mzinene Formation boundary (Aptian 15 %), but decreases up the succession to 10 % in the Cenomanian.

In Mozambique, recent collecting in the small exposures of the Maputo Formation SW of Maputo, and west of the Maputo River (Catuane, Powell's Camp, Chalala, and Mahube River) by Wachendorf (1967), Cooper (1974) and Förster (1975b) has produced a very similar assemblage to that described by Kennedy & Klinger (1975) from the Makatini and Mzinene formations of Zululand. There are some differences, however, and genera not recorded by Kennedy & Klinger (1975) include *Crioceratites* (Barremian), *Neosilesites* (Aptian), *Prohysteroceras, Cantabrigites* and *Goodhallites* (Albian).

Microfossils are the other Cretaceous fossil group that has received considerable attention in recent years, although from Zululand the Lower Cretaceous is less studied than the Upper Cretaceous. McLachlan & McMillan (1979) did not record planktonic foraminifera from the Makatini or lower Mzinene formations, but noted *Rotalipora apenninica* and *R.cushmani* from the upper part of the latter formation (Tab.28). These are international zonal forms, and *R.cushmani* confirms

Table 30. Ammonite genera from Barremian to Maastrichtian strata in Zululand-Natal (revised, after Kennedy & Klinger 1975).

Barremian (9)
*Acrioceras*
*Ancyloceras*
*Colchidites* (C)*
*Emericiceras**
*Eulytoceras*
*Heteroceras*
*Lytoceras*
*Phylloceras*
*Sanmartinoceras*

Aptian (20)
*Acanthoplites**
*Aconeceras*
*Adouliceras***
*Ancyloceras*
*Australiceras*
*Cheloniceras**
*Diadochoceras*
*Helicancyloceras***
*Lytoceras*
*Nonyaniceras***
*Phylloceras*
*Procheloniceras**
*?Protanisoceras*
*?Rossalites*
*Sinzovia*
*Theganeceras*
*Tonohamites*
*Toxoceratoides*
*Tropaeum**
*Valdedorsella*

Albian (43)
*Alopecoceras***
*Ammonoceratites*
*Anagaudryceras**
*Anisoceras**
*Arestoceras***
*Argonauticeras*
*Beaudaniticeras*
*Bhimaites***
*Cainoceras***
*Carinophylloceras***
*Cleoniceras*
*Deiradoceras*
*Desmoceras*
*Dipoloceras*
*Douvilleiceras**
*Durnovarites*
*Erioliceras***
*Eubrancoceras**
*Gaudryceras*
*Hamites**
*Hypengonoceras***
*Hypophylloceras*
*Hysteroceras*
*Idiohamites**
*Labeceras***
*Lechites*
*Lyelliceras**
*Manuaniceras** ***
*Mariella*
*Mojsisovicsia*
*Mortoniceras**
*Myloceras***
*Oxytropidoceras**
*Pictetia*
*Protetragonites*
*Puzosia*
*Rossalites*
*Sonneratia*
*Stoliczkaia*
*Tarfayites*
*Tetragonites*
*Umsinenoceras***
desmoceratids

Cenomanian (21)
*Acanthoceras**
*Anisoceras*
*Bhimaites***
*Borissiakoceras*
*Calycoceras**
*Desmoceras*
*Eucalycoceras*
*Flickia***
*Forbesiceras*
*Hypophylloceras*
*Hypoturrilites**
*Mantelliceras**
*Neostlingoceras**
*Ostlingoceras**
*Puzosia*
*Scaphites*
*Sciponoceras*
*Sharpeiceras**
*Stomohamites*
*Tetragonites*
*Turrilites**

Coniacian (29)
*Allocrioceras*
*Andersonites*
*Baculites**
*Basseoceras*
*Bostrychoceras*
*Eedenoceras***
*Fluminites*
*Forresteria**
*?Gauthiericeras*
*Hluhluweoceras*
*Hyphantoceras*
*Kossmaticeras*
*Lewesiceras*
*Miotexanites*
*Pachydesmoceras*
*Paratexanites*
*Peroniceras*
*?Praemunericeras*
*Proplacenticeras**
*Protexanites*
*Pseudoschloenbachia*
*Pseudoxybeloceras*
*Puzosia*
*Scaphites*
*Tetragonites*
*Texanites*
*Yabeiceras*
*Zuluiceras***
*Zuluites***

Santonian (16)
*Bevahites***
*?Eupachydiscus*
*Eutexanites***
*Gaudryceras*
*Hauericeras**
*Hyphantoceras*
*?Karapadites***
*Paratexanites*
*Plesiotexanites**
*Pseudophyllites*
*Pseudoschloenbachia*
*Reginaites*
*Saghalinites*
*Submortoniceras*
*Tetragonites*
*Texanites**
diplomoceratids

Campanian (22)
*Anapachydiscus*
*Australiella** ***
*Baculites**
*Bevahites***
*Bostrychoceras**
*Dalawarella*
*Eupachydiscus**
*Gaudryceras*
*Gunnarites***
*Hauericeras*
*Hoplitoplacenticeras*
*Maorites***
*Menabites**
*Neodesmoceras*
*Neogaudryceras*
*Nostoceras*
*Pachydiscus*
*Pseudophyllites*
*Pseudoschloenbachia*
*Saghalinites**
*Submortoniceras**
*Tetragonites*

Maastrichtian (6)
*Ephiphylloceras*
*Eubaculites**
*Hoploscaphites*
*Menuites*
*Neodesmoceras*
*Saghalinites*

* Abundant or diverse
** Endemic to Indo-Malagasy-Australian province

Table 31. Number of ammonite genera per stage and similarity between stages at generic level in Zululand (data from updated lists in Kennedy & Klinger 1975).

| Stage | No.of genera | % similarity | |
|---|---|---|---|
| Maastrichtian | 6 | | |
| | | 8 | |
| Campanian | 22 | | |
| | | 30 | due to Texanitinae |
| Santonian | 16 | | |
| | | 13 | due to Texanitinae |
| Coniacian | 29 | | |
| Turonian | – | 6 | |
| Cenomanian | 21 | | |
| | | 13 | |
| Albian | 43 | | |
| | | 2 | |
| Aptian | 20 | | |
| | | 12 | |
| Barremian | 9 | | |

the Upper Cenomanian age of the upper part of the Mzinene Formation. Benthic foraminifera have received only limited attention but Lambert & Scheibnerova (1974) have described an assemblage from the Albian of the Mzinene Formation NW of False Bay, which shows great similarities with faunas of a similar age from the Great Artesian Basin of Australia.

Siesser (1982) has reported that calcareous nannofossils from Makatini and Mzinene sediments are generally poorly preserved and sparse, with *Watznaueria barnesae* the most consistent taxon.

No details have yet been published of the ostracod faunas from the Lower Cretaceous of Zululand, but preliminary results (Dingle personal data) (Tab.24) indicate that the fauna contains elements that link it with the progonocytherid/schizocytherid/cytherurid dominated assemblages that characterized a Middle Jurassic to Cenomanian ostracod province in south Gondwanaland. The composition of the Zululand Aptian to Cenomanian fauna (no ostracods have so far been recovered from the Barremian strata which are heavily decalcified) is typical of a late assemblage from the south Gondwana faunal province, with no representatives of the earlier dominant *Progonocythere/Amicytheridea* group, and only small numbers of *Sondagella* (which is common in the Upper Jurassic and Neocomian of Argentina and the Agulhas Bank (Musacchio 1979)). It contains late representatives of *Majungaella* cf. *M.nematis*, as well as early trachyleberids (including *Isocythereis*) and shows close ties with Madagascar with *Pirileberis* spp. Although not fully documented, it is clear that the Zululand Lower Cretaceous ostracod fauna is relatively sparse, and taxonomically contrasts strongly with the assemblages that appear at the base of the overlying Coniacian St Lucia Formation. Only one cytheracean genus and none of the Lower Cretaceous species reappear after the Coniacian-Turonian hiatus, where the cytheracean fauna is trachyleberid/brachycytherid/schizocytherid-dominated. This dramatic change is ascribed to colonization by taxa from the

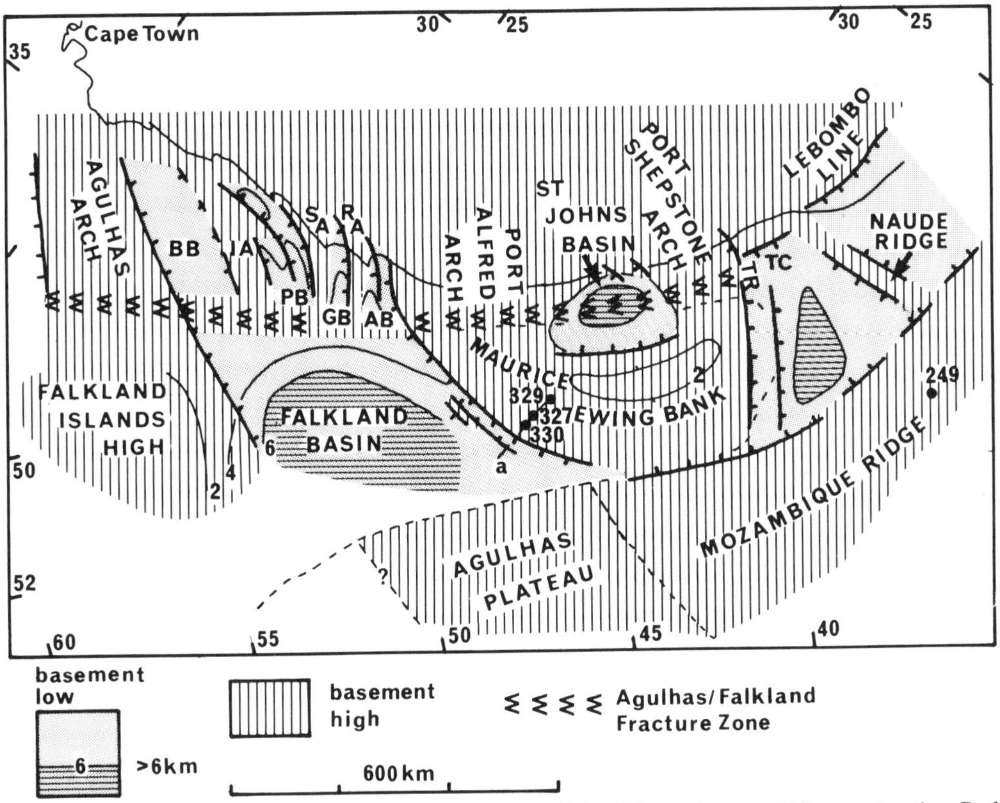


Figure 104. Main structural features of S.E.Africa and the Falkland Plateau in a pre-drift reconstruction. Dashed line around the Falkland Plateau is the present 3 km isobath. DSDP sites 249, 327, 329 and 330 shown by dots. Line 'a' is cross-section in Figure 110. Summary depths to basement (km) below present sea level are shown for some areas. Coordinates are present latitudes and longitudes. The Agulhas/Falkland Fracture Zone would not have been in existence in the pre-drift stage, but is shown for completeness. The refit is based on that of Simpson (1977), but involves 119 km less penetration into the Natal Valley, and is similar to that of Martin *et al.* (1982a) which was achieved independently. The southern Agulhas Plateau abuts the southern end of the Mozambique Ridge (following Tucholke *et al.* 1981). Abbreviations: BB = Bredasdorp Basin; IA = Infanta Arch; PB = Pletmos Basin; SA = St Francis Arch; GB = Gamtoos Basin; RA = Recife Arch; AB = Algoa Basin; TR = South Tugela Ridge; TC = Tugela Cone.

Table 32. Faunal composition of the Maputo Formation, Southern Mozambique (after Förster 1975b). Values are percentages of numbers of individuals.

| | Ammonites | Lamellibranchs | Gastropods | Echinoids | Others | No. of specimens |
|---|---|---|---|---|---|---|
| 1. Upper Albian | 78 | 16 | 4 | | 2 | 278 |
| 2. Upper Albian | 19 | 54 | 25 | | 2 | 123 |
| 3. Lower Aptian | 52 | 23 | 13 | 11 | 1 | 257 |

1. Lagoa Catuine, 26°48.98′S, 32°14.92′E.
2. Lagoa Catuine, shallow water assemblage, 26°48.77′S, 32°14.68′E.
3. Chalala, 26°18.00′S, 32°19.65′E.

South Atlantic which moved into the western Indian Ocean area in late Cenomanian-early Turonian times following palaeogeographic changes that allowed movement across the Walvis Ridge-Rio Grande Rise archipeligo from the Brazil and West Africa region (Dingle 1982a).

### 4.4.2.2 *Early Cretaceous strata in the northern Natal Valley and on the Mozambique Ridge*

The northern Natal Valley is a stratigraphically and structurally complex region that prior to continental separation between Africa and South America lay to the north of the Falkland Plateau (Fig.104). It is flanked on its east and west sides by large basement highs (the main continent, and the Mozambique Ridge) and is thought to be underlain by thinned and subsided continental crust. No boreholes have been drilled in the valley proper and all stratigraphic knowledge comes from seismic data, and borehole sites on peripheral basement highs: J(c)-1, DSDP 249 (both offshore) and

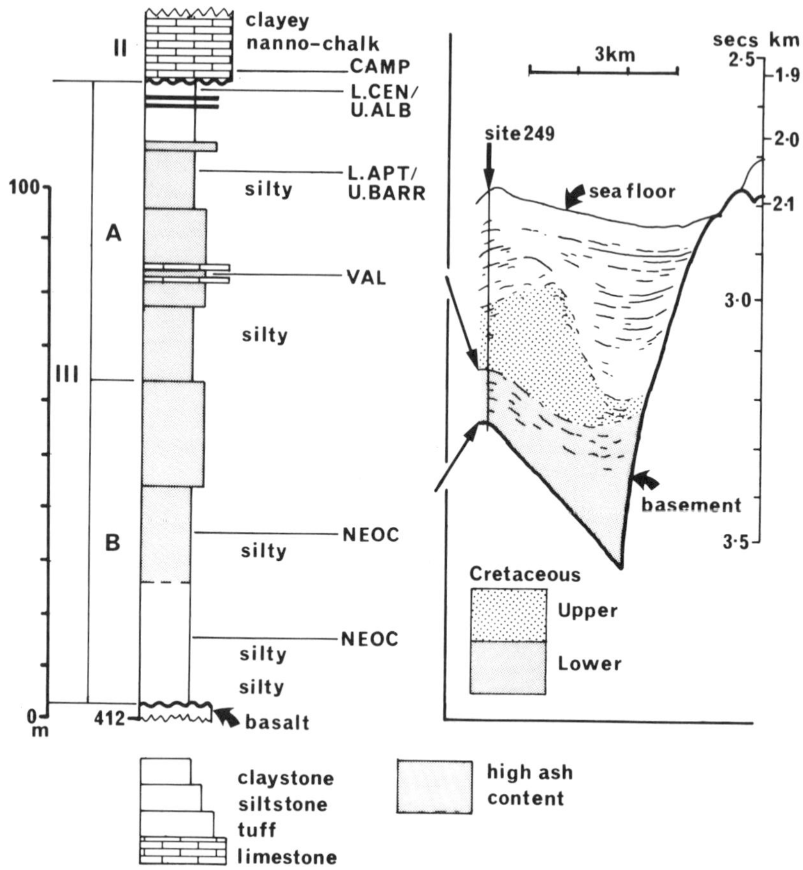


Figure 105. Details of Lower Cretaceous sediments on the Mozambique Ridge at DSDP site 249. Borehole log constructed from corelog descriptions in preliminary DSDP report. Cross-section shows drill site superimposed on seismic reflection profile, with vertical scale in sec and km below sea level. Km scale calculated at 1500 km/sec. All based on data in Simpson *et al.* (1974).

Zandamela (onshore, south Mozambique) (Fig.106). Dingle et al (1978) recognized two regionally-developed seismic reflecting horizons in the northern Natal Valley (McDuff and Angus) which they equated with mid-Cretaceous and mid-Tertiary sedimentary events (hiatuses, or facies changes), respectively (Fig.106). Figure 107 shows these horizons in relation to various non-sequences, unconformities and lithofacies changes in the Natal Valley and adjacent areas.

Reflector McDuff abuts basement on the western flank of the Natal Valley about 500 m below and 7 km east of the Cenomanian-Turonian base in J(c)-1 borehole, which overlies Palaeozoic sediments and granite (du Toit & Leith 1944) (Fig.106). Seismic records (e.g. Fig.171) show older (?Cenomanian-Turonian) sediments progressively lapping onto basement, indicating that McDuff is an important ?Cenomanian-Turonian sedimentary horizon. Dingle et al (1978) equate it with the early Cenomanian to Turonian hiatus at the top of the Mzinene Formation in Zululand, and the early Cenomanian to early Campanian non-sequence in the DSDP 249 borehole. McDuff's position on seismic records also indicates that it coincides with the refraction interface between 2 km/sec and 4 km/sec strata located by Ludwig et al (1968) at their site 160 on the eastern side of the Tugela Cone.

Nominally taking McDuff to represent the top of Lower Cretaceous sediments in the Natal Valley, it can be seen that pre-Cenomanian strata thicken under the Tugela Cone, where a large pre-McDuff sedimentary structure about 3.5 km thick lies along the western side of the Natal Valley; presumably a palaeo-Tugela Cone. Pre-Cenomanian strata apparently thin in an easterly direction and abut the western flank of the Mozambique Ridge below about 3.8 seconds. Reflector McDuff cannot be traced across the Mozambique Ridge, suggesting that most of the basement is covered by post-Cenomanian strata, but pockets of older sediment have been proved by DSDP site 249 (Simpson et al 1974), which lies on top of the ridge crest, in an asymmetric (?faulted) basin (Figs 105 & 106). Here, 121 m of Neocomian sediments overlie vesicular, glassy, weathered basalts. Although it was subdivided into two parts by the shipboard scientists (IIIa & IIIb in their terminology), the sequence is lithologically similar throughout and consists of laminated and locally low angle cross-bedded, olive black and medium grey silts and clays, often relatively rich in organic carbon (up to 1.6 %), and in reduced iron oxides and pyrite. With the exception of the upper and lower parts, and two thin limestones,

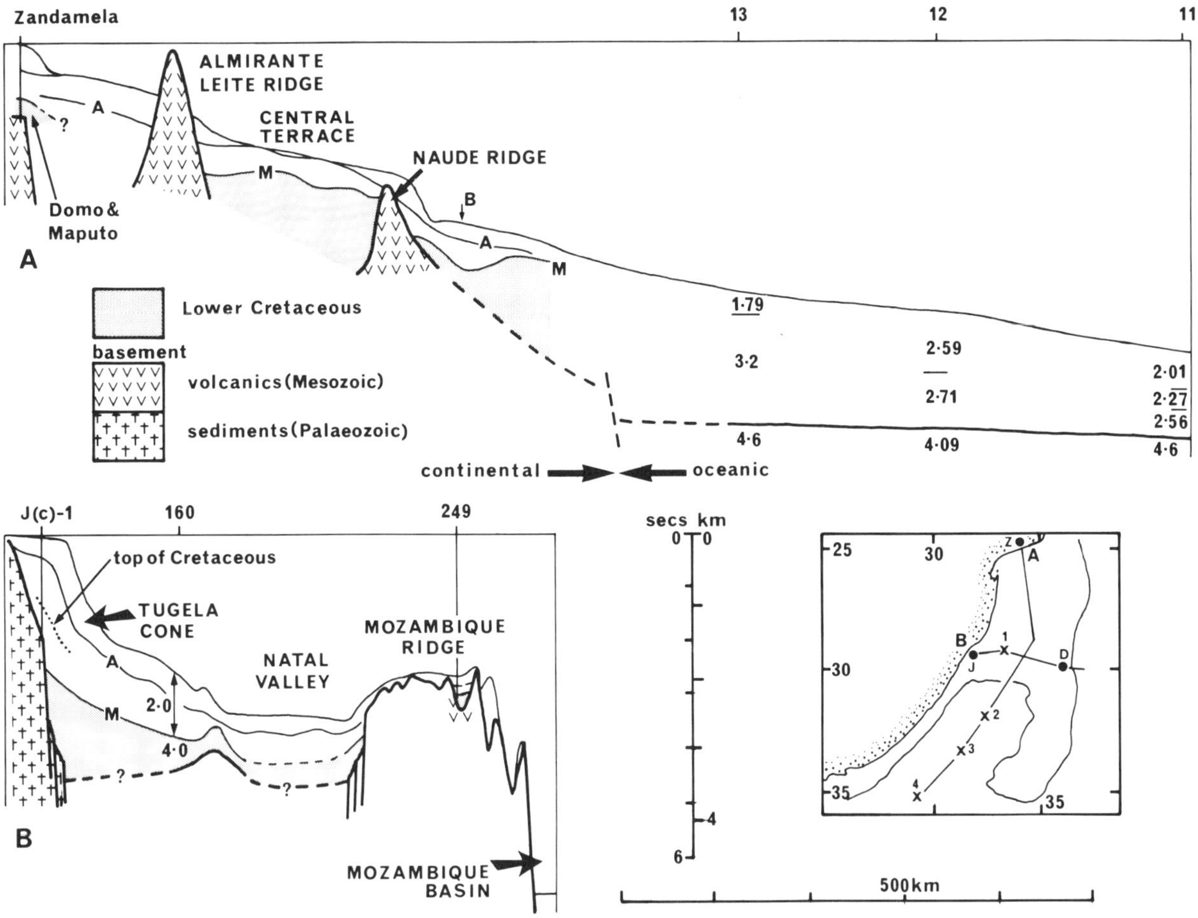


Figure 106. Sections along (A) and across (B) the Natal Valley. Strata designated Lower Cretaceous include Turonian in the Zandamela borehole, Albian/Cenomanian at DSDP 249, and may include Upper Jurassic in the deep basins N of about 30°S. Regional seismic reflectors McDuff (mid-Cretaceous) and Angus (mid-Tertiary) are used for correlation in the N.Natal Valley. In the south, subdivision is by seismic refraction data, although no consistent correlation can be achieved. Km scale only valid for water depths. Abbreviations: M = reflector McDuff; A = reflector Angus; Z = Zandamela; J = J(C)-1; D = DSDP 249; x = refraction site; 1 = 160; 2 = 13; 3 = 12; 4 = 11. Data from Ludwig *et al.* (1968a); Hales & Nation (1973); Flores (1973), du Toit & Leith (1974); Simpson *et al.* (1974), Dingle *et al.* (1978).

the whole succession has a high content of fine-grained volcanic debris (ash and glass) which locally constitutes 95 % of the sediment. Three thin limestones (composed of irregular silt and clay-sized carbonate fragments) occur in the upper unit (IIIa) and the silts locally have a calcite cement. Sparse detrital quartz and feldspar occur throughout, whilst glauconite, rare calcareous foraminifera and ostracods occur to within 2 m of the base. The age of these sediments is not precisely known. At the very top, core 23 (which is radiolaria-rich) produced a good planktonic foraminifera assemblage with *Hedbergella delrioensis, Hedbergella* (3 spp.), *Rotalipora balernaensis, Planomalina* sp. and *Ticinella* sp. which indicates an Upper Albian/Lower Cenomanian age (Simpson et al 1974). Below this level, positive identifications of age diagnostic taxa were few. Small planktonic foraminifera were encountered in core 26, and suggested an Upper Barremian/Lower Aptian age. They were globigerinids, but positive identifications were not possible. Ostracods similar to *Sondagella theloides* were also found in this core but this species has a range ?Portlandian to Albian and merely confirms the general position of the level. The presence of *Majungaella nematis* in cores 26 and 27 suggested a Valanginian age to Simpson et al (1974) (it probably has a range Portlandian to Cenomanian) and at two other levels (near the base) the benthic foraminifera were taken to indicate a Neocomian age. With this relatively poor palaeontological control it is neither possible to determine the age of the oldest sediments, nor to detect any breaks in the sequence. Nominally, therefore, a Neocomian (?pre-Valanginian) to ?Lower Cenomanian age is allocated for the lower part of DSDP site 249 (Fig.107).

Reflector McDuff can be traced northwards as far as the Almirante Leite Ridge (Dingle et al 1978). Assum-

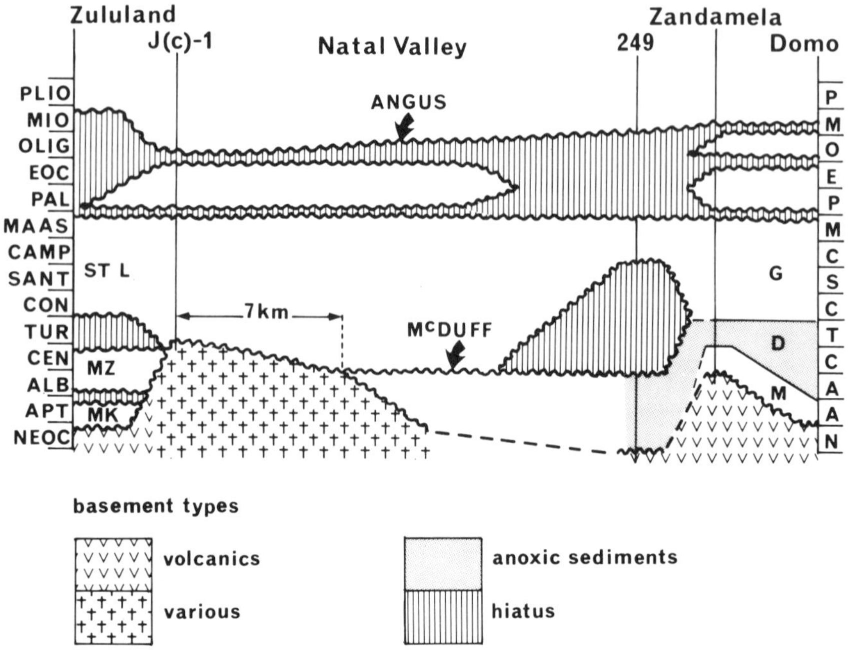


Figure 107. Stratigraphic relationships of regionally-developed hiatuses in the Natal Valley, Zululand, and S.Mozambique area. Not to scale. Abbreviations of formation names: STL = St Lucia; MZ = Mzinene; MK = Makatini; G = Grudja; D = Domo; M = Maputo. Data from Flores (1973), Simpson *et al.* (1974); du Toit & Leith (1974); Forster (1975a); Kennedy & Klinger (1975); Dingle *et al.* (1978); McLachlan & McMillan (1979). See Figures 98 & 106 for locations.

ing it is synchronous (for which there is admittedly no evidence) then both the Almirante and Naudé ridges are probably pre-Cenomanian in age and they subdivided the early Natal Valley sediment basin. Asymmetric elevations of McDuff to the north and south of the Naudé Ridge indicate that the ridge acted as an important sediment dam or current deflector during the pre-Cenomanian period of sedimentation, with over 2 km of pre-McDuff sediment to the north and nil or less than 500 m on the southern flank. Farther south, thicknesses increase, possibly as a result of lateral input across the Tugela Cone.

North of the Almirante Leite Ridge, McDuff has not been recognized beneath the present Limpopo Cone. The Zandamela borehole sits on top of a volcanic capped basement high, but correlation of its sedimentary succession (Fig.103) into the Natal Valley can be made only tentatively (Figs 106 & 107). It does not show a mid-Cretaceous hiatus, but there is a marked facies change from the Maputo Formation to the dark, organic shales of the Domo Formation (Fig.113). However, this occurs in Turonian/Coniacian sediments, which is significantly younger than the assigned age for McDuff – which may of course be diachronous, younging to the north. We have no way of testing these various possibilities and merely present Figure 107 as a working hypothesis. On Figure 106, therefore, the shaded sediments (pre-McDuff horizons) probably represent distal, deeper water facies equivalents of the Makatini, Mzinene, Maputo and the lower parts of the Domo formations, as well as the whole of unit III in DSDP 249. We will comment further on this in the section on palaeogeography.

#### 4.4.2.3 *Transkei Swell: Mngazana and Mbotyi outliers; St Johns Basin*

Two small outliers in the vicinity of Port St Johns on the Transkei coast (Figs 108 & 109) represent the onshore extensions of a small margin basin (St Johns Basin) which developed on the crest of the Palaeozoic basement high shown in Figure 97. A conjugate feature on the northern side of the Maurice Ewing Bank (Falkland Plateau) is thought to be the pre-drift eastern half of the St Johns Basin (Fig.104).

Both outliers lie on the southern side of large boundary faults and generally the sedimentary fill dips relatively steeply towards the fault planes (ca.15° at Mbotyi and 20-33° at Mngazana) i.e. they have a half graben configuration. The recently revised geological map of the area (Geological Survey of South Africa 1977) (Fig. 109) shows that the Cretaceous sediments are the youngest strata in a series of small E-W graben and half graben in the vicinity of Port St Johns which bring Upper Palaeozoic to Cretaceous rocks down against Lower Permian and Lower Palaeozoic strata, and that in the largest outlier (south of Mngazana) a reasonably complete Upper Permian to Middle Triassic sequence has been preserved. Although previously recorded in the literature, the most comprehensive description of the Cretaceous rocks of the two outliers is that recently given by McLachlan et al 1976b.

At Mngazana, the outcrop of Cretaceous rocks is confined to a small hill (Mqualeni) at the mouth of the Mngazana River estuary. Here, tough grey-green, coarse conglomerates, sandstones and grits, with minor pebble conglomerates, medium sandstones and rare calcareous

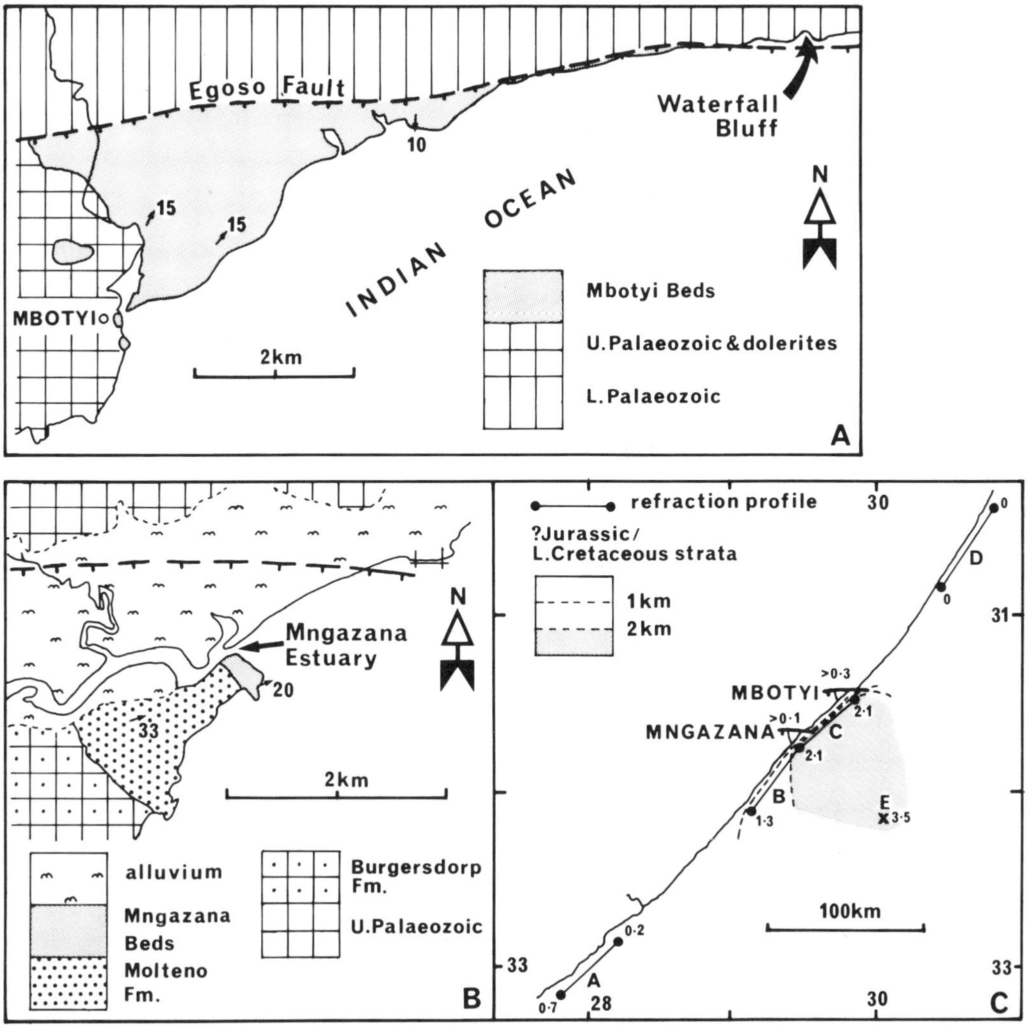


Figure 108. ?Jurassic-Lower Cretaceous strata along the SE coast of South Africa. A.Sketch map of Mbotyi outlier; B.Sketch map of Mngazana outlier; C. Thickness of strata on the continental shelf determined by seismic refraction lines: A = 154; B = 155; C = 156; D = 157, and E = 14. We have placed strata with velocities >2.0 and <5.0 km/sec in the category '?Jurassic-Lower Cretaceous', but the lower 2.32 km at site E (4.30 km/sec) may be pre-Jurassic 'basement'.
A & B are after McLachlan *et al.* (1976b), C compiled with data from du Toit (1920); Ludwig *et al.* (1968a), Chetty & Green (1977).

siltstones and shales overlie white gritty sandstones that du Toit (1912b, 1920) correlated with the Molteno Formation. The coarse conglomerates are composed mainly of dolerite boulders, but also contain fragments of Molteno sandstones, shale (? from the Ecca Group), lydianite, amygdaloidal basalt, and agates. The larger fragments (up to 4 m) are angular, but smaller clasts are well-rounded and often polished. Intercalated within the coarse conglomerates are fine conglomerates, greenish grits, brecciated and apparently slumped calcareous siltstones, and lignitic sandstones. The coarse beds are not confined to the basal parts of the sequence, but occur at intervals throughout it and shelly marine fossils occur in both the conglomerate matrix and within the rare calcareous siltstones. From the field relationships, lithology and fossils, it is clear that the Mngazana outlier represents a piedmont fan which was deposited directly into a shallow marine environment that at least periodically had access to the open sea. Du Toit (1920) estimated a total of about 100 m of exposed Cretaceous rocks. In spite of the rather unpromising lithofacies a surprisingly rich fossil assemblage has been obtained from the Mngazana Beds (Tabs 24 & 33). Cooper (1974), McLachlan et al (1976b) and Klinger & Kennedy (1979) have all assigned these beds to the Valanginian and equate them with the Lower Sundays River Formation. We have followed this correlation (Fig.67), although noting that in fact the palaeontological evidence for

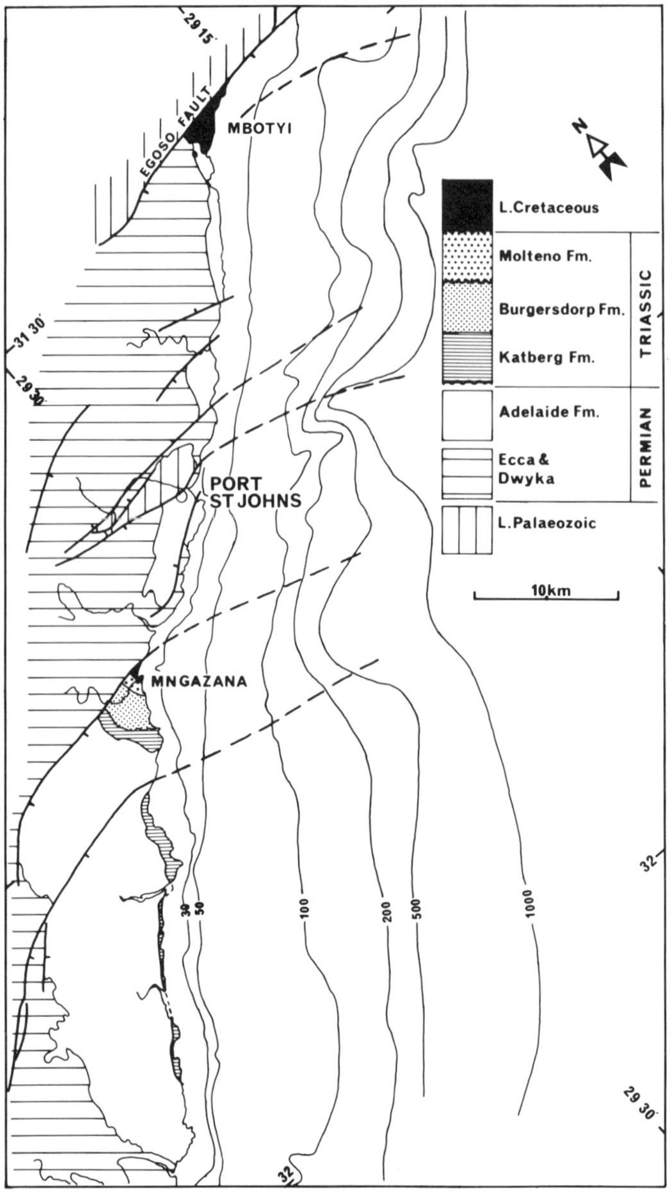


Figure 109. Mesozoic outliers on the collapsed crest of the Transkei swell. Discontinuities in the bathymetry of the adjacent continental shelf suggests extensions of some major faults. Geology is modified from Geological Survey (1977), and bathymetry is from unpublished charts (University of Cape Town).

synchroneity is not strong. For example, Klinger & Kennedy note that the ammonite fauna is 'unlike that known from the Uitenhage Formation', and the ostracods at Mngazana lack the important progonocytherid types which are the diagnostic elements in Gondwana Lower Cretaceous correlation. The types that are common to both areas (mainly Cytherellidae) are notoriously difficult to use and the lower ranges of the Sundays River Formation taxa have not been established, but extend at least into the Portlandian (McLachlan et al 1976a).

Mbotyi outlier (35 km to the north of Mngazana) appears to lie in a similar tectonic setting at the foot of a large fault scarp (Egoso Fault), along which the notable sea cliff features of Cathedral Rock and Waterfall Bluff occur. It unconformably overlies Karoo Supergroup rocks (Ecca and Lower Beaufort groups) and generally dips NE at about 15°. The outlier is triangular-shaped and mostly lies to the NE of the Mbotyi River, although small erosional outliers occur on the SW side of the stream. Good exposures are confined to the cliffed shore, and here the bulk of the sediments are coarse conglomerates with minor sandstones. The rudaceous components consist of poorly sorted cobbles and boulders (up to 3 m) which are predominantly dolerite, with subsidiary soft, Ecca-like shales, lydianite and Beaufort-like sandstones. Significantly, there are no quartzites unequivocally attributable to the Table

Table 33. Fossil assemblage from the Mngazana Beds (after du Toit 1912b, 1920, McLachlan et al. 1976b, Klinger & Kennedy 1979).

| | |
|---|---|
| Ammonites | *Lenticulina bifurcilla* |
| *Neohoploceras* sp. | *L.cultrata* gr. |
| *Umgazaniceras thieuloyi* | *L.nodosa* |
| *Bochianites* cf. *B.renevieri* | *L.subtilis* |
| Gastropods | *L.* spp. |
| *Pseudomelania suther-* | *Lingulina bettenstaedti* |
| *landi* | *L.* sp. |
| *Natica* sp. | *Marginulinopsis gracilissima* |
| Lamellibranchs | *Neoflabellina* sp. |
| *Trigonia shepstonei?* | *Nodosaria sceptrum* |
| *?Barbatia* sp. | *Nubecularia lucifuga* |
| *Pecten* sp. | *Patellina subcretacea* |
| *Panopea* sp. | *Planularia* cf. *madagascarien-* |
| *Corbula* sp. | *sis* |
| *Thetironia* sp. | *Planularia* spp. |
| Also unidentified corals and | *Pseudoglandulina humilis* gr. |
| belemnites | *Pseudopolymorphina* sp. |
| Plants | *Pyrulina cylindroides* |
| Ferns | *Quinqueloculina minima* |
| *Onychiopsis* sp. | *Reinholdella hofkeri* |
| *Cladophlebis* sp. | *R.valendisensis* |
| Cycads | *Saracenaria bronni* |
| *Nilssonia* sp. | *S.compacta* |
| *Otozamites* sp. | *Spirillina minima* |
| *Dictyozamites* sp. | *S.tenuissima* |
| Microfossils | *Tristix acutangulus* |
| Foraminifera | *T.insignis* |
| *'Allomorphina'* sp. | *Trocholina infragranulata* |
| *Ammobaculites* sp. | *Vaginulina* sp. |
| *Astacolus calliopsis* | *Webbinella subhemisphaerica* |
| *A.gibber* | Ostracods |
| *A.mutilatus* | *Acrocythere* sp. |
| *A.perobliqua* | *?Argilloecia* sp. |
| *A.* spp. | *Bairdiacypris* sp. |
| *Citharina sparsicostata* | *Cytherella algoaensis* |
| *Dentalina communis* gr. | *Cytherella inaequivalva* |
| *D.linearis* | *Cytherella schwarzi* |
| *D.* spp. | *Cytherella* sp.C |
| *Dorothia kummi* | *Eucytherura (V.)* aff. *tuber-* |
| *Eoguttulina anglica* | *culata* |
| *E.* cf. *inovroclaviensis* | *?Macrocypris* sp. |
| *E.* cf. *liassica* | *Metacytheropteron* sp. |
| *Epistomina caracolla* | *Monoceratina* sp. A |
| *Globulina prisca* | *?Paranotacythere* sp. |
| *Lagena apiculata neoco-* | *Procytherura maculata* |
| *miana* | Genus A *(?Paracytheridea)* |
| *L.hauteriviana hauter-* | Also unidentified radiolaria |
| *riviana* | |
| *L.* cf. *sulcata* | |

Mountain Group which crops out to the north of the Egoso Fault, yet the coarsest clasts generally occur in the vicinity of the fault plane. Wood fragments and lignite occur in minor sandstone intercalations, but no marine fossils have been found at Mbotyi so that their correlation with the Mngazana beds is purely on lithological similarity.

On the continental shelf adjacent to the Mngazana and Mbotyi outliers, two seismic refraction lines (155 and 156 of Ludwig et al 1968a) indicate a basin containing at least 2.1 km of medium velocity (2.5-4.24 km/sec) material (St Johns Basin, Fig.108). It lies between two basement highs (Port Alfred and Port Shepstone arches, Fig.104), and extends offshore for at least 60 km, where towards the base of the steep continental slope (in approximately 2.5 km of water), Chetty & Green (1977) have reported about 2.2 km of high velocity strata (4.3-4.84 km/sec), overlain by 1.3 km of medium velocity material. The nature of these strata are not known, but the presence of well-lithified Neocomian sediments on the coast suggests that much of the lower part of the offshore succession could be of this age or older – certainly if one applies the analogy of the southern Cape basins to this area, where locally velocities as high as 4.6 km/sec clearly equate with early graben fill (Dingle 1979). The apparent lack of high velocity material on the shelf (bearing in mind that the interval velocity is an average) suggests that the Mngazana/Mbotyi rudaceous facies may be confined to areas immediately adjacent to the fault plane extensions across the shelf, whilst a thick wedge of high velocity rudaceous facies may lie at the foot of and across the Agulhas Fracture Zone. How much of the medium velocity material can be ascribed to the Lower Cretaceous is not known.

Similarly, we have only the meagrest knowledge of the stratigraphy of the southward thinning medium velocity (2.27-3.2 km/sec) infill in the southern Natal Valley, and in the deep ocean basins to the southwest (Transkei Basin and Agulhas Passage (Figs 1 & 66), where between 1 and 3 km of medium velocity material rest on oceanic layer 2). A date of 127 my for the initiation of SW movement of the Falkland Plateau, suggests that the basal part of the layer progressively becomes younger in age from Valanginian in the vicinity of Durban to Albian at the southern tip of the Agulhas Bank. This implies that in the north, a greater part of the medium velocity layer is of Lower Cretaceous age than it is in the SW, but exactly how much cannot be determined without drilling. Also we have no knowledge of the lithofacies, although on the grounds of basin geometry and plate tectonic models, it can be predicted that the basal layer of sediments (i.e. those resting on ocean crust) will consist of shallow water marine carbonates and ridge crest debris and that these will be overlain by deep water clays deposited below the carbonate compensation level (Dingle & Camden-Smith 1979). Large quantities of terrigenous debris were dumped into the Natal Valley at its northern end (via the Tugela Cone) and similar, but smaller accumulations can be predicted adjacent to the Agulhas Fracture Zone. Also in the sector between 30 and 35°S, there is a strong likelihood that during Lower Cretaceous times the deeper parts of the basin had restricted circulation and may have accumulated anoxic sediments.

4.4.2.4 *Falkland Plateau*

The Falkland Plateau is an eastward prolongation of

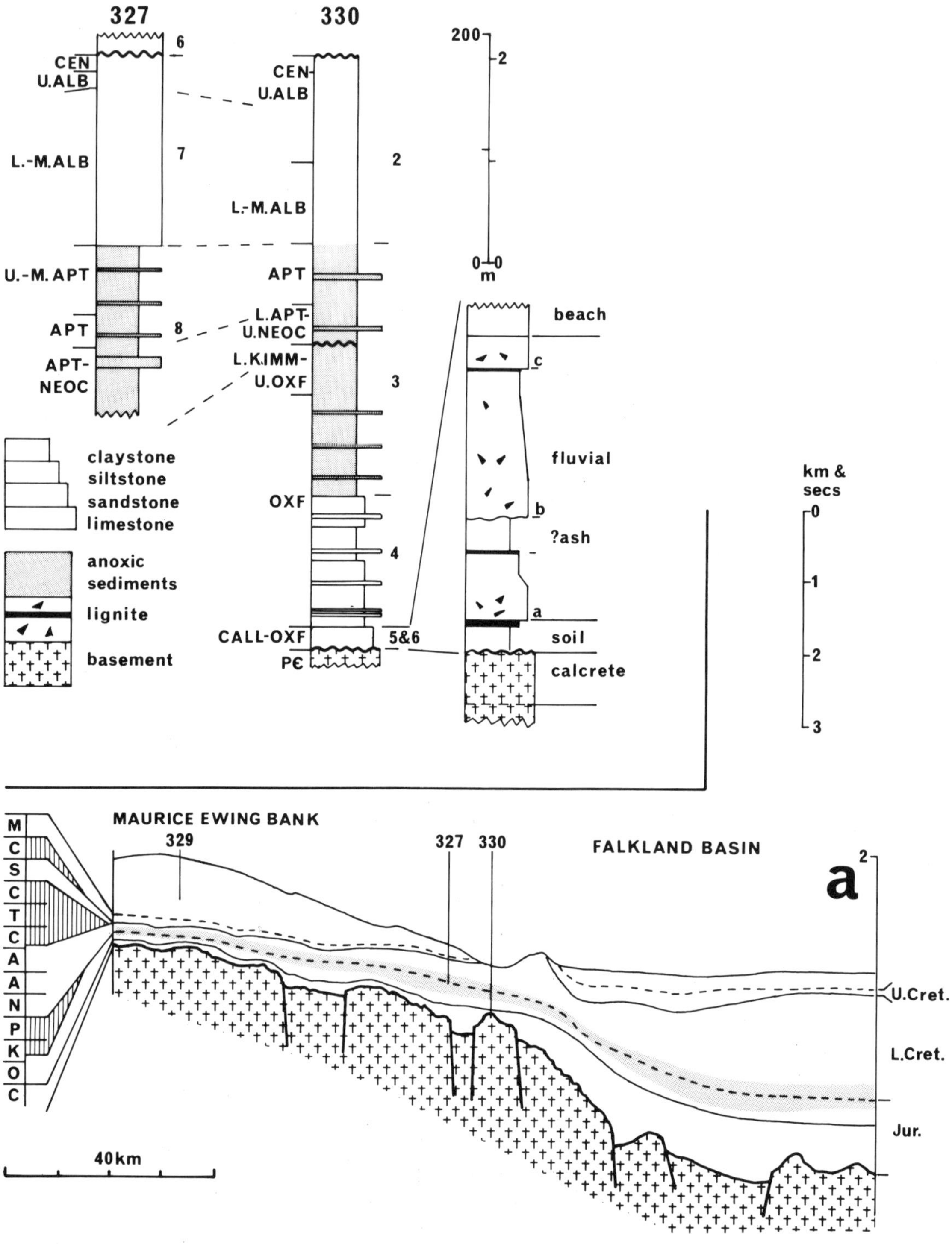


Figure 110. Borehole logs and seismic profile from the Falkland Plateau. Compiled and re-interpreted from data in Barker *et al.* (1977b), Barker (1977), and Thompson (1977). Vertical scale: left side for logs 327 and 330; right side for detail of 330. Units on logs numbered 6-8 and 2-6 are DSDP shipboard notations. Vertical scale for seismic section is in sec and km (assuming velocity of 2 km/sec), depth commences at 2 sec sub-sea level. Left side of profile indicates extent of hiatuses (vertical ruling) in the Callovian (C) to Maastrichtian (M) succession. Organic carbon-rich sequence is shaded. Locations on Figure 104.

the continental crust of South America, about 1 300 km long, which, prior to continental separation lay adjacent to SE Africa in what is now the southern Natal Valley. Between about 127 and 100 my, it sheared progressively past the continental margin of SE Africa along the Agulhas-Falkland Fracture Zone leaving a newly created ocean basin in its place. Because the Falkland Plateau was well clear of the western edge of the Agulhas Arch by Upper Cretaceous times, only its Jurassic and Lower Cretaceous structures and sedimentary history will be dealt with in this book.

Published details on the Falkland Plateau are relatively sparse but sufficient borehole (DSDP) and geophysical data (e.g. Ludwig et al 1968b, Urien & Zambrano 1973, Zambrano & Urien 1974 and Barker 1977) have been released to construct a preliminary structural and stratigraphic framework for the area east of the Falkland Islands.

4.4.2.4.1 *Structure.* Morphologically the Falkland Plateau consists of four main parts (from west to east) (Fig.104):

i) a western sector with water depths <1 km. The Falkland Islands are situated on this shallow portion (Falkland Islands High);

ii) a broad central depression with water depths about 2.5 km (Falkland Basin);

iii) a broad shallow bank with a crest depth in about 1.5 km of water (Maurice Ewing Bank);

iv) a complex eastern end which descends, via a small elevated section at 35°W, to water depths in excess of 5 km.

The northern edge of the Falkland Plateau is straight and precipitous, descending to water depths >6 km in the southern Argentine Basin. These morphological features correspond to underlying basement structures, which, as can be seen in the pre-drift palaeogeographic reconstruction in Figure 104, are continuations of the main structural elements along the eastern continental margin of SE Africa. The western bank on which the Falkland Islands are situated is underlain by shallow acoustic basement (<2 km beneath sea level) and lies adjacent to the Agulhas Arch whose NE faulted margin continues onto the Falkland Plateau and forms the eastern edge of the Falkland Islands High. Across this discontinuity basement descends rapidly to >6 km in the central Falkland Basin and locally (in the SW) lies at depths in excess of 10 km. In the palaeogeographic refit shown in Figure 104, this basin lies adjacent to the Outeniqua Basin and although few details are available of basement structures over most of its area, seismic profiles across the eastern margin show that here its floor is composed of small horsts and graben (Fig.110) What published data we have on its structure does suggest that the Falkland Basin has a similar internal geometry to the Outeniqua Basin, and that its eastern side consists of several large westward throwing boundary faults which are continuous with the basement discontinuity between the Port Alfred Arch and the Outer Algoa Basin.

The Maurice Ewing Bank is a major structural high with depths to basement below sea level over a large portion of its top <2 km. It lies adjacent to the Transkei Swell and in a median position on its northern flank, basement locally descends to 6 km in a small basin which lies opposite St Johns Basin on the African side. The least known portion of the Falkland Plateau is its eastern extremity and we have selected the 3 km isobath to fit against the Natal Valley embayment at about 30°S. The excellent fit of conjugate features across the trace of the Agulhas/Falkland Fracture Zone leads us to believe that this reconstruction is an improvement on that of Rabinowitz & LaBrecque (1979). A similar refit has independently been achieved by Martin et al (1981).

4.4.2.4.2 *Stratigraphy.* DSDP borehole and seismic data provide good litho- and biostratigraphic control over the area of the Maurice Ewing Bank/Falkland Basin transition zone (Barker et al 1977b). Three drilling sites were located on the western Maurice Ewing Bank (327, 329 and 330), but only two (327 and 330) penetrated into Lower Cretaceous sediments. Additional sites (511 and 512) have recently been completed, but no details have yet been published. Figure 110 shows borehole logs for sites 327 and 330, and a correlation between boreholes and seismic profiles across the Maurice Ewing Bank and the eastern Falkland Basin.

Site 330 penetrated to basement, which consists of high grade gneisses (quartz and biotite semipelites), overlying a coarse grained granite pegmatite. A small vein of intrusive microsyenite occurs below the pegmatite. The pelitic metamorphic rocks have been dated at 535 ± 10 my by Rb/Sr whole rock determinations (Beckinsale et al 1977). The top 70 cm of the basement has been calcretized and consists of metamorphic rock fragments cemented by hard calcite, presumably as a primary sedimentological phenomenon, because the immediately overlying sediments are not calcareous.

The overlying sedimentary succession has been divided into five lithological units (DSDP shipboard terminology) with a suspected non-sequence in the middle of unit 3 (Barker et al 1977a) (Fig.110).

The basal unit (5 & 6) is 3.1 m thick and consists of an assemblage of terrigenous silty sands and sandy silts with three thin interbeds (<3 cm) of lignite, and a 30 cm band of montmorillonite-rich clay. Thompson (1977) describes three upward fining cycles from this unit (a-c on Fig.110), which commence with yellowish-brown poorly-sorted medium to fine silty, angular to subangular sands, which grade upwards into olive grey sandy silts. Convolute bedding and mottling are common in the finer parts. Lignite fragments and angular quartz and K-feldspar fragments occur throughout,

Table 34. Mesozoic macrofossils from DSDP Sites 327 and 330 on the Falkland Plateau (from Jones & Plafker 1977).

| | Albian | Aptian | Aptian/Neocomian | Kimmeridgian/Oxfordian | Oxfordian |
|---|---|---|---|---|---|
| Lamellibranchs | | | | | |
| *Aucellina* sp. | x | – | – | – | – |
| *Inoceramus* sp. | x | – | – | x | x |
| *Malayomaorica* cf. *M. malayomaorica* | – | – | – | – | x |
| *Plicatula?* sp. | – | – | – | – | x |
| *Jeletzkiella falklandensis* | – | – | – | – | x |
| *Anomia* | – | – | – | x | – |
| Belemnites | | | | | |
| *?Hibolithes* | – | – | – | x | – |
| *?Belemnopsis* | – | – | – | x | x |
| Belemnite indet | – | – | – | x | x |
| Ammonite | | | | | |
| Ammonite idet | – | – | – | – | x |

but are more common in the coarser basal parts of each cycle. The provenance of the sands is thought to have been the local gneisses, and the sequence is non-marine; possibly fluviatile. The montmorillonitic clay may represent an ash band (Thompson 1977). No fossils were found, but unit 6 is reckoned to be Middle Jurassic in age (the age of the immediately overlying unit) and it is significant that volcanic horizons also occur at this level in Zuurberg Member of the Algoa Basin about 400 km to the west.

The non-marine beds of unit 6 are overlain by at least 2 m of olive-grey to medium grey, well-sorted subarkosic sandstone (unit 5). It is composed predominantly of quartz (up to 87 %), with minor K-feldspar, rock fragments, and traces of garnet, tourmaline, and zircon. Thompson (1977) considers this to represent a beach sand, and to mark the earliest Mesozoic marine transgression into the area. It is overlain by about 115 m of yellowish brown and olive grey silty clays, clayey silts, sands and limestones (unit 4). The whole sequence is laminated and moderately to intensely bioturbated. Sandy horizons are more common in the lower part of the unit, whilst thin (10-20 m) interbedded limestones are more common towards the top. Pyrite (1-7 %), plant fragments (charcoal and wood (1-15 %), traces of glauconite and heavy minerals (including garnet, zircon, tourmaline, apatite and siderite) occur throughout. Illite is the principle clay mineral. Marine fossils occur throughout, but are more common towards the base, where they are concentrated in drifted banks 3-5 m thick and include belemnite rostra, occasional lamellibranchs, rare coccoliths, and benthic foraminifera (mostly agglutinated forms). On palynological evidence, this unit has been dated as Oxfordian, with possible Middle Jurassic elements near the base.

The overlying sediments comprise about 220 m of dark, organic-rich marine clays and limestones, with a strong petroliferous odour (unit 3). Subordinate, thin (5-20 cm) porcellaneous limestones occur in the lower half, along with rare thin (<5 cm) montmorillonitic clays. The dominant lithology is olive-black laminated clay usually containing 3-4 % (up to 6 %) organic matter (Fig.113), with a restricted benthic fauna of thin shelled lamellibranchs and agglutinated foraminifera. Illite and montmorillonite are the dominant clay minerals. Limestones occur in the lower half, and are laminated, micritic beds with abundant radiolaria and organic carbon. Marine fossils occur throughout and include *Inoceramus,* belemnite rostra, lamellibranchs, onychites, radiolaria, benthic foraminifera and rare coccoliths. A major hiatus, representing a non- or condensed sequence occurs about half way up the unit, and spans Middle Kimmeridgian to Aptian (or possibly Lower Barremian) time. There is no nannofossil evidence for Berriasian-Barremian sediments, but palynological studies suggest the presence of one Barremian dinoflagellate genus (*Pyxidiella*) (Harris 1977).

The upper part of the Lower Cretaceous succession comprises 165 m of light brown and pinkish grey nannofossil-rich clay and ooze (unit 2), which has the following components: clay (45-55 %), nannofossils (20-30 %) and zeolites (5-15 %). A montmorillonite-rich layer occurs about midway up the unit, and the whole sequence is moderately to intensely bioturbated. Other marine fossils include *Inoceramus* and other lamellibranchs, and planktonic foraminifera. The organic carbon content of this unit is <1 % and indicates the establishment of open, oxygenated marine conditions after the lengthy period of anoxic sedimentation represented by unit 3. Unit 2 has been dated as early Albian to late Cenomanian, and at site 330 is overlain directly by Palaeogene sediments.

At site 327, the borehole penetrated only as far as Neocomian strata and stopped at an estimated 730 m above acoustic basement. The sediments at this site are almost identical to those at 330, comprising a lower, organic-rich clay unit (145 m) Aptian to Neocomian in age, and an upper nannofossil-rich clay (170 m) of Albian to Cenomanian age. A nonsequence occurs at the top of the nannoclay, which spans late Cenomanian to Santonian time.

Using the seismic profiles in Barker (1977) it is

possible to trace various reflectors from over the Maurice Ewing Bank into the Falkland Basin, and the DSDP sites allow some estimate to be made of the ages of the various horizons, assuming that they were synchronous – which of course they may not have been over long distances. In Figure 110 we have shaded the Oxfordian-Aptian anoxic sediment zone, using the control of sites 327 and 330, as well as a reflector near the bottom of the sequence, and one representing the top of the Cenomanian. From this it can be seen that of the approximately 3 km maximum thickness in this part of the Falkland Basin, 2.5 km are Cenomanian or older in age, and that in the basin, the Jurassic succession is at least 1.2 km thick (of which up to 900 m may have locally been non-marine). Barker (1977) shows that the Jurassic strata progressively lap onto and across the Maurice Ewing Bank, and that by Oxfordian times it was submerged. Clearly the Falkland Basin was a major depocentre by Middle Jurassic times. One interesting facet of the basin fill is the apparent lack of large offsets in reflectors, suggesting that synsedimentary faulting played no important role in basin development, certainly after marine conditions had been established by Oxfordian times. This is in contrast to the Outeniqua Basin, where large-scale differential movements between horsts and graben continued until Aptian times. It may transpire that the record interpreted in Figure 110 is atypical of the area (it penetrates only about 50 km across the basin) because a second record shown by Barker (1977, fig.3) taken somewhat to the NW shows some evidence of lateral discontinuity in horizons that are probably of Upper Jurassic or Lower Cretaceous age.

It is impractical to list the large microfaunas and floras that have been recovered from the Jurassic-Lower Cretaceous of sites 327 and 330, but we will briefly discuss some of the groups. Palynomorphs are relatively common throughout the Jurassic-Lower Cretaceous sequence, and it is on the basis of these microfossils and calcareous nannofossils that the Jurassic sections have been dated. Jurassic palynomorph assemblages are dominated by terrestrial spores and pollen, whilst in the Aptian, marine microplankton (dinoflagellate cysts, acritarchs and tasmanitids) dominate. Harris (1977) considered that the microfloral assemblages indicated a marginal marine environment and after comparing them with those from DSDP site 249 on the Mozambique Ridge, concluded that in the early Cretaceous the two areas were in close proximity. No Jurassic foraminifera have been described, but the Albian-Cenomanian sections are dominated by low diversity hedbergellid planktonic assemblages and a sparse benthic population of large nodosariids, agglutinated forms (such as *Ammodiscus, Dorothia, Gaudryina* and *Tritaxia*), *Gavelinella* sp. and *Gyroidinoidea* sp., plus minor types. Sliter (1977) regarded this as an austral fauna deposited in water depths of between 100 and 400 m.

One of the most interesting studies undertaken on the Falkland Plateau Jurassic-Lower Cretaceous fossils was that by Jones & Plafker (1977) who examined the macrofossils (Tab.34). They recovered six species of lamellibranchs, three species of belemnites, and an ammonite fragment. The assemblages were regarded as shallow water and deposited in depths no greater than 200 m. Although *Aucellina* has a wide geographical range, the lamellibranch *Malayamaorica* has been reported so far only from the austral province of the Antarctic Peninsula, New Zealand, New Guinea, etc., whilst the new genus *Jeletzkiella* has close relatives in New Zealand. An apparent total absence of these types from SE Africa and Madagascar, led Jones & Plafker (1977) to suggest that in fact the eastern Falkland Plateau did not originally lie adjacent to South Africa as shown in Figure 2, but far to the SW. This is certainly against the concensus view of most geophysicists (e.g. Rabinowitz & LaBrecque 1979) and is not in line with the palynomorph data of Harris (1977). So far, details of the benthic ostracod faunas from the Falkland Plateau have not been published but preliminary results by one of us (RVD) have shown that a typical South Gondwana genus *Arculicythere* which has previously been recorded from Madagascar (Portlandian-Valanginian), DSDP 259 and 263 off Western Australia (Albian) and the Outeniqua Basin (Albian) occurs in large numbers in Albian strata at sites 327 and 330. Five other species of ostracoda are common to SE Africa and the Falkland Plateau which suggests that the two areas were in close proximity at least during or immediately prior to Albian times (the Aptian rocks appear to be devoid of ostracods).

### 4.4.3 *Palaeogeography*

There are insufficient data to construct well-controlled palaeogeographies for specific time intervals over this large area; in particular we have no accurate stratigraphic detail from the Natal Valley nor for much of SE Mozambique. Nevertheless it is instructive to compile composite palaeogeographies which contain details from various non-contemporary but closely linked events and environments, and we have selected a pre-drift, and a post-drift situation to illustrate the main features of the complex area.

Figure 111 shows the Middle-Upper Jurassic predrift palaeogeography. Marine sediments are known only from the south where a transgression is first recorded on the southern flank of the Maurice Ewing Bank in Oxfordian times. These are marginal marine sediments which progressively lapped onto the basement high where they replaced swamp and estuarine

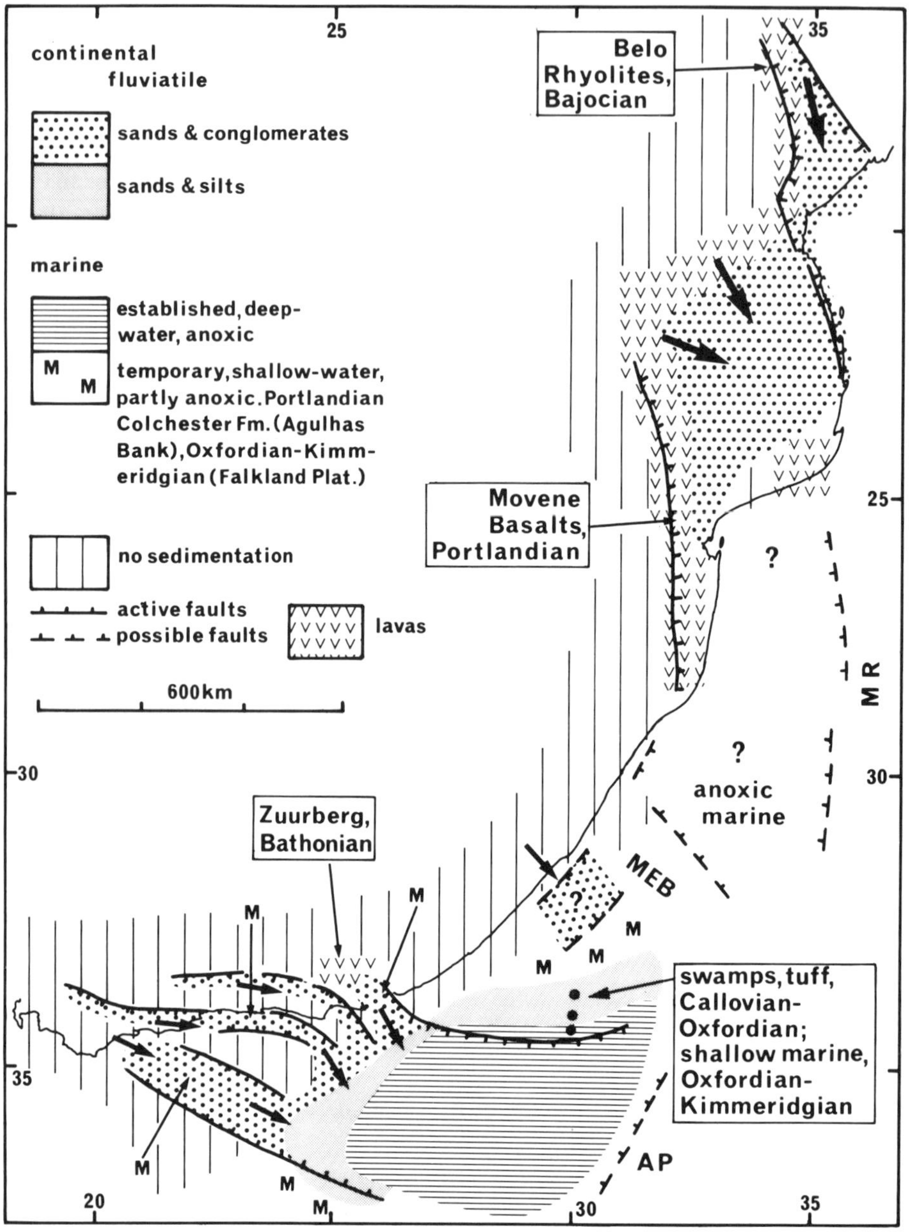


Figure 111. Middle-Upper Jurassic palaeogeography of S.E.Africa and the eastern Falkland Plateau. DSDP sites on the Falkland Plateau shown by dots, sediment dispersion routes with bold arrows. See text for discussion and data sources. Abbreviations: AP = Agulhas Plateau; MR = Mozambique Ridge; MEB = Maurice Ewing Bank.

conditions on an irregular granite/metamorphic basement surface. Oxygen levels in the bottom waters of this shallow sea rapidly decreased and by early Kimmeridgian times anoxic conditions obtained over the Falkland Basin and the flanks and perhaps intermittently over the crest of the Maurice Ewing Bank. Fluviatile sedimentation may have continued contemporaneously in the St Johns Basin area. Higher energy continental fluviatile sedimentation was proceeding farther west in the developing taphrogenic basins of the southern Cape/Outeniqua area and temporary incursions of the anoxic Falkland Sea are represented in this area by the Colchester Formation. There is some uncertainty as to whether the apparent lack of Upper Kimmeridgian to Middle Barremian sediments at DSDP site 330 indicates a non-sequence or whether the succession is complete (but condensed) and is present between cores 4 (Barremian) and 5 (Lower Kimmeridgian) which were cut approximately 20 m apart. Barker et al (1977a) assume the latter explanation, which implies a sediment accumulation rate of ca.1 m/my and suggests high current scour and/or a reduction in the rate of sediment supply. Contemporaneous sediment accumulation continued at a relatively high rate in the Falkland Basin, and resumed its previous rate over site 330 in Barremian times.

Over southern Mozambique widespread fluviatile sediments were accumulating in the large taphrogenic basins at the southern end of the East African Rift System. Large-scale Middle and Upper Jurassic volcanism accompanied crustal subsidence along major boundary faults and in the proximal areas at least, lavas and sediments interfingered. In addition, less intense volcanism also occurred in the Falkland/southern Cape area with the Zuurberg lavas and tuffs in the latter, and volcanic ash layers (possibly from the Zuurberg centres) in the lowermost (Callovian) sediments on the Maurice Ewing Bank. A major unknown factor in our reconstruction is the environments that obtained in the northern Natal Valley. The considerable thickness of pre-Cenomanian sediments, especially under the Tugela Cone, suggests that accumulation may have begun here in Upper Jurassic times, and if it did, palaeogeographic considerations require that the environment would have been non-marine or restricted circulation (anoxic) marine. At the moment we have no evidence to prove or disprove either of these suggestions. The Mozambique Ridge was apparently an area of non-deposition, and so may have been the southern Agulhas Plateau continental fragment, which lay at the SE end of the Falkland Basin.

A post-drift palaeogeography for late Barremian/early Aptian (nominally ca.115 my BP) times is shown in Figure 112. By this time, the Falkland Plateau and the attached Agulhas Plateau fragment had moved down the east coast of South Africa so that its eastern end lay in the vicinity of Port Elizabeth and the spreading centre just SSE of Port St Johns.

A change in oceanic circulation allowed the anoxic marine environment of the Falkland Basin to extend across the Maurice Ewing Bank in Barremian times and established an organic-rich sediment regime that is known to stretch from the SE Atlantic via the Falkland Basin to the eastern Maurice Ewing Bank. Accelerated crustal subsidence in the Outeniqua Basin area can probably be related to the onset of continental separation in Valanginian times, and resulted in the establishment of more extensive marine conditions than had hitherto prevailed (the Berriasian Infanta Formation is restricted to the distal parts of the sediment basins). During the accumulation of the locally mildly anoxic Lower Sundays River Formation, the Maurice Ewing Bank began its traverse across the SE end of the Outeniqua depocentre, and in our palaeogeography of Figure 112 (Upper Sundays River times) it lay across the length of the South African margin from the Agulhas Arch to the Algoa Basin, after which it progressively cleared the Outeniqua Basin from the NE. This could account for the persistence of Upper Sundays River lithofacies in the Bredasdorp Basin until Albian or Cenomanian times whilst the open water Alphard Formation was established in the Pletmos Basin in the Aptian. Anoxic sedimentation abruptly ceased over the Falkland Plateau and in the SE Atlantic with the establishment of permanent deep water connections between the latter and the SW Indian Ocean once the Plateau had cleared the tip of the Agulhas Bank at the end of Albian times (we estimate this at about 100 my). A further alteration in the oceanic environment in Cenomanian times, most likely an increase in current strength and/or a rise in the carbonate compensation level, terminated sediment accumulation on the Falkland Plateau and by the time it re-commenced (Santonian on the Maurice Ewing Bank) the Falkland Plateau lay well to the SW of our area.

The earliest marine sediments in Zululand (Barremian), are further evidence of a local Neocomian marine transgression (Figs 67 & 107) and indicate a muddy nearshore environment into which high energy alluvial fans were deposited directly from the adjacent Lebombo range, without the construction of fine grained deltas or extensive flood plains (Makatini Formation). This environment extended for a short distance into southern Mozambique, but was replaced NE by lower energy shallow marine sands and clays of the Maputo Formation. Late Aptian-early Albian sedimentary events in Zululand and southern Mozambique are synchronous with important lithofacies changes already noted in the southern area (change from anoxic to oxygenated bottom waters in the South Atlantic/Falkland Plateau/eastern Outeniqua Basin areas) and consist of a major non-sequence in Zululand and the Maputo area, and the establishment of anoxic conditions in the deeper parts of the marine basin of SE Mozambique (Domo Formation). A simple sea level rise or fall is unlikely to account for an event that had such contrasting results, and we speculate that it was caused by a radical readjustment of ocean current patterns related to the separation of the Falkland Plateau and southern Africa. Emergence in Zululand is not indicated, but possibly a shallowing and/or marked increase in bottom current strengths resulted in a widespread non-sequence and hard ground formation at the top of the Makatini Formation. Similar, though less widespread or prolonged fluctuations occurred throughout the Makatini and Mzinene formations, indicating a delicate balance between sediment accumulation and bottom current strengths in the shelf sea off Zululand during these times.

A further, even more widespread hiatus occurred during late Cenomanian to early Coniacian times. A break in the succession is recorded on the Falkland Plateau (by now well-separated from southern Africa), locally in the Outeniqua Basin, in the Orange Basin, on the Mozambique Ridge, and in Zululand and southern Mozambique, a distance of over 2 000 km. Only farther north in coastal Mozambique was sedimentation apparently continuous through the Domo Formation. This event effectively brought to a close the Jurassic to Lower Cretaceous sedimentary episode and its initiation was reasonably synchronous: late Albian to late Cenomanian in the Outeniqua Basin; early Cenomanian on

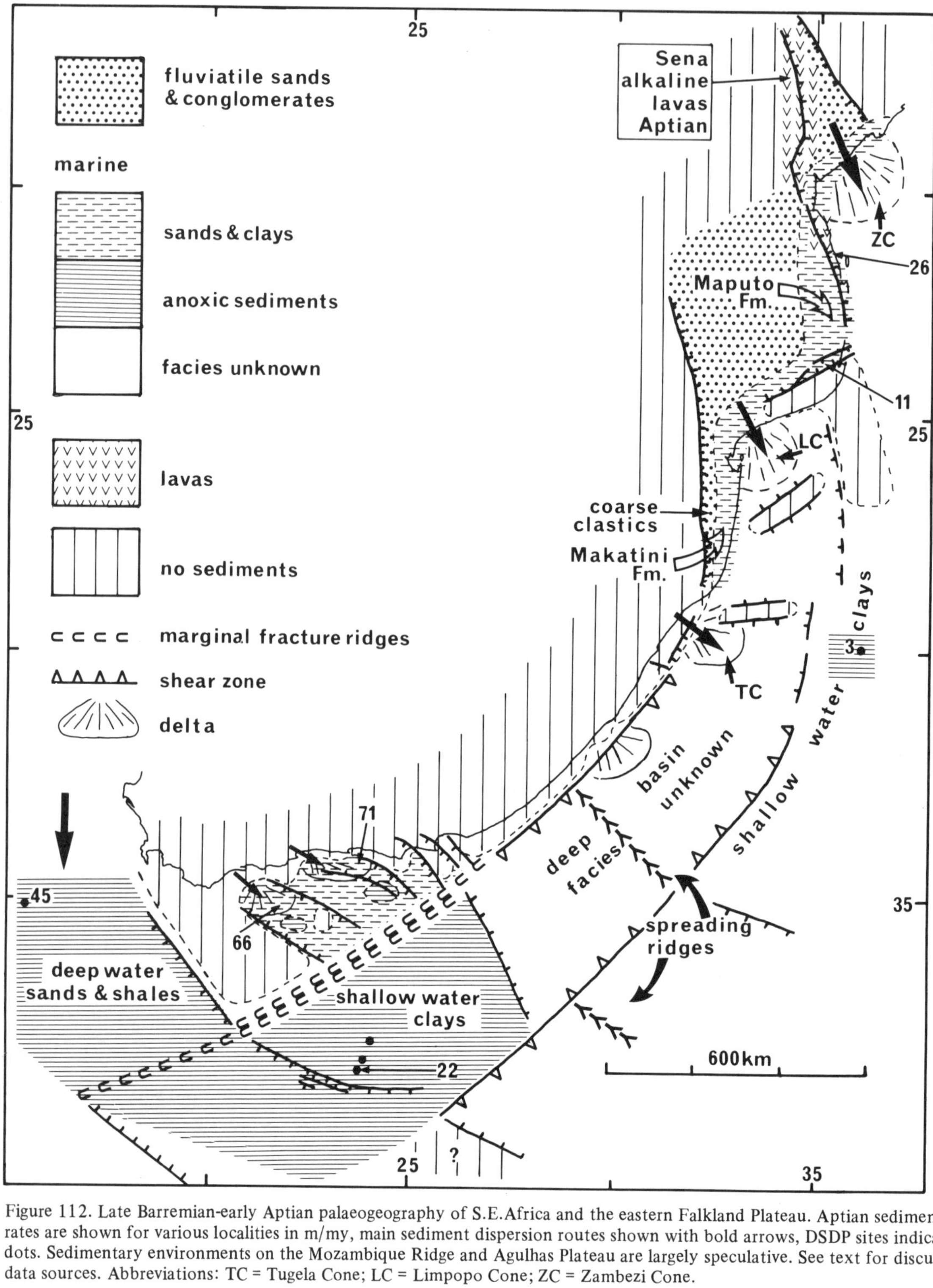


Figure 112. Late Barremian-early Aptian palaeogeography of S.E.Africa and the eastern Falkland Plateau. Aptian sedimentation rates are shown for various localities in m/my, main sediment dispersion routes shown with bold arrows, DSDP sites indicated by dots. Sedimentary environments on the Mozambique Ridge and Agulhas Plateau are largely speculative. See text for discussion and data sources. Abbreviations: TC = Tugela Cone; LC = Limpopo Cone; ZC = Zambezi Cone.

the Falkland Plateau and Mozambique Ridge; late Cenomanian in Zululand and southernmost Mozambique, ?late Cenomanian in the Orange Basin. There is a suggestion of a late Albian to early Cenomanian non-sequence in deep ocean basin sediments at DSDP site 361 in the SE Atlantic.

Synchronous with the deposition of marine sediments during this period, continental sedimentation continued in the Zambezi Graben, and over most of the Mozambique coast plain (Sena Formation). Fluviatile sands interfinger with the shallow-water marine facies along the vicinity of the present coastline, and it is obvious that the Limpopo-Save region either remained relatively buoyant during this period because of some underlying crustal (rheological) cause (which we consider unlikely), or that sediment input and accumulation was sufficient to keep pace with the various Neocomian marine transgressions that affected the areas surrounding it.

For the area between the Falkland Plateau and the northern Natal Valley, we have little precise data. The Mngazana marine rocks have been assigned a Valanginian age, but older sediments (which may include marine strata) probably occur in the deeper parts of the St Johns Basin, so we do not know at what stage the top of the Transkei Swell collapsed, although the fact that there is a conjugate basin on the northern Maurice Ewing Bank suggests that it took place in advance of continental separation. In view of this we have tentatively shown a basin with continental sediments in Figure 111. The tectonosedimentary environment of the Mngazana Beds was in many ways analogous to that of the Enon Formation the main difference being that the fan debris at the foot of the active fault scarp was dumped directly into the sea.

Evidence of volcanism during the period represented by Figure 112 is sparse. Aptian alkaline lavas are interbedded with Sena sandstones in the Zambezi Graben and lavas of a similar age possibly occur on the flanks of the graben at the Save River mouth (Mambone borehole). Elsewhere the evidence is limited to reports of thin tuff bands in the lower Sundays River Formation of the Algoa Basin, and the ash-rich silts and clays of Neocomian to Aptian/Albian age at DSDP site 249 on the Mozambique Ridge.

The abundance of plant material (comminuted, and as logs and stems) in sediments from all the areas along the east coast attests to a well vegetated hinterland, and the coarseness of much of the non-marine detritus indicates sharp local topographic discontinuities and high runoff rates. This is a conclusion also reached by Thompson (1977), on the basis of high kaolinite contents in the earliest marine and non-marine sediments at DSDP site 330. He did suggest, however, that the calcrete layer on top of the metamorphic basement surface represented a somewhat earlier (Lower to Middle Jurassic) drier climatic episode. This is in line with known Lower Jurassic arid conditions during Clarens Formation deposition.

### 4.4.4 *Distribution of organic carbon-rich sediments*

Sediments having relatively high organic carbon contents have been mentioned in our discussions of the Jurassic-Lower Cretaceous sediments, and it is instructive to summarize the distribution of the anoxic sedimentary environments that they represent in time and space (Fig.113).

Shallow water, anoxic marine environments were established in the south of the area (over the eastern Falkland Plateau) at least 18 my prior to continental separation, and the earliest substantial marine incursion into the Outeniqua Basin in Portlandian times indicates that they persisted over the Falkland Basin even though sedimentation stopped or was very slow on the higher parts of the Maurice Ewing Bank. This incursion gave rise to the Colchester Formation, which significantly has the highest average organic carbon values for the Uitenhage Group and has shown considerable small-scale petroleum potential in the Algoa and Outeniqua basins. Following the initiation of continental drift (Valanginian) anoxic conditions were maintained in the Outeniqua and Falkland areas until the time of final continental separation when there was a rapid switch to oxygenated sedimentation in both areas. This event was also registered in the SE Atlantic, where early organic carbon-rich, deep water sediment accumulation was terminated in Albian times. Synchronous with this Oxfordian-Albian anoxic sedimentary period in the south, the earliest sediments on the Mozambique Ridge (DSDP 249) also show moderate values of organic carbon (average of 0.8 %), but the upper limit here is formed by a non-sequence and no conformable passage to oxygenated sediments is seen.

Farther north, in Mozambique, the situation is somewhat different. Here the earliest marine sediments are well oxygenated (Maputo Formation), and anoxic environments were not established until Albian times, when the dark organic-rich Domo shales were deposited adjacent to the present coastline. This lithofacies persisted without a perceptable break until the late Turonian, and with the possible exception of the situation on the Mozambique Ridge, this northern development of organic-rich sediments is out of step with areas to the south (Fig.113). No organic-rich horizons are known from the Lower Cretaceous sediments of Zululand.

Speculation on the causes for such widespread anoxic events have been made by several authors in recent years (e.g. Jenkyns 1980) and three main hypotheses have been advanced:

i) the development of an oxygen-poor layer in the upper part of the water column, which, where it impinged upon shallow bathymetric features, produced anoxic sediment zones;

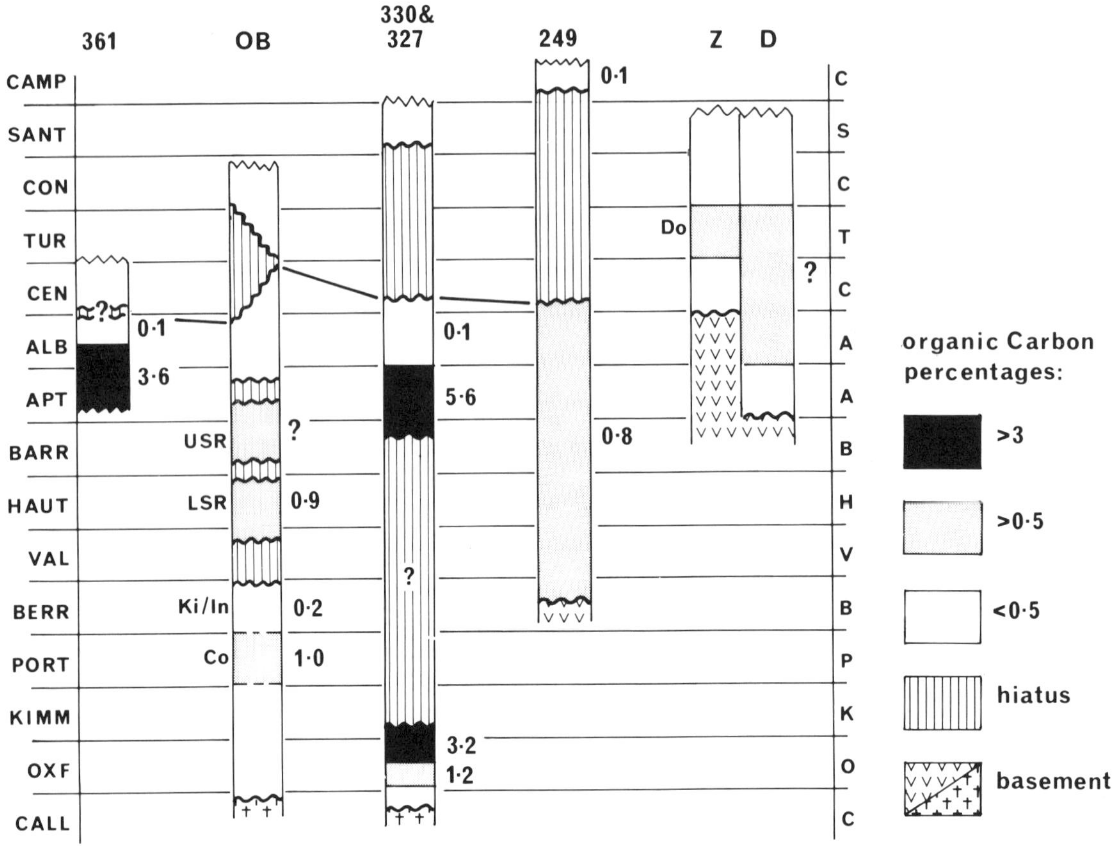


Figure 113. Distribution of organic carbon-rich Mesozoic sediments in S.E.Africa. Locations in Figures 98, 104, 112, 123: OB = Outeniqua Basin; Z = Zandamela borehole; D = Domo borehole; 249, 327, 330, 361 = DSDP sites. Values against stratigraphic sections are organic carbon percentages of total sediment: in 361 Aptian/Albian mean of 47 determinations; in 330 Aptian mean of 2, Oxfordian/Kimmeridgian mean of 17, Oxfordian a mean of 13; in 249 a mean of 31. Values in the Outeniqua basin are from the Algoa Basin. No values are available from Mozambique. Ranges of values are: 361, 0.1-14.6 %; 330, Aptian 5.3-5.8 %, Oxf/Kimm 1.9-5.8 %, Oxf 0.4-1.9 %. Data compiled from core log descriptions in Simpson *et al.* (1974), Barker *et al.* (1977a), Bolli *et al.* (1978). Algoa Basin data from Winter (1973). Abbreviations of formation names: USR = Upper Sundays River; LSR = Lower Sundays River; Ki = Kirkwood; In = Infanta; Co = Colchester; Do = Domo.

ii) zones of oceanic upwelling that led to a heavy rain of organic-rich debris on the underlying sea floor;

iii) restricted deep water circulation in ocean basins caused by topographic barriers.

With the exception of (ii), which cannot be assessed at the moment, both causes (i) and (ii) could have operated in our area, and in the case of DSDP site 361, isolated, stagnant, deep water 'trapped' between the Walvis Ridge, spreading ridge and the Falkland barrier is the only likely cause. Similar environments may have occurred in the Natal Valley to the north of the Falkland Plateau edge, where the new ocean basin was split into compartments by the spreading ridge which migrated southwards at half the rate of the motion between Africa and the Falkland Plateau. Although widespread Aptian oxygen poor layers have been reported from the Pacific Ocean (see Schlanger & Jenkyns 1976), and Barker et al (1977b) suggest that a worldwide mid-Cretaceous development of this is 'not unreasonable', it does not solve the question of the Upper Jurassic anoxic sediments on the Falkland Plateau. At present we believe that all the occurrences of anoxic rich sediments around southeastern Africa can be explained by restricted circulation in relatively small isolated basins.

Two areas of particularly high organic carbon preservation stand out on Figure 113: the deep water anoxic event in the early SE South Atlantic Basin (DSDP 361), and the long (?37 my) relatively shallow water episode on the Falkland Plateau. Average values of organic carbon in sediments from these areas exceed 3 %, and locally reach 14.6 % (Aptian DSDP 361), 5.8 % (Aptian DSDP 330), and 4.4 % (Kimmeridgian DSDP 330). Cromer & Littlejohn (1977) conclude that the Falkland Plateau sediments are potentially a rich oil source rock, but note that they are thermally immature. From their study of the organic-rich sediments of the Outeniqua Basin, Leith & Rowsell (1979) identified the Colchester Formation (organic carbon values < 1-4 %) and SR3 of the Upper Sundays River Formation (about 2 %) as the most promising potential oil source rocks. The former represents incursions directly from the Falkland Basin prior to the initiation of continental drift. Although having relatively high organic carbon values compared to open ocean marine sediments, the Lower Cretaceous sequence in DSDP 249 has maximum values of only 1.7 %, (mean 0.8 %). Timofeev & Bogolyubova (1981) identified the organic matter from sites 327, 330 & 361 as: 327 mainly sapropelic (autochthonous algae, higher plants, animals); 330 (and nearby 511) mixed humic and sapropelic; 361 (except Lower Albian) humic (allochthonous higher plants). Considered in conjunction with other Atlantic DSDP 'black shale' sites, these workers suggest that whilst all are immature, 361 is a potential gas source, and 327 and 330 potential oil source rocks.

## 4.5 WEST COAST (WALVIS RIDGE TO AGULHAS FRACTURE ZONE)

No occurrences of marine Jurassic or Lower Cretaceous strata are known on the west African coast between the vicinity of Benguela in Angola, and the southern Cape, a distance of nearly 2 500 km, although extensive outcrops of extrusive and intrusive basic, alkaline and granitic igneous rocks of this age occur in the Kaokoveld, Cape Cross, Hoachanas and Lüderitz areas. Non-marine sediments underlie the lavas in the Kaokoveld, and these have been correlated on faunal and lithological grounds with the upper part of the Stormberg Group (late Triassic-early Jurassic) of the main Karoo Basin (Keyser 1973) (Figs 41 & 42). They have been described in Chapter 2 because of their lithological similarity to, and stratigraphical relationships with the Triassic rocks of northern Namibia. The overlying Etendeka Lavas are Berriasian to Albian (132-108 my).

Between the Walvis Ridge abutment and the Agulhas Fracture Zone, two large marginal basins (Walvis and Orange) contain extensive developments of ?Jurassic-Lower Cretaceous rocks (Fig.66). One DSDP (361) and several commercial boreholes have encountered rocks of this age in the Orange Basin, and the former has allowed a major, regionally-developed seismic reflecting horizon (mapped and named 'Atlantis II' or 'A II' by Emery et al 1975), to be dated as Lower Albian. This horizon has been recognized in the Walvis Basin, and together with refraction interval velocity data, indicate that thick deposits of sediments older than the AII horizon of the distal Orange Basin (i.e. pre-Lower Albian) occur between the Walvis Ridge abutment and the vicinity of Walvis Bay. At DSDP site 361, which lies SW of Cape Town in a distal part of the Orange Basin, reflector AII corresponds to a major lithofacies change from restricted marine to open marine environments. It is possible that this change is diachronous over the 1 400 km that separate the Walvis Basin from site 361, but it is the only datum that we have, and provided discussions of the Walvis Basin are seen as no more than preliminary, this possibility should not serious detract from their usefulness. Also, there is reason to believe that south of the Walvis Ridge, the major lithofacies change represented by AII was effected rapidly over the whole area because of the very nature of the palaeogeographic reorganizations that caused it.

Additional evidence for the extensive distribution of well-lithified pre-AII strata are provided by seismic refraction and wide angle reflection data from the whole of the west coast margin by Bryan & Simpson (1971), Hoskins et al (1974), Emery et al (1975), Rabinowitz (1976), and by recently released commercial data (Gerrard & Smith 1980).

Continental separation north of the Agulhas Fracture Zone was effected by tensional rifting in which a clean break separation (i.e. with only minor shearing) took place between southern Africa and South America. Well-defined sea floor spreading magnetic anomaly patterns have been recognized in the Cape Basin, but interpretation of the width of the Upper Cretaceous quiet zone (i.e. between anomalies 34 and M0, 78-109 my) has given rise to some controversy in identifying the oldest anomalies adjacent to the west coast continental margin, i.e. there is no consensus on the age of the oldest zone of oceanic crust and hence the date of the onset of continental drift between SW Africa and South America. We have touched on this question in Chapter 1, but will continue further discussion here. The problem is threefold: (i) notation of pre-Quiet Zone anomalies, (ii) recognition of continent/ocean boundary, and hence location of the earliest anomaly and (iii) differences in correlating magnetic time scales to absolute time scales.

i) Much of the Cape Basin west of SW Africa is characterized by low amplitude, high frequency magnetic anomalies: the so-called Quiet Zone or Mercator Normal Interval of McElhinny & Burek (1971), and there is good agreement on recognizing its western boundary (about 1 660 km SW of Cape Town) at anomaly 34 (78 my, van Hinte 1976a) (Fig.1). Where controversy does arise, however, is in recognizing its eastern boundary. The Lamont school of geophysicists (e.g. Larson & Ladd 1973, Rabinowitz & LaBrecque 1979) assume that it is about 1 230 km wide SW of Cape Town, so that M0, the youngest anomaly in the Mesozoic sequence (the Serra Geral Mixed zone of McElhinny & Burek 1971) lies about 430 SW of Cape Town. Between M0 and the COB, Larson & Ladd (1973) recognized anomalies to M11, and Rabinowitz & LaBrecque (1979) to M13. Emery et al (1975) on the other hand recognized two quiet

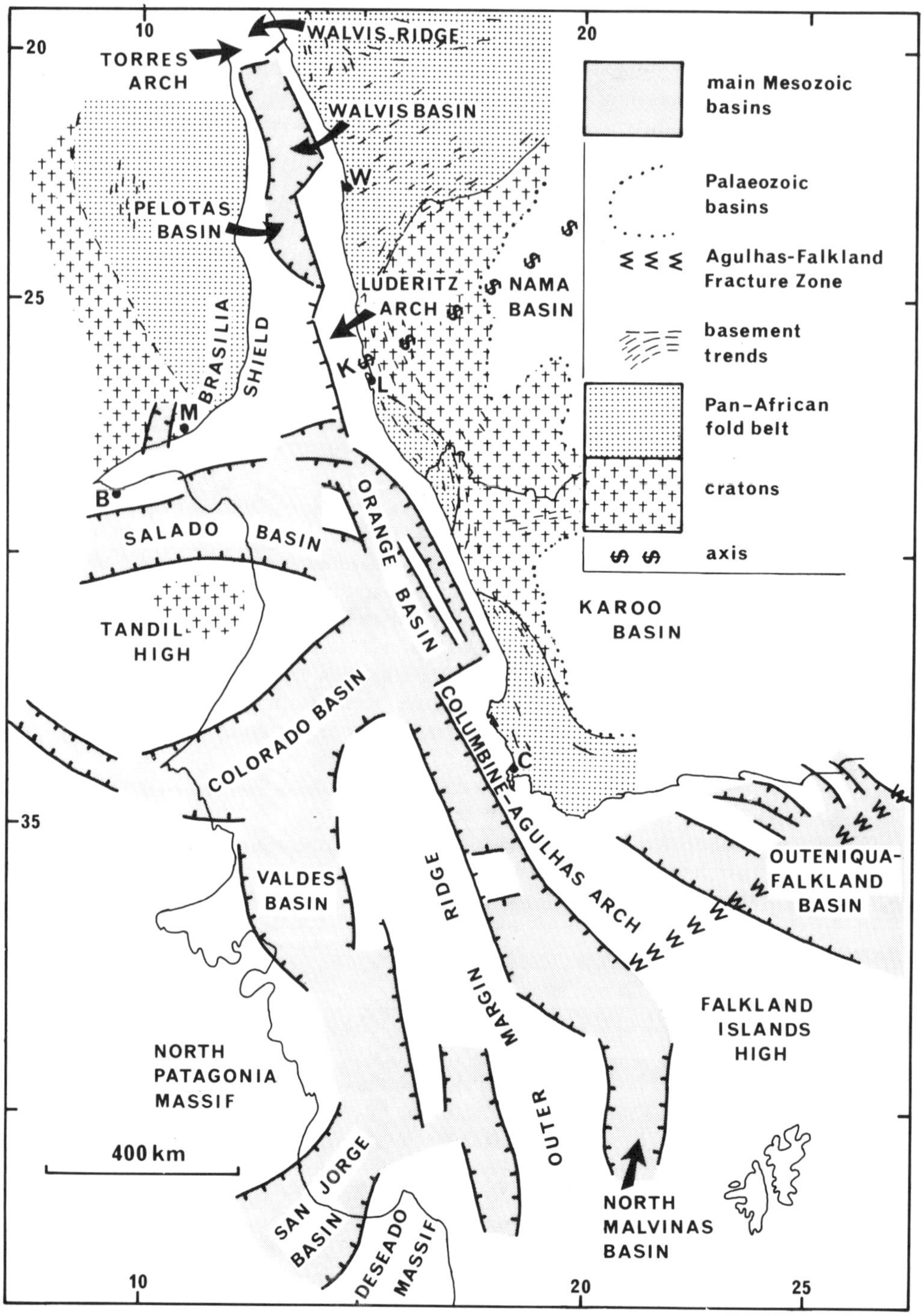


Figure 114. Main structural elements of S.W.Africa and S.E.South America in a predrift reconstruction. Data from: Martin (1973a, 1976); Urien & Zambrano (1973); Zambrano & Urien (1974); Soares & Landim (1976); Urien *et al.* (1976); Francisconi & Kowsmann (1976); Torquato (1976); Pretorius (1979); Dingle (1979, 1980a); Dalla Salda (1980). Abbreviations: W = Walvis Bay; L = Luderitz; K = Konkau radial upwarp; C = Cape Town; M = Montevideo; B = Buenos Aires.

zones (abyssal and marginal) and placed M1 (they did not identify M0) at 1 220 km SW of Cape Town. Between this point and the COB they identified anomalies to M19 (M11-15 falling within their marginal 'quiet zone').

ii) Unlike the the east coast margin, which has a marked topographic discontinuity, there is no obvious physical expression of the boundary between continental and oceanic crust along the west coast. In fact there may not be a sharp divide and a zone with transitional (intermediate) properties may be present. Several attempts to delineate the boundary in terms of geophysical parameters (magnetics, seismic refraction, and gravity) have been made, but models based on crustal thickness and density variations, for example by Emery et al (1975), and Scrutton (1976b) put the boundary up to 200 km seaward of that assumed by Rabinowitz (1976) and Rabinowitz & LaBrecque (1979). Under the circumstances, it is difficult to reach a consensus on what actually constitutes the earliest anomaly generated by oceanic crust, and both Emery et al (1975) and Rabinowitz & LaBrecque (1979) include anomalies in their oceanic sequences which have greater amplitudes and longer wave lengths than more typical oceanic anomalies farther west. This problem is mentioned further in our discussion of the Orange Basin (4.5.1).

iii) Finally there is the question of dating magnetic anomalies, although in relation to the other factors, this is far less of a problem. The time scale that we have adopted for this book is that of van Hinte (1976a), whereas other scales in use have been devised by Larson & Pitman (1972) and Larson & Hilde (1975), amongst others.

The differences of approach mentioned above, have given rise to various estimates of the age of the earliest oceanic crust in the Cape Basin, and although tentative, our conclusions on this dilemma were stated in Chapter 1.2, where we adopted the date of M10/11 (ca.127 my, Valanginian) of Larson & Ladd (1973).

A comparison of Figures 66 and 123 shows that pre-AII (i.e. pre-Lower Albian) sediments accumulated on continental crust, on the COB zone, and adjacent to the Walvis and Orange basins spread a considerable distance westwards onto oceanic crust. To some extent it is this relatively thick covering (>4 km in both main basins) that has made any basement expression of the precise location of the COB difficult to identify.

### 4.5.1 *Structural and basement features*

Figure 114 shows the major structural elements that controlled Jurassic to Lower Cretaceous sedimentation along the continental margins of SW Africa and eastern Argentina, and two main trends are evident: N-S graben and horsts; and south of a Lüderitz-Buenos Aires line, large E-W graben and horsts. A most striking feature of this region is the asymmetrical distribution of these taphrogenic structures, both before continental separation when E-W graben developed only in the west, and after separation when all except the N-S Walvis and Orange basins remained on the South American plate. Considering the scale of vertical subsidence that has taken place in the continental crust (>7 km in places), and the large tensional forces to which the area was subjected during continental separation, the lack of large-scale faulting and basin development onshore in SW Africa is quite remarkable.

The conclusion to be drawn from this pattern of fracturing is that the continental crust south of the Lüderitz-Buenos Aires line is heterogeneous, so that although areas to the east of the present African continental margin were, and have remained, particularly buoyant, the strength of the crust farther west was variable and it subsided significantly during the Jurassic to Lower Cretaceous phase of taphrogenesis. When continental separation took place it did so via the zone of greatest weakness which lay along the boundary between the buoyant basement in the east and the main N-S taphrogenic lineament adjacent to it. This line blocked out the new continental edges which correspond closely to the present coastline on the African side, but which are distant from the coastline on the South American side because considerable subsidence has taken place over the 'weak' old taphrogenic zone since continental separation.

The asymmetrical position of the line of separation can be further emphasized in reconstructed cross-sections of the Jurassic-Lower Cretaceous taphrogenic zone (Fig.115). Line a-b across the Colorado-Orange basins shows that in the former the deepest parts lie to the east, and that the two basins are separated by a basement high (Outer Margin Ridge). The line of continental separation lay between this ridge and the buoyant African block. In line c-d farther south, basins at either end of it are separated by a large complex basement high (Outer Margin Ridge), but here again the line of separation lay close to the eastern side of the whole zone which is marked by the Columbine/Agulhas Arch.

A further intriguing aspect of the crustal structure of this area is to account for the E-W course of the line of separation across the Agulhas Arch-Falkland Islands High, when an apparently weaker zone via the North Malvinas Basin would appear to have offered an 'easier' route.

Turning to the structures on the SW African margin, pre-AII isopachs (Fig.66) and basement structures (Fig.114) identify two negative and two positive areas of crust along the continental margin south of the Walvis Ridge abutment. At present, the oldest known Mesozoic rocks from the basins are of pre-Valanginian (?Upper Jurassic) age (Gerrard & Smith 1980), but in view of their probable relationship with the structures

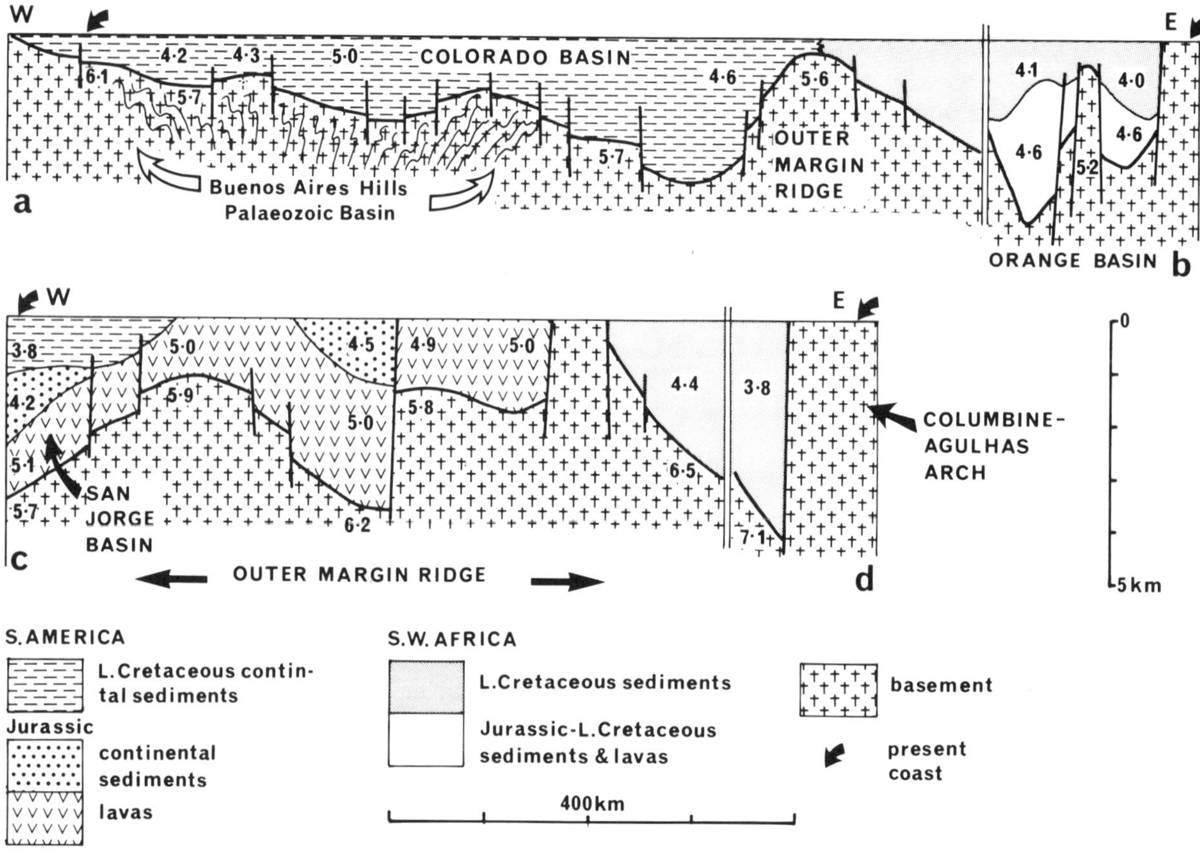


Figure 115. Sections across Mesozoic basins on the E.Argentina and S.W.African continental margins. The double vertical line nominally represents the COB. See text for discussion and Figure 123 for locations. P-wave interval velocities (km/sec) are given at various localities and aid correlation of lithostratigraphic units. Data mainly from Zambrano & Urien (1974), Hoskins *et al.* (1974), Urien *et al.* (1976), Rabinowitz (1976).

on the Argentine margin it seems reasonable to expect that older Jurassic lavas and/or continental sediments occur in their deeper parts.

The Orange Basin is the larger of the two Mesozoic west coast depocentres. It extends for about 500 km NNW-SSE on continental crust, but as a sedimentary feature has spilled over the COB and at the 3 km pre-AII isopach extends about 800 km NNW-SSE under the continental rise, where it merges with the narrow sediment prism west of Cape Town (Fig.66). It contains a maximum of at least 4.2 km of pre-AII strata, and in the southern half consists of two basinal areas under the continental shelf (Inner Graben complex and Main Basin) with a further depocentre (Outer Basin) about 350 km west of the coast (Fig.120c). The two basins under the continental shelf are probably both on continental crust, and are separated by a narrow horst (Central Ridge) that runs from near the southern edge of the basin NW for about 280 km. Very little data are available on the form of these shelf structures, but from seismic refraction data we conclude that the Inner Graben complex contains about 2.8 km of pre-AII strata and has a depth to basement of about 4.5 km. Gerrard & Smith (1980) state that in the Orange Basin, the Inner Graben complex lies seaward of about 45 km from the present coast, and that individual graben are 2-35 km in width. The Central Ridge probably has $<1$ km of pre-AII cover, is about 50 km wide, and rises to within 3 km of the sea floor. Its SW side descends steeply to an estimated depth to basement of about 7.5 km (ca. 6 sec) in the eastern part of the Main Basin, which has maximum thickness of pre-AII fill of about 4.3 km. These structures are separated by a broad basement high from the wide Outer Basin which contains about 3.5 km of pre-AII strata. On Figure 120c we have not attempted to identify the COB, but suspect that it lies to the west of the basement high crest, perhaps between 250 and 300 km from the eastern edge of the section. Rabinowitz (1976) prefers to place the COB at the discontinuity between the Central Ridge and the Inner Graben. The Central Ridge coincides with a prominent magnetic high of nearly 500γ peak to trough (Simpson & du Plessis 1968, Rabinowitz 1976), which suggests that it might be volcanic in origin, or that it is continental crust extensively intruded by basic material. Evidence from flexing in overlying Palaeogene sediments (Dingle 1973c) also suggests that it may have continued to act as a positive structure well into Tertiary times when it was the site of Lower Tertiary volcanism (Dingle 1979, Gerrard & Smith 1980). We have no evidence for the continuation of the Central Ridge north of about 28½°S.

Very little data have been published on the nature and extent of the Walvis Basin, and what data are available are difficult to interpret because of uncertainty about structures on the southern side of the Walvis Ridge. This dilemma is typified by Rabinowitz & LaBrecque (1979) who cannot even trace magnetic anomalies across the area adjacent to it. Interpretation of the seismic refraction data in Hoskins et al (1974) and Emery et al (1975), as well as reflection data published by the latter authors, leads us to propose the cross section in Figure 120c. A narrow(40 km), shallow basement ledge under the inner continental shelf descends rapidly to a depth of over 8 km at the eastern edge of the Walvis Basin, where a maximum of about 4.2 km of pre-AII strata are suspected. Westward the basement rises again to form an outer ridge ( 6 km beneath sea level) under the continental rise. North of the latitude of Walvis Bay, seismic data suggest that the basin is offset to the east and approaches to within 60 km of the coast. It is probable that only this northern part lies on continental crust, and that in section a on Figure 120 the COB lies near the foot of the steep basement escarpment.

The continental parts of the Orange and Walvis basins are separated by a buoyant basement high (Lüderitz Arch) which is about 500 km long north to south and extends about 100 km offshore. It has minimal Mesozoic cover and horizon AII abuts it well below the top of the outer edge (Fig.120b), beyond which, basement descends rapidly to between 6 and 7 km. The Lüderitz Arch may be a composite feature in which buoyant basement extends farther westwards in the southern part than it does in the narrower northern sector, and that the two areas are separated by an outer margin discontinuity at about 25°S. The northern edge of the northern sector of the Lüderitz Arch appears to lie along a SW extension of the Cape Cross Line, which as we will discuss later, is a volcanic lineament that was intermittently active during the period 195-115 my, and which is characterized by the alkaline nature of its igneous rocks. The 25°S discontinuity looks suspiciously like a SW extension of the southern edge of the Damara metamorphic zone which is a NE-SW striking arm of the Pan-African Fold Belt (ca.550 my) that separates the Angolan and Kalahari/Kaapvaal cratons. The location of the Walvis and Pelotas basins in Figure 114 is clearly linked in some way to the presence of this major structural feature, and in this respect the northern sector of the Lüderitz Arch is somewhat enigmatic – perhaps it is a transition zone between the wide southern part of the arch and the Walvis/Pelotas basin proper. The southern sector of the Lüderitz Arch lies adjacent to several analogous onshore geological and geophysical features whose inter-relationships are not clear at present, but which all presumably have a common cause. A small (40 x 60 km) alkaline igneous province lies 40 km south of Lüderitz. It was active for two short(?) periods during the Cretaceous (ca.130 my) and Lower Oligocene (35-37 my), and sits in the trough of a broad gravity low (see Pretorius 1979, fig.14, and Geological Survey S.Afr. 1:1 000 000 sheets) which deflects the 0 mgl bouguer anomaly isogal over the crest of the Lüderitz Arch. Just south of Lüderitz, the Pan-African Fold Belt re-crosses the coastline before making a long, but narrow incursion onshore in southern Namibia and northern Namaqualand. The region between the two mobile belt outcrops, which has its centre north of Lüderitz, coincides with one of Pretorius's (1979) fundamental subcontinental radial crustal upwarps (the Konkau line), and the extension of this axis projects precisely on to the crest of the southern sector of the offshore Lüderitz Arch. The core of the Lüderitz Arch may, therefore, coincide with a small westward extension of the Kalahari/Kaapvaal cratonic block, and the northern, southern and western edges (at least as accurately as we can delineate them) may in a rather imprecise manner mark the inner edges of the Pan-African mobile belt. This relationship is less clear at the southern end, where a major offset occurs in basement structures off the Orange River (ca.29°S) (see Fig. 122), but immediately to the north of which significant crustal subsidence did take place. A further transitional zone may occur between this point and 28°S where the Lüderitz Arch proper starts. The 29°S basement offset lies adjacent to two weak NE-SW gravity high axes onshore (see Pretorius 1979, fig.14), and with what Gerrard & Smith (1980) call the Kubuus line (a 200 km lineament of Pan-African aged quartz porphyry intrusions that runs NE from the coast across the Orange River). A further westward excursion of the craton occurs in central Namaqualand (between about 30 and 31.5°S), but this lies adjacent to the centre of the NW-SE Orange Basin graben complex, and no obvious relationship is suggested.

Following on from the above observations an apparent contradictory situation occurs in the south, where the Pan-African Fold Belt (which is intruded by numerous granitic bodies) lies adjacent to a major offshore structure: the Columbine/Agulhas Arch. This structure lies at the southern end of the Orange Basin, and stretches SE for about 700 km to the Agulhas Fracture Zone. It is a buoyant basement feature consisting of Cambrian granite plutons (553-522 my, Allsopp & Kolbe 1965, Schoch & Burger 1976) draped with Ordovician to Devonian sediments.

It has minimal Mesozoic and Tertiary sedimentary cover (0-3 sec, say 3 km) and gives rise to a shallow, narrow and rugged continental shelf, and a steep straight continental slope, where in the subsurface, acoustic basement descends rapidly via a few narrow fault zones to about 5 sec (ca.8 km) (Fig.121d).

The southern end of the Columbine/Agulhas Arch is truncated by the Agulhas Fracture Zone where there is a precipitous bathymetric discontinuity, but its northern end is less well-defined. It roughly coincides with the southern end of a large gravity high axis which obliquely cuts the coast between 33½ and 32½°S and continues inland to join the Griqualand-Transvaal axis of du Toit (1933) (Scrutton & Dingle 1974), whilst in the same area basement rises rapidly via a series of steps from the Orange Basin (Bryan & Simpson 1971). Gerrard & Smith (1980) report two narrow (3-10 km) N-S graben at the northern end of the Columbine/Agulhas Arch (about 35 km west of Saldanha Bay), which presumably represents a southerly extension of the Inner Graben complex of the Orange Basin. Scrutton & Dingle (1974) suggested that the boundary between the Orange Basin and Columbine Arch coincides with an original marginal fracture zone on the 61° small circle about Le Pichon & Hayes's (1971) predicted opening pole. No lateral offset in the COB was envisaged across this fracture zone, which in our refit in Figure 114 may coincide with a lineament along the southern side of the Colorado Basin in Argentina. The Columbine/Agulhas Arch itself has a conjugate structure in the Outer Margin Ridge which has a core of high velocity rocks (5.8-6.5 km/sec, Urien et al 1976) (Figs 114 & 115).

### 4.5.2 *Stratigraphy*

Our knowledge of the stratigraphy and lithofacies of the ?Jurassic-Lower Cretaceous rocks of the offshore basins is limited to regional seismic coverage (mostly Emery et al 1975), several boreholes in the proximal Orange Basin (McLachlan & McMillan 1979, Gerrard & Smith 1980) and DSDP site 361 (Bolli et al 1978). From these studies, two tectono-stratigraphic correlation schemes, based upon regionally recognizable seismic reflectors and deep boreholes have been devised, and locally it has been possible to correlate between them. In the ocean basin and outer margin areas, seismic reflectors AII and Davie can be recognized along most of the west coast margin, and DSDP site 361 has provided tentative ages for them of lowermost Albian and uppermost Maastrichtian/lowermost Palaeocene, respectively (Fig.116). In the Orange Basin, academic seismic refraction studies, and commercial seismic and drilling activity have provided evidence of a fairly complex stratigraphic sequence. Regionally-developed seismic reflectors have been notated T (bottom), R, P to L (top) by SOEKOR geologists (Gerrard & Smith 1980) (Fig.118), and these correspond approximately to major seismic P-wave interval velocity discontinuities recognized in earlier studies by Dingle (1979, 1980a) (Fig. 120). The deep oceanic reflector AII can be traced upslope and equated with SOEKOR reflector P (Fig.120), and in this chapter we will deal only with the pre-AII = pre-P stratigraphic interval, which consists of locally impersistent, high velocity (>4.3 km/sec) material at the base, and extensively-developed, regionally continuous medium velocity (3-4 km/sec) strata above (i.e. the T-R and R-P units, respectively, of Gerrard & Smith 1980).

#### 4.5.2.1 *Orange Basin*

Figure 66 shows the extent of the large Jurassic-Lower Cretaceous sediment lens that lies off the west coast margin adjacent to the Orange River. In the proximal areas (i.e. under the present continental shelf), the isopachs indicate considerable variation in thickness, which is a manifestation of irregular basement topography (e.g. Fig.120), whilst in distal parts less irregularities are evident, where sediment overlies oceanic crust. Because the distal parts of the Orange Basin merely represent sediment spread from a large input point (Orange River), there is no obvious western limit to the Orange Basin. Hence DSDP site 361 (which has only ca.400 m of pre-AII strata) does not appear on Figure 66 to lie at a significant location within the basin, being far to the SW on its feather edge. Nevertheless, a knowledge of the succession at 361 is crucial for a understanding of the stratigraphy and palaeogeography of the early SE South Atlantic, mainly because it is the only data point that we have in the distal parts of the Orange Basin, but also because the succession established there can be traced into the more proximal areas where the stratigraphy is more difficult to unravel.

We will first deal in some detail with the relatively simple stratigraphy in the deep oceanic area, and in particular that at DSDP site 361, before considering the more complex succession under the continental shelf and slope.

In the distal parts of the Orange Basin (i.e. on oceanic crust), horizon AII can be recognised <0.5 km above irregular oceanic basement. It rises towards the continent and at the same time basement descends rapidly so that under the continental rise SW of Cape Town, pre-AII strata reach a maximum of about 4 km and AII abuts basement about 3 km beneath sea level (Fig.121d). Traced NE into the main Orange Basin, AII can be followed under the continental shelf and apparently correlates with P-wave velocity discontinuities of generally 3.0 km/sec or less above, and >3 km/sec below (Fig.120c). At DSDP site 361 about 300 km SW of Cape Town, reflector AII is associated with a significant lithofacies change within Lower Albian strata from anoxic sediments below to relatively well oxygenated sediments above.

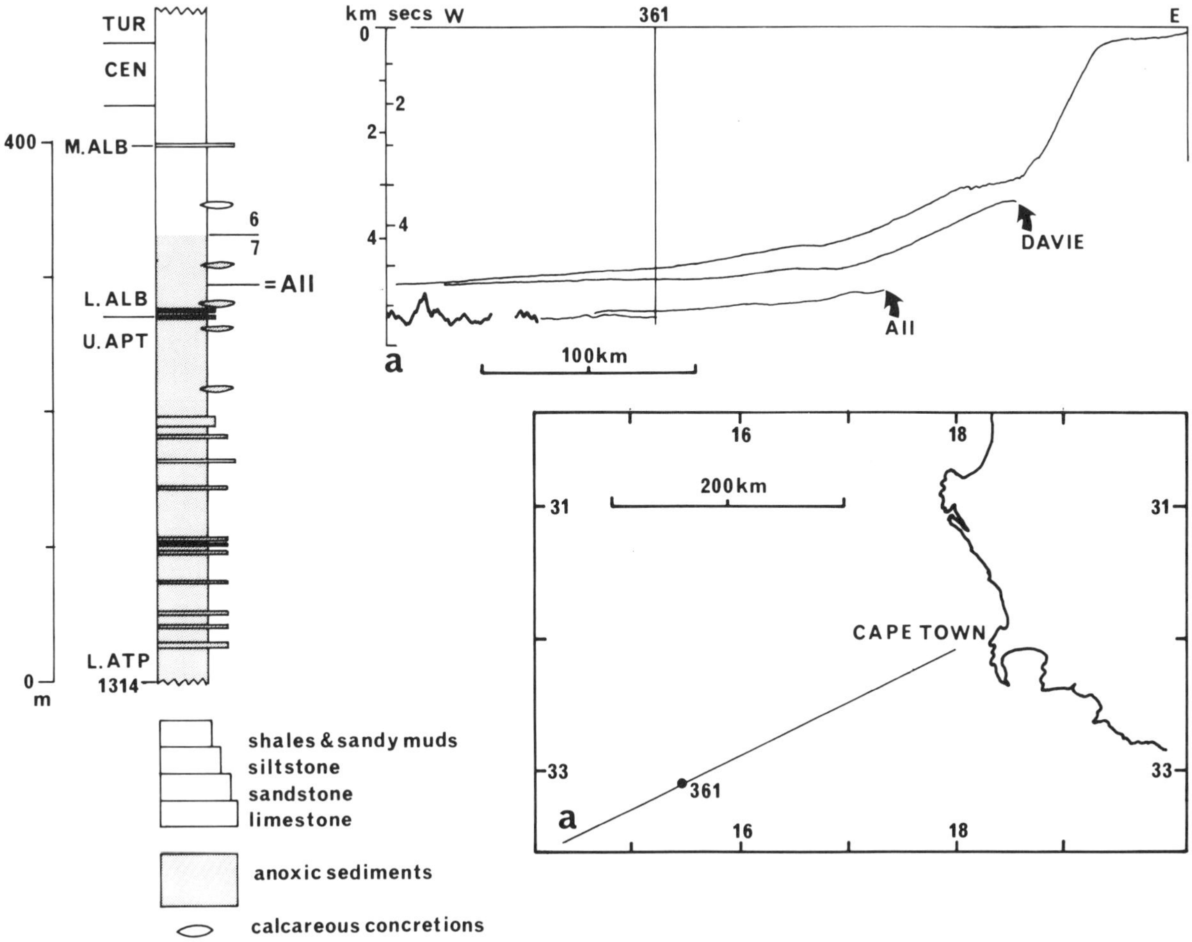


Figure 116. Lower Cretaceous sediments in DSDP borehole 361. Core log shows lithologies and their relationship to seismic reflector Atlantis II (AII) which can be traced towards the continent on the seismic record (a). Depth (m) at base of borehole is sub-sea floor, and vertical scale on seismic record is in sec and km (water depth). Data from Bolli et al (1978).

Bolli et al (1978) recognized two lithological units in the mid-Cretaceous section of DSDP 361 – units 6 and 7 (Fig.116). The lower Unit (7) includes all pre-AII rocks and its boundary with the overlying Unit 6 is about 35 m above the estimated position of AII. The base of the borehole was calculated to lie about 80 m above acoustic basement, so that at site 361 pre-AII sediments are about 380 m thick.

Unit 7 consists of shales, sandy mudstones and sandstones, with minor mudstones, siltstones and one 40 cm thick marly nannofossil chalk. The shales are typically black to dark grey, the mudstones greenish grey to greenish black, and the sandstones bluish grey to greenish grey. Bedding is generally massive, with thin sandy laminae and minor graded bedding. Cross-bedding is very rare and bioturbation is absent.

Sandstone units, which are well indurated by calcite and clay cement, are more abundant in the lower part, but shales and sandy mudstones are distributed throughout the sequence, which is characterized by abundant carbonaceous matter in all lithologies except the chalk (3.6 % organic carbon, mean of 47 determinations quoted by Bolli et al 1978). Pyrite laminae and small concretions are common throughout, and the carbonaceous detritus consists of both palynomorphs and wood fragments, the latter ranging from silt size grains to pieces 10 cm across. Occasionally, thin layers (up to 4 cm) contain such abundant carbonaceous debris as to warrant the description of coal. Methane gas from the bacteriological decay of the plant material occurs throughout the sequence.

From a study of the mineral assemblages and rock fragments in the sandstones (Tab.35), three distinct sediment provenances were proposed by the DSDP team (Bolli et al 1978);

i) a granitic and high grade metamorphic terrain. Undulose quartz, microcline, tourmaline, zircon and rose quartz suggest a granitic source, and schist and quartzite fragments, together with biotite, amphibole and muscovite suggest a high grade metamorphic terrain;

Table 35. Average modal analyses of lithologies in DSDP borehole 361, pre-AII sediments (unit 7) (data from Bolli *et al.* 1978), as percentages of total.

| | Quartz | Clay | Carb. debris | Pyrite | Feld. | Nannos. | Rock fr. | Cement calcite/clay |
|---|---|---|---|---|---|---|---|---|
| shales | 36 | 36 | 15 | 6 | 2 | 2 | 3 | 0 |
| sandy mudstone | 55 | 23 | 10 | 2 | 3 | 0 | 7 | 0 |
| sandstones | 55 | 8 | 4 | 2 | 3 | 0 | 2 | 26 |

Heavy minerals: up to 0.6 % in the sandy mudstones include: zircon, chlorite, garnet, sphene, apatite, opaques and tourmaline.
Mineral grains in trace amounts include: plagioclase, biotite, brown amphibole, muscovite, hypersthene, augite, ilmenite and glaucophane (one sample only).

ii) an acid volcanic terrain is suggested by the presence of small quantities of sanidine;

iii) a basaltic terrain is suggested by the presence of black rock fragments in the basal part of the sequence. These were interpreted as basalt chips.

The angularity of the bulk of the quartz and detrital volcanic grains suggests nearby sources for the components of the sands.

The boundary between lithological Units 6 and 7 is taken between cores 27 and 28 (i.e. between levels 962 and 1 000 m) (Fig.116). It is marked by a sharp downward increase in plant debris (from 0 up to 15 %) and pyrite (from trace up to 6 %) and a colour change from dark grey, greenish black and minor bluish grey to dark grey and greyish black. Both cores consist of shales but 27 (lower Unit 6) has higher sand size (average 8 vs 0 %) and silt size (33 vs 27 %) components.

With the exception of non-calcareous plant material, fossils are sparse in the pre-AII sediments of DSDP 361. No macro-invertebrates were recovered and most micro-invertebrates are poorly represented. Radiolaria and planktonic foraminifera are very sparse and poorly preserved and there are no benthic foraminifera. Calcareous nannofossils are also poorly preserved but sufficient forms were recovered to effect a palaeontological zonation. Three zones were recognized: *Prediscosphaera cretacea* (Middle Albian); *Parhabdolithus angustus* (Upper Aptian to Middle Albian); and *Chiastozygus litterarius* (Lower to Upper Aptian). Marine palynomorphs on the other hand are well-represented by a rich dinoflagellate and acritarch flora which indicates normal surface waters environments. Terrestrial plant debris is abundant in the sediments in the form of spores, pollen and wood fragments. Although large quantities of spores and pollen are present, they represent a low diversity assemblage, dominated by *Classopollis* which is the pollen from the conifer *Pagiophyllum* (also well represented by leaves, stems and cuticles). The flora is a mixed conifer/cycad assemblage which indicates a nearshore community possibly growing under unfavourable climatic conditions, or representing a restricted community. Wood fragments are abundant and pieces as large as 8 cm long were recovered.

Bolli et al (1978) correlate reflecting horizon AII with the top of a sequence containing numerous sandy mudstones in Unit 7, which occurs between cores 29 (with thick calcite cemented sandy mudstones and siltstones) and 28 (non-carbonate shales). The bottom of core 28 is at 1 010 m and the top of core 29 is at 1 029 m. Physical properties change significantly across this boundary with P-wave velocities exceeding 4.5 km/sec in core 29 compared to 2.25 km/sec in the shales of core 28. Assuming that the boundary occurs midway between these cores, and that the boundary between the carbonaceous Unit 7 and non-carbonaceous Unit 6 occurs midway between cores 27 and 28, then reflecting horizon AII lies about 35 m below the transition from anoxic to oxygenated sedimentary environments. On all but the highest quality seismic records this is not likely to be a resolvable difference (about 0.031 sec two way time at 2 250 m/sec) particularly in view of the rather diffuse and irregular nature of horizon AII. To all intents and purposes, therefore, AII can be equated with the boundary between both organic carbon-rich to poor, and shale/muddy sand to shale transitions. The boundary between lithological Units 6 and 7 (cores 27 and 28) is dated as Lower to Middle Albian, and that between cores 28 and 29 (AII) as Lower Albian – both within the *Parhabdolithus angustus* zone (Bolli et al 1978, tab.10).

The large quantities of terrestrial plant matter in the Unit 7 sediments attest to rapid burial, and Bolli et al (1978) estimate the sedimentation rate to have been between 45 and 70 m/my.

No detailed well logs have yet been published from the eight commercial boreholes in the Orange Basin that had been completed at the time of the report of Gerrard & Smith (1980), but information contained in their report, and the preliminary stratigraphy supplied by McLachlan & McMillan (1979) (Tab.37) allows us to construct provisional summary litho- and biostratigraphic logs and sections in the proximal Orange Basin, which can be compared with the more general cross-section of the basin constructed from seismic interval velocity data in Figure 120. We will consider the succession in terms of the important seismic reflectors identified by Gerrard & Smith (1980) as T, R and P (Figs.117, 118, 119 & 120).

Gerrard & Smith (1980) correlate reflector T with the top of an early rift sequence that we consider probably dates from mid-Jurassic times. Jurassic continental (lucastrine) sediments and acid lavas are widely dis-

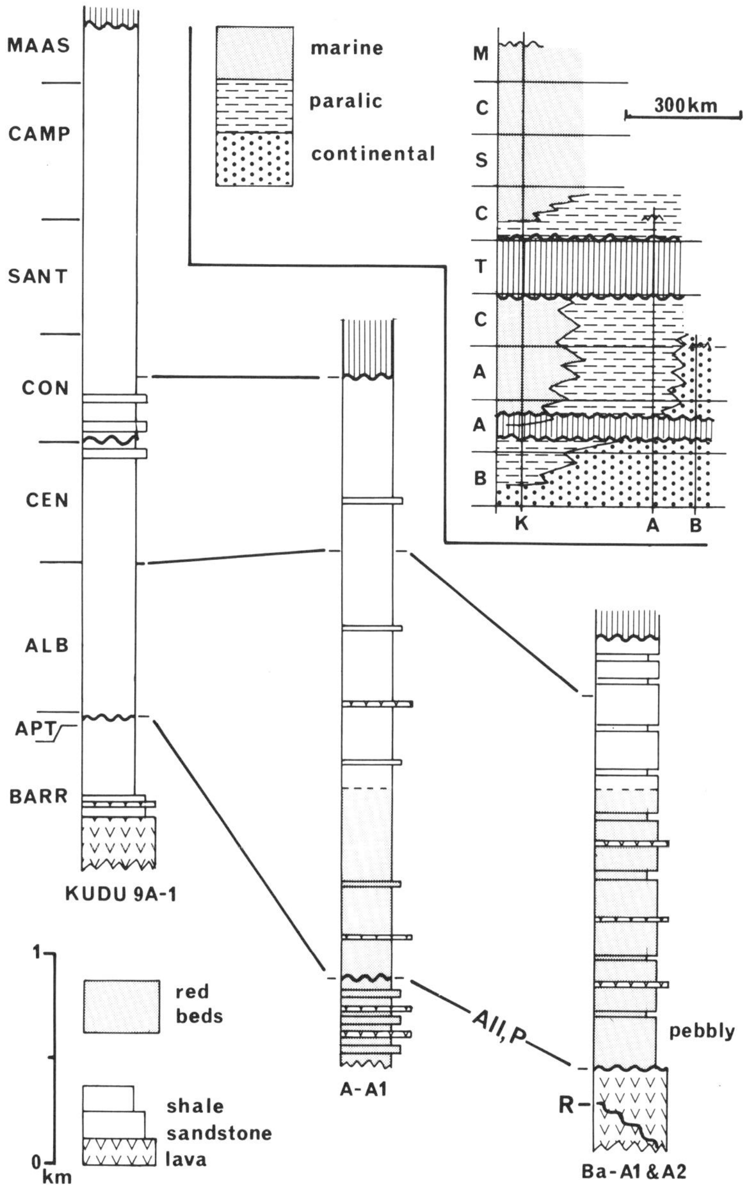


Figure 117. Mesozoic strata in boreholes in the Orange Basin. Constructed from sediment accumulation rate data given in McLachlan & McMillan (1979 figs.2 & 3), modified by data in Gerrard & Smith (1980). Kudu is reasonably accurate, but the others are partly diagrammatic and Ba-A1 & A2 is a composite log from two holes ca.10 km apart. Insert shows diachronous lithofacies relationships across the Orange Basin. See Figure 118 for locations and further regional stratigraphic data.

tributed in the taphrogenic basins and over horst tops south of the Colorado, and within the Salado Basins in the Argentinian offshore areas (Figs 114 & 122), and it is reasonable to suppose that similar facies are developed in the deeper parts of the Orange Basin and other areas along the west coast. These strata have suffered considerable faulting on the South American side, and were presumably laid down during and possibly prior to the main taphrogenic movements. This implies that the top of their presumed counterparts in the Orange Basin (i.e. reflector T) will occur at varying elevations, and will appear on seismic records as part of the faulted acoustic basement. The only rocks of comparable age onshore in SW Africa are the 161-173 my Hoachanas basalts in southern Namibia (Gidskehaug et al 1975). Gerrard & Smith (1980) consider that only borehole (A-C1) has penetrated into the pre-T Mesozoic strata, but that A-C2 and Ba-A1 may have done so. In all three, the rock types were acid lavas of unspecified petrological character. By analogy with Argentina, interbedded sediments may be expected in the pre-T succession, but so far none have been reported from the west coast offshore area. These presumed mid-Jurassic strata rest on pre-Mesozoic basement of generally unknown type. Figure 114 suggests that the Orange Basin area may be underlain by rocks that were involved in the Pan-African folding (ca.550 my), although a small extension of the Namaqualand cratonic basement can also be anticipated

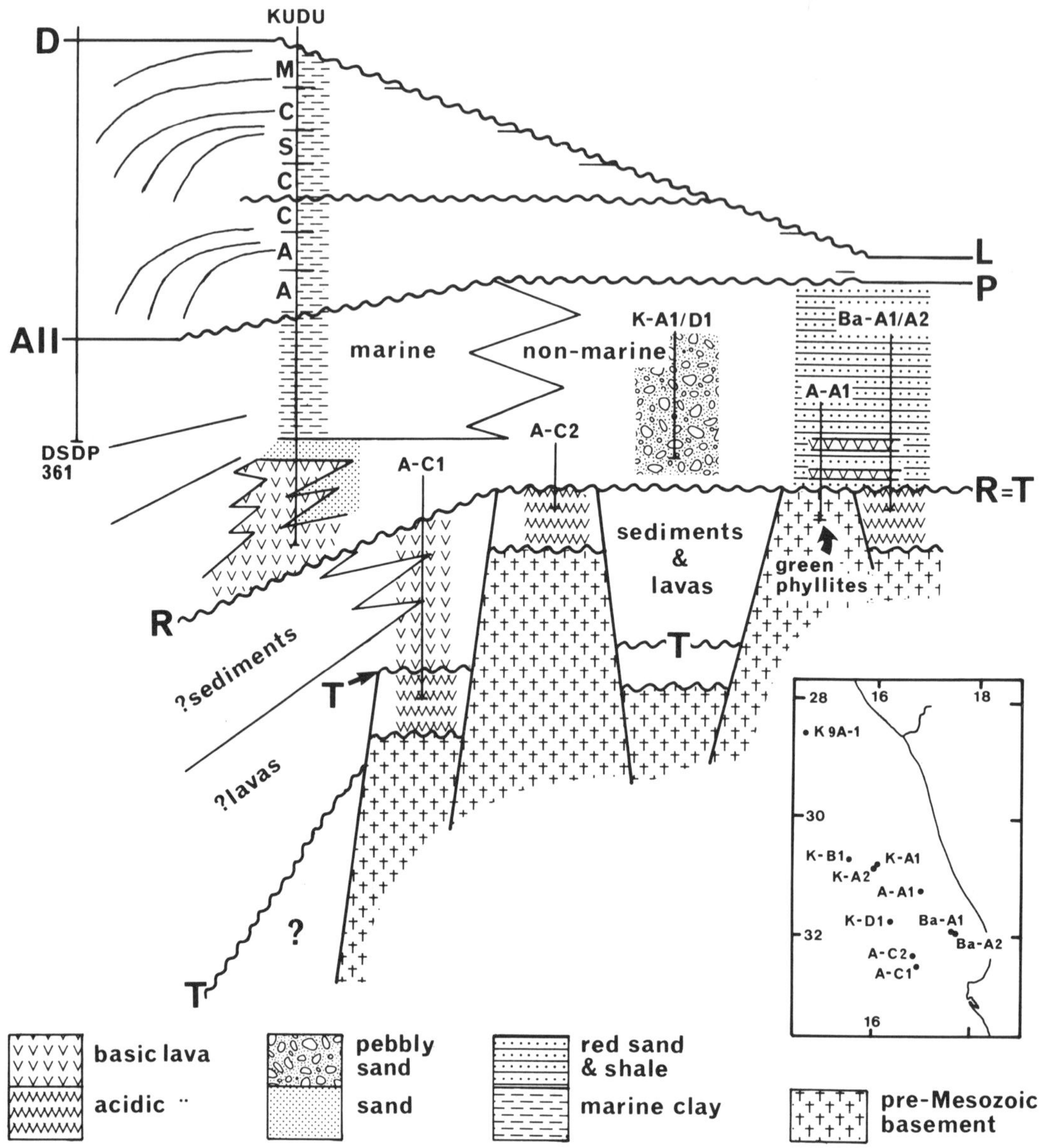


Figure 118. Schematic stratigraphic interpretation of borehole and seismic data in the Orange Basin. Based mainly on published SOEKOR results (McLachlan & McMillan 1979, Gerrard & Smith 1980). Not to scale. Selected borehole sections show specific phenomena and are not intended to be complete. Regional correlation is by reference to seismic reflectors T, R, P = AII, L = Davie, see text for discussion of their ages and significance. DSDP borehole 361 is out of context, but is shown for completeness. Insert shows location of SOEKOR boreholes mentioned by Gerrard & Smith (1980). Abbreviations: A-M = Aptian to Maastrichtian stages.

between about 30 and 31°S. One borehole (A-A1) is reported to have penetrated into a green phyllite that Gerrard & Smith (1980) equate with rocks in the Pan-African fold belt, whilst other evidence is provided by reports of pebbles and cobbles in Neocomian sediments, of orthoquartzites typical of the Table Mountain Group (Upper Cambrian-Lower Devonian) and quartz-feldspar porphyry similar to that occurring in the Saldanha area (553-522 my), suggesting nearby basement outcrops of these rock types in Lower Cretaceous times.

The T-R sequence of Gerrard & Smith (1980) (Fig. 118) corresponds to the high velocity graben fill in Figure 120. It has been identified in the Inner Graben complex, and as a unit 50-130 km from shore on the western flank of the Central Ridge. The former are bounded by basement highs, but the latter forms a generally westward thickening wedge. Some of the seismic reflectors within this sequence are conformable to reflector T, but locally they lap onto it. Reflector R at the top unconformably truncates the internal seismic reflecting horizons. Only one borehole (A-C1, to the west of the Central Ridge) is known to have

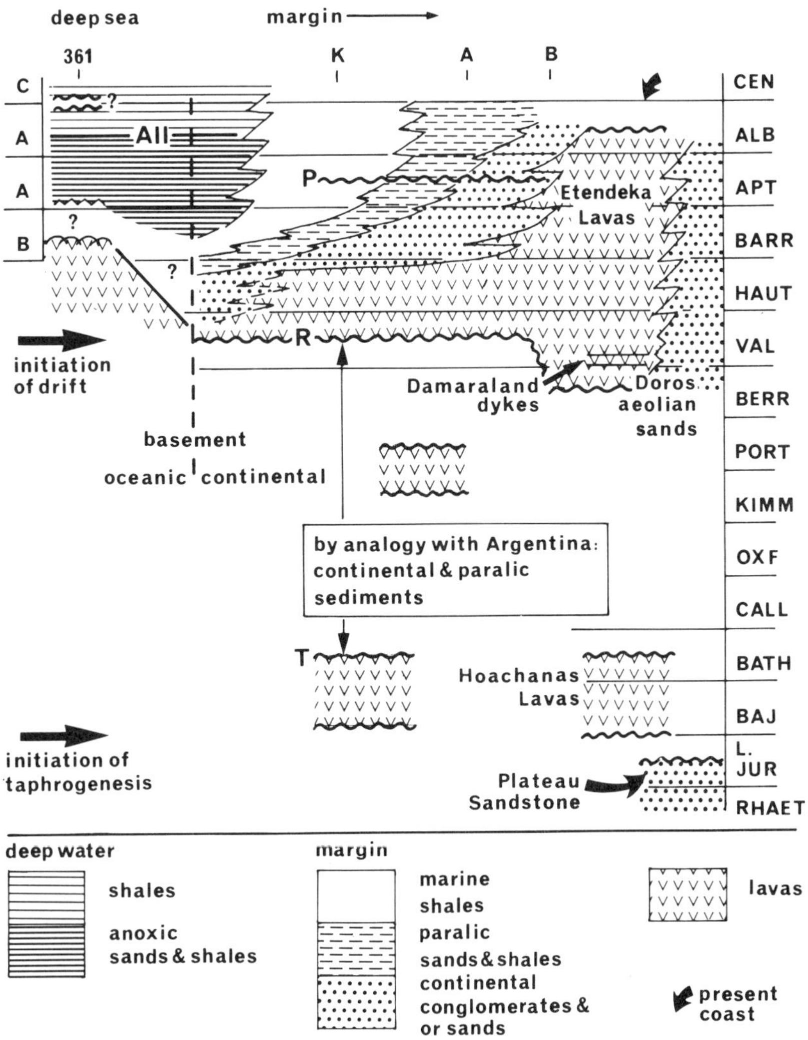

Figure 119. Schematic spatial and temporal distribution of Jurassic-Lower Cretaceous strata on the continental margin of S.W.Africa. Not to scale. Note that the aeolian sands at Doros (the 'Main Sandstone' in Figure 41) probably extend to the Rhaetian and are partly contemporaneous with the Plateau Sandstone. They may extend higher than the basal lavas (Berriasian) – say to the Albian. Abbreviations: 361 = DSDP 361; K = Kudu 9A-1; A = A-A1; B = Ba-A1, A2. Seismic reflectors AII, P, R, are those shown in Figures 118 and 120.

penetrated into the T-R sequence, and no data are available on the succession with the Inner Graben complex. In A-C1, Gerrard & Smith (1980) report 690 m of basic lavas with minor alkaline lavas and subordinate sediments or pyroclastics. The basic lavas have an average density of 2.67 g/cc, which is rather low, suggesting a high percentage of amygdaloidal types. They have an average P-wave velocity of 5 km/sec. The relatively thin T-R sequence in A-C1 is to be expected considering its position on the western flank of the Central Ridge, and nearly 4 km of strata belonging to this interval are suspected farther west. Gerrard & Smith (1980) report a commonly-developed intra-unit seismic reflector, which they suspect may represent a sediment/lava contact, suggesting that the thin sediment or pyroclastic interbeds found in A-C1 merely represent the inner edge of a more extensive upper sediment-pyroclastic sequence farther west (Fig.118). By analogy with the Argentine margin, these sediments can be expected to be fluviatile/lacustrine, laid down in an area of active synsedimentary faulting and rugged topography in the proximal parts of the Orange Basin. Within the Inner Graben, internal drainage might be expected. As already mentioned, reflector R cuts unconformably across the T-R graben fill, and locally coincides with T where it planes the top of horsts and basement structures. It is overlain by a sequence of rocks that fill in and smooth out the remnants of the irregular horst and graben topography, and has as its upper surface another regionally-developed reflector:P. The R-P interval corresponds to the medium velocity layer on Figure 120, and reaches a thickness of about 0.75 sec (ca.1.5 km) under the continental shelf. Reflector P is developed as an unconformity near to the present coast, but may be disconformable, or even conformable under the present continental slope. SOEKOR geologists (e.g. Gerrard & Smith 1980, p.12) have dated reflector P as mid-Aptian and have equated it with reflector AII of the deep ocean basin succession.

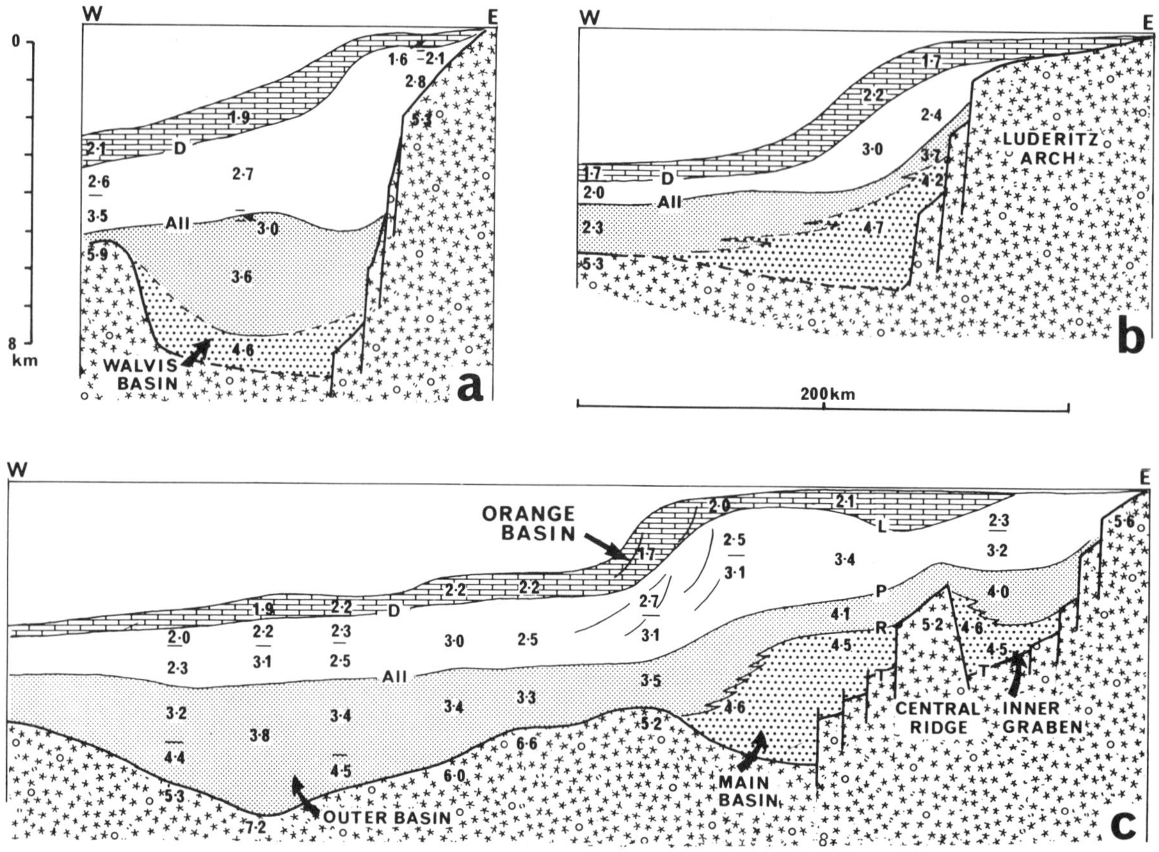


Figure 120. Cross-sections of the continental margin of S.W.Africa. a. Walvis Basin; b. Luderitz Arch; c. Orange Basin. Locations and key on Figure 121. Datum is sea level. Constructed mainly from seismic refraction data, and P-wave interval velocities (km/sec) are shown as control points. These have been integrated with regional seismic reflectors AII, D, T, R, P and L. Vertical scale (km) has been corrected for velocity variations. Modified from Dingle (1979, 1980a) and based mainly on the following sources: Bryan & Simpson (1971); Hoskins *et al.* (1974); Emery *et al.* (1975); Rabinowitz (1976); Lehner & de Ruiter (1977); Bolli *et al.* (1978); Gerrard & Smith (1980).

Mid-late Aptian is the age commonly quoted for AII, but as mentioned earlier, close inspection of the DSDP records (Bolli et al 1978) suggests to us that this date is incorrect, and that this prominent reflector is in fact earliest Albian. Assuming that the SOEKOR data are well-founded and not based on an *a priori* correlation of P and AII, then it is clear that AII and its lateral equivalent are diachronous, being older in the proximal part of the basin, and slightly younger in the distal areas. That is to say, the lithofacies change that it represents took place first in the shallow continental margin areas and spread later but rapidly into the deep ocean basin. Gerrard & Smith (1980) follow McLachlan & McMillan (1979) in assuming a Barremian age for reflector R, based on the dating of the lowest marine sediments in borehole Kudu 9-A1. All the boreholes reported by Gerrard & Smith (1980) (as well as those mentioned by McLachlan & McMillan 1979) penetrate into the R-P succession, so we have reasonably good knowledge of its lithology. Internally, there is no evidence of re-activation of the taphrogenic boundary faults, so syn-sedimentary faulting and large contemporaceous topographic discontinuities cannot be expected during its deposition. Gerrard & Smith (1980) report that west of a 'hinge line' that lies approximately over the outer edge of the buried Central Ridge, the R-P sequence thickens rapidly. It might also be noted here that this lineament is also recognized as a control for Palaeogene volcanism, and as a weak tectonic 'high' structure as late as Miocene times. It also coincides with an important facies change during deposition of the R-P interval: to the east the sediments are relatively thin, continental red beds; to the west they are thicker grey-coloured restricted marine or lacustrine. These relationships are illustrated in Figures 117, 118 and 119, with Kudu 9-A 1 typical of the most distal marine location, and A-A1, and Ba-A1/A2 lying in progressively more proximal positions. In addition, there is a significant N-S facies change across the Orange Basin, with the marine rocks apparently restricted to the northern areas and coarser sands and conglomerates, and continental red shales and minor sands to the south. The sands are reported by Gerrard & Smith (1980) to be 'radioactive'.

All the localities reported in the R-P sequence have basal lavas: the thickest being in Kudu, where 130 m of basic flows underlie, and interfinger with gas-bearing sands of Barremian age (McLachlan & McMillan 1979). They are age equivalents of the Kaokoveld (Etendeka) lavas of NW Namibia, which have a Berriasian-Albian age (132-108 my), but the Barremian age of the Orange Basin extrusives suggest that they are slightly younger than the climax age of the Kaokoveld suite (between 122 and 115 my compared to 125 my according to Siedner & Miller 1968). They also are of comparable age to Lower Cretaceous lavas interbedded with red beds that have been reported from the Colorado Basin by Zambrano & Urien (1974) and Urien et al (1976). In the more proximal locations, thinner lava flows interfinger with continental red beds (e.g. Ba-A1/A2). Reference to Figure 117 shows that in all three localities represented by the boreholes illustrated, there is a lower (Barremian) sequence of continental sandstone interbedded with lavas, and that in the proximal localities, these are of a red bed facies (pebbly in Ba-A1/A2). In the more distal parts of the Orange Basin, continental sands with some shales (red in the lower half), and thin interbedded lavas (or sills) make up the whole of the Barremian to ?Cenomanian succession, but in a NW direction, a brackish water (paralic) facies replaces the continental strata across the Aptian/Albian hiatus (reflector P). Here the rocks are dominantly argillaceous, with only minor sands and possible thin lava interbeds. At the most distal site (Kudu 9-A1), situated on the present-day Orange Banks (on the NW edge of the present continental shelf) marine rocks appear above the Aptian/Albian hiatus, and with the exception of a temporary return to a paralic lithofacies (shales and sandstones) across the Cenomanian-Coniacian hiatus (the Turonian is missing), the whole of the Albian-Maastrichtian succession is represented by marine shales. In Lower Cretaceous times, therefore, the whole of the proximal Orange Basin was affected by volcanism interspersed with continental (?fluviatile) sedimentation, but marine conditions were established in the more distal areas by Albian (possibly late Aptian) times. At present we do not have complete information on the lateral and lower extent of the lava/continental sediment sequence, but Figures 67 and 119 diagrammatically place the three boreholes in a regional context with known volcanic and continental rocks onshore and DSDP in the adjacent deep water basin. These show that distally, the continental sediments pass via paralic mudrocks and sands into shallow marine shales, which in turn pass in to deep-water dark anoxic sandy mudrocks (below AII) and greyer shales (above AII). The fact that horizon AII can be traced under the continental slope suggests that in this region the deep-water, organic carbon-rich lithofacies interfingers with shallower-water facies of the Kudu 9-A1 type with good prospects for hydrocarbon generation and accumulation.

#### 4.5.2.2 *Walvis Basin and Lüderitz Arch*

Very little can be said about the stratigraphy and lithofacies of pre-AII sediments to the north of the Orange Basin because there have been no offshore boreholes and very little published detailed geophysics from either region. Horizon AII has been recognized in the area (Emery et al 1975), and in Figure 120a, b, it generally coincides with the top of sequences having P-wave velocities >3 km/sec. Assuming that it still represents a Lower Albian horizon, then thick pre-Lower Albian sediments lie along the western flank of the Lüderitz Arch (ca.3 km) and in the Walvis Basin (> 4 km). In both areas the lower part of the succession consists of high velocity strata (4.2-4.7 km/sec). This may represent a basal lava sequence or highly indurated sediments, but if the latter, we are not in a position to speculate as to whether they are marine or non-marine (the earliest sediments deposited on the new oceanic crust need not necessarily have been marine). However, the very presence of horizon AII suggests that the upper part of the shaded areas on Figure 120a and b is anoxic marine. Horizon AII does not transgress across the Lüderitz Arch, indicating that in Valanginian to Albian times the arch stood as a wide rocky continental shelf with a precipitous descent into the adjacent deep ocean basin. Because the position of the COB is not accurately known, the possibility of predrift volcanic or continental sediments occurring along this northern offshore sector is difficult to assess. Certainly, the extent of any such deposits adjacent to the main part of the Lüderitz Arch is likely to be very limited (perhaps lapping against fault blocks on the outer edge), but in view of the extensive coastal lava outcrops in the Kaokoveld, the possibility seems high for the Walvis Basin.

#### 4.5.2.3 *Onshore volcanic rocks*

The oldest Mesozoic volcanic rocks in SW Africa are the Hoachanas lavas in south central Namibia (Bajocian-Bathonian 173-161 my) and Jurassic dates in the Cape Cross lineament on alkaline plugs and dykes (Siedner & Miller 1968, Gidskehaug et al 1975, Siedner & Mitchell 1976) (see Chapter 7).

Large outcrops of intrusive and extrusive volcanic rocks occur in the coastal area of northern Namibia, to the north of Walvis Bay. The largest and most extensive of these lie to the north of the Ugab River in Kaokoland, where the Etendeka lavas form high terrace mountains and cover an area of approximately 18 000 km$^2$ between 21 and 18°S. They are composed of massive and amygdaloidal andesites and basalts in widespread flows that are usually <60 m thick (Haughton 1969). The present-day uppermost layers are rhyolitic, but as Haughton (1969) has pointed out, these must originally

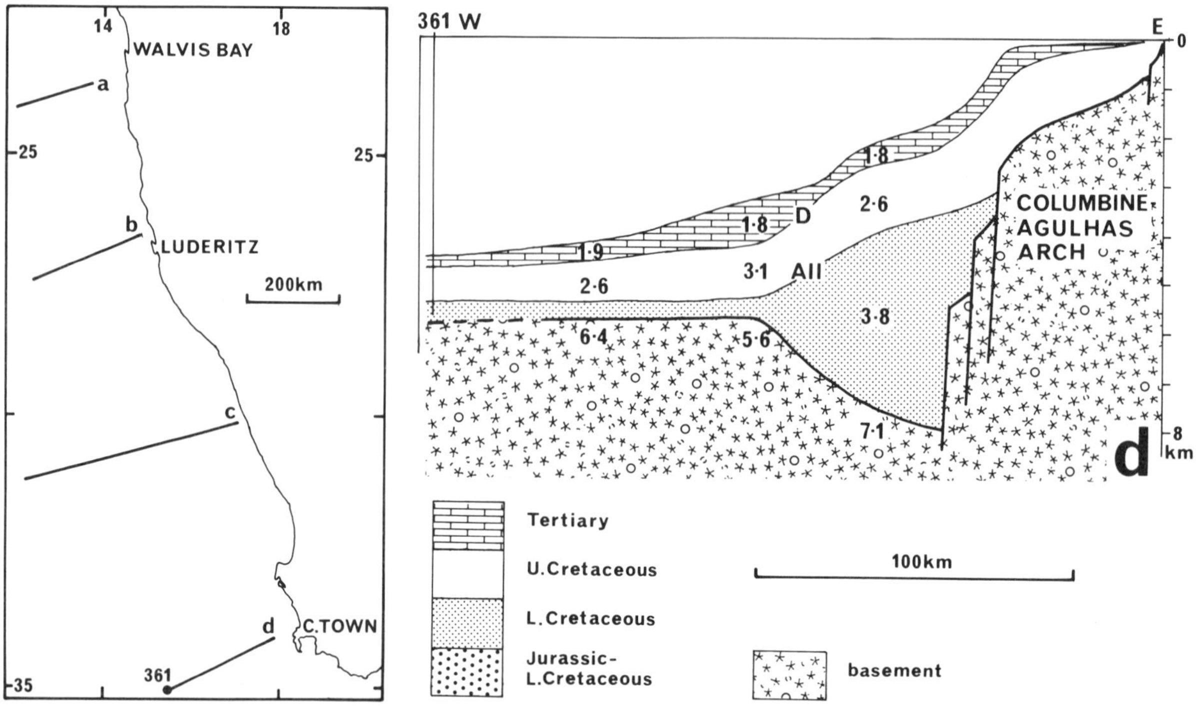


Figure 121. Cross-section of the continental margin SW of Cape Town. DSDP site 361 is used as stratigraphic control in the west. Insert shows location for this and Figure 120. See Figure 120 for further details.

have been some way down in the sequence. The total thickness preserved is about 800 m (Martin 1976). Radiometric dates on the Etendeka lavas have a range of 132-108 my (Tab.36), and at their base they interfinger with the 'main sandstone' at Doros, which is probably Rhaetian at its base. Small outcrops of Etendeka lavas occur south of the Ugab River NE of Cape Cross (1 300 km$^2$), where Siedner & Miller (1968) have obtained a radiometric date of 132 my.

Running east from Cape Cross (140 km NW of Walvis Bay) for a distance of about 350 km, and in a zone up to 120 km wide, is a series of large intrusive bodies associated with basaltic lava and ringdykes. This is the line of Cape Cross alkaline intrusives which consists of circular plug and vent-like bodies up to about 20 km in diameter. A large number of rock types are found in these plugs but most have alkaline affinities: alkaline granites, anorthosite, syenite, quartz prophyry foyaite, tinguaite, essexite, olivine gabbro, and two occurrences of carbonanite (Haughton 1969). Siedner & Miller (1968) and Siedner & Mitchell (1976), dated several of the lavas and basaltic ring dykes associated with these intrusives and found one to be Hettangian (188 my) (Klein Spitzkop), one to be Bathonian (164 my) (Okonjeje), one Portlandian (136 my) (Paresis), and two Hauterivian (Messum and Doros, 123 and 125 my respectively).

Finally, a large number of dolerite dykes cut the area of the Cape Cross line (Botha & Hodgson 1977). They are often found in en echelon distribution within swarms, with individual dykes seldom exceeding 3 m in width but continuing for distances of up to 20 km. They have a general northerly trend, but Botha & Hodgson (1977) stress that this is invariably oblique to the regional basement structural grain, which is usually more east of north than the dyke trend, and is also not parallel to the coastline which is here NW-SE. The most common of the dyke types is olivine dolerite, and mineralogically they are similar to those of the main Karoo Basin except that they lack pigeonite and have some quartz-dolerite varieties SW of Brandberg. Siedner & Miller (1968) radiometrically dated nine of these dykes and although there is a marked threefold distribution in time they appear to be randomly distributed in space: two dykes are Upper Triassic (196 my), two are Bathonian (163 and 165 my) and the rest (five) are Lower Cretaceous (Berriasian to Aptian/Barremian) (Tab.36).

Clearly, the Cape Cross line was intermittently active for about 80 my (Fig.193), with earliest activity in Upper Triasssic times (contemporaneously with the main Karoo volcanism in the south) (dykes), in the Middle Jurassic (dykes and plugs) and at its climax in late Jurassic-Lower Cretaceous times (dykes, plugs and lavas). In contrast, the coastal zone between Cape Cross and 18°S was active only during the late Jurassic-Lower Cretaceous period with the extrusion of the Etendeka lavas.

The three alkaline complexes south of Lüderitz

Table 36. Radiometric dates on Jurassic and Lower Cretaceous volcanic rocks in the Kaokoveld, Cape Cross and Luderitz areas.

| Kaokoveld & Cape Cross | | range |
|---|---|---|
| A. Lavas not directly associated with intrusives | | |
| Etendeka plateau [1,2,4] | 116, 114, 111, 109, 124, 120 116, 108 | 124-108 |
| Cape Cross[1] | 132 | |
| B. Lavas and ring dykes associated with alkaline intrusives | | |
| Paresis[1] | 136 | 181-123 |
| Messam[1] | 123 | |
| Doros[1, 4] | 125, 128, 131 | |
| Okonjeje[1] | 164, 181 | |
| C. Dykes and sills[1,4] | 196, 196, 165, 163, 134, 133, 121, 129, 128, 128, 115 | 196-115 |
| Luderitz | | |
| Granitberg nepheline syenite[3] | 130 | |

1 – Siedner & Miller (1968)
2 – Gidskehaug *et al.* (1975) (averaged)
3 – Marsh (1975)
4 – Siedner & Mitchell (1976)

show some significant petrological contrasts. The two Lower Cretaceous intrusions (Granitberg dated at 130 ± 2 and Drachenberg assumed) are formed of feldspathoidal (nepheline syenites and foyaites) and saturated and over-saturated syenites respectively, whereas the Oligocene (presumed) Pomona complex is again feldspathoidal (nepheline syenites). Marsh (1976) concludes that two distinct magmas must have been available for these intrusions, but the relative closeness of the rock types, and the reactivation of the province indicates a tectonic control, similar to that of the Cape Cross line.

### 4.5.3 *Palaeogeography*

Because we have relatively few control points, we will present composite palaeogeographies to represent pre-drift and post-drift situations, even though the data on each may not be strictly contemporaneous.

Figure 122 shows the pre-drift palaeogeography, nominally for late Jurassic times, during which extensive graben development took place in the south and SW, e.g. Valdes, San Jorge and Salado basins on the South American side. In these areas, Triassic to Jurassic volcanism is recorded on a massive scale with main lava fields known to extend as far north as the Valdes Basin and locally exceeding 2 km in thickness. These are the Chon Aike and Pampa de Agnia acid extrusives and pyroclastics (161 my Zambrano & Urien 1974), and are overlain, at least in the south, by Jurassic lacustrine sediments and pyroclastics (La Matilde and Canadón Asfalto formations). Jurassic lavas are also known from the Hoachanas Basin in southern Namibia, the southern Orange Basin, the Salado Basin, and the Algoa Basin (Zuurberg volcanics) to the east. At least two periods of volcanism are indicated by the Orange Basin lavas – an earlier, acid phase, whose products were displaced by further large-scale block faulting, during which a second phase of amygdaloidal basalts and minor alkaline lavas were extruded. In Upper Jurassic times, we envisage a series of large graben and horsts in the south and central region associated with widespread volcanic activity and towards the close of the period, elongate lakes fed by internal drainage, at least in the south. Marine conditions obtained to the east in the Falkland-Outeniqua Basin, and in the west in the Neuquen Basin. The ostracod faunas of these last two named areas are very similar and indicate free connections between them (Musacchio 1979, Dingle 1982a), but at present, there is no evidence to confirm that this was via a proto-south Atlantic seaway. Arid conditions probably prevailed in the far north, because it seems likely that the 'Main Sandstone' at Doros is Upper Triassic to Lower Jurassic or even Lower Cretaceous in age.

The second palaeogeography is for Lower Cretaceous times, nominally Aptian (Fig.123). By then, continental separation had been underway for about 15 my and a narrow oceanic basin had formed in the South Atlantic between the Walvis Ridge-Rio Grande Rise barrier and the Falkland Plateau, which still abutted the Agulhas Bank. The latter barrier had, however, subsided sufficiently to allow the proto-South Atlantic to become flooded. Large-scale taphrogenic basin subsidence continued on the Argentine margin where the Salado, Colorado and San Jorge cross-margin basins and their linking north-south graben were accumulating thick (nearly 3 km in the Colorado Basin) continental and paralic clastics and tuffaceous clastics, with red beds in the Salado and Colorado basins. No marine incursions have been reported from these areas, though they must be expected in the distal regions. The Outer Margin Ridge appears to have remained a buoyant feature with no sediment cover at this time. Volcanics are known from the Colorado Basin, but at the northern end of the proto-southern Atlantic massive lava fields developed in late Jurassic to early Cretaceous times over the Walvis Ridge-Rio Grande Rise barrier and on the adjacent continental margins: Serra Geral in Brazil and the Etendeka Lavas in Namibia, as well as lavas and alkaline intrusions along the Cape Cross line. This activity effectively brought to a close the continental (aeolian) sedimentation in the Doros area, and its equivalent the Botucatu sandstone in southern Brazil.

A consideration of the texture and components of the pre-AII sediments at DSDP site 361 shows that the following factors need to be considered in predicting the sedimentary environments in the SE Atlantic at this time:

i) lack of normal benthic faunas (e.g. molluscs, benthic foraminifera, ostracoda, etc.) and any sign of bioturbation;

ii) abundance of authigenic pyrite;

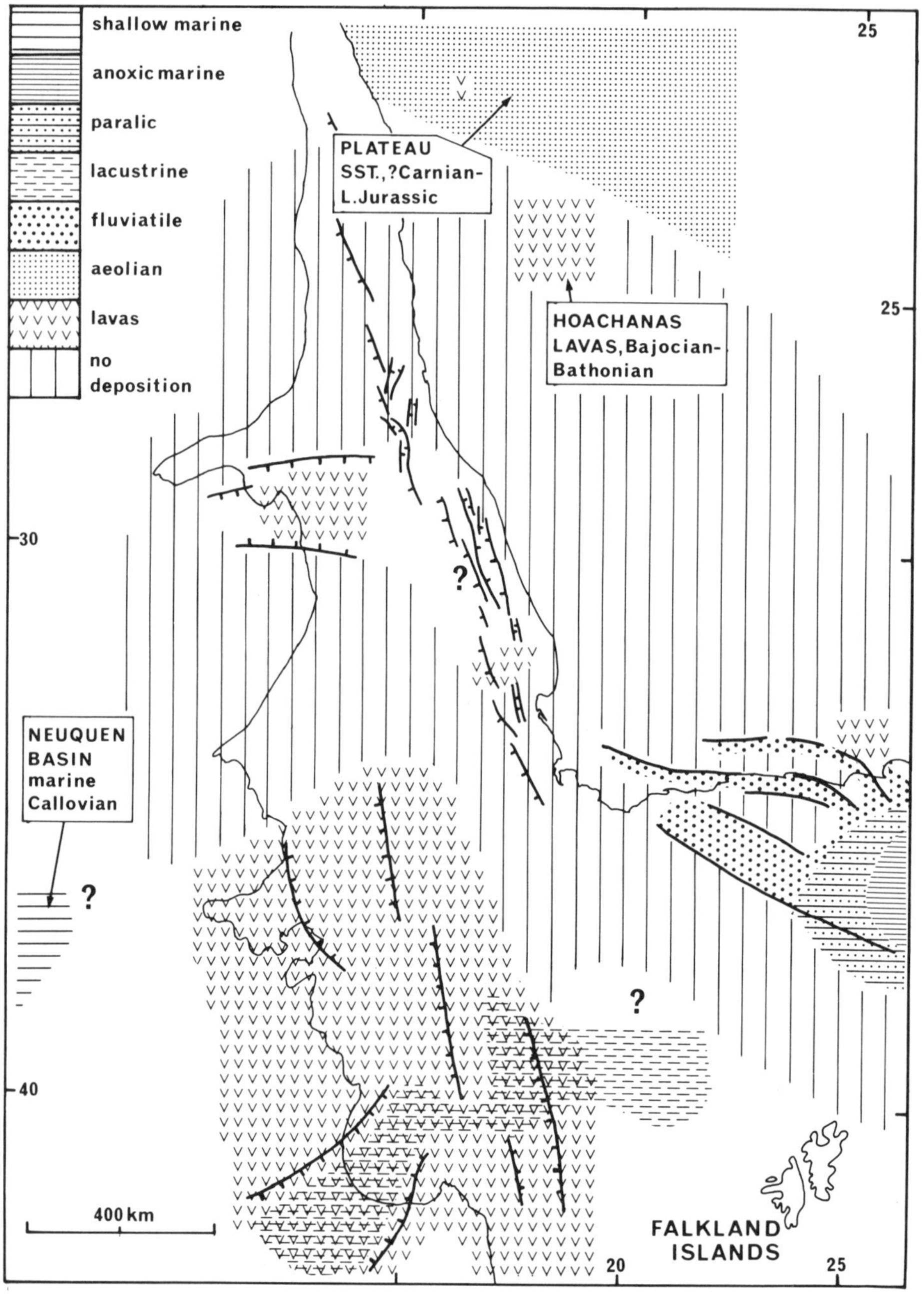


Figure 122. Jurassic palaeogeography of S.W.Africa and E.Argentina. See text for discussion. Data from numerous sources cited in the text, but particularly Urien & Zambrano (1973); Zambrano & Urien (1974); Gerrard & Smith (1980). See Fig.114 for locality names.

iii) abundance, and good preservation of carbonaceous matter;

iv) large quantity and angularity of quartz grains, and abundance and good preservation of feldspars in the sand facies;

v) presence of calcareous nannofossils throughout the shales.

The abundance of pyrite, preservation of the carbonaceous debris, lack of a benthic fauna, and preservation of calcareous nannofossils indicate that the environment of deposition was in an anoxic marine basin, and a consideration of the Parsons & Sclater (1977) age versus depth curve for oceanic crust suggests that the basin had a maximum depth to basement of 3.5 km and that the DSDP site 361 had a depth to basement of 2.5-3 km. Proximity to a rapidly eroding high hinterland is suggested by the obviously short transporation routes of the sand grains and the relatively high sedimentation rates (up to 70 m/my). The DSDP team (Bolli et al 1978) discarded the possibility of margin-

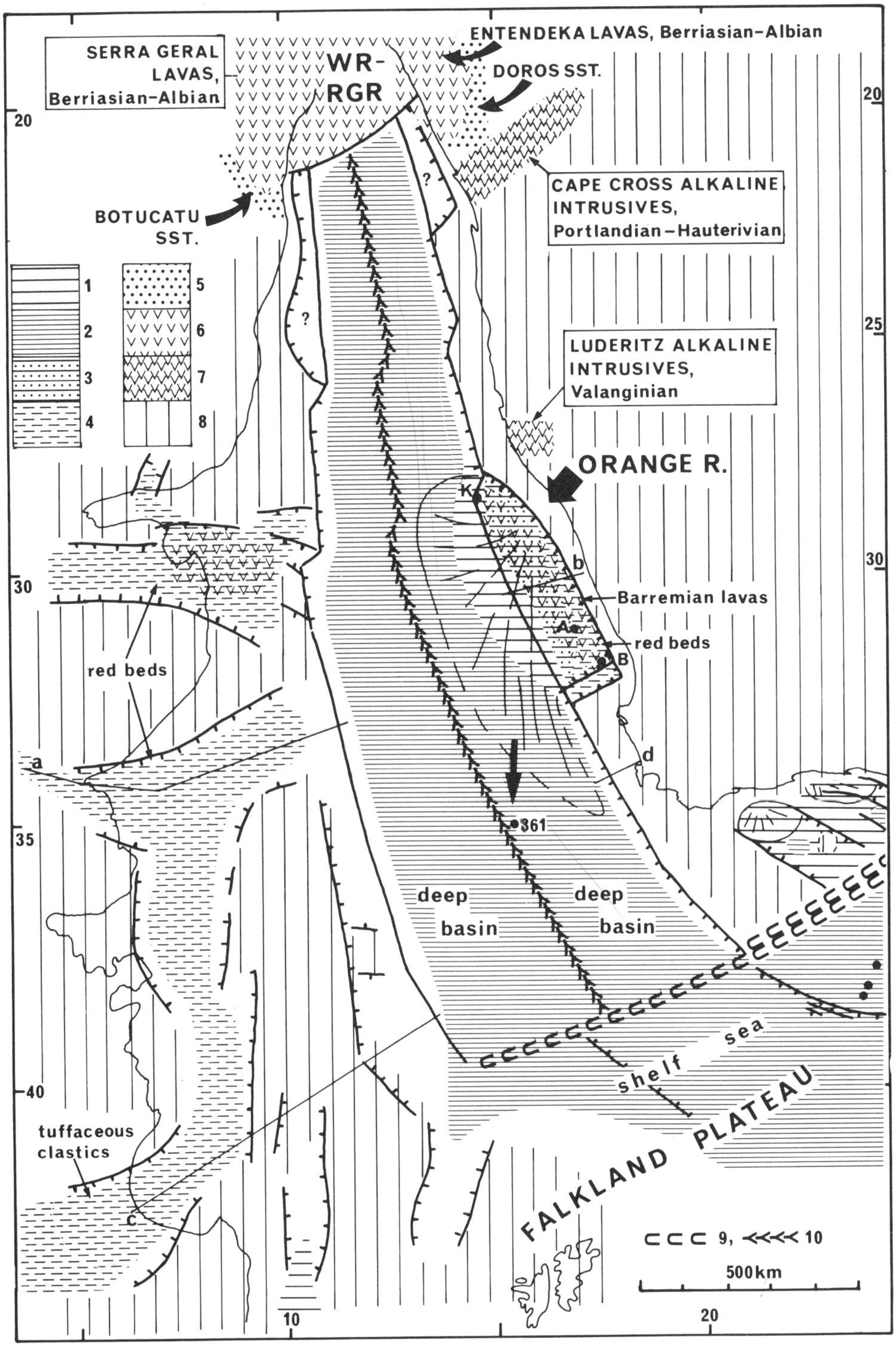


Figure 123. Aptian palaeogeography of S.W.Africa and E.Argentina. See text for discussion. Data from numerous sources (see text and Figure 122). Cross-section lines a-d are those for Figure 115. DSDP sites on the Falkland Plateau shown by dots. Abbreviations: WR-RGR = Walvis Ridge/Rio Grande Rise; K = Kudu 9A-1; A = A-A1; B = Ba-A1/A2; 361 = DSDP 361. Key: 1 = normal marine shales; 2 = anoxic marine shales; 3 = paralic sediments; 4 = lacustrine/fluviatile sediments; 5 = aeolian sediments; 6 = extrusive volcanics; 7 = intrusive volcanics; 8 = no deposition; 9 = marginal fracture ridge; 10 = ocean spreading ridge.

derived turbidity current transport down the continental slope (because of the lack of an allochthonous shallow water biota) in favour of a deep sea fan forming at the foot of a submarine canyon into which a large river (Orange) directly discharged its detritus. Dispersion on the deep sea floor was, presumably, by turbid flow. The anoxic state of the ocean floor indicates that at this stage there was no deep-water connection with the world ocean, and that the elongate marine basin had stagnant bottom water which supported no benthic communities. A combination of rapid sedimentation and lack of oxygen accounted for the preservation of land-derived plant material.

The surface waters of this narrow, isolated ocean basin, which was separated from its south American counterpart by the spreading ridge with sill depths of about 2 km, were, at least periodically, and possibly permanently, connected to the world ocean by surface water overflow across the Falkland Plateau. This is shown by their support of abundant cosmopolitan calcareous and non-calcareous phytoplankton, the latter contributing to the overall high organic content of the bottom sediments. Widespread volcanism accompanied the subsidence of the Orange Basin in Lower Cretaceous times: basaltic lavas over the whole basin area itself and alkaline intrusives along the southern edge of the Lüderitz Arch. The proximal parts of the depocentre were above sea level and accumulated fluviatile pebbly sands and shales in mainly red bed facies – probably analogous to the Kirkwood Formation of the Outeniqua Basin. Shallow marine, and paralic environments prevailed in the more distal areas with marine conditions on what is now the outer shelf region. The Orange River was probably the main input point and constructed a large prograding delta across the COB which spread as a deep sea fan into the adjacent deep, stagnant ocean basin and provided sediments for the more distal parts of the basin such as DSDP site 361.

Sediment components in the DSDP 361 sands allow some predictions to be made of the nature of the coastal hinterland of SW Africa. The large quantity of woody matter suggests that it was at least locally well vegetated albeit with a restricted flora, and that the proto-Orange River drainage area included granitic and high grade metamorphic terrains – presumably the Namaqualand area. In addition, acid volcanic rocks were also locally being eroded and these are probably the ?mid-Jurassic acid lavas reported from the southern Orange Basin. Bolli et al (1978) considered that the basaltic glass sand grains in the basal part of Unit 7 in DSDP 361 were locally eroded from oceanic basalts which formed the irregular floor to the basin on which the sediments were deposited. The high sedimentation rates experienced in the distal parts of the Orange Basin suggest that runoff from the hinterland was high, and that topographic discontinuities were locally large. Regionally the South American and SW African coastlines presented strong contrasts. Both regions would have possessed relatively rugged stretches of coastline adjacent to basement and lava-covered horsts, but whereas these dominated the coastline on the African side with very long straight stretches, and only two sedimentary embayments, the south American side coastline was broken by several long wide straight sided embayments with extensive alluvial or intertidal flats, which stretched several hundred kilometres inland.

# 5

# Upper Cretaceous

At the opening of Upper Cretaceous times (Cenomanian), we are dealing with a subcontinent surrounded on all its present day coasts by oceanic areas: the proto Atlantic SW of Cape Town was about 1 600 km wide, whilst the Natal Valley, Transkei Basin and Mozambique Channel/ Basin areas were all established geomorphological features.

Crustal subsidence and sedimentary basin development on the continental margins of southern Africa underwent a fundamental change in style in Aptian/ Albian times, which was approximately synchronous with similar rheological events described from many other parts of the world, including the whole of the west and east African margins (e.g. Kent 1974, 1977, Dingle 1982b). This change involved a switch from horst/graben basement dislocation, characterized by large boundary faults and relatively narrow, but deep sediment basins (taphrogenic style), to regional downwarping with the production of saucer-shaped depocentres that overstep the older boundary faults (epeirogenic, miogeosynclinal). The establishment of these large marginal basins, with typically no major coast parallel basement discontinuities resulted in accelerated oceanward progradation and shift of depocentres, the effects of which are well-illustrated along the west coast and in the Outeniqua Basin (compare Figs 66 & 124). Because of the more complicated palaeogeography of the east coast area, the effects here are less obvious, but an examination of the stratigraphic record reveals coeval lithofacies discontinuities and depocentre shifts which are related to a change in crustal subsidence patterns. On a continental scale, this major rheological event is marked by two phenomena: changes in lithofacies and tectonic style. The latter, characterized by a cessation of movement along graben boundary faults, is variously dated as Aptian/Albian (Kent 1977). In southern Africa the main change occurred at the end of Aptian times. As a consequence, Upper Cretaceous (plus Albian) and subsequent Cenozoic sedimentation took place largely in elongate depocentres on the continental margin that tended to prograde oceanwards, rapidly extending beyond the confines and destroying the surface identity of most of the older taphrogenic basins. The notable exception is the outer part of the Outeniqua Basin, which remained dammed behind the Agulhas Marginal Fracture Ridge. A consequence of this is the vestigial or non-existent continental rise SE of the present day Agulhas Bank, which is in strong contrast to the well-developed rise prism along the west coast and in the Natal Valley.

The base of the post-taphrogenic sediment sequence is locally an unconformity, disconformity or facies change, and its onset is usually marked by a transgression. These changes are best documented on the south and east coasts, where the earlier episode of sedimentation came to a close in Aptian times, and the subsequent transgression onlapped progressively onto the continental block (usually with a late Cenomanian-Turonian intermission) until, reaching its highest point in Campanian times, it topped the outer edge of the Transkei Swell. In offshore basinal areas (e.g. Outeniqua Basin, offshore Natal, and SE Mozambique), sedimentation continued with only a minor unconformity, disconformity, or, in the case of the latter, no apparent break at all. The situation is somewhat more complex off the west coast, where a strong facies change occurred in the distal parts of the basin in early Albian times, and was followed by a Cenomanian transgression onto the Lüderitz Arch. Because the switch from taphrogenic to epeirogenic crustal subsidence took place about 10 my before the beginning of the Cenomanian, the earliest phase of sedimentation in the newly developing epeirogenic basins (Albian) in some areas has already been dealt with in the previous section (e.g. Mnzinene and Domo formations), whereas others (early Alphard Formation and post-AII sediments on the west coast) will be dealt with here. This is because in some areas we have subdivided the succession into major lithological units, and in others have no data other than seismic horizons with which to effect a satisfactory subdivision.

In all areas around southern Africa, the top of the Mesozoic is marked by a regression. Its effects are noted earlier (Campanian/early Maastrichtian) over basement highs (e.g. the Transkei Swell), but by late Maastrichtian times it had affected all parts of the continental shelf, and even in the more distal parts of the basins there is usually a marked lithofacies change.

Present day outcrops of Upper Cretaceous strata are confined to the Natal-Zululand-Mozambique coastal basin, isolated outcrops on the east and west coast, and

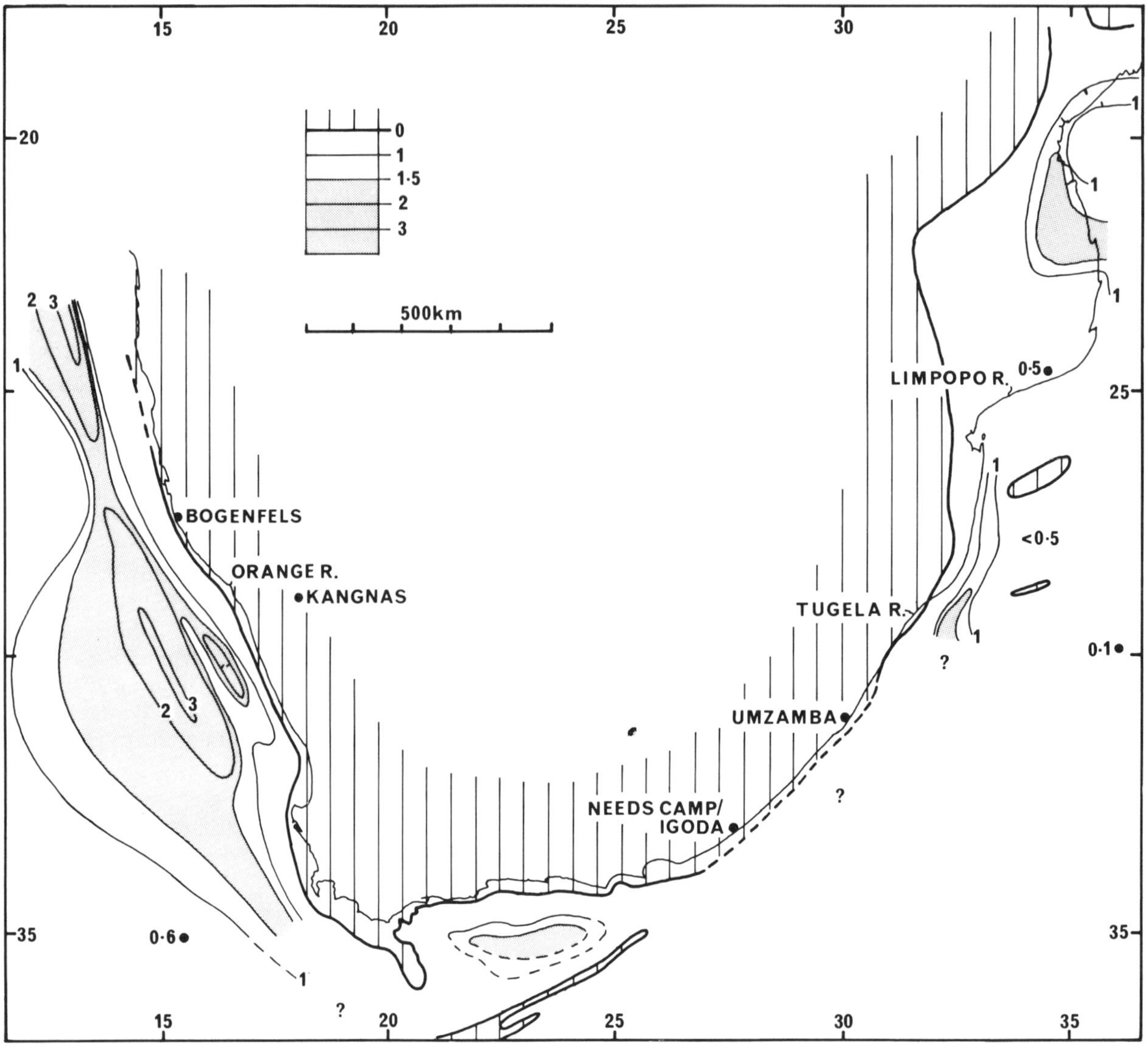


Figure 124. Isopachs (km) of Upper Cretaceous sediments on the continental margins of southern Africa. Lowermost Palaeocene have been included in S.Mozambique (Grudja Formation). Contour interval 1 km (with addition of 1.5 km). Data sources cited in text.

major sediment basins on the continental margins (Figs 67 & 124). In the latter, maximum accumulations occur in the Orange Basin (>3 km), Outeniqua Basin (>1.5 km), under the Tugela Cone (1.6 km), and in southern Mozambique (>1.7 km). A comparison of Figures 67 and 124 shows that sedimentation in this developing epeirogenic phase was no longer controlled by the major fault lines that operated during Jurassic and Lower Cretaceous times, and that by the end of the Mesozoic era sediment dispersal from the main terrigenous input points on the south and west coasts had become widespread, both transverse and coast parallel. Despite the great aereal extent of Upper Cretaceous sediments, known exposures are not large. On the west coast they are limited to two very small occurrences onshore (Bogenfels and Kangnas), and a narrow nearshore outcrop, whilst on the south coast no onshore outcrops are known, and only a narrow offshore zone occurs along the north edge of the Outeniqua Basin. Small, discontinuous outcrops have been found along the shelf break of the Agulhas Bank, where tensional fissures produced by the late Cenozoic slumps have exposed the underlying strata. Small outcrops occur along the coast of Transkei and southern Natal (Umzamba area), and only along the western side of the Zululand coastal plain is a continuous Upper Cretaceous outcrop found. Here, however, individual exposures are poor, with deep weathering, alluvium, and thick scrub restricting outcrops to man-made cuts and occasional river banks. Despite this, the Zululand/Mozambique Upper Cretaceous sequence is well-known, and forms the standard for comparison with other areas in South Africa and the various components of Gondwanaland. Its molluscan and ostracod faunas have recently been studied in detail, and it is the only part of southern Africa for which a good biostratigraphic zonation is available.

Boreholes in the Outeniqua Basin provide the best available potential for a well-zoned offshore sequence, but little systematic work has yet been published on the samples. The J(c)-1 borehole off Natal contains only poorly-developed microfossil assemblages and is not a suitable standard for comparison. Figure 67 summarizes the spatial and temporal distribution of Upper Cretaceous strata around Southern Africa.

## 5.1 OUTENIQUA BASIN

In the Outeniqua Basin, the change from taphrogenic to epeirogenic crustal subsidence coincides approximately with the boundary between the Upper Sundays River and Alphard formations (there is some diachroneity to this boundary in the west), so in this section we will deal with uppermost Aptian, Albian as well as Upper Cretaceous (Cenomanian-Maastrichtian) strata.

By the close of Lower Cretaceous times, movements along the main boundary faults of the taphrogenic basins had all but ceased. Although some of the large onshore fractures are still active (e.g. the Worcester Fault and related faults in the Ceres area which were responsible for the 1969 earthquake), those on the Agulhas Bank became dormant over a short period of time: of the 33 faults with a throw of more than 100 milliseconds (say 200 m) that were active during the Hauterivian (uppermost Lower Sundays River Formation), six were active in the Lower Aptian (uppermost Sundays River Formation), and only one (on the western side of the Recife Arch) was active in Turonian times (mid-Alphard Formation) (data from maps in SOEKOR 1976). Because this rapid reduction in tectonic activity began during deposition of the Upper Sundays River Formation, most of the original taphrogenic basin irregularities and basement topographic highs were smoothed out early during the deposition of the Alphard Formation, and certainly before the start of the Upper Cretaceous (i.e. base Cenomanian). The cessation of large-scale fault movement allowed the remaining sub-basin depressions to fill rapidly, with the result that the change from the taphrogenic to epeirogenic sedimentary environments is marked by a major facies change, in particular the destruction of localized restricted sedimentary environments, and the establishment of regionally efficient sediment dispersal routes and mechanisms.

The surfaces onto which Albian sedimentation commenced were relatively smooth, having locally suffered some planation, and dipped seaward with a 'shelf break' along the outer edge of the prograding Upper Sundays River wedge. In simple terms, subsequent Upper Cretaceous (and Cenozoic) sedimentation took the form of a large up- and outward prograding sediment lens which advanced the shelf edge oceanwards, although advancement was not symmetric and depended upon the local balance between subsidence rates and sediment supply. The main axis of sedimentation lay approximately E-W stretching from the western end of the Bredasdorp Basin to the edge of the Port Alfred Arch (Fig. 126). Of all the older basins, only the Bredasdorp, which had never developed strongly faulted margins or significant asymmetry, retained its identity, although the area of sedimentation progressively extended farther onto the flanks of the Agulhas and Infanta arches than had been the case hitherto (Fig.73A,B). Locally buoyant basement features such as the St Francis and Recife arches were covered by only a thin veneer of Upper Cretaceous sediment (less than 400 m), and the Agulhas and the Port Alfred arches were probably never completely covered.

### 5.1.1 *Alphard Formation*

Open shelf, post-Upper Sundays River Formation sediments of the Outeniqua Basin are represented by the Alphard Formation which overlies dark lignitic shales (SR3 and SR5) in the distal, and lignitic sandstones (SR4 and SR6) in the proximal parts of the older basins (Fig.92). Locally, the Alphard Formation oversteps onto basement, and occasionally onto Lower Uitenhage Group. Because the change from restricted to open shelf sedimentary environments was diachronous, the base of the Alphard Formation varies in age from Upper Aptian to Cenomanian (McLachlan & McMillan 1979). Generally, the change took place earlier over basinal and eastern areas, where the base of the Alphard Formation is Upper Aptian, and later in the western region (Bredasdorp Basin) and on originally more buoyant areas (such as the crest of the Superior ridge). In the latter, the basal Cenomanian sediments are locally still lignitic and indicate a restricted marine environment (e.g. in borehole G(a)-A/2).

The term Alphard Group was introduced by Dingle (1973a) for the sediment sequence on the Outeniqua Basin that lies between the Lower Cretaceous graben infill, and the basal Tertiary unconformity. At that time, only three sediment samples had been collected from the unit, but it was obvious that it ranged in age from at least early Senonian to late Maastrichtian. Du Toit (1976) formally defined the Alphard Formation,

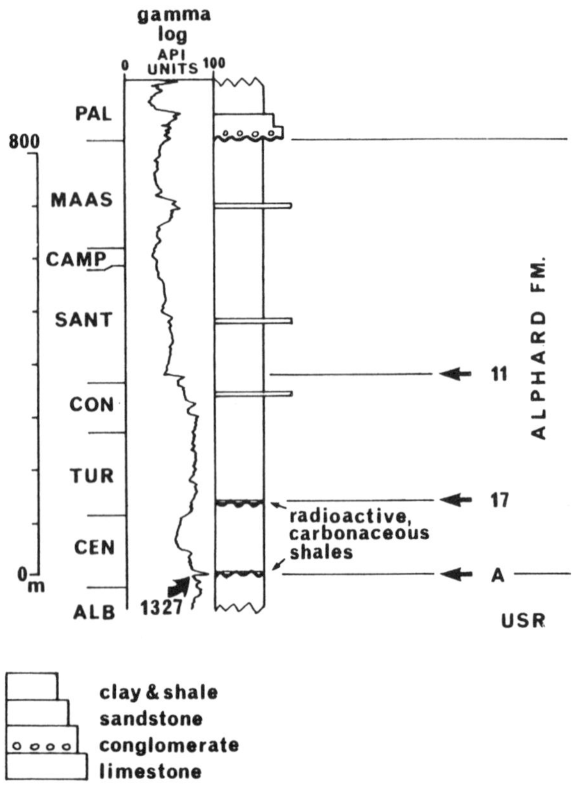


Figure 125. Type section of Alphard Formation in borehole G(a)-A/2. Modified slightly from du Toit (1976) with data in SOEKOR (1976) and McLachlan & McMillan (1979). A, 17, and 11 are regional seismic reflectors shown in Figures 72 & 92. Depth (m) at lower left side of log is sub-sea level. Reflectors 17 and A are equated with small hiatuses and thin carbonaceous shales.

and used the succession in borehole G(a)-A/2 as his stratotype (Fig.125). In the type section, the Alphard Formation consists of 820 m of monotonous soft shales with subordinate thin limestones and sandstones. *Inoceramus* prisms are common throughout, but lignite is rare with concentrations being confined to two thin black shale horizons near the base. Downhole geophysical logging indicates two horizons at which significant lithological changes occur, and these horizons have been used as rough data for lateral correlation with other boreholes, where they correspond to prominent reflecting horizons:

i) beneath 969 m, the gamma ray and sonic logs indicate a decrease in porosity and a decrease in P-wave velocity, respectively, suggesting that the sediment beneath 969 m is more argillaceous. This is seismic reflector 11 of SOEKOR (1976);

ii) at 1 212 m, a further sudden downhole decrease in sonic velocity is indicated, but a corresponding slight increase in porosity suggests a somewhat more arenaceous lithology. This is reflector 17 of du Toit (1976), which he equates with a regionally-developed erosion surface, that away from the type area (e.g. in the Bredasdorp Basin, and in boreholes G(a)-B/1 and G(b)-G/1) locally forms the base of the Alphard Formation (Fig.92). It coincides with a thin layer of radioactive carbonaceous shale.

The base of the Alphard Formation in the type section corresponds to reflector A of SOEKOR (1976) which is a regionally developed erosion surface that is equated with a thin layer of radioactive carbonaceous shale. Away from the type section, reflector A locally merges with horizon 17 (in the Pletmos Basin), but in the distal parts of the Bredasdorp Basin it locally forms the base of the Alphard Formation, and in the proximal parts underlies the SR5 and SR6 cycles of the Upper Sundays River Formation. This correlation is a revision of SOEKOR (1976) and du Toit (1976), following revised palaeontological dating by McLachlan & McMillan (1979).

Away from the G(a)-A/2 borehole, which is situated on the crest of the Superior Ridge, du Toit (1976) records few lithological changes, and the Alphard Formation maintains its rather monotonous character over a wide area. However, two important facies changes have been recognized in the boreholes (du Toit 1976, SOEKOR 1976, McLachlan & McMillan 1979) (Fig.128). The most widespread is a 50 m thick well-sorted glauconitic sandstone, with minor brownish limestones, which occurs along the north side of the Bredasdorp Basin, and the crest of the Infanta Arch. Both du Toit (1976) and SOEKOR (1976) suggest that this sandstone (which has been dated as Coniacian in three boreholes, and Upper Turonian in a fourth borehole) is equivalent to horizon 11. The presence of glauconite suggests a low sedimentation rate, and the well-sorted texture suggests an increase in energy of the sedimentary environment indicative of a minor regression or increased bottom current strengths.

A second, less well-developed facies change is recorded by du Toit (1976) from the eastern part of the Outeniqua Basin, where a 'conspicuous sandstone' horizon occurs in the Santonian of the G(b)-Gemsbok/1 borehole, and a thick (125 m) lignitic sandstone of ?Santonian/Campanian age underlies the typical shaly Alphard Formation (Maastrichtian) in the H(b)-Hartebeest/1 borehole. Coarse horizons from other boreholes in the eastern region can also be equated with this facies change (Fig.128).

The Alphard Formation has been sampled from surface outcrops on the Agulhas Bank and the following lithologies have been recorded (see Fig.126 for locations): grey clays (510), dark greenish-grey siltstones (510, 4 492), and fine grained well sorted quartz sandstones (4 510), (all Lower Coniacian from the western end of the Bredasdorp Basin); and yellowish sandy lime-

Table 37. Microfaunal zonation for Outeniqua Basin Alphard Formation (after McLachlan & McMillan 1979).

| Stage | Species | Zone |
|---|---|---|
| Maastrichtian | *Rugoglobigerina rugosa* <br> *Globotruncana arca* | UC14 |
| | *Globotruncana tricarinata* | ? |
| Campanian | *Globotruncana ventricosa* | UC9 |
| Santonian | *Globotruncana coronata* <br> *Globotruncana carinata* | UC7 & 8 |
| Coniacian | *Globotruncana primitiva* <br> *Globotruncana renzi* | UC6 |
| | *Gavelinella plummerae* | ? |
| | *Gaudryina 'agulhasensis'* | ? |
| Turonian | *Praeglobotruncana stephani* | UC3-LC19 |
| | *Praeglobotruncana helvetica* | UC5 |
| | *Lingulogavelinella turonica* | ? |
| Cenomanian | *Rotalipora reicheli* | UC2/UC3 |
| | *Rotalipora apenninica* | LC19-UC1 |
| Albian | *Majungaella pyriformis* | ? |
| | *Tritaxia tricarinata* | ? |
| | *Hedbergella washitensis* | LC18-LC19 |
| | *Clavulina gaultina* | ? |
| Aptian | *Marginulinopsis pristipellis* | ? |
| | *Bilingulogavelinella australis* | ? |

stones (1 336) and olive glauconitic clayey sands (with pyrite, shell fragments and fish teeth) (818) (Maastrichtian from St Francis Bay and the upper continental slope, respectively) (Dingle 1971b, Klinger et al 1976, 1980).

Figure 126 shows the distribution and thickness of the Alphard Formation in the Outeniqua Basin. The present northern limit mostly lies hidden under the feather edge of the Tertiary, but can be traced on seismic profiles. It locally crops out on the sea floor: SE of Mossel Bay; in St Francis Bay; over a large area at the western end of the Bredasdorp Basin around the Alphard Banks (after which the unit was named and from where it was first sampled); and on the upper continental slope where slumping exposes Alphard Formation between faulted masses of Tertiary strata. Upper Cretaceous sediments have been intersected in all the offshore boreholes quoted by du Toit (1976), except those close to the coast, and the maximum recorded thickness is 1 410 m on the northern side of the Bredasdorp Basin south of Mossel Bay (borehole F-1). From seismic records, it can be seen that the Alphard Formation forms a seaward prograding lens, which reaches its maximum known development under the present continental slope. The axis of this depocentre runs approximately E-W and at least 1 500 m of sediments occur in it south of the present shelf break (assuming a P-wave velocity of 2 258 km/sec, calculated from the G(b)-Springbok/1 borehole). Westwards, the succession thins as the depocentre becomes a symmetric basin confined between the Agulhas and Infanta arches. Here the Upper Cretaceous sediments lap onto the flanks of both these highs, and overstep directly on Palaeozoic basement. Dips into the Bredasdorp Basin are very low (< 1°), and the basin's axis plunges at about 2°SE. Some of the few examples of post-Lower Cretaceous faulting occur at the western edge of the Bredasdorp Basin, where the Alphard Formation is gently folded and faulted along lines apparently controlled by older structures. It is along these lineaments that early Tertiary volcanics intruded the Upper Cretaceous sediments to form the extensive complex of lava flows and plugs centred on the Alphard Banks (6.1.1).

At the eastern end of the Outeniqua Basin, the Port Alfred Arch remained buoyant and was probably never covered by Upper Cretaceous sediments, although a thin veneer would be beyond the resolution of the low frequency commercial records. SOEKOR (1976) shows Alphard Formation at the southern end of the Algoa Basin, but there is no evidence to suggest that the basin edge lies north of the Coega Fault. Both the St Francis and Recife arches are covered by thin cappings of Upper Cretaceous strata (500 and 340 m respectively) (SOEKOR 1976).

Du Toit (1976) gave details of stage thicknesses in the Alphard Formation and these have been plotted as an E-W section across the Outeniqua Basin in Figures 127 and 128. They show that although the basin as a whole underwent epeirogenic downwarping, the areas centred on the old Bredasdorp and Pletmos basins subsided considerably more than those over the Infanta Arch and Superior Ridge areas where the whole of the Upper Cretaceous is represented by about 500 m of sediment. Although stages are locally absent (e.g. Cenomanian is missing in F-B/1, and the Coniacian is missing in G(a)-A/1), the thinness is clearly due to slow sediment accumulation and not major non-sequences. Figure 128 shows an analysis of individual stage thicknesses and a graphical presentation of hiatuses and sandy horizons in time and space. At the stage level, only the Turonian and Coniacian show strong thickness variations coupled with a patchy distribution, and it is significant that these levels also contain sandy and locally glauconitic lithofacies. Du Toit (1976) reported that the Coniacian planktonic foraminiferal faunas were sparse and that the full zonal sequence had not been recognized, in addition to which, reflector 11 is shown as locally being an erosion surface. This evidence points to a Turonian/Coniacian regression with increased bottom current strength in the Outeniqua Basin, an event that correlates with a major Upper Cenomanian-Lower Coniacian non-sequence in Zululand (Kennedy & Klinger 1971). Cenomanian strata show some lateral thickness variations, which might be accounted for by unevenness in the pre-Alphard Formation floor, but the Santonian

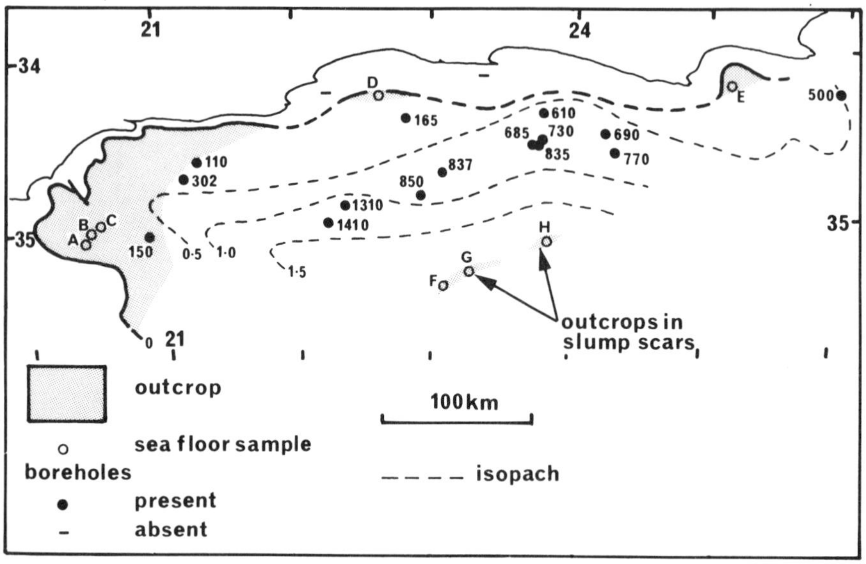


Figure 126. Distribution and thickness of Cenomanian-Maastrichtian sediments (mainly Alphard Formation) in the Outeniqua Basin. Isopachs in km, borehole values in m. Thickness data from du Toit (1976), sea floor sample data from University of Cape Town. Sample notations: A = #510 – Coniacian; B = #4492 – Coniacian; C = #4510 – Lower Coniacian; D = #690 – Cenomanian-Maastrichtian; E = #1336 – Maastrichtian; F = #4450 – Upper Maastrichtian; G = #4439 – Upper Maastrichtian; H = #818 – Upper Maastrichtian. For details of age determinations see text, and Tables 37, 38 & 39.

and Campanian maintain remarkable conformity of thickness across the whole area of the Outeniqua Basin. There is, however, considerable variation in the thickness of the Maastrichtian sediments, with a maximum of about 425 m in the area over the northern Bredasdorp Basin and about 70 m over the Superior Ridge. Faunal evidence cited by du Toit (1976) indicates that the top zones of the stage are missing and that the upper sediments are of a sandy lithofacies. This evidence, taken in conjunction with seismic data which shows truncation of Maastrichtian reflectors by the base of the Tertiary, indicates a major non-sequence with regression and erosion at the top of the Upper Cretaceous. These correlations raise an interesting point because they indicate that whilst crustal subsidence was relatively slow and regionally even during periods of overall sedimentation, it temporarily increased under the old taphrogenic basins prior to periods of regression that led to non-sequences in Turonian/Coniacian, and Upper Maastrichtian times.

Palaeontological information from the Alphard Formation is limited to a generalized microfossil zonation scheme by McLachlan & McMillan (1979) and descriptions of the small faunas of molluscs and ostracods dredged from surface outcrops. McLachlan & McMillan (1979) have recognized a succession of mostly planktonic foraminiferal assemblages (Tab.37) which we have tentatively equated with the international zonal schemes of Postuma (1971) and van Hinte (1976a) (Tab.28). Several of the local zones cannot be correlated with the international scheme because they are based on benthic foraminifera and in one case an ostracod, and also because no details of the faunas have been released. Du Toit (1976) mentioned that the Lower Maastrichtian, Campanian, and Santonian have abundant planktonic foraminiferal assemblages, but that the Upper Maastrichtian and Coniacian faunas were sparse. An assessment of the planktonic foraminifera from sample 818 (locality H in Fig.126) suggests that the highest zone represented on the Agulhas Bank is UC14 (*Globotruncana gansseri*) of van Hinte (1976a) (Dingle 1971b), whilst the ostracods indicate an age of Maastrichtian III of Kennedy & Klinger's (1975) Zululand ammonite scheme. This date is corroborated by Siesser (1982) on the basis of the calcareous nannofossil *Nephrolithus frequens*. The sparseness of the Coniacian assemblage is reflected by McLachlan & McMillan (1979) being forced to utilize benthic foraminifera for zonation in the lower part of the stage. In fact the whole of the UC3-UC5 period of van Hinte (1976a) is rather uncertain (Tab.37).

A rich ostracod assemblage has been described by Dingle (1971b, 1981) from the Upper Maastrichtian (III) of the Alphard Formation (locality H, Fig.126). It is dominated by trachyleberid types (6 spp., 96 % of individuals) (Tab.38) which show some interesting associations with Upper Cretaceous faunas elsewhere: offshore Natal (borehole J(c)-1) (*Dutoitella mimica, ?Phacorhabdotus anomala*), Tanzania (*Dutoitella, ?Pharcorhabdotus anomala*) and Western Australia (*Apateloschizocythere*). The fauna suggests a quiet, shallow water environment dominated by open ocean influences on the outer shelf and upper slope.

The macrofauna of the Alphard Formation is poorly known, but dredged assemblages have been described by Klinger et al (1976, 1980) (Tab.39). Three species of ammonite and two species of *Inoceramus* have been recovered from Coniacian strata and these have the following associations: *Yabeiceras manasoaense* has been recorded from the Lower-Middle Coniacian of Japan, and the Middle Coniacian of Madagascar and Zululand;

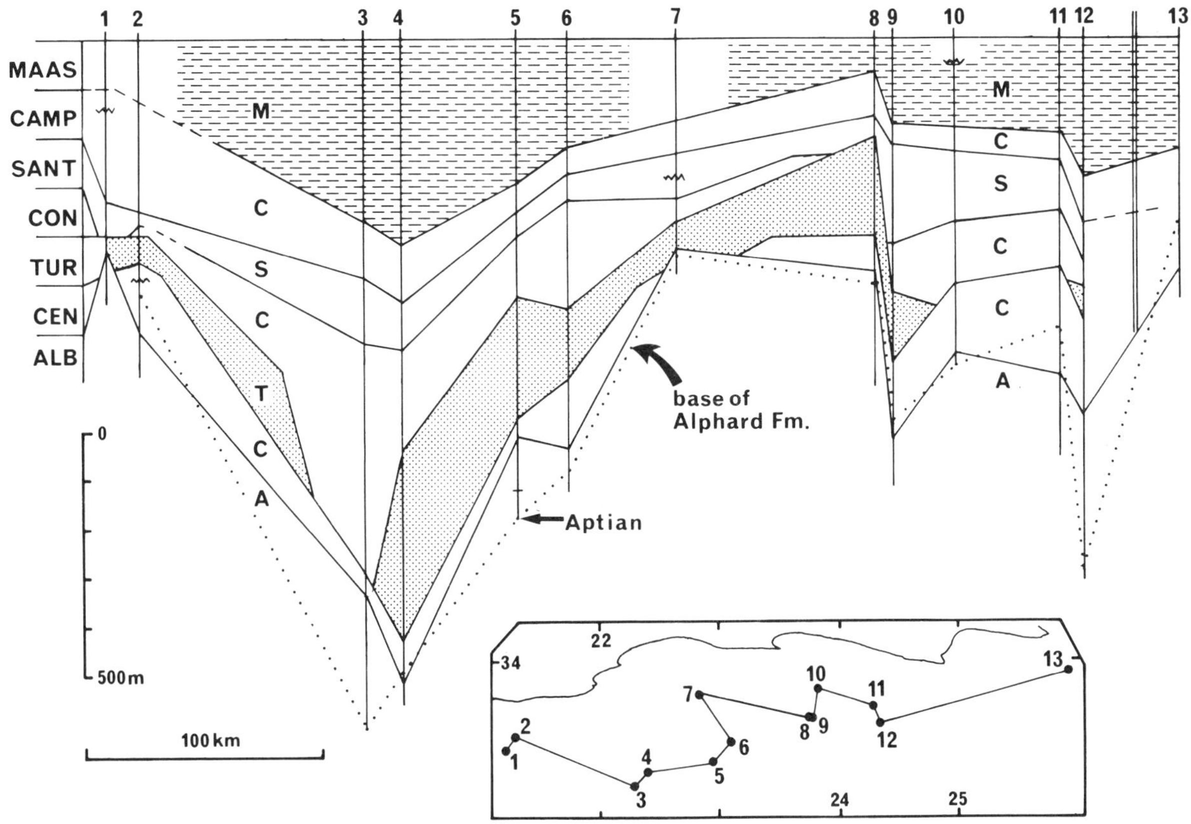


Figure 127. Thickness variations in Upper Cretaceous stages across the Outeniqua Basin. Base of Alphard Formation is marked with a dotted line and is clearly diachronous (Aptian to Cenomanian), although micropalaeontological control is apparently not good. Note condensed horizontal scale between sites 12 and 13. Data from du Toit (1976). Shaded horizons are for clarity and have no geological significance. Insert shows borehole locations: 1 = E-B/1; 2 = E-1; 3 = F-1; 4 = F-E/1; 5 = F-D/1; 6 = G(a)-C/1; 7 = F-B/1; 8 = G(a)-A/1; 9 = G(a)-A/2 (type section for Alphard Formation); 10 = G(a)-B/1; 11 = G(b)-Gemsbok/1; 12 = G(b)-Springbok/1; 13 = H(b)-Hartebeest/1.

*Proplacenticeras kaffrarium* has been recorded from the Lower Coniacian of Zululand. Both *Inoceramus* species are cosmopolitan, and *I.ernsti* has been recorded from the Coniacian of Europe, North America, Madagascar, North and South Africa, and Pacific USSR; with *I.frechi* from North America, ?Japan, East Russia, North and West Europe, and Mediterranean Tethys. Klinger et al (1980) note that the *Inoceramus* population is dominated by left valves and conclude that bottom currents have preferentially removed the smaller right valves. An inner shelf, slow sedimentation rate with moderate bottom current strengths is indicated for sample 4510 (locality C on Fig.126). Maastrichtian ammonites have been found at site 1336 (locality E on Fig.126) (Dingle 1973a, Klinger et al 1980). *Eubaculites latecarinatus* has previously been recorded from the Maastrichtian of Western Australia, Madagascar and Zululand (Maastrichtian I), whilst *Gunnerites* is typical of the subantarctic islands and Grahamland. A species comparable to *G.*cf.*kalika* occurs in the Maastrichtian of Zululand.

### 5.1.2 *Palaeogeography*

Generally well-represented planktonic foraminifera, the dominance of argillaceous lithofacies, and seaward prograding sedimentary structures on seismic profiles indicate that for the most part the Alphard Formation was deposited as a seaward prograding lens into a relatively deep marine environment which maintained direct connection with the open ocean. The bulk of the sediments intersected in the boreholes on the Agulhas Bank represent middle or outer shelf environments, an inner shelf facies having been encountered only at certain horizons or at outcrop close to the modern shoreline. From Figures 73, 74 and 128, it is clear that except during regressive episodes, the old horsts sank at approximately the same rate as the old basins, so that for most of the period represented by the Alphard Formation (i.e. Upper Aptian to Upper Maastrichtian) water depths in an E-W direction across the basin were similar, in strong contrast to earlier times when there were distinctive N-S or NW-SE discontinuities.

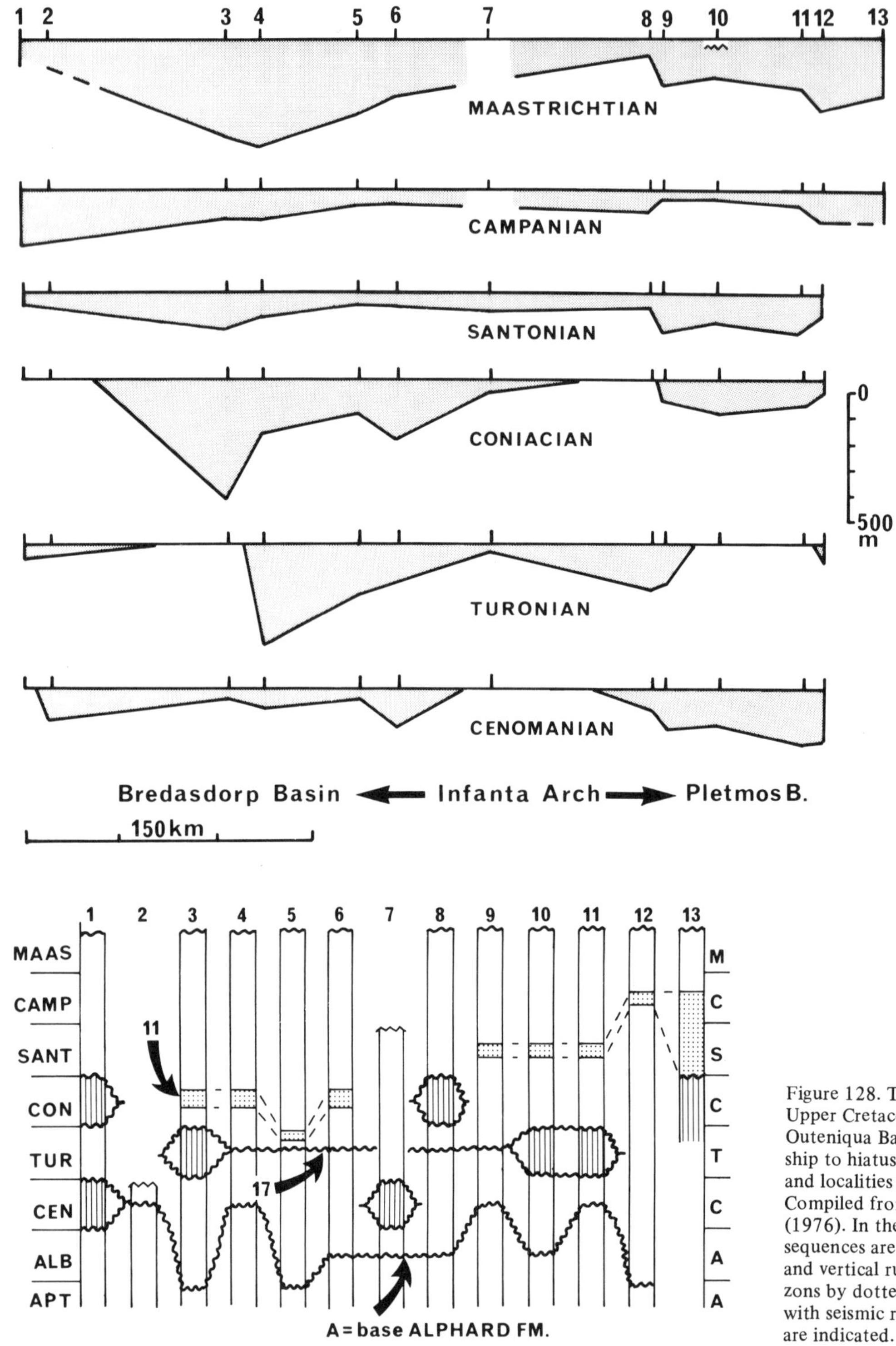


Figure 128. Thickness variations in Upper Cretaceous stages across the Outeniqua Basin and their relationship to hiatuses. Borehole notations and localities as in Figure 127. Compiled from data in du Toit (1976). In the lower diagram non-sequences are shown by wavy lines and vertical ruling, and sandy horizons by dotted screen. Correlations with seismic reflectors A, 17 and 11 are indicated.

We will consider two palaeogeographic situations (Fig.129), one representing a transgressive phase, and one a regressive phase. Major transgressions commenced in Albian-Cenomanian and Upper Coniacian-Santonian times and in Figure 129 the high point of the latter (Upper Campanian/Lower Maastrichtian) is shown, although it probably applies equally to the earlier sea level high. In this situation, the coast lay a short dis-

Table 38. Coniacian-Maastrichtian ostracods in South Africa (after Dingle 1982a). Data from numerous sources mentioned in the text.
Umz. = Umzamba; I.NC = Igoda and Needs Camp; C/M = late Campanian/early Maastrichtian. ● = presence; ○ = cf.

| | Zululand St Lucia Formation | | | | JC-1 borehole | | Outeniqua Alphard Fm. | | Transkei Swell Umz. | Transkei Swell I.NC |
|---|---|---|---|---|---|---|---|---|---|---|
| | Con. | Sant. | Camp. | Maas. | Camp. | Maas. | Maas. | Con. | Sant. | C/M |
| *Cytherella* sp. | ● | ● | ● | ● | | ● | | | ● | ● |
| *Cytherelloidea newtoni* | ● | ● | | | | | | | | |
| *C.umzambaensis* | ● | ● | ● | | | | | | ● | |
| *C.gardeni* | | ● | | | | | | | ● | |
| *C.griesbachi* | | ● | ● | | | | | | | |
| *C.contorta* | | | ● | | | | | | | |
| *C.mfoloziensis* | | | ● | ● | | | | | | |
| *Platella africana* | | | ● | ● | | | ● | | | |
| *Bairdoppilata andersoni* | | ● | ● | ● | | | ● | | ● | ● |
| *B.africana* | | | | | ○ | ○ | | | | ● |
| *B.* sp.A | | | | | | | | | | ● |
| *Bythocypris richardsbayensis* | ● | ● | ● | ● | | | ● | | | ● |
| *B.* sp. | | | | | | ● | | | | |
| *Paracypris umzambaensis* | | ● | ● | ● | | | | | | ● |
| *P.zululandensis* | | ● | ● | ● | | | ● | | | |
| *P.* sp.A | | | | | | | | | | ● |
| *P?* sp. | | | | | | ● | | | | |
| *Pontocyprella nibelaensis* | | | ● | ● | | | ● | | | |
| *P.* sp. | | | | | | | | | | ● |
| *Krithe nibelaensis* | | | ● | ● | | | ● | | | |
| *K.* sp. | | | | | | ● | | | | |
| *Xestoleberis luciaensis* | | | ● | ● | | | ● | | | ● |
| *Klingerella aranearius* | | | | ● | | | | | | |
| *Pariceratina hirsuta* | | | | ● | | | | | | |
| *Pondoina sulcata* | | ● | | | | | | | ● | |
| *P.igodaensis* | | | | | | | | | | ● |
| *Buntonia?* sp. | | | | | | | | | | ● |
| *Paraphysocythere thompsoni* | | | | | | | | | ● | |
| *Hutsonia?* sp. | | | | | | | | | | ● |
| *Brachycythere longicaudata* | ● | ● | ● | ● | | | | | ● | ● |
| *B.pondolandensis* | | ● | | | | | | | ● | |
| *B.rotunda* | | | | | | | | | ● | |
| *B.agulhasensis* | | | | | | | | ● | | |
| *B.sicarius* | | ● | ● | ● | | | | | | |
| *Pterygocythere lanceolata* | | | | ● | | | | | | |
| *Ponticulocythere biremis* | | | | ● | | | | | | |
| *Eucytherura? pyramidatus* | | | ● | | | | | | | |
| *Pedicythere fragilis* | | | | ● | | | | | | |
| *Cytheropteron brenneri* | | | | ● | | | | | | |
| *C.* cf. *westaustraliense* | | | | ● | | | | | | |
| *Amphicytherura tumida* | | ● | ● | | | | | | ● | |
| *A.zululandensis* | | | ● | | | | | | | |
| *A.armatus* | | | ● | ● | | | | | | |
| *A.* sp.A | | | ● | ● | | | | | | |
| *A?* sp. | | | | | | | | ● | | |
| *Cnestocythere?* sp. | | | | | | | | | ● | |
| *Apateloschizocythere laminata* | | | | | | | ● | | | |
| *A.mclachlani* | | | ● | | | | | | | |
| *Cythereis klingeri* | ● | ● | ● | ● | | | | | | |
| *C.transkeiensis* | | ● | | | | | | | ● | ● |
| *C.* spp. | ● | | | | | | | | | |
| *Veenia obesa* | | | | | | | | | ● | |
| *Haughtonileberis haughtoni* | ● | ● | ● | | | | | | ● | |
| *H.vanhoepeni* | | ● | ● | | | | | | | |
| *H.fissilis* | | ● | ● | ● | | | | | ● | |
| *H.nibelaensis* | | | ● | | | | | | | |
| *Australileberis stangerensis* | | | | | | ● | | | | |
| *Oertliella* sp.A | | ● | ● | | | | | | | |
| *O.pennata* | ● | ● | ● | | | | | | ● | |
| *O.africana* | | | ● | ● | | | | | | |
| *O.maastrichtia* | | | ● | ● | | | | | | |
| *Hermanites kennedyi* | | | ● | | | | | | | ● |

Table 38 (continued).

| | Zululand St Lucia Formation Con. | Sant. | Camp. | Maas. | JC-1 borehole Camp. | Maas. | Outeniqua Alphard Fm. Maas. | Con. | Transkei Swell Umz. Sant. | I.NC C/M |
|---|---|---|---|---|---|---|---|---|---|---|
| *H? arcus* | | | | ● | | | | | | ○ |
| *Trachyleberis minima* | | | ● | | | | | | | |
| *T.zululandensis* | | | ● | ● | | | | | | |
| *T.schizospinosa* | | | | | | ● | | | | |
| *Unicapella reticulata* | | | ● | | | | | | | |
| *U.sacsi* | | | ● | ● | | | | | | |
| *U.stragulata* | ● | ● | | | | | | | | |
| *Dutoitella dutoiti* | | | ● | ● | | | | | | |
| *D.mimica* | | | | | ● | ● | ● | | | |
| *Gibberleberis africanus* | ● | ● | | | | | | | ● | |
| *G.elongata* | | ● | ● | | | | | | | |
| *G.* sp.A | | | ● | | | | | | | |
| *Parvacythereis monziensis* | | | ● | ● | | | | | | |
| *P.spinosa* | | | | | | | ● | | | |
| *Rayneria nealei* | ● | ● | ● | | | | | | ● | |
| *Curfsina monziensis* | | | ● | | | | | | | |
| *Cativella? dubia* | | | | ● | | | | | | |
| *Phacorhabdotus? anomala* | | | | | | ● | ● | | | |
| *P?* | | | | | | ● | | | | |
| *Agulhasina quadrata* | | | | | | | ● | | | |
| *Paraplatycosta reticulata* | | | | | | | ● | | | |

Table 39. Fossils from the Alphard Formation (from Dingle 1971b, 1981, Klinger *et al.* 1976, 1980) (University of Cape Town sample number in brackets).

| | |
|---|---|
| Macrofossils | |
| Lower Coniacian | *Yabeiceras manasoaense* Collignon (4492) |
| | *Proplacenticeras kaffrarium* (Etheridge) (510) |
| | *Scaphites (?Otoscaphites)* sp.(510) |
| | *Inoceramus (I.) ernsti* Heinz (4510) |
| | *I.frechi* Flegel (4510) |
| Maastrichtian | *Eubaculites latecarinatus* (Brunnschweiler)(1336) |
| | *Gunnerites* cf. *G.kalika* (Stoliczka) (1336) |
| Microfossils | |
| Lower Coniacian and Maastrichtian | Ostracoda – see Table 38 |
| Upper Aptian-Albian | Ostracoda – see Table 24 |
| Albian-Maastrichtian | Foraminifera – see Table 37 |

stance south of the present shoreline and most of the depocentre was covered by moderate water depths with a smooth, mud covered continental shelf and slope. No islands or major offshore shoals are envisaged. The rapidity of the early transgression is indicated by the fact that with the exception of the crest of the Infanta Arch, the whole of the basin was an area of sediment accumulation during Cenomanian times. During the regressions represented by erosion across reflector 17 (Turonian) and the associated sandy facies in the Coniacian (horizon 11), and the shallow water late Maastrichtian facies, the coastline retreated southwards, and differential subsidence appears to have taken place along the length of the basin. This resulted, in Coniacian times for example, in distinct lithofacies changes in E-W directions (Fig.129): glauconitic shallow water sands and sandy limestones over the old Infanta Arch and northern Bredasdorp Basin, with quieter (?deeper) water clays and calcareous silts to the south. Erosion in the Turonian may have been subaerial, but was probably caused by increased sea bottom scour by currents. Similar conditions set in towards the middle Maastrichtian, and by the end of the Upper Cretaceous the present day continental shelf area was probably dry land. Earth movements occurred in the western Bredasdorp Basin where the Cretaceous sediments are gently folded and faulted in the vicinity of the early Tertiary Alphard Banks igneous intrusions.

During Upper Cretaceous times, the continental shelf edge slowly advanced southwards, so that by the end of the Maastrichtian it lay up to 30 km south of its present position, a maximum advance of about 35 km in about 45 my (778 m/my).

By Upper Cretaceous times, the Falkland Plateau had separated from southern Africa and a deep-water ocean basin lay beyond the SE margin of the continent. During separation a large ridge complex (composition unknown) was emplaced along the line of the Agulhas Fracture Zone: the Agulhas Marginal Fracture Ridge (Scrutton & du Plessis 1973), and this effectively dammed Upper Cretaceous sediments behind it and restricted sedimentation to areas underlain by continental crust. Apart from these observations, we know

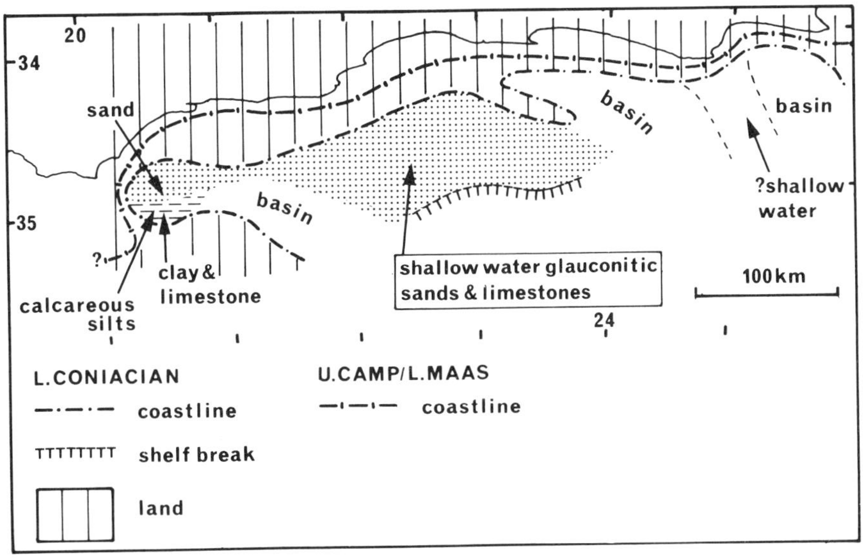


Figure 129. Lower Coniacian palaeogeography in the Outeniqua Basin. Compiled to show lithofacies at the level of reflector 11 (see Figure 128). A speculative Upper Campanian-Lower Maastrichtian shoreline is also plotted to illustrate a transgressive situation.

nothing of the sedimentary environments between the present shelf break and the marginal ridge, and can only surmise that they were deep water areas in which clays were deposited in front of the advancing shelf lens. Conceivably, restricted deep water environments could have occurred in this area.

## 5.2 EAST COAST AND ADJACENT OFFSHORE AREAS

Upper Cretaceous sequences are widely distributed and relatively complete over the Natal-Zululand-Mozambique coastal plain, while even thicker successions occur in the offshore areas (Fig.124). Over the Transkei Swell, however, only thin, small and widely scattered outcrops of late Cretaceous rocks occur (Figs 67 & 124).

By late Lower Cretaceous times, the present-day sedimentary and oceanic basin geometry of the east coast region was well established, with the Natal Valley-south Mozambique depocentres effectively separated from the Zambezi Basin by various basement highs. Consequently, when the major mid-Cretaceous hiatus affected the southern area (either through subaerial exposure or increased strength of bottom currents), sedimentation was continuous across the Albian/Cenomanian/Turonian boundaries in the Zambezi Basin. The effects of this event are most noticeable in the present-day onshore parts of Zululand-south Mozambique where they are represented by a widespread non-sequence, but are also recorded in lithofacies changes throughout the northern Natal Valley across a major regionally-developed reflecting horizon (McDuff), and as an important hiatus on the Mozambique Ridge at DSDP site 249. The presence of this mid-Cretaceous event in areas of differing crustal structure suggests that it was not occasioned primarily by crustal uplift, and the fact that no effects were registered in the Zambezi depocentre indicates that eustatic sea level movements were probably not involved. A radical reorganization of oceanic circulation patterns caused by altering palaeogeographies during the South America/Africa separation is the only cause that can account for this event which (as we have discussed above) also affected the Outeniqua Basin. Local crustal instability is possibly indicated however in the region of the Tugela Cone where, in contrast to the Cenomanian/Turonian hiatus registered onshore, rocks of the same age offshore are transgressive over a basement surface. Further evidence of small-scale earth movements will be discussed later. A major transgression commenced in late Turonian times and onlapped southwards, topping the Transkei Swell in the late Cretaceous (Fig.132). This event was eustatically controlled.

Maximum thicknesses of Upper Cretaceous rocks along the east coast occur under the Tugela Cone (1.6 km), and SW of the Zambezi Cone (1.7 km) (Fig. 124). Extensive deposits of unknown thickness also probably occur in the southern Natal Valley.

### 5.2.1 *St Lucia Formation of Zululand*

In Zululand, the Upper Cretaceous is represented by the St Lucia Formation which was formally named and described by Kennedy & Klinger (1975), from sections in the Mzinene and Munywana rivers around the Skoenberg Hill about 20 km north of Hluhluwe (locality 60, Fig.133). These had been made well-known by Anderson (1907), Crick (1907a), Etheridge (1907) and particularly van Hoepen (1926 et seq.) from their studies of the rich ammonite faunas. The St Lucia Formation can be traced at outcrop as far north as the Mkuze River

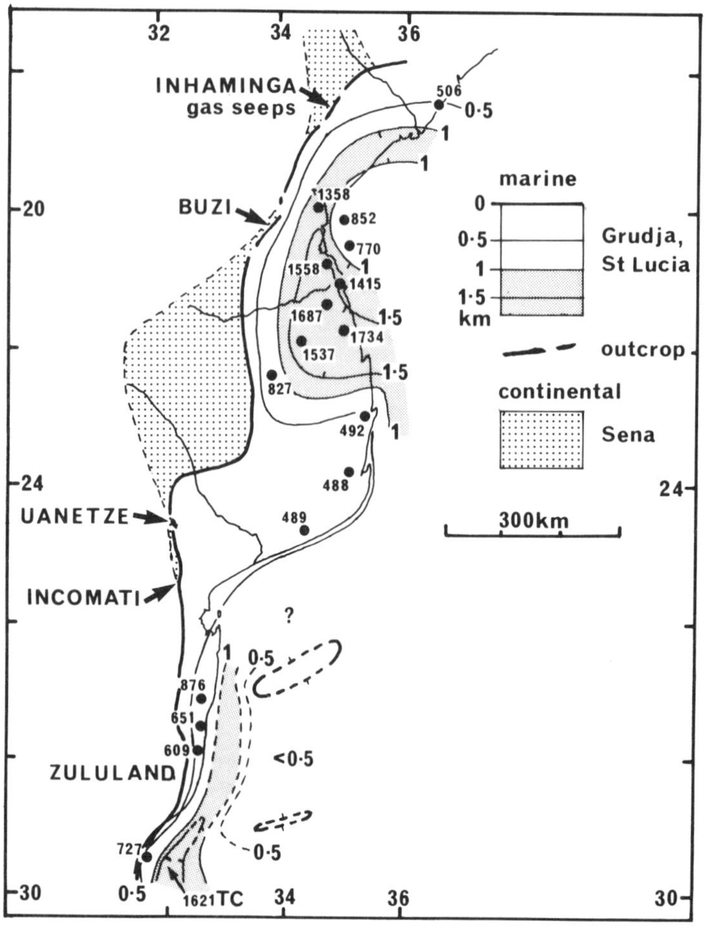


Figure 130. Distribution and thickness of Upper Cretaceous sediments in Zululand and S.Mozambique. In Zululand these rocks range Turonian/Coniacian to Maastrichtian (St Lucia Formation), offshore (J(c)-1) they are ?Cenomanian to Maastrichtian, and in Mozambique Lower Coniacian to Upper Palaeocene/Lower Eocene (Grudja Formation). Continental Sena sandstones may range up to Santonian.Boreholes are shown by dots, and thicknesses are in m. Offshore isopachs in km compiled using an average P-wave velocity of 2948 k/sec (calculated from the seismic profile of du Toit & Leith 1974). Data from Flores (1973); du Toit & Leith (1974); Forster (1975b), Dingle *et al.* (1978); McLachlan & McMillan (1979).

(Fig.133) beyond which it has been intersected in boreholes in northern Zululand (McLachlan & McMillan 1979), and is represented by a shallower water facies in SW Mozambique (Mandajene Formation) (Fig.131). Thickness estimates for the whole formation are difficult to make at outcrop, but borehole records (Fig.130) show that it thickens northwards from the type area (ca.600 m) to 876 m in northern Zululand, and offshore to 727 m on the Tugela Cone.

The St Lucia Formation consists of richly fossiliferous buff and greenish grey glauconitic silts and fine sands, with interbedded large calcareous concretions. The fauna is diverse, consisting of both in- and epifaunal, and planktonic elements: lamellibranchs, gastropods, ammonites, nautiloids, echinoids, ostracods, and benthic and planktonic foraminifera. Plant fossils consist of drifted logs, comminuted debris and calcareous nannofossils. There is abundant evidence of bioturbation in the silts and current bedding is common, although horizontal bedding is dominant. The succession is very similar to that in the underlying Mzinene Formation, in particular the presence of numerous upward fining sedimentary units (Fig.135). These reach thicknesses of up to 3.5 m and typically consist of the following sequence (1 (bottom) to 4 (top)):

1. a basal concretionary limestone (ca.40 cm) crowded with large mollusc fragments (especially *Inoceramus* and ammonites). Its upper surface is bored by *Lithophaga* and often encrusted with oysters and serpulids;
2. upward fining, glauconitic, current aligned shell lag (ca.15 cm);
3. glauconitic sand with few broken shells (ca.5-10 cm);
4. fine glauconitic sands grading into glauconitic silts with scattered molluscs, both drifted and in situ. This unit is 2-3 m thick and is usually bioturbated.

These cycles clearly represent alternating high (scour) and low (accumulation) energy bottom current activity.

In the type area (locality 60 on Fig.133), the base of the St Lucia Formation belongs to the Coniacian I zone of Kennedy & Klinger (1975), which is typified by *Kossmaticeras theobaldianum* (Stoliczka) (Fig.134). It consists of cross-bedded fine sandstones and silts resting on a 28 cm *Pterotrigonia* conglomerate containing pebbles of agate, igneous rocks, lamellibranch fragments, and rare *Proplacenticeras*. This in turn rests on

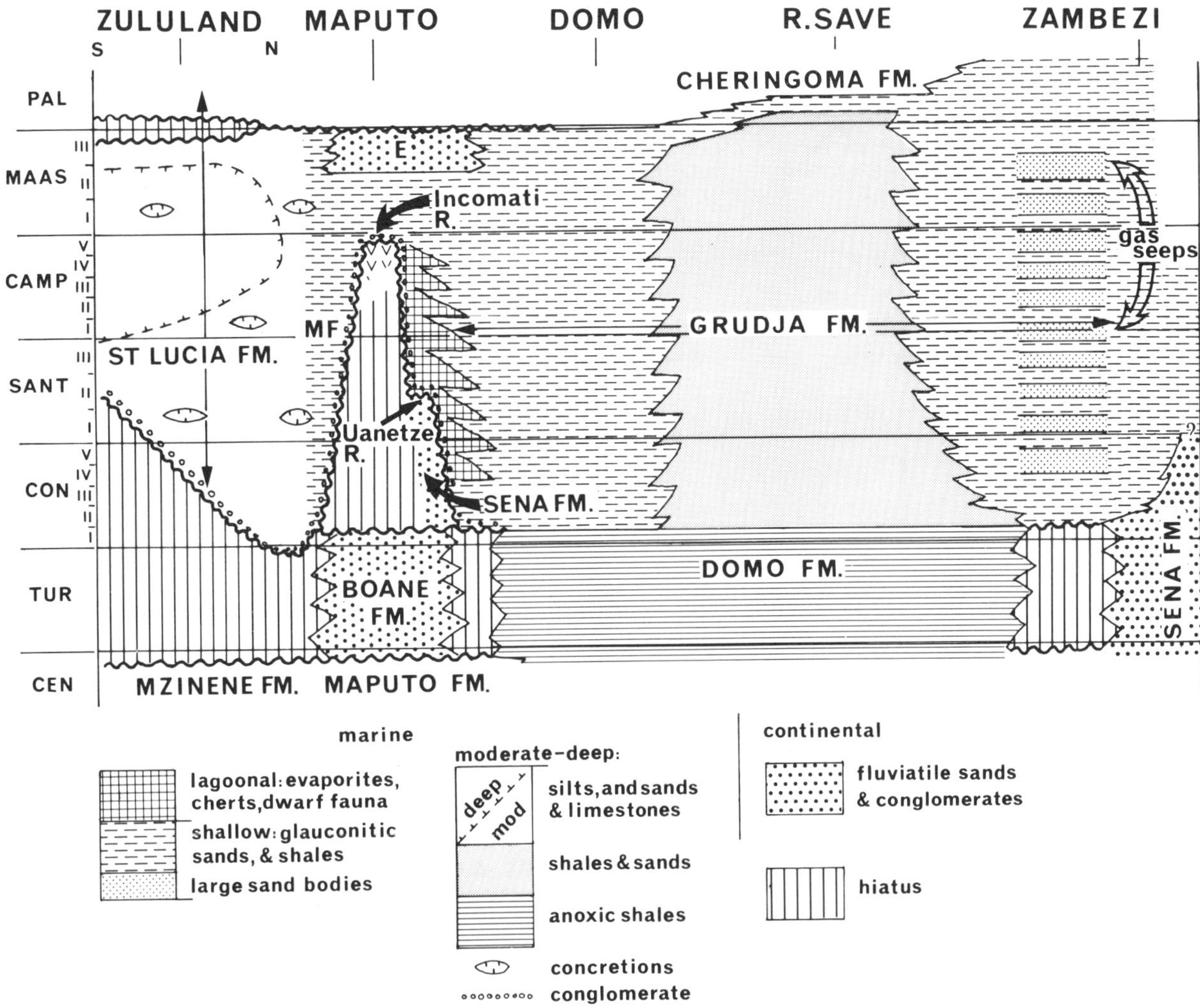


Figure 131. Schematic lithostratigraphic correlation of Upper Cretaceous strata between Zululand and the Zambezi Graben. Approximate length of section is 1200 km. Not to scale. Based mainly on the work of Flores (1973), Forster (1975a, 1975b). The relationships around Maputo have been simplified (the Upper Cretaceous here rests on Sena Formation and basement), and the large sand bodies in the Grudja Formation are shown diagrammatically. Basal volcanics in the Incomati River are probably Jurassic rhyolites. MF = Mandajene Formation (its position follows Cooper 1974, quoting Soares & Henriques da Silva 1970); E = Elephantes Formation.The Sena may range into the Santonian.

a Cenomanian hard ground with bored concretions at the top of the Mzinene Formation.

At Matubatuba, 60 km to the south, on the north bank of the Mfolozi River, and on its south bank around Umkwelane Hill (localities 7-10, Figs 133 & 134), Coniacian II/III, characterized by various species of *Proplacenticeras, Peroniceras, Forresteria* occur at the base of the St Lucia Formation, which here rests on strata older than the Mzinene Formation. To the north of the river (localities 7 and 8 on Figs 133 & 134), about 2 m of fine calcareous, pebbly sandstone with scattered molluscs (Coniacian II fauna) rest on about 25 m of interbedded conglomerates and sandstones of the Makatini Formation. The conglomerate contains boulders and cobbles of Lebombo volcanics, Karoo dolerites, quartzite, granite, banded ironstones, and abundant, silicified logs. The sandstones are cross-bedded with comminuted plant matter and Aptian ammonites have been recorded from this outcrop by Frankel (1960). To the south of the river (locality 10 on Figs 133 & 134), the base of the St Lucia Formation (Coniacian II) consists of 80 cm of hard, silty limestones with scattered quartz and quartzite pebbles, abundant oysters, other molluscs and echinoid spines (the 'oyster conglomerate') resting directly on deeply weathered Lebombo lavas. It is overlain 20 m of deeply weathered silts and very hard shelly concretionary linestones. A further 3 km south, at locality 9 (Figs 133 & 134) a basal conglomerate consisting of large (up to 30 cm) boulders of granite and schist set in a bioclastic lime-

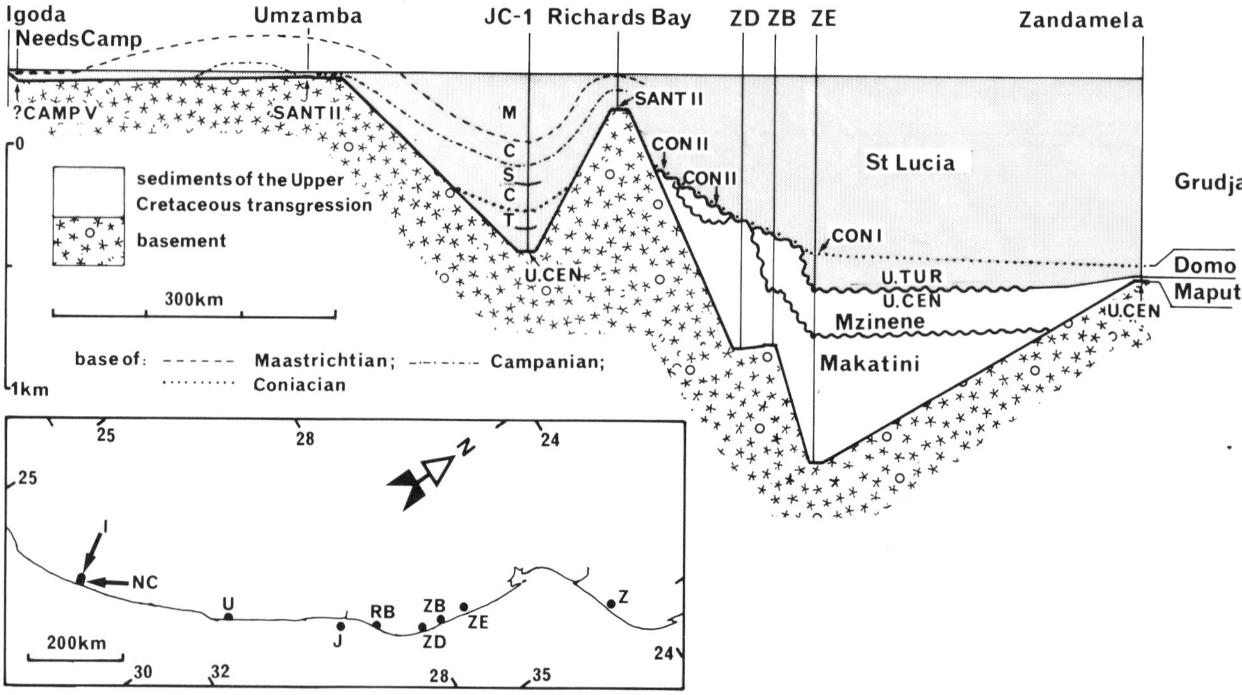


Figure 132. Distribution and thickness of Upper Cretaceous stages along the continental margin of S.E.Africa. Datum is top of Maastrichtian except in the Zandamela borehole where the Grudja Formation ranges into the Palaeocene. Stages in J(c)-1 are denoted by their initials. Insert shows locations, with sites denoted by their initials from the cross-section. ZD, ZB, and ZE are SOEKOR boreholes in N.Zululand. Data from Flores (1973); du Toit & Leith (1974); Kennedy & Klinger (1975); Maud & Orr (1975); Klinger & Kennedy (1977, 1980); Klinger & Lock (1978); McLachlan & McMillan (1979).

stone is overlain by 3 m of hard shelly limestones and interbedded soft shelly lenses. The limestones have a Coniacian fauna and rest on metamorphic granite basement.

Eastwards, in outcrops along the Nyalazi, Hluhluwe, and Mzinene rivers, and around the cliff sections of False Bay and Lake St Lucia, an apparently complete sequence from Coniacian to Lower Maastrichtian can be traced. The succession is richly fossiliferous throughout, particularly on the Nibela Peninsula at the east end of False Bay, where the bases of excellent cliff exposures in Campanian strata are frequently littered with beautifully preserved ammonites. Dips are low to the east (ca.3°) so shore exposures give easy access to much of the succession. It is in this area that the cyclic sediment units shown in Figure 135 (locality 110) are particularly well-developed. The highest parts of the succession crop out on the west shore of the central part of Lake St Lucia, where Maastrichtian III buff silts are packed with large *Inoceramus* fragments and prisms, and rare oyster and polyzoa, boring sponges, and cirripeds.

South of locality 9 (Fig.133), natural exposures do not occur, but engineering works at Richards Bay, on the Nyokaneni River (north of Empangeni locality 6, Fig.133), and and at Durban have revealed Santonian-Campanian strata. At the Nyokaneni River site (28.687°S, 32.022°E, locality 6 of Kennedy & Klinger 1975) grey/green glauconitic and lignitic silts with calcareous concretions containing a Santonian II ammonite fauna characterized by *Plesiotexanites stangeri* (Baily) rest directly on Pre-Cambrian basement, and similar relationships have been recorded from boreholes in the Richards Bay area (Klinger & Kennedy 1977). These boreholes further indicate that the upper part of the Maastrichtian is missing and that there is an hiatus between the Upper Cretaceous and Danian sediments. The hiatus includes the upper part of the *Globotruncana mayorensis* zone (uppermost Maastrichtian) and the whole of the *Globorotalia pseudobulloides* zone (lower Danian) (Stapleton 1975). The Mesozoic/Cenozoic contact is not exposed. In the vicinity of Durban, Kenney & Klinger (1975) record Upper Santonian-Lower Campanian soft glauconitic silts, sands, and calcareous concretions with wood fragments, and although depths to basement are not known it is thought that the regional southward younging of the basal Upper Cretaceous unconformity continues into this area (Fig. 133).

From the foregoing discussion, the picture emerges of an Upper Cretaceous transgression that proceeded southwards from the St Lucia area of Zululand (where it is Coniacian I in age) to Durban (probably Upper Santonian) and westwards across the continental shelf

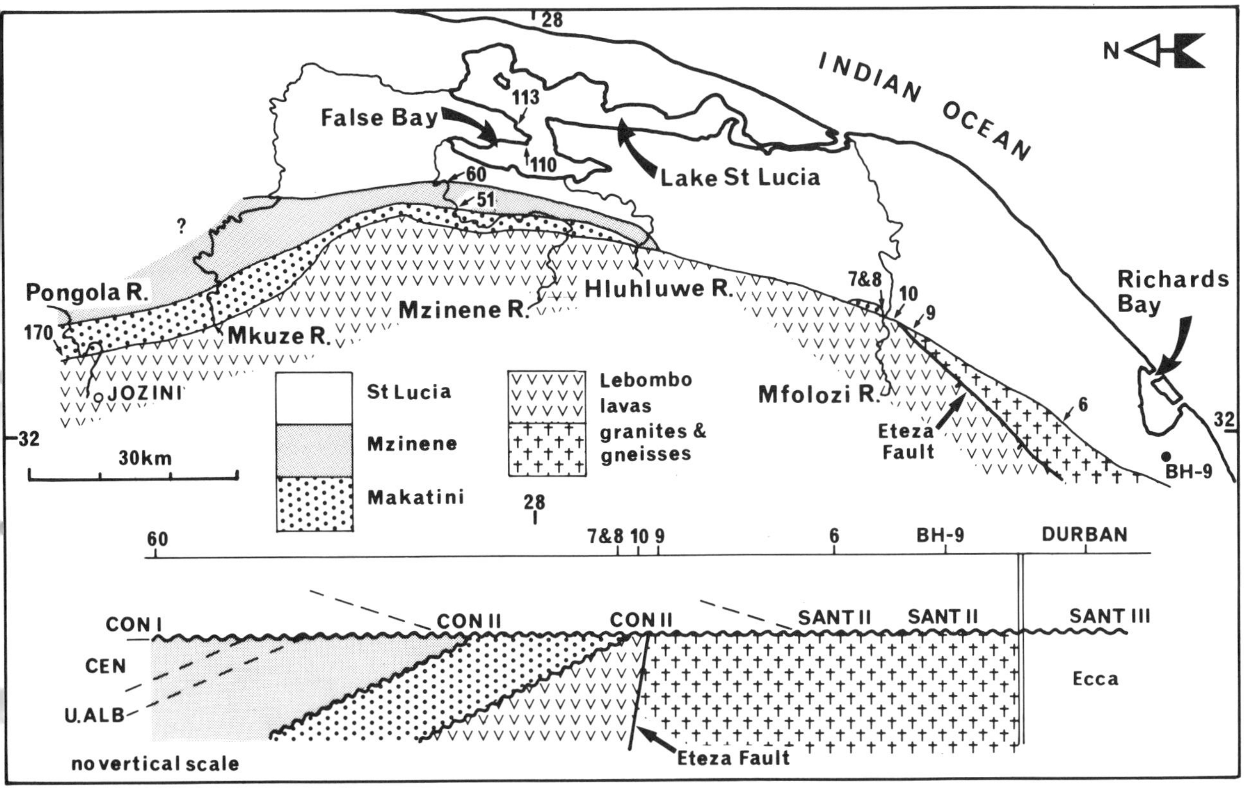


Figure 133. Distribution of Coniacian to Maastrichtian sediments in Zululand. Locality numbers are those of Kennedy & Klinger (1975). Lower diagram is a schematic N-S cross-section between Mzinene River and BH9 (Richards Bay), which is extended (not to scale) 140 km farther south to Durban. It shows the southward onlap of successively younger St Lucia Formation strata. Section and map have same horizontal scale. The map is compiled from Lambert & Scheibnerova (1974), and Kennedy & Klinger (1975). Erratum: for locality 170 read 158.

(under the present Tugela Cone) where it is Cenomanian in age about 10 km east of borehole J(c)-1 (Fig. 132).

There are no outcrops of St Lucia Formation in north Zululand on the dense bush-covered Makatini Flats, but several boreholes have proved up to 876 m of it. In this area the base of the formation is Upper Turonian (*Praeglobotruncana helvetica,* UC5 of van Hinte 1976a) (McLachlan & McMillan 1979) and the underlying Mzinene Formation is Upper Cenomanian (Fig.132). This non-sequence is considerably shorter than that found farther south, and emphasizes the southward younging of the transgression already discussed.

Over the border, in southernmost Mozambique (SW of Maputo), the earliest marine Upper Cretaceous sediments are Coniacian in age and consist of up to 500 m of fine sandstones and silts with limestone concretions (Mandajene Formation) (Cooper 1974) (Fig.131). The rocks are clearly lateral equivalents (possibly a slightly shallower-water facies because it is poor in ammonites) of the St Lucia Formation. Immediately west of Maputo, they rest on a fluviatile sandstone/conglomerate sequence (Boane Formation) which is up to 100 m thick. This formation has been dated as Turonian-Coniacian (Flores 1973) and presumably represents a continental facies laid down during a Turonian-Coniacian regression that produced the non-sequence in Zululand. A very restricted provenance from the adjacent Lebombo mountains is indicated for the Boane Formation by pebble (>1 cm) counts: rhyolite – 97 %; agate – 1 to 1.5 %; Pre-Cambrian basement – 1 %; basalt – 0.5 % (Förster 1975b).

5.2.1.1 *Palaeontology of the St Lucia Formation*

In common with the underlying Makatini and Mzinene formations, the St Lucia Formation contains an abundant and well-preserved macrofauna which is dominated by lamellibranchs, gastropods and cephelopods. Only the last of these, however, has been revised and biostratigraphically analysed in recent times. Other taxa are relatively poorly represented, although we have no quantitative data such as that provided by Förster (1975b) for the underlying sediments. Kennedy & Klinger (1975) report that corals, belemnites, echinoids, and brachiopods are all sparse, although the intense bioturbation of most of the St Lucia Formation indicates that it supported an abundant infauna. Likewise,

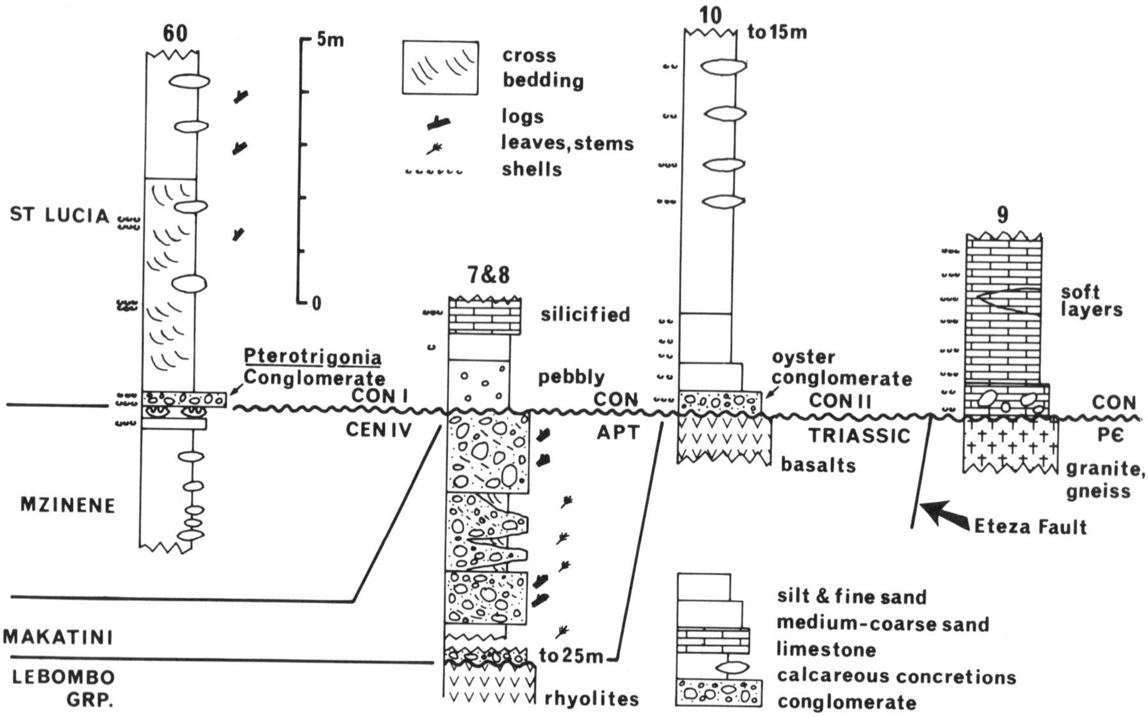


Figure 134. Measured sections across the basal beds of the St Lucia Formation in Zululand. They show overstep across progressively older basement. Location of sites shown on Figure 133, following numbering system of Kennedy & Klinger (1975, and in press).

reptile remains were only noted at a few horizons in the Santonian to Lower Campanian strata. Plant remains are very abundant, with logs, stems and comminuted debris. Logs are not as common as they are in the underlying Makatini and Mzinene formations, but lignite chips are an important sedimentary component as high up as the Campanian.

The rich ammonite fauna has been revised by Kennedy & Klinger (1975 et seq.), who have subdivided the stages into zones based on the local assemblages (Tabs 29 & 30). The Coniacian (29) and Campanian (22) stages have the most genera on record, but there is a sharp decline into the Maastrichtian (six genera), with the uppermost subdivision (Maastrichtian III) devoid of ammonites, and characterized only by abundant *Inoceramus* remains. Generic similarity is relatively low across the Turonian hiatus with only three genera (6 %) common to the Cenomanian and Coniacian strata (Tab.31), but it increases rapidly upwards to reach 30 % between the Santonian and Campanian faunas. Numerically, the Upper Coniacian to Middle Campanian is dominated by the Texanitinae, and at the generic level the Tetragonitidae (*Tetragonites, Saghalinites,* and *Pseudophyllites*) represent long ranging linkages across stage boundaries.

Kennedy & Klinger (1975) have identified a sparse fauna of Collignon's (1965) Middle Coniacian *Kossmaticeras theobaldianum* and *Barroisiceras onilahyense* zone of Madagascar from the lowermost Coniacian rocks above the mid-Cretaceous hiatus, which is equivalent to the UC6 (late Turonian-early Coniacian) *Globotruncana renzi-G.sigali* planktonic foraminifera zone of van Hinte (1976a). In boreholes farther north in Zululand, McLachlan & McMillan (1976) have recognized the *Praeglobotruncana helvetica* UC5 zone (Middle-Upper Turonian) above the hiatus, which indicates that Coniacian rocks older than Kennedy & Klinger's (1975) Coniacian I are present in Zululand (Tab.28).

Ostracods are the other fossil group that have recently received detailed attention, and although studies of the Coniacian and Santonian faunas are not complete, sufficient information exists to recognize the main trends. To date, 62 species belonging to 39 genera have been recognized in the Zululand Coniacian-Maastrichtian rocks (Tab.38) and although the Cytherellidae and Bairdiacea are locally numerically dominant, the character of the faunas is determined by their cytheracean components. These are diverse, with particularly vigorous trachyleberid evolution in the Campanian. As previously mentioned, the composition of the Cytheracea in the Lower Cretaceous and Cenomanian was Cytheruridae/Progonocytheridae/Schizocytheridae dominant, but the faunas that re-colonized Zululand after the mid-Cretaceous hiatus show an entirely different make-up at the generic level and are Trachylebe-

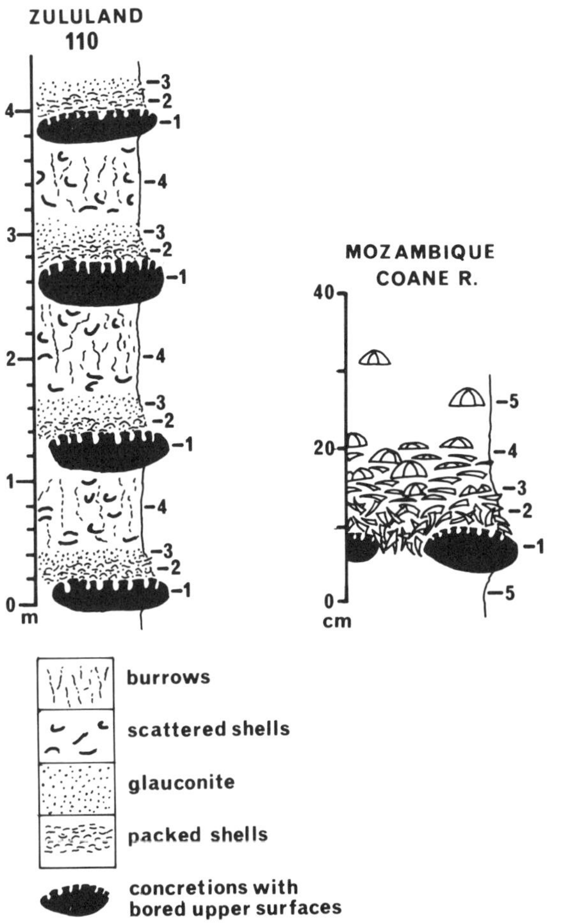


Figure 135. Current-generated sediment cycles in the St Lucia and Maputo formations.

Locality 110 is shown in Figure 133 and exposes Campanian fine sands and silts on the Nibela Peninsula. The four units of each cycle consist of: 1 – bored hard ground with the ammonites *Menabites, Baculites* and various pachydiscids and inoceramids with bored upper surfaces; 2 – glauconitic packed shell horizon with *Pycnodonte, Baculites, Menabites,* inoceramids, heterodont lamellibranchs, nautiloids, and pectinids; 3 – glauconitic sand with diminishing upward fauna of unit 2; 4 – burrowed silts with scattered *Pycnodonte,* inoceramids and spatangoid echinoderms.

The Mozambique locality is on the Coane River (32.299°E, 26.661°S) in Cenomanian sediments. The five units recognized in each cycle consist of: 1 – concretions bored by lithophagus bivalves with epifaunal encrustations; 2 – irregular bedded echinoderm test fragments; 3 – parallel-bedded echinoderm test fragments; 4 – complete but mostly crushed echinoderm tests; 5 – scattered uncrushed echinoderms in argillaceous silts and sands. The echinoderms are predominantly *Hemiaster* cf. *H.zululandensis.*

Figures are modified after Kennedy & Klinger (1975 fig.9) and Forster (1975b fig.11).

rididae/Brachycytheridae/Schizocytheridae dominated. In particular, the Lower Coniacian is marked by the appearance of *Brachycythere,* which migrated from the West African and Brazilian area, and *Haughtonileberis* and *Gibberleberis.* These types are joined higher up by a large number of new (especially trachyleberid) types, but the first two genera continue through into the Maastrichtian. The specific diversity (%) of the cytheracean populations steadily increases from seven (provisional estimate) in the Coniacian, 23 (Santonian), 32 (Campanian) to 35 in the Maastrichtian, whilst the number of extant genera follows a similar trend (Tab. 41).

Comparison of the Campanian and Maastrichtian ostracods of Zululand (including J(c)-1 borehole offshore Natal, and sample 818 from the Agulhas Bank) with other Gondwanide localities at the generic level (Tab.42) indicates relatively strong similarities with Western Australia and East Africa (Tanzania), but only relatively weak links with West Africa and Argentina (the 14 % similarity between South Africa and the Middle Maastrichtian of Argentina is misleading because it is solely in cosmopolitan types). The two most significant links that the Zululand Upper Cretaceous ostracod faunas do have with South Atlantic areas is the importance of *Brachycythere* in the Brazilian and West African assemblages, and the deeper water elements of the Rio Grande Rise/Sao Paulo Plateau (Dingle 1981).

### 5.2.2 *Grudja Formation of Mozambique*

NW of Maputo, in the Incomati and Uanetze regions (Fig.130) Sena Formation sandstones of Coniacian-?Santonian age are overlain by and interfinger with paralic sediments; lacustrine and marine cherts and evaporites with a dwarf fauna. Farther east over the Zandamela High, these shallow water facies pass into thicker sequences whose main depocentre lay farther north in an elongate trough just west of the present day coast (Fig.130). These rocks form the Grudja Formation and they range in age from Coniacian to Palaeocene, although in the vicinity of the Incomati River the transgression did not arrive until Maastrichtian times, where the Grudja Formation rests directly on weathered Lebombo lavas. The Grudja Formation is about 500 m thick over the Zandamela-Domo region, but northwards thickens to over 1 700 m in the main basin. In the deeper parts of this depocentre, the lithology of the formation is predominantly shaly, but in the proximal areas it becomes silty and sandy with abundant glauconite, and in the region of Beira it contains ten prominent sandstone units which have produced good gas shows during hydrocarbon exploration (Flores 1973) (Fig.131). Gas seepages occur at outcrops NW of Beira (Inhaminga).

In the Lower Zambezi Valley, and along the eastern edge of the Save River region, the lower part of the Grudja Formation interfingers with, and grades into the upper part of the continental Sena Formation, which locally may range into the Santonian (pre-*Globotruncana arca,* Flores 1973, fig.3) and consists of

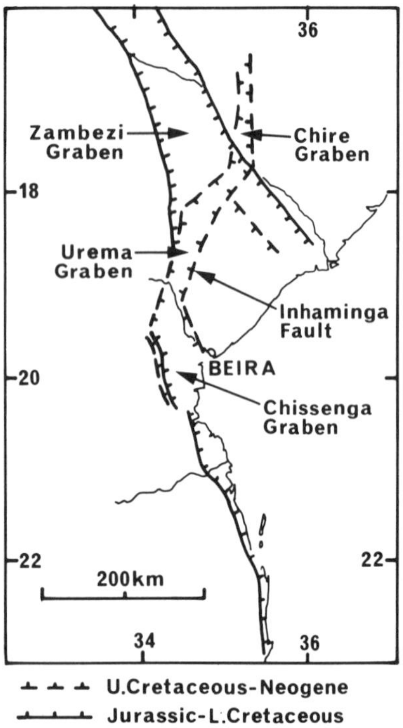


Figure 136. Taphrogenic trends that affected mid-late Mezosoic and early-middle Cenozoic sedimentation in the Zambezi/Beira area. After Flores (1973).

poorly-sorted, cross-bedded reddish arkoses, sandstones and conglomerates. East of the present coast off Save River, and between 23°S and the area around the Zandamela and Domo boreholes, marine sedimentation was continuous during Cenomanian-Turonian-Coniacian times, and there are no stratigraphic breaks between the Maputo, Domo and Grudja formations, so that dark Domo shales conformably pass upwards into the better oxygenated environment represented by the Grudja Formation (Fig.131). In this area, therefore, the Lower/Upper Cretaceous boundary is marked by a facies change only. However, the area of sedimentation expanded as the sea spread progressively west of the Save River mouth, so that by Maastrichtian times it covered most of the present coastal plain, and in the SW lay along the foothills of the Lebombo range (Förster 1975a). This was the first occasion on which marine conditions had penetrated across this area during Mesozoic times. It finally brought to a close the deposition of the Sena Formation which since at least the Middle Jurassic had accumulated continental deposits, and in the lower parts, interbedded lavas.

During Upper Cretaceous times a transverse graben commenced development at the SW end of the Zambezi Valley: Urema-Chire Graben (Flores 1973) (Fig. 136), whilst at the same time, reactivation of the western edge of the SW extension of the Zambezi Graben may also have taken place south of Beira. The boundary faults of the Urema graben cut the Sena Formation, and an angular unconformity separates Sena and Grudja strata north of Beira. The youngest rocks to be affected by the faulting are lower Miocene sandstones. Initiation of this phase of tectonic activity seems to have been synchronous with the Cenomanian-Turonian non-sequence in Zululand, and caused the Zambezi River to be diverted to the SW, where it flowed down the new graben, to debouch in the vicinity of Beira. Downwarping along the coastal strip was responsible for the location and narrowness of the main Grudja depocentre (compare Figs 130 & 136).

### 5.2.3 *Tugela Cone, J(c)-1, Natal Valley and Mozambique Ridge DSDP 249*

In the Natal Valley, the mid-Cretaceous hiatus is probably represented by seismic reflector McDuff (Figs 106 & 107). No large break in sedimentation is suspected across this horizon because Cenomanian-Turonian sediments occur in borehole J(c)-1 immediately above McDuff (which, 7 km to the east, laps against the basement under the Tugela Cone) (Figs 106 & 171). In the borehole, they rest upon 52 m of poorly-sorted, pebbly, muddy sandstone, which in turn rests on hard quartzites that are correlated with the local representatives of the Palaeozoic Table Mountain Group (Fig. 137) (du Toit & Leith 1974). Although the age of the pebbly mudstone is unknown, the SOEKOR geologists consider it to be pre-Cretaceous in age because the contact with the underlying quartzite is gradational, whereas it is unconformably overlain by the Cretaceous sediments. The Upper Cretaceous sequence in J(c)-1 consists of an apparently uninterrupted succession of monotonous light grey marine claystones with subordinate fine grained sandstones. Numerous thin (<65 cm) hard, white to brown limestones are present, particularly near the base. This lithological unit has a total thickness of 773 m and an erosional non-sequence at the top. Du Toit & Leith (1974) originally considered the passage into the Palaeogene to be conformable, but recent micropalaeontological work by McLachlan & McMillan (1979) has detected a break which includes the *Globotruncana mayorensis* zone (UC17 – topmost zone) (Tab. 40). A similar break was detected in the Richards Bay area of Zululand by Stapleton (1975). Du Toit & Leith (1974) recorded moderately abundant planktonic foraminiferal faunas at some horizons, but the benthic fauna is somewhat impoverished with large numbers of agglutinated benthic foraminifera and only poor to very sparse ostracod assemblages. Pyrite is common and fresh-water charophyte algae occur over a 60 m zone in the Middle Maastrichtian. Dingle (1981) concluded that during the late Cretaceous, water depths on the Tugela Cone top temporarily increased to about 200 m on at least two occasions, but that local bottom conditions were mildly anoxic, with a large fresh water

Table 40. Planktonic foraminiferal zonation of Upper Cretaceous rocks on the Tugela Cone (J(c)-1) and onshore Zululand (after McLachlan & McMillan 1979).

| | J(c)-1 | Zululand | Standard zones |
|---|---|---|---|
| | *G.contusa* and *G.stuarti* | *G.contusa* and *G.stuarti* | UC 16-11 |
| Maastrichtian | *G.fornicata* | *G.fornicata* | UC 10 |
| | *G.tricarinata* | *G.tricarinata* | ? |
| Campanian | *G.ventricosa* | *G.ventricosa* | UC 9 |
| | *G.elevata* | *G.elevata* | UC 9 |
| Santonian | *G.coronata* | *G.coronata* and *G.carinata* | UC 8 |
| | *G.concavata* | *G.concavata* | UC 7 |
| Coniacian | no zonation | *G.renzi* and *G.primitiva* | UC 6 |
| Turonian | *P.stephani* | *P.helvetica* | UC 5 |
| | | non-sequence | |
| Cenomanian | *? R.* sp. | *R.cushmani* | UC 3 |
| | basement | *R.apenninica* | UC2-LC19 |

*G.* = *Globotruncana*
*P.* = *Praeglobotruncana*
*R.* = *Rotalipora*

and sediment supply nearby. This inhibited the establishment of diverse benthic faunas even though access to open ocean conditions at the surface water levels was good. These data have been incorporated in a late Cretaceous sea level fluctuation curve for Zululand that will be discussed in the palaeogeography section (5.2.5). One aspect of the ostracod faunas of the late Cretaceous on the Tugela Cone that is somewhat unexpected is that several of the cytheracean types have close relatives on the Agulhas Bank (1 000 km away), but not particularly close links with southern Zululand sites only 130 km to the north.

Landward, the Upper Cretaceous strata thin as they overstep across the basement and probably pinch out close to the present coast (Fig.171). Seaward of J(c)-1 they thicken rapidly as a prograding sequence, and 15 km east of the borehole are probably 1 500 m thick. Because there is no obvious lithological change across the Mesozoic/Cenozoic boundary on the Tugela Cone, and hence no regionally-developed seismic reflector which corresponds to it, it has not been possible to trace the top of the Cretaceous succession into the deeper water environments of the Natal Valley. From their seismic data, Dingle et al (1978) deduced that this boundary falls between the two major reflectors in the Natal Valley; McDuff & Angus (mid Cretaceous and base Oligocene, respectively) (Figs 106, 107 & 171). The intervening sediment unit reaches its maximum development (ca.2 sec, ?2 km) under the Tugela Cone, just east of the present shelf break, and thins to about 0.6 sec (600 m) in the deep ocean basin farther east (Fig.106). On the top of the Mozambique Ridge, post-Cenomanian sediments intersected by DSDP at site 249 (Simpson et al 1974) consist of 120 m of brown and grey, intensely bioturbated foraminifera-bearing clay-rich nannochalk (Fig.137). The clay content varies from about 25 % in the lower part, to 45 % near the top, and the other non-carbonate components include quartz (up to 10 %, but usually ca. 1%), K-feldspar, mica, and palygorskite. Abundant planktonic foraminifera and nannofossils allow the sequence to be accurately dated as late Campanian (*Globotruncana calcarata,* UC12) to late Maastrichtian (*Globotruncana mayorensis,* UC17), showing that the mid-Cretaceous hiatus here encompasses mid-Cenomanian to mid-Campanian time, far longer than at any of the other locations for which we have data (Fig.107). The Upper Cretaceous sediments at site 249 are separated from the overlying Miocene rocks by a further massive hiatus.

Northwards into the Natal Valley, the interval between reflectors McDuff and Angus is generally <0.5 sec (ca.500 m). Bearing in mind that the Upper Cretaceous represents only a part (an unknown part) of this total, it is obvious that Cenomanian to Maastrichtian sedimentation was quite slow in the deeper parts of the northern Natal Valley. This suggestion is strengthened by the fact that the Grudja Formation (Coniacian to Palaeocene) in the Zandamela borehole (situated at the north end of the basin) is only about 490 m thick (Figs 103, 106 & 130). During this interval in time the two main basement ridges in the Natal Valley (Almirante Leite and Naudé ridges) continued to act as major barriers to sediment dispersion.

### 5.2.4 *Umzamba Formation and Igoda Beds*

South of Durban, Upper Cretaceous rocks crop out in two areas: Pondoland (Transkei/Natal border area), and near East London. Both represent remnants of sediment cover along the edge of the buoyant Transkei Swell, and mark high points in the Upper Cretaceous transgression along the eastern coast of South Africa (Figs 67, 124 & 132). The former (centred on Umzamba) lies at the southern end of the Port Shepstone Arch and the latter (Needs Camp/Igoda) on the Port Alfred Arch, and they are separated by the crest of the Transkei Swell over a distance of nearly 300 km. At present, there is no evidence that Upper Cretaceous sediments were deposited in the intervening areas, even within the Lower Cretaceous Mngazana and Mbotyi basins.

The Pondoland outcrops have been studied by numerous workers since they were first reported by Baily in 1855. They locally contain a rich, well-preserved invertebrate fauna, and form numerous small outcrops along the coast between the Umnyameni (Trafalgar Beach) and Mtentu estuaries (ca.40 km) (Fig.138). The main outcrop occurs in the vicinity of the Umzamba River

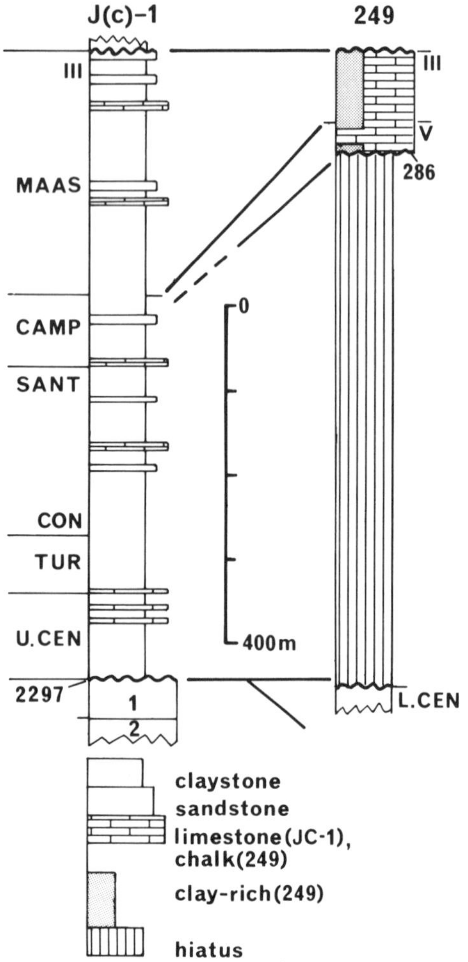


Figure 137. Upper Cretaceous sediments in the J(c)-1 and DSDP 249 boreholes. In J(c)-1 the Upper Maastrichtian contains *Globotruncana contusa* and is equivalent to Kennedy & Klinger's (1975) Maastrichtian III. In DSDP 249 only Campanian V (with *Globotruncana calcarata*) to Maastrichtian III occurs. 1 = pre-Cretaceous pebbly mudstone, 2 = quartzites of the ?Table Mountain Group. Compiled from data in du Toit & Leith (1974), and Simpson *et al.* (1974).

mouth and in the cliffs about 1.5 km to the north; the so-called Fossil Head and Petrified Forest outcrops. The best general account of the area is that by Plows (1921), although various aspects of the geology of the area have been discussed by Baily (1855), Griesbach (1871), Rogers & Schwarz (1902b), du Toit (1913, 1920a), and Gevers & Little (1946) amongst others. Modern revisions of the ammonite faunas have been undertaken by Spath (1953), van Hoepen (1965), Kennedy & Klinger (1975), and Klinger & Kennedy (1980), whilst microfossils have been described by Chapman (1904, 1923), Dingle (1969b, 1980c, 1981) and Makrides (1979). The distinctive lithology and extent of the outcrops have led to these rocks being referred to as the Umzamba Formation which has been recognized as far north as Durban (Kennedy & Klinger 1975).

Measured sections of the Umzamba Cliff have been published by Plows (1921), Dingle (1969b), and Klinger & Kennedy (1977), but the most detailed and complete is that by Klinger & Kennedy (1980) who compiled a composite section from locality C on the south side of the Umzamba estuary where the basement is seen, and locality A at the main cliff for which they allocated bed numbers as used on Figure 138. Stage subdivisions are correlations with Kennedy & Klinger's (1975) ammonite zonation for Zululand. The base of the sequence consists of conglomerates with water-worn pebbles of sandstone and lydianite (up to 20 cm) with abundant bored silicified logs (up to 12 m long), sharks' teeth, and chelonian and reptile bones (bed C1). These are seen to rest directly on Table Mountain Group rocks 1 km south of the Mtamvuna estuary, and in the Umzamba estuary. No maximum thickness for these conglomerates has been recorded (at least 1.5 m according to Klinger & Kennedy 1980), but they are overlain by at least 21 m of glauconitic fine sands and silts (up to 2 m thick) and discontinuous, cross-bedded hard shelly or concretionary glauconitic, sandy limestones (up to 50 cm). The silts and fine sands are bioturbated and contain small concretionary cross-bedded shell lenticles (up to 1 m) and abundant lamellibranchs (both fragments and in situ) and logs (up to 1 m). When weathered, the sands and silts are relatively soft and buff coloured, but fresh they are bluish grey and hard. The conspicuous hard bands are mostly shelly limestones (cross bedded), representing scour fill, or shell banks, and are discontinuous along the cliff sections. The faunas are diverse, but are dominated by thick shelled varieties, and preservation varies from rounded to well-preserved. Several layers are dominated by particular faunal associations; e.g. turritelid gastropods, pectins, giant *Inoceramus* (up to 1 m long in bed A3). The lowest fossiliferous beds with a datable ammonite fauna are C4 (*Baculites capensis*) and these are overlain by horizons with *B.capensis, Gaudryceras varicostatum* and *Texanites umzambaensis* (C6). Klinger & Kennedy (1980) compare this fauna to the uppermost Lower to Middle Santonian assemblages in Madagascar (Collignon 1966) and to the Santonian I or II (most probably the latter) of Zululand (Kennedy & Klinger 1975), whilst the presence of *Baculites sulcatus, Hauericeras madagascariense* and *Scaphites* cf. *aequigramensiformis* in A14 and A15 indicates that the top of the section is probably Campanian II in age (equivalent to uppermost Lower Campanian faunas in Madagascar). An analysis of the ammonite faunas recorded by Klinger & Kennedy (1980) reveals some interesting differences between the Umzamba area and Zululand. *Hauericeras gardeni* (A3, A4, A8) and *Pseudoschloenbachia* spp. (C9, C11, A3, A7) are both abundant at Umzamba but only rare to common farther north, and *Plesiotexanites stangeri* (A6) which is common at Umzamba is rare in Zululand.

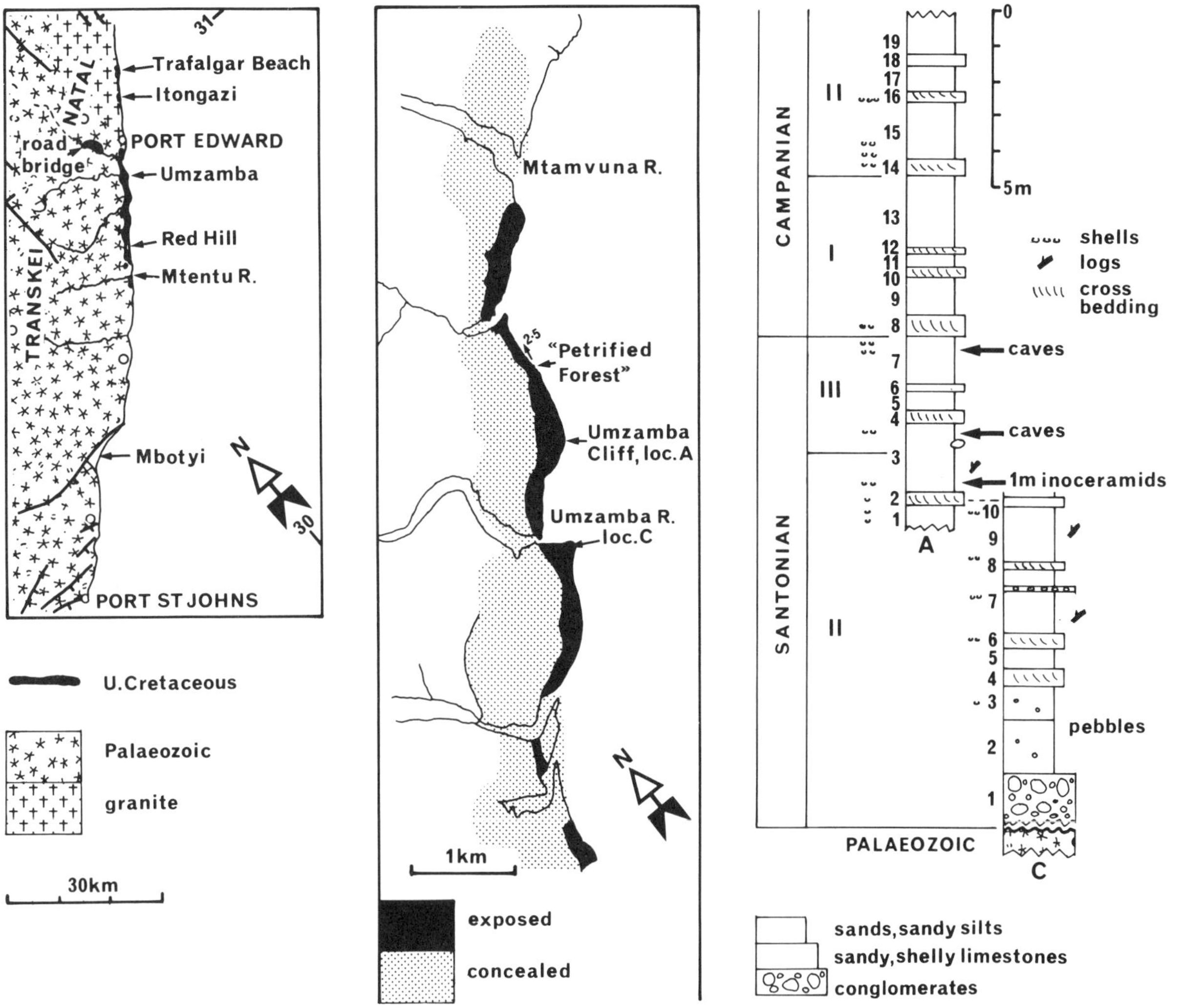


Figure 138. Distribution of Santonian-Campanian Umzamba Formation in the Transkei and S.Natal coastal zone. Regional map is mainly from Geological Survey (1970) with additions from Gevers & Little (1946), and Makrides (1979). Centre map shows outcrops on the N.Transkei coast with type localities N of the Umzamba River mouth. After du Toit (1913) with modifications from Makrides (1979). Localities A and C are those of Klinger & Kennedy (1980) from which the measured sections were taken. Left side of the composite section shows ammonite zones, with bed numbers 1-10 and 1-19 following the notation of Klinger & Kennedy (1977, 1980). The dip value of 2.5° on a bearing of 010° in the vicinity of the 'Petrified Forest' was reported by Plows (1921).

On the other hand, two species of *Baculites (B.capensis, B.bailyi)* which are abundant in Zululand are only occasionally found at Umzamba. The fact that 16 of the 40 Umzamba species have also been recorded from Zululand precludes the presence of a barrier between the two areas and Klinger & Kennedy (1980) suggest that the shallow water, relatively high energy environment in the south was preferred by some species, whilst the deeper water environments of Zululand was preferred by others. Compositional differences can also be recognized in the ostracod faunas which Dingle (1980c) compared between Umzamba and Richards Bay. In the 28 species identified from the Santonian sections in the two areas there was 61 % similarity: four Umzamba species were not found at Richards Bay and seven species vice versa. Dingle (1980c) also noted that although several species were common to the two areas, they were differently represented. For example: *Cythereis transkeiensis* is abundant in the south and very rare in the north, whilst *C.klingeri* has a reversed role; also *Pondoina sulcata,* which is a characteristic element of Umzamba Santonian III assemblage is only sparsely represented at Richards Bay. Because both areas were, in Santonian III times, relatively shallow, high energy environments (with only trace numbers of *Bairdoppilata* and *Cytherella*) these compositional differences cannot be attributed in any great measure to differences in sedimentary environments, and must reflect other,

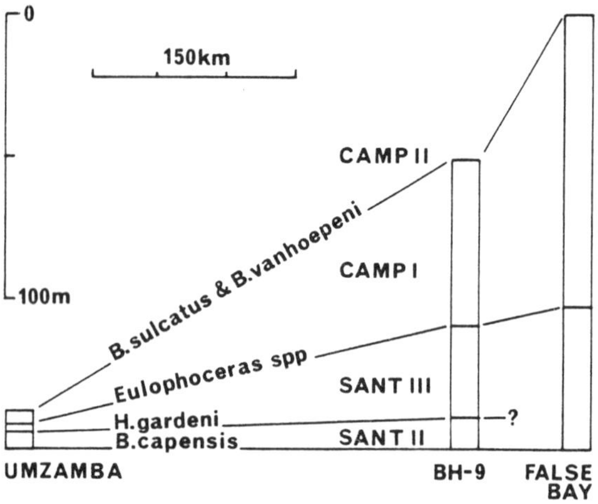


Figure 139. Thinning in Upper Cretaceous stages along the Transkei-Zululand coast. Completeness of the succession indicates differential sedimentation rates (Umzamba taken as unity):
Santonian II/III-Umzamba = 1, BH-9 = 4.5, False Bay = 5
Campanian I-Umzamba = 1, BH-9 = 11.6, False Bay = 20
Ammonite horizons serve as markers for correlation. B = *Baculites*, H = *Hauericeras*.
Data from Klinger & Kennedy (1977, 1980).

possibly climatic factors (e.g. water temperature). Significantly, the Umzamba ostracods have strong similarities with those from Igoda and Needs Camp farther south.

Twenty kilometres north of Umzamba in the vicinity of Trafalgar Beach, conglomerates overlain by glauconitic silts and fine sands, and concretionary shelly limestones rest on basement gneisses. These outcrops have been described by Crick (1907b), Anderson (1907), du Toit (1920), Gevers & Little (1946), and Kennedy & Klinger (1975). The conglomerates are glauconitic and contain boulders (up to 60 cm) which mostly consist of Karoo-type dolerites, with minor quartzites, gneisses, and sandstones, with rolled bones and molluscs. The overlying sands and silts are lithologically similar to those at Umzamba and from them the following ammonites have been recorded: *Natalites* sp., *Baculites sulcatus* and *Pseudoschloenbachia umbulazi,* which Kennedy & Klinger (1975) interpreted as Campanian.

Smitter (1955) examined foraminifera in samples taken slightly south of Trafalgar Beach, from a locality to the north of the Itongazi River. He recorded nine species, one of which was planktonic (*Globotruncana marginata*), and assigned a Middle Maastrichtian age. Several of the species reported by Smitter (1955) were also found at Umzamba by Makrides (1979), along with a more diverse planktonic foraminiferal assemblage. On the basis of SOEKOR's correlation of the local *Globotruncana coronata* zone with an Upper Santonian age (Tabs 40 & 43) (e.g. McLachlan & McMillan 1979), Makrides (1979) equates the presence of this species in beds A2 to A5 with Upper Santonian strata, whilst she dates beds A7 to A10 as Lower Campanian on the presence of *Epistomina pondensis. Rugoglobigerina rugosa* was recorded from beds A15-A17, which, according to Postuma (1971), has a range Lower Campanian (UC9) to Upper Maastrichtian (UC17). Makrides (1979) places these top beds in the Upper Campanian, which is at variance with the ammonite evidence of Klinger & Kennedy (1980). To us there seems no compelling reason on the foraminifera content why a Lower Campanian (Campanian I-II) should not be preferred.

A comparison of the thicknesses of various stage subdivisions and ammonite zones between Zululand and Umzamba (Klinger & Kennedy 1977) shows a marked convergence of the isopachs in a southerly direction (Fig.139). In the False Bay area, the Santonian II to Campanian I sequence is about 155 m thick, and at Richards Bay (BH9 borehole) this has reduced to ca. 100 m. About 330 km south, at Umzamba, this has reduced still further to ca.18 m, but the presence of a complete ammonite sequence at all three localities indicates that sedimentation was continuous and the thinning was the result of slow sedimentation, rather than non-sequences. It was probably caused by a combination of lower sediment input and higher energy of sedimentary environment in the south. Average sediment accumulation rates were approximately: False Bay 37 m/my, Richards Bay 24 m/my and Umzamba 4.3 m/my.

Finally, it should be noted that all available geological evidence suggests that the Umzamba Formation outcrops are not downfaulted outliers as was suggested by Griesbach (1871).

Upper Cretaceous sediments in the East London area differ from those in Pondoland on several points:

i) are calcareous, with only minor sands;
ii) are slightly younger (probably late Campanian/ early Maastrichtian);
iii) occur well above present sea level.

They occur at two localities: Needs Camp, and in the vicinity of the Igoda River estuary (Fig.140). The former area, in the Lower or East Quarry on the road between Kingwilliamstown and East London at Needs Camp, was first mentioned by Lang (1908) and Woods (1908), who described the polyzoa, echinoids, brachiopods and lamellibranchs, whilst microfossils have since been recorded by Chapman (1916), McGowran & Moore (1971), King (1972) and Dingle (1981). No measured sections have been published of the outcrop, which consists of a few metres of off-white polyzoan limestones with ostracods and benthic foraminifera, occasional *Inoceramus* fragments and other macroinvertebrates, and rare rounded quartz and glauconite grains. It occurs as a thin veneer resting on an erosion surface cut across a resistant 'Karoo' dolerite still at 367 m above sea level (King 1972). The most detailed account of the microfauna of the Lower Quarry outcrop was given by McGowran & Moore (1971) who listed 17

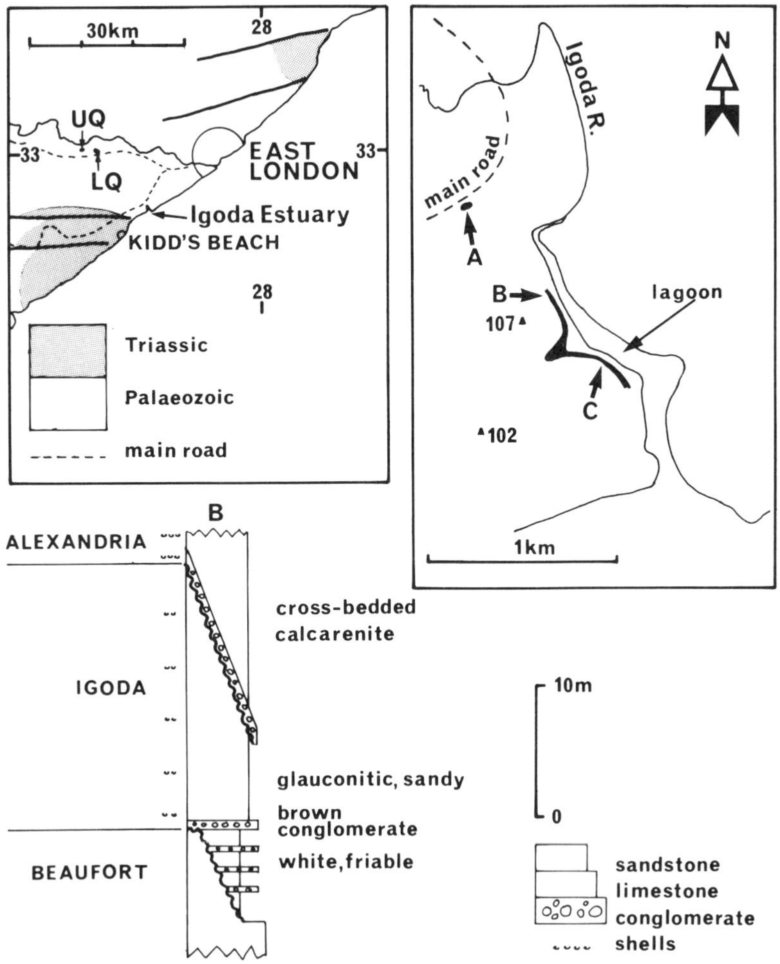


Figure 140. Upper Cretaceous outliers in the vicinity of East London. Top left map shows locality of the Igoda Estuary and Lower Quarry Needs Camp (LQ). Both have been dated as late Campanian/early Maastrichtian. The Upper Quarry Needs Camp (UQ) is Palaeogene. Localities A-C at Igoda (top right) are those described by Klinger & Lock (1978), although A was first recorded by Mountain (1974). The section measured at B is the type for the Igoda Formation, and is composite from two outcrops 30 m apart (extent of exposures shown in black). Spot heights on the top right map are in m.

species of foraminifera, all but two of which were benthic types. The two planktonic forms they identified as *Rugoglobigerina* cf. *rugosa* and *Hedbergella?* sp. and on the basis of these they dated the sediments as late Campanian/early Maastrichtian. They also recorded sharks' teeth (*Isurus, Carcharias*), and a plesiosaurid tooth. Klinger & Lock (1978) compared the Needs Camp Lower Quarry fauna with that at Igoda, and concluded that they were the same age on the basis of the common presence of *Eolacazella affine.* Finally, Dingle (1981) has re-examined the ostracod fauna including Chapman's (1916) material. The fauna is sparse, dominated by *Bairdoppilata* (> 90 %) and of the four species recorded, only *Hermanites?* cf. *arcus* was considered a significant correlative, occurring at Igoda and in Zululand. There is insufficient published evidence for a high resolution age determination and McGowran & Moore's (1971) date, which has recently been confirmed by Siesser & Miles (1979) quoting unpublished dates of Siesser's, is the best available. The dominance of benthic elements suggests that the area was not in connection with the open ocean when the Upper Cretaceous sea reached this point, its farthest penetration onto the Transkei Swell, 15 km from the present coast. The nearby Upper (West) Quarry at Needs Camp exposes Eocene strata, not Upper Cretaceous as suggested by King (1972), but Klinger & Lock (1978) state that further small outcrops of the Upper Cretaceous polyzoan limestone lithofacies have been found about half way between the two (3.2 km apart).

Evidence for other Upper Cretaceous outcrops in the East London area was found in 1974, when Mountain recorded a few centimetres of glauconitic, sandy limestone on the coast road about 14 km SW of East London. Subsequent field work revealed a relatively extensive outcrop of similar lithofacies in cliffs on the SW side of the Igoda River estuary (Klinger & Lock 1978), where the Igoda Formation has been formally designated (Fig.140). Here, a basal sequence up to 3 m thick of white pebbly sands and a prominent brown coloured conglomerate rest on lower Beaufort Group sandstones and mudstones, and are overlain by up to 20 m of sandy limestones. The lower sands and pebbly sands are unfossiliferous, but the brown conglomerate,

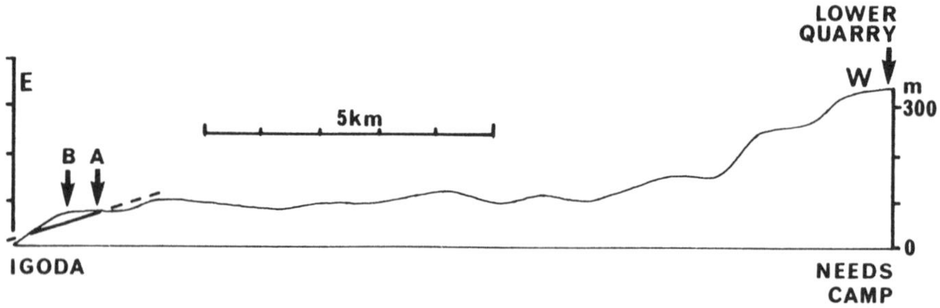


Figure 141. Smoothed topographic section (20 points) between the Igoda and Lower Quarry Needs Camp outcrops. The heavy line is the surface upon which the Igoda sediments rest. It tilts at 1 in 28.57 (to E). Datum is sea level. See Figure 140 for localities.

which is matrix-supported with well-rounded (up to 2 cm) small pebbles of quartzite, has a glauconite/calcite matrix, and contains small shell fragments. It passes upwards into the arenaceous limestones and calcareous sandstones which are both highly glauconitic with a shelly fauna dominated by osterids. These glauconitic beds are overlain by a conglomerate at the base of the Tertiary Alexandria Formation.

Klinger & Lock (1978) describe quite a varied macrofauna from the Igoda Formation, which includes corals, echinoderms, three species of ammonite, gasteropods, six species of lamellibranch, polyzoa, and four species of brachiopod. The lamellibranchs are the same as those found at Mountain's (1974) exposure, and the ammonites *Baculites subanceps* Haughton, *Saghalinites* sp. and *Eupachydiscus* sp. indicate a Campanian/Maastrichtian age. The brachiopod *Eolacazella affine* (Bosquet) also occurs at the Lower Quarry of Needs Camp. The ostracod fauna is likewise much more diverse than that from Needs Camp (Dingle 1981) and so far 16 species have been recognized with the cytheracean taxa dominant. Some species (e.g. *Hermanites kennedyi, ?H.* cf. *arcus*) provide links with late Campanian/early Maastrichtian faunas in Zululand, but significantly the three most abundant cytheraceans (*Cythereis transkeiensis, Pondoina igodaensis,* and *Brachycythere longicaudata*) indicate closest ties with the Umzamba fauna.

Although the Needs Camp and Igoda lithologies are somewhat different, they do have in common the fact that they are both carbonate rocks and within the limits of faunal determinations, are approximately contemporaneous. In the Igoda cliff, the surface on which the Upper Cretaceous strata rest dips seawards at 1 in 28.57. This surface projected inland to the vicinity of the Lower Quarry at Needs Camp gives a height well in excess (600 m) of the height at which the outcrop occurs (367 m) (Fig.141), suggesting that there has been flexing of the present coastal region since Upper Cretaceous times.

### 5.2.5 *Palaeogeography*

The main Upper Cretaceous depocentres of the east coast lay in the Natal Valley and Zambezi Graben/Mozambique Basin, and were supplied with sediment by the Tugela and Limpopo, and Zambezi rivers, respectively. These two areas were effectively separated by the Mozambique Ridge, which acted as a buoyant axis and a submarine barrier under the present day onshore and offshore areas, respectively. Proximal parts of these basins extended over present day land areas because the extent of the shallow shelf seas was very sensitive to any eustatic, tectonic or oceanographic fluctuation, as was the crest of the Mozambique Ridge. In addition, local areas continued to experience differential crustal subsidence which led to subtle facies and thickness changes across and around them. Micropalaeontological studies in south Zululand (Dingle 1980c, 1981) have enabled the late Cretaceous history of one such area to be documented (Figs 142 & 143). This lies at the southern end of the Lebombo line, where an arcuate fault zone crosses the coast east of Empangeni in the vicinity of Richards Bay. Archaean basement gneisses and granites separate the downfaulted Mesozoic volcanics (Lebombo Group) to the north from the Tugela Cone offshore basin which is underlain by Palaeozoic sediments. Except for a period of subsidence in the Campanian, this Archaean block (Richards Bay Arch, Fig.145) remained relatively buoyant and exhibited shallower water facies and thinned sedimentary sequences compared to areas lying north and south (Fig.143). Specific examples will be mentioned below. Similar areas are suspected across the Zandamela high in southern Mozambique, but insufficient data are available to document their history accurately.

Despite the differing sedimentary environments that prevailed in various parts of Zululand, microfossil data (Dingle 1981) have also allowed a local and regional Upper Cretaceous sea level curve to be constructed for the northern Natal Valley and these will be used to illustrate the discussion that follows.

To describe Upper Cretaceous palaeogeographies in this area we have chosen two periods which typify regressive and maximum transgressive situations: late Cenomanian-Turonian and late Campanian, respectively. During the earlier period, Zululand and the Mozambique Ridge (Fig.145) experienced non-deposition whilst distal areas of the Natal Valley underwent an

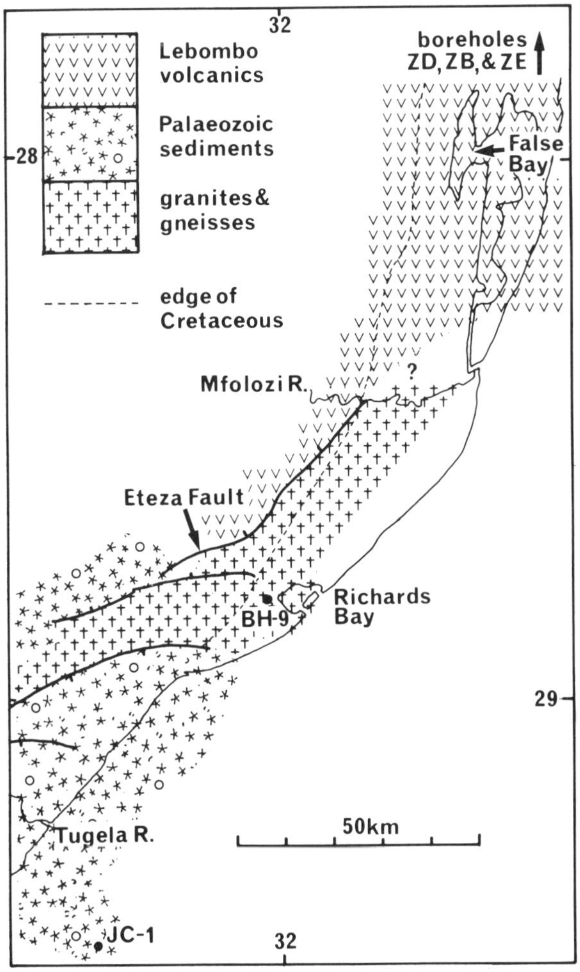


Figure 142. Pre-Cretaceous basement rock types between J(C)-1 and False Bay, Zululand. The Richards Bay Pre-Cambrian metamorphic block is separated from the Lebombo volcanics by the Eteza Fault, and from the Palaeozoic sediments of the Tugela coastal area by several faults, the largest of which crosses the coast in the vicinity of the major coastline deflection (29°S). Borehole J(c)-1 penetrated into ?Dwyka diamictite, and Table Mountain Group quartzites.

environment change that is recorded in the regionally developed seismic reflector McDuff. During this episode, which commenced in late Cenomanian times, there is little evidence for large scale erosion of underlying sediments, but rather the establishment of vigorous bottom currents which prevented sedimentation, and in Zululand led to the development of extensive hard grounds. This may have been caused by a slight lowering of sea level which created a very shallow shelf sea over Zululand. Local crustal instability is suggested for the area south of the Richards Bay Arch (Fig.145), because in the J(c)-1 borehole there is a Cenomanian-Turonian transgression across the nearshore Palaeozoic shelf area (Figs 132 & 143a). South of Durban, the coastline probably lay in the vicinity of the present continental edge, where a straight rocky coastline extended to the vicinity of the Algoa Basin, a distance of about 700 km. Although we have equated reflector McDuff with hiatuses in Zululand and on the Mozambique Ridge (Fig.107), we in fact do not have substantive evidence that the physical discontinuities (e.g. reflection coefficient) across this horizon necessarily represent an erosive, or hard ground effect. They could have resulted from a temporary increase in sedimentation brought about by shelf currents sweeping shallow water detritus into a quieter, deeper depocentre.

Although these erosive/non-depositional situations affected the SW of Mozambique (SW of Maputo) and the lower Zambezi Graben, they appear to have had no noticeable influence on the Domo depocentre that lay along the eastern side of the Mozambique Ridge extension. In Figure 145 we have shown a narrow zone of shallow water sediments (Maputo Formation) separating the Domo basin from the areas to the west. Flores (1973) is ambiguous on the question, but it seems unlikely that continental (lacustrine and fluviatile) Sena and Boane lithofacies pass directly into deep water anoxic Domo shales. Because there are no outcrops, only borehole data can provide the evidence, and this has not been published in detail. In Figures 99 and 131 we have queried the connection. A further possibility for the nature of reflector McDuff is that the Domo lithofacies originally extended into the northern Natal Valley during Albian to Cenomanian times, fostered by poor circulation in the basin compartments created by the Almirante Leite and Naudé ridges. Increased aeration in the water column caused by the intensification of current activity in the northern Natal Valley in late Cenomanian to Turonian times (and perhaps accompanied by increased terrigenous sediment input) brought this sedimentary environment to a close. If this was the case, then McDuff is analogous to horizon AII in the SE Atlantic (although considerably younger). The likelihood of shallow water sands being injected into the northern Natal Valley from the Limpopo and Tugela cones leads us to believe that the mid-Cretaceous sediments of this area are interesting hydrocarbon prospects.

In deeper waters off SE Africa, we have no data on Turonian lithofacies, and can only predict from ocean crust sinking curves and published carbonate compensation level (CCL) fluctuations, that the whole of the southern Natal Valley probably lay in deep water below the CCL, but that the Agulhas Plateau lay above it (Dingle & Camden-Smith 1979) (Fig.145).

An eustatic sea level rise had commenced by late Turonian times (Fig.144), with the result that a regional transgression advanced west and SW into Zululand (working its way progressively southwards onto the Transkei Swell), and also over the present Mozambique coastal plain (Figs 132 & 144). The effect of this was

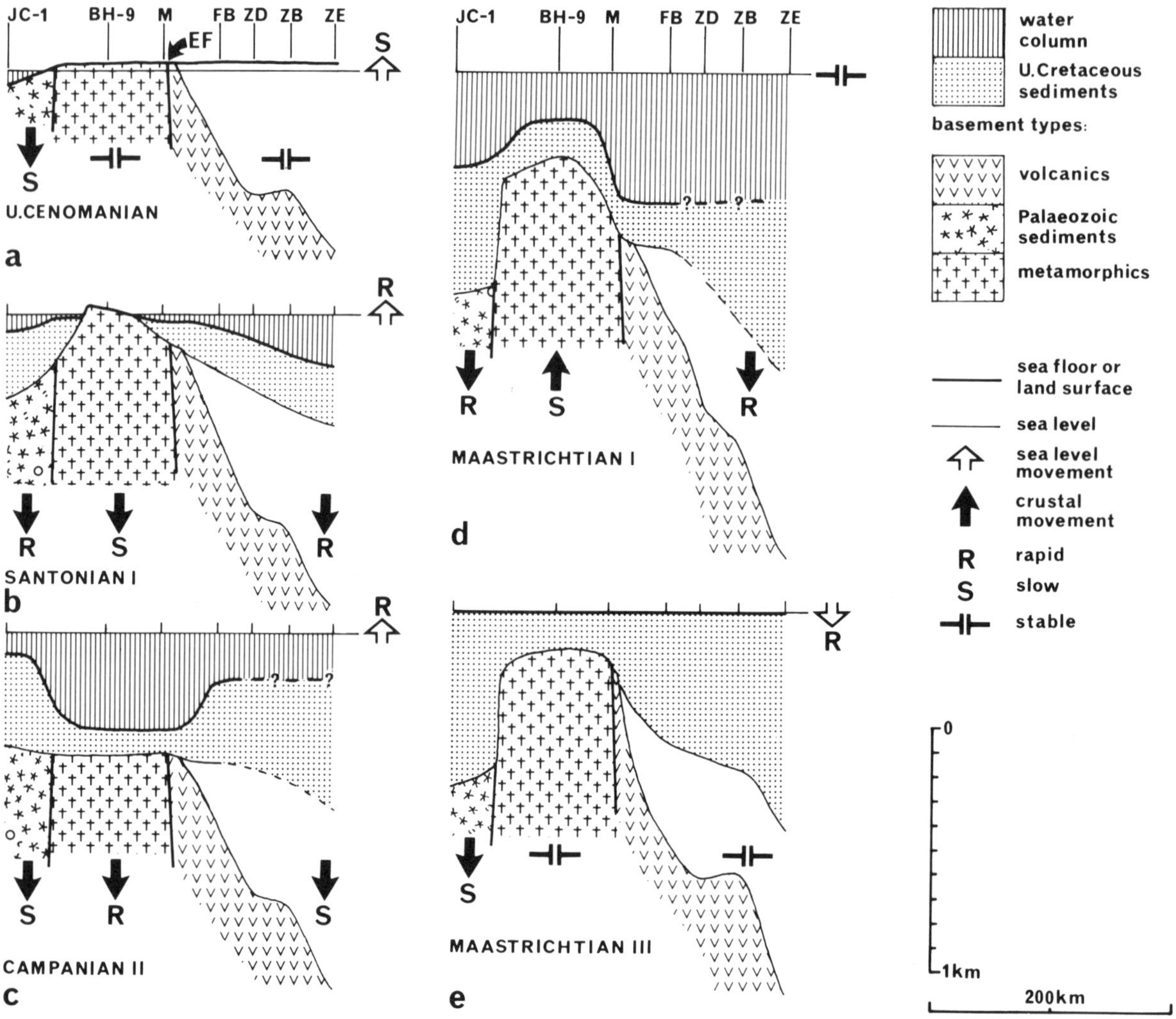


Figure 143. Schematic and partly speculative relationship between sea level and crustal movements during Upper Cretaceous sedimentation in S.Zululand. In a-d the datum is local sea level, in e it is top Maastrichtian. The various configurations of the top of the Richards Bay basement high denote tilting. The sequence of events is discussed in the text, but in summary they indicate that relative crustal movements controlled the local sedimentary environments. Figures 132 and 142 show localities: EF = Eteza Fault; M = Mfolozi; FB = False Bay; JC-1, BH-9, ZD, ZB, ZE = boreholes. Based on data from Dingle (1981), and McLachlan & McMillan (1979).

to alter the hydrodynamics of the area to the extent that sedimentation could again proceed in Zululand (St Lucia Formation), and with its progressive rise, on the Mozambique Ridge (Campanian) (Fig.146). At the same time, more efficient bottom water circulation in eastern Mozambique dispelled the anoxic conditions responsible for the dark Domo shales, which were replaced by well aerated Grudja sediments. Differential subsidence along the Chissenga Graben, at the SW end of the Zambezi Graben (Fig.136) resulted in the diversion of the Zambezi River and high sedimentation rates in the main Grudja trough, where argillaceous lithofacies dominated, whilst sandier sediments obtained to the north and over much of the present Mozambique coastal plain (Fig. 146).

Only moderate sedimentation rates occurred over the Zululand shelf, where periodic strong bottom currents gave rise to cyclic hard ground formation. Maximum sedimentation in this area lay in a zone along the present day continental shelf edge. Farther east, in the deep basin, rates are unknown, but probably low. Apart from the main Grudja trough (83 m/my) the highest known sediment accumulation rates in the east coast area were under the Tugela Cone (54 m/my). Several spurts in the sea level rise have been recognized (late Coniacian/early Santonian, and early and middle Campanian) (Fig.144), and the transgression reached its high point in late Campanian/early Maastrichtian times when the Pondoland and East London areas were temporarily inundated. In the former, a relatively high energy sedi-

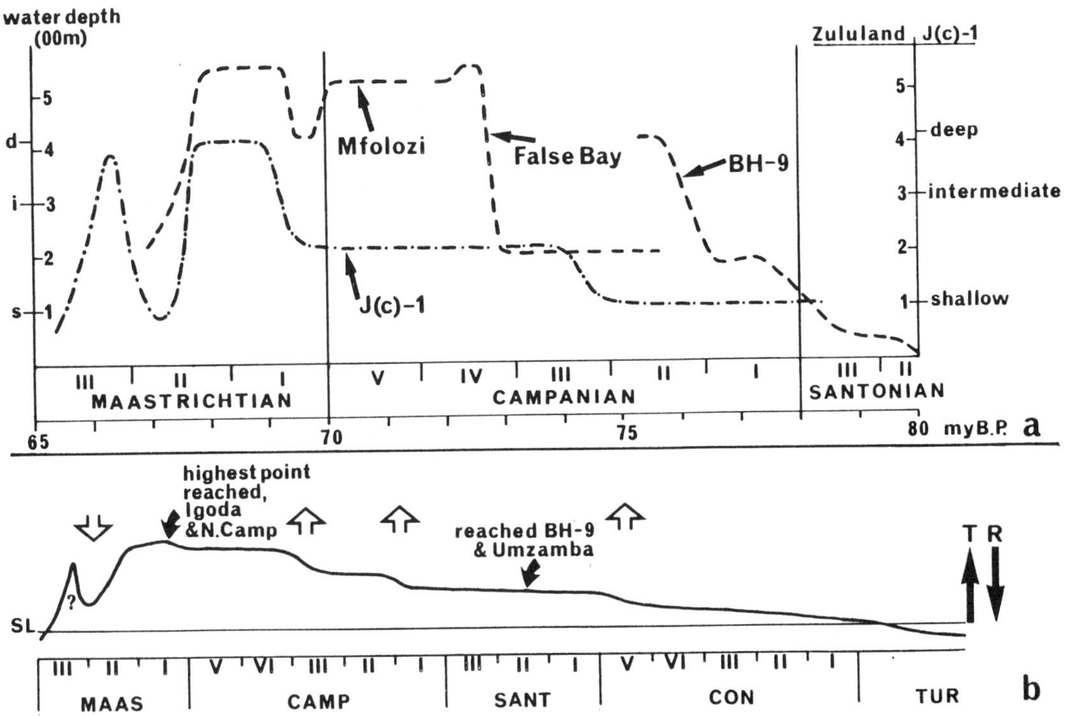


Figure 144. Upper Cretaceous sea level movements in Zululand.
A. Water depth fluctuations for various localities based on changes in ostracod faunas.
B. Integrated sea level curve for S.E.Africa during the Upper Cretaceous transgression. Bold arrows indicate regression (R) and transgression (T), open arrows indicate main pulses of sea level movement. Sea level (SL) is for present day. There is no vertical scale but the highest point reached by the transgression (above present) may have been between 330 m (present height of Needs Camp) and 460 m (water depth of 550 m above depth to base of Campanian in BH9, −87m).
Data mainly from Dingle (1980c, 1981).

mentary environment was established, producing a coarse basal conglomerate with terrestrial vertebrate remains and logs, followed by muddy sand and shelly limestones with shell banks and pebbly horizons. Open connections to the sea are attested by the abundant planktonic foraminifera and ammonites. Farther south, a quieter environment formed around East London, where carbonate sediments were laid down: polyzoan arenites nearshore, and sandy chalks farther away from the coast at Igoda. Again, the presence of ammonites indicates a connection between the Umzamba area and the open sea.

Sea levels began to fall at the end of Maastrichtian I times (Fig.144) and dropped dramatically during Maastrichtian II and III, although microfaunal evidence from the J(c)-1 borehole suggests a temporay rise in early Maastrichtian III (Dingle 1981). This massive regression is registered throughout the area in non-sequences and coarse lithofacies, even at the moderately deep water J(c)-1 and Mozambique Ridge (DSDP 249) sites, although we have not been able to equate any well-developed reflecting horizon in the northern Natal Valley with its effects. The resultant changes in coastal configuration probably altered surface and bottom water circulation patterns, and Stapleton (1975) claims that a marked cooling of sea water temperatures is reflected across the Maastrichtian-Danian hiatus in the Richards Bay area.

Details of the interplay caused by the progressive sea level rise during the Upper Cretaceous transgression, and differential subsidence athwart the Richards Bay Arch can be examined by comparing successive microfossil assemblages from the Richards Bay borehole which lies on the crest of the arch, with those in the J(c)-1 borehole and at outcrops in Zululand (Dingle 1981). The Archaean arch and areas to the south (sites of BH9 and J(c)-1 respectively) had remained buoyant features during earlier phases of Cretaceous sedimentation in Zululand, but whereas the Palaeozoic basement under J(c)-1 had been flooded by the sea in Cenomanian-Turonian times, the earliest marine sediments to be deposited over the Richards Bay Arch were Santonian II. These are overlain by 80 m of silts and clays that can be divided into five assemblages based on their ostracod faunas, each one representing a different sedimentary environment. The lower three assemblages indicate shallow water conditions (probably < 100 m), the bottom two of which represent somewhat restricted (poor cir-

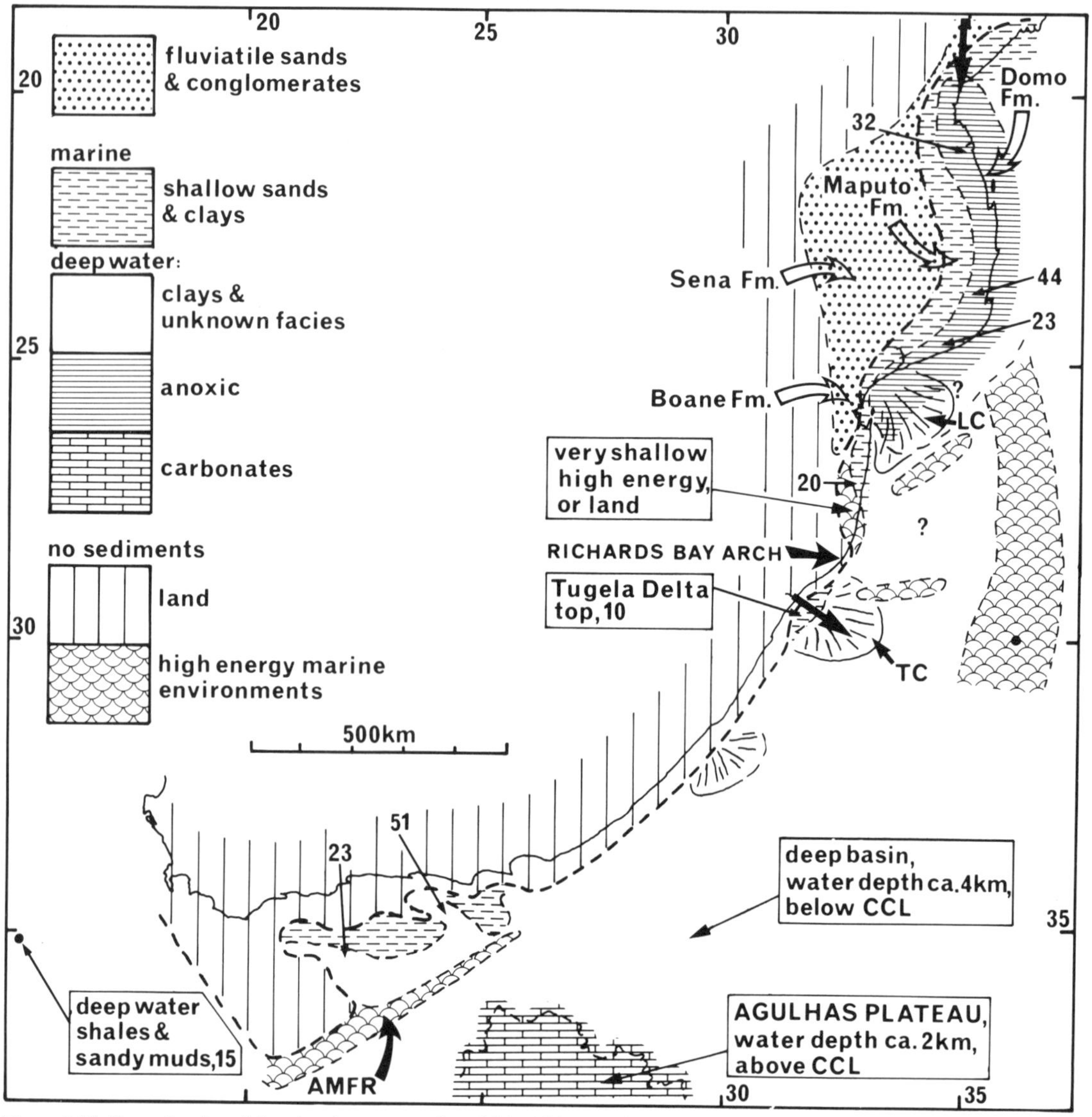


Figure 145. Turonian (say 86 my) palaeogeography of S.E.Africa. Sedimentation rates at various sites are in m/my, coastline is heavy dashed line. Richards Bay Arch refers to the basement high identified in Figures 142 and 143. DSDP sites are shown by dots. See text for discussion. Data from numerous sources cited in the text. Abbreviations: AMFR = Agulhas Marginal Fracture Ridge; TC = Tugela Cone; LC = Limpopo Cone; CCL = carbonate compensation level.

culation) environments. Assemblage 4 (Campanian I) reflects a distinct deepening of the water column which progresses into assemblages 5 (Campanian II) that has typical deep water (200-500 m) elements characterized by complex shell architecture and blindness. A particularly interesting genus (*Unicapella*) in this outer shelf assemblage is a close relative of a deep water (1 500 m) genus (*Paleoabyssocythere*) that has only previously been recorded from Santonian sediments on the continental margin of the Sao Paulo Plateau in the western South Atlantic. Ostracod assemblages of a similar age (Campanian II) on the Tugela Cone, and farther north in Zululand, represent somewhat shallower water environments than those on the Richards Bay Arch, which can only mean that the arch had sunk relatively faster than the adjacent areas, even assuming they experienced higher sedimentation rates (Fig.143c). Three short-lived fluctuations in the faunal makeup indicate the temporary re-establishment of shallow-water, higher energy environments, but overall the trend is of a progressive increase in water depth as the transgression moved SW. In the Richards Bay area, therefore, we have evidence that between Santonian II and Campanian III the sedimentary environment developed from sub-littoral to outer shelf (i.e. 0 to 200-500 m) over a period of about 5 my (i.e. a net sea floor subsidence rate somewhere between 40 and 100 m/my), which was faster than the local sediment supply. The thinness of the Maastrich-

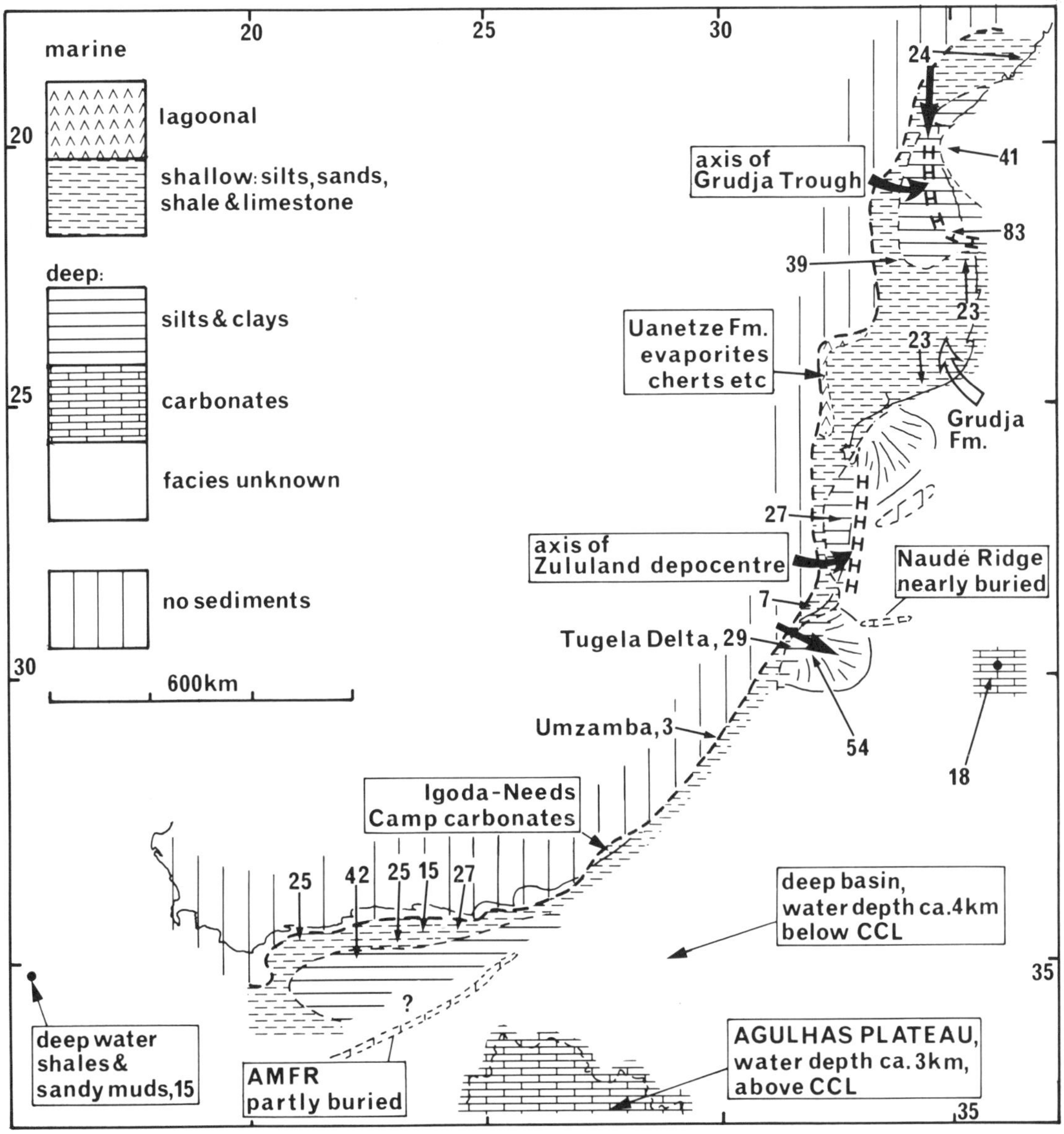


Figure 146. Upper Campanian (Camp V, 70 my) palaeogeography of S.E.Africa. Sedimentation rates at various sites are in m/my, coastline is heavy dashed line. DSDP sites are shown by dots. See text for discussion. Data from numerous sources cited in the text. Main depocentres offshore Zululand and in the Grudja Trough are shown with H. Abbreviations: AMFR = Agulhas Marginal Fracture Ridge; CCL = carbonate compensation level.

tian sediments at BH9 indicates a substantial slowdown in sedimentation rates, which we interpret as a relative shoaling in Maastrichtian I. Adjacent areas, however, show increased water depths and sedimentation rates for Maastrichtian I (e.g. 22 m/my Campanian to 47 m/my Maastrichtian at J(c)-1, McLachlan & McMillan 1979), but they all experience a shallowing in Maastrichtian II as the sea levels began to fall in the end Mesozoic regression.

Our data are rather scanty for a consideration of Turonian palaeoclimatic conditions. Upward construction of the Tugela Cone, with expected high sedimentation rates in distal areas, as well as the coarseness of fluviatile sediments of the Boane Formation indicates high runoff rates, and an ample supply of clay size debris suggests deep weathering. For Campanian times we have more data. A continued ample supply of fine detritus suggests a well-watered, deeply weathered hinterland, and the abundance of plant matter and logs in both the St Lucia and Umzamba rocks indicates extensive vegetation on the coastal strip. The low sedimentation rates suspected in the northern Natal Valley suggest that most of the mud brought down by the rivers was deposited adjacent to the wide, shallow shelf over the

Table 41. Vertical distribution of cytheracean ostracod genera in the Coniacian-Maastrichtian of Zululand.

| | Coniacian | Santonian | Campanian | Maastrichtian |
|---|---|---|---|---|
| *Krithe* | | | ● | ● |
| *Xestoleberis* | | | ● | ● |
| *Klingerella* | | | | ● |
| *Pariceratina* | | | | ● |
| *Pondoina* | | ● | ● | ● |
| *Brachycythere* | ● | ● | ● | ● |
| *Pterygocythere* | | | | ● |
| *Ponticulocythere* | | | | ● |
| *?Eucytherura* | | | ● | |
| *Pedicythere* | | | | ● |
| *Cytheropteron* | | | | ● |
| *Amphicytherura* | ● | ● | ● | ● |
| *Apateloschizocythere* | | | ● | |
| *Cythereis* | ● | ● | ● | ● |
| *Haughtonileberis* | ● | ● | ● | ● |
| *Oertliella* | ● | ● | ● | ● |
| *Hermanites* | | | ● | ● |
| *Trachyleberis* | | | ● | ● |
| *Unicapella* | ● | ● | ● | ● |
| *Dutoitella* | | | ● | ● |
| *Gibberleberis* | ● | ● | ● | |
| *Parvacythereis* | | | ● | ● |
| *Rayneria* | ● | ● | ● | |
| *Curfsina* | | | ● | |
| *Cativella* | | | | ● |
| | 8 | 9 | 18 | 20 |

present day Zululand-Mozambique coastal plain, although the high clay content of the Campanian/Maastrichtian strata on the Mozambique Ridge indicates that a good deal of fine terrigenous material was carried offshore by surface currents. Only in the south (around East London) is there not evidence for high runoff rates and sediment loads, but rather quiet, shallow to medium water depth rocky marine substrata, backed by a low hinterland.

## 5.3 WEST COAST (WALVIS RIDGE TO AGULHAS FRACTURE ZONE)

Epeirogenic subsidence of the continental margin off SW Africa during Upper Cretaceous and Cenozoic times affected the whole area and resulted in the development of a large sediment lens between the Walvis Ridge and the Agulhas Fracture Zone. During the Upper Cretaceous its axis shifted slowly seawards as material built upwards and prograded away from the coast, with important depocentres marking the sites of the old Walvis and Orange basins. The regional scale of sedimentation however, blurred the boundaries of the original depocentres, with the result that the 2 km isopachs, which define their thickest parts, almost join, and extend much farther west than their Jurassic-Lower Cretaceous equivalents (compare Figs 66 & 124). From Figures 120, 121 and 124, it is clear that localized tectonic control played little or no part in determining the geometry and development of the Upper Cretaceous depocentres, beyond initial infill of the previous irregular basin outlines.

For the purposes of this presentation, Upper Cretaceous strata on the west coast have been defined as those lying between the two regionally-developed reflecting horizons Atlantis II and Davie, which were mapped by Emery et al (1975), and dated by the DSDP at site 361 as Lower Albian and top Maastrichtian, respectively (Bolli et al 1978).

### 5.3.1 *Stratigraphy*

Substantive stratigraphic control for the west coast Upper Cretaceous stratigraphy is limited to DSDP site 361, the commercial boreholes in the Orange Basin area, and two dated outcrops between the Orange River and Lüderitz (Bogenfels onshore and a shallow borehole just off the coast). The locality at Kangnas consists of poorly-dated continental sediments.

P-wave velocity discontinuities from seismic refraction and wide angle reflection work (Hoskins et al 1974, Rabinowitz 1976), together with Emery et al's (1975) mapping of AII and Davie allow a crude acoustic stratigraphy to be developed for the west coast offshore that shows that the Upper Cretaceous strata form a regionally asymmetric lens-shaped body, with its maximum thickness in the vicinity of the present shelf break. It thins gradually westwards, where it rests directly on ocean crust (beyond the limits of sections in Figures 120 and 121), and thins rapidly eastwards where it underlies the continental shelf. Over the old buoyant highs (Lüderitz and Columbine/Agulhas arches), Upper Cretaceous strata rest directly on basement and progressively lap across it, and at one location (Bogenfels), 150 km NW of the Orange River, extend onto Pre-Cambrian rocks onshore. In the original taphrogenic basins (Walvis and Orange), they overstep the Lower Cretaceous rocks to rest directly on basement close to the present coast, where they form a narrow sea floor outcrop between a coastal rocky belt and the eastern feather edge of the Tertiary cover. South of Cape Columbine, Tertiary sediments overstep the Cretaceous, so that Mesozoic outcrops occur at only two places between here and Cape Agulhas.

Maximum thicknesses for Upper Cretaceous sedi-

Table 42. Similarity at generic level between ostracod faunas at various localities in Gondwanaland and South Africa (from Dingle 1981).

| | Campanian | Maastrichtian |
|---|---|---|
| Argentina | non-marine | Lower - 5 %, Middle - 14 % |
| West Africa | 9 % | 7 % |
| East Africa | 14 % | 17 % |
| West Australia | 27 % | no data |

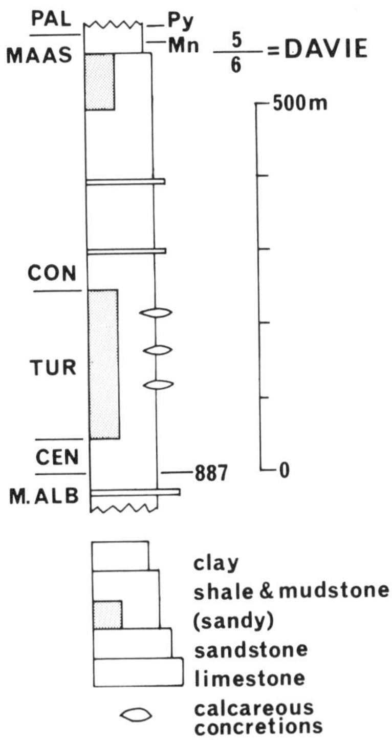


Figure 147. Upper Cretaceous sediments at DSDP site 361. Compiled from core log descriptions in Bolli *et al.* (1978). Boundary between lithological units 5 and 6 (shipboard terminology) corresponds to the regional seismic reflector Davie (D) of Emery *et al.* (1975). The Coniacian-Maastrichtian section was not subdivisible palaeontologically. Abbreviations: Py = pyrite; Mn = manganese nodule. See Figure 121 for location.

ments occur in the Walvis Basin (estimated 4.5 km) and Orange Basin (>3 km), whereas along the western edge of the Lüderitz and Agulhas/Columbine arches about 2 km is the maximum.

5.3.1.1 *Orange Basin*

Upper Cretaceous sediments in the Orange Basin constitute a wide, thick lens that straddles the COB, and whose 1 km isopach extends over 500 km SW of the Orange River mouth. The geometry of this large deposit is more regular and its isopach gradients smoother than its Jurassic-Lower Cretaceous counterpart (compare Figs 66 & 124), indicating a more extensive dispersal of detritus from the main input point and onlap of significant sedimentation onto parts of the Lüderitz and Columbine/Agulhas arches. Nevertheless, the main depocentres remained in the Walvis bight, and adjacent to the Orange River where it extends as an elongate zone which runs approximately parallel to the coastline between 70 and 100 km westward of the outer edge of the Central Ridge.

In common with the Jurassic-Lower Cretaceous succession in the Orange Basin, the Upper Cretaceous rocks have been studied in both the distal (deep ocean) and proximal (continental margin) locations. In the former, at DSDP site 361, Bolli et al's (1978) work has allowed the dating of regional seismic reflectors AII and Davie (earliest Albian, and late Maastrichtian/early Palaeocene, respectively), and it is the sediments that lie between these two seismic events that we take as constituting Upper Cretaceous in this area. Both AII and Davie can be traced onto the continental margin (Fig.116), where they have been equated by SOEKOR geologists (Gerrard & Smith 1980) with regionally-developed reflectors P and L, respectively. Both are unconformities: the former cutting across early Aptian red beds in the most proximal areas; and the latter cutting across progressively older horizons eastward to rest on Albian near the present coast. Several additional, but less important reflectors have been identified, the most significant of these representing a Turonian non-sequence (reflector N of Gerrard & Smith 1980).

Figures 120 and 124 show a maximum thickness of just over 3 km for Albian-Maastrichtian sediments in the Orange, and 4.2 km in the Walvis basins, with the 1.7 km isopach outlining a continuous deposit along the length of the west coast. Recently-published SOEKOR data show considerably more than this in the Orange Basin (3.5 sec, ca.5 km), and somewhat less (2.0 sec, ca.3 km) in the Walvis Basin, with the main depocentre in the latter displaced to the north of that in Figure 124. Figures 120 and 124 were constructed mainly from seismic refraction and wide-angle reflecting interval velocity/depth calculations, and clearly rocks of higher velocity lower in the sequence have been assigned to the Upper Cretaceous by Gerrard & Smith (1980) than in Figure 124. Additionally, we have designated over 1.5 km of the upper sediment column in the Orange Basin to the Tertiary, whereas Gerrard & Smith (1980) place horizon L at about 900 m sub-seafloor. This accounts for about 600 m of the discrepancy. Obviously their data are more complete, and presumably more reliable, and in Figure 148 we have included for comparison, a conversion of Gerrard & Smith's (1980) isochron map for the reflector P-L interval.

We will first deal in detail with the deep-water succession before considering the continental margin stratigraphy. A complete Upper Cretaceous sequence occurs at DSDP site 361, where ca.660 m of shales with minor sandy mudstones (Unit 6) rest conformably on the anoxic Lower Albian to Lower Aptian sediments of Unit 7 (Figs 116 & 147). This lower contact lies about 35 m above seismic reflector AII, whilst the Mesozoic/Cenozoic boundary occurs within an Upper Maastrichtian/Lower Palaeocene deep water clay which corresponds to seismic reflector Davie of Emery et al (1975) (Bolli et al 1978). The Upper Cretaceous shale sequence (Unit 6) contains minor sandy mudstones (near the top), one marly nannofossil chalk (near the base), and the two beds of sandstone (Fig.147). The shales are

Table 43. Microfaunal zonation for Kudu 9A-1 borehole, Orange Basin (amended after McLachlan & McMillan 1979 from data in Gerrard & Smith 1980).

| | |
|---|---|
| Maastrichtian | *Globotruncana arca*<br>*Rugoglobigerina rugosa*<br>*Globotruncana tricarinata* |
| Campanian | *Globotruncana ventricosa* |
| Santonian | *Globotruncana coronata*<br>*Globotruncana carinata* |
| Coniacian | *Globotruncana renzi*<br>*Valvulineria* sp. |
| Cenomanian | *Praeglobotruncana helvetica*<br>*Coronotalites aptiensis* |
| Albian | *Tritaxia tricarinata*<br>*Osangularia californica* |
| Aptian | *Epistomina chapmani* |

alternatively grey or greenish black and red, a character which Bolli et al (1978) attributes to environmental or diagenetic cyclicity, and are distinctly parallel or cross-bedded with fine sand and silt laminae, with minor convolute and contorted layers. Calcareous, and rare pyrite concretions and pyrite-rich laminae occur in the lower half of the succession, and there appears to be an antithetic relationship between the number of concretions and the pyrite content and associated laminated bedding. Evidence of bioturbation is scattered throughout the section. An average modal analysis of the shales is clay – 69 %; quartz – 21 %; zeolite – 2 %; iron oxides – 1 %; trace components – 7 % (Bolli et al 1978). The trace components (usually less than 1 % each) include: zircon, epidote, garnet, tourmaline, opaques, monzanite(?), barite(?), apatite, piedmonite, leucoxene, sphene, pyrite, biotite, chlorite and staurolite(?). This suggests a predominantly granitic provenance with some admixture of basaltic and/or andesitic components.

Although the organic carbon values (over 0.1 %) of Unit 6 (Lower Albian to Maastrichtian) indicate significantly better oxygenated bottom conditions than obtained during the deposition of Unit 7, there is palaeontological evidence that the sea floor may have still been slightly anoxic, and was certainly below the carbonate compensation level. Autochthonous planktonic foraminifera have not been recorded, because the two assemblages noted by the DSDP team at 366 and 955 m (Campanian mixed with Aptian, and Lower Aptian, respectively) were considered reworked (Bolli et al 1978). Calcareous nannofossils are represented only by rare, small specimens of *Watznaueria*. The basal 400 m of Unit 6 (Lower Albian to ?Coniacian) has a primitive, sparse fauna of dwarfed agglutinated foraminifera which suggests that bottom conditions were not conducive to the establishment of diverse communities. Taxa recognized are: *Bathysiphon, Ammodiscus, Glomospira* and *?Trochammina.* In contrast, spores, pollen and phytoplankton are abundant and varied, with the terrestrial element indicative of a favourable climate for a diverse vegetation. It is on the basis of the phytoplankton that the Upper Cretaceous was provisionally zoned (Fig.147) although subdivision was not confident enough to assess fluctuations in sedimentation rates, and an average value was calculated at 15 m/my for the whole of the Upper Cretaceous, a significant decrease compared to that of Unit 7 (Bolli et al 1978).

DSDP site 361 lies on the distal edge of the Upper Cretaceous sediment lens where, according to seismic records, stratification is approximately horizontal. Moving eastwards towards the Orange Basin, individual beds gradually thicken and under the present-day continental slope and rise they pass into strongly prograding sets on the outer edge of a delta-like structure which built upwards and outwards on the ocean side of the Lower Cretaceous sediment wedge (Fig.120c). In the Orange Basin, a maximum thickness of >3 km can be measured from seismic data for the sediments between reflectors AII and Davie. Farther east, they thin rapidly and smooth out irregularities in the basin floor inherited from the underlying taphrogenic fill, before pinching out against the inshore basement platform, which is usually <10 km wide (it varies from about 40 km in St Helena Bay, to <4 km northwest of the Olifants River).

Under the present continental shelf, Upper Cretaceous strata have been intersected in several SOEKOR boreholes (McLachlan & McMillan 1979, Gerrard & Smith 1980) where they form a large composite prograding lens between reflecting horizons P and L (see Fig.118). The latter is a Palaeocene hiatus which cuts across Maastrichtian to Albian strata when traced eastwards. Upper Cretaceous sediments reach a thickness of ca.2.9 km in borehole Kudu 9A-1 (Figs 117 & 118), within a total marine succession (Upper Aptian to Maastrichtian) 3.25-5 km thick (seismic refraction, and reflection/borehole data, respectively). The whole sequence consists of grey shales and clays, except across an Upper Cenomanian-Lower Coniacian hiatus which has several thick sandstones, and probably represents a marginal marine facies (no planktonic foraminifera were available here to effect a zonation) (Tab.43). Farther SE, borehole A-A1 probably has Coniacian at the top, and is in a paralic facies for the whole sequence above the Aptian/Albian red beds. In the southernmost part of the basin on the present continental shelf, the upper part of composite borehole Ba-A1 & A2 may be as young as Cenomanian, and consists of continental sandstones and minor shales. These boreholes are important because they show that normal marine conditions had not been established in the SE part of the present continental shelf area by Coniacian times, and

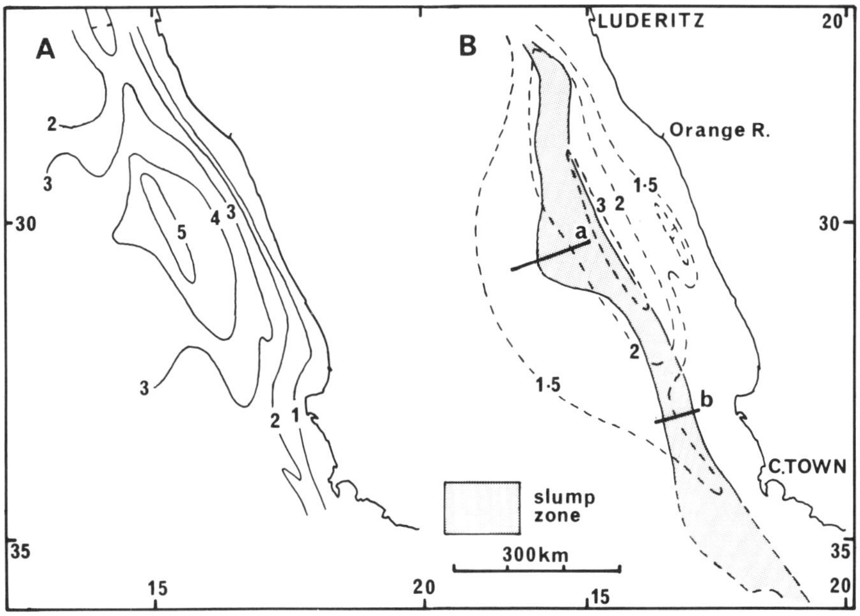


Figure 148. Isopachs for Upper Cretaceous strata in the Orange Basin.
A. Values converted from isochron data in Gerrard & Smith (1980) using a P-wave velocity of 2.97 km/sec.
B. Values from seismic refraction data used to construct cross-sections in Figure 120. Shaded zone is affected by large-scale slump structures. Lines show cross-section locations for Figure 149. Data from Emery *et al.* (1975), Lehner & de Ruiter (1977), Dingle (1979, 1980b).

that the Turonian-Coniacian regression recorded in the Outeniqua Basin, and along the east coast also affected the proximal Orange Basin.

McLachlan & McMillan (1979) give no details of the microfossil faunas of these boreholes, but planktonic foraminifera suitable for zonal purposes appeared in the Cenomanian, and the presence of *Rugloglobigerina rugosa* indicates that all but the top of the Maastrichtian is present under the Palaeocene unconformity (Tab.43). They imply that there is a regression at this level, but give no details of the Upper Maastrichtian lithofacies.

A feature of the depocentre SW of the Orange River is the large number of slumps and other post- and possibly synsedimentary structures on the outer (western) face of the Upper Cretaceous sediment lens (schematically shown in Figure 120c) (Dingle 1980b). These can be traced from seismic profiles in a swathe up to 120 km wide that stretches from south of Lüderitz in the north, to at least 36°S, a distance of at least 1 000 km (Fig.148). Seismic records show that slumping has occurred at two main horizons between the regional reflectors AII and Davie, with the oldest resting on horizon AII, presumably affecting sediments of immediately post-lowermost Albian age, whilst the younger is overlain by undisturbed reflector Davie (late Maastrichtian/early Palaeocene) (Fig.149). The base of the upper unit rests on a smooth surface that appears to have been deposited after the earlier episode of movement had been completed. Although these relationships do not allow a very refined dating scheme to be applied, we can say at a first approximation that the older unit is lower Upper Cretaceous and the younger is upper Upper Cretaceous in age. The boundary between these two sequences may correspond to the Turonian non-sequence recorded by Gerrard & Smith (1980). Both have a maximum thickness of about 1.5 sec (at least 1.5 km) and consist of huge cascade-like structures with large-scale down slope over thrusting onto the adjacent fold. Distances between individual folds vary from 3.5 to 15 km. At present it is not possible to say how much of this movement took place in a synsedimentary fashion (for example in the form of growth faults or diapiric structures) or as instantaneous slumps, but similar features have been described (on a somewhat smaller scale) from the modern Mississippi delta where extremely high sedimentation rates of 80 000 m/my have been reported (e.g. Coleman & Garrison 1977). Consequently we favour trigger mechanisms that were related to slope instability caused by rapid sedimentation rates and resultant high internal pore fluid pressures, which were caused by rapid dumping of terrigenous debris by the Orange River on its outer delta front. As we will discuss later, these slumps are but the earliest in a sequence indicative of seafloor instability that continued to ?Pleistocene times.

Very little data on the sedimentary sequences have been published from the Walvis Basin area. Deep penetration seismics have located reflector AII (Emery et al 1975) which, combined with seismic refraction results (Hoskins et al 1974) suggests that about 4.2 km of Upper Cretaceous strata occur between Walvis Bay and the bight of the Walvis Ridge. By the end of Upper Cretaceous time, the outer edge of the coastal basement platform was completely buried and a shelf edge had become established in approximately its modern position. No large slumps have so far been recorded from this area, which is joined by a wide, thinner (2 km)

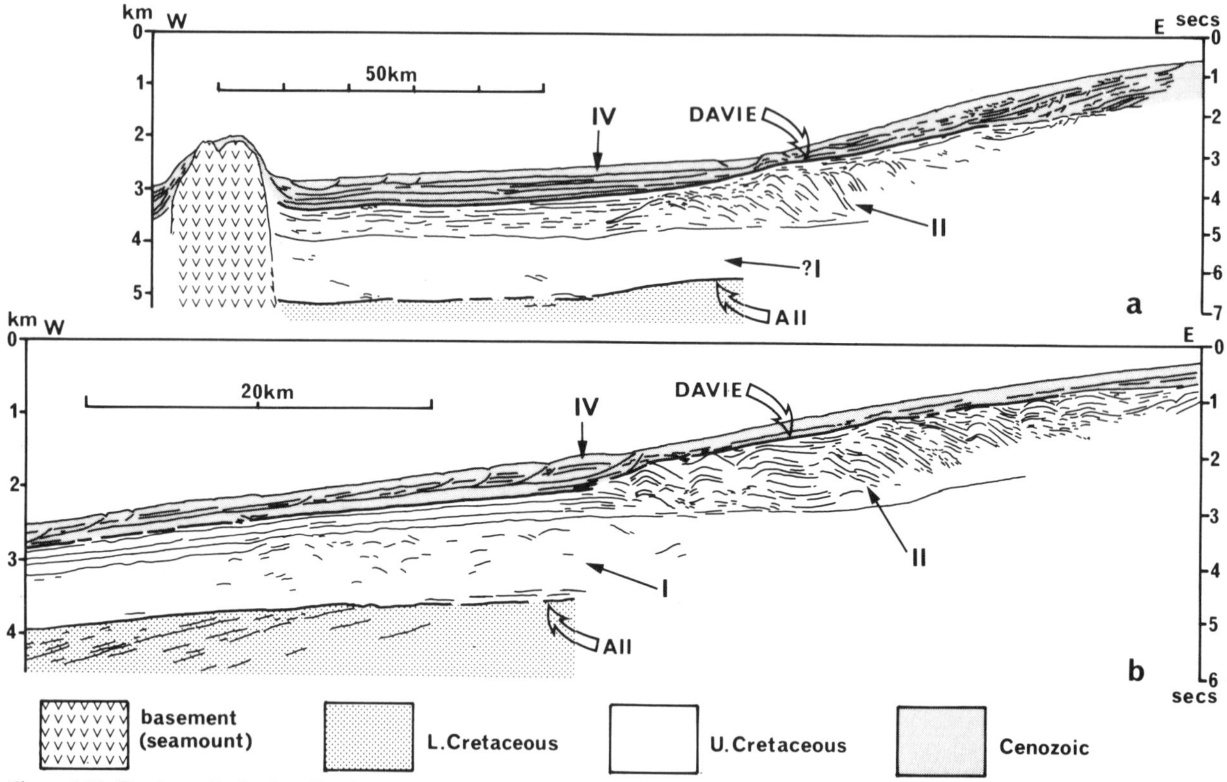


Figure 149. Tracings of seismic reflection profiles across slumped Upper Cretaceous and Tertiary strata NW of Cape Town. See Figure 148 for locations. Two periods of sedimentation/slumping are evident between regional seismic reflectors AII (early Albian) and Davie (uppermost Maastrichtian). These are designated I and II (see Table 66 for further information on west coast slump stratigraphy). Vertical scale in sec (right) and km (water depth) (left). Datum is sea level.

prism (adjacent to the Lüderitz Arch) to the Orange River depocentre.

5.3.1.2 *Bogenfels and Kangnas*

Upper Cretaceous sediments occur onshore in two areas: marine strata near Bogenfels on the south side of the Lüderitz Arch, and non-marine strata at Kangnas, about 100 km south of the Orange River (Figs 124 & 150).

The earliest record of Upper Cretaceous sediments from the Orange River-Lüderitz coastal belt was by Merensky (1910) who recorded Senonian molluscs presented to him by Mr Schettler, a mining company representative. The locality from which they were collected was not mentioned. In 1930 (a & b), Haughton described a small outcrop of Upper Cretaceous sediments from the vicinity of Bogenfels, and on the evidence of one ammonite, *Proplacenticeras merenskyi* (Haughton), suggested a Coniacian age. The vagueness of the locality data, and subsequent mining operations for diamonds which have substantially altered the surface morphology of the area, resulted in several unsuccessful attempts before Haughton's outcrop was finally re-located in 1975. Klinger (1977) has subsequently described the outcrop and fauna in detail. Although it is very small, this is the only substantiated onshore exposure of marine Upper Cretaceous strata south of Moçamedes in Angola, a distance of over 2 000 km, which further emphasizes the buoyant nature of the onshore west coastal margin.

The Bogenfels outcrop occurs on the northern side of a low, sand-filled depression (Langetal) about 2.4 km north of the abandoned mining village of Bogenfels in remarkably desolate and featureless terrain at 27.3842°S, 15.4056°E (Fig.150). It lies at about 60 m above sea level and consists of approximately 20 cm of material, which crops out over about 5 $m^2$. Klinger (1977) records that in the upper part of the sequence there is a hard, calcified shell bed crammed with valves of *Rhynchostreon suborbiculatum,* both fragmented and in life position, overlying a yellowish brown silty sand with heterodont lamellibranchs and *Rhynchostreon* near the top. The *Rhynchostreon* shell bed is overlain by several metres of unfossiliferous olive green silt. Haughton's (1930b) ammonite apparently came from the upper shell bed, and its good state of preservation suggests that it was not reworked. Klinger (1977) and Cooper (1974), on the basis of the ammonite and the similarity of the oysters with assemblages in Angola, suggest that the outcrop is Cenomanian in age (late Cenomanian according to Cooper). It is apparently overlain by Eocene strata, which crop out in the vicinity, but their

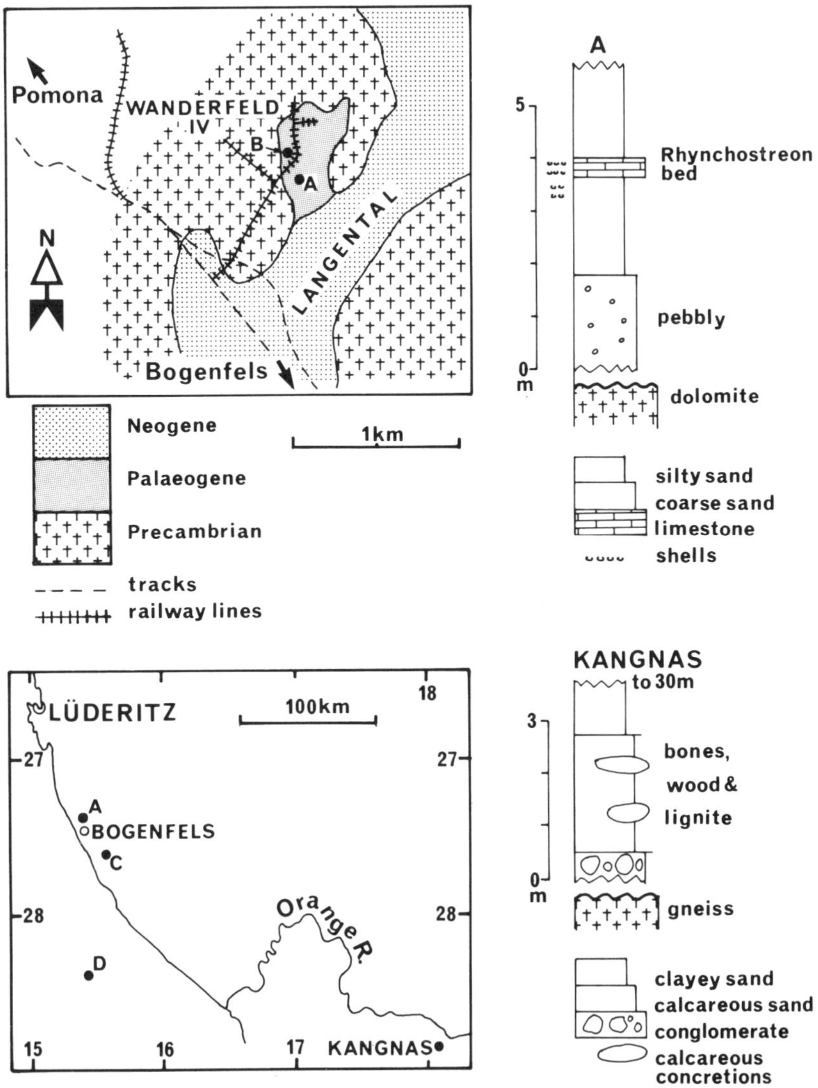


Figure 150. Upper Cretaceous outcrops in the coastal zone of S.W.Africa. Location map (lower left) shows outcrops at: A = Cenomanian of Wanderfeld IV near Bogenfels; C = suggested locality of Merensky's (1910) mid-Santonian-early Campanian assemblage; D = diamond exploration borehole with mid-Santonian-early Campanian assemblage (28°23'S, 15°25'E).

The Wanderfeld IV locality (A on map at upper left, measured section at upper right) is at 27°23.05'S). The locality map is after Klinger 15° 24.33'E (1977), and Siesser & Salmon (1979), and the section is after Klinger (1977). The Kangnas section has been drawn from Rogers' (1915) original account of the excavation. The dinosaur bones recorded by Haughton (1915) were recovered from the calcareous sand with concretions above the basal conglomerate. Kangnas is a farm at an altitude of 1 050 m, ca.80 km SE of the locality map (at 29°34.2'S, 18°21.0'E).

Locality B in the map at the top left is the type locality for the Langental beds (Eocene) (see Figure 167).

relationships are not accurately known, nor is the status of the feldspathic grit which immediately overlies the fossiliferous Cretaceous beds. Klinger (1977) formally designated the Upper Cretaceous strata at this locality the Wanderfeld IV beds after the name of the old mining locality in which it occurs.

Merensky's (1910) original fauna has not been redescribed (its whereabouts are not known), but Haughton (1930a) assumed that it was collected from outcrops overlying the ammonite-bearing strata at Wanderfeld IV, whereas Klinger (1977) cites Merensky's reference to Plumpudding Island to indicate that the collection was made from a locality south of the Bogenfels cliff on the coast. By comparison with the assemblages from Angola, Cooper (1974), who accepted Merensky's identifications of *Protocardia hillana* (Sowerby), *Turritella (Zaria) bonei* Baily and *T. (Haustator) meadi* Baily, suggests this 'lost' fauna is of Santonian age. There is a possibility, therefore, that Santonian sediments occur in the Bogenfels area, either directly overlying the Cenomanian strata, or from a locality nearby.

Sediments of mid-Santonian to late-Campanian age are also suspected offshore. Klinger (1977) reports that samples containing a specimen of *Sphenoceramus* aff *S.schmidtii* have been drilled (presumably during the course of diamond prospecting) at 28.383°S, 15.417°E, 49 km offshore and 95 km south of Bogenfels (locality D on Fig.150).

Non-marine Upper Cretaceous sediments are known from one locality on the west coast. At Kangnas, 140 km inland and 100 km S of the Orange River, dinosaur bones and teeth were found in an excavation about 32 m beneath the surface in sands and clays with calcareous concretions above a sub-angular breccia (Haughton 1915, Rogers 1915) (Fig.150). The fossils were not rolled and were found along with calcified and silicified wood fragments and thin (<2 cm) lignite streaks. Haughton (1915) erected a new genus on the basis of this material (*Kangnasaurus coetzii*) whose affinities with other Ornithopoda suggest a late Cretaceous age (confirmed by Dr Cluver, personal communication 1980). The Cretaceous strata occur as fill in a river channel that was cut into Namaqualand gneiss as part of a drainage system originally flowing northwards into the Orange River (via the present-day Henkries valley).

### 5.3.2 *Alkaline volcanism*

Upper Cretaceous volcanic activity was limited to the intrusion of large numbers of plugs of alkaline affinities along the west coast area (Namaqualand), and particularly in the interior (coinciding in the main to the Karoo Basin as previously defined) (Fig.168). These intrusions were part of a sequence of Lower Cretaceous-Tertiary emplacements, whose temporal relationships are reviewed in Chapter 7. The rock types involved can, in simplest terms, be classified as olivine melilitites, and kimberlites, but several field examples (the 'pseudo-kimberlites' of Moore 1979), as well as experimental evidence suggests that the two are genetically related and that kimberlitic magmas can be transformed into melilititic magmas by loss of volatiles such as $CO_2$ and $H_2O$ (see Moore 1979 for a summary). This suggests that the various rock types involved in the Upper Cretaceous activity may be the result of emplacement of a common parent magma under differing tectonic circumstances (e.g. Moore 1976).

Olivine melilitite is an alkaline ultrabasic rock composed of olivine (20 %), melilite (10 %), perovskite (5 %), magnetite (15 %) and ground mass 49 % (average of ten determinations from Namaqualand pipes by Moore 1979). The ground mass is dominantly interlocking clinopyroxene laths or needles with interstitial nepheline. Kimberlite is a widely used term for ultra basic rocks consisting of (in order of importance) olivine, enstatite, diopside, phlogopite, Mg-garnet, ilmenite and accessories which typically include diamond. The rock usually includes large xenoliths of ultramafic rocks such as lherzolite, saxonite and eclogite, all set in a fine ground mass of perovskite, apatite, chromite and magnenite. Although intermediate varieties between olivine melilitite and kimberlite occur, typical 'end' members show some distinct mineralogical differences, which have been summarized by Moore (1979) as:

i) diamonds have never been reported from olivine melilitites;

ii) ultramafic xenoliths are far more common in kimberlites than they are in olivine melilitites;

iii) melilite is absent from kimberlites;

iv) different ground mass assemblages; clinopyroxene and nepheline dominate in olivine melilitites, and phlogopite, calcite, serpentine diopside, montechellite and garnet dominate in kimberlites.

Moore (1979) suggests that the absence of diamond in olivine melilitite may be due to the relative scarcity of ultramafic xenoliths, because recent work that he cites has indicated that the diamonds in kimberlites may ultimately be derived from the xenoliths.

Figure 168 shows that olivine melilitites are concentrated along the west coast margin, and Dawson (1971) has noted that diamondiferous kimberlites are confined to the central areas. A further point for consideration is that a histogram of age determinations on kimberlites and olivine melilitites collated by Moore (1979) has three distinct peaks: 93-83 my (Upper Cenomanian-Middle Coniacian); 68-63 my (Lower Maastrichtian to Lower Palaeocene) and 38-33 my (Upper Eocene to Lower Oligocene). A fourth and earlier phase, peaking at 128-123 my (Upper Valanginian to Upper Hauterivian) is also seen if the alkaline granites and syenites of the Lüderitz and Cape Cross lines are included, which are coeval with the Etendeka lavas of the Kaokoveld. Aside from this latter category, which is clearly related to continental separation, distribution of the kimberlite-olivine melilitite phase in space shows a distinct west and SW younging – the older diamondiferous kimberlites lying over the central and eastern Karoo, with the younger, and in the main olivine melilitite and phonolitic rocks progressively closer to the COB along the west coast. It should be noted that kimberlites do fall into this latter category (Davis 1977).

In summary, Upper Cretaceous plugs occur in the central Karoo Basin, in central Namaqualand, and at two localities in the southern Cape, near Sutherland and at Heidelberg (Duncan et al 1978). This distribution will be further reviewed in Chapter 7.

### 5.3.3 *Palaeogeography*

As discussed in the Lower Cretaceous palaeogeographies, the SE Atlantic during its earliest stage of development was a narrow, deep basin, connected to the open ocean by a shallow shelf sea over the Falkland Plateau. This arrangement, which resulted in dark anoxic muds and sandy muds being deposited in the Cape Basin, persisted until the basin geometry altered sufficiently to allow deep-water connections with the open ocean. The change took place in the Lower Albian a short time (?5 my) before the Falkland Plateau cleared the southern tip of the Agulhas Bank, as a result of which the anoxic black muds were rapidly replaced by grey shales with a low pyrite and carbonaceous content, and the establishment of an infauna (i.e. bioturbation). At this stage of development, Bolli et al (1978) envisaged a deep-water environment for the post-AII sediments at site 361, which was supplied with terrigenous debris via turbidites from the Orange River source and lateral contour current transport. Deep water connections were not established with the Angola Basin until Tertiary times so that although the Upper Cretaceous Cape Basin was moderately

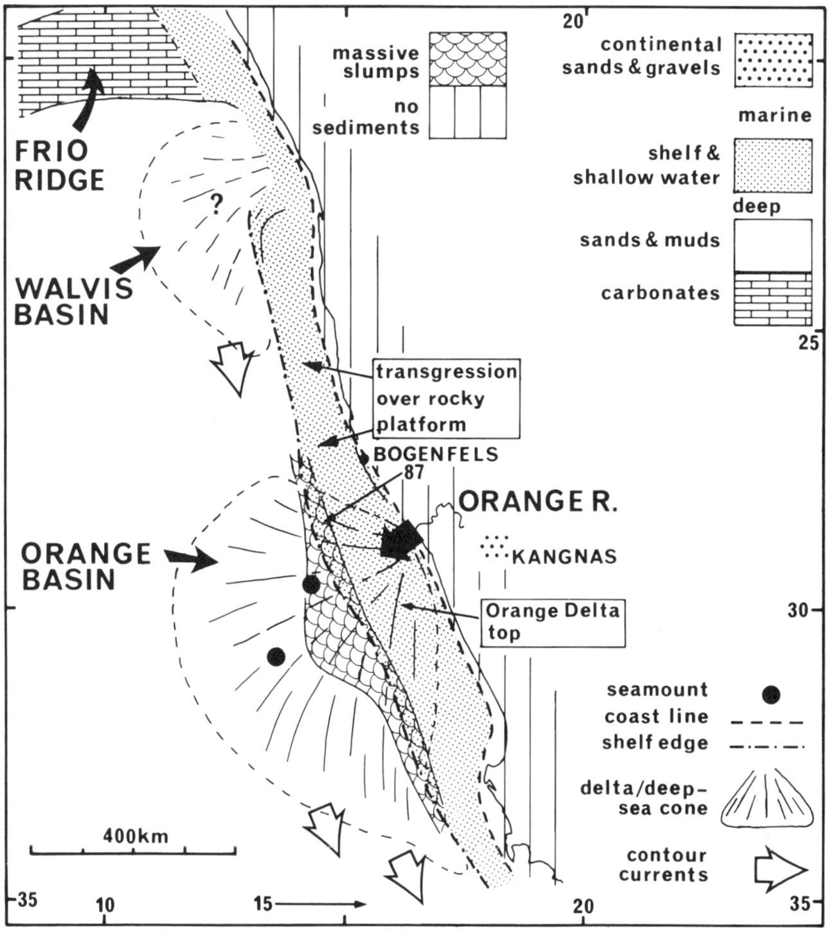


Figure 151. Upper Cretaceous palaeogeography of S.W.Africa. The sedimentation rate for the Kudu 9A-1 borehole (N.W.Orange River mouth) is in m/my for the period Cenomanian to Maastrichtian. Data from numerous authors cited in the text.

well-oxygenated at all levels, anoxic sediments continued to accumulate north of the Walvis Ridge until early Santonian times.

We present one palaeogeography (Fig.151), which represents the situation in Cenomanian times, when the Upper Cretaceous transgression had reached its maximum extent. It would, however, be applicable to most of Upper Cretaceous time, with only minor amendments. The Orange River was the main sediment source along the west coast and a large cone continued to develop over the site of the early Cretaceous depocentre. Large-scale crustal subsidence and high sediment supply established a major depocentre SW of the present river mouth in a position straddling the COB. The resultant sediment wedge advanced into deep water by a combination of progradation and allochthonous injection – the latter probably being dominant. From the evidence of the delta front structures we envisaged a Mississippi-type situation, which advanced the shelf break about 30 km (750 m/my) during Upper Cretaceous time west of the Orange River. In addition, SE distribution of terrigenous debris by contour currents took place along the foot of the adjacent continental slope and constructed a large continental rise prism south of Cape Columbine. The establishment of these currents resulted from a reorganization of ocean circulation in the SE Atlantic following the separation of Africa and the Falkland Plateau and accounts for the rapid facies change between Units 6 and 7 at DSDP site 361: the virtual dissipation of the anoxic bottom conditions, and the finer grain size of Unit 6 sediments. These oceanic changes may have also caused an amelioration in the climate of the Orange River hinterland, because the terrestrial plant matter in Unit 6 indicates the establishment of a more diverse vegetation.

In our Cenomanian reconstruction (Fig.151), the coastline probably lay close to its present position, but is known to have advanced 4 km east of this in the Bogenfels area at the height of the transgression. There is no evidence that it maintained this high stand for very long, but equally, it probably did not retreat much farther west until the general withdrawal in Upper Maastrichtian times. Apart from the possible temporary salients opposite the Orange River mouth (delta top emergences), we envisage a relatively straight coastline, with hyposaline and paralic delta top flats for the shelf

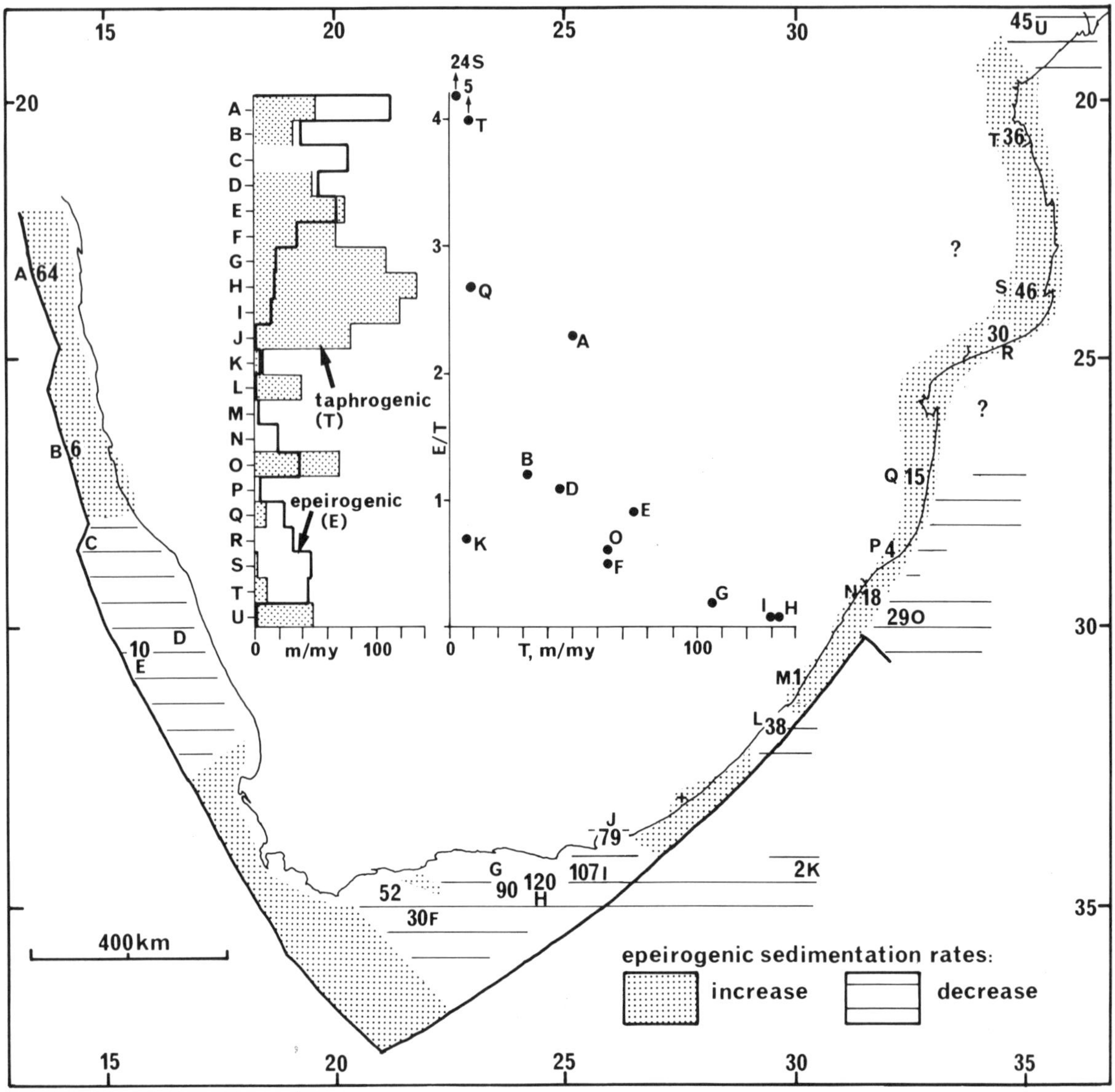


Figure 152. Comparison of Jurassic-Cretaceous taphrogenic and epeirogenic sedimentation rates. Taphrogenic period taken as 162-109 my, and epeirogenic as 108-65 my. Main map shows epeirogenic rates relative to taphrogenic rates, with difference in m/my at various sites. Histogram at left shows actual rates during the two periods for the same sites (A-U). Graph plots epeirogenic/taphrogenic against taphrogenic rates and shows that sites with an intially high taphrogenic rate continued with a relatively low rate in the following period, and vice versa. See text for discussion and for the numerous data sources.
Abbreviations: A = Walvis Basin; B = Luderitz Arch; C = Kudu borehole; D = Orange Basin (Inner Graben); E = Orange Basin (main basin); F = Bredasdorp Basin; G = Pletmos Basin (south of PB-A/1); H = Pletmos Basin (near Gemsbok borehole); I = Gamtoos Basin; J = Algoa Basin; K = Falkland Plateau (DSDP 330); L = Mngazana; M = Umzamba; N = proximal Tugela Cone (J(c)-1); O = distal Tugela Cone; P = BH9 (Richards Bay); Q = N.Zululand; R = Zandamela borehole; S = Domo borehole; T = Save River Mouth; U = lower Zambezi Graben. Solid line is approximate COB. Cross at Igoda is unquantified increase.

area between the vicinity of Chamais Bay and Cape Columbine. Although we have no substantive evidence, a similar, though smaller-scale situation may have obtained in the Walvis Bay area which, in addition possibly also received some coarse, shallow marine debris from the eastern end of the Walvis Ridge (Goslin et al 1974).

Data on Upper Cretaceous climates are not very abundant. The high clay content and relatively rapid sediment accumulation in the distal parts of the basin (17 m/my), as well as the large delta front slumps indicate a continued high sediment input rate which suggests a well-watered, warm climate hinterland. This is reinforced by the diverse terrestrial vegetation indicated by spore and pollen counts in the site 361 sediments, and the circumstances of the Upper Cretaceous terrestrial vertebrates and wood fragments in the valley fill at Kangnas.

A late Cretaceous (Maastrichtian) regression affected the west coast continental margin, and judging from the angular discordance between the uppermost Mesozoic and early Tertiary rocks on the continental shelf much of the latter was exposed to either subaerial or shallow marine (high energy current) erosion. The increased coarse fraction shedding that resulted from this event is clearly seen in the sandy texture of the Maastrichtian sediments of Unit 6 at DSDP site 361 (Fig. 147). An earlier regressive episode is also indicated by coarse horizons in Coniacian-Turonian sediments at DSDP site 361, and in the Kudu 9A-1 borehole which lay over the outer part of the Orange delta front. Both these regressions correlate with regionally developed hiatuses in the Outeniqua Basin and along the east coast (Fig.67).

Apart from the small-scale kimberlite and olivine melilitite intrusives along the Namaqualand coast and hinterland (Fig.168), no igneous activity is known from the west coast margin in Upper Cretaceous times. The two seamounts in the Orange Basin, and those that lie on the continental slope and rise farther south are probably pre-AII in age and had little effect on sediment dispersion except possibly in impeding the local development of slump structures.

## 5.4 FURTHER ASPECTS OF JURASSIC-CRETACEOUS SEDIMENT BASIN DEVELOPMENT

In the foregoing two chapters we have detailed the litho- and biostratigraphy of the Jurassic and Cretaceous strata around the continental margins of southern Africa, and have stressed the contrasts in sediment basin shape that resulted from the taphrogenic and epeirogenic styles of crustal subsidence operating during mid-Jurassic to late Aptian-early Albian, and post early Albian times, respectively. These differences are not only manifest in basin geometries and disposition, but also in lithofacies and thickness variations.

Although detailed borehole information has not been published for the west, and most of the east coast areas, sufficient biostratigraphic data of one sort or another are available to allow us to make a further preliminary comparison of the two tectono-sedimentary styles by investigating, on a regional scale, the history of crustal subsidence as it is reflected in sediment accumulation rates. For this sort of study to be done properly a refined biostratigraphy is required for numerous localities, and corrections made for sediment compaction in borehole sections. Insufficient data to fulfil either of these requirements have been released by SOEKOR, but some preliminary observations are still worthwhile.

For the general statement that sediment accumulation rates directly reflect crustal subsidence rates, to be true, then one or other of the following conditions must be met:

i) sedimentation rate must remain constant, in which case fluctuation in depths of deposition can be used to monitor sea floor movement;

ii) the basin must be full all the time, in which case very shallow marine conditions will be observed throughout the fill sequence and sedimentation rates can be used to monitor sea floor movement.

If these conditions are not met, as is presumably often the case, then the quality of data available to us will not allow the problem to be addressed in any but a rudimentary fashion. Further, item (i) also requires a level of palaeoecological input that is presently not possible, so we have had to make the assumption that sedimentation at all the locations we discuss here took place under shallow marine or paralic conditions. Whilst this is unlikely to be true, it is not such an outrageous assumption to make as a first approximation for at least the continental margin sites, in the light of the sedimentary history of the region.

In Figures 152 and 153, we have plotted sediment accumulation rates for various localities around the continental margins and have tried to illustrate two aspects; a comparison of gross rates for taphrogenic and early epeirogenic (i.e. Upper Cretaceous) subsidence (Fig.152), and stage by stage comparisons for selected localities to show fine-scale irregularities through time (Fig.153). Whilst the quality of our data militates against authoritative pronouncements, several trends that are worth further investigation do appear to emerge:

i) taphrogenic rates are higher in the Outeniqua Basin (mean of 93.1 m/my) than in the Orange Basin (mean of 68.5 m/my) and the northern Natal Valley (mean of 58.2 m/my);

ii) epeirogenic rates are higher in the Orange Basin (mean of 53.8 m/my) than in the northern Natal Valley (mean of 29.2 m/my) and the Outeniqua Basin (mean of 25.3 m/my);

iii) relatively speaking, the Orange, Outeniqua, St Johns, and Tugela depocentres show a decrease in sedimentation rates during the epeirogenic compared to taphrogenic phase. In the case of the Orange Basin this is probably a debatable conclusion in view of the thick sequences assigned to the Upper Cretaceous by Gerrard & Smith (1980);

iv) all of the relatively buoyant areas show an increase during the epeirogenic phase because they received little or no sediment during earlier times;

v) areas which experienced high absolute rates during the taphrogenic phase have relatively low values for the epeirogenic phase, and vice versa.

Much of this has already been implied during the earlier descriptions of regional stratigraphy, but Figure 152 quantifies some of these observations. They may

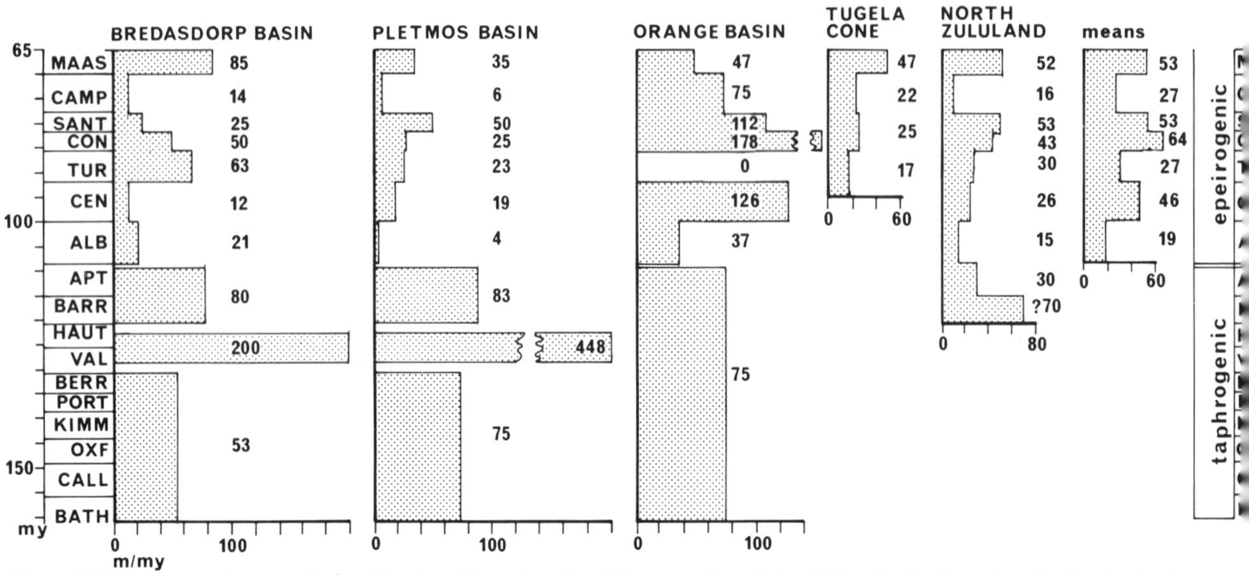


Figure 153. Sedimentation rates (m/my) by stage for various localities around southern Africa. In the Outeniqua Basin, the Bredasdorp site is for borehole F-E/1, and the Pletmos site is for borehole G(a)-A/3. Orange Basin rates have been based on Kudu 9A-1 and the main basin average for the pre-Upper Cretaceous Tugela Cone is from J(c)-1, and N.Zululand calculated from ZE borehole. Data from du Toit & Leith (1974), du Toit (1976), McLachlan & McMillan (1979), Gerrard & Smith (1980), Figure 120.

be summarized by the general statement that during the taphrogenic phase of sedimentation absolute accumulation rates were locally high, and that whilst there was an overall reduction in maximum rates during the epeirogenic phase, sedimentation affected a larger area.

Looked at on a finer time scale (Fig.153), both phases show possible cyclicity, although with so few data points we must be aware of the danger of generalizing from specific examples. There appears to be a short period of maximum sediment accumulation in Neocomian times (Valanginian to Barremian) when, in the Outeniqua Basin values as high as 582 m/my are recorded (we show two somewhat lower values in Fig. 153). For the few calculations that we have made, these maxima are between 3.7 and 8.5 times higher than mean values for the pre-Neocomian period. This spurt was followed by regionally lower values which reached minima during the Albian when they were between 1 and 49 % of their former values. This corresponds to the period when the switch from taphrogenic to epeirogenic basin formation was generally taking place.

During the epeirogenic phase, all our histograms show a bimodal distribution, even if the peaks are not strictly synchronous. There is a period of rapid sediment accumulation during ?Turonian/Coniacian/Santonian time, a distinct decrease in most areas during the Campanian, and a further short period of rapid accumulation during the Maastrichtian before the end Mesozoic regression. The dating of the earliest of these high periods is locally enigmatic because in most areas there is a Turonian hiatus and yet we seem to have a peak at this time in the Bredasdorp Basin. The problem will almost certainly resolve itself when the biostratigraphy of the offshore area is clarified.

In terms of crustal subsidence, the above indicate a lengthy period (at least 50 my) of localized rapid collapse, with at least one particularly short (?<10 my) sharp episode. This period occupied the time prior to (at least 35 my) and immediately after (ca.15 my) the onset of continental drift, and corresponds to Falvey & Middleton's (1981) 'rift' and 'breakup' phases, respectively. In view of our previous discussions on the timing of drift onset around southern Africa (see Chapter 1) it is tempting to correlate the Neocomian high subsidence episode with this event at ca.127 my, and note in particular that there is suspicion that the highest rates occurred in the Outeniqua sub-basins that lay closest to the Agulhas/Falkland Fracture Zone (Algoa, Gamtoos and SE Pletmos).

The Upper Cretaceous epeirogenic phase covers early to middle post-drift time, and whilst it is generally true to say that overall crustal subsidence values were lower than during the taphrogenic phase, episodes during Coniacian-Santonian and early-mid Maastrichtian time do show moderately high values. Whether these apply only to areas that were original taphrogenic basins is not clear, but such aberrations are not predicted in the simple thermal model for cooling lithosphere curves propounded by Steckler & Watts (1978). Although crustal inhomogeneities might be expected to modify such empirical predictions, the apparently regional scale of these Upper Cretaceous episodes around southern Africa suggests some other causes. This leads to the question of why sediment accumulation generally decreased during the epeirogenic phase (and falls still further during the Tertiary). Clearly, in terms of isostatic equilibrium there is a limit to the amount of subsidence that the continental crust will undergo as it adjusts to heat loss, stretching, basal erosion, or whatever combination of phenomena are the driving forces that cause continental margin subsidence. The basin formed by the depressed crust will ultimately become filled with sediment as subsidence ceases and will be unable to accommodate more – a case of sedimentary constipation. Additional sediment must bypass this depocentre and accumulate beyond it. If a progressive decrease in crustal subsidence is anticipated during the epeirogenic phase (and data from the Tertiary strata supports this contention), then models which combine progressive (?exponential) decay linked to crustal cooling (e.g. Steckler & Watts 1978) and sediment loading (e.g. Beaumont & Sweeney 1978) may provide satisfactory explanations. It should be noted, however, that the latter workers found an upper theoretical limit to basin size of 400 km wide and 2.5 km deep with an original 100 km wide graben, and to enlarge this it may be necessary to appeal to factors such as lower crust metamorphism, as proposed by Falvey & Middleton (1981).

# 6

# Tertiary

Until relatively recently, the paucity of easily-dated marine rocks onshore, and the low economic potential of Tertiary strata in general, had militated against their systematic stratigraphic study and the development of regionally applicable correlation schemes. Several factors have now altered this state of affairs to the point that during the last decade the traditional, and inadequate lithostratigraphic classifications have been in a state of flux from which no consensus has emerged. These factors include: the application of micropalaeontological studies (especially planktonic foraminifera and calcareous nannofossils) to the onshore outcrops; the recognition, classification and mapping of extensive, relatively thick and comparatively complete Tertiary sequences on the continental shelf; and the realization that the Tertiary may not be as economically barren as previously thought. On the last point, minerals worthy of economic assessment include glauconite ($K_2O$), phosphatic rocks ($P_2O_5$), diamondiferous and metal enriched sediments (particularly along the west coast), as well as manganese-rich nodules in some of the deep sea areas. In addition drilling for hydrocarbons in the pre-Tertiary has provided considerable incidental knowledge on the overlying cover.

With the exception of the central Kalahari Basin (non-marine), and the wide Zululand-Mozambique coastal plain, all the extensive deposits of Tertiary strata occur on the continental shelf and slope, where cover is complete apart from nearshore zones in the south and west Cape. In fact, the only large area of sea floor where pre-Tertiary rocks crop out any distance from shore is SE of Cape Agulhas. The distribution and thickness of Tertiary rocks in southern Africa are shown in Figure 154, where it should be noted that over the continental margin and adjacent ocean basins, isopachs have invariably been derived from seismic reflection records. It is usually not possible to resolve or identify Quaternary cover, so that in reality the map shows Cenozoic isopachs although where borehole or other evidence are available, it has been found that the Quaternary cover is rarely >50 m thick, at least on the continental shelves. The thickest and most extensive deposits of Tertiary sediments occur under the outer shelf and continental slope of the west coast, and as a sinuous, linked series of depocentres under the shelf, slope and deep sea basin from the Zambezi Cone to the southern part of the Natal Valley off the east coast. The west coast Tertiary forms a vast, coast parallel lens along the whole margin from the Walvis Ridge to the Agulhas Fracture Zone. It is generally between 1 and 1.5 km thick but sub-depocentres lie off the Orange River (>1.5 km) reflecting the continued, but diminished importance of this terrigenous input point during Tertiary times, and on the Walvis Ridge abutment where seismic and borehole data indicate over 4 km. This abnormal thickness is probably due primarily to intense biological production in an upwelling Benguela Current. Figures 120, 121 and 154 clearly illustrate the virtual obliteration of the outlines of the original taphrogenic highs and basins along the west coast. This reflects the cessation of differential crustal subsidence, and the progressively subordinate role played by terrigenous sedimentation since late Cretaceous times, with the concomitant increase in the importance of biogenic and authigenic sediment production. The bulk of these Cenozoic sediments accumulated as outbuilding lenses on surfaces that had an initial high gradient (i.e. onto the late Cretaceous continental slope) with the result that there has been large-scale mass movement downslope (slumping, debris flows, etc.) and the present-day meso- and micro-topography of the sea floor has been largely generated by these huge allochthonous masses.

Tertiary sediment distribution along the east coast was, by contrast, still dominated by major terrigenous input points: Zambezi, Limpopo, and Tugela rivers. Each of these constructed a significant Tertiary cone into adjacent deep sea areas, although the rate of construction changed through time. Tertiary sediments extend widely across the Mozambique coastal plain, but it is only along the present-day continental margin that accumulation exceeds 1 km, and south of the Limpopo, the main depocentre was severely restricted to the western edge of the basin by strong bottom currents which swept the central parts of the northern Natal Valley clear of sediments. The thickest sediment accumulation in this area lies under the Tugela Cone (>2 km) SW of which a wide, thick swathe of sediment projects along the western side of the southern Natal Valley. Although much of this material probably originated from the Tugela, the other southern Natal and Transkei rivers

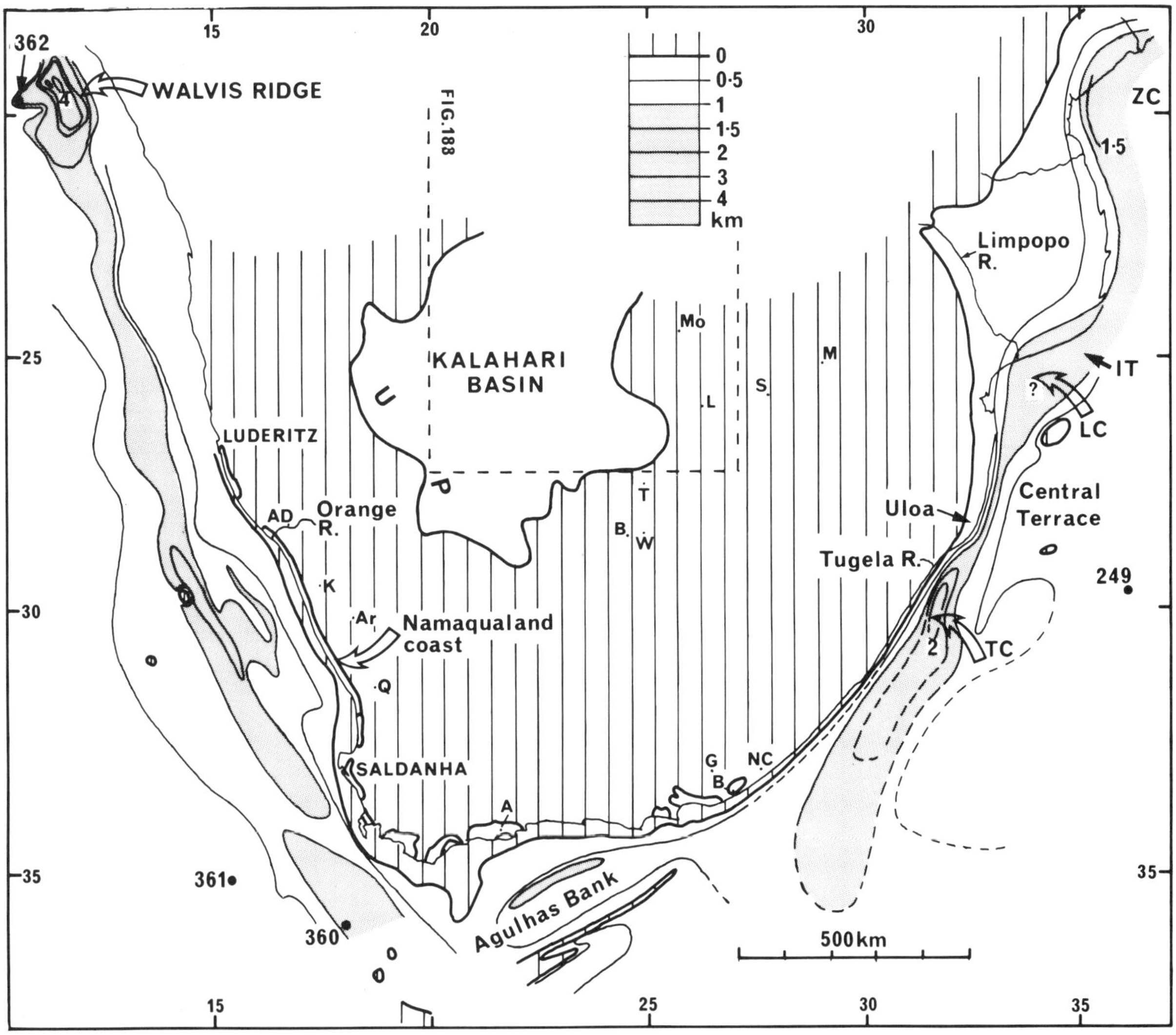


Figure 154. Isopachs (km) of Tertiary sediments on the continental margins of southern Africa. Heavy line shows landward limit of marine sequences, with Kalahari Basin indicating approximate extent of interior non-marine succession. DSDP sites 249, 360, 361, and 362 are also shown. Abbreviations: AD = Arris Drift; K = Kamaggas; Ar = Arnot (or Banke); Q = Quaggaskop; A = Albertinia; G = Grahamstown; B = Birbury; NC = Needs Camp; TC = Tugela Cone; LC = Limpopo Cone; IT = Inharrime Terrace; ZC = Zambezi Cone; UP = Urananib Plateau; B = Barkly West; W = Windsorton; T = Taung; L = Lichtenburg; S = Swartkrans, Sterkfontein and Kromdraai; M = Makapansgat; Mo = Molepolole. Dashed outline is location of Figure 183.

must have played a significant role in the construction of this continental rise prism, aided by the sweeping action of the Agulhas Current which today carries large quantities of terrigenous debris beyond the shelf break. At the southern end of the Natal Valley, Tertiary fill decreases to $<500$ m in the Transkei Basin and Agulhas Passage, reflecting the geographical isolation of this area from major terrigenous sources and the screening effect of the marginal fracture zone along the outer edge of the Agulhas Bank.

The Agulhas Bank itself (site of the Outeniqua Basin) is mostly covered by Tertiary rocks, but except for a narrow zone under the present shelf break, thicknesses are $<1$ km. This reflects relatively low terrigenous input rates, particularly in the Neogene. Large-scale slumping has modified the distribution of the Cenozoic strata on the outer margin in this area. It has exposed narrow irregular outcrops of older strata on the upper slope, and is responsible for the present day arcuate asymmetry of the outer margin, and the highly irregular topography along the upper slope.

With the exception of the Zululand-Mozambique coastal plain, marine Tertiary rocks are restricted to narrow, discontinuous outcrops along the coastal fringes, which are usually overlain and partly obscured by Neogene and Quaternary non-marine limestones and sands, the latter often unlithified and mobile. The largest Cenozoic depocentre onshore in southern Africa is the

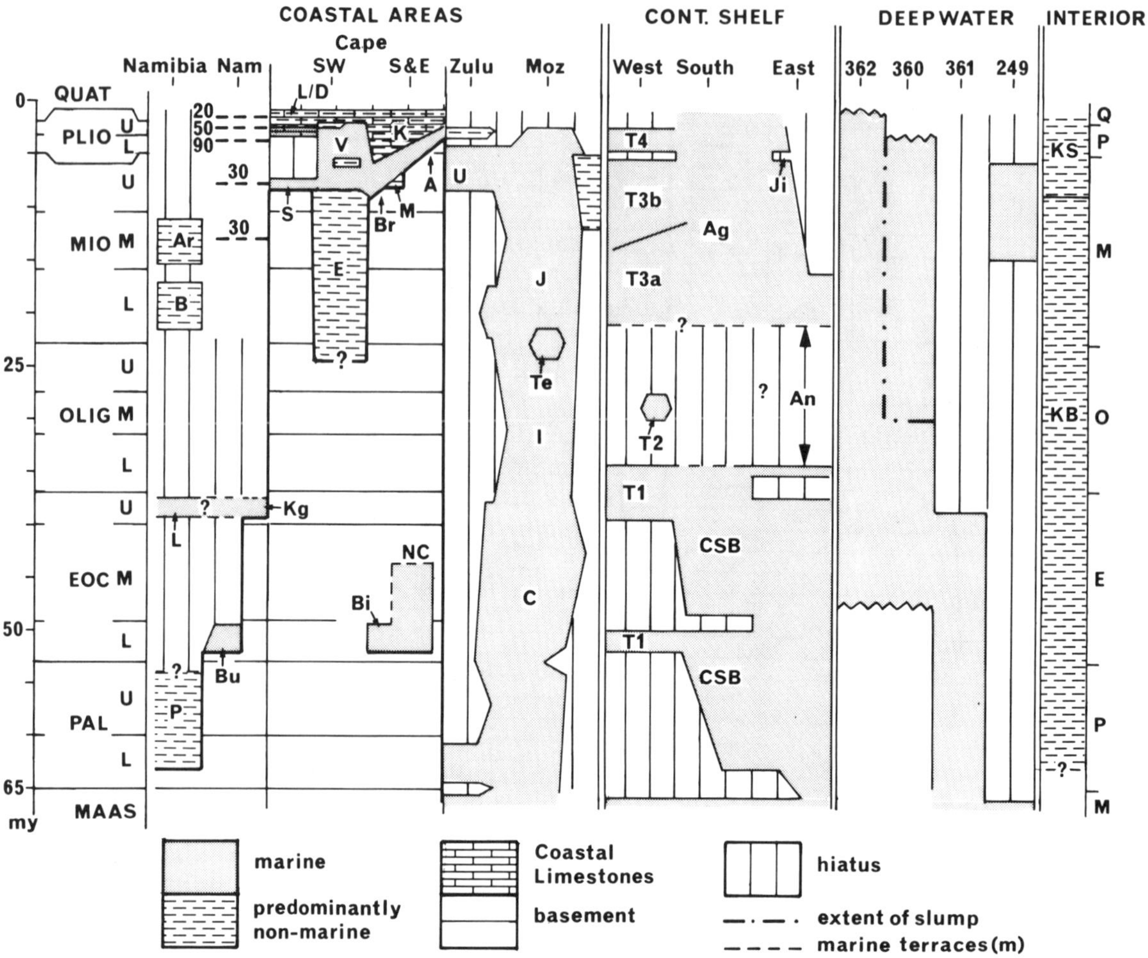


Figure 155. Summary regional correlation of Tertiary sediments around southern Africa. The sections showing coastal areas and continental shelf are displayed west to east, with the former subdivided into Namibia/N.W.Cape; SW, S and SE Cape; and Natal/Zululand/Mozambique. Deep water sites show DSDP borehole records. For clarity sediments are shown either 'marine' or 'non-marine', although locally this will result in oversimplification of the succession. Hiatuses indicated by vertical lines, basement devoid of Mesozoic strata has a bold border and horizontal lines (corresponding to stage and system boundaries to aid correlation across the chart). Timescale is that used by Vail *et al.* (1977). Namaqualand terraces are identified by elevations (m) above present sea level. Stratigraphic position of many sequences is uncertain, see text for discussions.
Abbreviations
Column headings: Nam = Namaqualand; Zulu = Natal/Zululand; Moz = S.Mozambique.
Localities, formations, etc. (from left to right): Ar = Arris Drift; B = Bogenfels, Elizabeth Bay, Plant 4 etc.; L = Langental beds; Kg = Kamaggas (40 km W of Springbok); P = Pomona beds; Bu = Buntfeldschuh, Advokat Bake, Eisenkieselklippenbake; L/D = Langebaan and Dorcasia limestones; S = Saldanha Member; V = Varswater Formation; E = Elandsfontyn Formation; Br = Bredasdorp Formation; M = Martingdale beds; K = Knysna beds; A = Alexandria Formation; Bi = Birbury; NC = Upper Quarry Needs Camp; U = Uloa; J = Jofane Formation; Te = Temane evaporites; I = Inharrime Formation; C = Cheringoma Formation; T1-T4 = acoustic stratigraphy; Ag = Agulhas Formation; CSB = Cape St Blaize Formation; Ji = reflector Jimmy; An = reflector Angus; KS = Kalahari sands; KB= Kalahari beds.

Kalahari Basin which is centred on Botswana, but which extends into Namibia, and South Africa, with further extensions into Angola and Zimbabwe. This is a vast shallow intracontinental depression lying at about 1 000 m above sea level which acts as an internal drainage basin with main input from the Okavango River and its tributaries into the Okavango Graben and Makgadikgadi Depression. Exposures of the Cenozoic strata are not good and because of the low dips, the oldest sediments are known only from boreholes which have proved over 200 m of non-marine Cenozoic sediments. Most of the area occupied by the Kalahari Basin is semi-arid grass and woodland, and is separated by the Great Escarpment from the numerous thin and irregular deposits of non-marine and shallow marine Cenozoic strata that lie along the coastal fringes. Figure 155 shows our correlation of the various onshore and offshore localities and illustrates the difficulties in fitting the many small,

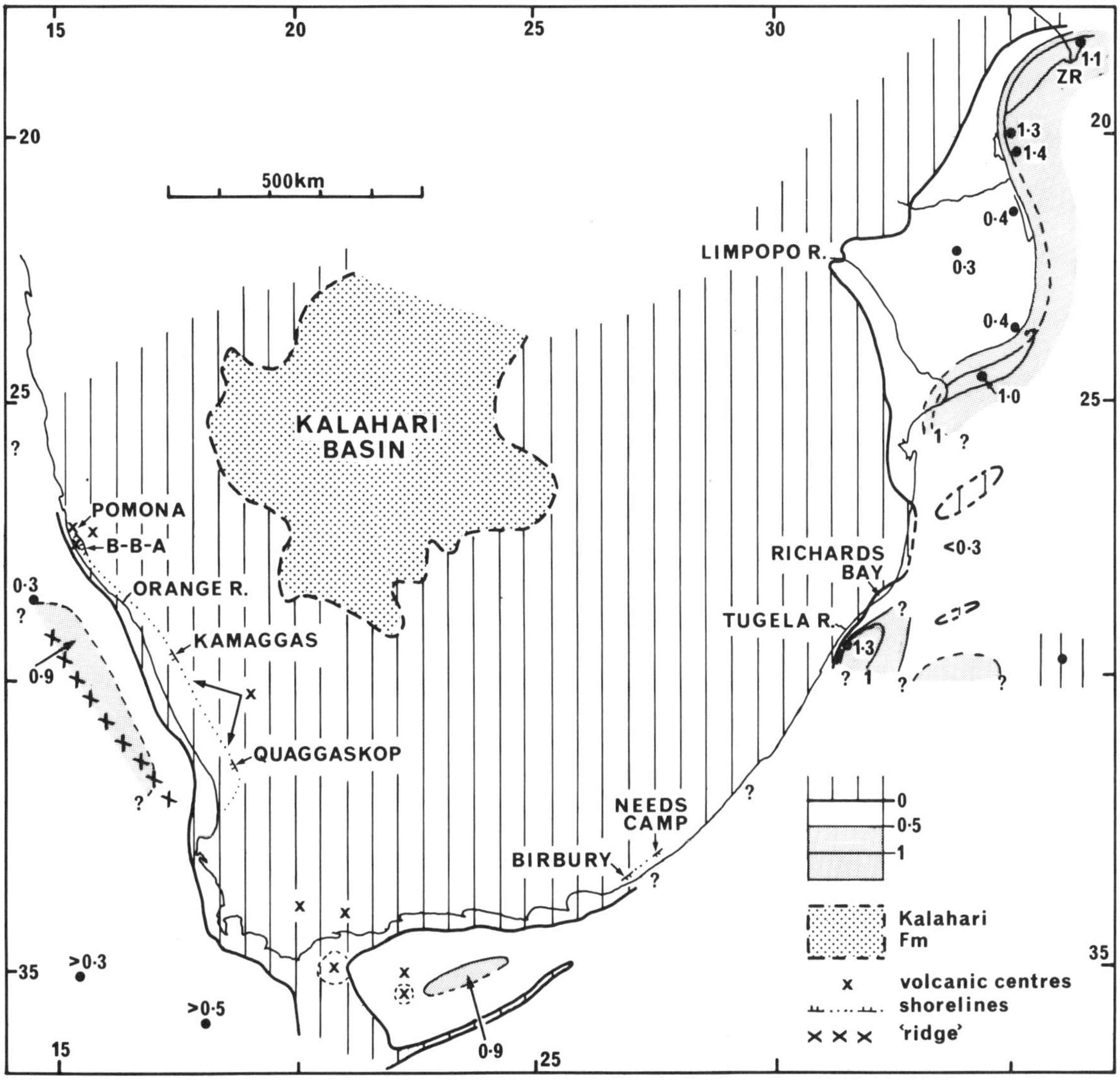


Figure 156. Distribution and thickness of Palaeogene sediments around southern Africa. Isopachs are in km, with borehole sites shown as dots. Shorelines are: Bogenfels area – Lower and Upper Eocene; Kamaggas/Quaggaskop – possibly Eocene; Birbury/Needs Camp – Lower Eocene. Abbreviations: B-B-A = Bogenfels-Buntfeldschuh-Advokat Bake; ZR = Zambezi River.

scattered and often poorly-fossiliferous sections into a regional scheme.

The best-known and most complete Cenozoic sequences are on the Agulhas Bank, in the Natal-Mozambique basins (onshore and offshore), and at certain localities along the west coast, and with the exception of deep water sites, the succession in all these areas has two lithostratigraphic subdivisions which are Palaeogene and Neogene in age. In the proximal areas (i.e. on the continental shelves) these two units are separated by an hiatus which spans parts of Oligocene to Lower Miocene time, and on seismic records the event is usually marked by an unconformity. More distal areas often exhibit no hiatus (e.g. DSDP sites off the west coast), but there are regional lithofacies changes (e.g. northern Natal Valley). Although onshore outcrops are mostly small erosional outliers, and usually represent very short time intervals, they do fit into this broad twofold division, whilst several other, usually less well-developed hiatuses and lithofacies changes occur within the Cenozoic succession and serve as useful stratigraphic markers. Prominent amongst these are Lower-Middle Eocene and late Miocene-early Pliocene events.

The Mesozoic/Cenozoic boundary around southern Africa is marked by a major non-sequence which locally spans parts or all of uppermost Maastrichtian and Lower Danian time. In proximal areas, it was accompanied by emergence and, locally (e.g. on the NE Agulhas Bank and in the SW Cape) by mild faulting, folding and/or igneous activity. In the distal areas important facies

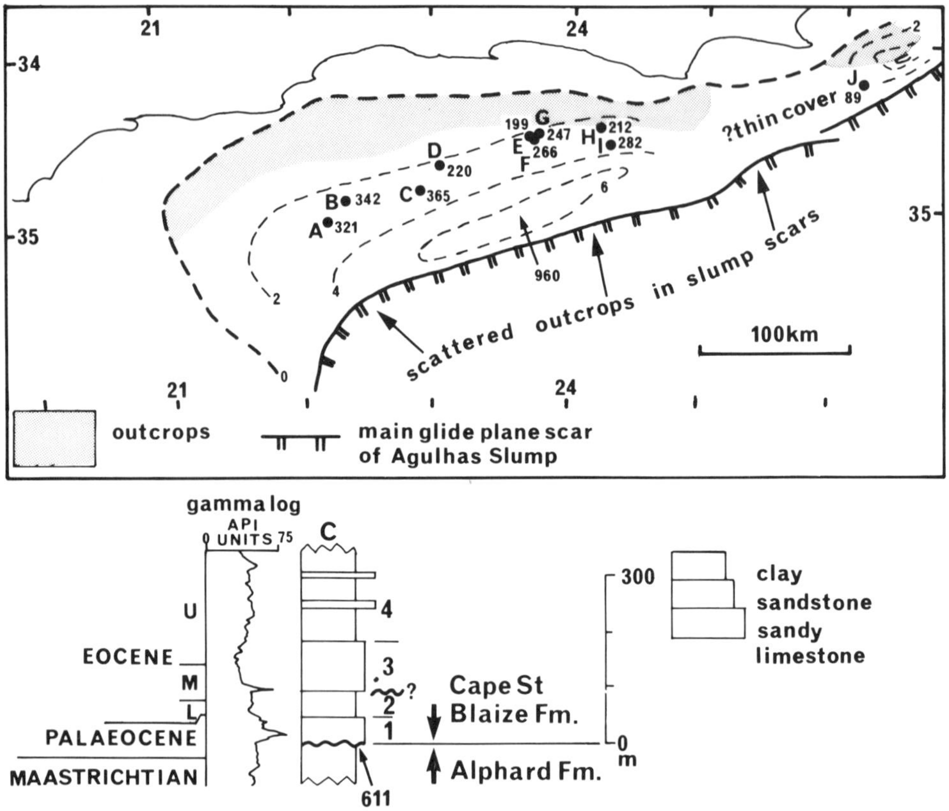


Figure 157. Isopachs of the Palaeogene Cape St Blaize Formation on the Agulhas Bank. Borehole log C is the Palaeogene stratotype designated by du Toit (1976) (which he called the Alexandria Formation, but this name is not available). Units 1-4 are after du Toit (1976) and form the basis for correlations in Figures 158 and 159. Thicknesses in boreholes are shown in m, and isopachs are x 100 m. Borehole locations: A = F-1; B = F-E/1; C = F-D/1; D = G(a)-C/1; E = G(a)-A/1; F = G(a)-A/2; G = G(a)-A/3; H = G(b)-Gemsbok/1; I = G(b)-Springbok/1; J = H(b)-Hartebeest/1. Data mainly from Dingle (1973a), du Toit (1976).

changes occurred, particularly in the Cape Basin, where major oceanographic and sediment supply characteristics altered. This event clearly correlates with the world-wide late Cretaceous eustatic regression.

## 6.1 PALAEOGENE

Although it has been possible to compile isopach maps for total Tertiary sediments, the lack of borehole data frequently inhibits subdivision of the Tertiary sequence as seen on seismic records. Consequently, it is not possible to present Palaeogene isopach data for most deep sea areas off the west coast and in the southern Natal Valley. Nevertheless, what data we do have (Fig.156) suggest that in most areas the bulk of the total Tertiary sediments are Palaeogene in age. This is particularly so off the Orange River (ca.900 m), Tugela Cone (1 300 m), SW of the Zambezi River (>1 400 m), and on the Agulhas Bank (ca.900 m), and emphasizes that the post-Cretaceous slowdown in sediment accumulation rates became more acute in Neogene times. The main exceptions to this are on the Walvis Ridge Abutment and on the Limpopo Cone, where Neogene sediments have accumulated faster than their Palaeogene counterparts.

Eocene high sea levels are recorded as scattered exposures along the east and west coasts, but otherwise, the only extensive marine outcrops are in southern Mozambique where the wide coastal plain was covered by the transgression.

### 6.1.1 *Agulhas Bank (Cape St Blaize Formation)*

Palaeogene sediments crop out at several places on the Agulhas Bank, particularly in the nearshore zone south of Plettenberg Bay (Fig.157). They form a large, oval deposit on the mid and outer shelf, but appear to be missing from the crest of the Agulhas Arch in the west, and probably from the Port Alfred Arch in the east. Du Toit (1976) reported Palaeogene sediments in ten of the 17 SOEKOR boreholes from which data had been released, and seismic data show that they form a seaward dipping wedge (1-2°) which reaches its maximum thickness of about 0.9 sec (ca.900 m) over the outer shelf, where outcrops are locally and irregularly exposed in the head region of the Agulhas Slump. On the basis of

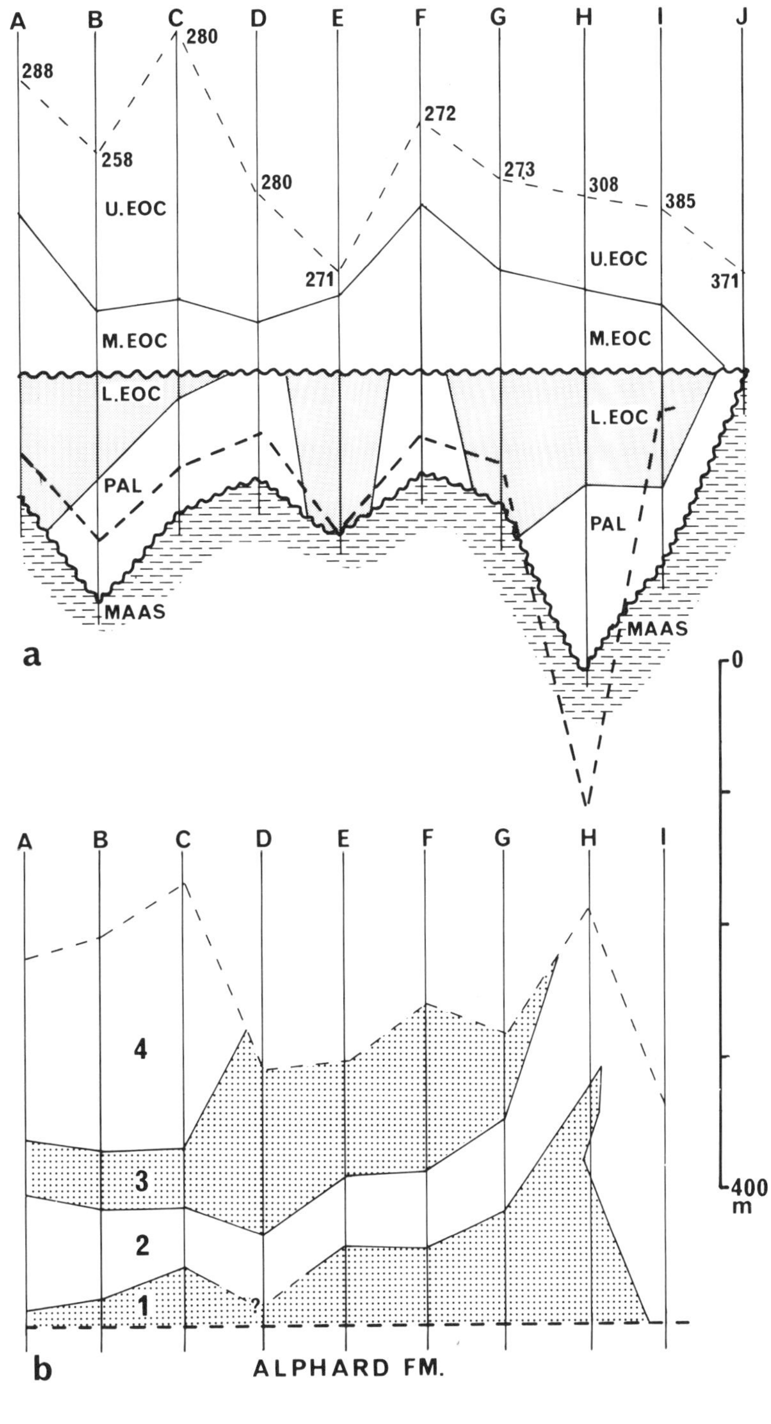


Figure 158. Thickness variations in Palaeogene series and lithofacies across the Agulhas Bank. Shown as W-E cross-sections with borehole locations (A-J) on Figure 157.

a. Series thicknesses, datum is base of Middle Eocene. Non-sequences shown as wavy lines, thick dashed line is base of Cape St Blaize Formation, thin dashed line is top of borehole record. Shading is for clarity in correlation with no geological significance. Depths (m) to top of borehole are below sea level.

b. Main lithofacies of Cape St Blaize Formation expressed as units 1-4 (see borehole log C, Figure 157). Sandstones are dotted, clays and shales blank. The scheme follows du Toit (1976) in the west, but is provisional in the east.

All data compiled from du Toit (1976).

seismic records and surface samples, Dingle (1973a) described this lithostratigraphic unit and its relationship to the two bounding unconformities and named it the Cape St Blaize Group (herein referred to a formation) and considered it to be Palaeocene-Eocene in age. Du Toit (1976) presented more data on this unit which he called the Alexandria Formation, an unfortunate choice because of the obvious nomenclatorial conflict with the Neogene Alexandria Formation that crops out on-shore. Du Toit (1976) selected borehole F-D/1 to be the stratotype for the Agulhas Bank Lower Tertiary succession which we follow and use as the type section for the Cape St Blaize Formation (Fig.157). Here, approximately 345 m of strata can be divided into two

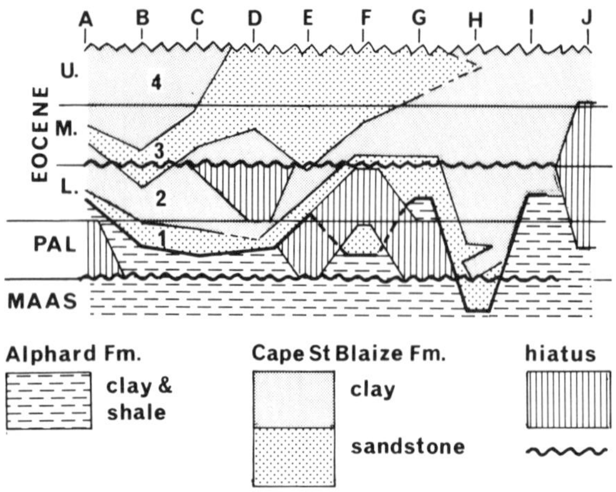


Figure 159. Distribution of Agulhas Bank Palaeogene lithofacies in time and space. Base of Cape St Blaize Formation shown by heavy solid/dashed line. Units 1-4 are those in Figures 157 and 158.

upward fining cycles that consist from the base (1) to top (4) of:

4. light greenish grey, fossiliferous, calcareous, glauconitic clay with subordinate siltstones, sandy limestone and marl – 155 m;

3. fine to very fine, well-sorted sandstone, with abundant glauconite, shell fragments and foraminifera – 86 m;

2. light-medium grey fossiliferous calcareous clays – 45 m;

1. fine to coarse sandstone with occasional scattered pebbles – 45 m.

These subdivisions have been recognized over most of the Agulhas Bank, except for the central-northern area, where Unit 4 is missing, probably because of erosion or non-retention during drilling. The other argillaceous member (Unit 2) maintains a fairly constant thickness across the area (ca.50 m) but the two arenaceous members (Units 1 and 3) are generally thickest in the east (Fig.158). When plotted against SOEKOR's biostratigraphic time scale (du Toit 1976) (Fig.159), these various members are seen to be strongly diachronous, with arenaceous lithofacies persisting for much of the Palaeocene-Eocene in the central area, but argillaceous lithofacies representing longer periods of time in the west and east. The implication of Figure 159 is that only one cycle (Units 1 and 2) developed in the east, contemporaneously with two cyles (Units 1-4) in the west.

It is not possible to make a final assessment of the distribution of the lithostratigraphic members of the Cape St Blaize Formation in time and space until the faunas of the boreholes have been released, because up to now only a few general remarks (du Toit 1976) and a provisional zonation scheme (McLachlan & McMillan 1979) have been published. Du Toit (1976) indicated a hiatus between the Lower and Middle Eocene on the Agulhas Bank, and whilst not specified by McLachlan & McMillan (1979), their zonal scheme is clearly condensed in the Middle Eocene section (Tab.44). Using the division and stage thicknesses given in du Toit (1976) (Fig.158a), the Middle Eocene is seen to overstep onto Palaeocene in some boreholes whilst the Lower Eocene varies greatly in thickness. Also, the Palaeocene is missing at several sites. This evidence indicates an upper Lower or lower Middle Eocene hiatus (probably the latter). With the published data it is also difficult to assess the nature and age of the Alphard-Cape St Blaize formation boundary. All the evidence points to it being a non-sequence and probably erosional, but the age of the base of the Cape St Blaize Formation as described by du Toit (1976) ranges from Lower Eocene (boreholes F-1, G(a)-A/1, G(a)-A/3, G(b)-Springbok/1) to Upper Maastrichtian (G(b)-Gemsbok/1). In most boreholes, however, it lies in the lower part of the Palaeocene and it is possible that contamination in ditch cuttings could account for some of the discrepancies. Nevertheless, McLachlan & McMillan (1979), with more data to hand, still suggested a range of Lower Palaeocene to Upper Palaeocene. They also indicate a carbonate sequence at the top of the Cape St Blaize Formation which ranges into the Lower Oligocene. This presumably refers to the Unit 4 (which contains subordinate sandy limestones) of du Toit (1976). Seismic evidence (e.g. Dingle 1973a) indicate that the upper boundary of the Cape St Blaize Formation is an erosional unconformity below the Neogene Agulhas Formation which spans upper Lower Oligocene to Lower Miocene time.

Palaeogene igneous rocks have been recorded from three places on the Agulhas Bank, all within the Mesozoic Bredasdorp Basin (Fig.160). The largest covers

Table 44. Planktonic foraminiferal zonation of the Cape St Blaize Formation, Agulhas Bank (after McLachlan & McMillan 1979).

| | | | |
|---|---|---|---|
| Oligocene | | *Gl.yeguaensis* | P18-19 |
| | Lower | *G.centralis* *Globigerapsis index* | P16-17 |
| Eocene | Upper | *Globigerapsis mexicana* | P15 |
| | | *Truncorotaloides rohri* | P14 |
| | Middle | *G.bullbrooki* *Tritaxilina colei* | P12-13 |
| | | *Gl.boweri* | P10-11 |
| | Lower | *Gl.collactea* | P8-9 |
| | | *G.rex* | P6-7 |
| Palaeocene | Upper | *G.aequa* | P5-6 |
| | | *G.pseudomenardii* | P4 |
| | | *G.angulata* | P3 |
| | Lower | *Gl.* cf. *pauciloculata* | P2 |
| | | *Globoconusa daubjergensis* | P1 |

*Gl.* = *Globigerina*
*G.* = *Globorotalia*

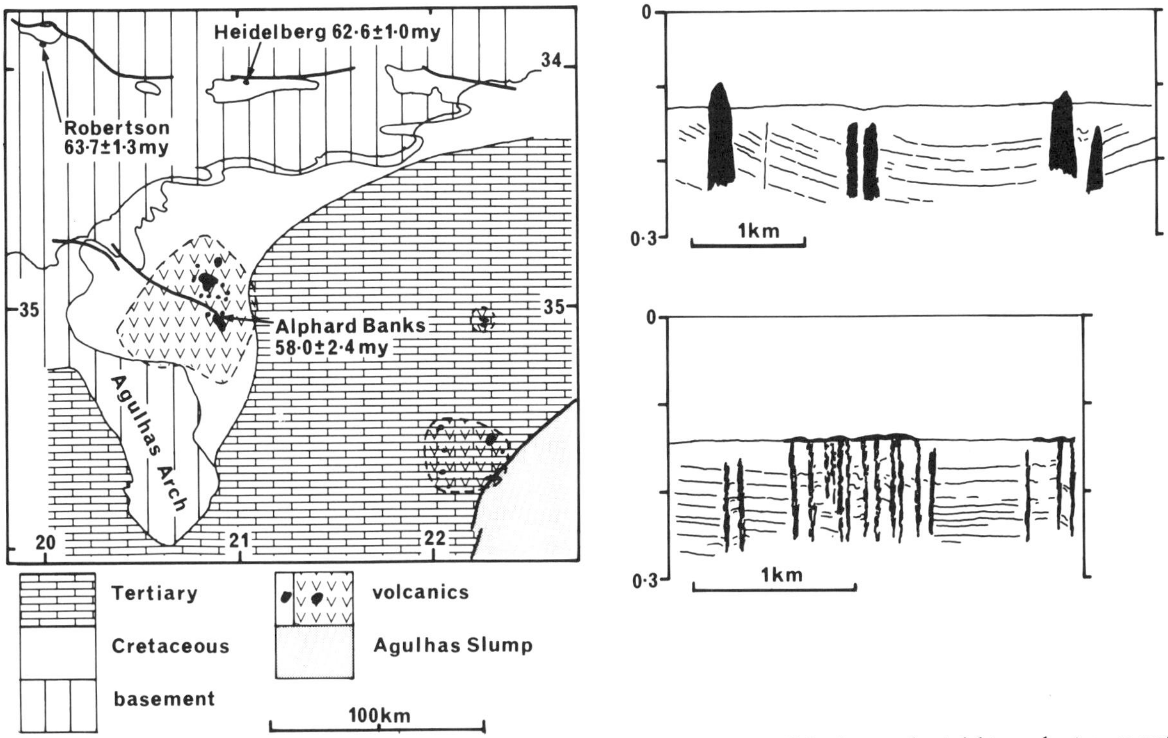


Figure 160. Tertiary igneous rocks on the Agulhas Bank and adjacent southern Cape. Seismic records at right are short cross-sections in the vicinity of the Alphard Banks, and show plugs intruding gently folded and faulted Cretaceous sediments, with lavas at the sea floor. Vertical scales are in sec. Data from Dingle & Gentle (1972) with additions from SOEKOR (1976) and Duncan *et al.* (1978).

about 2 900 km² centred on the Alphard Banks (about 50 km SE of Cape Agulhas), a cluster of small jagged pinnacles that protrude to within 18 m of the sea surface. Here, tuffs, trachybasalts, aegerine-augite trachytes and aegerine-augite phonolitic trachytes dated at 58 ± 2.4 my (K/Ar) have been intruded into and extruded onto Upper Cretaceous sediments and the Maastrichtian-Palaeocene erosion surface (Dingle & Gentle 1972). The area lies at the NW end of the Bredasdorp Basin, mostly on fault-bounded basement blocks to the north of the main boundary fault of the Agulhas Arch. This suggests that reactivation of Jurassic-Lower Cretaceous fault lines in late Cretaceous-early Tertiary times caused gentle folding and faulting of the older sediments facilitating the upward movement of the alkaline magmas.

A similar igneous centre lies about 120 km farther SE at the eastern end of the Bredasdorp Basin. It covers about 1 200 km² and is located on the F-F/2 structure (see Fig.68). No samples have been collected from the area, but small pinnacles representing post-Cretaceous plugs protrude to within short distances of the sea floor, and are buried by Tertiary sediments. A similar, but much smaller centre lies about 50 km due north. The general tectonic and stratigraphic setting of these two centres suggest that they are of a similar age (i.e. Palaeocene) to the Alphard Banks volcanics.

Alkaline intrusions (olivine melilitites) are known from three areas onshore in the southern Cape: Robertson, Heidelberg, and Sutherland/Saltpetrekop (Figs 160 & 168), and various aspects of their geology have been summarized by Rogers & du Toit (1904a,b), Taljaard (1936), and Gerrard (1958). All are plug-like bodies and the first two intrude Mesozoic sediments along the Worcester-Pletmos graben line. Radiometric dating of these by Duncan et al (1978) indicates that the Alphard Banks volcanics represent the climax of alkaline magma generation in the SW Cape that spanned Campanian to Palaeocene times in which strong, but localized tectonic control is evident in all areas except Sutherland/Saltpetrekop, where field relationships are not clear. The main characteristics of the alkaline igneous rocks in this area are summarized in Table 45.

### 6.1.2 *East coast and adjacent offshore areas*

It is only in the eastern part of the subcontinent that marine Palaeocene, Eocene and Oligocene strata are known both onshore and offshore (Figs 156 & 161), and apart from small outcrops on the southern end of the Transkei Swell (Birbury and Needs Camp), all the localities lie north of 30°S. The Richards Bay-Tugela Cone area, which contains a maximum thickness of 1.3 km of Palaeogene sediments, is separated from the extensive south Mozambique deposits by a wide zone in northern Zululand which is devoid of Lower Tertiary strata. North of 27°S the area of coastal plain underlain

Table 45. Late Cretaceous-early Tertiary alkaline igneous rocks in the southern Cape.

| Locality | Nature of bodies | Main rock types | Setting | Size | Age my |
|---|---|---|---|---|---|
| 1. Alphard Banks | plugs, lavas, dykes & sills | plugs: aegerine-augite trachyte, aegerine-augite phonolite, trachyte<br>lavas: tuffs, olivine basalt, trachybasalt | intrudes folded Cretaceous sediments on a horst | 2 900 $km^2$, plugs up to 200 m across | 58 ± 2.4 (Palaeocene) |
| 2. F-F/2 structure | plugs & ?extrusives | unknown | intrudes Cretaceous sediments on a horst | 1 200 $km^2$, size of plugs unknown | not known |
| 3. Unnamed, 50 km north of F-F/2 | plugs | unknown | intrudes Cretaceous sediments, north side Bredasdorp Basin | unknown | not known |
| 4. Heidelberg | small plug | olivine melilitite, $SiO_2$ & $TiO_2$-rich | intrudes Enon Formation in half graben | 200 m across | 62.6 ± 1 (Palaeocene) |
| 5. Robertson | small plug | olivine melilitite, $SiO_2$ & $TiO_2$-rich | intrudes Palaeozoic sediments just south of Mesozoic half graben | 200 m across | 63.7 ± 1.3 (Palaeocene) |
| 6. Sutherland commonage | composite intrusion with main vent & several cone sheets | olivine melilitite, MgO $P_2O_5$, CaO-rich | no obvious tectonic control, intrudes Permian sediments | several hundred metres across | 74.9 ± 1.3 (av.2 determinations) (Campanian) |
| 7. Saltpetrekop | numerous plugs, & dykes around a central vent filled with tuff & breccia. Other plugs in vicinity | olivine melilitite, trachyte, carbonatite & melilitite basalt | no known tectonic control, intrudes Permian sediments | total area at least 80 $km^2$. Individual plugs several metres across | 74.7 ± 1.3 (av.2 determinations) (Campanian), plug from Matjesfontein |

by Palaeogene rocks widens steadily to reach 400 km at about 23°S, although over most of this area thicknesses are less than 400 m and it is only in the coastal zones, and particularly adjacent to the Limpopo and Zambezi rivers, that thickness exceeds 1 km (Fig.156).

6.1.2.1 *South Mozambique*

The Palaeogene sediments of south Mozambique constitute a complex of shallow water carbonates, and argillaceous and sandy rocks (Palaeocene-Eocene) which are disconformably overlain by somewhat deeper water argillites, and at the top by a local evaporite development (Oligocene). The temporal and spatial distribution of these various lithofacies is summarized in Figure 161.

In the proximal parts of the Grudja depocentres (Fig. 130) there is a non-sequence across the Maastrichtian/Palaeocene boundary (Flores 1973), but once re-established, Palaeocene sedimentation in these areas (in the north, and SW) was of an identical lithofacies (glauconitic sands and shales) to the Cretaceous Grudja Formation, although the sequence is characterized by numerous hiatuses and local facies changes. One of the more important of these is represented by the fine-grained calcareous sands and silts of the Middle Palaeocene Mangulane Formation (Förster 1975a). Flores (1973) quotes a K/Ar date of 60 ± 6 my from glauconites from these Lower Tertiary Grudja lithofacies, which borehole data show stretched from the region of Beira almost into Zululand. In the main Grudja depocentre, there is apparently no sedimentary break across the Mesozoic/Cenozoic boundary, and in the lower Palaeocene these argillaceous rocks were replaced by a carbonate lithofacies known at outcrop in the north of the area, from which locality it is named: Cheringoma Formation. This is an algal-oolitic limestone sequence which becomes sandier and marly towards the east. At the top, where it is overlain by the argillaceous Oligocene rocks, the Cheringoma Formation is represented by coquinas of large foraminifera (*Nummulites*). Flores (1973) reports thicknesses of up to 100 m for the Cheringoma Formation in the type area, but this increases to at least 500 m in the subsurface beneath the Zambezi delta.

In the vicinity of Maputo, the Lower Palaeocene Cheringoma Formation interfingers with the Grudja Formation and its sandier equivalents, but in the Upper Palaeocene these are replaced by a fine grained, foraminiferal phosphatic and glauconitic limestone lithofacies called the Salamanga Formation by Frankel (1972) and Flores (1973). Flores (1973) suggests that the

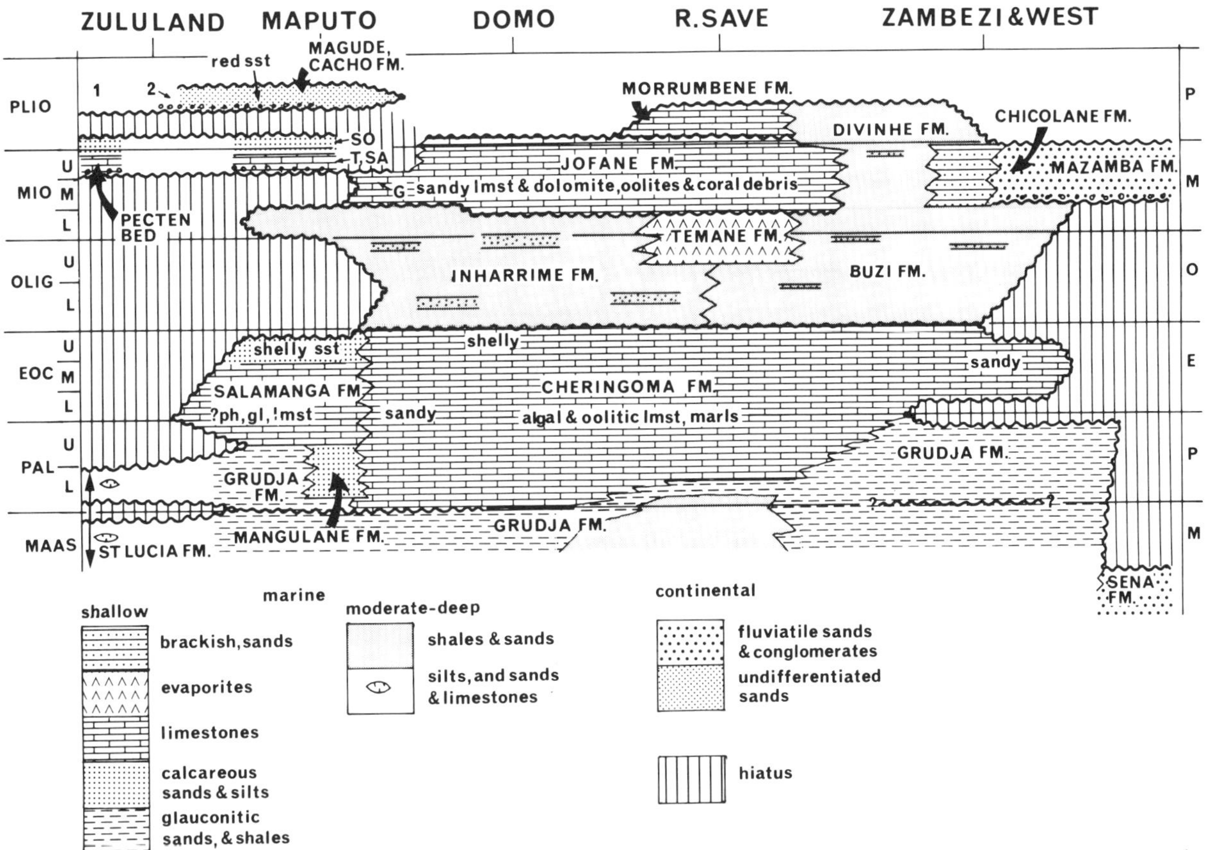


Figure 161. Schematic lithostratigraphic correlation of Tertiary strata between Zululand and the Zambezi area. Approximate length of section is 1200 km. Not to scale. For cartographic convenience the Sena Formation is placed above the Santonian to illustrate the unconformable relationship between it and the Mazamba Formation. Data mainly from du Toit (1954); Frankel (1972); Flores (1973), Forster (1975a). Abbreviations: SO = Santiago Formation; T = Tembe Formation; SA = Santaca Formation; G = Govuro Formation; ph = phosphatic; gl = glauconitic; 1 = Umkwelane Hill, Uloa; 2 = Lakeview.

Grudja Formation passes conformably into the Salamanga Formation, but south of Maputo the latter rests unconformably upon the Maputo Formation. The upper part of the Salamanga Formation consists of coarse shelly sands.

The main depocentre for Palaeocene-Eocene rocks in Mozambique lay under the Zambezi Cone, where over 700 m of sandy Grudja and Cheringoma rocks have been drilled (Flores 1973) (Fig.162a). These thin rapidly westwards, and apart from a large salient of the 200 m isopach at about 23° S, are generally less than 200 m thick over most of the coastal plain.

There was a shallowing of the sedimentary environment in late Eocene times when the Salamanga and Cheringoma formations were represented by coarser and sandier lithofacies. Oligocene sediments are absent from most of the Mozambique coastal plain, but relatively thick deposits do occur in the coastal region and these are the only known marine Oligocene rocks onshore in southern Africa (Flores 1973). They consist of a northern clay/sand/limestone lithofacies (Buzi Formation) and a southern sandy clay/marl/calcareous sand lithofacies (Inharrime Formation). The Buzi Formation reaches a maximum known thickness of 701 m at the southern end of its distribution, whilst the Inharrime Formation is 760 m thick in the Zandamela borehole. Both have a rich planktonic foraminifera fauna characterised by *Globigerina oligocaenica* and *G. tripartita topurensis*, and Flores (1973) reckons them to represent relatively deep water environments. In the south, the Inharrime Formation ranges up into the Lower Miocene, whilst in the north the Buzi argillaceous facies is continuous into the Pliocene, where Flores (1973) calls the top 182 m the Divinhe Formation although he gives no details as to how it can be recognized as a separate lithostratigraphic unit in subsurface intersections.

In the vicinity of the Save River mouth, the top of the Oligocene is marked by a local development of dark grey clays with interbedded gypsum/anhydrite beds and gypsiferous limestones and sands. This lithofacies is only known from boreholes in which it has been mapped over 37 500 km$^2$ and reaches a maximum thickness of 173 m south of the Save estuary. Flores

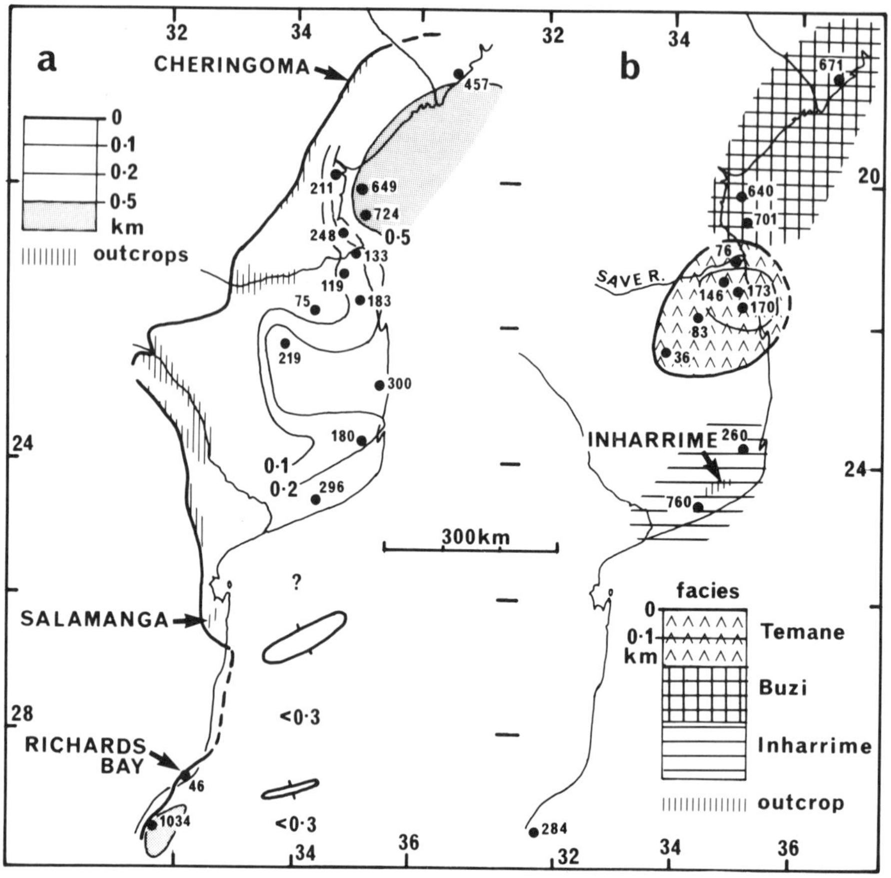


Figure 162. Distribution and thickness of Palaeogene sediments in Zululand and S.Mozambique.
a. Palaeocene and Eocene. In Mozambique isopachs exclude the Tertiary part of the Grudja Formation, and in the Natal Valley all strata below Angus (mid-Tertiary) are included. Borehole values are in m, isopachs in km.
b. Oligocene. Natal Valley rocks of this age are included on map (a).
Data from Flores (1973), du Toit & Leith (1974), Stapleton (1975), Forster (1975a, 1975b), McLachlan & McMillan (1979), and Figure 106.

(1973) has called it the Temane Formation, and although no recognizable fossils have been found in it, *Streblus becarri* is abundant above and below it, placing the formation somewhere between Upper Oligocene and Lower Miocene.

### 6.1.2.2 *Richards Bay, Tugela Cone and northern Natal Valley*

The only known Palaeocene rocks onshore in South Africa occur in boreholes and excavations at Richards Bay (Fig.163), where 46 m of dark olive-grey silts with thin lenses of clay and hard, sandy limestone concretions dip seaward at 2-3° (Maud & Orr 1975). Although they are lithologically similar to, and are included within, the St Lucia Formation, there is a non-sequence across the Cretaceous/Palaeocene boundary (Fig.161). Stapleton (1975) has in fact recognized two hiatuses, one at the top of the *Globotruncana gansseri* Zone (UC14 of van Hinte 1976a), and a larger one spanning the top of the *Globotruncana mayorensis* to the base of the *Globorotalia triloculinoides* zones. Maud & Orr (1975) report that the siltstones below the boundary have a load-equivalent of 260 m, whilst those above have a value of 140 m suggesting that 120 m of sediment were removed from the top of the Cretaceous sequence prior to deposition of the Middle Danian, and that 95 m have been removed from the top of the present Palaeocene sequence. Figure 163 shows that the upper surface of the Palaeocene is cut across by a pre-Upper Miocene planation, and that post-Miocene erosion has carved deep channels in both the Cretaceous and Tertiary sediments.

Moulds and casts of lamellibranchs and gastropods are the most common fossil remains in the Richards Bay Palaeocene, although nautiloids, corals, echinoderms, decapods and fish remains have also been found (Orr & Chapman 1976). Micro- and nannofossils are particularly abundant with over 100 species of benthic foraminifera, ten species of planktonic foraminifera, several species of ostracods, and 15 species of calcare-

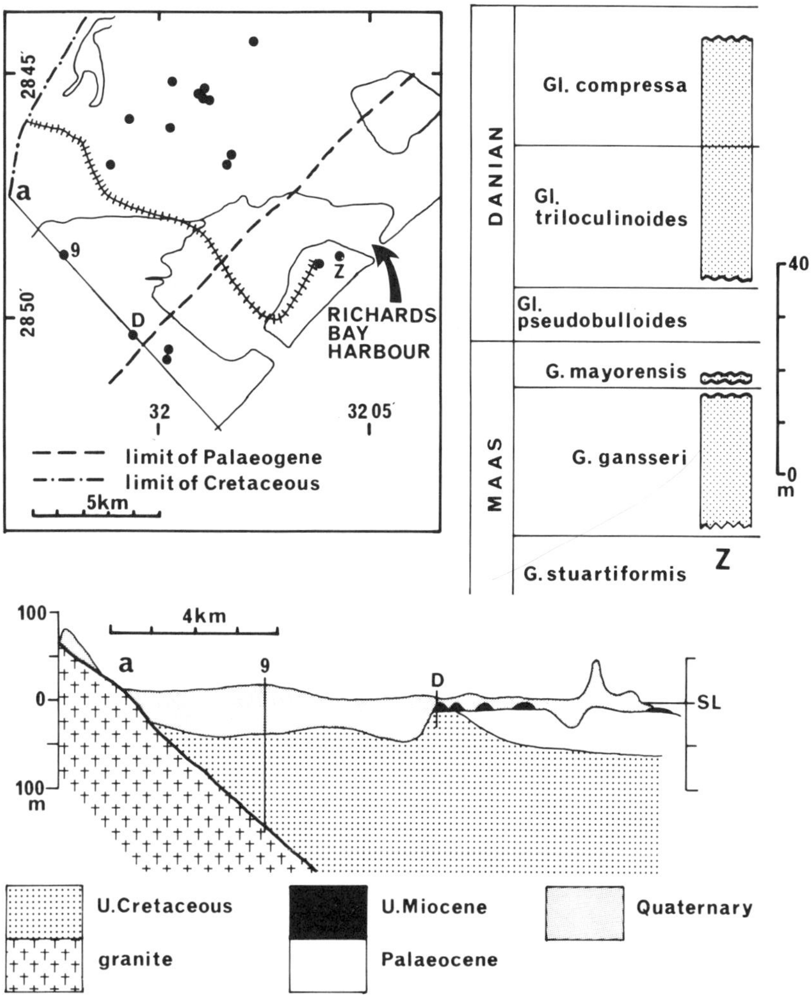


Figure 163. Palaeogene sediments in the Richards Bay area. Map at top left shows location of cross-section (a) and borehole log Z. The latter is plotted at top right and shows non-sequences across the Cretaceous/ Palaeocene boundary. In the cross-section, the Quaternary sediments include the red Berea Sands, Port Durnford beds, and alluvium. G = *Globotruncana,* Gl = *Globorotalia.* Data from Maud & Orr (1975) and Stapleton (1975).

ous nannofossils (Orr & Chapman 1974; Stapleton 1975). On the basis of the relative abundances of the planktonic foraminifera, and the generic composition of the benthic foraminifera population, Orr & Chapman (1974) suggested a mid-continental shelf environment for the Richards Bay Danian sediments.

Seismic and borehole data indicate that the Palaeogene rocks in Zululand and southern Mozambique dip and thicken offshore into the Natal Valley, and probably reach their maximum development under the Tugela Cone (Figs 106 & 162). Borehole J(c)-1 (du Toit & Leith 1974) intersected 1 034 m of Palaeocene-Eocene seaward prograding sediments which thicken to about 1 300 m 10 km farther east (Figs 106 & 171). In the borehole there is a non-sequence across the Mesozoic/Cenozoic boundary (McLachlan & McMillan 1979), but this does not seem to coincide with a prominent reflecting horizon on seismic records across the top of the Tugela Cone (Fig.171). Palaeogene sediments in the J(c)-1 borehole can be divided into three lithological units (Fig.164), modified by McLachlan & McMillan (1979), after the work of du Toit & Leith (1974). The lower unit (4) has a basal 180 m of light grey clays with numerous thin sandstones and sandy limestones which are separated by a 6 m undercompacted clay from 646 m of monotonous light grey-greenish grey clay and minor sandstones and sandy limestones. These clays have small amounts of shell fragments, glauconite, pyrite and lignite, and they are overlain by 228 m (unit 3) of Upper Lutetian-Priabonian buff, sandy, argillaceous limestones with abundant shell fragments and rare glauconite grains. Individual sandstone bodies become better sorted upwards and are considered by du Toit & Leith (1974) to be typical of regression-laid sands. An erosional hiatus separates these Upper Eocene limestones (with *Globorotalia cerroazulensis* and *Cribrohantkenina inflata* (Priabonian) ) from the overlying Oligocene sediments (with *Globigerina sellii, G.angulisutura* and *G.tripartita topuriensis* (Lattorfian)), and this horizon can be seen on seismic records as a strong unconformity.

The Oligocene section is 285 m thick and consists

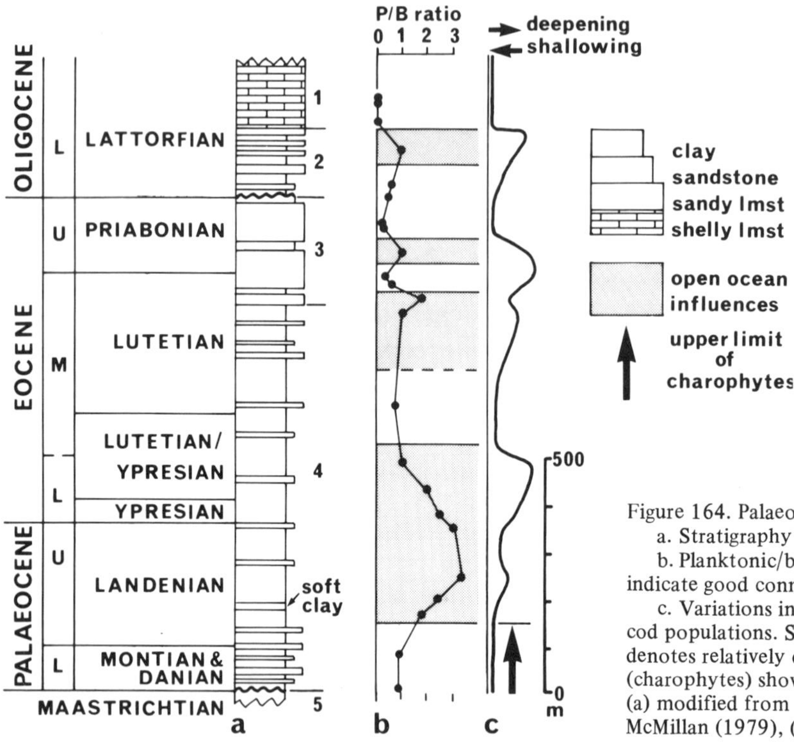


Figure 164. Palaeogene succession in the J(c)-1 borehole.
a. Stratigraphy and lithofacies of units 1-4.
b. Planktonic/benthic foraminiferal ratios. Values of 1 or >1 indicate good connections to oceanic waters (shaded sections).
c. Variations in depth of deposition (no scale) based on ostracod populations. Significant presence of *Bairdoppilata* and *Krithe* denotes relatively deep water. Upper limit of fresh water algae (charophytes) shown by bold arrow.
(a) modified from du Toit & Leith (1974) after McLachlan & McMillan (1979), (b) & (c) after Dingle (1976).

of a lower unit (129 m, Unit 2) of glauconitic, calcareous clays and quartzose and shelly limestone with abundant fragments of *Pecten* and reworked Eocene foraminifera (du Toit & Leith 1974). The coquina is considered by du Toit & Leith (1974) to be a diachronous lateral equivalent of the Uloa *Pecten* bed, implying a slow landward transgression beginning in early Oligocene times that did not reach coastal Natal until the late Miocene.

A study of the benthic microfossils (foraminifera and ostracods) from borehole J(c)-1 (Dingle 1976) shows that the Danian to early Landenian section was deposited under conditions similar to those that had obtained during the late Cretaceous, i.e. shallow restricted marine, with large influxes of terrigenous debris, including freshwater algae (charophytes). Above these levels, the planktonic/benthic foraminifera ratio increases sharply and the diversity and abundance of ostracods increases, with an accompanying decrease in the percentage of agglutinated benthic foraminifera. Faunas indicative of open ocean and more restrictive environments alternate up the borehole, presumably reflecting sea level changes brought about by either eustatic events, or alterations in the local configuration of the Tugela delta top (Fig.164). There is a sudden decline in the planktonic/benthic foraminifera ratio between lithological Units 1 and 2 in the Oligocene. This coincides with the disappearance of deep water ostracod species such as *Krithe* and across which boundary there is less than 7 % similarity at the species level in the ostracod fauna. These changes are taken to indicate a regression and are similar, but more marked, than those noted across the Upper Eocene-Lower Oligocene hiatus.

Figure 171 shows a correlation of these major lithostratigraphic events with the acoustic stratigraphy on top of the Tugela delta. Although the Maastrichtian/Palaeocene boundary is prominently displayed on the figure by shading, it is not in fact an horizon that can be traced with any certainty down the cone and into the Natal Valley (Fig.106), whereas the Upper Eocene-Lower Oligocene hiatus shows as a major unconformity and reflecting horizon that has been recognized over the whole of the adjacent deep sea area – horizon Angus of Dingle et al (1978). This means that unfortunately we have no way of confidently identifying the Palaeogene (or the Upper Cretaceous) sediment sequence other than on the outer edge of the Tugela Cone itself. Here the Palaeocene to Eocene (i.e. pre-Angus) succession is about 1.1 sec thick (at least 1 100 m), but it must thin rapidly east and northwards because in the Natal Valley east of the cone the whole mid-Cretaceous to Lower Oligocene section is no thicker than 0.5 sec (500 m), and between the Naudé and Almirante Leite ridges 0.5-0.28 sec (500-280 m). These figures imply that whilst Lower Tertiary (Palaeocene to Eocene) sedimentation rates on the Tugela Cone were relatively high (39 m/my), they were probably between about 9

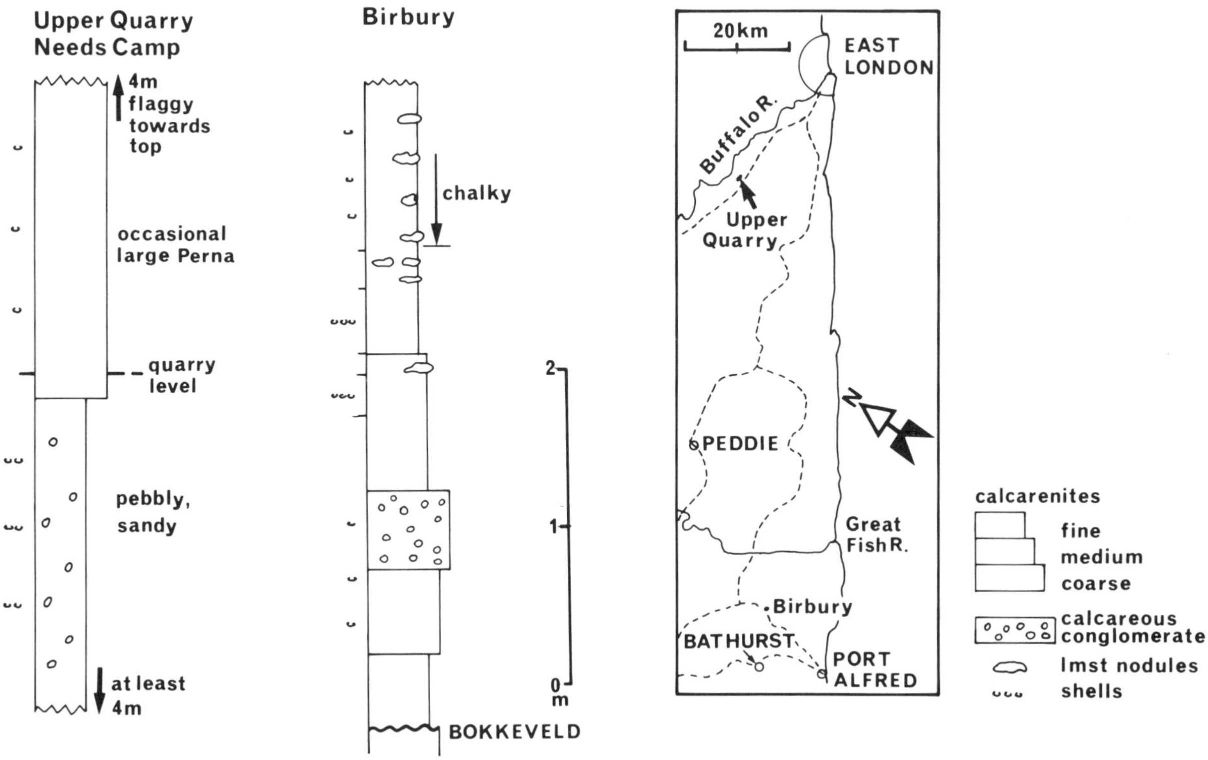


Figure 165. Palaeogene outcrops in the Eastern Cape. Map at right shows locations, roads are dashed lines.
Upper Quarry Needs Camp section is diagrammatic and based on the descriptions given by Lock (1973), and Siesser & Miles (1979). Pebbles are composed of limestone (some glauconitic), sandstone, dolerite, and chert. The rocks are probably Eocene in age (see Table 47 for microfossil lists).
Birbury section was measured by Siesser & Miles (1979). Pebbles are composed of quartz, quartzite and siltstone. The rocks are Lower Eocene in age (see Table 46 for microfossil lists).

and 5 m/my in the deep water areas, with the lower rates prevailing over the Central Terrace (between the two ridges). Thicker pre-Angus sediments are suspected north of the Almirante Leite Ridge but no values other than about 1 000 m of Palaeogene in the Zandamela borehole are available. Seismic evidence of sediment thickness and bedform changes indicate that pre-Angus sediment starvation (i.e. during the Palaeocene-Eocene interval) was due to a combination of shielding by basement ridges, and high bottom currents, although the relative importance of these factors is difficult to gauge at present. The thick deposits in the Tugela Cone indicate that it was not due to an overall drop in sediment input. High sea floor currents are also considered the cause of non-deposition of sediment on the Mozambique Ridge DSDP site 249 during Upper Maastrichtian to Miocene time (Fig.107).

6.1.2.3 *Birbury and Upper Quarry Needs Camp*
Palaeogene rocks are poorly represented on the Transkei Swell. No Palaeocene sediments are known onshore, and none have so far been sampled offshore, although they are suspected in the deep water distal Natal Valley. Two small outliers of Eocene strata occur SW of East London at Birbury and in the Upper Quarry at Needs Camp (Fig.165). At Birbury (205 m above sea level) 1 to 1.5 m of sparsely fossiliferous, poorly-consolidated calcareous conglomerate and medium-coarse calcarenite rest on weathered Bokkeveld basement, and are overlain by about 3 m of nodular, fine to medium calcarenites (chalky at the top) (Siesser & Miles 1979). Sharks teeth, molluscs, polyzoa and echinoid fragments are scattered intermittently throughout the section but are most abundant in the medium calcarenite above the conglomeratic unit. It should be noted that the general lithology of the Birbury Beds is dissimilar to the Neogene Alexandria Formation widely exposed farther west, and to the Upper Quarry Needs Camp strata to the SE.

On the basis of the benthic foraminifera *Discocyclina pratti,* which occurs in abundance along with *D. varians,* Chapman (1930) assigned an Upper Eocene (Bartonian = Priabonian) age to the Birbury beds, and this was generally accepted until Bourdon & Magnier (1969) reexamined the outcrop. They reported an abundant but poorly preserved Eocene assemblage of benthic foraminifera (at least 20 species), mixed with a sparse, but well preserved Miocene assemblage of five tentatively identified benthic foraminifera. On the basis of this composite assemblage they assigned a Miocene age to

Table 46. Planktonic foraminifera and calcareous nannofossils from the Birbury beds (early Eocene) (data from Siesser & Miles 1979).

| Calcareous nannofossils | Age range |
|---|---|
| *Braarudosphaera bigelowi* | Palaeocene-Pliocene |
| *Chiasmolithus solitus* | Lower-Middle Eocene |
| *C.* sp. | |
| *Coccolithus eopelagicus* | Palaeocene-Pliocene |
| *C.formosus* | Eocene-Oligocene |
| *Cruciplacolithus* sp. | Palaeocene-Eocene |
| *Cyclococcolithus gammation* | Palaeocene-Eocene |
| *Discoaster* sp. | – |
| *?Lophodolithus nascens* | Palaeocene-Eocene |
| *Pontosphaera* spp. | – |
| *Reticulofenestra coenura* | Eocene-Oligocene |
| *Transversopontis pulcher* | Eocene |
| *Zygodiscus plectopons* | Palaeocene-Eocene |
| *Zygrhablithus bijugatus* | Eocene-Oligocene |
| Planktonic foraminifera | |
| *Acarinina esnaensis* | Palaeocene-Lower Eocene |
| *A.nitida* | Palaeocene-Lower Eocene |
| *A.primitiva* | Upper Palaeocene-Middle Eocene |
| *A.pseudotopilensis* | Palaeocene-Lower Eocene |
| *A.soldadoensis soldadoensis* | Palaeocene-Lower Eocene |
| *Morozovella subbotinae* | Lower Eocene |
| *Subbotina eocaena* | Lower Eocene-Upper Oligocene |

these rocks, suggesting that the diverse, abundant, and confidently identified Eocene forms were reworked. Siesser & Miles (1979) have made a further study of the Birbury microfossils and identified at least 14 species of calcareous nannofossils and six species of planktonic foraminifera. These are mostly Palaeogene forms, plus a few longer-ranging types, but none are specifically Neogene. The overlapping age ranges of several of the nannofossils and foraminifera (Tab.46) especially *Morozovella subbotinae* and *Chiasmolithus solitus* indicates an early Eocene age for the Birbury beds (planktonic foraminifera zones P6-8).

At Needs Camp, marine limestones crop out in two quarries at an elevation of about 400 m. They are separated vertically by about 17 m and the Lower Quarry (or East Quarry) contains Upper Cretaceous strata (see 5.2.4). A total of about 8 m has been recorded at various times from the Upper Quarry, although only 4 m occur above the present quarry floor (Fig.165). The lower part consists of pebbly and sandy fine calcarenites and the upper part coarse calcarenites which become flaggy towards the top. The rocks are generally hard, recrystallized and yellowish grey in colour, whilst the pebbles in the fine calcarenites consist of limestone, sandstone, dolerite and chert. Siesser (1971) reports that in thin sections the rock (above the quarry floor) is a skeletal grainstone rich in molluscs, cirripeds, polyzoa and coralline algae cemented by microspar. Large *Perna* valves are a conspicuous constituent of the rock at outcrop. These rocks were originally considered to be lateral equivalents of the Alexandria Formation (Bullen-Newton 1913, du Toit 1954), which was itself thought to be Eocene in age by correlation with the Birbury beds (Chapman 1930, Haughton 1925). Lock (1973) reviewed the literature, but despite the presence of Eocene benthic foraminifera, was wary of assigning a Palaeogene age, probably in view of the allegedly reworked Eocene foraminifera at Birbury (reported by Bourdon & Magnier 1969). Siesser & Miles (1979) have extracted a sparse assemblage of calcareous nannofossils (Tab.47) from some of the softer layers near the quarry floor which consists of a mixture of Tertiary and reworked Cretaceous taxa. The ratio is about 1:10, respectively. No identifiable planktonic foraminifera were recorded, but the ranges of the Tertiary nannofossils indicates an Eocene (probably Lower Eocene) age. This suggests that the Birbury and Upper Quarry Needs Camp outcrops were deposited during the same high sea level stand. The heavy reworking also indicates that the Eocene strata in the Upper Quarry physically overlie the late Cretaceous Lower Quarry beds below the level to which excavations in the Upper Quarry have been taken.

### 6.1.3 *West Coast (Walvis Ridge abutment to Agulhas Fracture Zone)*

Palaeogene sediments underlie most of the continental margin off the west coast (Fig.156) but exposures are limited to a narrow inner zone, and relatively little is known about the lithofacies and internal stratigraphy of the succession. Under the continental shelf, Lower Tertiary strata form a seaward thickening wedge which rests with a strong unconformity on the underlying Mesozoic sediments, and may locally overstep onto the basement (Fig.173). In some areas under the continental slope, Palaeogene sediments can be seen resting paraconformably on the underlying Cretaceous (Fig.149) but elsewhere large-scale slumping has disrupted the sediment pile to such an extent that it is not possible to confidently locate the Mesozoic/Cenozoic boundary. Details of the lithology of the Palaeogene rocks are limited to DSDP boreholes in the deep ocean basin (sites 360, 361 and 362), borehole Kudu 9A-1, and

Table 47. Tertiary calcareous nannofossils from the Upper Quarry at Needs Camp (Eocene) (data from Siesser & Miles 1979).

| | Age range |
|---|---|
| *Braarudosphaera bigelowi* | Palaeocene-Pliocene |
| *Chiasmolithus* cf. *C.solitus* | Lower-Middle Eocene |
| *Coccolithus eopelagicus* | Palaeocene-Pliocene |
| *C.* cf. *C.formosus* | Eocene-Oligocene |
| *Cyclococcolithus gammation* | Palaeocene-Eocene |
| *?Cyclicargolithus floridanus* | ? |
| *Reticulofenestra coenura* | Eocene-Oligocene |
| *Toweius* sp. | not found above Lower Eocene |
| *?Zygrhablithus bijugatus* | Eocene-Oligocene |

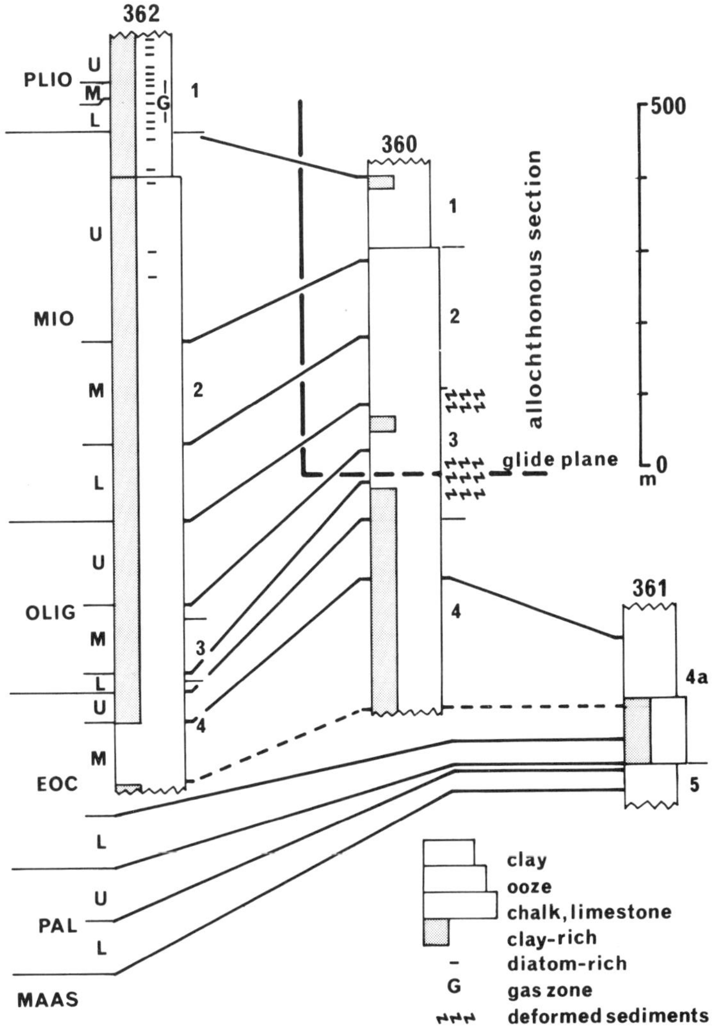


Figure 166. Tertiary succession in three DSDP boreholes off S.W.Africa. The dashed line (Middle Eocene) equates synchronous horizons at the level of *Nannotetrina fulgens* zone. Upper part of the sequence at site 360 (S.W. Cape Town) is allochthonous, where it has been involved in slumping. Site 362 (Frio Ridge) shows distribution of diatomaceous and gas-charged sediments. Numbering sequence at each site is the DSDP shipboard notation. Compiled from core log descriptions in Bolli *et al.* (1978). Data on slump after Dingle (1980b).

surface sampling sites in the Cape Canyon,on the continental shelf, and scattered small outcrops in the coastal zone between Lüderitz and about 31°S.

6.1.3.1 *DSDP sites 360, 361 and 362 (Cape Basin and Walvis Ridge)*

A composite section through the Palaeogene sediments of the deeper oceanic areas is provided by DSDP boreholes 360, 361 and 362 (Fig.166). In the Cape Basin SW of Cape Town (site 361), grey Upper Maastrichtian shales are abruptly overlain by 23 m of brown pelagic clay rich in zeolites and fish remains (Unit 5 of Bolli et al 1978). Its average composition is: clay – 77 %, quartz – 11 %, zeolite – 4 %, fish remains – 2 %, iron oxides – 3 %, nannofossils – 1 %, trace components – < 1 %. A minor 80 cm bed of marly nannofossil chalk occurs near the top of the clay sequence, whilst euhedral authigenic barytes is locally abundant in the coarse fractions (up to 20 %) and manganese nodules occur at one horizon. The exact position of the Mesozoic/Cenozoic boundary cannot be located within this clay, but it probably ranges in age from uppermost Maastrichtian to the lower part of the Upper Palaeocene, and represents a period of slow sedimentation beneath the carbonate compensation level. This clay unit corresponds with the regionally developed seismic reflecting horizon Davie of Emery et al (1975) which can, therefore, be used to approximately mark the Cretaceous/Palaeocene boundary (assuming a P-wave velocity of 2 km/sec it would appear only 0.023 sec thick on seismic records). It is overlain by at least 340 m (a composite thickness from sites 360 and 361, Fig.166) of mud, calcareous mud and marly nannofossil ooze (site 361) which passes laterally into greenish grey marly nannofossil chalk at site 360. Table 48 gives details of these lithofacies, which are characterized by a relatively high clay content (35-55 %). As might be expected, the series appear to thicken towards the continent (compare the thickness of Middle Eocene sediments at sites 360 and 361, Fig. 166). Eocene strata lie at the sea floor in DSDP 361

Table 48. Lithology of Palaeogene sediments (Units 3 and 4) at DSDP Sites 360 and 361 (data from Bolli *et al.* 1978).

| | Colour | Composition % Nanno-fossils | Clay | Quartz | |
|---|---|---|---|---|---|
| Oligocene (part of Unit 3) | | | | | |
| 360 | | | | | |
| nannofossil chalk with subordinate marly horizons | light-medium grey, olive-grey laminae | 75 | 20 | minor | |
| Minor amounts of: foraminifera, quartz, pyrite, heavy minerals, fish fragments, ostracods, glauconite and mica | | | | | |
| Palaeocene-Eocene (Unit 4) | | | | | |
| 361 | | | | | |
| a) Marly nannofossil oozes and chalks | light brown, yellowish-grey | 55 | 35 | 7 | <1 % trace components |
| b) Muds and calcareous muds | light brown, brown-greenish grey | 8 | 50 | 35 | 3 % heavy minerals, 2 % feldspar |
| Heavy minerals include: zircon, garnet, epidote, opaques, chlorite, tourmaline and sphene | | | | | |
| Also present: glauconite, fish remains | | | | | |
| 360 | | | | | |
| Marly nannofossil chalk | greenish-grey | 40 | 55 | 3 | ~2 % trace components |
| Trace amounts include: foraminifera, pyrite, glauconite, fish remains, authigenic carbonate, plagioclase, microcline, zircon, tourmaline, garnet, epidote, chlorite and muscovite | | | | | |

although shear strength tests indicate that at least 500 m of overburden have been removed from this site (Bolli et al 1978). Eocene at site 360 is overlain by ca.160 m of Oligocene nannofossil chalks with subordinate marly horizons which make up most of Unit 3 of Bolli et al (1978) (see Tab.48 for details of lithology). Unit 3 is characterized by two zones of deformed sediments, the lower one of which Dingle (1980b) correlates with the main glide plane of a large allochthonous sediment mass on the continental rise (Fig.187). This feature will be discussed in more detail in the section on Neogene strata, but it can be noted here that the intense soft sediment deformation noted by DSDP shipboard scientists include matrix supported clasts, and normal and reversed faults. The main zone of deformation is about 66 m thick and coincides with the boundary between sediments having different physical properties, which are summarized in Table 49. The slump glide plane separates Middle Oligocene sediments (above) from Lower-Middle Oligocene sediments (below) and coincides with a zone that Bolli et al (1978) consider may represent a 'brief undetected hiatus'. There is an apparently conformable transition into the Neogene at site 360, but according to borehole logs reported by Bolli et al (1978) the boundary falls within a second zone of sediment deformation, which Dingle (1980b) suggests could be a subsidiary glide plane within the main allochthonous mass.

Palaeogene sediments at DSDP site 362 on the Frio Ridge comprise ca.380 m of carbonate sediments which Bolli et al (1978) subdivided on the percentage of clay minerals and carbonate components. The succession appears to be complete from Middle Eocene to across the Mio-Oligocene boundary (which falls within DSDP Unit 2) and comprises Units 4, 3 and the lower third of 2. There are: marly limestones characterized by coarse calcarenites and slumps (20-30 % clays); brownish-greenish grey marly nannochalks interbedded with soft white *Braarudosphaera* chalks (15-30 % clays); and brownish-greenish grey marly chalks (foraminiferal and nannofossil rich) (10-40 % clay). Bolli et al (1978, p.234) interpreted the upward fining calcarenite/calcareous siltstone horizons of the Middle to Upper Eocene (unit 4) as turbiditic sequences and speculated that the decrease in the extent of foraminifera preservation in the Oligocene (it becomes more pronounced in the Miocene) reflects high elevations for the carbonate compensation level, which were caused by the establishment of cold water current systems in the South Atlantic. Of particular interest are the Oligocene *Braarudosphaera* chalks. Similar blooms of this genus have been recorded at numerous other DSDP sites in the South Atlantic (14, 17, 19, 20, 22, 331, 358, 359). A small contradiction appears in the report of Leg 40 (Bolli et al 1978) at this point. *Braarudosphaera* chalks are reported to also occur at site 360 (Bolli et al 1978, p.235) and to correlate with those at site 362, yet on page 57 of the same report 'near-shore indications such as *Braarudosphaera* . . . are totally absent'. Notwithstanding this uncertainty, the widespread distribution of this characteristically nearshore (reduced salinity) taxon appears to indicate 'periodic South Atlantic-wide reduction in surface-water salinity . . .' (Bolli et al 1978, p.235). Steinmetz (1981) considers the presence of *Braarudosphaera* to indicate a period of environmental stress which he equates with the intrusion of cool, dense and nutrient-rich Antarctic Bottom Water into the South Atlantic.

### 6.1.3.2 *Continental shelf*

Although reflector Davie of Emery et al (1975), which we nominally take to mark the Maastrichtian/Palaeocene boundary, cannot unequivocally be traced under the present continental shelf, it can be followed part

Table 49. Physical and sedimentological parameters across the base of the Cape Town Slump at DSDP Site 360 (at about 475 m subbottom) (data compiled by Dingle 1980b from Bolli *et al.* 1978).

| | above glide plane | below glide plane |
|---|---|---|
| 1. P-wave velocity | 1.89 km/sec | 2.21 km/sec |
| 2. Coring rate (indicating relative lithification and hardness) | 37 m/hour | 16 m/hour |
| 3. $CaCO_3$ content | >70 % | ~60 % |
| 4. Sediment porosity | 20 % | 30 % |

of the way up the slope at several localities where mid-Tertiary slumping (cycle III, see Tab.66) has not severely disrupted it. Seismic profiling (Dingle 1973c) and borehole data (Kudu 9A-1, McLachlan & McMillan 1979) indicate an erosional unconformity at the base of the Palaeogene (horizon L of SOEKOR, Gerrard & Smith 1980) (see Fig.118), but to date, no Palaeocene sediments have been recorded. The upper boundary of the Palaeogene is also an erosional unconformity, and the sediments between these two bounding reflectors on seismic records have been referred to as T1 by Dingle (1971a, 1973a). They reach a maximum thickness of about 900 m on the upper part of the continental slope SW of the Orange River, which is the general locality of the only published borehole data through T1 (Kudu 9A-1 west of the Orange River McLachlan & McMillan 1979) (Fig.174). Here a thin (?<20 m) Lower Eocene sandstone rests unconformably on Upper Cretaceous (Maastrichtian) sediments and is itself unconformably overlain by an estimated 260 m of Upper Eocene clays and sandstones and ?60 m of Lower Oligocene clays and limestones (Tab.50). It is probable that on seismic records the thin basal Lower Eocene sandstone would not be resolved (<0.02 sec thick) and that it is included within the reflectors marking the base of T1. The regional reflector which coincides with the hiatus at the top of the Lower Oligocene represents an unconformity at the base of the Neogene succession, and on the continental shelf west and NW of Cape Town is associated with large-scale erosion features in the Cape Canyon (Fig.173b). On the basis of seismic data Dingle (1971a) placed these between acoustic stratigraphic Units T1 (Palaeogene) and T3 (Neogene) and concluded that they were Oligocene. This dating has been strengthened by the results from Kudu 9A-1. The present canyon is the youngest erosional feature along the site of what appears to have been an early succession (T2) of cut and fill structures (at least 200 m deep) affecting a zone up to 18 km wide, which were eroded into underlying Eocene clays and locally may have even reached the Upper Cretaceous (Dingle 1971a). Large channel-like, and other irregular features are common along the outer (buried) edge of the shelf and upper slope (e.g. western side of Fig.173b) where they are cut into T1 strata. Dingle (1973c) attributed all these to erosional phenomena, as did van Andel & Calvert (1971) with similar features off Namibia, which they suggested may have been caused by periods of particularly vigorous flow by the Benguela current. Whilst such processes may have operated, and been locally dominant, we suggest that the irregularities along the outer shelf and upper slope may also have been partly the result of fissuring in the head regions of the huge slumped masses that lie along the whole west coast margin (Fig.187), some of which are thought to have formed in mid-Tertiary times (Dingle 1980b). There may even be a link between the Cape Canyon lineament and incisions, and the massive glide plane scars farther south, though the nature of this relationship is rather obscure with present data.

Remarkably rapid thickness variations and locally thick Lower Tertiary sediments occur on the continental shelf under the Walvis Ridge Abutment Plateau. Here, Eocene sediments rest directly on block-faulted Mesozoic (?locally Lower Cretaceous) strata east of DSDP site 362, which penetrated a horst-like feature at the western end of the Plateau. Nearer the continent, asymmetric graben-like basins contain up to 2 km of presumed ?Palaeocene-Eocene (Fig.154). The attitude of some of the reflecting horizons in these basins (Lehner & de Ruiter 1977) suggests that the faulting was synsedimentary, but that it had ceased by Oligocene times. We have no further data on this period of early Tertiary tectonism on the Walvis Ridge, nor on the lithofacies involved, except to note that sediments of presumed Lower Tertiary age nearer the coast form a strongly seaward prograding wedge.

### 6.1.3.3 *Namibia and Namaqualand*

Small scattered outcrops of marine and non-marine beds are known from the coastal zone of southern Namibia, while problematical outcrops occur farther south in Namaqualand (Figs 155, 156 & 167).

Stratigraphic relationships in the numerous small outcrops south of Lüderitz (Fig.167) are not well-established, but there appear to be three distinct associations: a non-marine sequence; a high elevation, poorly fossiliferous and dominantly rudaceous marine sequence; and a lower, fossiliferous, calcareous marine sequence.

Table 50. Palaeogene zonation in borehole Kudu 9A-1 (after McLachlan & McMillan 1979).

| | | | |
|---|---|---|---|
| Oligocene | Lower | *Globigerina yeguaensis* | P18-19 |
| | | *Globorotalia centralis* *Globigerapsis index* | P16-17 |
| Eocene | Upper | *Globigerapsis mexicana* | P15 |
| | Lower | *Globorotalia rex* | P6-7 |
| Maastrichtian | | | |

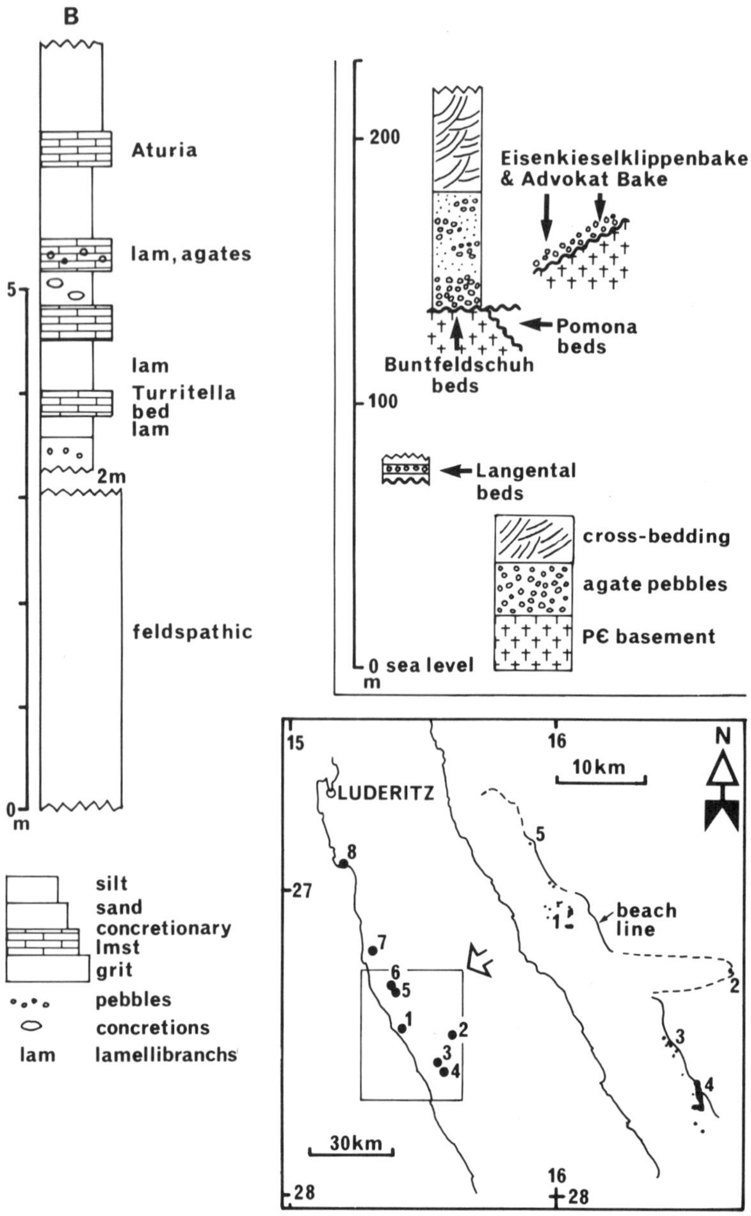


Figure 167. Palaeogene localities in S. coastal Namibia. Map at lower right shows locations of various outcrops. Section at left is that at Bogenfels which is site 1 on location map, and site B in the Langental Valley shown in Figure 150. It was measured by Klinger (1977) and designated as the type section for the Langental Beds by Siesser & Salmon (1979). Measured sections at top right are for various localities arranged relative to present sea level (scale in m on left side). Map also shows late Palaeocene/early Eocene beach line identified by Martin (1973b). Localities: 1 = Bogenfels area; 2 = Eisenkieselklippenbake; 3 = Advokat Bake; 4 = Buntfeldschuh; 5 = Granitburg; 6 = Luteritz Krater; 7 = Pomona; 8 = Elizabeth Bay.

The oldest strata are the Pomona Beds of Kaiser (1926), and although they are included here, they could conceivably be any age from Cretaceous to Eocene. They crop out 15-20 km north of Bogenfels and typically consist of a basal pebble conglomerate which is overlain by clayey and calcareous sandstones with occasional pebble stringers. The only fossils so far recorded are of the land snail *Dorcasia antiqua* (by du Toit 1954). The succession is up to 35 m thick, and the sandstones, which are probably mostly fluviatile, are capped by dolomitic silcrete and calcrete. The latter diagenetic horizons indicate a period of aridity prior to 37 ± 1 my (Siesser 1978), when they were overlain and metamorphosed by alkaline lavas in the Klinghardt Mountains (Martin 1973a). In addition, in the Bogenfels area, Stocken (1962) reports that Buntfeldschuh Beds overlie and contain pebbles and boulders of Pomona Beds (Kaiser 1926, Haughton 1930a). This evidence suggests a pre-late Eocene age for the Pomona Beds, and Stocken (1962) has assigned them a Cretaceous age, noting their development on an 'end Cretaceous' erosion surface. We have tentatively placed them in the Palaeocene.

The oldest marine Tertiary rocks are associated with a beach line that Kaiser (1926) and Martin (1973a) have mapped for about 50 km between Buntfeldschuh in the south and Granitberg in the north. The most extensive and best-known exposures occur at Buntfeldschuh

Table 51. Microfossils from the Langental Beds (Upper Eocene) near Bogenfels (data from Siesser & Salmon 1979).

| Calcareous nannofossils | % | Range |
|---|---|---|
| *Braarudosphaeara bigelowi* | 1 | |
| *B.discula* | trace | |
| *Chiasmolithus* spp. | 1 | |
| *Coccolithus eopelagicus* | 28 | |
| *C.formosus* | 6 | |
| *Discoaster saipanensis* | 5 | Middle to Upper Eocene |
| *D.tani* | 3 | |
| *Isthmolithus recurvus* | 3 | Upper Eocene to Lower Oligocene |
| *Reticulofenestra bisecta* | 37 | |
| *R.coenura* | trace | |
| *R.umbilica* | 12 | |
| *Zygrhablithus bijugatus* | 4 | |
| Benthic foraminifera | | |
| *Textularia* sp. | | |
| *Astacolus* sp. | | |
| *Lenticulina simplex* | | |
| *L.* cf. *L.pseudo-mamilligerus* | | |
| *L.* cf. *L.oblonga* | | |
| *L.subalata* | | |
| *L.* sp. | | |
| *Siphoglobulina* sp. | | |
| *Glandulina* sp. | | |
| *Elphidium* cf. *E.crispum* | | |
| *E.* sp.A | | |
| *E.* sp.B | | |
| *Nonion costiferum* | | |
| *N.sloanii* | | |
| *Valvulineria aegyptina* | | |
| *Ammonia* cf. *A.beccarii* | | |
| *A.* sp. | | |
| *Pararotalia inermis* | | |
| *Cibicides pseudoungerianus* | | |
| *C.* sp. | | |
| *C.* spp. | | |

(Beetz 1926) where they rest on a wave cut platform on Pre-Cambrian basement at 120-140 m above sea level (Martin 1973a). Here, about 45 m of marine conglomerates and clayey fine sands are overlain by 40 m of aeolian sandstones. The conglomerates contain pebbles of chalcedony, agate and jasper, and above the basal bed, are found interbedded in the sandstones. They are diamondiferous, and Beetz (1926) reports gemstones as large as 2.5 cts. The marine sandstones are mostly greyish green in colour, fine grained and are composed of quartz (dominant), feldspar, rock fragments and opaques. They are overlain by brown cross-bedded aeolian sandstones which are capped by about 1 m of calcrete with diagenetic ooids and intraclasts (Siesser & Salmon 1979). No microfossils were found in the Buntfeldschuh Beds by Siesser & Salmon (1979), and the only positively identified macrofossils have been marine fish teeth (Böhm 1926, Stocken person communication, quoted in Siesser & Salmon 1979), and burrows (Siesser & Salmon 1979).

Siesser & Salmon (1979) assigned a late Palaeocene-early Eocene age to the Buntfeldschuh Beds on admittedly equivocal evidence, but points in favour of this are: their relatively high elevation (135 m above sea level); and the fact that on lithological grounds (their content of Pomona lithologies) they appear to be older than the marine beds at Bogenfels (e.g. Haughton 1930a, 1963). Deposits similar to those at Buntfeldschuh occur along the length of the beach line shown in Figure 167: at Eisenkieselklippenbake (between 150-165 m above sea level), Advokat Bake (ca.165 m), Lüderitz Krater; and Granitberg (Beetz 1926, Kaiser 1926, Martin 1973a, 1973b), and we equate all these to the same late Palaeocene-early Eocene high sea level stand (Fig. 186).

Fossiliferous marine strata occur 3-5 km north of Bogenfels on either side of the Langental in the same area as the Wanderfeld IV Cenomanian beds (Figs 150 & 167). Siesser & Salmon (1979) have described the microfossils from the outcrop B (location on Fig.150), whilst the general geological setting and measured sections have been published by Beetz (1926) and Klinger (1977). The marine beds are about 4 m thick and consist of a pebbly sandstone overlain by various thin concretionary calcareous sandstones, and marly sandstones and silts (section B, Fig.167). These rocks are separated by 3-4 m of unfossiliferous feldspathic sandstones from the underlying Cretaceous rocks (Wanderfeld IV beds), although no section across the boundary has been recorded. The whole upper marine sequence contains agate pebbles and macrofossils, and is generally calcareous (in contrast to the Buntfeldschuh Beds which are not). Böhm & Weissermel (1913) originally assigned a Miocene age to these beds on the basis of molluscs and fish teeth, but this date was later revised to Middle-Upper Eocene by Böhm (1926). A large number of macrofossils have been reported from this outcrop, including: lamellibranchs (nine new species by Böhm 1926), gastropods (13 new species by Böhm 1926), polyzoa, a nautiloid (*Aturia*), cirripeds, a crab, corals, hydrozoa, fish teeth (13 new species by Böhm 1926), and *Callianassa* burrows. Shells of *Turritella* are especially abundant, and Haughton (1930a) called these the '*Turritella* beds'. Siesser & Salmon (1979) have renamed them the Langental Beds, and using floral lists made by Siesser (1977b) have firmly established an Upper Eocene age for these strata on the basis of calcareous nannofossils and benthic foraminifera (Tab.51).

Two other possible Palaeogene (Eocene) localities on the Namaqualand coastal plain have been suggested by Hendey (1981a). Both are equivocal palaeontologically, but we have included them here because their relatively high elevation accords well with the Palaeogene outcrops in Namibia – they could, however, conceivably be Neogene. At Kamaggas on the Buffels River about 37 km west of Springbok (Fig.156) Hendey (1981a) reports a shark's tooth locality in beach gravels (elevation ca.140 m) which he equates with a shark's tooth

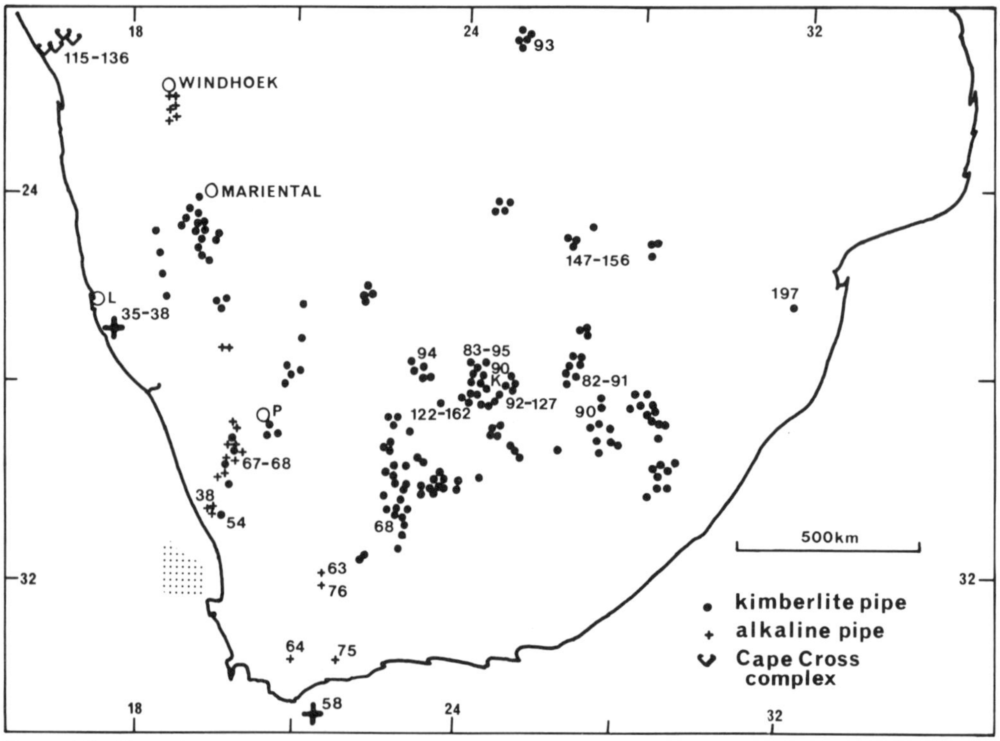


Figure 168. Kimberlites and associated Cretaceous and Tertiary alkaline volcanic rocks in southern Africa. Data mainly from Moore (1979). Individual 'pipes' are in several instances clusters. The category 'alkaline pipes' are mostly olivine melilitites, but phonolites (south of Luderitz) and phonolitic and trachytic types (offshore Cape Agulhas) have been included. Ages and age ranges in my are shown beside certain localities. Large crosses indicate extensive developments of small pipes. Shaded area off the W.coast shows location of Lower Tertiary intrusions of unknown affinities in the S.Orange Basin. K = Kimberley, P = Pofadder, L = Luderitz.

occurrence mentioned by Lamont (1947) at Quaggaskop 26 km north of Vanrhynsdorp. Lamont's (1947) report was in fact a personal communication from Dr A.W. Rogers who 'gave the author two photographs of capping deposits in the Knersvlakte in which he had found marine fossils including sharks' teeth'. Lamont could find no more sharks' teeth but he located several outcrops of the calcareous conglomerate mentioned to him by Rogers, in which he found a '?coral', and from which local farmers had reported shells (Lamont considered these might have been oysters). These rocks, which rest directly on Pre-Cambrian basement, lie at between 150 and 230 m above sea level.

Finally, mention can again be made to the Palaeogene igneous rocks which occur in the coastal zone between Lüderitz and the Agulhas Bank (Fig.168). As discussed elsewhere (5.3.2) these form part of the Lower Cretaceous to Palaeogene alkaline phase of volcanism that includes the kimberlites of the interior and the scattered olivine melilitites, kimberlites, phonolites and nephelinites along the west coast. Intrusives specifically dated as Palaeogene are:

i) Klinghardt Mountain phonolite lavas – 37 ± 1 my and 35.7 my (Lower Oligocene) (Kröner 1973);

ii) Olivine melilitite plug in south Namaqualand – 38 my (Upper Eocene) (Kröner 1973);

iii) a kimberlite plug close by (ii) – 54.1 my (Upper Palaeocene) (Davis 1977).

On the basis of petrological similarity and tectonic setting, Marsh (1975) assumed that the olivine melilitite and nephelinite plugs in the vicinity of Pomona and Bogenfels are the same age as the nearby Klinghardt phonolites (i.e. Lower Oligocene). As can be seen from Figure 168 most of these intrusions form a distinctly younger set that the Palaeocene alkaline intrusives that lie farther south; with the Upper Palaeocene kimberlite plug dated by Davis (1977) as the most northerly example of the latter province. As such, the south Namibia-Namaqualand alkaline intrusives are the youngest known examples of volcanic activity on the southern African continental block.

Several of the plugs in Namaqualand contain sediment, some of which comprise blocks that collapsed into the throats of the pipes (e.g. on the farm Riembreek in the Gamoep district, Moore 1979), and others which contain later fill. Of the latter, one pipe at Arnot (or Banke as it is alternatively known) (Fig.169) has been described by Haughton (1931a) (amongst others). Here, at least 70 m of clays and sands have accumulated in the pipe crater and from these a collection of leaves, spores, wood, and frog skeletons have been recorded. Haughton (1931a, 1969) favoured a late Cretaceous age

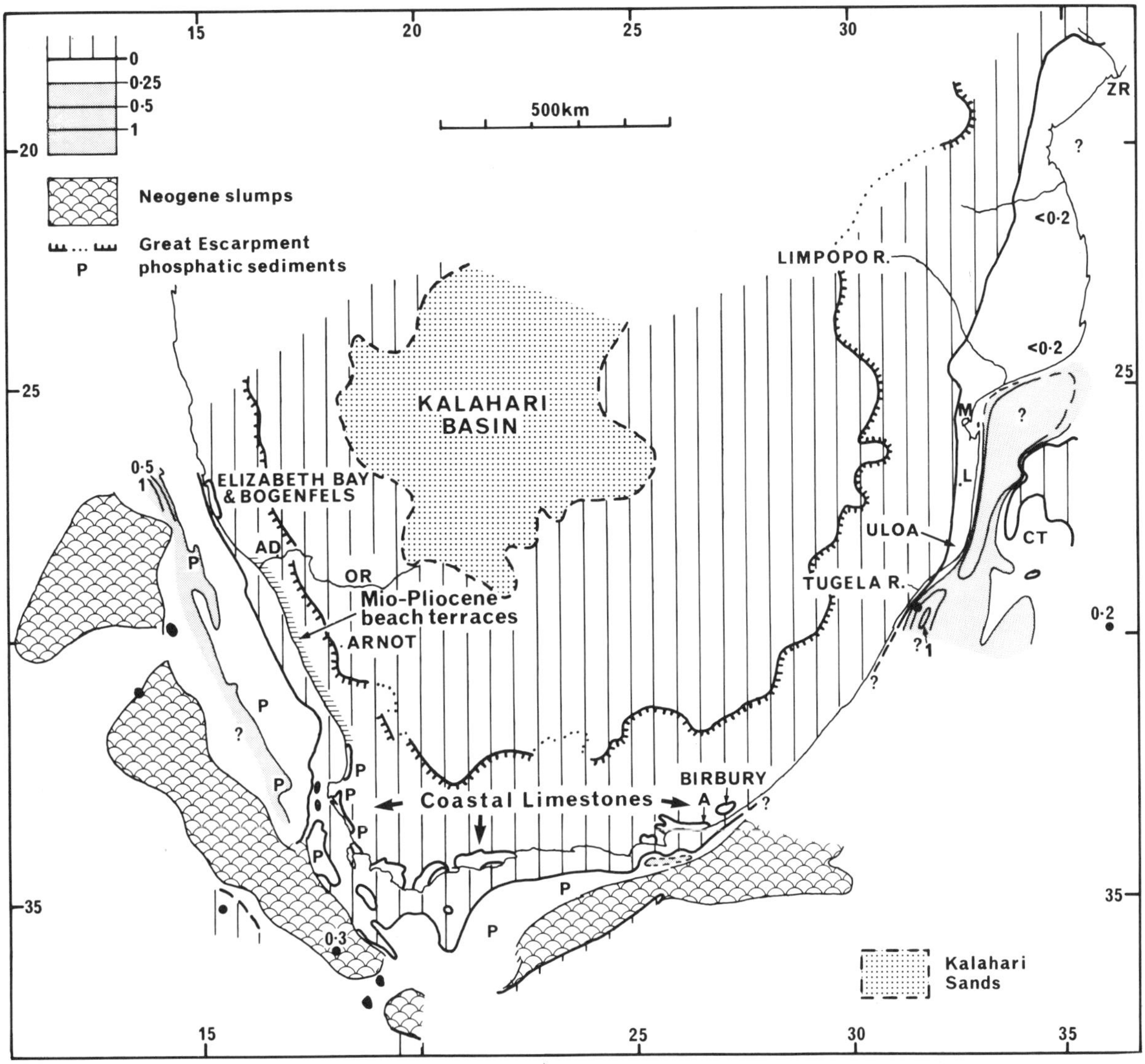


Figure 169. Distribution and thickness of Neogene sediments in southern Africa. Inner edge of coastal and offshore outcrops indicated by thick line. Differentiation from Quaternary strata is often uncertain, particularly on seismic records, and in the descriptions of boreholes from Mozambique. Where available, data suggest that the younger cover is thin (<200m offshore). DSDP boreholes are shown by dots. Data from numerous sources, but particularly Frankel (1968), Siesser (1972a), Flores (1973), Dingle (1973c, 1980b), Dingle *et al.* (1978). Abbreviations: AD = Arris Drift; OR = Orange River; A = Alexandria; L = Lakeview; CT = Central Terrace; M = Maputo; ZR = Zambezi River.

for this material, but its age has not been unequivocally established and could be anywhere between Upper Cretaceous to Miocene. Palynological studies are most likely to resolve the age of the infill – which of course may be considerably younger than the volcanic intrusion beneath.

Palaeogene igneous intrusions occur along the southern part of the 'hinge line' of Cretaceous sedimentation in the Orange Basin (i.e. along the outer edge of the magnetic Central Ridge), especially in the vicinity of the A-C boreholes, as well as sporadically elsewhere (e.g. around the Ba boreholes) (Dingle 1979, Gerrard & Smith 1980). To our knowledge, no radiometric dates or illustrations of the form of these structures have so far been published, and they are dated merely on stratigraphic relationships. Gentle flexing of Lower Tertiary sediments, presumably in late Palaeogene times, has been reported along a 200 km zone coinciding with the outer edge of the Central Ridge (Dingle 1973c). This movement may have been caused by reactivation of older tectonic lineaments, and was possibly synchronous with the igneous activity, on the other hand the folding could have been caused by updoming associated with the intrusions.

We will review the significance of the Namaqualand intrusions in time and space in Chapter 7.

## 6.2 NEOGENE

Miocene and Pliocene sediments occur extensively over the whole of the continental margin (Fig.169), although on the shelves the latter are probably confined to large erosional outliers. Neogene thicknesses vary from less than 250 m over most of the continental shelf, to locally > 1 km under the upper continental slope (e.g. west of Lüderitz and on the Tugela Cone). Over large areas of the outer margin (particularly off the south and west coasts) thicknesses are unknown because massive slumping has rendered uncertain the correlation of the various acoustic stratigraphic units in the upper part of the sediment column with the Neogene subdivisions on the shelves. In fact, until the age of the slumping has been better established it is possible that significant portions of the thick allochthonous slope and rise units may be Quaternary in age. Elsewhere, on the shelves, Quaternary sediments are generally thin (except for narrow, elongate nearshore mud and sand prisms) and consist of veneers of Holocene, and relict and residual Pleistocene biogenic and terrigenous sands, and local coarse biogenic rudites.

Knowledge of Neogene sediment distribution in deeper water is very patchy, but at some localities at least (e.g. SE Cape Basin, and in the Agulhas Passage) they are missing and thin Quaternary cover rests directly on Eocene strata. In contrast, relatively thick Neogene sequences are suspected in the northern Natal Valley (Fig.169) particularly adjacent to the Zululand continental margin and south of the Limpopo River.

The most extensive Neogene deposits onshore occur under the Mozambique coastal plain (Fig.169). These narrow southwards into Zululand where the western feather edge passes onto the continental shelf over the Richards Bay coastline offset. Elsewhere, coastal deposits are confined to small scattered outcrops in the SE Cape Province (in the Birbury, Bathurst, Algoa Bay area), and more extensive, but scattered outliers between the Still Bay area on the south Cape coast and the Varswater-Saldanha area on the west coast. Numerous very narrow but long beach terrace and occasional terrestrial deposits occur scattered along the Namibia-Namaqualand coast between Elizabeth Bay and Lamberts Bay. The diamondiferous and phosphatic nature of some of these latter sediments and their SW Cape equivalents have made them economically important and locally they have been intensively studied.

Marine Neogene rocks are characterized by the extensive development of biogenic sediments. They are frequently rich in authigenic minerals and, locally, have high organic carbon and opaline silica contents, particularly off the west and south coasts. Aeolian sediments occur in most coastal Pliocene sections, and in the large Kalahari Basin, the whole of the Neogene (and probably Palaeogene, although differentiation cannot be readily effected) is of continental facies, with aeolian sands forming a dominant element.

With the exception of Mozambique (north of about 25°S) and the two DSDP sites 360 and 362, Neogene strata rest disconformably or unconformably on Palaeogene or older rocks (Fig.155), and except in the coastal fringe, Kalahari Basin, and possibly some deep water sites, are separated from Quaternary sediments by major erosion surfaces.

### 6.2.1 *East coast and adjacent offshore areas*

Figure 169 shows that the main Neogene depocentres along the east coast lay off the major rivers (Tugela, Limpopo, and Zambezi although no thicknesses are available for the last area), whilst only thin deposits were laid down over the Mozambique and Zululand coastal plains (< 200 m).

#### 6.2.1.1 *South Mozambique*

Over much of southern Mozambique, Miocene sediments rest conformably upon Oligocene strata where the boundary between the two lies in the upper parts of the argillaceous Inharrime and Buzi formations. In the area of the Save River mouth, the upper part of the Temane evaporites is also probably of Lower Miocene age (Fig.161). Flores (1973) reports a regional hiatus between the Jofane Formation and the underlying argillaceous/evaporitic beds, with the Middle and Upper Miocene sediments transgressing far to the west where they overstep the Palaeogene to rest directly on Mesozoic formations.

In the west, the Middle and Upper Miocene is represented by coarse sands and conglomerates (Chicolane Formation of Forster 1975a) which are similar at outcrop to the underlying Sena Formation. Eastwards these grade into brackish water micaceous purple or violet coloured silty, fine-grained sandstones with poorly-preserved crustacea, and dwarfed molluscs. This continental/brackish facies is called the Mazamba Formation by Flores (1973) (type area on the Mazamba Creek near Beira) and it occurs extensively throughout southern Mozambique as far east as about 30½°E. East of this longitude the Middle-Upper Miocene is represented by a clear, warm-water shallow marine facies (Jofane Formation) which consists of oolitic and coral limestones and dolomites with rich faunas of polyzoans, echinoids and orbitoid foraminifera, but relatively poor planktonic assemblages. Forster (1975a) recognizes small offshore bars composed of detrital limestones interfingering with sandstones and channel lags of coarse sandstone and conglomerate, patch reefs, and nodular detrital algal limestones. The Jofane Formation reaches a maximum thickness of about 180 m in offshore boreholes, and Flores (1973) reports that northwards it passes into an argillaceous facies which he equates with a late Miocene development of the underlying Buzi Formation. He calls this the Divinhe Formation, but gives no details as to how it may be distinguished from the underlying formation (Fig.161). South of about 22°S, the Jofane Formation is represented by sparse outcrops of glauconitic marine sandstones with shale and marl interbeds (Govuro Formation) (Frankel 1972, Flores 1973). These may be equivalent to the early Miocene Delagoa Formation referred to by Henriques da Silva (1967) (which Frankel 1972 places in the Upper Miocene), but are probably in part also correlatable with several formations mentioned by Frankel (1972) that crop out along the Uanetze River about 120 km NNE of Maputo (see Fig.130 for location). These were originally described by Oberholzer (1963), and consist of (i bottom, iii top): i) about 11 m of mollusc-rich calcareous sandstone overlain by 4 m of pebbly, *Pecten* bearing limestone; ii) 15 m of arkosic, dolomitic limestone with interbedded limestones and foraminifera-bearing calcarenites, overlain by 1.5 m of chert with replaced fossils; iii) 3 m of sandstones, sandy arkoses and conglomerates with few fossils. Frankel (1972) dates these rocks as Middle to Upper Miocene, and reckons that they represent shallow marine (locally estuarine conditions). Units (i) & (ii) combined are variously called the Tembe and Santaca Formations and unit (iii) the Santiago Formation (Fig.161). Frankel (1972) equates the widely developed *Pecten*-bed within them with the Uloa *Pecten* bed in Zululand. The uncertainty in correlation of some of these lithofacies is caused by the paucity of reliable palaeontological dates, and the considerable local lateral facies variations, which in turn give rise to differences of opinion on the ages of the Miocene non-sequences. For instance: Flores (1973) believes a 'limited' Middle-Late Miocene transgression was followed by a Late Miocene-Pliocene regression in Mozambique, whereas Förster (1975a, p.272) indicates a Lower Miocene transgression followed by an abrupt regression. Förster (1975a) also maintains that only short, local 'ingressions' occurred from Middle Miocene to Pliocene times, in contrast to Frankel (1972) and Flores (1973), who would place the greater part of the marine Miocene sedimentation during middle and late parts of the epoch. He cites, as examples, Upper Miocene-Pliocene deposits along the Tembe River (30 km SW of Maputo), as well as the Morrumbene Formation (small outcrops of oolitic, sandy, fine grained limestone, at least 15 m thick east and south of the Domo borehole).

Other Pliocene strata in Mozambique are the argillaceous Divinhe Formation, which according to Flores (1973) indicates continuous deposition of a mudrock lithofacies from early Oligocene to Pliocene times in the offshore area NE of the Save River, as well as various red sandstones west and south of Maputo. These coarse red sands are non-marine sediments and frequently lie, with a basal conglomerate up to 1 m thick on the Upper Miocene to Lower Pliocene Santiago and Santaca/Tembe formations (for example at Santaca on the Pongola River just north of the Zululand/Mozambique border). They have been variously called the Magude and Cacho Formations (Frankel 1972) and reach a maximum thickness of about 30 m at Santaca. Frankel (1972) considers these red sands to be late Pliocene in age and traces them into Zululand as far south as Lakeview (Fig. 169), 20 km to the north of which, at Ndumu they rest directly on Cretaceous sediments.

Flores (1973, p.101) suggests that locally the non-marine Mazamba Formation may also be Pliocene in age and he likens it to the Kalahari sands.

#### 6.2.1.2 *Zululand*

Neogene sediments are poorly represented in Zululand compared to Mozambique, and consist of scattered outcrops of Miocene marine calcareous rocks and minor clays, locally overlain by the Pliocene red sands mentioned in the previous section. The best outcrops of the Miocene strata are at Uloa, Sapolwana, Warners Drain, Umkwelane Hill, The Jungle and Lakeview. They may reach their thickest development (about 24 m) at Lakeview (Frankel 1972), although Truswell (1977) has reported that up to 65 m have been recorded somewhere in Zululand. Figure 170 illustrates the development of the marine Miocene in the Mfolozi Valley of Zululand, with the important section in the Uloa area taken as the type. Here, the basal sediments occur in a narrow, central, N-S zone and comprise the Nodule bed of Frankel (1966). This is a locally ironed stained, light brown to reddish brown friable sandy clay with pebbles, brown shiny phosphatic nodules, and remarnié fossils in pockets near the base. The reworked fossils contain a wide variety of Cretaceous molluscs, including several ammonite species, wood fragments, fragments of cetacean bone and Tertiary sharks' teeth, whilst the pebbles include micro-perthitic granite, quartzite, basalt and Cretaceous siltstones. The phosphatic nodules are up to 5 cm long and are mostly conglomeratic with phosphatized Tertiary and Cretaceous sediment rock fragments set in a carbonate apatite matrix. Some of the fragments consist of phosphatized and partially phosphatized micrites, whilst others include algal nodules, glauconitic sandstone, and calcareous siltstone. Reworked Cretaceous planktonic foraminifera (*Globotruncana*) occur in the collaphane matrix of these nodules, as well as glauconite and reworked Tertiary foraminifera (*Victoriella*). Frankel (1966, p.222) quotes a date of 55 my (Upper Palaeocene) on the glauconite from one nodule and a whole rock $P_2O_5$ content of 16.2 % for one nodule analysis.

The Nodule bed, which reaches a maximum thickness of about 1 m at Uloa, is overlain by up to 4.6 m of unbedded calcirudite – the so-called *Pecten* bed. This overlies the irregular upper surface of the Nodule bed, and thins rapidly, but irregularly northwards. To the east and west of the shaded area on Figure 170, the *Pecten*-bed rests directly on the Cretaceous, but is itself absent a few kilometres west of Uloa. The *Pecten*-bed is a coquina-like limestone composed dominantly (95 %) of

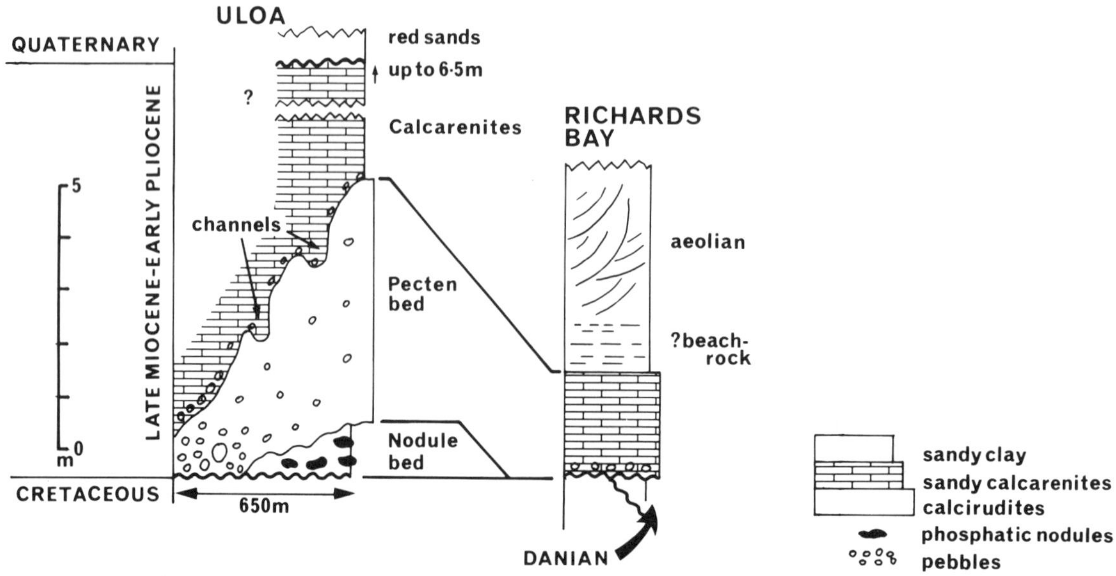


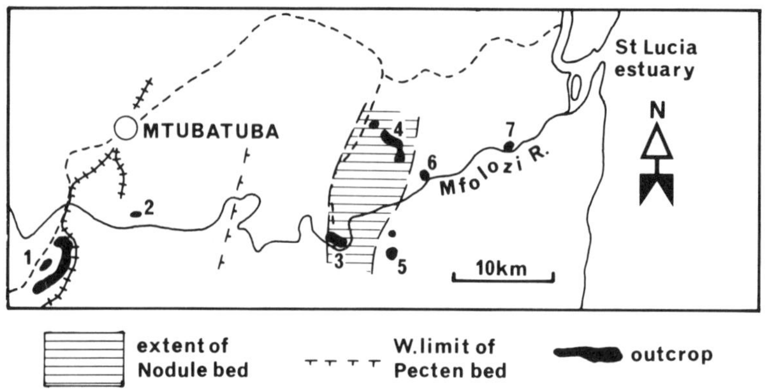


Figure 170. Neogene sediments in S. Zululand. Top shows a composite section for the Uloa area in the Mfolozi Valley and correlation with a borehole sequence at Richards Bay. Map shows the numerous outcrops and lateral distribution of various lithofacies in the Mfolozi Valley. Localities: 1 = Umkwelane Hill (including the outcrop named 'The Jungle' by Frankel 1968); 2 = Peaston; 3 = Uloa; 4 = Lot U178; 5 = Sapolwana; 6 = Warners Drain; 7 = *Pecten* bed locality. Data from Frankel (1966, 1968), and Maud & Orr (1975).

calcareous biogenic debris with minor grains of quartz, microcline, perthite, glauconite and mica flakes, together with small pieces of phosphatized foraminiferal mudrock. Small boulders and pebbles of granite, pegmatite, quartzite, sandstone, dolerite, basalt, and foraminiferal limestone are sparsely scattered throughout the unit, but locally they form a basal conglomerate.

The main part of the *Pecten*-bed consists of an abundant shelly invertebrate fauna set in a matrix of fragments of broken and comminuted coralline algae, crinoid ossicles, echinoid fragments, polyzoa, and foraminifera. Locally, the rock is an algal limestone. This richly fossiliferous rock has yielded over 100 species of macrofossils and 45 species of microfossils (King 1970), but is dominated by the lamellibranch *Aequipecten uloa* (King), followed by *Conus haughtoni* King and *C. proliferatus* (King).

The thickest (up to 6.7 m) and most extensively developed unit of the Uloa sequence is an upper calcarenite layer which consists of alternating 5-15 cm, flaggy well-cemented fine calcarenites and softer, coarser cross-bedded calcarenites up to 30 cm thick. These rocks rest on an eroded surface cut across the *Pecten*-bed, and at their base there are local concentrations of pebbles of older rock types. In the western part of the Mfolozi valley (i.e. south of Matubatuba), the calcarenites rest directly on the Cretaceous. These calcareous rocks are overlain by yellow, red and brown sands with minor clays which Frankel (1966) considers Quaternary in age. There has been considerable discussion as to the age of the Mfolozi Valley outcrops (see Stapleton 1977c for a review). King (1953 et seq.) dates the Nodule and *Pecten* beds as early Miocene (based on macro- and microfossil evidence) and recognizes an unconformity between these units and the overlying calcarenites, which he dates as Pliocene on the basis of mollusc fragments. Conversely, Frankel (1968) prefers a Middle or Upper Miocene to Pliocene age for the whole sequence (based on sharks' teeth and foraminifera) and recognizes intra-formational channelling at the top of the *Pecten*-bed, but no major non-sequence. Most of the uncertainty has been generated by the lack of high resolution age-diagnostic fossils, but two recent publications dealing with planktonic foraminifera and calcareous nanno-

Table 52. Microfossils from Mio-Pliocene beds in Zululand (after Siesser & Miles 1979).

| Calcareous nannofossils | Range |
|---|---|
| *Reticulofenestra pseudoumbilica* [1] | Middle Miocene-Lower Pliocene |
| *Coccolithus pelagicus* | |
| *Discoaster surculus* | Upper Miocene-Upper Pliocene |
| *D.* sp. | |
| Planktonic foraminifera | |
| *Globigerina* cf. *G.praedigitata* | |
| *Globigerinella aequilateralis* [2] | |
| *Globigerinoides conglobatus* | |
| *G.obliquus extremus* [2] | |
| *G.trilobus* [3] | |
| *Globoquadrina altispira altispira* | |
| *G.altispira globosa* | |
| *Neogloboquadrina humerosa* | |
| *Orbulina universa* | |

All species were found in the Pecten bed at Uloa, and in addition: 1 – Lakeview, Morrisvale, Lot U-178; 2 – Lakeview; 3 – Lot U-178.

fossils give better discrimination than was previously available. Stapleton (1977c) found an assemblage of 13 species of planktonic foraminifera in a sample collected at the base of the Uloa *Pecten*-bed from which he concluded that its age is latest Miocene, or at the Miocene/Pliocene boundary. He also re-assessed published information on the planktonic foraminifera in the overlying calcarenites and concluded further that they can be little younger than the *Pecten*-bed, and are probably approximately the same age. Siesser & Miles (1979) examined samples from the outcrops of the Mfolozi Valley, as well as other *Pecten*-bed localities farther north. They found that although material from Sapolwana, Warners Drain, Umkwelane Hill, The Jungle, Mkuze and Mtoti Pan were barren of calcareous nannofossils, and were barren or contained unidentified planktonic foraminifera, eight species of planktonic foraminifera were found in samples from Lakeview, Lot U-178 and the *Pecten*-bed at Uloa. All ranged from late Miocene through the Pliocene. In addition, samples from Lakeview, Morrisvale and Lot U-178 contain rare to sparse poorly preserved specimens of the calcareous nannofossil *Reticulofenestra pseudoumbilica,* and a soft pocket in the *Pecten*-bed just south of the railway line at Uloa yielded a sparse, but moderately well-preserved nannofossil assemblage including *R.pseudoumbilica* (middle Miocene to early Pliocene) and *Discoaster surculus* (late Miocene to late Pliocene). The presence of these taxa indicate a late Miocene-early Pliocene age (within Martini's (1971) calcareous nannofossil zones NN11-NN15) for the Uloa *Pecten*-bed and its equivalents in Zululand, with the overlying calcarenites falling within the same age range. Siesser & Miles (1979) consider the planktonic foraminifera to be indicative of warm-temperate water. The list of taxa identified by Siesser & Miles (1979) is given in Table 52.

Sediments of a similar age and lithology have been reported from boreholes in the Richards Bay area by Maud & Orr (1975) (Figs 163 & 170), where a basal coquina is overlain by calcarenites. Here there is no lower phosphatic nodule bed. The coquina is up to 2 m thick and rests on both Palaeogene and Cretaceous rocks. It consists of a highly leached, iron stained shell bed from which many of the originally aragonitic molluscs have been removed, with pebbles of reworked indurated glauconitic, ferrugenous late Cretaceous siltstone and a matrix that contains reworked glauconite grains, Cretaceous molluscs and foraminifera, and Tertiary molluscs. Macrofossils from the shell bed include lamellibranchs (especially *Pecten*-like valves), gastropods, polyzoa, as well as ostracods and well preserved foraminifera. The calcirudites are sharply overlain by sandy, hard, light grey coloured calcarenites up to 4 m thick. The lower part of this unit is well-bedded with coarse sand that suggested a beach rock lithofacies to Maud & Orr (1975), whilst the upper part has aeolian textural and steep-cross-bedding characters. Its upper surface suggests subsequent karst-like weathering. The sporadic distribution of these rocks at Richards Bay (see the section on Fig.163), led Maud & Orr (1975) to propose a post-Miocene, pre-Middle Pleistocene period of solution weathering, karst development and erosion. They reported seven species of benthic and five species of planktonic foraminifera from the calcarenite of these sections and proposed a late Miocene age for them (Tab. 53). They specifically excluded a Pliocene age, and although the fauna found within the lower coquina is not as age diagnostic, it was considered that the similarities between the two units were sufficient to also indicate a late Miocene age for the coquina. The faunas indicate deposition in or adjacent to shallow water (15 m or less) with access to the open ocean (Maud & Orr 1975).

Table 53. Neogene foraminifera from Richards Bay, data from Maud & Orr (1975) refers specifically to their borehole B.

| Benthic foraminifera |
|---|
| *Noniella* sp. |
| *Elphidium* sp. |
| *Rotalia beccarii* |
| *Discorbis* sp. |
| *Eponides* sp. |
| *Triloculina* sp. |
| *Quinqueloculina* sp. |
| Planktonic foraminifera |
| *Globigerina nepenthes* |
| *Globigerinoides obliquus* |
| *G.trilobus** |
| *G.bolli* |
| *Globoquadrina acostaensis* |

* Common to Uloa Pecten bed.

In summary, Maud & Orr (1975) proposed a late Miocene age for the *Pecten*-bed and overlying calcarenites, whilst Siesser & Miles (1979) date the lateral equivalents in the Mfolozi Valley as late Miocene-early Pliocene. This evidence can be interpreted in two ways: either the various lithofacies are diachronous and are slightly older in the Richards Bay area, so that early Pliocene elements were extant by the time the Uloa type section was deposited, or that Siesser & Miles's (1979) more cautious and wider age range can in fact be refined to indicate a late Miocene age. Whatever the case, the recent datings tend to corroborate Frankel's (1968) interpretation of a late Miocene age for the Nodule and *Pecten* beds with a possibility of an earliest Pliocene age for the calcarenites in the Uloa area, and confirm his interpretation of a Middle or Late Miocene transgression followed by a latest Miocene-earliest Pliocene regression.

Other sediments in Zululand that have been dated as Pliocene are probably the local equivalents of the Cacho Formation of southern Mozambique. The most recent detailed description of these rocks has been by McCarthy (1967) from the Durban area, where he demonstrated the lateral continuity of the Bluff Beds and Berea Red sands (herein referred to as the Bluff and Berea Beds) which had previously (e.g. King & Maud 1965) been considered separate formations of different ages. The Bluff and Berea Beds consist of two basal conglomerates and an overlying sequence of aeolianites. The former (30 cm and 60 cm to 1.5 m thick, respectively) consist of boulders and cobbles of Table Mountain Group quartzites, Dwyka Tillites and dolerites, usually 15-23 cm across, but occasionally over 1 m, along with reworked Upper Cretaceous shells, calcareous grits and mudrocks, *Inoceramus* fragments and sharks' teeth. In weathered sections of the Bluff and Berea Beds (i.e. in the Berea Red Sands), these large clasts are also intensely degraded by weathering, but elsewhere they are fresh, although the basal conglomeratic unit is missing altogether from the easternmost outcrops under the Bluff Ridge area. The overlying sequence consists of up to about 120 m of cross-bedded, calcareous (locally up to 50 % $CaCO_3$) yellowish sands (Bluff beds) which grade laterally and vertically into clayey red and brown sands (Berea Red Sands, weathered facies). The Bluff beds are well-sorted aeolianites, whilst their weathered equivalents contain abundant kaolinite intermixed with goethite as weathering products of original feldspar, ferromagnesians and magnetite in the dune rock. In the Durban area, beach boulders overlain by regressive coastal dune sequences extend to a height of about 91 m above present sea level.

Rocks of a similar lithology extend along the Zululand coastal zone to Mozambique (King 1966) (lateral equivalents of the Cacho Formation), as well as southward along the Natal coast where they form a degraded coastal barrier and associated offshore features such as Aliwal Shoal (Belderson 1961, McCarthy 1967), which are analagous to the Durban Bluff (rising 107 m above sea level). McCarthy (1967) has re-emphasized the similarity of these deposits to the lithified calcareous dune sands that occur around the southern Cape coastal fringe as far west as Saldanha Bay, and drew attention to the observations of earlier workers that the red clays overlying the Alexandria Formation aeolianites were residual weathering products (e.g. Schwarz 1908, Haughton 1928).

McCarthy (1967) favoured a Pliocene age for the Bluff and Berea Beds, on the basis of foraminiferal studies by Carter (1966) (quoted in King 1966), but other workers (e.g. Frankel 1967, Maud 1968) prefer to regard these deposits as Pleistocene. We have included them in the Neogene because the more substantive (albeit equivocal) evidence is in favour of a Pliocene age, although the question is still open.

6.2.1.3 *Natal Valley and Tugela Cone*

The J(c)-1 borehole on the upper Tugela Cone did not systematically sample Neogene sediments, but one sample of sandy limestone was collected 41 m below the sea floor. This contained the following foraminifera which du Toit & Leith (1974) considered to be a Middle to Upper Miocene assemblage: *Globoquadrina dehiscens, G.altispira, Globigerinoides* sp., *Quadrilobatus trilobus, Lepiclocycline* sp. and *Miogypsina* sp. Consequently, the thickness of the Neogene strata in the borehole is not known, but post-Angus sediments (i.e. post Lower Oligocene) are 428 m thick, and dip and thicken steadily eastwards to 750 m, 11 km farther east (Fig. 171). Post-Angus sediments reach a maximum thickness under the shelf break on the Tugela Cone (at least 1 284 m at 2 140 m/sec which is the average P-wave velocity for these rocks at J(c)-1) but thin rapidly eastwards into deep water and are about 350 m thick in the Natal Valley south of the Naudé Ridge (Section B on Fig.106, Fig.107).

North of the Naudé Ridge, they show even greater thickness variations (Section A, Fig.106). On the western side of the Central Terrace, post-Angus sediments probably exceed 500 m (Fig.169), but over the greater part of the area between the Almirante Leite and Naudé ridges, reflector Angus lies at or close to the sea floor. Seismic records and sea floor sampling show that Neogene sedimentation has been strongly influenced by bottom currents under the major western boundary current system (Agulhas Current) which has reduced sediment accumulation rates in some areas and produced impressive moats and constructional bedforms in the vicinity of the Almirante Leite and Naudé ridges (the latter being buried by a large sediment prism) (Dingle et al 1978, Martin et al 1982b).

Reflector Angus (probably Lower Oligocene) clearly marks a major alteration either in the Agulhas Current

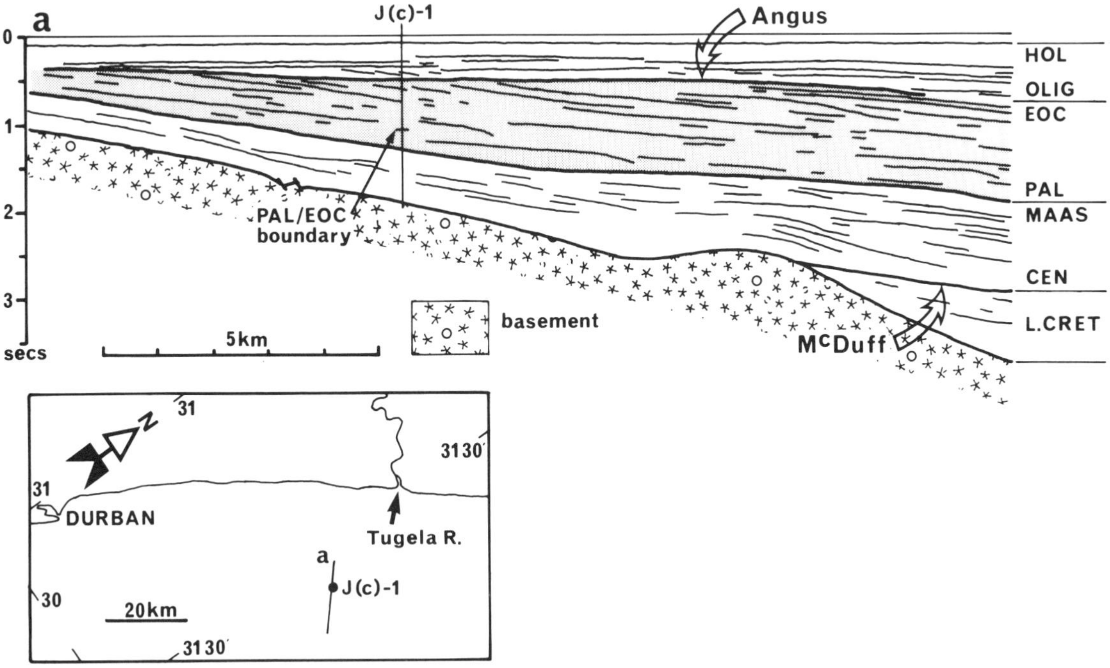


Figure 171. Seismic reflection profile showing the mid-Tertiary unconformity on the proximal Tugela Cone. This horizon is correlated with reflector Angus in the Natal Valley. The mid-Cretaceous horizon McDuff abuts basement SE of borehole J(c)-1. The Palaeocene-Eocene sequence is shaded. Interpretation of a seismic profile presented by du Toit & Leith (1974).

strength or its main core location, since which time it has flowed strongly in a SW direction obliquely across the Central Terrace of the northern Natal Valley. Martin et al (1982b) correlate this event with a mid-Tertiary increase in Antarctic Bottom Water flow and the establishment of the psychrosphere, which stimulated Indian Ocean surface and near surface circulation (the northern Natal Valley presently lies at depths shallower than 2 000 m). They also recognize a younger, regionally developed reflecting horizon (Jimmy) which they date, using sea floor samples, as late Miocene-early Pliocene. This reflector crops out widely over the Central Terrace and its generation is thought to have been linked to new Antarctic Bottom Water production associated with changes in the extent of Antarctic ice cover, and early Pliocene Arctic glaciation. Like its predecessor Angus, Jimmy was produced during a period of major local (and worldwide) regression (Figs 155 & 188).

Post-Jimmy (post late Miocene-early Pliocene) sedimentation in the Central Terrace has been minimal, with large areas exposing coarse Miocene and Pliocene biogenic sands, but farther north post-early Pliocene sediments thicken rapidly under the distal Limpopo Cone where they reach a known maximum of about 0.8 sec (at least 850 m at 2 140 m/sec). In this area Martin et al (1982b) recognized a fourth reflecting horizon (L) which is post-Jimmy in age and may be late Pliocene or even Quaternary. When traced under the proximal Limpopo Cone, and north of the Almirante Leite Banks (i.e. under the Inharrime Terrace) it is seen that in these two areas maximum Neogene sedimentation is in fact post horizon L in age.

In summary, sedimentation patterns in the northern Natal Valley have undergone major changes on three occasions since Eocene times:

i) the establishment of a vigorous Agulhas Current in early Oligocene times dramatically reduced sedimentation rates over the whole area, and especially over the Central Terrace (initiation marked by reflector Angus);

ii) a modified sedimentary regime was established with increased scour over the Central Terrace, but increased sediment accumulation under the distal Limpopo Cone in late Miocene-early Pliocene times (initiation marked by reflector Jimmy);

iii) and finally, in either late Pliocene or during Quaternary times a further shift in sediment depocentres took place (producing reflector L) since when maximum accumulation has occurred under the proximal Limpopo Cone. Upwelling over the Inharrime Terrace has produced an authigenic mineral-rich (apatite and glauconite) sediment lens on the outer shelf where the Agulhas Current impinges against it.

On the Mozambique Ridge (DSDP site 249), Neogene sediments are separated by a large hiatus from underlying Upper Maastrichtian strata (Fig.107). They comprise Unit 1 of the DSDP shipboard terminology (Simpson et al 1974) which consists of 172 m of white to light grey pelagic ooze that is predominantly foraminifera-rich nanno-ooze, which passes below about 159 m into chalks. The unit is streaked, contains small black pyrite specks throughout, and has 4-10 % clay in the lower 152 m. The basal part of this unit (ca.3.5 m) is of Middle Miocene age (planktonic foraminifera zone N12), with the bulk Upper Miocene (ca.153.5 m) that encompasses planktonic foraminifera zones ?N13 to N18. There is a thin (< 1 m) Pliocene sequence and about 1.7 m of Pleistocene material (all the same lithology at the top). DSDP shipboard scientists report that the Lower Pleistocene and most of the Pliocene is missing at site 249 (Simpson et al 1974), so that although Unit 1 is lithologically continuous there is a probable hiatus near the top of 3-4 my. They also note that Palaeocene and Maastrichtian planktonic foraminifera and nannofossils are reworked into the Middle Miocene at the base of Unit 1, and that the sediment accumulation rate of 60 m/my for the Upper Miocene is so much more rapid than might be expected at the site that they postulate 'technical mixing'. By this we assume they imply some form of sequence 'elongation' or duplication by the drilling process. Sedimentation rates in the lithified Middle Miocene is 8 m/my (Simpson et al 1974). The cause of the 50-55 my hiatus between the Neogene and Maastrichtian strata is presumably to be found in bottom current phenomena, because site 249 was too elevated to be significantly influenced by carbonate compensation level fluctuations. The worldwide eustatic regressions at the end of the Maastrichtian and the accompanying ocean circulation changes must have remained as an effective sedimentation prophylactic until the equally widespread Middle-Upper Miocene transgressions (that gave rise to the *Pecten*-bed and its equivalents in Zululand) restored a bottom current regime conducive to sediment accumulation, with, throughout most of the subsequent period, a minor influx of fine terrigenous debris.

We have no other dated material from the deep water areas adjacent to SE Africa, and although a continuous Cretaceous-Cenozoic sedimentary sequence is anticipated in the main Natal Valley, erosion and non-deposition (associated with massive current-generated bedforms along the flanks of the Mozambique Ridge, Agulhas Plateau and adjacent to the continental margin in the Agulhas Passage) occurred around the periphery of the deep ocean basin (Dingle & Camden-Smith 1979). In this connection, Salmon (1979) has reported the succession in a piston core from the western Agulhas Passage (36.1817°S, 23.4883°E) as consisting of 25 cm of Quaternary foraminiferal sand overlying Middle Eocene clays (planktonic foraminifera zone P14).

### 6.2.2 *South and west coasts*

Neogene sediments are widely distributed on the Agulhas Bank and the west coast continental shelf (Fig.172) and form several medium sized outliers in the coastal zone between Bathurst and the Varswater-Saldanha area. They rest with a strong unconformity on the Palaeogene strata where only local Oligocene rocks have been reported (e.g. Kudu 9A-1, Cape Canyon area, and 100 km south of Plettenberg Bay). Present-day distribution of Neogene rocks suggests that they were originally more extensive and that their outcrop has been modified by late Neogene (late Pliocene) or Quaternary erosion. Sediments on the outer continental margin have been affected by large-scale slumping.

Seismic and sampling surveys (Dingle 1971a, 1973c, Dingle & Siesser 1977) allow a twofold subdivision of the continental shelf succession along the west coast into $T_3$ (Miocene) and $T_4$ (Pliocene) units, but although both Miocene and Pliocene rocks have been identified on the Agulhas Bank, no confident subdivisions can be made in this area on seismic records, and Dingle (1971a) was only able to recognize a Palaeogene ($T_L$) and Neogene ($T_u$) acoustic stratigraphy. Age determinations on the Mio-Pliocene sediments onshore are neither precise nor numerous enough to effect a satisfactory correlation between most of the sections, with the result that there is still considerable uncertainty in interpreting the Neogene sedimentary history of the Namaqualand to south Cape coastal zone. However, recent work in the Varswater-Saldanha area, notably by Hendey (e.g. 1981a, 1981b), has opened the way for a fresh appraisal of the relationship of the various marine, and non-marine sediments, and beach terraces with the continental shelf succession.

#### 6.2.2.1 *Continental shelf*

Neogene sediments form a thin cover over the mid and outer Agulhas Bank where they reach a maximum thickness of about 250 m just north of the present shelf break. High resolution seismic records show (e.g. section d, Fig.173) flat lying or very gently seaward dipping, and locally seaward prograding units that rest unconformably on the Palaeogene. The outer edge of this outcrop is truncated by the glide planes of the proximal part of the Agulhas Slump (Fig.187). Here, large blocks and tilted units lie in a jumbled sequence along and below the shelf break where rocks ranging in age from Neogene to Upper Cretaceous are locally exposed. Detailed mapping of this zone has not been carried out. On the shelf, adjacent to the main glide plane scars the upper surface of the Neogene sediments is cut by numerous fissures (incipient faults) which give the sea floor a highly indented micro-relief. Dingle (1973a) proposed the name Agulhas Group for these Neogene rocks and recognized several subunits, although because the extent and significance of them has not yet been assessed it is proposed to refer to the whole sequence as the Agulhas Formation (Fig.155). Dingle (1971a) and Siesser (1977a) identified several Miocene and Pliocene rock samples from the Agulhas Formation, as well as a large number

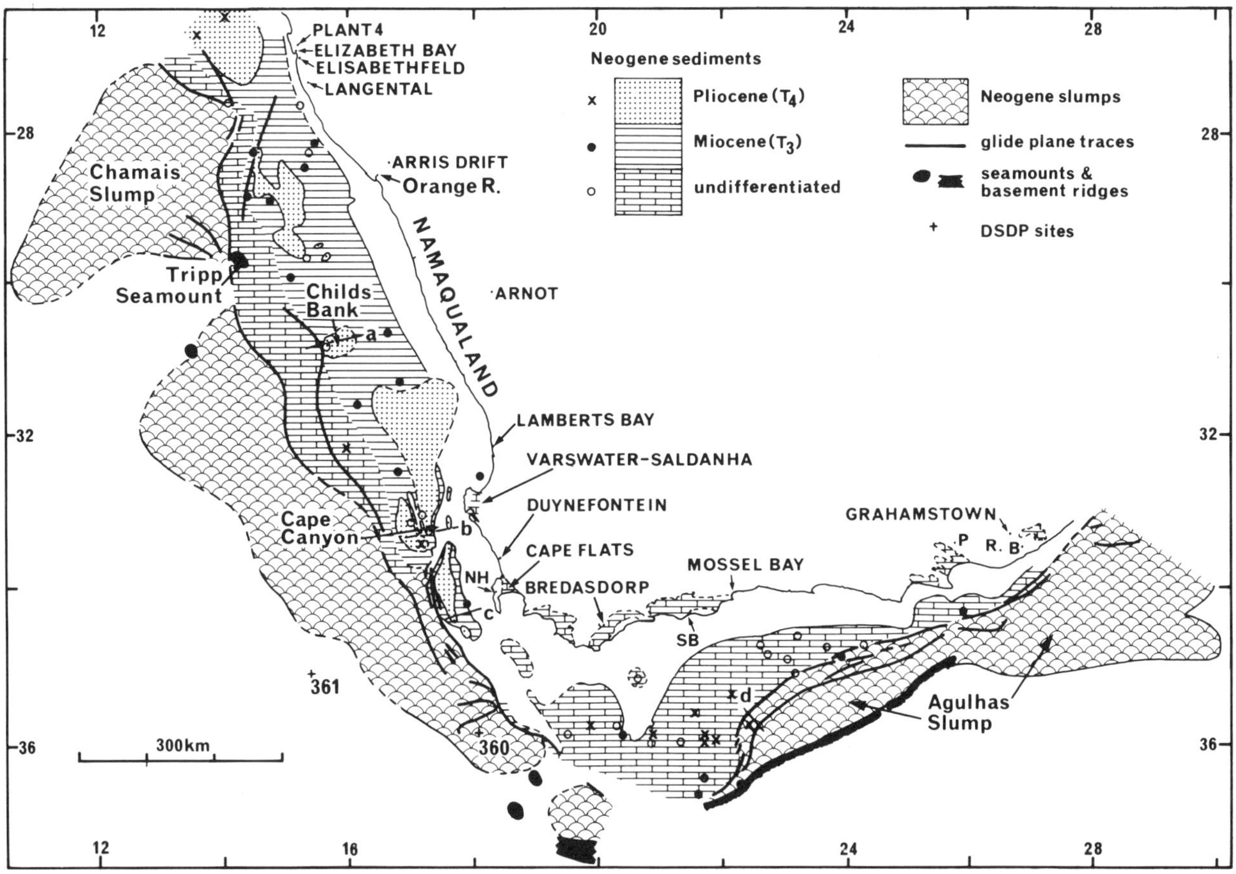


Figure 172. Distribution of Neogene strata in the S. and S.W.Cape and adjacent offshore areas. Dated samples offshore are mainly after Siesser (1977a) and the geology updated from Dingle (1971, 1973a, 1980b). Data sources for onshore areas are numerous and are cited in the text. Cross-sections a-d are shown in Figure 173. Abbreviations: SB = Still Bay; R = Rooikrans; P = Paterson; B = Bathurst; +360, 361 = DSDP sites, NH = Noordhoek.

of undifferentiated Mio-Pliocene outcrops (Fig.172), but so far no consistent pattern in either distribution or lithology has been established to form the basis of a subdivision of the Agulhas Formation. Figure 172 shows that all the positively identified Pliocene samples lie in the southern and central areas (i.e. between about 19.5°E and 22.5°E) but the presence of numerous undifferentiated samples makes this distribution only preliminary.

Siesser (1971, 1972a) has described the texture and mineralogy of the surface rock samples collected from Agulhas Bank and shows that the Agulhas Formation is predominantly a limestone sequence where the main rock type consists of moderately to loosely-consolidated yellowish grey and grey lime wackestones and packstones containing up to 90 % skeletal matter (mainly planktonic foraminifera and polyzoa), up to 10 % glauconite, and never more than 5 % terrigenous matter. Up to 80 % of the cementing matrix is micrite. Occasional calcareous clays have also been found. Phosphatic concretions (slabs, boulders and cobbles) occur abundantly within the Agulhas Formation (Parker 1971) in rocks of both Miocene and Pliocene age (Dingle 1974, Siesser 1978a).

Along the west coast continental margin, the Neogene sediments can be divided into two well-defined sequences: a lower, relatively extensive sequence of flat lying, or gently seaward dipping units, and an upper generally thinner series of flat, irregularly bedded or steeply seaward dipping units that occur as outliers along the middle or outer shelf (Figs 173 & 174). Dingle (1973c) referred to these as $T_3$ and $T_4$ respectively, and tentatively dated them as Miocene ($T_3$) and Pliocene ($T_4$). The subsequent large number of age determinations carried out by Siesser (1977a) on sea floor samples has supported, but not unequivocally established this scheme. The lower unit ($T_3$) covers most of the continental shelf and upper slope, and generally forms a seaward thickening sequence (maximum on mid shelf of about 250 m) of complexly bedded subunits (prograding banks, channel and fill sequences, and flat lying deposits), which locally at least, are subdivided by an erosion surface ($T_3$a below, $T_3$b above) (Dingle 1973c). The most prominent sedi-

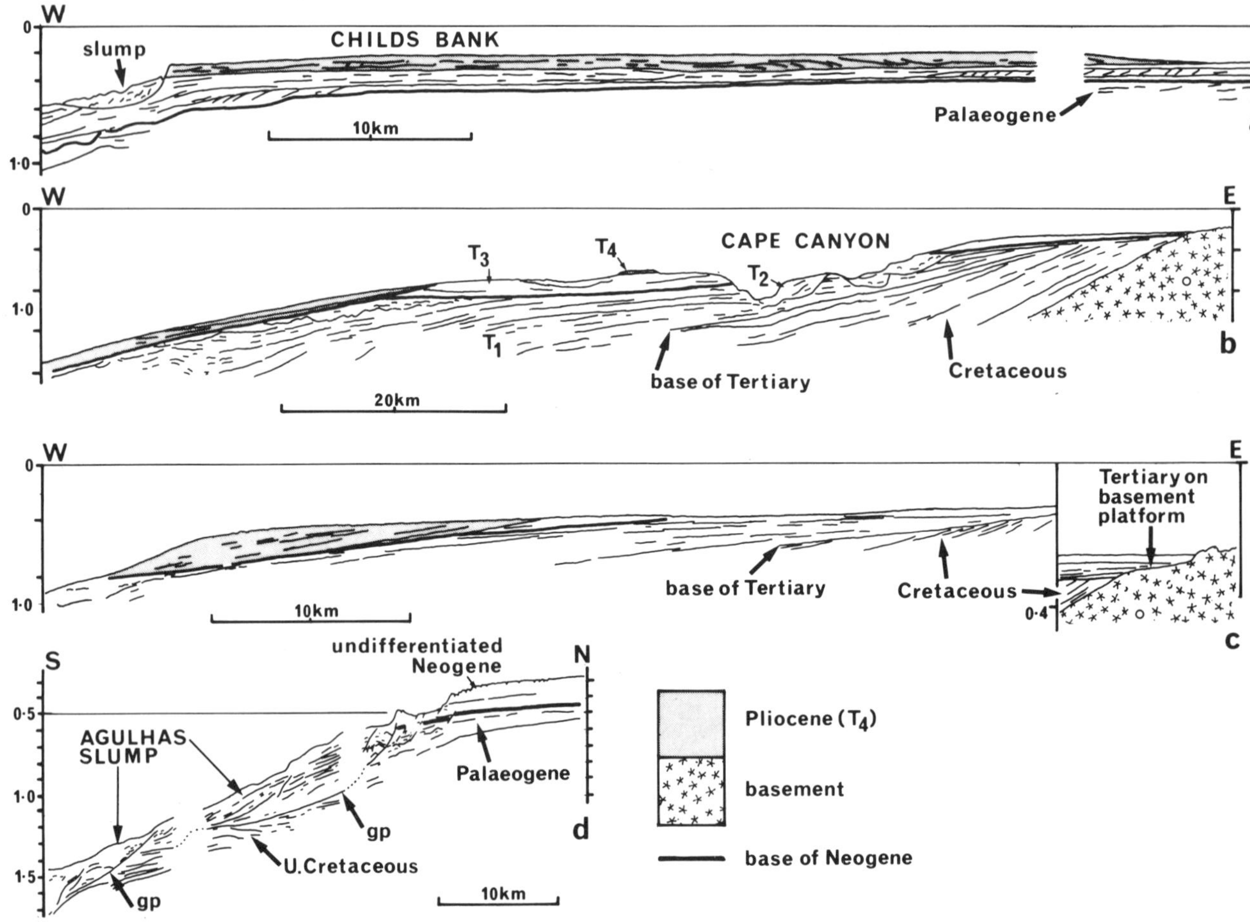


Figure 173. Seismic reflection profiles showing Tertiary sediments on the continental shelf off S. and S.W.Cape. Locations are on Figure 172.

a. SW of the Orange River towards the outer edge of the shelf showing the Pliocene (T4) outlier of Childs Bank resting on a practically horizontal erosion surface across Miocene (T3) strata. The outer edge of the bank is cut by a small slump.

b. W of Saldanha Bay across the Cape Canyon. Neogene strata (T3 and T4) are relatively thin and rest unconformably on Palaeogene, Cretaceous and Palaeozoic/Pre-Cambrian basement. The Cape Canyon is a complex polyphase feature partly filled with T2 (?Oligocene) sediments, and may be associated with the extensions of some of the large slumps shown in Figure 187.

c. W of Cape Peninsula. Neogene strata are apparently confined to the outer shelf, with the Pliocene (T4) perched as a lens-shaped body of seaward prograding sediments over the shelf break. Inshore, the Tertiary (?Palaeogene) rests on a gently seaward dipping platform cut across the pre-Mesozoic basement.

d. N-S profile across the outer Agulhas Bank S of Mossel Bay. Tertiary strata form a seaward thickening wedge which has been truncated by the proximal part of the Agulhas Slump. Cretaceous outcrops are locally exposed in the glide plane scars.

Datum for all profiles is sea level, and vertical scales are in sec. Modified from Dingle (1971a, 1973c, 1977).

mentary structure in the $T_3$ (?Miocene) sediments is a wide seaward prograding bar-like feature which lies on the outer continental shelf and can be traced on seismic records for over 300 km between about 32 and 29°S. It lies on top of, or seaward of, a topographical ridge which may have been formed by flexure in the underlying Palaeogene sediments, and a combination of the two presumably formed a shallow water barrier along the Miocene outer shelf. The sedimentary bar is less well-developed in the youngest units of $T_3$ (i.e. $T_3$b), but under the eastern end of Childs Bank (section a, Fig.173) is represented by a wide sedimentary complex which exhibits internal bedding that dips seaward and landward about a mid point situated at the eastern end of section a in Figure 173. All these features suggest sedimentation under shallow water, relatively high energy conditions (oceanic or tidal currents). On the upper continental slope, the $T_3$ sediments are cut by and lose their identity within the massive west coast marginal slumps.

Miocene sediments have been recovered from the walls of the Cape Canyon (clays), between the Cape Canyon and Bogenfels (mostly foraminiferal wackestones and calcareous sandstones), and north to the

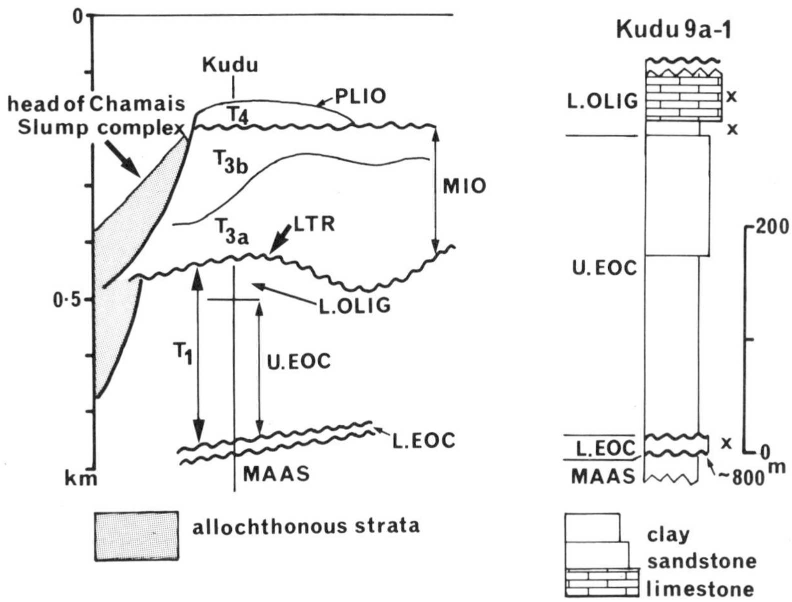


Figure 174. Schematic Tertiary stratigraphy of the continental shelf W of Orange River. Acoustic stratigraphy ($T_1$-$T_4$) is after Dingle (1973c), and the bio- and lithostratigraphy of borehole Kudu 9A-1 has been compiled from data in McLachlan & McMillan (1979). Datum is sea level and the vertical scales are in km and m. No Neogene has been reported from Kudu 9A-1. Exact thicknesses could not be calculated with the published data for horizons marked 'X' on the borehole log. LTR = Lower Tertiary Ridge.

Kunene River (calcareous clays), and whilst most are of Middle Miocene age, a few give Lower and Upper Miocene dates (Siesser 1977a) (Fig.172).

The scattered nature of the $T_4$ (?Pliocene) outliers suggests that they have suffered severe post-Neogene erosion, but it is possible that their present configuration is partly due to their original deposition as bank-like features on the mid-shelf and steeply dipping progrades over the shelf break. The $T_4$ outliers are relatively thin (maximum about 0.2 sec, 180 m) and exhibit a variety of internal sedimentary structures including prograding, horizontal and ?cross-bedding. The mid-shelf outliers were probably deposited under shallow-water, relatively high energy environments, but the significance of the perched, seaward dipping shelf edge lenses has yet to be satisfactorily explained. They rest with a marked angular discordance on the underlying $T_3$ and Palaeogene strata and give rise to small scale unevenness where individual bedding surfaces intersect the sea floor. It is possible that their internal geometry is in some way connected with the outer margin slump features, but such a connection remains obscure at present.

All Pliocene samples so far recovered have been calcareous clays. They come from the western wall of the Cape Canyon, NW of Cape Columbine, and off Lüderitz and Walvis Bay, and where well-resolved ages can be assigned, all are early Pliocene (Siesser 1977a). Erosional surfaces cut both the $T_3$ and $T_4$ sedimentary units of the west coast shelf (e.g. within the Cape Canyon, on Section b, Fig.173). indicating widespread denudation (either subaerial, or high energy submarine) in post-Pliocene times. In addition, $T_4$ strata have been affected by slumping (e.g. on the outer edge of Childs Bank, Fig.173), indicating that the west coast outer margin slumps are post-Pliocene in age.

The wide distribution of authigenic apatite and glauconite in these Neogene rocks (Bremner 1975, Siesser 1978a, Birch 1979a), together with their thinness, and complexity of internal sedimentary structures indicates deposition under shallow marine, locally high energy environments, with low terrigenous inputs. A dominant factor in this sedimentary regime was the strong upwelling by the Benguela current system, which resulted in high oceanic biological activity, and aridity of the adjacent coastal hinterland. We return to these phenomena in Section 6.2.2.3.4.

### 6.2.2.2 *DSDP sites 360 and 362*

The Neogene succession is relatively thick and complete at the DSDP sites in the Cape Basin and on the northern Walvis Ridge (Frio Ridge). At the former (site 360), Mio-Pliocene sediments occur as part of a large allochthonous mass that has moved down slope onto the upper continental rise, and although there is no strong palaeontological evidence for a break across the Oligocene-Miocene boundary it coincides with a zone of soft sediment deformation and lamination (Bolli et al 1978) (Fig.166). Here the Miocene sequence appears to be complete and is about 306 m thick in light grey nannofossil oozes (most of Unit 1) that pass downwards into light grey nannofossil chalks (Unit 2). The oldest Miocene sediments (Lower Miocene) occur within the uppermost 2 m of the deformed and laminated nannofossil chalks of Unit 3. We believe that this zone of deformation coincides with a subsidiary glide plane within the slumped mass. The bulk of the Miocene succession (Unit 2) consists of sediments with an average composition of 75 % nannofossils and 12-15 % clay, with minor to trace amounts of planktonic foraminifera, sponge spicules, pyrite, quartz, mica, glauconite, fish fragments, echinoid fragments, radiolaria, pteropods, ostracods and heavy minerals. This lithology grades upwards into the nannofossil oozes which contain a slightly higher clay content (ca. 20 %), less fish fragments, and more abundant pyrite. The upper part of the sequence comprises 9.5 m of Lower Pliocene nannofossil ooze of the same lithology found in the Miocene. Like the nearby site 361, no Quaternary sediments were recovered from the sea floor. Bolli et al (1978) recognized two sediment accumulation maxima within the Neogene succession at site 360: at 6.5 my (at least 30 m/my Upper Miocene), and 13.5 my (about 45 m/my Middle Miocene) set against a background Neogene rate of about 10 m/my. Compared with Palaeogene sedimentation parameters, the Neogene succession shows a higher $CaCO_3$ accumulation rate (expressed in $gm/cm^2/my \times 10^2$) is approximately the same as it had been during Oligocene times (ca.6-8) and was lower than during the Eocene (ca.35-40). This identifies a lower terrigenous contribution during Mio-Pliocene times as the prime cause for a reduction in bulk accumulation rates from Eocene to Neogene (55 m/my to 10 m/my), but an input decrease may not have been the sole reason for lack of fine supply because Bolli et al (1978) detected increases in sand size components between periods of high Neogene accumulation rate. These suggest intensified winnowing by bottom currents.

At DSDP site 362 on the Frio Ridge, there is an expanded Neogene succession (ca.670 m) which consists of nannofossil chalks (Lower to Upper Miocene) and nannofossil oozes (Pliocene to Pleistocene) (Bolli et al 1978) (Fig.166). One important contrast between the lithologies at site 362 compared to those in the Cape Basin, are the relatively high clay contents and the appearance of abundant siliceous microfossils in the upper part of the sequence. There is an apparently conformable passage between Oligocene and Miocene sediments in marly nannofossil chalks although the Lower Miocene is characterized by somewhat higher percentages of planktonic foraminifera compared to older and younger sediments (Tab.54). The greenish grey marly chalks of Unit 2 continue up into the Upper Miocene, but a significant change in the late Miocene is the first appearance of siliceous microfossils (1-4 %). These increase still further in numbers (6-20 %) in the Lower Pliocene olive-dark olive grey sandy and silty nanno oozes where pyrite, organic matter and $H_2S$ gas and accompanying deformation structures are in evidence. The sequence culminates in uppermost Pliocene and Pleistocene dark organic rich sediments which are strongly deformed by gas structures and have high siliceous microfossil contents, particularly diatoms (Tab. 54). As we will discuss later, these changes are related to the establishment of cells of strong upwelling in the

Table 54. Variation in major components (%) of Neogene sediments at DSDP Site 362 (Frio Ridge) (after Bolli *et al.* 1978).

| | Nanno-fossils | Clay | Dia-toms | Planktonic foraminifera | Radio-laria |
|---|---|---|---|---|---|
| Nanno ooze | | | | | |
| Pleistocene | 15-60 | 10-60 | 5-40 | 1-5 | 1-4 |
| Lower Pliocene | 50-60 | 25-30 | 5-20 | 1-2 | 1 |
| Nanno chalk | | | | | |
| Upper Miocene | 50-70 | 20-30 | 1-3 | 1-5 | <1 |
| Middle Miocene | 35-30 | 30-60 | – | 0.5-5 | – |
| Lower Miocene | 50-60 | 10-40 | – | 15-60 | – |

Benguela Current System, and accompanying sharp increases in biological productivity. Seismic records published by Lehner & de Ruiter (1977) show that about 85 km NE of DSDP site 362 on the abutment plateau, the Oligo-Quaternary succession has thickened to about 2.8 km.

6.2.2.3 *Coastal region*

The Neogene rocks of the coastal region consist of various marine and non-marine sediments, which can be conveniently described under three headings: the carbonate rocks of the SE, south and SW area (coastal limestones); the non-carbonate (locally phosphatic rocks) of the SW Cape (Elandsfontyn and Varswater formations); and the non-marine, and diamondiferous beach terraces of Namaqualand and southern Namibia.

6.2.2.3.1 *Coastal Limestones.* Limestones of probable Miocene-Pliocene age are found intermittently along the south and SW Cape Province coast in a narrow belt, usually about 10 km wide (Fig.155 & 172) (Truswell 1970). The limestones rest on a fairly even surface that reaches a maximum height above sea level of about 305 m near Bathurst and Paterson (on the Grassridge Terrace of Ruddock 1973, King 1972), 153 m in the Riversdale district (30 km north of Still Bay), and 76 m near Bredasdorp (Haughton 1925, 1963). This surface descends seaward in a series of wide benches in the eastern Cape (Ruddock 1973), but falls seaward more evenly farther west. Exposures of the limestones are generally poor because of extensive sand and calcrete cover, but several good sections are found in deep valleys and quarries. Thicknesses are variable, probably reaching a maximum of 150 m or more in the Riversdale district (Haughton 1963), whilst dips are low, usually 1-2° to the south. Although Mio-Pliocene carbonate rocks occur on the Agulhas Bank, Siesser (1971) emphasizes that the limestones of the coastal region are lithologically dissimilar to the various Agulhas Bank lithofacies. However, within themselves, the coastal Mio-Pliocene limestones show remarkably little petrographic, mineralogic or geochemical variation when the great lateral extent of the formation is considered.

In general the formation consists of a thin marine

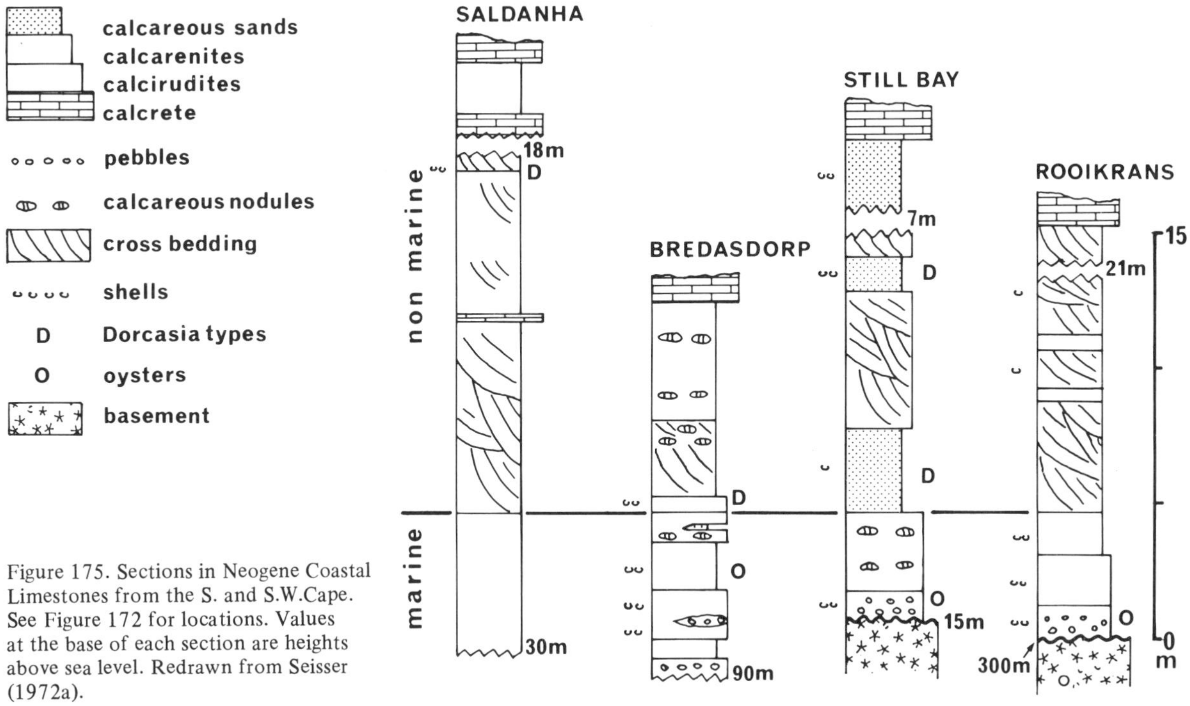


Figure 175. Sections in Neogene Coastal Limestones from the S. and S.W.Cape. See Figure 172 for locations. Values at the base of each section are heights above sea level. Redrawn from Seisser (1972a).

limestone, usually less than 10 m thick, overlain by an aeolian limestone/sandstone section as much as 150 m thick in places. Figure 175 shows measured sections at various localities along the length of the outcrop. The lowermost part of the marine bed commonly consists of a pebble- or broken-shell-bearing fine calcirudite, usually less than 1 m thick, resting unconformably on Palaeozoic and Mesozoic rocks. This basal bed is not always present, but where it does occur, it is conformably overlain by a massive or thick-bedded calcarenite, which may itself contain scattered pebbles or pebble lenses. The calcarenite, which usually has the texture of a skeletal lime grainstone, contains a variety of whole and fragmented marine fossils of which molluscs (lamellibranchs and gastropods) are the most prominent, and *Ostrea* locally abundant. Other taxa include: polyzoa, echinoids, sharks' teeth, and benthic foraminifera. Massive marine calcarenites are also occasionally found lying directly on older rocks, without an underlying pebble bed. The contact between the marine calcarenite and the overlying aeolian calcarenite and calcareous sandstones is often indistinguishable except where aeolian cross bedding is developed, although frequently there is a change in lithology with the appearance of *Trigonephrus* (*'Dorcasia'*) bearing sediments (e.g. at Still Bay, Fig.175). Siesser (1972a) has published detailed textural, mineralogic and geochemical data from the coastal Mio-Pliocene rocks and summarizes the limestones as pale or yellowish orange, moderately to well-consolidated, and consisting mostly of sand-size terrigenous grains and comminuted skeletal fragments of shallow-water organisms set in an intergranular fabric dominated by sparry calcite and pseudospar. Thin section analysis of over 100 representative samples gave the following component averages: intergranular fabric 25 %, terrigenous grains 22 %, and fossil fragments 49 %.

Previous studies of these rocks have usually referred to the outcrops in three separate areas and there has been a tendency for lithostratigraphic terminology to be applied to each, somewhat in isolation of the others. These areas are: the Alexandria-Bathurst region of the Eastern Cape; around Bredasdorp; and the Saldanha Bay area of the west coast. The earliest studies of this group of limestones were made in the eastern Cape by Atherstone in 1850 (unpublished field notes, quoted in Schwarz 1908), who divided the strata into three units: a lower Blue clay, a middle Bathurst limestone, and an upper Pecke limestone. Schwarz (1908, p.110) proposed the name Alexandria Formation for the two upper beds, but did not include the Blue clay which he realized must belong to the underlying Lower Cretaceous. Moulle (1885, quoted in Mountain 1946, p.36) apparently called the same Tertiary sequence the Albany Formation, but this name has been generally discarded. All workers have not adhered strictly to the term Alexandria Formation. The same beds have been more loosely called the Alexandria Series (Wybergh 1920, Haughton et al 1937), and Alexandria Beds (Haughton 1928). The wind-blown calcarenite (probably Atherstone's Pecke limestone) overlying the marine limestone has been referred to separately as

Dune Rock by Mountain (1946) who mentions that it is also called 'aeolianite' or 'consolidated dune' (Mountain 1962, p.10).

For the outcrops farther west, Wybergh (1919) designated the limestones in the Bredasdorp district the Bredasdorp Formation, but this usage has also not been uniform with, for example, Spies et al (1963, p.15) who refer to the same rocks as the 'Bredasdorp Beds'. From around Saldanha Bay, du Toit (1917, p.13) referred to the limestones as the 'Dorcasia Limestone', but although he was certainly referring to the aeolian calcarenites exposed there, it is not entirely clear whether he also meant to include the underlying marine limestones in his classification. More recently, Visser & Schoch (1973, p.49) formally proposed the name 'Langebaan limestone' for the recent aeolian limestones of this area, but make no mention of the underlying marine limestone. It seems probable that they would correlate the Langebaan limestone with the upper part of the Bredasdorp Formation, as does Rogers (1980) in his account of the west coast Cenozoic sediments. Further aspects of the late Cenozoic carbonate rocks from the SW Cape have been described by Knox (1977), De la Cruz (1978) and Flemming (1980a).

Siesser (1972a) presented a series of measured stratigraphic sections between Bathurst and Saldanha Bay (Fig.175), including detailed petrographic, geochemical and mineralogic data. Comparison of the sections and other exposures along the coast shows a remarkable lithologic similarity among the Langebaan ('Dorcasia') limestones, Bredasdorp Formation and Alexandria Formation, and Siesser (1971, 1972a) following earlier usage by Wybergh (1919) therefore suggested the informal term 'Coastal Limestones' for all the Tertiary and Pleistocene carbonate rocks exposed along the Cape coast from Saldanha Bay to East London. This designation was not meant to replace the two names most commonly used, i.e. the Bredasdorp Formation and Alexandria Formation (Haughton 1963), but was proposed merely as a term of convenience to include *both* the marine and aeolian parts of the Alexandria Formation, the Bredasdorp Formation and the Langebaan limestone (Fig.155). It should not include the unlithified sands overlying the formation, nor the calcretes, which are considered only as a secondary modification of the formation. A formal lithostratigraphic name will, no doubt eventually be assigned to all these rocks and because the best exposures are in the eastern Cape Province, a stratotype will probably be designated there. The already entrenched name in that region 'Alexandria Formation' would seem to be most appropriate for this unit, as it also has historical priority (Schwarz 1908). The marine and aeolian facies of the formation can often be distinguished and may be suitable for designation as members.

Notwithstanding the extensive palaeontological collections that have been described from these rocks by Newton (1913), Mountain (1946), Engelbrecht et al (1962), Spies et al (1963), Siesser (1971), Ruddock (1973) and Visser & Schoch (1973), relatively few age determinations have been made on the Coastal Limestones. Newton (1913) assigned a Miocene-Pliocene age to some low-lying parts of the Alexandria Formation close to the present shore-line, but King (1953, p.64) and Haughton (1963, p.312) believed that these rocks are probably Pliocene. Spies et al (1963) have further shown that the Miocene-Pliocene faunas identified by Newton (1913) are practically identical to those in the Bredasdorp Formation, and consider that the two formations are thus of the same age. Parr (1958) has identified Pleistocene to Recent foraminifera from rocks at the coast south of Bathurst, and Singer & Fuller (1962) identified fossil bones in an Upper Pleistocene aeolianite on the northern shore of False Bay. Pliocene aeolianites have been reported from the Natal coast (McCarthy 1967), and King (1973) has recorded Pliocene benthic foraminifera from marine rocks of the Alexandria Formation at Paterson in the eastern Cape at about 305 m above sea level. It appears, therefore, that the Coastal Limestones span a considerable part of the Cenozoic, generally progressing from older to younger beds in a seaward direction. The marine facies is probably Miocene-Pliocene in age, with Pliocene rocks predominating closer to the coast. The aeolian facies may be Miocene or Pliocene farther inland (Siesser, 1972b shows aeolian calcarenites at more than 300 m above sea level) but is probably almost entirely Pleistocene near the coast.

The Coastal Limestones have been interpreted as deposits laid down by regressive seas (Haughton 1925, Siesser 1972a). Where present, the thin basal part of the formation is a beach and nearshore deposit which is overlain by lithified coastal dunes (aeolianites) whose thickness varies considerably. Outcrops nearer the coast have probably gone through several regression-accumulation, transgression-destruction phases during the late Tertiary, and thus represent a younger accumulation than those farther inland, which later transgressions did not reach.

Several local lithological variants have been described from the coastal limestone sequences. At Martindale, in the Bathurst area, Mountain (1946) found stratified, highly ferruginous sandstones, which in places contain well-rounded pebbles and boulders of quartzite near the bottom of the unit. He suggested the name Martindale Beds for these rocks which reach a thickness of just over 3 m and have a slight seaward dip. The only apparent indication of their age is that they contain pebbles of silcrete, which is believed to be of early or middle Tertiary age in this area. Mountain (1946) makes no suggestion as to the age of these beds (other than to indicate they are 'Tertiary'), although the stratigraphic column on his map locates them immediately above the silcretes and below the Alexandria Formation. In

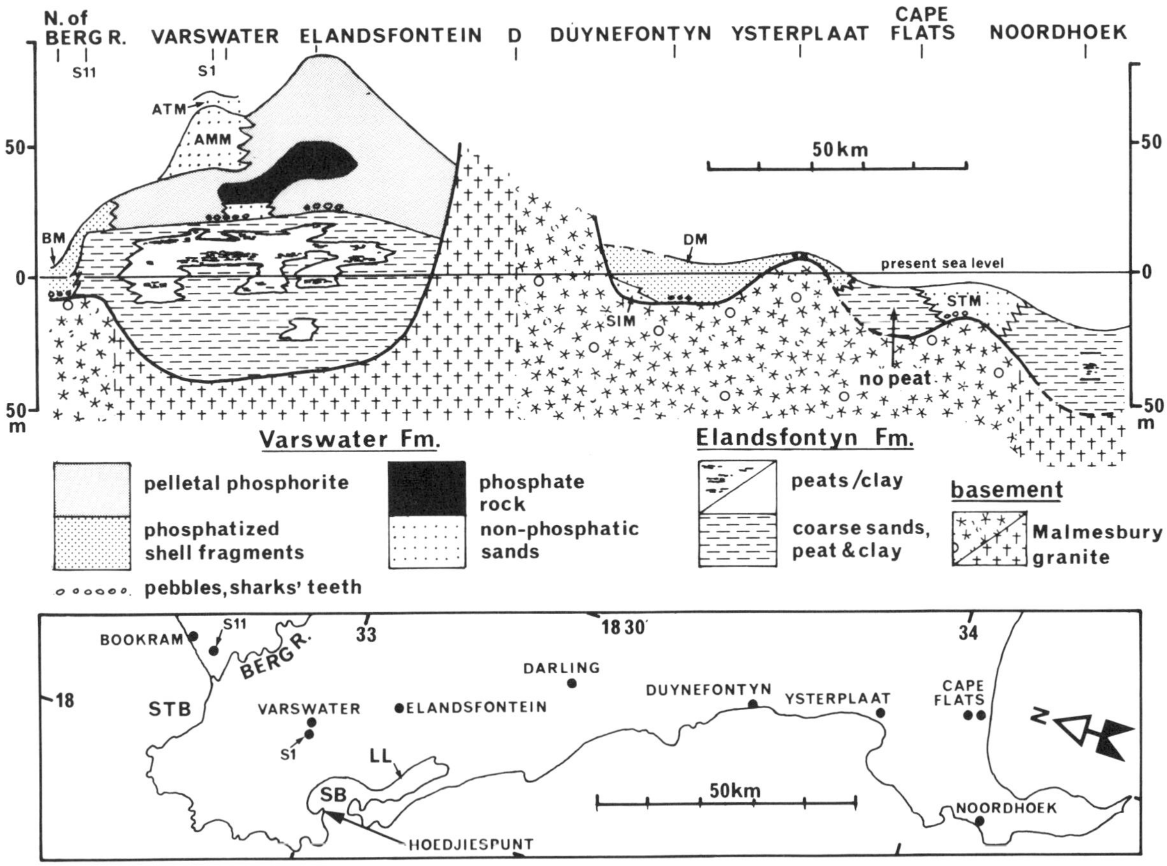


Figure 176. Distribution of Neogene sediments in the S.W.Cape. Datum is present sea level. Note spelling: the Elandsfontyn Formation is named after the original farm which is near the present settlement of Elandsfontein. Data mainly from Tankard (1975b), Rogers (1979, 1980), Hendey (1981a,b). Abbreviations: BM = Bookram Member; ATM = Anyskop Terrestrial Member; AMM = Anyskop Marine Member; SIM = Silverstroom Member; DM = Duynefontyn Member; STM = Strandfontein Member; STB = St Helena Bay; SB = Sandanha Bay; LL = Langebaan Lagoon; D = Darling. S1 and S11 are boreholes mentioned in Rogers (1980).

Figure 155 we have placed them in a nominal Upper Miocene position.

The Knysna Beds (du Toit 1954) are deposits of slightly clayey, quartzose sand cropping out near Knysna which may also be associated with local beds of clay and lignite. The sands rest unconformably on Enon conglomerate and Table Mountain Sandstone, and are fine grained, moderately well indurated, massive to cross bedded and white, yellowish or reddish brown in colour. A ferricrete horizon up to 3 cm thick and a pebble layer separate the sands from the overlying soil in some places. Haughton (1963) notes that the beds are up to 67 m thick and rest on a platform over 430 m above sea level. The only fossils found are those of plants which still live in the vicinity today: *Widdringtonia* (Cape cedar), *Podocarpus* (yellowwood), *Conioma,* and *Curtisia.* Du Toit (1954) speculates that the Knysna Beds may correlate with some of the soft sandstones of the Alexandria Formation (?Pliocene).

Ruddock (1973) has summarized the macro- and microfossils reported from the marine Mio-Pliocene coastal limestones of the SE and SW Cape (Alexandria, and Bredasdorp plus Dorcasia sequences, respectively). Lamellibranchs form the most diverse group (28 spp., 18 genera SE; 14 spp, 10 genera SW) with oysters the single most important taxa (at least six species), followed by gastropods (27 spp., 21 genera SE; 10 spp., nine genera SW). Within each of these groups 8 spp. and 4 spp., respectively, are common to the two areas. Other macro invertebrates are relatively poorly represented: 8 spp. all from the SE (scaphopods, brachiopods, echinoids, barnacles and corals). On the other hand, sharks' teeth are well-represented, with 15 spp. reported from the SE Cape and two species from the western outcrops, though none are noted as common to both areas. Of the microfossils, only foraminifera have received much attention and Ruddock (1973) lists 25 spp. and 29 spp. from the SE Cape and SW Cape respectively (five are shown as common), with specimens occurring in both the marine and non-marine (aeolianite) litho-

Table 55. Previous classifications of the Elandsfontyn and Varswater formations in the S.W.Cape.

<table>
<tr><th colspan="2">Tankard (1974b, 1975a)</th><th colspan="2">Dingle, Lord & Hendey (1979)</th><th colspan="2">Rogers (1980)</th><th colspan="2">Hendey (1981a) (modified from 1974 and 1978)</th><th colspan="2">This work</th></tr>
<tr><td>Pelletal phosphorite Member</td><td rowspan="6">Varswater Formation</td><td>A<br>B<br>C<br>D</td><td>Upper Varswater Fm.</td><td>Phosphorite Member</td><td rowspan="3">Varswater Formation</td><td>Pelletal Phosphorite Member</td><td rowspan="4">Varswater Formation</td><td>Pelletal Phosphorite Member</td><td rowspan="4">Varswater Formation</td></tr>
<tr><td>Fluvial Sand Member</td><td>E</td><td rowspan="7">Lower Varswater Formation</td><td>Quartzose Sand Member</td><td>Quartzose Sand Member</td><td>Quartzose Sand Member</td></tr>
<tr><td rowspan="2">Beach Gravel Member</td><td rowspan="2">F</td><td>Gravel Member</td><td rowspan="2">Gravel Member</td><td rowspan="2">Gravel Member</td></tr>
<tr><td colspan="2">Saldanha Formation</td></tr>
<tr><td>Kaolinitic Clay</td><td rowspan="4">G</td><td colspan="2" rowspan="4">Elandsfontyn Formation</td><td colspan="2" rowspan="4">Pre-Gravel Member deposits</td><td rowspan="4">Various sands, peats and clays</td><td rowspan="4">Elandsfontyn Fm.</td></tr>
<tr><td>Fine Quartzose Sand</td></tr>
<tr><td>Basal Bed</td><td rowspan="2">Saldanha Fm.</td></tr>
<tr><td>Mottled Silty Clay</td></tr>
<tr><td colspan="10">Basement</td></tr>
</table>

facies. Only one planktonic species (*Globigerina* sp.) has so far been reported from the western outcrops (Haughton et al 1937).

6.2.2.3.2 *Elandsfontyn and Varswater formations.* A complex of non-marine sands, clays and peats, and marine sands and gravels is developed in the SW Cape between the Cape Flats (south of Cape Town) and the Lamberts Bay area (north of the Berg River) (Figs 172 & 176). Early geological studies in this area (e.g. du Toit 1917, Haughton 1932) were prompted by, and concentrated upon, the commercially exploitable apatite (phosphate) deposits that occur in the marine sands, whilst in recent years attention has also focused on the remarkably varied and well-preserved fossil assemblages (especially the vertebrates) that have been recovered during mining operations. This research has led to several attempts at litho- and biostratigraphic correlation of the various rock units in the regions, and although neither has been unequivocally established, the work of Hendey (1970 et seq.), Tankard (1974a,b & 1975a, b) and Rogers (1980) in particular, allow a satisfactory regional picture to be presented (Figs 176 & 177). The importance of the vertebrate fauna in the Mio-Pliocene rocks of the Varswater Quarry cannot be over stated and it has been emphasized by Hendey (1976) that 'certainly no other recorded African fossil occurrence of Pliocene age has produced so large an assemblage of specimens representing so wide a variety of species'.

The Neogene sediments of the SW Cape are best developed in the Varswater-Saldanha embayment (Fig. 176), where a maximum thickness of at least 120 m overlie granite basement which descends to about 40 m below sea level. The sequence thins north and south, and is separated from that in the Duynefontyn-Cape Flats region by a zone of basement high in the hills around the village of Darling. Outcrops in the southern region are relatively poor, and the succession thin (ca. 30 m or less), resting on a basement floor that generally lies above the −20 m level (Noordhoek being the exception). Because stratigraphic relationships are better known in the Varswater-Saldanha area, and it is from here that all the main lithostratigraphic type sections have been established, we will take this as our standard for comparison.

The succession in the Varswater-Saldanha embayment consists of a lower non-marine sequence (Elandsfontyn Formation), and an upper sequence characterized by the presence of apatite-bearing marine sands and lithified sands, but also including a complex of non- or poorly-phosphatic estuarine and fluviatile sands and minor clays (Varswater Formation). These rocks are capped by various calcareous sands and calcrete within which some of the upper parts of the Varswater Formation may still be erroneously included. The type sections for this succession are in excavations and boreholes in the Varswater (E or New) Quarry at Langebaan Road, and at boreholes in the vicinity of Elandsfontein.

The Elandsfontyn Formation was formally designated by Rogers (1980) from a borehole (G30878) in the Elandsfontein district, about 16 km SE of the Varswater Quarry. The succession consists of about 60 m of coarse, angular, gravelly sands, with subordinate medium and fine sands, interbedded with three main clay horizons that contain two peat layers. Argillaceous rocks constitute about 30 % of the succession in the type borehole, where seven individual clayey horizons were intersected, the thickest being about 12 m with a basal peat 4 m thick. Rogers (1980) notes that the lithology of the

Elandsfontyn Formation is very variable, with the proportion of argillaceous material altering considerably between the various boreholes in the Varswater-Saldanha embayment, but the characteristics of the formation are the presence of clays and peat, and coarse, angular sands. No natural outcrops of the Elandsfontyn Formation are known, but details of its lithology were given by Tankard (1974b, 1975a), who noted sands, clays (kaolinitic, and peaty) and a phosphatic horizon. In his classification of the succession Tankard (1974b, 1975a) placed the upper two members of Roger's (1980) Elandsfontyn Formation within what is now taken as the Varswater Formation (Tab.55), whilst the phosphatic horizon (his Basal Bed) (and presumably the underlying mottled silty clay) were equated with the Saldanha Formation. As we will discuss later, the latter correlation is considered unproven, and in fact the presence of the Basal Bed has never been reconfirmed. Rogers (1980) intersected the Elandsfontyn Formation in his borehole S1 on the southern edge of the Varswater Quarry (on Anyskop), and proved 62 m of sands (including angular, coarse and gravelly sands), with a 10 m thick peat/clay member 10 m below the top of the formation.

Correlation of various members within the Elandsfontyn Formation has not yet been achieved, but preliminary work by both Rogers and Hendey (personal communications 1981) shows that the argillaceous horizons with their local developments of peat occur at well-defined levels across the Varswater-Saldanha embayment, and that these may prove valuable markers for understanding the sedimentary history of the succession. In the Varswater Quarry, Tankard's (1975a) borehole record represented a composite section from two closely-spaced borings, and re-examination of the original logs shows two peaty clay horizons at each site (as does a new borehole in the quarry taken by Hendey, personal communication 1981), in contrast to the one in Roger's S1 borehole. Clearly there is a complex interfingering relationship between argillaceous and arenaceous lithofacies in this area because at all three sites the clays with peats are confined to a zone that lies at an elevation of between 2-3 m and 20-22 m above sea level. In the Elandsfontein district, Rogers (1980) records clays with peats at similar elevations, and his other borehole data show that there is continuity of this zone across the embayment to the north of the Varswater area (Fig.176). A further, lower level (−25 m to +2 m) series of locally-developed clays and peats (interbedded with sands and gravelly sands) suggests that this lithofacies was established in the southern part of the embayment somewhat earlier than it was in the Varswater area, where the lower part of the succession consists solely of arenaceous sediments.

If the peats represent depositional environments at or very close to sea level (i.e. lagoonal, estuarine, or lower flood plain), then the +2-22 m deposit may provide a suitable datum about which to construct sedimentary models when further sedimentological data are available.

The Elandsfontyn Formation is overlain (disconformably, possibly unconformably) by the Varswater Formation whose type sections are in the Varswater (E or New) Quarry at Langebaan Road. Hendey (1970) originally proposed the name Varswater Beds (later revised to formational status, Hendey 1974a), but the first formal descriptions of the sequence were given by Tankard (1974b). Table 55 shows that Tankard's definition included strata that have since been referred to the Elandsfontyn Formation (i.e. his kaolinitic clays and fine Quartzose sand), and the classification that we adopt here is that revised by Hendey (1981a), in which the base of the formation is taken as the Gravel Member. This differs slightly from that recently proposed by Rogers (1980), who places the Saldanha Formation between the Elandsfontyn and Varswater formations (see Tab.55) on the assumption that Tankard (1974b) correlated his Beach Gravel member with the Saldanha Formation, whereas the latter author plainly stated (Tankard 1974a tab.1, 1974b tab.1, p.269 & 1975a) that it is the Basal Bed that is correlated with the Hoedjiespunt phosphatic rocks (type section of Tankard's Saldanha Formation).

The base of the Varswater Formation is taken as the Gravel Member. This is a 0.5-2 m phosphatic, locally silicified, conglomeratic sandstone which shows evidence of at least two episodes of re-working. It occurs as a horizon of rounded, discoidal, and slab-like blocks set in a partially lithified phosphatic sand. These fragments have clearly been water-worn, and consist themselves of water-rounded cobbles and pebbles of phosphatic sandstone set in a phosphatic matrix. These relationships indicate at least three phases of phosphate-rich sediment formation separated by two intervals of erosion. The rock has the appearance of a hardground horizon and must represent an hiatus or condensed sequence at the top of the Elandsfontyn Formation. Furthermore, it indicates that the Elandsfontyn sedimentary episode was brought to close by a marine transgression which resulted in the deposition of phosphate-rich sands that were subsequently lithified then eroded to produce the oldest clasts in the Gravel Member conglomerate. Tankard's (1975a) record of 'phosphate' in the Elandsfontyn Formation (which he referred to as the Basal Bed) has not yet been reconfirmed by an examination of the original borehole material (Rogers, personal communication 1981), but if substantiated, indicates that the Elandsfontyn Formation consists of no less than two transgressive sedimentary cycles terminating in marine strata, and that the original concept of the formation by Rogers (1980) as one non-marine sequence needs revision.

In the Varswater Quarry area, the Gravel Member occurs at an elevation of about +23 m in the south,

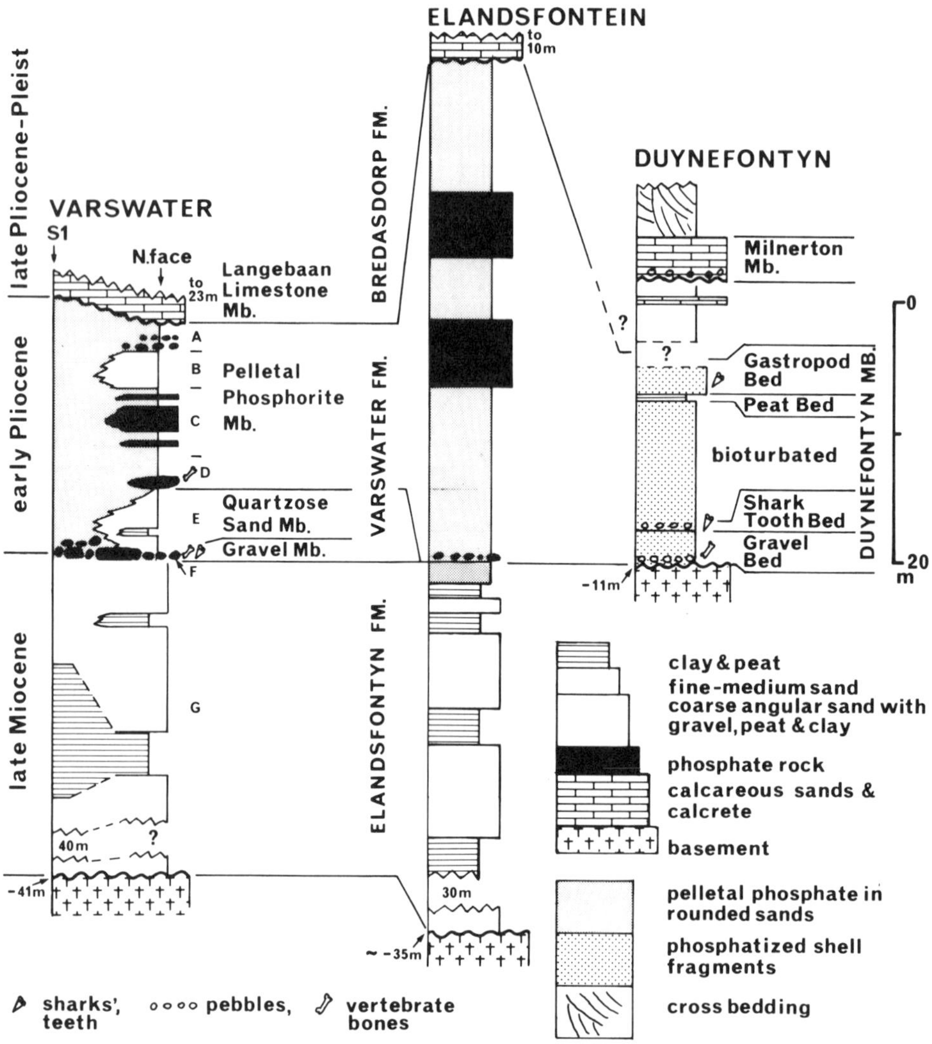


Figure 177. Measured sections and boreholes in the Mio-Pliocene strata of the S.W.Cape. At left is the type section of the Varswater Formation in the New Varswater Quarry, Langebaan Road. It is composite from quarry exposures, Geological Survey borehole S1, and quarry boreholes (quoted in Tankard 1975a). Notation A-G is after Dingle *et al.* (1979). In the centre is the type section of the Elandsfontyn Formation in Geological Survey borehole G30878 at Elandsfontein, and at right a composite section in the vicinity of excavations for the Koeberg nuclear power station at Duynefontyn. Lithostratigraphic terminology has been slightly revised from original descriptions. Values at the left base of each section are heights (m) below present sea level. Datum is taken at the level of the Gravel Member or its equivalent. Data mainly from Tankard (1974b, 1975a), Dingle *et al.* (1979), Rogers (1979, 1980), Hendey (1981a, 1981b).

and rises to about +26 m in the north. Rogers (1980) recorded probable correlatives at +28-30 m in the Elandsfontein district, so here perhaps we have another useful datum for interpreting the Varswater-Saldanha embayment. Rogers (1980) also recorded various conglomerates with sharks' teeth at several other localities in the SW Cape, but their relationship to the Gravel Member is far from clear.

The gravels representing the younger of the erosional events in the Gravel Member in the Varswater Quarry contain a molluscan fauna indicative of intertidal, rocky environments which contain elements that suggest a warmer water environment than prevails off the SW Cape coast today (e.g. *Cellana capensis* and *Turbo sarmaticus*) (Kensley 1972). Table 56 summarizes the number of species in major taxa as reported by Hendey (1981a). For age determinations, a significant record is that of *Hipparion* cf. *H.primigenium* by Hendey (1976) which places a maximum age on the Gravel Member at 12.5 my (Middle Miocene), the earliest date at which *Hipparion* could have entered Africa (Hendey 1981a).

In the Varswater Quarry, the Gravel Member is overlain by a complex assemblage of phosphatic and non-

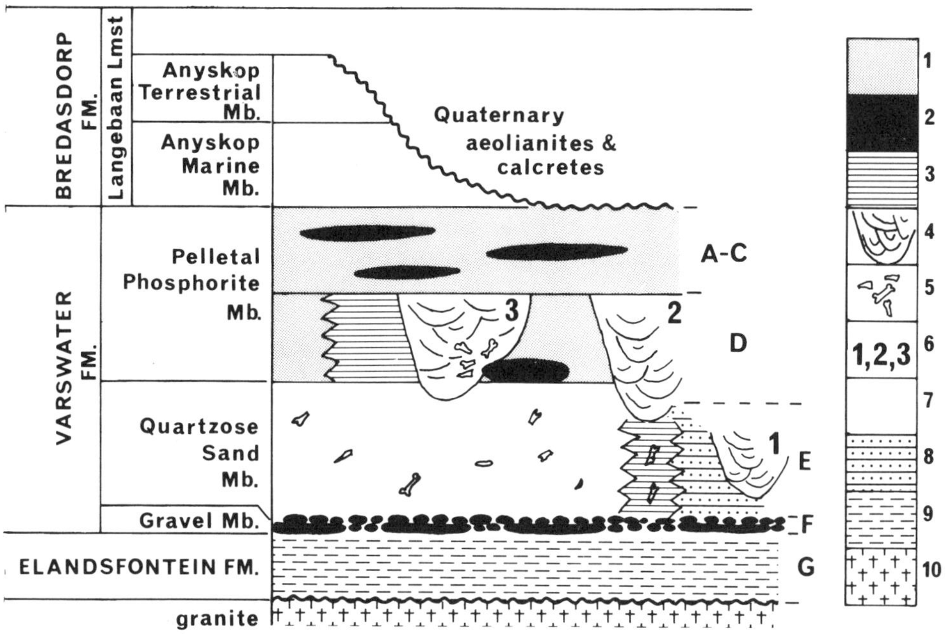


Figure 178. Lateral lithofacies relationships in the Varswater Formation as exposed in the New Quarry, Varswater. Classification and data are after Hendey (1978b, 1981a), Dingle *et al.* (1979), Rogers (1980). See text for discussion. Key: 1 = pelletal phosphorite sands; 2 = phosphate rock; 3 = peat and clay; 4 = river channel deposits; 5 = vertebrate bone accumulations; 6 = successive positions of the Berg River channel; 7 = quartzose sands; 8 = intertidal clays; 9 = various clays, peaty clays, coarse angular sands, gravels, silts; 10 = granitic basement.

phosphatic sands, with subordinate clays, peaty clays, and locally, thick phosphatic rocks. Hendey (e.g. 1970, 1973, 1981a,b) has interpreted this succession as a fluvial-estuarine-shallow marine transgressive sequence in which the channel of the Neogene Berg River can be inferred or identified in at least three successive positions (Fig.178). Lithofacies variations are rapid within the quarry, but the lower units are essentially non-phosphatic and consist of: river channel sands; intertidal muds; peaty clays; and flood plain sands. These sediments constitute the Quartzose Sand Member (Tab. 55), and the last-named lithofacies are the most widespread, consisting of rounded, iron stained, quartzose, moderately-sorted, interbedded coarse-medium and fine sands (Tankard 1974b). They contain a rich mixed terrestrial and aquatic vertebrate fauna (Hendey 1981a), the most characteristic elements of which are a tortoise (*Chersina* sp.), a francolin (indet sp.), a rhinoceros (*Ceratotherium praecox*), a pig (*Nyanzachoerus* cf. *pattersoni*), and a boselaphine (*Mesembriportax acrae*). The last-named, along with the tortoise being the most abundant form (often in very large numbers). Two primate teeth, which Grine & Hendey (1981) consider close to *Parapapio jonesi*, a taxon recorded in the Makapansgat and Sterkfontein caves, have also been found in the flood plain facies. Occasional juvenile seal bones (*Homiphoca capensis*) occur in this lithofacies, but are more abundant (adult forms) along with other marine taxa (molluscs and amphineura) in the estuarine peats and clays, which are typically dark, sandy, carbonaceous, waxy-looking clays with a high iron mineral concentration. Their fossil content has not yet been thoroughly analysed, but Hendey (1981a) states that they are rich in both pollens and vertebrates. The intertidal clays of the Quartzose Sand Member contain an abundant invertebrate assemblage, but are poor in vertebrate remains, which are mainly allochthonous small bone fragments, but include birds eggs (Hendey 1981a). Kensley (1977) reports that the invertebrate fauna includes terrestrial, fresh water, estuarine and marine species typical of an intertidal mud flats assemblage.

Table 56 shows that birds (74 spp.) and mammals (59 spp.) are the most abundant types in the Quartzose Sand Member, and of the latter 19 (32 %) are carnivores.

The Quartzose Sand Member peats occur at an elevation of about +24 m in the Varswater Quarry, and may be correlatives of peats reported by Rogers (1980) at between +37 and +54 m in the Elandsfontein area.

The Quartzose Sand Member passes laterally and vertically into the Pelletal Phosphorite Member, which in the S1 and Elandsfontein boreholes rests directly on the Gravel Member (Rogers 1980) (Fig.177). This distribution strengthens Hendey's (1981a) assumption that the Quartzose Sand Member was deposited on the flood plain and within the estuary of the Berg River (i.e. it was a strictly local lithofacies). The Pelletal Phosphorite Member, on the other hand, has a wide distribution although one of its characteristic elements in the type area (the phosphate rock is more restricted. In the Varswater Quarry, the Pelletal Phosphorite Member exhibits four main lithofacies: a non, or poorly phosphatic quartzose sand channel-fill unit with abundant vertebrate bones; an adjacent peat/clay bed; pelletal phosphate-rich sands; and tough lenticular phosphate rock horizons (Fig.178). Dingle et al (1979) reported that the upper part of the member consists of phosphate sands and rock layers, whilst the lower part contains only one known phosphate rock layer, but contains two river channel fills and phosphatic sands. It is the pelletal phosphate sands that are of commercial importance in the Varswater Quarry. Tankard (1974b) quotes a maximum thickness about 25 m for the Pelletal Phosphorite Member in the Varswater area, but about 38 m occur in the Elandsfontein borehole (Rogers 1980). The phosphatic sands, which form the bulk of this member, are reddish brown moderately sorted fine phosphatic, quartzose sands, with a high percentage of rounded grains. Coarse and medium size fractions form only minor components. Tankard (1974b) found that the sands are generally bimodal and that the pelletal phosphate fraction is somewhat coarser (2-2.5 ∅) than the quartz fraction (3-3.5 ∅) which is rounded to well-rounded and polished. Phosphorite pellets smaller than 2.5 ∅ are also well-rounded and polished. Tankard (1974b) records the $P_2O_5$ values in the Pelletal Phosphorite Member in boreholes from the Varswater Quarry and shows that they are generally higher than 12 % and locally reach 22 %. Although the phosphatic sands are frequently compacted, they are usually friable, but particularly in the northern part of the present quarry, they pass laterally into tough greyish brown lithified phosphate rock layers up to 2 m thick. Some of these can be individually traced for at least 300 m laterally, but they often vary in thickness from a few centimetres to over 1 m within a distance of 3-4 m. The cement of these rock units is phosphatic and they clearly developed by progressive in situ phosphatization of pre-existing pelletal phosphate sands. That phosphatization occurred a short time after the deposition of any individual sand horizon and was not the product of a single late stage event is suggested by the impingement of the Berg River channel at one point against what was obviously a hard rock layer.

Although vertebrate fossils do occur occasionally in the main body of the pelletal phosphate sands, the bulk of the specimens have been recovered from the two channel-fill units in this member(marked 2 and 3 in Fig.178). These contain yellowish or white friable sands with relatively low pelletal phosphate contents compared to the adjacent main phosphatic sands, and locally have rich accumulations of bones in small banks which were formed against older phosphate rock horizons. Hendey (1981a) points out that whilst these river channel deposits in the Pelletal Phosphorite Member have many species in common with the Quartzose Sand Member, their faunas are in some respects distinct and the conditions of specimens and relative numbers of individuals of any one taxon may be very different. The most characteristic taxa of these beds are: the seal (*Homiphoca capensis*); giraffids (*Giraffa* sp., *Sivatherium hendeyi*); alcelaphine antelopes (*Damalacra neanica*, *D.acalla*) and a reduncine (*Kobus subdolus*).

Tankard (1974b) states that sharks' teeth, molluscs, echinoid spines, fish teeth and phosphatized foraminifera (mainly *Elphidium*) occur throughout the pelletal

Table 56. Number of species of major taxa in the Mio-Pliocene Varswater Formation. Summarized from Hendey (1981a) – full details given in Appendix.

| | Gravel Member | Quartzose Sand Member | Pelletal Phosphorite Member | |
|---|---|---|---|---|
| | | | *river channel facies* | *main sand body* |
| echinoids | 1 | | | |
| brachiopods | 1 | | | |
| cirripeds | 1 | | | |
| ?decapods | 1 | | | |
| gastropods | 19 | 20 | | |
| lamellibranchs | 4 | 2 | | |
| sharks | 17 | ?1 | | indet. |
| other fish | 5 | | 2 | 1 |
| amphibia | | 1 | 1 | |
| tortoises | 1 | | 2 | |
| chameleons, gekkos, lizards, snakes | | 5 | 4 | |
| penguin | 1 | 4 | 2 | |
| other birds | | 70 | 40 | |
| moles and shrews | | 5 | 3 | |
| bats | | 1 | | |
| monkeys | | 1 | | |
| pangolins | | 1 | | |
| aardvarks | | 1 | 1 | |
| fox, bears, badgers, etc. | | 1 | 4 | |
| seals | 1 | 1 | 1 | |
| mongooses, etc. | | 5 | 4 | |
| hyaenas | | 4 | 5 | |
| cats | | 5 | 5 | |
| indet. carnivores | | 3 | | |
| elephants, etc. | | 2 | 1 | |
| dassies | | 1 | | |
| horses | 1 | 1 | 2 | |
| rhinoceras | | 1 | | |
| pigs, peccaries | | 1 | 2 | |
| hippopotamus | | | 1 | |
| giraffes | | 2 | 3 | |
| bovids | | 7 | 11 | |
| hares | | 1 | 1 | |
| rodent moles | | 2 | 1 | |
| porcupines | | 1 | 1 | |
| rats, mice, etc. | | 12 | 1 | |
| cetacea | indet. | indet. | indet. | |

phosphate sands and indicate a shallow marine setting between an offshore bar and the beach.

Figure 177 shows that the vertical and lateral distribution of the various lithofacies within the Pelletal Phosphorite Member is variable: Rogers (1980) did not report any phosphate rock in the S1 borehole, a short distance (ca.1 km) from the outcrops described by Dingle et al (1979), whereas he did note at least two thick (ca.5 m) beds in the Elandsfontein borehole.

At Varswater Quarry, the Varswater Formation is unconformably overlain by various sands, calcareous sands and calcretes which Visser & Schoch (1973) placed in the Langebaan limestone. Tankard (1974b) considered all these as Pleistocene in age, whilst Rogers (1980) regarded the Langebaan limestone as one of the younger members of the Bredasdorp Formation which he believed to be late Pliocene to Holocene in age. We have tentatively followed Rogers' (1980) classification, but accept that at least two of the units in what Rogers identifies as Langebaan limestone are most likely lateral facies equivalents of the youngest part of the Varswater Formation. The beds in question cap the Anyskop, a low rounded hill immediately south of the Varswater Quarry, and consist of calcareous sands with abundant benthic foraminifera (mainly *Anomalina* and *Elphidium*) and phosphatized echinoid spines. The upper part of this sequence has a sparser marine fauna (likely to be allochthonous), and the hillside of Anyskop is littered with land snails *Trigonephrus globulus* and an indeterminate species. Tankard (1974b) accounts for the marine fossils by aeolian transport, but as Hendey (1981a) points out, this origin is inconsistent with the large number of fresh (i.e. not wind-abraided) specimens in the assemblage. The latter author suggests that the lower part of the Anyskop sequence may in fact be a non-phosphatic marine sand deposited as a bar to the west of the Berg River mouth during early Pliocene times, after which the younger terrestrial strata were deposited on a surface that closely followed the contours of the present hill during a late Pliocene regression.

Two other locations within this northern depocentre are significant for an understanding of the various lithostratigraphic relationships shown in Figure 176. Two boreholes reported by Rogers (1980) from north of the Berg River (Bookram and S11) show that in this area, the Neogene succession is reduced to less than 30 m and that the basement surface lies at an elevation of about −10 m. At S11, a lower sequence of coarse-medium sands and clays (ca.25 m), correlated with the Elandsfontyn Formation, is overlain by about 5 m of fine sand with polished phosphatized shell fragments. Although this unit lacks pelletal phosphate, its stratigraphic position above probably non-marine sediments and general lithology suggest a correlation with the Varswater Formation. Farther north at Bookram, a similar fine sand (14 m), rests, with a basal conglomerate of vein quartz pebbles, black, polished phosphatized shell fragments, and sharks' teeth, directly on basement. Rogers (1980) refers to this northern phosphate-bearing fine sand as the Bookram Member of the Varswater Formation which obviously represents a marginal, relatively phosphate-starved facies of the thicker, higher grade phosphatic rocks of the Varswater area. As we will note below, a similar marginal facies occurs south of the Varswater-Saldanha embayment. At Bookram the Varswater Formation is overlain by 12 m of muddy sand which Rogers (1980) correlates on what seem to be rather equivocal evidence with the Bredasdorp Formation (his Papkuils Member). Other isolated sites at which phosphatic rocks have been recorded north of the Berg River are: around Graauwe Duynen and Elands Bay (friable phosphate rock, strewn phosphorite cobbles); and Lamberts Bay (phosphorite cobbles beneath unconsolidated quartzose sands) (Fig.181) (Birch 1977a).

The second locality of interest in this area is on the Hoedjiespunt Peninsula at Saldanha, where phosphatic rocks crop out in the Bomgat and at the base of the cliffs behind the fishing quay (sites 1 and 2, respectively, Fig.179). Of these, the Bomgat is the most important because it was chosen by Tankard (1975b) as the type section for a lithostratigraphic unit: Saldanha Member of the Varswater Formation (originally designated a formation). Here the succession consists of two phosphatic layers resting directly on a highly irregular weathered granite surface. The lower layer (1 in Fig. 179) fills in a large hollow in the basement and reaches a maximum thickness of about 70 cm. It is a reddish-brown hard, brittle, resinous, highly phosphatic rock (34-36 % $P_2O_5$) that consists of phosphatized calcareous biogenic debris (shell fragments, polyzoa and minor foraminifera) set in a collophane matrix. As pointed out by Birch (1977a), it is unique among the onshore phosphate rocks of the western Cape in being practically non-quartzitic. This layer is bedded, in contrast to the overlying layer 2, which consists of unbedded phosphatized shell debris and collophane matrix with numerous rounded granite boulders (up to 30 cm). Layer 2 oversteps the bedded unit and rests directly on granite away from the immediate vicinity of the Bomgat Cave. The phosphate-rich sequence is unconformably overlain by semi-consolidated and friable shell limestones with rounded phosphorite cobbles (layer 3) and bedded shell banks (up to 4 m) with a large granite boulder basal conglomerate (layer 4). The sequence is capped by up to 30 m of aeolianite. Tankard (1975b) designated the phosphate-rich layers (1 & 2) at the Bomgat as the type section of his Saldanha Formation, which he was unable to date at this locality. In his discussion of this rock unit (1975b) he identifies correlatives in the Varswater Quarry, at Ysterplaat, and on the continental shelf, but bases this correlation on the fact that all the rocks are phosphatic. Subsequent

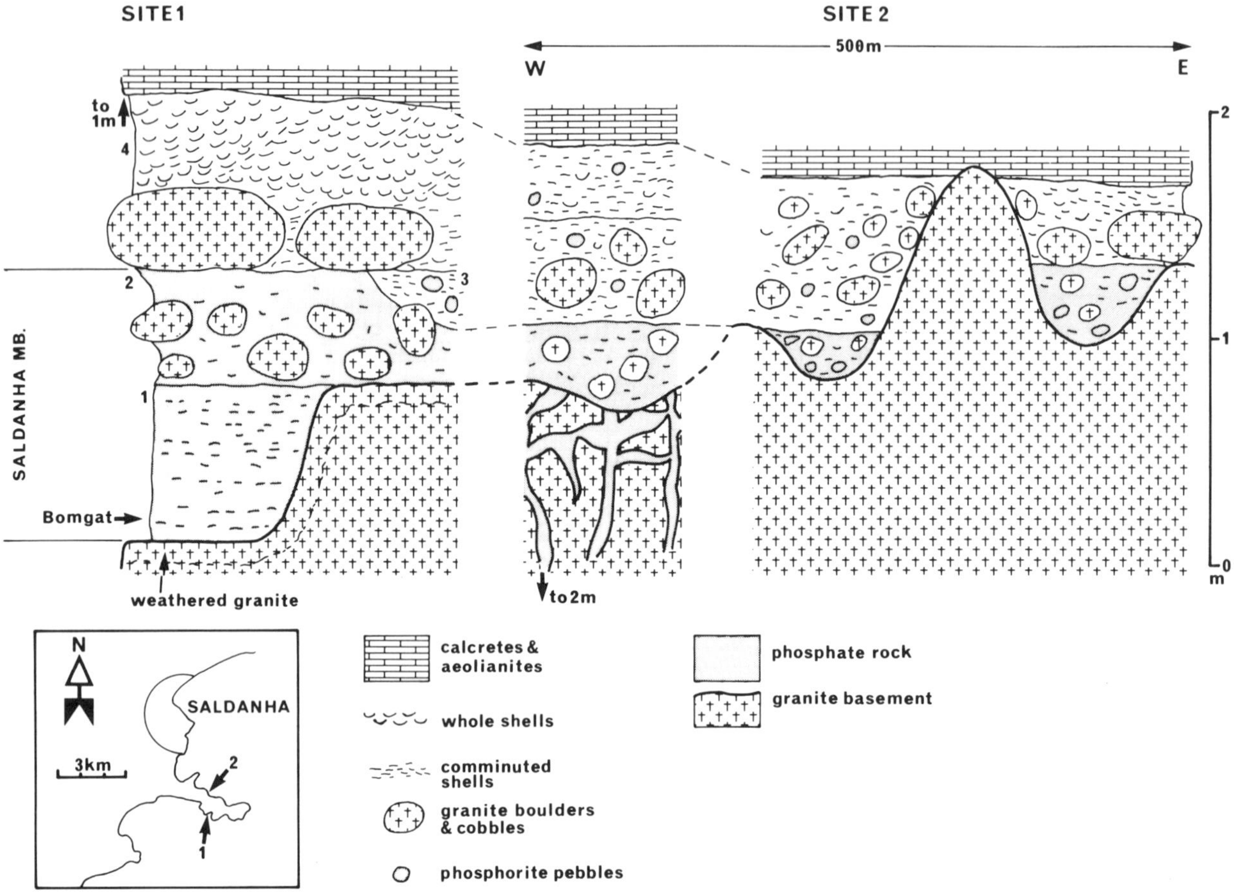


Figure 179. Sections in the phosphatic and associated sediments on the Hoedjiespunt Peninsula S of Saldanha. Site 1 is the Bomgat Cave, which is the type section of Tankard's (1975b) Saldanha Formation. Site 2 is a cliff section behind the fishing quay in which granite with apatite-filled joints is overlain by a sequence of shelly rudaceous phosphatic sediments that can be correlated with the Bomgat succession. Layers 1 and 2 are probably Mio-Pliocene in age, whilst Tankard (1975b) assigned a Pleistocene age to layers 3 and 4.

work (e.g. Siesser 1978a, Birch 1978, Dingle et al 1979, Rogers 1980) has shown, however, that numerous phosphatic rock units occur in the Neogene of the continental margin of SW Africa and the Agulhas Bank, and that this criterion alone is insufficient to prove lithostratigraphic continuity. There is no substantive palaeontological or mineralogical evidence, for example, which, if any, of the phosphatic horizons at the Varswater Quarry are a lateral equivalent of the Bomgat sequence. Under these circumstances, the regional significance placed by Tankard (1975b) in the Saldanha Formation is unjustified and we consider is better viewed as a minor (although mineralogically interesting) member of the Varswater Formation. Taking into account the nature of the outcrops on the eastern side of the Hoedjiespunt (site 2, Fig.179) it is clear that the deposit accumulated locally in 'pools' on an irregular granite surface, and that in places, where the granite was fractured or deeply weathered, phosphatic solutions penetrated into the bed rock as veins up to 2 m long.

Marine and non-marine Neogene sediments have been intersected in several boreholes, excavations and small natural exposures along the coastal zone between about 33.5°S and the northern shores of False Bay (southern Cape Flats). The thickest and stratigraphically most significant of these are in the excavations for the Koeberg nuclear power station at Duynefontyn (Rogers 1979), and in a borehole at Noordhoek (Coetzee 1978a, b, 1981) (Fig.176). At Duynefontyn, Rogers (1979) has described about 16 m of marine sediments which are overlain by strata that are more fully exposed in coastal outcrops to the north and south (Fig.177). The marine part of the sequence has been allocated to the Duynefontyn Member of the Varswater Formation (the upper boundary has been slightly modified from Rogers' original account), and overall consists of fine, well-sorted, polished quartzose sands with minor black phosphatized shell fragments. There is a basal Gravel Bed (0.4 m) consisting of well-rounded to angular pebbles and cobbles of Malmesbury basement and vein quartz set in a grey quartzose sand matrix rich in phosphatized shell debris, which rests on pholad bored and gullied Malmesbury basement. The Gravel Bed is overlain by two grey quartzose sand units (the upper one strongly bioturbated) with minor phosphatized shell fragments, and muddy and gravelly layers, which are separated by a ca.1 m layer of grey, gravelly, coarse quartzose sand with decomposed basement pebbles and abundant sharks' teeth, fish and whale bone debris (Shark Tooth Bed). At the top of the sequence a coarse laminated sand with shell fragments, vein quartz granules, sharks' teeth, and phosphatized gastropod casts (Gastropod Bed, ca.3 m) is separated from the grey bioturbated sands by a thin (10-15 cm) peaty clay. These rocks are overlain by various non-phosphatic sands (Springfontyn Member), calcareous (Milnerton Member) and aeolian sands of the Bredasdorp Formation. The Milnerton Member was first described by Kensley (1975) (quoted in Rogers 1979) who recorded a basal conglomerate of shelly quartz sand, with cobbles of calcrete and Malmesbury basement. We follow Rogers (1980) in placing the Duynefontyn Member in the Varswater Formation, and draw attention to its similarity, both lithologically and spatially to the Bookram Member. No precise dating has so far been possible on the fossils from Duynefontyn, but Simpson (1979) recorded two species of penguin, one of which was a new taxon *Nucleornis insolitus* (the other was indeterminate). Because of some unusual characteristics, he was unable to make a close comparison with other penguin types and the suggested age of Miocene for this outcrop seems to be based mainly on the fact that Miocene rocks are suspected nearby at Ysterplaat. Other elements of the rich vertebrate assemblages which includes selachid fish, teleost fish, whale, dolphin and seal bones and teeth, and bovid, hare, snake and tortoise bones, have not been studied,

Table 57. Late Cenozoic pollen zonation of the Noordhoek borehole (from Coetzee 1978a).

| Pollen zone | Vegetation | Climate | Suggested ages |
|---|---|---|---|
| M | present fynbos | present | Quaternary |
| Lvii | first strong development of fynbos | colder, drier | Pliocene |
| Lvi | forest: coniferae Casuarinacea Cupaniedites | cool, wet | late Miocene |
| Lv | palmae | sub-tropical, tropical | middle Miocene* |
| Liv | restionaceae swamp | ?temperate, locally wet | early Miocene* |
| Liii | forest: coniferae first compositae | cool, wet | |
| Lii | palmae | subtropical, tropical | |
| Li | forest: coniferae | cool, wet | late Oligocene* |

* May all be middle-late Miocene.

although Rogers (1979) quotes an identification of a giant shark *Megaselachus megalodon* by Hendey (personal communication).

Outcrops of a similar lithofacies have been reported by Tankard (1975b) from a quarry at Ysterplaat where about 70 cm of coarse sands are set in an apatite cement with occasional phosphate pellets and phosphatized shell fragments. The sand contains a relatively rich fossil assemblage of gastropods (2 spp.), lamellibranchs (9 spp.), polyzoa, echinoids, as well as a shark (*Oxyrhyna hastalis*) and two penguins *?Palaeospheniscus huxleyorum* and *P.* sp. (the last two identified by Simpson 1973). No precise age can be given to these sediments, but Simpson (1973) suggested that the penguins indicated a pre-Pliocene, probably Miocene age. Rocks of a similar age are suspected to lie in water below normal wave base (ca.10 m) off Milnerton Beach (2.5 km NE of Ysterplaat) because storm waves frequently deposit fossils on the beach, which include sharks' teeth (*Megaselachus megalodon,* and *Oxyrhyna hastalis*). We correlate these deposits with the Duynefontyn Member of the Varswater Formation, which in turn we consider analogous to the Bookram Member. All represent marginal facies of the main phosphate rich Varswater depositional environment and are probably of late Miocene or early Pliocene age. Phosphatic sediments have not been reported south of Ysterplaat and Rogers (1980) records various coarse and fine sands from the Cape Flats, and peaty sands at Noordhoek. He has correlated the former with the Elandsfontyn and Varswater formations respectively, on rather equivocal lithological grounds (Fig.176), but no age determinations are available even to substantiate a Tertiary age, and the two lithologies are apparently laterally equivalent.

At Noordhoek two boreholes have been sunk, and in S20 Rogers (1980) reports at least 27 m of angular, coarse, gravelly sands with thick (up to 10 m) peat horizons beneath 23 m of decalcified dune sand (the last being his Noordhoek Member of the Bredasdorp Formation). So far, no fossils have been described from this borehole, but Coetzee (1978a, 1981) has given a preliminary report of the palynology of an earlier nearby (500 m) borehole from which no lithological data have yet been published. In addition, no sample depths were included in Coetzee's (1978a) stratigraphic table so it has not been possible to establish a correlation between her biostratigraphic (Tab.57) and the regional lithostratigraphic successions. The flora in the lower levels of the borehole contain abundant pollens comparable with the most primitive angiosperms. In particular, Coetzee (1981) notes the presence of *Clavatipollenites* cf. *C.hughesii* Couper and a taxon referred to *?Pseudowinterapollis* sp. The former is similar to forms in the extant genus *Ascarina* which now occurs only in the tropics, but not on the African continent, though it is found in Madagascar. Similar genera to the latter occur today in South America, Madagascar, and the South

Table 58. Chronology of Neogene succession, SW Cape (modified from Hendey 1981a).

| Formation, Member | Sedimentary environments | Sea level cycle | Age |
|---|---|---|---|
| 7. Baards Quarry deposits | fluviatile | regression | early Pleistocene (<1.9 my) |
| 6. Anyskop Terrestrial Member | sandy coastal plain | regression | late Pliocene |
| 5. Anyskop Marine Member | offshore barrier complex | transgression | early Pliocene |
| 4. Pelletal Phosphorite Member | fluviatile/shallow marine | transgression | early Pliocene |
| 3. Quartzose Sand Member | estuarine/salt marsh/flood plain/ river channel | early transgression | early Pliocene (5 my) |
| 2. Gravel Member | beach, very shallow marine | regression | late Miocene (⩽12.5 my) |
| 1. Elandsfontyn Fm. | terrestrial/estuarine/marine | transgression | ?early to middle Miocene |

5, 6, 7 constitute part of the Langebaan Limestone Member of the Bredasdorp Formation as defined by Rogers (1980).

Pacific area. Coetzee (1981) concludes from these distributions that both taxa survived in southern Africa until the late Miocene, when climatic changes caused their extinction on an inter-continental scale, so that their surviving relatives have a much more limited distribution.

Coetzee (1978a) recognizes several vegetations that suggested to her alternating periods of subtropical/ tropical (with palms) and cooler, wetter climate (with coniferous forest). Stratigraphically important events are marked by the appearance of the Compositae (?early Miocene), and the establishment of fynbos and related shrubland vegetation (typified by Proteaceae, Restionaceae, ericacea, etc.) (Pliocene). Coetzee (1978a) has dated the Noordhoek borehole as late Oligocene at the base to Quaternary, and whilst she does not claim that the record is continuous and acknowledges that it consists of assemblages from various peaty horizons up the core, we feel that additional data are needed to unequivocally establish that a record stretching for possibly 25 my is contained in the 33 m of borehole. In particular, whether the first appearance of the Compositae is a phylogenetic event, or may have been occasioned by local changes in the climatic or sedimentological milieu. Similarly, it seems conceivable that rather than reflecting wholesale climatic changes, the alternating record of palm and conifer-dominated forests, as well as the intervention of a swamp vegetation, could also be explained by local changes in distribution of forest types, and stream and pond location, especially since palm pollen was present in all the Tertiary samples studied.

Notwithstanding these uncertainties, Coetzee's zonal scheme (1978, fig.2) gives for the first time a potential regional biological correlative for the SW Cape outcrops. In this connection, Coetzee has studied samples from a Geological Survey borehole programme and preliminary details have been reported by Rogers (1980). At both boreholes S1 (Anyskop) and Elandsfontein (G30878), peats from the Elandsfontyn Formation have been placed in her zones LVII and LVI (Late Miocene-Pliocene), whilst peats from the Quartzose Sand Member and the Elandsfontyn Formation in the Varswater Quarry are considered to be no older than early Miocene (Coetzee personal communication 1979).

The age of the Elandsfontyn and Varswater formations has been the subject of considerable debate in the literature. By and large, the invertebrate faunas have proved of little use and most emphasis has been placed on vertebrate taxa, especially sharks' teeth and mammals. Hendey (1981a) provides a thorough review of previous attempts and concludes that 'the principle difficulty with faunal dating has been that broadly contemporary faunas elsewhere in Africa . . . are not well known'. Hendey (1981a) suggests a 5 my age (early Pliocene) for the Quartzose Sand, Pelletal Phosphorite, and Anyskop Marine members on the basis of comparison with the 6 my Lothagan 1 and 4 my Kanapoi mammal assemblages in Kenya. The Varswater assemblages are intermediate in evolutionary development between these two, specifically with regard to the pig *Nyanzachoerus* cf. *pattersoni,* the bovid *Simatherium demissum,* the elephant *Mammuthus subplanifrons,* and in particular the hyenas *Ictitherium preforfex, Hyaena abronia* and *Hyaenictitherium namaquense.* This dating fits well with the worldwide early Pliocene eustatic sealevel rise, since the Quartzose Sand-Pelletal Phosphorite-Anyskop Marine series is thought to be a transgressive sequence. Dating the youngest Neogene sediments is more subjective and depends primarily upon the presence of *Equus* in the Baard's Quarry (adjacent to Varswater Quarry) deposits which are probably younger than 1.9 my (the earliest date of entry of *Equus* into Africa). This places the Baard's Quarry fauna in the early Pleistocene, indicating that the bulk of the Langebaan limestone is younger.

There is no substantive fossil evidence for an accurate date of the pre-Quartzose Sand Member strata, but the presence of *Hipparion* cf. *primigenium* suggests a late Miocene age for the Gravel Member (Hendey 1981a). The original phosphate rock from which the eroded and recemented gravel member is derived was obviously older, but by how much is not known. Hendey (1981a,b) concluded that it represented the marine phase of a Miocene transgression which reached its highest point in Middle or lower Upper Miocene times, and that erosion and the second phase of phosphatization (of the matrix) was effected during a late Miocene

Table 59. Miocene mammals from the southern Namib desert (after Hendey 1978a).

| | Elisabethfeld | Plant 4 | Unspecified locality south of Lüderitz[1] | Langental | Unspecified, south Namib desert |
|---|---|---|---|---|---|
| *Myohyrax oswaldi* | ● | ● | ● | | |
| *Protypotheroides beetzi* | | | ● | ● | |
| *Pterodon kaiseri* | ● | | | | |
| Creodonta/Carnivora indet. | ● | | | | |
| *Prohyrax tertiarius* | | | | ● | |
| *Gomphotherium* sp. | | | | | ● |
| *Brachypotherium heinzelini* | | | | ● | |
| Artiodactyla indet. | ● | ● | | | |
| *Xenochoenus africanus* | | | | ● | |
| *?Propalaeochoerus* sp. | | | | ● | |
| *Brachyodus africanus* | | | | | ● |
| *Dorcatherium* sp. | | | ● | | |
| Pecora indet. | | | ● | | |
| *Propalaeoryx austroafricanus* | ● | ● | | ● | |
| *Austrolagomys inexpectatus* | ● | | | | |
| *Kenyalagomys simpsoni* | | | ● | | |
| *Parapedetes namaquensis* | ● | | | | |
| *Bathyergoides neotertiarius* | | ● | | ● | |
| *Paraphiomys pigotti* | | ● | ● | ● | |
| *P.stromeri* | | | ● | | |
| *Phiomyoides humilis* | | ● | | | |
| *Diamantomys luederitzi* | | | | ● | |
| *Pomonomys dubius* | | | | ● | |
| *Paracryptomys mackennae* | | | | | ● |
| Rodentia indet. | | ● | ● | ● | |

1 – Hopwood (1920)

regression (the evidence of *Hipparion* and the warm water mollusc assemblages favour this date over an early Pliocene age). If these dates are correct, and the fossil evidence does not preclude it, then the oldest parts of the Elandsfontyn Formation in the type areas are presumably Middle (or even early) Miocene in age. As we will discuss later these dates fit well with the known Tertiary sea level curves, and have important implications for dating terraces on the Namaqualand coast. There remains the problem of the age of the phosphatic Bookram and Duynefontyn members. What fossil evidence there is (from Duynefontyn and Ysterplaat) has been interpreted as indicating a Miocene age (Tankard 1975b, Simpson 1973, 1979), which suggest a correlation with the Middle or lower Upper Miocene transgression high point and phase of phosphatization that ended the Elandsfontyn cycle of sedimentation. In this case the subsequent late Miocene regression and following early Pliocene transgression must be present in the Duynefontyn Member profile. Although the Sharks' Tooth Bed indicates a regressive or higher energy phase, Rogers (1979) did not report definite evidence of an intra-member erosion surface although he does mention an 'undulating boundary' at its base. The alternative explanation is for a wholly Pliocene Duynefontyn Member with complete removal of previous (Miocene) sediments. At present, we have no unequivocal evidence for either suggestion. Table 58 summarizes our preferred chronology for the Neogene succession of the SW Cape.

6.2.2.3.3 *Namaqualand and South Namibia.* A variety of small but fossil-rich non-marine outcrops occur along this coastal zone: Elizabeth Bay area and Arris Drift (Miocene), and Kleinzee (Pliocene), together with the geomorphologically and commercially important (diamondiferous) Neogene beach terraces.

Miocene vertebrates were first described from the southern Namib desert in a series of publications by Stromer (1922 to 1926). His localities were at Elisabethfeld, and a borehole near Plant 4 of the Kolonial Bergbau Gesellschaft in the Elizabeth Bay area, and from the Langental near Bogenfels (Fig.172). Further localities (Fiskus and Grillental in the Elisabethfeld area) have been identified by Greenman (1966), and the fauna from all these sites, plus additional material were re-assessed by Hamilton & van Couvering (1977) (Tab.59). The sediments in the Elizabeth Bay area were placed in the Elizabeth Bay Formation by Greenman (1966) and consist of up to 15 m of slightly dolomitic clays and coarse sandstone. At their seaward edge these may be shallow marine, but they grade landward into a fluviatile lithofacies. Similar fluviatile marly sandstones occur in the Langental area (Haughton 1969).

In their assessment of the vertebrates from these localities, Hamilton & van Couvering (1977) considered the fauna as a whole to represent a Lower Miocene (Burdigalian) assemblage, but Hendey (1978a) believes that there may be slight differences between the various sites which suggest that they might not be coeval. In particular, a comparison of *Pterodon kaiseri* with east African forms led him to tentatively suggest an Aquitanian (> 20 my) age for the Elisabethfeld fauna, whilst a comparison of *Austrolagomys* (Elisabethfeld) and *Kenyalagomys* (Langental) indicates that the former is more primitive, and the Langental assemblage is slightly younger (Burdigalian) than that at Elisabethfeld. The occurrence of *Protypotheroides beetzi* in both the Langental and material collected by Hopwood (1929) (Tab.59) suggests that these faunas may be the same age (Burdigalian). The fossils at Plant 4 are similar to those from Langental, but also have elements in common with Arris Drift, which suggests that the sediments at Plant 4 may be younger (Middle Miocene) than the other deposits from the southern Namib desert area (Hendey 1978a).

Arris Drift (16.783°E, 28.250°S) is the site of the largest Miocene vertebrate assemblage from a single locality in southern Africa (Corvinus & Hendey 1978, Hendey 1978a). It lies 900 m north of the Orange River, about 30 km from its mouth, in terrace deposits, which consist of up to 35 m of gravels with a feldspathic sand matrix, and interbedded sands and muds, resting on an irregular bedrock surface. In this area, terrace elevations are at 56, 45, 30 and 20 m above the mean river level, which at this point itself lies about 7 m above mean sea level. The fossils, which occur in well-lithified material that had to be extracted with power tools, were found in Pit 2, AD8 at about 33 m above the river where the terrace cover is between 1 and 4 m thick.

Hendey (1978a) records 28 vertebrate species from Arris Drift. Mammals (22 species) are the most important group, and include three new species and at least eight genera not previously reported from southern Africa. Amongst the more interesting finds are the abundant remains of a giant dassie (*Prohyrax*), what may be the first African record of a bear-dog, (*Amphicyon* cf. *steinheimensis*), and a rare deinothere (*Prodeinotherium hobleyi*) (of which only two other species have previously been recorded from Africa). Hendey (1978a) considers that eight of the 22 mammals indicate a Middle Miocene age when comparison is made with the East African assemblages, whilst ten of the remainder provide no direct evidence, but are not inconsistent with a date betwen 18 and 14 my. The closest assemblage is that from Maboko in Kenya (e.g. Bishop 1967).

Non-mammalian taxa are dominated by a species of crocodile, with minor birds, fish and tortoises. The only invertebrate recorded so far is a serpulid polychaete, which is a typical estuarine taxon, and led Hendey (1978a) to postulate that the accumulation of this assemblage took place during a period of high sea level (40-50 m above present sea level) with the sea no more than a few kilometres away. The presence of two proboscideans, as well as browsing types such as *Prodeinotherium*, gomphotheres, rhinoceras and *Climacoceras* indicate a forested riverine setting (gallery forest), backed by dense bush adjacent to the river (vide the small herbivores) (Hendey 1978a). Similar environments have previously been postulated for the other Miocene sites in the Namib (e.g. Stromer 1926, Hopwood 1929), indicating that wooded river valley settings extended at least from the Lüderitz area to the present Orange River in Miocene times (Hendey 1978a).

Pliocene non-marine deposits have been reported from Kleinzee where a small assemblage of mammalian fossils has been described by Stromer (1931a,b). Hendey (e.g. 1970, 1974a) considered these to be broadly contemporaneous with those from the Quartzose Sand and Pelletal Phosphorite members of the Varswater Formation (i.e. early Pliocene).

The study of marine Neogene sediments along the coast of southern Namibia and Namaqualand has been

Table 60. Miocene vertebrates from Arris Drift Pit 2/AD 8 (after Hendey 1978a).

| | |
|---|---|
| Osteichthyes | gen. et sp(p). indet. |
| Amphibia | gen. et sp. indet. |
| Reptilia | |
| Squamata | gen. et sp. indet. |
| Crocodilia | *?Crocodylus niloticus* |
| Chelonia | gen. et sp(p). indet. |
| Aves | gen. et spp. indet. |
| Mammalia | |
| Insectivora | |
| Macroscelididae | *Myohyrax* cf. *oswaldi* |
| Carnivora | |
| Amphicyonidae | *Amphicyon* cf. *steinheimensis* |
| Amphicyonidae or Ursidae | gen. et sp. indet. |
| ?Ursidae | ?Hemicyoninae gen. & sp. indet. |
| ?Felidae | *?Metailurus* sp. |
| Mustelidae | *?Ischyrictis* sp. |
| indet. | gen. & sp. indet. |
| Hyracoidea | |
| Procaviidae | *Prohyrax* n.sp. |
| Proboscidea | |
| Gomphotheriidae | gen. et sp. indet. |
| Deinotheriidae | *Prodeinotherium hobleyi* |
| Perissodactyla | |
| Rhinocerotidae | *Dicerorhinus* sp. |
| Artiodactyla | |
| Suidae | gen. et sp. indet.<br>*Lopholistriodon moruoroti* |
| Tragulidae | *Dorcatherium* cf. *pigotti* |
| Palaeomerycidae | *Climacoceras* sp.nov. |
| Bovidae | gen. et sp. indet. |
| Pecora | gen. et sp. indet. |
| Lagomorpha | |
| Ochotonidae | *Kenyalagomys* sp.nov. |
| Rodentia | |
| ?Bathyergidae | *?Bathyergoides* sp. |
| Thryonomyidae | *Paraphiomys pigotti* |
| indet. | gen. et spp. indet. |

closely linked with the prospecting for, and exploitation of diamondiferous deposits. Diamonds were discovered at Kolmanskop (east of Lüderitz) in 1908 and in subsequent years intensive prospecting was undertaken along the whole coastline to the north and south. Several small diamonds were found at Alexander Bay (immediately south of the Orange River) in 1909 (Hallam 1964) but the prospect was considered too poor to warrant exploitation and it was not until 1925 that several worthwhile discoveries were made in Namaqualand, culminating with the proving of large reserves at the Orange River mouth in 1927. During, and since this early period, a large amount of data have been acquired by various mining concerns on the nature of the diamond bearing deposits, and although most of this has not been publicly released, several important publications have appeared which summarize the location, sedimentology and fossil content of some of the areas that have been exploited, including prospecting and mining operations offshore the Namibia and Namaqualand coasts: Hallam (1964), Murray (1970), Murray et al (1970), O'Shea (1971) and Keyser (1972). In addition, academic geologists have contributed to an understanding of the area, mainly through accounts of the geomorphology and palaeontology of the diamond-bearing strata: e.g. Krige (1927), Haughton (1931b, 1969), Carrington & Kensley (1969), Tankard (1975b), whilst Hendey (1981b), proposed a revised dating and correlation scheme for the diamondiferous terraces.

Diamonds have been discovered, and locally commercially exploited at one time or another along almost the whole of the coastal strip between the Kaokoveld and Olifants River, including offshore operations between the Lüderitz area and various parts of Namaqualand. They occur in sediments of both marine (beach and nearshore gravels and sands) and alluvial (river terraces and valley fill) origin, but the most important occurrences are in beach terraces. These lie in a narrow coastal zone (up to 8 km wide) and show strong positive associations with river mouths (both modern and buried) that have been summarized by Hallam (1964) and Keyser (1972) as:

i) the largest concentrations of diamonds occur immediately north and south of river mouths;

ii) at the mouth of any particular river, the deposits are more extensive and contain larger stones on its north side but there is a progressive deterioration in size and number away from the river;

iii) the largest rivers have the largest reserves.

From this distribution it seems clear that the diamonds were originally brought to the coast by rivers and that they were subsequently transported mainly northward along beaches by long shore drift, although erosion and re-working of older beach terraces may have resulted in individual stones going through several episodes of exhumation and deposition. In addition to primary concentration around river mouths, other shoreline features have been found to preferentially concentrate the heavy minerals (Hallam 1964):

i) on the north side of south opening bays;

ii) on the south side of westerly projecting headlands,

iii) in pocket bays.

Beach terraces are cut across Pre-Cambrian bedrock, which usually has a complexly-gullied upper surface whose form is controlled by the dominant joint and foliation directions. Small cliffs, sea stacks and potholes are common, and are buried by beach detritus which frequently has a storm beach profile with laterally graded cobbles and pebble mounds. This sequence is locally overlain by unconsolidated aeolian sands and calcrete. It is popularly held that diamonds are preferentially concentrated in bedrock depressions and in well-sorted gravel layers, but Keyser (1972) after considerable experience in observing mining operations in the Alexander Bay area remarks that '. . . concentrations of diamonds . . . occurs in all types of marine deposits, irrespective of the composition or physical characters . . .' In the northern areas (Kaokoveld, and around Elizabeth Bay) many of the terraces had little sediment cover and had been deflated by the high prevailing winds, resulting in some locally remarkable concentrations of diamonds which were, naturally, the easiest and first to be collected. By and large, mining activities in the northern areas have now been abandoned and the main extraction effort is concentrated around Oranjemund just north of the Orange estuary, whilst several mines are operating along the Namaqualand coast (e.g. Kleinzee Annex, Koingnaas, Klipvlei/Geewal all associated with river or dried river mouths: Buffels, Swartlintjies and Olifants, respectively). Statistics on grades, reserves and distribution of diamonds have not been widely published and Hallam (1964) and Keyser (1972) remain the only modern sources. Hallam (1964) states that immediately north of the Orange River, the largest diamonds are found in the highest terraces and quotes an average size of 1 carat/stone, locally reaching 3 carats/stone. Ninety per cent of the stones in this area are pure white to yellow, and 95 % are of gem quality (with only minor amounts of small, dirty cracked industrial grade). Some idea of the importance of these deposits can be gauged from the following statistics:

i) at 1970, the Oranjemund reserves (to about 70 km north of the Orange estuary) were reckoned to be $25 \times 10^6$ carats;

ii) $5.35 \times 10^6$ carats were extracted from the now exhausted Alexander Bay State Alluvial diggings immediately south of the Orange estuary;

iii) the area around the Buffels estuary is estimated to have held $3.8 \times 10^6$ carats ($1.8 \times 10^6$ already extracted).

The distribution, and age of the beach terraces – with which these diamond-rich sediments are associated, have also been of geological interest because of the be-

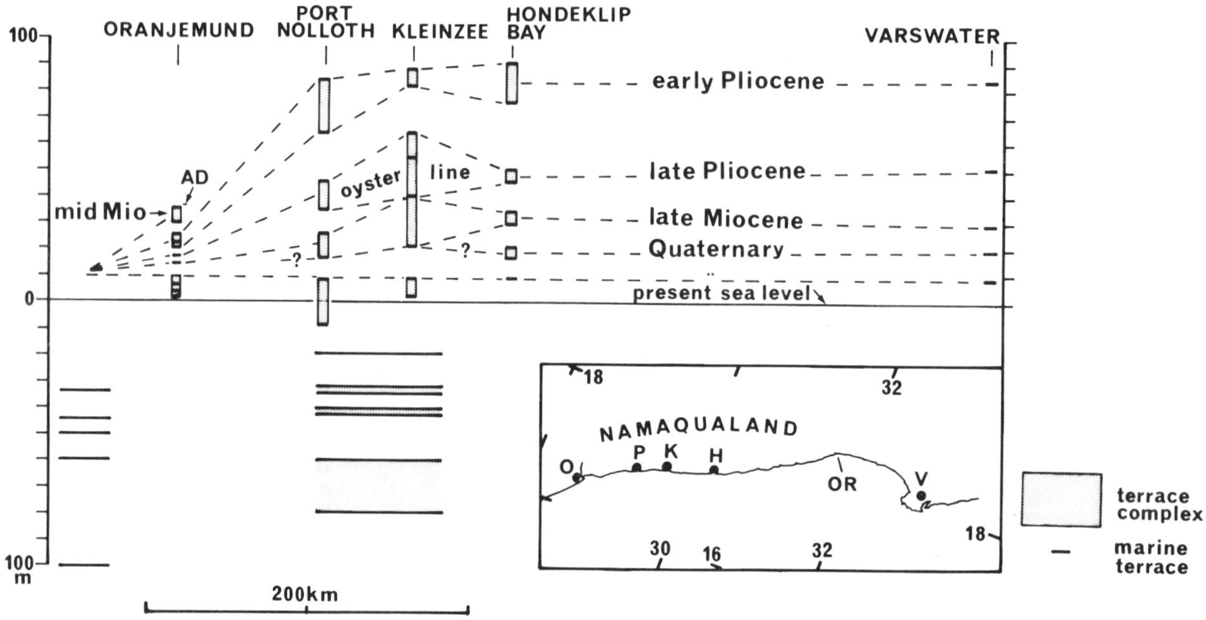


Figure 180. Vertical and lateral distribution of diamond-bearing beach terraces along the Namaqualand and S.Namibia coasts. The dating of these terraces follows the suggestions of Hendey (1981b). Submarine terraces have been identified during mining and prospecting operations by bathymetry and seismic profiling. Sharks' teeth have been reported from the base of the Oyster Line (50 m terrace) on the Olifants River by Haughton (1969) who suggested a Mio-Pliocene age for it. See Table 61 for further details of terrace elevations. Data from Hallam (1964); Carrington & Kensley (1969); Murray *et al.* (1970); O'Shea (1971); Keyser (1972). Abbreviations: AD = Arris Drift; O = Oranjemund; P = Port Nolloth; K = Kleinzee; H = Hondeklip Bay; OR = Olifants River; V = Varswater Quarry.

lief that they were mainly Pleistocene in age and that the variation in terrace elevation was evidence of differential crustal warping in Quaternary times (e.g. du Toit 1933, Haughton 1969). Hendey's (1981b) re-assessment of the ages of these terraces indicates that the higher ones are in fact Neogene in age so that although differential Neogene crustal warping has to be appealed to, the previously-postulated high Quaternary crustal movement rates are no longer necessary to explain the terrace elevations.

Between the Oranjemund area (Orange estuary) and the Kaokoveld, Neogene terraces are less well developed than to the south. In the Kaokoveld itself, Hallam (1964) identifies beach terraces at 4.5, 9, 15, 23 and 61 m, with the higher three only poorly-developed. Farther south, between this area and Oranjemund only a 4 m terrace is well-developed, and this only locally (e.g. Elizabeth Bay and Bogenfels), but south of the Orange River along the Namaqualand coast, a series of well-defined and easily correlatable terraces occurs (e.g. Hallam 1964) (Fig.180, Tab.61). Although they fluctuate slightly, a remarkable continuity is evident with mean heights at: 90, 50, 30, 20 and 9-0 m, whilst a series of submarine terraces has also been located at −20, −30, −40, −60 to −80 and −100 m (Murray et al 1970, O'Shea 1971). There are no published age data on the submarine features, but there is general agreement that the +20 and +9 to 0 m terraces on shore, which contain modern, cold water fossil assemblages, are Quaternary in age (Haughton 1931b, Hallam 1964, Carrington & Kensley 1967, Hendey 1981b), and when these are traced northwards to the Oranjemund area there is only a small amount of convergence in levels (Fig.180).

The higher terraces are more complex because they frequently occur as a series over an elevation range that may be as much as 20 m. Nevertheless, in the three areas south of the Orange River in which they have been well-studied (Port Nolloth, Kleinzee and Hondeklip Bay) they form distinct groupings.

Hendey (1981b) has used the Mio-Pliocene succession in the Varswater-Saldanha area to date the Namaqualand terraces, and although palaeontological confirmation is needed, his correlations are plausible. The late Miocene Gravel Member lies at 30 m and represents an erosional platform that can be correlated with the 29-34 m shoreline at Hondeklip Bay, and terraces at similar altitudes farther north, which are thin and discontinuous (this is the Middle Terrace of Hallam 1964 and Keyser 1972). As Hendey (1981a,b) points out, this poor development sets it apart from the other Namaqualand shorelines and suggests that it may have been degraded by subsequent reworking during later high sea levels. In the Varswater-Saldanha area, the late Pliocene marine Varswater Formation is found at maximum elevations of ca.90 m (Mabbutt 1956, Rogers 1980) and Hendey (1981b) correlates this with the 75-90 m terraces in Namaqualand (the Grobler Terrace of Keyser 1972, Highest Terraces of Hallam 1964).

The most poorly-defined of the terrace features in the Varswater-Saldanha area is that correlated with the 34 to 60 m complex in Namaqualand. Hendey (1981b) equates it with a 50-55 platform identified by Tankard (1974b) which may be the same feature as the 45-60 m erosion levels noted by Davies (1972) in the False Bay-Gansbaai area. At Varswater, this level is associated with a late Pliocene Anyskop mixed marine and fresh water assemblage, and in Namaqualand it constitutes the well-known 'oyster line' (Middle Terrace of Hallam 1964) because its fauna is dominated by *Ostrea prismatica* (= *Crassostrea margaritacea*) (Haughton 1969, p.460). Other important molluscs in this assemblage are *Donax rogersi* and *Chamelea* spp., and overall it has similarities with the fauna of the Alexandria Formation along the south coast. Because of this, Haughton (1931b, 1969, p.527) had suggested a Miocene or Pliocene age for the 'oyster line', although this view has generally not found support (e.g. Hallam 1964).

Between Port Nolloth and Oranjemund, there is a rapid decrease in the elevation of the Neogene terrace levels, and according to Hallam (1964) they have all merged at a height of 9-10 m above sea level 50 km north of Oranjemund (Fig.180). This is the elevation of the lower Pleistocene set of terraces in Namaqualand, so that whilst there is strong evidence for crustal movement at the northern end of the section in Figure 180 during the period Miocene-?lower Pleistocene, there is no evidence for movement in late Pleistocene-Holocene times. In addition, provided the correlation shown in Figure 180 is correct (i.e. the Neogene terraces of Namaqualand = the Upper Terraces Group, and the Quaternary terraces of Namaqualand = Lower Terraces Group of Hallam at Oranjemund) then the convergence in the Oranjemund area can be accounted for by upward crustal movement (ca.50 m) between late Miocene and early Pliocene times, followed by a progressive subsidence to a position about 16 m below its late Miocene level by the beginning of the late Pleistocene (i.e. a fall of 66 m since early Pliocene time). These movements were of greater magnitude to the north, and rapidly diminished to the south, although the Port Nolloth terrace levels are slightly lower than those at Kleinzee. This suggests a boundary between the stable and unstable areas just south of the Orange River, which coincides with the old lineament marking the boundary of the Lüderitz Arch and Orange Basin. If such a correlation is permissible, it is interesting to note that the stability characteristics of the two areas have reversed since Jurassic-Lower Cretaceous times.

6.2.2.3.4 *Phosphatic rocks.* Neogene sediments on the continental shelves off the south and west coasts, and onshore along the SW Cape coast, are rich in authigenic and diagenic apatite (microsphorite and lithosphorite of Birch 1978, respectively) (Fig.181). Fragments of lithosphorite rock (so-called phosphorite nodules) were first recovered from the continental shelf south of Cape Town by HMS Challenger, in 1873. These were described by Murray & Renard (1891), who proposed for their formation, *in situ* precipitation of apatite from phosphate-rich colloidal suspensions in the interstices of glauconitic mud, along with subsidiary replacement of calcium carbonate by calcium phosphate. They also noted the association of phosphate formation with zones of strong oceanic upwelling and attendant high biological activity. Further collections of phosphatic-rocks from offshore were described and discussed by Collet (1905), Cayeaux (1934) and Haughton (1956), and from onshore by du Toit (1917), but considerable progress has been achieved since 1967 following systematic surface sediment sampling on the continental shelves by South African and Russian institutes. Detailed accounts of mineralogical, geochemical and petrographic relationships have been given by Parker (1971, 1975), Parker & Siesser (1972), Summerhayes (1972, 1973a), Baturin & Dubinchuk (1974), and Birch (1975), whilst additional descriptions, as well as accounts of phosphate rock distribution and formation have been presented by Baturin (1969, 1970), Parker

Table 61. Elevations of diamond-bearing terraces at various localities along the Namaqualand and S.Namibia coasts. Data mainly from Hallam (1964) and Hendey (1981b). See Figure 180 for localities. Height in m above present sea level.

| | A | B | C | D | E | F |
|---|---|---|---|---|---|---|
| Orangemund | | | | 35, 30, 25, 22 | 20, 17, 14, 9 | 6, 4, 2 |
| Port Nolloth | 84-64 (Grobler) | | 47-34 (Upper) | 26-17 (Middle) | 9- -1 (Lower) | |
| Kleinsee | 88-82 | 64-52 | 52-40 (Upper) | 40-21 (Middle) | | 4-3 (Lower) |
| Hondeklip Bay | 90-75 | | 50-45 | 43-29 | 21-7 | |
| Langebaan[1] | 84 | | 50 | 30 | 20 | |

A-F = terrace levels (A– highest, F – lowest)

1 – not diamondiferous

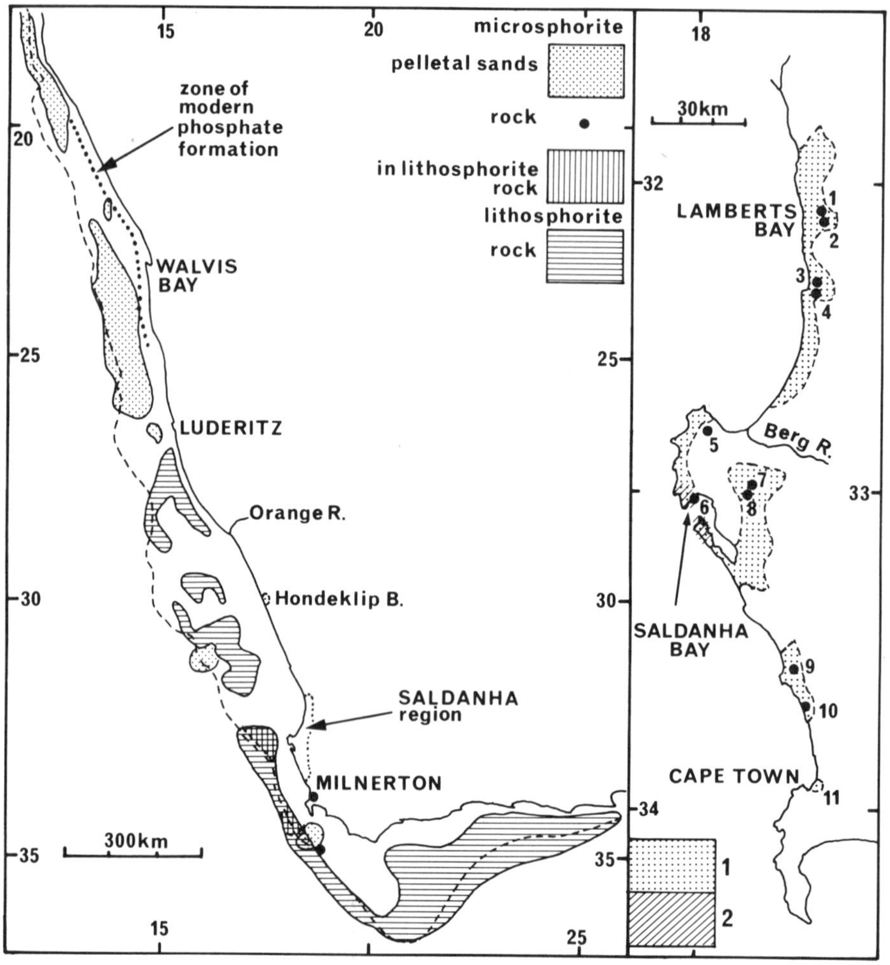


Figure 181. Distribution of phosphatic Neogene sediments on the continental shelf around S. and S.W.Africa. Left map shows distribution of authigenic (microsphorite and diagenic (lithosphorite) types and the narrow zone of modern authigenic production in the Walvis Bay area (from Rogers 1977; Birch 1978; Bremner 1978a). The coastal locality at Hondeklip Bay is the site mentioned by Tankard (1974b). Right map shows localities in the Saldanha region of the S.W.Cape: 1 & 2 = Lamberts Bay; 3 = Gaauwe Duynen; 4 = Elands Bay (at 100 m); 5 = Skoongesig; 6 = Bomgat Cave; 7 & 8 = Varswater Quarry area; 9 = Crywagenskraal (at 80 m); 10 = Duynefontyn; 11 = Milnerton. Key: 1 = phosphate-rich sediments; 2 = aluminium phosphate of Saldanha Bay. Data from Tankard (1974a,b); Birch (1977a, 1978); Bremner (1978b); Rogers (1979).

& Simpson (1972), Summerhayes (1973b), Summer hayes et al (1973), Bremner (1975, 1978a, 1978b, 1980a, 1980b) and Birch (1977b, 1979a,b). General reviews of results of this research have been given by Dingle (1974) and Fuller (1979). Less work has been done on the onshore deposits, but Visser & Schoch (1973), Tankard (1974a,b 1975a) and Birch (1977a) have given detailed accounts of the Varswater Formation sediments.

Following the classifications of Birch (1978) and Bremner (1978b), the Neogene phosphate-rich sediments of the west and south margins can be placed in three main groups:

i) lithosphorites (usually in the form of consolidated rock);

ii) microsphorites (usually as pelletal grains);

iii) concretions (which constitute the only known modern phosphatic sediments).

Lithosphorites generally occur as lithified rocks which are found as partly eroded pavements or scattered cobbles and slabs – the so-called phosphorite nodules, or less-commonly, as sand-sized grains in superficial sediments on the continental shelf. The rocks are restricted to the Agulhas Bank and the western continental shelf south of Lüderitz, and locally form extensive cover on the sea floor. Parker (1971, 1975), Parker & Siesser (1972), and Birch (1979a) have given detailed descriptions of the petrology and mineralogy of these rocks and have shown that they comprise a heterogeneous group that includes both conglomeratic and non-conglomeratic varieties (Tab.62). The non-conglomeratic types comprise phosphatized and partially phospha-

Table 62. Classification of Neogene lithosphorite rocks ('phosphorite nodules') from the continental shelf off south and southwestern Africa (from Birch 1979a).

*Non-conglomeratic phosphate-rich rocks*

| No. of samples dredged | | Foraminiferal (NI) (30) | Goethite-rich (NII) (3) | Glauconitic (NIII) (23) |
|---|---|---|---|---|
| Colour | | yellow-grey to dusky brown | moderate yellow brown | greyish-green; olive grey and yellow-grey |
| Shape and size | | rounded or tabloid pebbles & cobbles | ovoid cobble | irregular & rounded cobbles & pebbles |
| Components | Dominant | Foraminifera 5-60 % (plank. > benth.) Commonly filled with glauconite or collophane. Calcitic shells frequently altered to glauconite | foraminifera 5-25 % (plank. & benth.) | glauconite 5-75 % medium sized, rounded, mostly variegated or oxidized. Apatitic rims |
| | Minor | glauconite < 5 %<br>very fine quartz 0-20 %<br>bryozoan fragments<br>coraline algae | glauconite < 1 %<br>very fine quartz < 1 %<br>echinoderm fragments<br>bryozoan fragments | macrofossils<br>pelletal phosphorite<br>very fine quartz 5-40 %<br>feldspar<br>jasper<br>chalcedony |
| Cement | | fluorapatite<br>minor glauconite and micrite | Fe-rich (?goethite) | fluorapatite<br>minor glauconite and micrite |
| Texture | | packstone | packstone | packstones/wackstones |

*Conglomeratic phosphate-rich rocks*

| No. of samples dredged | | Glauconitic (CI) (15) | Goethite-rich (CII) (6) |
|---|---|---|---|
| Matrix colour | | dusky yellow-orange and yellow grey | orange or dark yellow-brown |
| Shape and size | | ovoid and tabloid cobbles and boulders | cobble and boulder slabs |
| Components | Dominant | glauconite 10-25 % medium sized, rounded, variegated, apatitic rims | foraminifera 10-20 % (plank. > benth.). Usually filled with Fe-rich collophane |
| | Minor | quartz ~5 %<br>foraminifera < 5 % | macrofossils<br>bone fragments<br>quartz < 5 %<br>glauconite < 5 %<br>coraline algae |
| Cement | | fluorapatite | Fe-rich (?goethite). Minor glauconite |
| Pebble inclusions: | Type A: | identical to foraminiferal non-conglomeratic phosphorites. Mainly foraminifera (5-40 %), minor glauconite (< 5 %) and quartz (< 5 %). Cement collophane, minor glauconite and micrite | similar to ferruginous, non-conglomeratic phosphorites. Mainly foraminifera (plank. > benth.), minor glauconite (< 1 %) and quartz (< 1 %). Cement Fe-rich goethite or collophane |
| | Type B: | similar to glauconitic, non-conglomeratic phosphorites. Mainly glauconite (10-20 %), minor quartz (5-15 %) and foraminifera (< 5 %). Collophane cement | |

tized limestones which show varieties rich in foraminifera, goethite, and glauconite, whilst the conglomeratic types contain a variety of clast and matrix components that can be divided into two broad categories, although as Birch (1979a) points out, they form a continuous series. These categories are: those with abundant pebbles (often NI type fragments) set in a matrix of quartz, glauconite microfossils and micrite/apatite cement; and those with a significantly lower glauconite content, but large numbers of macrofossils and a high goethite content. Table 63 shows that $P_2O_5$ values in these rocks vary from 17.83 % on the west coast (13 analyses) to 14.97 % on the Agulhas Bank (17 analyses). The Hoedjiespunt phosphatized coquina, and the phosphatic phase of the Bookram and Duynefontyn members belong in this category as they are clearly apatite replacement of original shell debris. In the former, analysis of the resinous Bomgat deposit shows a particularly high percentage of $P_2O_5$ (33.82 %) (Tab.63, Column 5) because of the very low quartz content of the rock. The manner in which apatite veins have penetrated the weathered granite basement at the exposures behind the fishing quay testify to the diagenic nature of the Hoedjiespunt occurrences, but a distinction can be drawn between these outcrops and other, more extensive deposits along the western side of Saldanha Bay. These have been described in detail by du Toit (1917), Hutchinson (1950) and Tankard (1974a) and consist of

Table 63. Major element composition of Neogene phosphate rocks from the South African continental margin and of phosphorites from other localities.

| | West Coast | | | | | | | Agulhas Bank | | | | Onshore West Coast | | | | N.W. Africa | Phosphoria Fm |
|---|---|---|---|---|---|---|---|---|---|---|---|---|---|---|---|---|---|
| Authors | 1 | | | | | | 2 | 3 | | | | 4 | 5 | 6 | 7 | 8 | 9 |
| Rock type | NI | NII | NIII | CI | CII | Average | | NCI-III Fe-poor | Fe-rich | CI | CII | | | | | | |
| No. of analyses | 4 | 1 | 4 | 3 | 1 | 13 | | 3 | 3 | 8 | 4 | 6 | 1 | 1 | 2 | 11 | 60 |
| $SiO_2$ | 4.42 | 12.06 | 24.81 | 19.76 | 13.94 | 13.44 | 4.97 | 6.20 | 3.45 | 15.30 | 13.79 | 55.64 | 4.12 | 4.13 | 62.30 | 6.73 | 11.90 |
| $TiO_2$ | 0.08 | 0.12 | 0.13 | 0.11 | 0.13 | 0.11 | – | 0.06 | 0.04 | 0.10 | 0.11 | – | – | – | – | 0.08 | 0.10 |
| $Al_2O_3$ | 1.12 | 2.05 | 2.45 | 1.89 | 1.47 | 1.70 | 1.12 | 1.13 | 0.92 | 2.14 | 2.12 | 3.29 | 1.51 | 1.73 | 7.55 | – | 1.70 |
| $Fe_2O_3$-FeO | 1.69 | 11.11 | 6.06 | 5.17 | 5.24 | 5.55 | 2.29 | 1.40 | 25.80 | 5.58 | 7.34 | 1.92 | 0.33 | 1.77 | 4.66 | 4.30 | 1.10 |
| MnO | 0.02 | 0.04 | 0.01 | 0.01 | 0.05 | 0.03 | – | 0.01 | 0.06 | 0.01 | 0.02 | – | – | – | – | 0.06 | <0.01 |
| MgO | 1.04 | 1.17 | 1.69 | 1.46 | 1.72 | 1.29 | 0.83 | 0.97 | 1.49 | 1.34 | 1.45 | 0.17 | 0.69 | 0.28 | – | 3.29 | 0.03 |
| CaO | 46.83 | 37.67 | 31.33 | 33.98 | 40.82 | 39.27 | 46.93 | 47.02 | 33.42 | 37.04 | 36.54 | 18.21 | 47.46 | 46.97 | 0.05 | 31.16 | 44.00 |
| $Na_2O$ | 1.09 | 0.96 | 0.88 | 1.18 | 0.67 | 0.98 | 0.50 | 0.62 | 0.34 | 0.78 | 0.73 | 0.19 | 1.02 | 0.68 | – | – | 0.60 |
| $K_2O$ | 0.44 | 0.57 | 2.50 | 2.01 | 0.91 | 1.19 | 0.38 | 0.43 | 0.39 | 1.57 | 1.51 | 0.26 | 0.20 | 0.26 | 0.54 | 0.73 | 0.50 |
| $P_2O_5$ | 18.76 | 15.42 | 17.86 | 18.92 | 15.41 | 17.83 | 28.04 | 14.82 | 10.26 | 17.89 | 16.82 | 14.49 | 33.82 | 33.16 | 11.91 | 16.96 | 30.50 |
| S | – | – | – | – | – | – | 2.31 | 0.31 | 0.21 | 0.46 | 0.45 | – | – | – | – | – | – |
| F | 2.57 | 2.10 | 2.06 | 2.54 | 2.22 | 2.46 | 3.20 | 2.12 | 1.42 | 2.21 | 2.04 | 1.31 | 1.66 | 3.44 | Tr | – | 3.10 |
| L.O.F. | 12.49 | 19.02 | 11.94 | 14.07 | 20.48 | 17.86 | – | 25.52 | 23.24 | 16.15 | 16.89 | – | – | – | – | – | – |
| Total | 101.70 | 101.41 | 101.72 | 101.09 | 103.06 | 101.99 | 93.27 | 100.61 | 101.04 | 100.56 | 99.81 | 95.74 | 91.14 | 92.42 | 87.01 | 63.31 | 93.54 |
| $C_{org}$ | 0.52 | 0.31 | 0.54 | 0.64 | 0.11 | 0.43 | 2.70 | – | – | – | – | 0.26 | 0.33 | – | – | – | – |
| $CaO/P_2O_5$ | 2.55 | 2.44 | 1.76 | 1.80 | 2.65 | 2.23 | 1.67 | 3.17 | 3.26 | 2.07 | 2.17 | 1.14 | 1.40 | 1.42 | 0.004 | 1.84 | 1.44 |

1 – Lithosphorite rocks offshore (Birch 1979a)
2 – Microsphorite pellets, offshore Namibia (Birch 1978)
3 – Lithosphorite rocks offshore (Parker 1971, Parker & Siesser 1972)
4 – Average, localities 1, 3, 5, 7 and 9 on Figure 181 (Birch 1977)
5 – Hoedjiespunt phosphatized shell bed (Birch 1977)
6 – Pelletal phosphorite, Varswater Formation (Tankard 1974a)
7 – Average two phosphatized porphyry basement Aluminium phosphate-Saldanha-Bay (Tankard 1974a)
8 – Summerhayes 1970
9 – Gulbrandsen 1966

scattered outcrops of phosphatized bedrock (weathered granite and limestones) on both sides of the entrance to Saldanha Bay, mainly on small hills (Fig.181). The phosphate rock varies from hard, yellowish brown, to greyish brown clay, and the deposits take the form of a phosphatized weathered surface of the underlying bedrocks which varies in thickness from 0.5 to 12 m. It occurs at altitudes which vary from 40 to 100 m above sea level, although Tankard (1974a) remarks that the lower levels are due to percolating solutions from higher deposits. The largest deposit is on Konstabelskop (on the south side of the bay entrance), where commercial exploitation has been carried out. Analysis of the phosphate-rich rock shows a characteristically high $Al_2O_3$ content (5.66-31.33 % in four analyses quoted by Tankard 1974a, hence his term aluminium phosphate for these deposits) with $P_2O_5$ values ranging from 0.92 % in limestone bedrock to 18.31 % in weathered granite basal deposits. In thin section, the rock is composed of phosphatized clay, with quartz grains. The geological setting and geochemistry of these aluminium-rich phosphates indicate that they were formed subaerially by leaching solutions from thick sea bird guano cover of coastal rock outcrops.

Microsphorite is the name given by Birch (1978) to any of the phosphate rock species which has formed authigenically (i.e. by direct precipitation, interstitial growth, or accretion). It is mostly found as resinous dark brown and reddish-brown ovate or spherical pellets which are well rounded, very well sorted fine to medium sand size. They usually have nuclei of quartz, feldspar or biogenic debris and an oolitic internal structure (Birch 1978), which indicate accretionary growth under conditions of constant agitation. Locally, pelletal microsphorite has been cemented by an apatite cement to form lithified rock layers, which pass laterally into uncemented pelletal sands in the Varswater Quarry. Whilst microsphorite rock is relatively abundant in the Varswater-Saldanha area it has only been reported at two localities offshore. In contrast, pelletal microsphorite occurs extensively on the mid-shelf areas south of the Kunene River, and between Walvis Bay and Lüderitz (Bremner 1975, 1978a), but is restricted to isolated patches on the mid-outer continental shelf south of the Orange River, and to the Varswater-Saldanha embayment onshore (Fig.181).

Although strictly speaking belonging to Birch's (1978) microsphorite category, we here draw a distinction, and separate a third class of apatite rich sediment which has been described by Bremner (1978a,b) from the continental shelf of Namibia; namely, the Quaternary concretionary deposits. These sediments are intimately associated with the diatomaceous muds off Walvis Bay and are forming at present: one of only two such areas reported (the other is off Chile-Peru). We mention these sediments here because the environment in which they occur may have been more widespread in Neogene times, and an understanding of it helps to explain the production of the older microsphorites and lithosphorites. They occur as pebble to fine sand size particles in a wide variety of shapes ranging from dark lithified irregular lumps with abundant fish scales and vertebrae to soft yellowish friable granules with abundant diatom frustules (Bremner 1978b, Baturin 1969, Veeh et al 1974). They are confined to the diatomaceous mud belt that extends as a narrow, discontinuous strip (18-27 km wide) a short distance from the coast for over 700 km north (and a short distance south) of Walvis Bay. This coincides with areas where large quantities of coarse fish debris are concentrated as a lag deposit by storm-wave induced bottom currents (Bremner 1978b).

Mineralogical studies of the Neogene phosphate rocks (Parker 1971, Bremner 1978a, Birch 1979a) indicate that the phosphate phase in the lithosphorite rocks (conglomeratic and non-conglomeratic), microsphorite (pelletal and rock) and the modern concretionary phosphate is francolite, a carbonate fluorapatite with $> 1$ % F, whilst that in the Hoedjiespunt lithosphorite is dahllite, a carbonate apatite with $< 1$ % F (Tankard 1974a). Mineralogically, the west coast and Agulhas Bank deposits are very similar, with the dominant components: fluorapatite, calcite, quartz, and glauconite. The calcite component is greatest in the partly phosphatized limestones, whilst glauconite is important in some of the pelletal sands which are glauconite/apatite mixtures (Birch 1975). A very low quartz (as well as illite and feldspar) content in the Quaternary concretions indicates a negligible terrigenous input into the present-day zone of precipitation (Bremner 1978b). Differences in composition between the various phosphate rock types can, therefore, be explained mainly on the basis of variation in sedimentary environment, and not in terms of the phosphate mineral-forming processes which were regionally controlled by common oceanographic and climatic factors.

During the course of investigation of these rocks, several workers have attempted to define sedimentary models which take into account their considerable sedimentological and structural varieties. These include the mixed associations of shallow and deep water microfossils and the rudaceous nature of the conglomeratic lithosphorites, as well as the contrasts in clearly diagenic partly phosphatized limestones and the oolitic nature of the authigenic pelletal sands. Earlier attempts (e.g. Murray & Renard 1891, Cayeaux 1934, and Parker 1971, 1975) have been summarized by Dingle (1974) and Fuller (1979), and whilst they were successful in explaining individual types, they lack an overall appreciation of the relationships of the various phosphate rock species in terms of the regional oceanographic and climatic setting. A successful synthesis has recently been achieved by Birch (1978, 1980) with amplification of certain settings by Bremner (1978b).

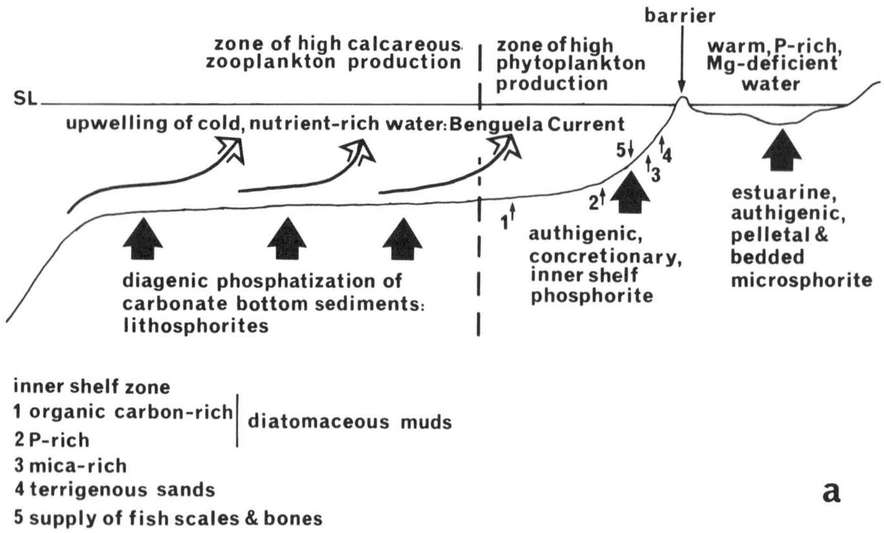


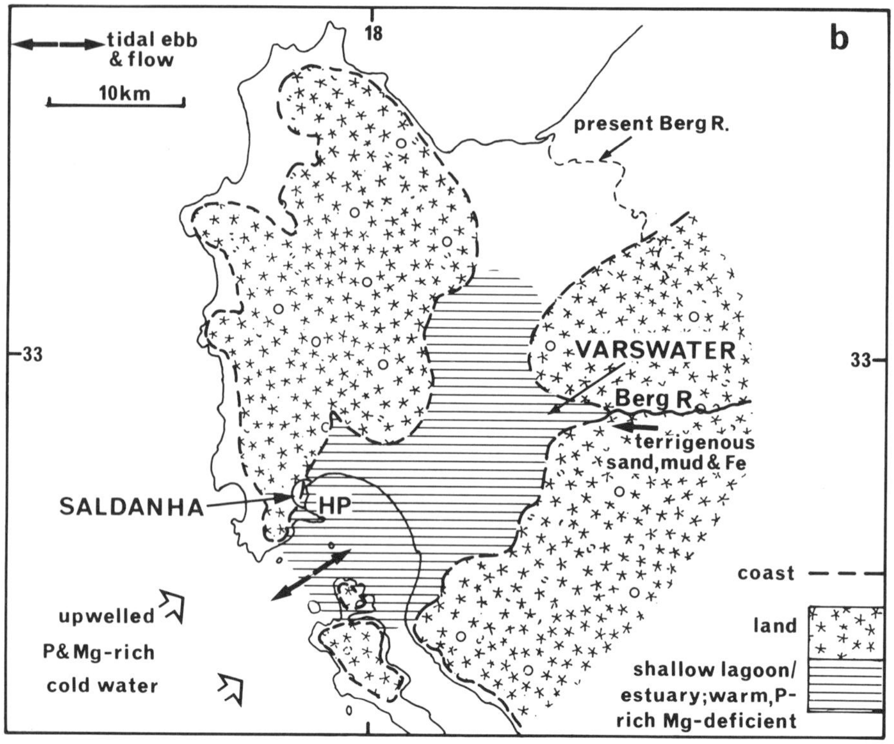


Figure 182. Suggested mechanisms and palaeogeographies for Mio-Pliocene phosphatic sediment production along the coast of S.W.Africa.

a. Schematic model for production of diagenic and authigenic phosphate on the continental shelf, inner shelf zone and lagoonal/estuarine environments. It combines the schemes proposed by Birch (1977a), Bremner (1978b) and Bremner (unpublished, cited in Flemming 1980).

b. Suggested palaeogeography for phosphate production in the Saldanha/Varswater Embayment. This represents the right side of (a) and is a modification of Bremner's unpublished model for Walvis Bay. See text for discussion. HP = Hoedjiespunt.

We have combined their schemes in our summary diagram (Fig.182) which shows Birch's model for the continental shelf, nearshore zone, and coastal lagoon, with a detail of the Varswater-Saldanha embayment modified from Bremner's (1978b) model for Walvis Bay.

Birch (1978, 1980) considers lithosphorite production (diagenic replacement of lime mud) as a continental shelf phenomenon, with microsphorite production (authigenic precipitation) confined to coastal lagoonal settings. The present west coast-continental shelf areas are subjected to periodic upwelling of cold nutrient-rich water within the Benguela Current System and as we will discuss in a later section, this has been an established oceanographic phenomenon for much of Neogene time. Birch postulates that on the mid shelf areas, the upwelling and attendant high zooplankton productivity results in the deposition of abundant calcareous debris on the sea floor (Fig.182a). Lower sea levels than present resulted in intensive upwelling with an increase in zooplankton productivity. Bacterial breakdown of the zooplankton organic matter resulted in the concentration of $PO_4^{-3}$ ions in bottom sediment pore waters to the extent that replacement of carbonate (micrite, lime mud) by a carbonate fluorapatite complex was initiated (0.1 ppm $PO_4^{-3}$ in sea water is sufficient for the action to proceed according to Ames 1959, quoted in Birch 1978). Continued lowering of sea level resulted in re-working of the semi-lithified phosphatized and partly phosphatized sediment to form conglomeratic textures, which were recemented by apatite cements during a subsequent sea level rise. The finely disseminated glauconite component was produced following Fe-enrichment of the sediment by the biological input and a contemporary glauconitization of clay particles.

In the nearshore zones, the high biological productivity is expressed by large-scale phytoplankton production which results in high organic-carbon, P-rich zones on the sea floor in a narrow nearshore zone of diatomaceous muds (Bremner 1978b). This highly productive zone supports a prolific fish fauna, whose bones accumulate in large quantities on the sea floor in this zone, particularly at the inner edge of the mud belt where wave action winnows the fish debris. $PO_4^{-3}$ ions released by bacterial decay of the organic matter on the sea floor are prevented from escaping into the overlying water column on the nearshore side of the diatomaceous mud belt by an overlying 'barrier' of iron-rich illite clays in suspension (land derived debris). These clays adsorb the $PO_4^{-3}$ ions, and ultimately result in a build up of phosphorus in the interstitial waters of the bottom muds to values 26 times higher than found in the overlying water column (Calvert & Price 1971a). At those concentrations authigenic apatite precipitation proceeds around nucleii of fish debris. This process is occurring at present along the inner edge of the diatomaceous mud belts in the Walvis Bay area (Bremner 1978b, 1980a,b). It might also be noted that the diatomaceous muds in this area are enriched in trace metals (Cu, Ni, Pb, U and Zn) (Calvert & Price 1971b, Veeh et al 1974).

In a lagoonal/estuarine situation adjacent to such a zone, the upwelled P and Mg-rich cold water, could, with suitable barriers and/or coastal bedrock geometry, be channelled into very shallow environments with significant terrigenous sediment input (such as Walvis Bay, or the Neogene Varswater-Saldanha embayment –Fig.182b). High solar heating in such shallow embayments raised the water temperature, encouraging precipitation of apatite by reducing its solubility. In addition, Fe in clay minerals brought down by rivers (e.g. the Berg River which entered the sea from the east) allowed a $Fe^{3+}$ exchange for $Mg^{2+}$ in the sea water, further enhancing the conditions for apatite precipitation ($Mg^{2+}$ inhibits apatite precipitation, Bremner 1978b). Constant agitation by tidal or wave action in the estuary and lagoon resulted in a dominance of pelletal and oolitic structures for the apatite grains. Where quartz sand concentrations were sufficiently high, apatite precipitation proceeded in the intergranular pores, whilst the local development of finely bedded, varve-like pure phosphate layers indicate that precipitation and lithification took place rapidly (Birch 1977a, 1978). As Birch (1978) points out, offshore wind action is most intense in the summer months, so that upwelling and a P-rich water supply would have been greatest at the same time as solar heating was strongest.

North and south of the Varswater-Saldanha embayment, conditions were not conducive to large-scale pelletal sand production, but the high biological productivity associated with upwelling along the whole coastal area was sufficiently intense to induce apatite production on a small-scale in the form of replacement of carbonate fragments and mammalian bone and fish teeth. Hoedjiespunt was a localized and peculiar development caused by the incursion of phosphate-rich waters into depressions (large pools) in the irregular basement surface on the edge of the lagoon. These periodically became isolated from the sea, suffered intense solar heating, with resultant replacement of the shell accumulations in these depressions, and localized replacement of the weathered granite along joints and fractures.

The large shelf deposits of microsphorite pelletal sands presumably formed during periods of low sea level, when very wide lagoons developed behind long outer continental shelf bars (Dingle 1973c).

The lack of microsphorite along the south coast (i.e. on the Agulhas Bank) indicates that although oceanic upwelling did operate in this area during Neogene times (lithosphorite is widespread), it was not as intense as along the west coast. Siesser (1978a) has suggested that differences between the two areas may have been the result of two styles of upwelling, with 'dynamic upwelling' along the south coast (caused by the flow of a strong current parallel to the coast on its right side), in contrast to 'oceanic upwelling' along the west coast (caused by wind-driven water movement). Clearly, conditions on neither of the shelves (with the exception of the Walvis Bay area) are conducive to production of apatite at present.

Early workers on the Agulhas Bank lithosphorites (e.g. Murray & Renard 1891, Collet 1905) believed them to be Recent or sub-Recent deposits, and it was not until Cayeaux (1934), who noted the presence of *Aturia*, that a Miocene or even older age was suspected. Haughton (1956) examined fossils from the rocks dredged off Cape Point and Cape Columbine and suggested a Miocene (possibly early Miocene) age, which was supported by Miller & Furnish (1956) who identified *Aturia lotzi* (which is now known to range Palaeocene-Miocene). Further dating attempts on the Agulhas Bank material by Kolodny (1969) and Kolodny & Kaplan (1970) using uranium isotope methods indicated ages greater than 800 000 years, which is supported by the presence of phosphorite fragments in late Pleistocene limestones offshore (Siesser 1971).

On west coast material, Rogers (1977) reported conglomeratic phosphorite attached to an eroded surface of early-middle Miocene limestones, implying a maximum age for the phosphorite. Siesser (1978a) has made age determination on 33 phosphorites from the west and south coast margins using the planktonic foraminifera and calcareous nannofossils in the lithosphorite rocks (both conglomeratic and non-conglomeratic varieties). Siesser was able to date ten of these samples as definitely pre-Pliocene (most likely early-middle Miocene), and 11 as definitely Pliocene or younger. The distribution of these rocks was: all the Pliocene-dated rocks were from the Agulhas Bank; 9 out of 10 of the Miocene rocks were from the west coast shelf. The remaining 12 samples gave age ranges of Mio-Pliocene and these were distributed: Agulhas Bank 9; west coast 3. Siesser (1978a) emphasized that his dates referred to the age of the original carbonate debris, and that phosphatization could have taken place (in the lithosphorite varieties) at any later stage provided the carbonate remained semi-lithified and close to the sea floor. It should be noted that of 21 definite pre-Pliocene, or Pliocene dates, only one (on the Agulhas Bank) was made on a conglomeratic variety, whilst on the rocks dated as Mio-Pliocene four were non-conglomeratic and eight conglomeratic. From this we conclude (noting Siesser's caution):

i) that phosphatization commenced in Miocene times (probably Middle Miocene according to Siesser 1978a) and continued into Pliocene times, stopping before the start of the Pleistocene;

ii) that there were three distinct phases with respect to production of the lithosphorite species – (a) a Miocene phase of lime mud phosphatization, (b) a late Miocene-early Pliocene phase of conglomeratic lithosphorite production, (c) a Pliocene phase of lime mud phosphatization.

Phases (a) and (c) correlate with relatively higher sea levels compared to (b) which may relate to the late Miocene-early Pliocene regression. Significantly, it is during regressive phases that Birch (1978a) postulates conglomeratic lithosphorite formation. Unfortunately, we have no well-dated samples of microsphorite from the continental shelf, but the Varswater Formation is definitely early Pliocene in age (Hendey 1981a)

Finally, it should be noted that Birch et al (1983) have recently obtained Pleistocene dates (62-76 ky) for some lithosphorite varieties from the continental shelf SW of Cape Town. They consider the question of phosphatization, particularly along the south coast (i.e. in typically non-upwelling western boundary current situations), in terms of processes related to bacterial action. These researches may open the possibility of recognizing contemporaneous polygenetic (organic and inorganic) mechanisms that have operated around southern Africa for most of Neogene time and offer alternative explanations for phosphate-rich sediments at other east coast sites: for example, the Nodule Bed at the base of the Mfolozi Valley Miocene strata.

### 6.2.3 *Non-marine Neogene deposits*

#### 6.2.3.1 *Kalahari Formation*

The central part of southern Africa consists of a huge shallow depression (Kalahari Basin) into which sediments have been moved by internal drainage and aeolian agencies, possibly since early Tertiary, or even Cretaceous times. Because of poor exposure, relatively little is known about the detailed stratigraphy of the sediment fill, but in the past it has been practice to follow du Toit's (1954) classification of a lower sequence of 'Kalahari beds' (primarily mostly lithified) and overlying 'Kalahari sands' (unconsolidated sands). In a recent study in the northern Cape, Smit (1977) established a local stratigraphic succession using over 2 000 borehole records, in which he placed all the sediments in the Kalahari Formation, and subdivided it into four units (top 4, base 1):

4 aeolian sands, rarely > 20 m thick;

3 fine-grained clayey and calcareous sandstones. Silcrete and calcrete are common in the sandstones. Up to 80 m thick;

2 red calcareous clay containing layers of fine gravel or coarse sand, with sandstone lenses. Up to 100 m thick;

1 clayey gravel and gravelly clay. Up to 100 m thick. The maximum observed thickness at one place was 180 m.

Smit (1977) noted that the pre-Kalahari Formation drainage flowed northwards, and postulated that during Tertiary times the river valleys became choked with gravel, clay and sand eroded from nearby highlands by periodic floods. Blockage of the old valleys eventually forced streams to cut new channels. Calcretization began after the valleys had been choked, and flooding had become less intense. Progressive aridification brought desert conditions in which fine sands were spread over the whole area by winds blowing from the NE.

Similar sequences can be observed elsewhere. The three lower units of Smit's classification correspond to du Toit's (1954) Kalahari beds. These are widely distributed, and the Urananib Plateau sequence of Namibia (east of Keetmanshoop) (Fig.154) is typical. Here, the basal beds are conglomerates and coarse calcareous sandstones grading upwards into finer, more calcareous sandstones. These are overlain by a hard, partly silicified limestone crust (calcrete?) from which layers of ochrous sandstone and purplish shales and clays are known in boreholes (Haughton 1969). Du Toit (1954) interprets this sequence as fluvial (although some aeolian influence is not excluded), and it is thought to be over 150 m thick in places.

In the Lake Ngami area (Fig.183) Passarge (1904) described the basal units as calcareous sandstones, chalcedonic sandstones, silcretes and laterites. He called this sequence the 'Botletle beds', a name which has enjoyed only limited usage. Other chalcedonic rocks (including the 'Pipe Sandstone' and 'Gwampa Chalcedony') considered to be equivalents of the Botletle beds (Haughton 1963), are known from Victoria Falls south to Bulawayo. These contain freshwater molluscs and *Chara* seeds and stems. Other exposures are found west of Molepolole, where yellowish shale beds with fibrous limestone bands are overlain by massive reddish sandstones, and in Zimbabwe, where massive red sandstones are known as the 'Gwai Reserve beds' (Haughton 1963). It should be noted, however, that these basal beds found towards the margins of the basin may not be coeval with basal deposits nearer the centre where boreholes show that marls (up to 67 m thick, Haughton 1969) become more common.

The age of the Kalahari beds has not been satisfactorily established. Passarge (1904) suggested that they were Eocene, whereas Mabbutt (1955) favoured Pliocene on the basis of one mammal (*Notohipparion namaquense*). Cahen & Lepersonne (1952) divided the Kalahari beds of the Congo Basin into three units (1 bottom, 3 top): (1) the Kamina Series (sands, sandstones and gravels) which is truncated by a Late Cretaceous (Gondwana) erosion surface; (2) the overlying Middle Series (sands, silicified sandstones and chalcedonic limestones), cut by a middle Tertiary (African) erosion surface; and (3) the Upper Series (ochrous sands), which is cut by an end-Tertiary erosion surface, over which lie Recent sands. The lithological and stratigraphic similarity of this sequence to the beds in the Kalahari Basin is apparent, although corresponding erosion surfaces have not been proven. The Middle Series with its chalcedonic limestones (grès polymorphes) is considered to be correlative with the Botletle beds (du Toit 1954), and van Zinderen Bakker (1975, p.72) assumes an Oligocene age for the grès polymorphe, based on an oceanographic-climatic model.

The Kalahari sands (of du Toit 1954) form a widespread blanket of quartzose sand that rests unconformably on both the Kalahari beds and on older rocks and correspond to Smit's (1977) unit 1. The intervening hiatus is marked by the presence of ferricrete and silcrete immediately below the sands (du Toit 1954). Outliers of the Kalahari sands indicate a much wider distribution in the past, reaching far into the Orange Free State, Mozambique and possibly western Namibia. Underlying rock units have been eroded to produce the sands, whose colour varies from deep red, to pink, yellow or grey, with grain sizes normally 0.15-0.40 mm (fine sand). They are usually not well-sorted or rounded, and frequently contain diatoms, presumably derived from fresh-water sources (du Toit 1954). They are known to be as much as 67 m thick, but may be much thicker in places. Earlier studies in the Kalahari region of Botswana (e.g. Boocock & van Straten 1962, Grove 1969, Wright 1978) have recently been supplemented by an interpretation of Landsat imagery of the whole of Botswana, resulting in a synthesis of the regional distribution and stratigraphic relationships of the late Tertiary deposits of this area, that clarifies the situation considerably (Mallick et al 1981). We will summarize this work because although the Kalahari sands are at least partly early or middle Pleistocene in age (indicated by enclosed stone artefacts) some are probably a good deal older (?Pliocene) (Haughton 1963).

Mallick et al (1981) identify three elements in the upper part of the Kalahari sands: duricrusts (mainly calcretes); aeolian sands; and lacustrine and alluvial sediments, which they mapped over the whole of east, north and central Botswana (Fig.183). Tentative stratigraphic relationships of these various rock types are also shown in Figure 183, where the crucial factors in establishing a relativity for the various elements are the recognition of two phases of duricrust formation (Boocock & van Straten 1962), two separate episodes of dune formation (indicating periods of increased aridity), and a period of widespread lacustrine environments and large-scale river systems. The oldest elements in this sequence are the old duricrust horizons which underlie the Kalahari sands. We have previously referred to these as forming extensive outcrops on the Urananib Plateau in eastern Namibia, from where they dip gently eastwards under the Kalahari basin as a duricrusted erosion surface. These duricrusts are mainly calcrete, but locally they contain abundant silicreted breccias and silcalcretes (Wright 1978), and rest both on pre-Tertiary bed rock, and various parts of the Kalahari beds.

In Botswana, these older duricrusts crop out extensively in the major old river channels (particularly the Molopo) which cut through the mid-Tertiary erosion surface, and also lie close to the present ground surface in the southern and central parts of the country, where they are exposed around the edges of some of the major pans. Here, they form cliffs up to 10 m high. Locally, the calcretes have a basal layer of up to 35 m of silcreted material.

The older duricrusts are overlain by two dune systems which occur in the north and south of Botswana. The former are the oldest, and have been identified as alab or seif sand ridges that were formed longitudinally to the prevailing wind directions, which were easterly. They consist of large parallel-sided features about 1.7 km apart that occur in a wide zone from western Zimbabwe, across Botswana into northern Namibia and southern Angola. Across this region their alignment swings from 260° in the east to about 280° in the west. Although they are often badly degraded and vegetated, Mallick et al (1981) were able to easily identify them in satellite

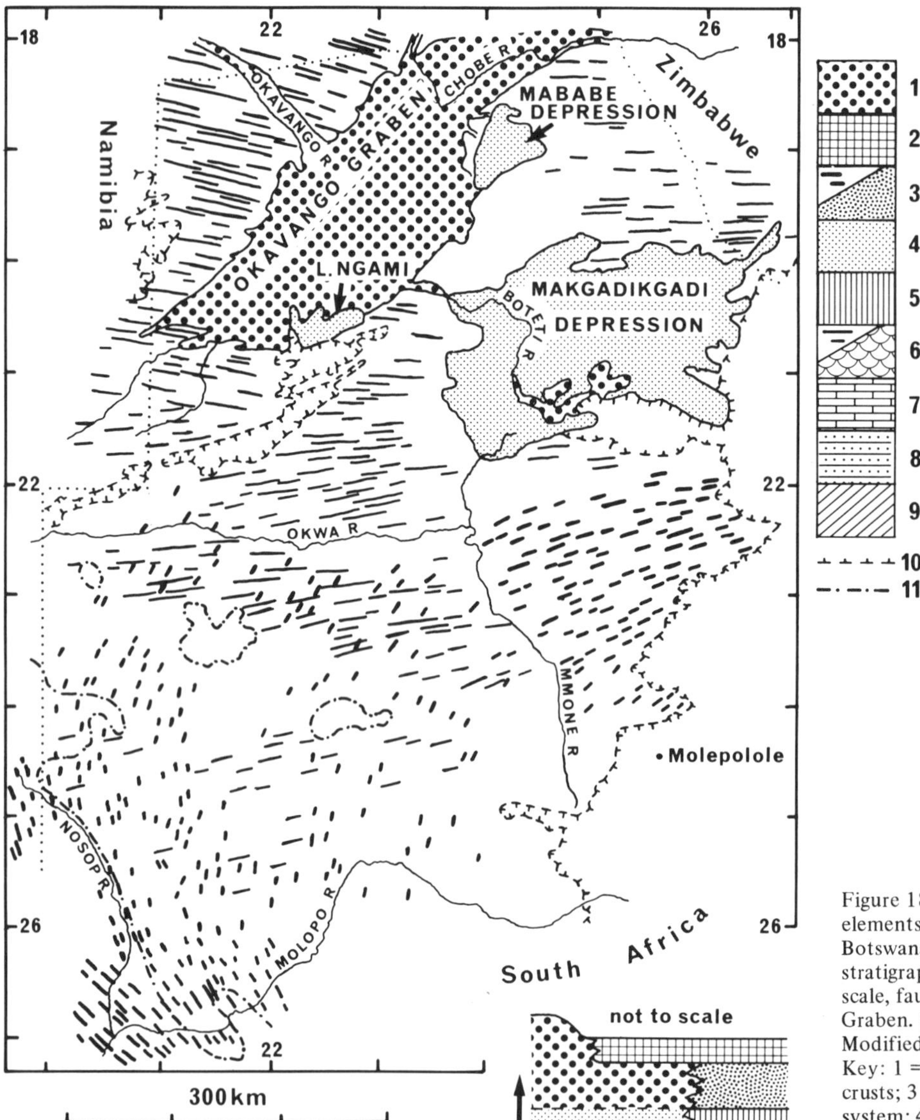


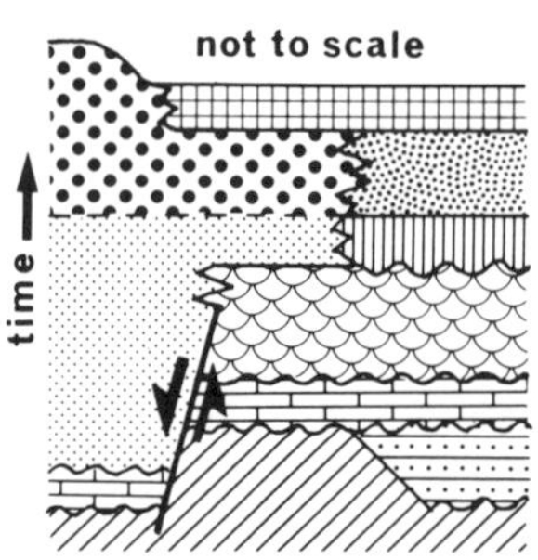


Figure 183. Distribution of various elements of the Kalahari sands in Botswana. Insert shows diagrammatic stratigraphic relationships and is not to scale, faults relate to the Okavango Graben. See Figure 154 for location. Modified from Mallick *et al.* (1981). Key: 1 = alluvium; 2 = younger duricrusts; 3 = younger (southern) dune system; 4 = lacustrine sediments; 5 = hiatus; 6 = older (northern) dune system; 7 = older duricrusts; 8 = Kalahari beds; 9 = pre-Tertiary strata; 10 = extent of pre-Tertiary outcrop (minor inliers not shown); 11 = extent of red (Namibian) sands of the younger dune system.

photographs because the interdune areas are finer grained (silty sands) with dark grey-brown soils that support a different vegetation to the dune ridges, possibly a reflection of the presence of calcretes (the younger duricrusts) close to the surface in the slight depressions.

The other dune system is younger and occurs mainly in the south of Botswana, although in the central area they are seen crossing the older alab dune lineaments. This system is characterized by large regional variations in alignments: 250-260° in the east; to 160-180° in the west and centre; to 130-150° in the SW. Mallick et al (1981) related this pattern to two prevailing wind directions by identifying this younger system as also being longitudinally formed and detecting continuity between the regionally curved pattern in the west, and a separate E-W set in the east. They also recognized several different categories, with the eastern parts consisting mainly of isolated, well-preserved dunes, degraded hummocky topography in the central areas, and well-preserved, possibly locally active dunes in the west and SW. A characteristic of the westernmost areas is the occurrence of brick red coloured dunes which are found particularly in Namibia and South Africa (the Namibian variety), that merge eastwards into yellowish brown sands (the Botswana variety). Locally, the latter overlie the former. Mallick et al (1981) speculate on the provenance of the red sands, and suggest that they may be re-worked alab dunes, or that the coloration represents different bedrock types because it has been found that sand in the dune systems may be locally-derived. For example, buried kimberlite pipes have been located by studying the mineralogy of the covering dunes.

The younger duricrusts typically occur in interdune hollows, particularly in the southern system, and there is good correlation between soil colour and the presence of duricrusts at shallow depth beneath the surface. For example, grey brown soils have calcretes 1-2 m below the surface.

Stratigraphically, the extensive lacustrine and alluvial deposits of northern Botswana partially intervene between the older and younger dune systems (Fig.183). There are three major former lakes in northern Botswana: Lake Ngami; the Mababe Depression; and the Makgadikgadi Depression, which are clearly outlined by shoreline features and soil types on satellite photographs (Mallick et al 1981). All three are partly fault-bounded, with the first two associated with movements of boundary faults on the southern side of the Okavango Graben that caused ponding of rivers in depressions against southern basement rock walls, and the latter partly fault-bounded in the SE. Lake Ngami, which historically has varied greatly in size (Grove 1969), was probably initially fed by rivers draining from the SW, and possibly the Okavango. Judging by the distribution of lacustrine deposits mapped by Mallick et al (1981), at its maximum extent the lake had a surface area of at least 2 100 km$^2$, but desiccation of the SW rivers, and encroachment of Okavango alluvium have greatly reduced its size, with retreat clearly shown by a set of recessional shorelines. The present lake in the Mababe depression occupies the central part during the wet season, but originally it also had a greater surface area (at least 3 200 km$^2$), and may have received water from the Okavango and Chobe rivers, which built small deltas into it. At its western end the Mababe depression has a particularly prominent shoreline feature or offshore bar: the Magibwe Sand Ridge.

The largest of these late Tertiary/Quaternary lakes lay in the Makgadikgadi Depression, where Mallick et al (1981) recognize shoreline and offshore bars that delineate an area of about 43 000 km$^2$ (Fig.183). They suggest in fact that the lake may originally have been as large as 80 000 km$^2$ (or even bigger), by extending northwards to the Mababe Depression. The relationship of the alab dunes and possible shoreline features in the north indicated to Mallick et al (1981) that the lake may have originated at least during the later stages of the older dune system, implying that it was in existence in late Miocene/Pliocene times. Clearly at their maximum extent, these lacustrine features related to periods of relatively high rainfall which intervened between the deposition of the older and younger (i.e. the north and south) dune systems, because sands from the western part of Lake Makgadikgadi provided material for the younger dune fields. Modern deposition is on a much smaller scale, as most of the former river courses which fed the lakes are now dry.

Tectonic activity in the Okavango Graben postdates the alab dunes, but has probably continued to recent times, so that the top of the Okavango Graben fill is partly contemporaneous with the younger dune system. As mentioned earlier, these faults are re-activations of much older lines of movement, and Permo-Triassic sediments also occur in the graben, in addition to Palaeozoic and even Pre-Cambrian sediments. Detritus has been brought down from the Angolan highlands by the Okavango and Chobe rivers, which today are still perennial, although in fairly recent times the latter has been deflected into the Zambezi. These two rivers have constructed large alluvial fans which have spread into the Okavango Graben, and formerly fed the Boteti which transferred alluvium into the Makgadikgadi Depression.

Because this sediment sequence has not been accurately dated, the portion of the Kalahari sands that belong to the Tertiary is not known, but if the older duricrusts do relate to the mid-Tertiary erosion surface, then the alab dune system is probably Mio-Pliocene in age, with the other parts of the succession possibly late Pliocene or Pleistocene. From a palaeoclimatic point of view, at least two periods of high winds and aridity, with a major intervening episode of high rainfall, as well as possible other minor wet phases, are indicated in the Kalahari sands succession of Botswana. Additional evidence for periods of higher rainfall is provided by Mallick et al's (1981) recognition of elongate pans lying between the northern (older) alab dunes, which are remnants of interdune-restricted drainage patterns, as well as pans formed by ponding behind linguoid dunes as they blocked drainage channels in wetter periods in the red sand areas of SW Botswana and eastern Namibia.

#### 6.2.3.2 *Duricrust deposits*

The calcretes, ferricretes and silcretes (duricrusts) mentioned above deserve further comment, as they are not solely associated with the Kalahari sands, but are known to sporadically cover rocks of all ages over much of southern Africa.

Calcretes (indurated caliche) are probably the most widespread of the three rock types, being found from Cape Agulhas in the south (calcretes have been dredged from as deep as 60 m on the continental shelf) to beyond the northern limits of the area under review, and from the Atlantic to the Indian Ocean coasts. In general, calcretes are common over most areas west of 27°E and south of 17°S, and are well developed in arid and semi-arid areas where rainfall is less than 550 mm/yr (see Netterberg 1969, 1971 for a detailed discussion of calcrete distribution).

Netterberg (1980) defines calcrete as 'almost any material of almost any consistency and carbonate content formed within the regolith by in situ cementation and/or replacement of a pre-existing material by carbonate precipitation from the soil water or ground water' and that 'calcretes may be of pedogenic or non-pedogenic origin or both'. For the sake of convenience, materials such as 'speleothems, spring tufa, aeolianite and beach rock' are usually excluded. As Netterberg (1980) points out, the calcretes of southern Africa have remained until recently relatively poorly known, but increasing demand for roadbuilding material, cement, lime, limestone, uranium, and agricultural development have prompted research, particularly related to engineering activities in the arid and semi-arid areas of the sub-continent. At the time of writing, only the first part of a comprehensive study of calcretes by Netterberg has been published (Netterberg 1980), but a significant increase in understanding of the formation, distribution and stratigraphic value of calcrete deposits can be expected with the publication of subsequent parts.

Netterberg's (1980) proposed classification of calcretes is a descriptive one based largely on secondary (chemical) structure and sequence of development, and his following categories are arranged in the order in which they are considered to form: calcareous soil; calcified soil; powder calcrete; glaebular calcrete; calcrete pseudotubules, etc.; honeycomb calcrete; hardpan calcrete; and calcrete boulders and cobbles. He considers that calcretes form by two main processes:

i) by precipitation in a host soil above a perched or permanent groundwater table: groundwater or non-pedogenic calcrete;

ii) by downward leaching of carbonate from the upper soil horizons by percolating waters which carry it in solution to lower levels where it is precipitated: pedogenic calcrete.

The rock formed will thus be rich in the inorganic components of the soil: clay; fossil fragments; quartz; feldspar and other terrigenous grains, which are cemented by calcium carbonate in the form of calcite (dolomitic calcretes have been reported by du Toit 1954). It normally forms over (or within, Siesser 1972c) carbonate-bearing rocks such as aeolianites, marine limestones, dolomites, calcareous shales, tillites and basic igneous rocks (du Toit 1954). Netterberg (1969) has assigned southern African calcrete formation to pre-Pliocene, Pliocene, Pleistocene and Holocene times, and suggests Pliocene and Pleistocene as the most important because of widespread climatic deterioration during these periods. It should be noted, however, that pre-Pliocene and Pliocene calcretes are 'dated' simply by their stratigraphic inclusion in the sequence of Kalahari beds, which themselves are not well-dated.

Ferricretes (lateritic ironstones) are pedogenic accumulation of iron oxides that differ from true laterites (found only in Zimbabwe, Mozambique and possibly Natal) chiefly in that they are high in silica and low in aluminium oxides. Ferricretes have a widespread but sporadic distribution in South Africa, Zimbabwe and Mozambique, but are not common in Namibia. In South Africa they form over and cover a wide variety of rock types: slate, granite, sandstone, basalt and dolomite (to mention the best-known). Ferricrete is formed by the deposition of iron oxide in the soil zone through the action of ground water, so that the rock is thus mostly made up of clay, quartz and other terrigenous grains cemented by iron oxide. A pisolitic texture is a common feature. Netterberg (1975) states that the iron oxide may be leached from an overlying, underlying, or lateral iron-bearing source, and as with calcrete, may take the form of individual nodules in soil, nodules cemented together ('honeycomb ferricrete'), or a well-indurated layer of rock locally more than 1 m thick (hardpan ferricrete). Netterberg (1975) believes that ferricrete can form in most climates where definite wet and dry seasons occur. Some ferricrete is of Pleistocene age (indicated by enclosed stone age tools, du Toit 1954), but its presence in the Kalahari beds suggests some earlier periods of formation, probably much the same age as the calcretes.

Silcretes (surface quartzites) are pedogenic accumulations of silica in the soil zone which occur frequently, but sporadically, along the coastal areas of southern Africa from the Kunene River to East London. Inland deposits are more scattered, but are known in the Transvaal and Zimbabwe, and somewhat more extensively in Botswana. Silcrete varies considerably in colour (white, grey, yellow, brown, red and green varieties are common, du Toit 1954) but consists overwhelmingly of quartz grains with occasional feldspar and other terrigenous fragments tightly cemented by silica. These rocks sometimes contain small, irregularly shaped cavities, and in a few cases, relict rhombs of calcite and dolomite have been noted, suggesting replacement of an original carbonate cement (du Toit 1954). The silica is most often in the form of chalcedony or opal, especially in the silcretes of the interior. Layers of silcrete up to 8 m thick are known, although deposits are usually much thinner.

Silcretes have been studied most thoroughly in the Grahamstown area (Frankel & Kent 1937), and near Albertinia (Frankel 1952)(Fig.154). Those from around Grahamstown were found to have formed from colloidal silica (chalcedony) derived from underlying clay, but du Toit (1954) cautions that underlying rocks may not everywhere be the silica source, as silcrete is also found on basalt. Other suggestions for the silica source are diatoms (Passarge 1904) and siliceous grasses (phytolithariens) (du Toit 1954). Passarge (1904) suggests an arid climate, interrupted by periods of greater rainfall, as the environment necessary for producing silcrete. Presumably, during the arid period alkaline conditions prevail which are favourable for precipitating calcium carbonate and taking silica into solution; during the wetter, acidic conditions, calcium carbonate would be dissolved and silica would be precipitated, both in the soil zone and as a replacement of limestone.

Little is known about the age of southern African silcretes but their occurrence in the pre-late Eocene Pomona beds indicates that they range well back into the Tertiary. Frankel & Kent (1937) suggest an early-middle Tertiary age, based on geomorphic and other evidence, and Haughton (1969) notes that pebbles of silcrete have been found in the basal limestones of the Alexandria Formation, implying that silcretization was pre-Miocene or pre-Pliocene. Moreover, the widespread occurrence of silcretes at the base of the Kalahari sands suggests a Pliocene age for many of them.

6.2.3.3 *Alluvial deposits*

Alluvial deposits in southern Africa extend in age from at least the Pliocene (and possibly before) into the Pleistocene. The best known of these are the 'High-level Gravels' (du Toit 1954) which occur on terraces at elevations of 700 m or more and are best developed in the region between Grahamstown and Mossel Bay, although they extend sporadically much farther east and west. The degree of induration of the 'gravels' (actually boulder- to pebble-size) varies from loose to hard, silica-cemented conglomerates, and they are often underlain by clays, some of which were formed in situ, but most of which are of alluvial origin. These deposits slope coastward or towards valley centres and must represent valley and scree deposits (Haughton 1969), but their age is not precisely known, as no diagnostic fossils have yet been recovered.

Farther inland, the so-called 'Low-level Gravels' are found. These are best known along the Vaal River at Windsorton and Barkly West (Fig.154), where Söhnge et al (1937), Cooke (1947), van Riet Lowe (1952), Partridge & Brink (1967) and Butzer et al (1973), among others, have contributed important studies. The Low-level Gravels are divided into 'Older Gravels' (late Pliocene-early Pleistocene) and 'Younger Gravels' (middle Pleistocene) (Söhnge et al 1937, Butzer et al 1973). The Older Gravels are mostly cobble conglomerates set in a matrix of calcretized, silty sand, and occur 3-30 m above modern flood level (Butzer et al 1973). They are not well-consolidated and are seldom more than 5 m thick (a maximum of 18 m has been reported). Their age is not well-established, but Partridge & Brink (1967) estimate a Pliocene age for Older Gravels at Windsorton (based on geomorphic evidence) and Butzer et al (1973) assume late Pliocene age for Older Gravels from the Lower Vaal (based on mastodonts). Remains of a supposed Pliocene elephant, *Mammuthus subplanifrons,* have been found in fluvial sands and gravels (the 'Virginia beds') in the Orange Free State (Butzer 1973), but Hendey (1978b) has since suggested that the age range of this mammoth is still not certain, and that the Pliocene age assignment should be treated with caution.

The 'Low-level Gravels' along the Vaal River and its tributaries are famous for their diamond content, but other diamondiferous fluvial deposits do occur, the best known being the 'Lichtenburg gravels'. These gravels are found in solution sinkholes in dolomite and have been attributed to the late Tertiary, although there seems to be no substantive palaeontological or stratigraphic evidence for this date (Haughton 1969).

6.2.3.4 *Cave deposits*

Certain caves in the Transvaal (Kromdraai, Makapansgat, Sterkfontein, Swartkrans) and a quarry in the northern Cape Province (Taung) are well-known for their deposits containing early hominid remains (locations on Fig.154). These deposits are now believed to span the Pliocene-Pleistocene boundary. Description from Brain (1958, 1976) of the Swartkrans cave will serve as an example of the rock types involved.

The deposits are stratified, calcareous breccias, often rich in bone debris. They originated by the infilling of subsurface fissures and caves developed in dolomite into which sediment and bone were carried and cemented during two cycles of deposition. The older cycle comprises a pink to orange breccia, rich in owl pellets and blackened bones, whilst the younger breccia is brownish, with blackened (by MnO) pebbles, but unblackened bones. The two breccias are separated by an erosional unconformity. The most important fossil remains to come out of these cave sites are those of the Australopithecine, the first of which (*Australopithecus africanus*) was discovered by R.A.Dart in 1924 from Taung. More specimens of this species were later discovered in the Transvaal caves, as was *A.robustus,* a much larger form, whilst very crude stone tools are found in some of the breccias (Brain 1976).

Recently, Hughes & Tobias (1977) have reported finding at Sterkfontein the skull of a hominid very similar to *Homo habilis.* It was found, with associated stone tools, in Member 5 (2.0-1.5 my) of the Sterkfontein Formation, which is underlain by Member 4 (3.0-2.5 my) containing *Australopithecus africanus,* but no stone tools. Hughes & Tobias (1977) believe that this supports the view that *A.africanus* is ancestral to *Homo habilis* and to later species of man. Much effort has gone into trying to date these deposits but so far radiometric techniques have been generally unsuccessful so that H.B.S.Cooke (quoted in Tobias 1973) attempted to date the South African hominid deposits by correlating the faunas with radiometrically dated East African assemblages. He tentatively suggested that the Makapansgat and Sterkfontein breccias were in the vicinity of 2.5-3.0 my, and that Swartkrans was at least 1.8 my. Butzer (1974) suggested that Taung was much younger than the other cave deposits.

Partridge (1973), using geomorphological evidence, dated the opening of the cave, and thus established a maximum age for the deposits. His technique depends on an assessment of rates of cyclic erosion relative to the age of erosion surfaces ('datum levels') established by King (1967), and calculates mean rates of nickpoint recession and extrapolates the age of the arrival of the oldest nickpoint opposite the site. The ages deduced in my BP are: Makapansgat 3.67 my; Sterkfontein 3.26 my; Swartkrans 2.57 my; Taung <0.87 my. Partridge's technique and results have been strongly criticized by de Swardt (1974), but his estimate at Makapansgat has gained support from recent palaeomagnetic work by Brock et al (1977). They found that the second-oldest unit in the cave (Member 2) is between 2.8 and 3.3 my. Most of the *Australopithecus africanus* remains have been found immediately above this unit, in Member 3. Other dates for the Transvaal cave deposits are Vrba's (1974, 1975) estimates based on bovid remains which suggest that the fauna in the older breccia has a 1.5-2.5 my age range, and the younger breccia is probably <0.5 my old.

Several other primate taxa have been found at the Makapansgat and Sterkfontein sites in members 1-4, which Szalay & Delson (1979) and Howell (1978) have dated as covering the period 3-2.5 my BP. These have been referred to: *Cercopithecoides williamsi; Parapapio jonesi; P.broomi; P.whitei; P.*cf.*izodi; P. hamadryas robinsoni; Theropithecus darti; Cercocebus* sp. (Szalay & Delson 1979, Eisenhart 1974). As noted previously, the primate teeth found in the Quartzose Sand Member at the Varswater Quarry were considered closest to *P.jonesi* by Grine & Hendey (1981).

Two workers have recently proposed formal lithostratigraphic names for these cave deposits. Partridge (1975, 1978) has proposed a Makapansgat Formation and a Sterkfontein Formation, both with Members 1-6. These are lithologically distinct units, but are not mappable away from their respective caves. Butzer (1976) has described a Swartkrans Formation, consisting of Member I (the 'older breccia') and Member II (the 'younger breccia'). These formational names are now widely used amongst workers on these deposits and although the formal naming of isolated, unmappable rock units is a break with accepted stratigraphic practice, it is perhaps justified in this case by the unique palaeontological importance of the deposits.

Finally, Hendey (1974b) has proposed a formal system of mammal-age chronology based on relative ages of mammal assemblages in Southern Africa. The ages are: the Namibian, Langebaanian, Makapanian, Cornelian and Florisian. Namibian faunas are of Miocene age (e.g. the fauna found near Arris Drift, Bogenfels and Elizabeth Bay), Langebaanian fauna are of Pliocene age with no extant species (e.g. those in the Varswater Formation), Makapanian fauna probably span the Pliocene-Pleistocene boundary and record the first extant species (e.g. represented by the Australopithecine sites of the Transvaal). Cornelian and Florisian fauna are of Pleistocene age.

## 6.3 FURTHER ASPECTS OF TERTIARY SEDIMENTATION

In this section we will examine the rôle of various environmental factors such as climate, sea level movements, sediment supply, and oceanic circulation in determining the regional trends in Tertiary sedimentation in both time and space. Any repetition under the different headings will serve to emphasize the interrelationships that exist between these factors and their imprint on the stratigraphic record.

### 6.3.1 *Accumulation rates and lithofacies variations*

Tertiary sedimentation on the continental margin represents the youngest part of the epeirogenic phase which commenced in mid-Cretaceous times, and an assessment of facies variations and sedimentation rates needs to include comparison with Upper Cretaceous strata for long term trends to be fully appreciated. As previously described (Fig.154), the main Tertiary depocentres lay under the present continental slope and rise along the west coast, on the Agulhas Bank, and under the Tugela and Limpopo cones and their southern extensions off the east coast. A comparison with Figure 124 shows some distinct changes from their Upper Cretaceous positions.

On the west coast, the zone enclosed by the 1 km Tertiary isopach is considerably narrower off the Orange Basin than its Upper Cretaceous counterpart (maximum of 350 km compared to 470 km), but is wider SW of Lüderitz (200 km compared to 80 km). This results in a relatively symmetric swathe of Tertiary sediment along the whole outer western continental margin compared with a strongly asymmetric distribution of Upper Cretaceous sediments. This contrast is further emphasized by the fact that the Tertiary isopachs reach a maximum thickness of just over 1.5 km (and are generally between 1.0 and 1.5 km) compared to Upper Cretaceous values which vary from <2.0 km SW of Lüderitz to >3 km in the Orange and Walvis basins. Only on the Walvis Ridge abutment (Frio Ridge) where over 4 km are suspected, is there evidence of a sizeable disparity in Tertiary accumulation rates along the western margin. On the Agulhas Bank, both the Upper Cretaceous and Tertiary depocentres are narrow and elongate, but whereas the former reaches a thickness of > 1.5 km, the latter is only just over 1.0 km. Lack of data on the distribution of Upper Cretaceous strata in the southern Natal Valley inhibits a meaningful comparison in this area, but one significant difference farther north is the relative thickness of the Tertiary depocentre between the Limpopo River and the Almirante Leite Bank.

The main trends in Cenozoic sedimentation rates and lithofacies changes are summarized in Figures 184 and 185, and Table 64 using a 3-fold time division: early Tertiary (Palaeocene-Eocene, 28 my); middle Tertiary (Oligocene, 14 my); and late Tertiary (Miocene-Pliocene, 21 my). Values are averages for these periods. Lithofacies data in Figure 184 are from boreholes, onshore outcrops and sea floor samples. In gross terms, our data show that Tertiary sedimentation rates were significantly lower than those during the Upper Cretaceous, and that lithofacies changed from terrigenous- to biogenic-dominated. Whilst these trends were to some extent progressive, there are large local fluctuations, particularly in sediment accumulation rates. In addition, two areas (Walvis Ridge abutment-Frio Ridge, and the Limpopo-Almirante Leite Bank sector of the Natal Valley) show strong divergences from the regional picture.

Figure 184a summarizes the Palaeocene-Eocene lithofacies and average accumulation rates. The latter are highest on the mid-shelf areas and locally exceed 20 m/my for this 28 my period around the subcontinent. With the notable exception of the Frio Ridge, however, these values represent significant decreases in the Upper Cretaceous rates that range from 88 % (Orange Shelf: Kudu 9A-1) to 19 % (Tugela Cone: J(c)-1) of their former values. On the Frio Ridge, early Tertiary sediments accumulated at 169 % of their Upper Cretaceous rate in boreholes 362/363, and at probably much greater rates closer to the continent where a remarkable 71 m/my is estimated from seismic data. Figure 185 shows that similar, and in some cases (Orange Shelf – 100 %, Tugela Cone – 46 %), even larger reductions are evident across the Mesozoic-Cenozoic boundary when a more refined time scale (Maastrichtian-Palaeocene) is used. These figures will, of course, be misleading with regard to actual accumulation rates in areas where erosion has removed sediments, but in practice this can be expected to have *decreased* the disparity by apparently reducing Maastrichtian sediment thicknesses. With the exception of the Frio Ridge, therefore, a large, real reduction in sediment accumulation rates is noted on the continental margins across the Mesozoic-Cenozoic boundary, which, as we will discuss later, can be attributed to a major low sea level stand in early Palaeogene times.

The trend in falling sedimentation rates on the shelves was arrested in Eocene times, when several areas (e.g. Tugela Cone and Orange Shelf), show significant increases, although rates did not return to their Maastrichtian levels (Fig.185). On the Frio Ridge, Eocene sedimentation was at least at 183 % of the Maastrichtian value.

We have only one site in deep water with which to make comparisons (DSDP 360/361), but here similar trends are evident. An 18 % overall reduction in early Tertiary sediment accumulation compared with Upper Cretaceous rates can be broken down into a sharp 71 % decrease across the Maastrichtian-Palaeocene boundary, followed by a large 360 % increase in Eocene times to 23 m/my which is 135 % of the Maastrichtian rate, and is the only site (except for the Frio Ridge) which shows such a trend.

As noted above, Tertiary lithofacies show significant and progressive changes compared with Upper Cretaceous sediments, but although some of these are evident in Palaeocene-Eocene strata, rocks of this age can best be considered as transitional to the more impressive regional lithofacies alterations that occurred later. Figure 184a shows that the bulk of continental shelf sedimentation was dominated by terrigenous debris, with only local (e.g. East London and Zandamela-Domo) regions of shallow water carbonate accumulation. The deep water areas SW of Cape Town do, however, herald the shift to biogenic sedimentation with Eocene chalks and limestones (locally clay-rich) overlying the thin deep water Palaeocene clay at site 361. These lithofacies contrast with the Upper Cretaceous sequence which is a shale and mudstone sequence with only minor limestones. Clearly a significant decrease in the supply of fine grained terrigenous debris to the deep ocean basin along the west coast is indicated by this record. The decrease coincided with the large sea level drop at the end of Maastrichtian times (Fig.186), but was not reversed during the subsequent late Palaeocene-early Eocene rise. The latter event was, however, coeval with the lowering of the carbonate compensation level in the area, which allowed carbonate debris to accumulate at DSDP sites 360 and 361, although at the latter, Middle Eocene strata are again clay-rich (Fig.166), possibly due to lysocline fluctuations that had a greater effect on the deeper water site.

From this evidence it appears that during Lower Tertiary times, with the exception of the Frio Ridge, all the marginal areas around southern Africa suffered a decrease in terrigenous sediment supply and a slow down in crustal subsidence rates (there is no evidence for deep water facies on the shelves) compared with the Upper Cretaceous situation. Despite this, terrigenous components still dominated most areas of shelf sedimentation, whereas biogenic carbonates replaced clays in the deep sea, showing that the detritus that

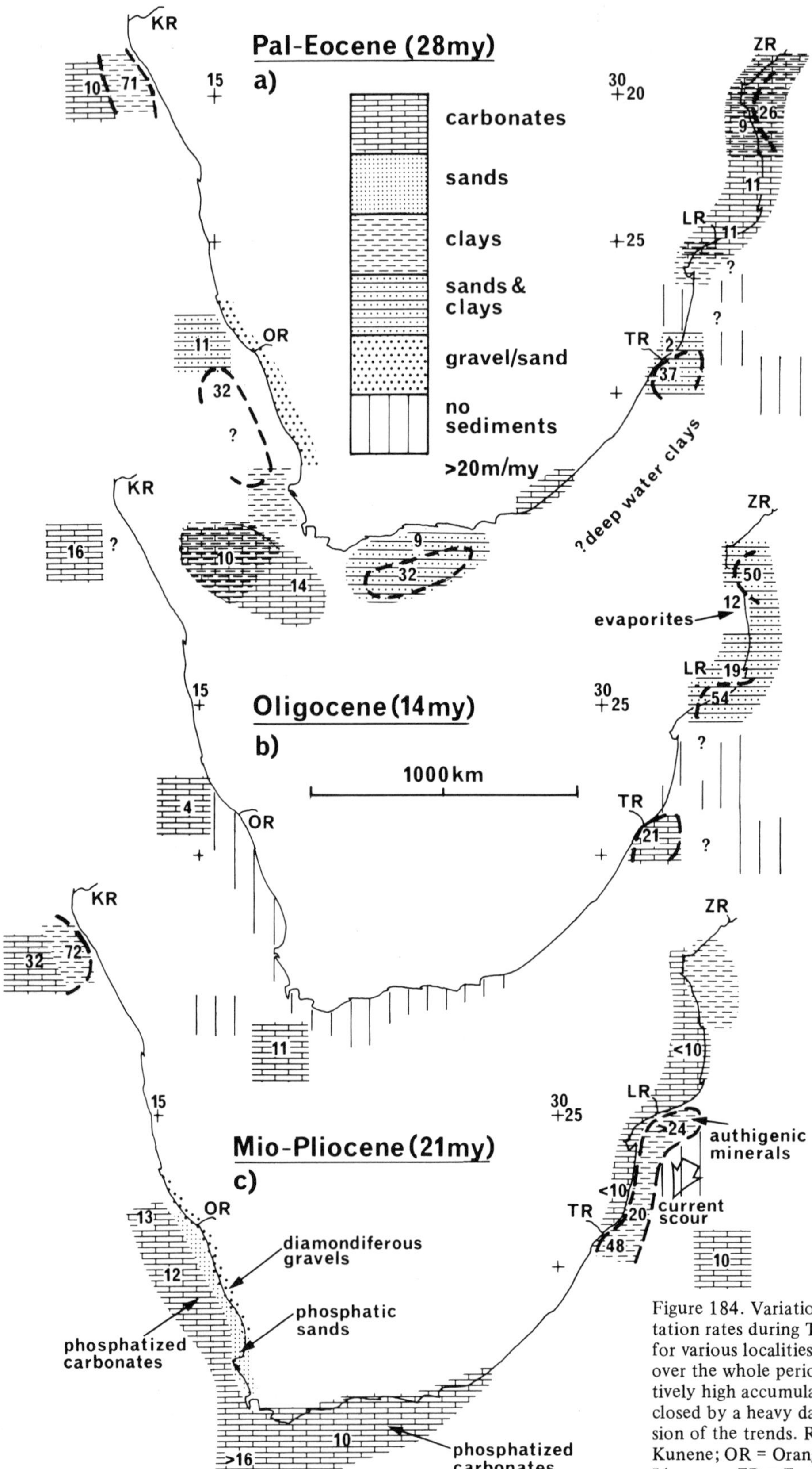


Figure 184. Variations in lithofacies and sedimentation rates during Tertiary time. Rates are shown for various localities in m/my and are averages over the whole period specified. Areas of relatively high accumulation (>20 m/my) are enclosed by a heavy dashed line. See text for discussion of the trends. Rivers are shown: KR = Kunene; OR = Orange; TR = Tugela; LR = Limpopo; ZR = Zambezi.

rivers did bring to the estuaries was largely retained in the continental margin environment.

Figure 184b shows that during Middle Tertiary (Oligocene) times sedimentation was maintained at a significant level only in the NE part of the subcontinent. This period coincides with a major sea level fall which resulted in only thin and isolated patches of sediment on the south and west coast margins that overall indicate nil or very low rates of accumulation in carbonate lithofacies. The sedimentation rate in the deep water areas apparently maintained its early Tertiary levels (11 m/my), and although the original status of the Oligocene succession at DSDP site 360, which is allochthonous, is not precisely known there is little doubt that it accumulated in a continental slope environment.

Even the relatively high Oligocene accumulation rate of 13 m/my at J(c)-1 on the Tugela Cone top represents a 25 % decrease of the Palaeocene-Eocene value, with the youngest-preserved Oligocene beds consisting of shallow-water carbonates. This contrasts with the situation that obtained in coastal southern Mozambique where Oligocene sedimentation rates locally show sharp increases compared to early Tertiary values: 390 % off the Limpopo River, and 92 % SE of the Zambezi River. Local regressions were also evident, with development of the Temane evaporites. Whilst it is possible that such high accumulation rates were solely the result of increased sediment discharge from the large rivers, a more likely explanation is thought to be local crustal downwarping associated with renewed movement along boundary faults in the Chissenga Graben (Fig.136), and undefined structures in the northern Natal Valley. Strong bottom currents appear to have maintained a sediment-free region on the Central Terrace between the Limpopo and Tugela Cone depocentres in the Natal Valley.

The return to widespread continental margin sedimentation off the west and south coasts in late Tertiary (Mio-Pliocene) times confirmed the establishment of the two trends that were evident before the Oligocene hiatus, viz: the subordination of terrigenous sedimentation; and relatively low sedimentation rates. Again with the exception of the Frio Ridge, late Tertiary sedimentation rates on the Orange and Agulhas shelves were as low, or in some cases lower, than their early Tertiary counterparts. Generally, the sediments from these areas are poor in terrigenous debris, but rich in biogenic carbonate and, in Upper Miocene and younger rocks, authigenic phosphate and glauconite, as well as, locally, fish debris and diatom frustules. Clearly, a combination of low terrigenous input, low crustal subsidence and oceanic upwelling combined to produce an unusual sedimentary environment which obtains in several essential features at present. The deposition of diamondiferous gravels in the littoral zone was probably most strongly influenced by the very low mid-Tertiary sea level stands, and Neogene distribution represents reworking of the original Oligocene deposits. Figure 184c emphasizes the dominance of biogenic sedimentation in both shallow water and deep water environments off the west coast in late Teritary times. This occurred at long term rates that were marginally higher than early Tertiary values (>16 m/my vs 14 m/my), although data from site 360 suggest a sharp decrease (to 3 m/my) during the Pliocene interval.

Although sedimentation rates on the Frio Ridge continued to be out of step with the regionally low values along the rest of the west coast margin (an increase of 220 % over the early Tertiary rate at DSDP site 362) the establishment of upwelling in the Benguela Current system, as traced in biogenic and related components, can be seen *par excellance* in this region (Fig.166, Siesser 1978b, Diester-Haas & Schrader 1979). It is the climatic deterioration occasioned by this oceanic phenomenon that played an important role in determining the sedimentary environments of the west coast areas and we will discuss it in more detail under the palaeoclimate section (6.3.4).

Mio-Pliocene sedimentary environments in southern Mozambique and Natal also show some disparities in both lithofacies and accumulation rates (Fig.184c). Whilst nearshore and present day coastal regions indicate slow, discontinuous carbonate-rich marine and aeolian sedimentation during the late Tertiary, the Limpopo and Tugela rivers appear to have discharged terrigenous detritus at similar, or even high rates than during earlier Tertiary times. This is particularly so of the Limpopo, where seismic data indicate increased accumulation in Pliocene or younger times. Purely oceanic factors influenced the lithofacies and accumulation patterns by promoting authigenic mineral production to the east of the Limpopo Cone, whilst at the same time continuing to sweep the Central Terrace clear of any significant sedimentation. The result has been the development of a rapidly growing detrital wedge along the base of the Zululand slope that links the Mio-Pliocene parts of the Limpopo and Tugela cones. One unknown factor at present is the quantity of Quaternary sediments involved, especially in the case of the Limpopo Cone.

Summarizing Tertiary sedimentation, two overall trends emerge:

i) a progressive decrease in sedimentation rates on the continental shelves, so that by the end of Tertiary times they were between 20 and 60 % of their Upper Cretaceous values. Anomalous areas lie on the Walvis Ridge abutment (at least 442 % increase) and in the northern Natal Valley (at least 60 % increase). In the Cape Basin, sediment accumulation upslope of site 360, on the continental rise, was at similar rates during Upper Cretaceous and late Tertiary times (17, and at least 16 m/my respectively), but in more distal areas (site 361) there has been no net post-Eocene sedimen-

Table 64. Tertiary sediment accumulation rates.

| | Sedimentation rates m/my | | | | Changes compared to U.Cret. | | | | | | Rate of change | | | | | |
|---|---|---|---|---|---|---|---|---|---|---|---|---|---|---|---|---|
| | U.Cret. | e.Tert. | m.Tert. | l.Tert. | eT/ U.Cret. | % Change | mT/ U.Cret. | % Change | lT/ U.Cret | % Change | eT/ U.Cret. | % Change | mT/ eT | % Change | lT/ mT | % Change |
| Frio Ridge[1] | 5.9 | 10 | 16 | 32 | 1.69 | **+69** | 2.71 | +171 | 5.42 | **+442** | 1.69 | **+69** | 1.60 | + 60 | 2.00 | **+100** |
| Cape Basin[2] | 17.0[3] | 14 | 11 | >16 | 0.82 | *−18* | 0.64 | − *36* | 0.94 | − *6* | 0.82 | *−18* | 0.79 | − *21* | 1.45 | **+ 45** |
| Orange Shelf[8] | 85.7[3] | 10.3 | 4 | 17 | 0.12 | *−88* | 0.05 | − *95* | 0.20 | − *80* | 0.12 | *−88* | 0.39 | − *61* | 4.25 | **+325** |
| Agulhas Bank[7] | 42.9 | 32 | 0 | 10 | 0.75 | *−25* | 0 | *−100* | 0.23 | − *77* | 0.75 | *−25* | 0 | *−100* | inf. | |
| Tugela Cone[4] | 21.4 | 17.4 | 13 | 13 | 0.81 | *−19* | 0.60 | − *40* | 0.60 | − *40* | 0.81 | *−19* | 0.75 | − *25* | 1 | 0 |
| Limpopo River mouth[5] | 15.0[6] | 11 | 54 | >24 | 0.73 | *−27* | 3.60 | **+260** | >1.6 | **> 60** | 0.73 | *−27* | 4.90 | **+390** | <0.44 | < *56* |

bold = increase; italics = decrease; 1 = Maastrichtian value taken at DSDP 363; 2 = DSDP 360 + 361; 3 = Cen.-Maas (35 my); 4 = J(c)-1; 5 = Zandamela; 6 = L.Coniacian-U.Palaeocene (Grudja Fm); 7 = Outer Agulhas Bank; 8 = Kudu borehole + seismics. e = early; m = middle; l = late; T, Tert. = Tertiary; U.Cret. = Upper Cretaceous. '% Change' indicates percentage of former value.

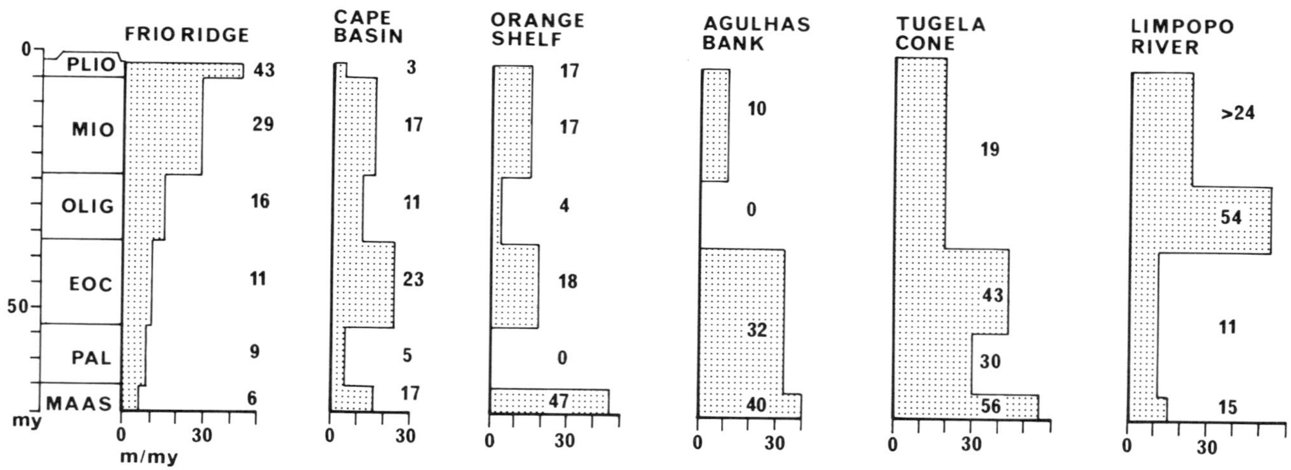


Figure 185. Comparison of Tertiary sedimentation rates for various localities. Rates are shown in m/my, and Maastrichtian values are added for comparison. Frio Ridge rates are calculated at DSDP site 362, and in the Cape Basin a composite histogram combines Palaeocene from DSDP 361 and Eocene to Pliocene from DSDP 360. Orange Basin rates are from Kudu 9A-1 for the Maastrichtian to Lower Oligocene and seismic data for the Neogene. Agulhas Bank rates are maxima for the area and have been gauged from seismic thicknesses in units TL and TU. Maximum values from seismic sections adjacent to J(c)-1 have been used for the Tugela Cone, and for the Limpopo area the Zandamela borehole and seismics in the adjacent Natal Valley were used.

tation. This indicates either no actual post-Eocene accumulation, or later erosion. Physical property tests by the DSDP team (Bolli et al 1978) suggest the latter explanation;

ii) a progressive change from terrigenous to carbonate and/or authigenic-rich sediments in all areas, with the possible exception of the Limpopo Cone. This trend is evident earlier in the deep basin environments.

The main factors which determined these trends, and upon which we will briefly comment are:

i) rate of terrigenous sediment input onto the continental shelves;

ii) rate of continental shelf subsidence and the creation of basins;

iii) reorganization of ocean current patterns.

The rate of terrigenous input onto the shelves depends on the hinterland's climate (which controls precipitation/runoff, and rock weathering) and topography. King (1967) has identified three erosional cycles during Upper Cretaceous to Recent times (post Gondwana, African, and post-African) which have been tentatively dated as representing mid-Cretaceous, late Cretaceous/early Tertiary, and early Miocene periods of relative sea level movements, respectively. The tectonic significance of these has been challenged (e.g. de Swart & Bennet 1974), but their validity is generally accepted by geomorphologists, although no modern quantitative studies have been carried out on the problem. Crustal uplift of the kind envisaged by King, or alternatively, eustatic sea level changes would have created major topographic discontinuities in the coastal hinterland, leading to particularly vigorous erosion and sediment shedding in Palaeocene and Miocene times. As we will discuss in more detail below, palaeoclimatic data indicate warmer, wetter conditions during the early Tertiary, with marked aridity beginning (particularly in the west) in Miocene times. A combination of these two lines of evidence suggest that the terrigenous supply to the continental margins would have been higher in early Tertiary than in late Tertiary times. This in fact is what appears to have occurred, with carbonates largely replacing terrigenous deposition after mid-Tertiary times, although high terrigenous supply may have continued through the Neogene on the NE coast. In absolute terms, we can quote only modern supply rates as a basis for comparisons (Tab. 65) although there is considerable disparity in the figures available. Dingle & Scrutton (1974) made estimates based on the data provided by Midgley & Pitman (1969), who assumed a 25 % bedload component, and gave a figure of 60.5 x $10^7$ $m^3$ per annum as the terrigenous input to the sea for the southern African river system. 40 x $10^7$ $m^3$ (66 %) was calculated as the contribution of the east coast margin, but a more recent evaluation by Flemming (1981) based on the data of Rooseboom (1978), gives the latter as 10 x $10^7$ $m^3$ (he does not discuss the south or west coasts). For the purposes of the present discussion we will use the figures of Flemming (1981) for the east coast, and the figures of Dingle & Scrutton (1974) for relative values on the south and west coasts. Currently, the west coast continental margin between the Walvis Ridge and Cape Agulhas receives only 30 % of the river borne sediment load eroded from the southern Africa land surface, with the Orange River (16.5 x $10^7$ $m^3$ pa) as the only significant source. Despite this quite considerable input at one point, the modern Orange River 'delta' is a very small feature (1 500 $km^2$) (Hoyt et al 1969), and studies of the sediments of the adjacent continental shelf (Rogers 1977) show that the bulk of the sand fraction brought down the river is moved northwards by long shore drift and most is blown back from the beaches into the Namib desert. Fine material (clays) is carried south by nearshore currents and deposited between the Orange and Olifants rivers. This dispersion pattern is a direct result of the shallowness of the wide continental shelf and the accompanying high energy water column. It follows, therefore, that lack of Neogene terrigenous sediment accumulation on the shelf off the Orange River is at least partly due to the fact that the shelf has subsided only slightly since mid-Tertiary times and is not solely a consequence of low

Table 65. Modern sediment loads of southern African rivers ($m^3$ x $10^7$ p.a.).

| | Run off | Terrigenous detritus | | | | % | |
|---|---|---|---|---|---|---|---|
| | a | a | | b | | a | |
| Orange River | 1 030 | 16.5 | | | | 27 | 30, west coast |
| S.W.Cape (mainly Berg & Olifants) | 91 | 1.5 | | | | 3 | |
| S.Cape (Breede, Gouritz, Gamtoos, Sundays) | 465 | 2.5 | | | | 4 | south coast |
| S.E.Cape, Transkei & Natal | 1 350 | 29.0 | | 3.4 | | | |
| Tugela | 475 | 6.0 | 40 | 0.4 | 10 | 66 | east coast |
| Zululand & S.E.Mozambique } Limpopo | 1 523 | 5.0 | | 1.3 | | | |
| | | | | 4.9 | | | |

a – From Dingle & Scrutton (1974), based on the data of Midgley & Pitman (1969).

b – From Flemming (1981), based on the data of Rooseboom (1978), using a 5 % bed load component. Flemming estimates biogenic input at 15 % of terrigenous value i.e. 1.5 x $10^7 m^3$.

river-borne terrigenous input. Wind-borne terrigenous input probably plays a significant role off the west coast, where red/brown silt and clay size particles frequently give rise to extensive low level dust layers many kilometres out to sea off Namaqualand and Namibia, but no systematic studies have yet been made to quantify this process (Chester & Johnson 1971).

These modern sediment patterns, and their climatic controls along the west coast were initiated by early Late Miocene times (see next section). The resultant lack of dilution by large quantities of terrigenous matter, together with the geochemical environment attendant on the active upwelling of the Benguela Current have produced ideal conditions for the formation of authigenic minerals (glauconite and apatite) and carbonate-rich sediments.

A somewhat similar, sediment-starved environment obtains along the south coast, where the rivers between Cape Agulhas and Algoa Bay carry only 2.5 $m^3 \times 10^7$ of detritus per annum (4 % of the subcontinent total), in addition to which, there is no significant aeolian input. Modern mud and sand are retained in the nearshore area as narrow, elongate prisms and swathes off the main river mouths, whilst the mid and outer shelf are current swept with relict and residual sediment cover derived from late Tertiary and Pleistocene sediments. Modern material largely consists of biogenic debris. Following the accumulation of moderately thick clastics in Lower Tertiary times, crustal subsidence under the Agulhas Bank slowed, creating a shallow, flat-topped feature with favourable conditions for Neogene authigenic- (glauconite and apatite) and biogenic-rich sediment formation.

The east coast (Cape Padrone to Limpopo) currently receives about 66 % of the river-borne material from southern Africa, 34 % of which enters the sea south of the Tugela (Tab.65). Flemming (1978, 1980b) has shown that the narrow continental shelf off Natal and Transkei, which is swept by the strongly flowing Agulhas Current (up to 4 kt), moves large quantities of sand parallel to the coast in extensive dune fields. These display a wide variety of bed forms (the largest dunes reach a height of 17 m with a wave length of 700 m) and rest on a surface of scoured late Tertiary carbonate gravels. Replenishment of the dune field is by sediment migration from the nearshore sand prism reservoirs by meandering of the current. The reservoirs are constantly topped up from the rivers, and once in the main sand stream the sediment is carried along the continental shelf until it is ultimately funnelled down the continental slope, presumably via canyons. Despite the very large quantities of material being supplied to the shelf, little or none permanently accumulates because of the high energy of the current regime. A sediment budget for the fine material brought down by the rivers has not yet been worked out, but this fraction is not accumulating on the shelf, though in the Durban area there is some evidence that locally it is carried northwards. Off the two largest input points (Tugela $0.44 \times 10^7$ $m^3$ pa and Limpopo $4.9 \times 10^7$ $m^3$ pa) late Tertiary sediment accumulation has maintained relatively high rates (48 m/my and >24 m/my, respectively).

Looking at the question of sediment input in a regional context identifies a problem of 'over supply'. Taking the effective area of Tertiary sediment accumulation off SE Africa (Fig.154) as 214 000 $km^2$ (130 000 $km^2$ south of the Tugela, and 84 000 $km^2$ north of the Tugela), and assuming an average thickness of 1.5 km for Cenozoic strata, a volume of 321 000 $km^3$ of sediment has accumulated in the Natal Valley north of Cape Padrone during the last 65 my (ca.23 m/my). At the rate of present supply (0.1 $km^3$/year), this volume represents only 3 210 000 years input. Even a theoretical increase in the sedimentation area (for instance, to include the Central Terrace) and a doubling of the average sediment thickness produces only a volume that could be provided by 12 my supply at the present rate. Three possibilities present themselves:

i) the discharge estimates are wrong;
ii) that most (95 %) of the input is lost from the system and is deposited outside the Natal Valley;
iii) that modern sediment input rates are much larger than they were during most of Tertiary time.

Possibility (i) must be conceded, but it is unlikely that errors as large as twenty times have been made. Similarly, whilst some of the fine sediment brought to the Natal Valley by rivers may be carried out of the area, we have no evidence for routes by which the bulk of modern input is lost, and the studies of Martin et al (1981) suggest that the modern circulation patterns have been operative since at least mid Tertiary times.

We conclude, therefore, that modern erosion rates are significantly higher than the Tertiary values, which must have produced an average sediment input of about $43 \times 10^5$ $m^3$/year (or $185 \times 10^5$ $m^3$/year for the alternatively larger volume discussed above). Speculation on the reasons for such relatively high modern values could range from climatic differences to agricultural activity, and cannot be resolved without further study.

A major influence on regional sedimentation is the oceanic climate: it controls the biogenic productivity of the water column, as well as playing an important role in the run off and weathering characteristics of the adjacent hinterland. We will examine the question later from a palaeoclimatic viewpoint, but here will mention some of the geochemical and sediment dispersion changes that were attendant on the establishment of the modern surface current patterns.

The major current affecting the west coast and its continental margin is the Benguela, which is defined 'as the region of cool upwelled water found within

100 nautical miles of the west coast between 15 and 34°S' (Shannon 1966, Hart & Currie 1960). This is both a surface and a deep current, flowing northward with a surface velocity of 50-100 cm/sec. Today, it is a zone characterized by subsurface upwelling over the shelf break and spasmodic surface upwelling close to shore. Upwelling is caused when the wind blows parallel to the shore, as it normally does along the west coast, and produces a leftward (seaward) divergence of water, that results in an inshore deficiency of surface water. Atlantic Central (or, less likely, Antarctic Intermediate) water rises to replenish the shortage and during maximum periods of upwelling, the current's temperature and salinity are as low as 9°C and 34.6‰, respectively.

Upwelled waters are rich in nutrients, especially nitrates, phosphates and silicates, which are vital to the growth of marine phytoplankton. Today the vast numbers of phytoplankton in the Benguela Current area provide the initial food link which makes this area one of the world's richest fishing grounds. The most important sediment-forming plankton are diatoms and calcareous nannoplankton (phytoplankton), and planktonic foraminifera (zooplankton). It is believed (e.g. Siesser 1978b, Diester-Haas & Schrader 1979) that initiation of major Benguela upwelling resulted in a considerable expansion in production of these organisms, with a concomitant increase in supply of mostly calcareous (the nannoplankton and foraminifera) sedimentary components.

We consider this to have been the main factor in Neogene increased sediment accumulation on the Walvis Ridge abutment, and large-scale generation of authigenic minerals on the continental shelf, for which massive supplies of organically-derived phosphorus and potassium were necessary. At the present time, locally intense upwelling in the vicinity of Walvis Bay has resulted in large areas of the sea floor being blanketed in a thick sludge rich in organic carbon (> 5 %), phytoplanktonic and fish debris, and fine aeolian material (Bremner 1978a, 1980a,b). These unusual sediments have high trace element concentrations: Cu > 120 ppm, Ni > 300 ppm, Pb > 16 ppm, Zn > 100 ppm (Calvert & Price 1971b) and U >10 ppm (Veeh et al 1974). Because upwelling has been intense along the whole of the west coast since late Miocene times (Siesser 1978b, Diester-Haas & Schrader 1978), similar organic-carbon and trace element-rich sediments have probably accumulated at several places during the last 10 my. Their economic potential has not been assessed, but may prove considerable.

Van Zinderen Bakker (1975) has suggested that the Benguela Current originated in early Oligocene times, initiating aridification of the west coast of southern Africa and creating, inter alia, the Namib Desert. This timing is largely based on isotopic evidence from Shackleton & Kennett (1975), who found that deep-water temperatures in high southern latitudes were near the present low values by the beginning of the Oligocene, but not before. Van Zinderen Bakker (1975) also believes that only by mid-Tertiary (Oligocene) times did the South Atlantic have sufficient longitudinal width for the establishment of its circular wind and ocean current systems.

Siesser (1977c, 1978b) has presented a variety of data including corrected sediment accumulation rates, diatom frustule abundances, planktonic foraminifera and calcareous nannofossil temperature preferences, primary productivity (expressed in $C_{org}$) and phosphorus incorporation in calcareous skeletons plotted against time for a DSDP Site (362) under the direct influence of the Benguela Current. These sedimentological, palaeontological and geochemical data suggest weak, spasmodic introduction of cool, upwelled waters along this coast from middle or late Oligocene until middle Miocene times. In the early late Miocene, conditions changed markedly, suggesting intensification of upwelling which brought cold, nutrient-rich waters to the surface. This can be seen in Figure 166 where diatom-rich layers have been identified at DSDP Site 362.

The effects of the Agulhas-Mozambique currents on sedimentation are less easy to surmise. In the Indian Ocean, the Trade Wind Drift (South Equatorial Current) hits the African coast and is deflected both north and south. The southward-flowing branch is split by Madagascar. One branch, consisting of numerous discontinuous eddies, flows between the mainland and Madagascar and eventually rejoins the Mozambique Current south of Madagascar. The rejoined current becomes the strongly flowing Agulhas Current (e.g. Lutjeharms & Valentine 1981). This current is both a surface and a deep (at least to 2 500 m) feature. It flows above and parallel to the shelf break off Natal at velocities up to 4 kt (Bang & Pearce 1976). About 300 km south of Cape Agulhas, the major part of the current veers to the south, then east to become the Return Agulhas Current. A minor branch of the Agulhas Current continues westward, eventually rounding Cape Point.

The surface waters of the Agulhas Current are poor in nutrients and have a low biologic productivity. Only weak, spasmodic surface upwelling occurs in the Agulhas Current today, and it is localized and seasonal (Shannon 1966). However, sub-surface upwelling (deep oceanic water is brought up and over the shelf edge) is a common feature of the Agulhas Current. This is known as 'dynamic upwelling' and occurs in the Southern Hemisphere when a strong current flows with the coastline on its right side. This sub-surface upwelling brings phosphate-rich waters onto the Agulhas Bank. It has been suggested (Siesser 1978a) that this mechanism was the source which provided phosphorus for the abundant phosphorite rocks that now litter the Agulhas Bank. As the dated phosphorites are probably all Miocene-Pliocene in age, subsurface upwelling may have

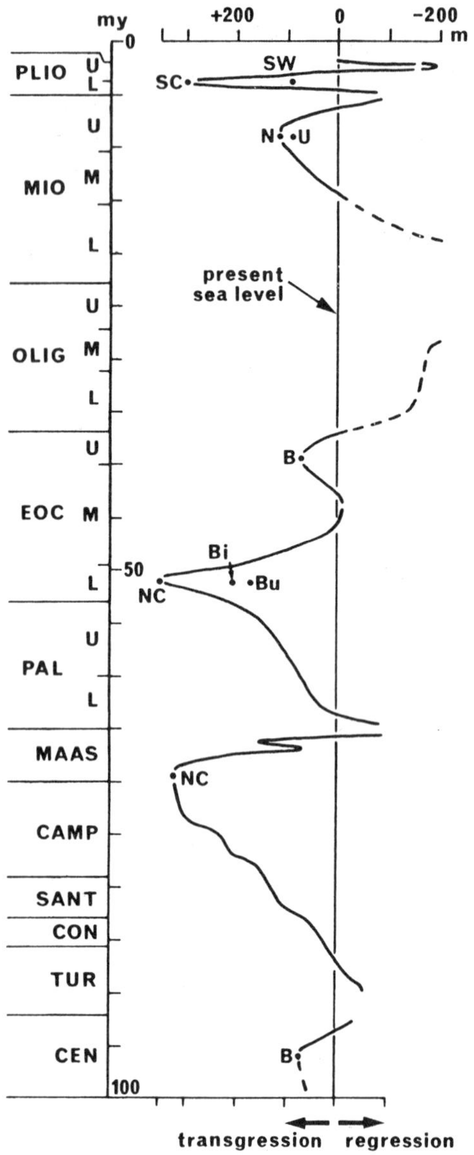


Figure 186. Upper Cretaceous and Tertiary sea level movements plotted in relation to present sea level. Slightly modified from Siesser & Dingle (1981). See text for discussion. Localities: B = Bogenfels; NC = Needs Camp; Bu = Buntfeldschuh; Bi = Birbury; N = Namaqualand coastal terraces; U = Uloa; SC = south coast; SW = S.W.Cape localities.

been more intense during those epochs. Martin et al (1982b) have shown that the Agulhas Current has played an important role in sweeping the Central Terrace of the Natal Valley clear of sediment during much of Neogene time, implying that it was initiated by at least the middle Tertiary.

### 6.3.2 *Tertiary sea level movements*

Numerous authors (e.g. King 1967, Haughton 1969, Truswell 1970) have discussed the problem of Tertiary sea level movements around southern Africa, but probably the first attempts to summarize the whole Tertiary on a regional basis were made by Dingle (1971a, 1973c) and Dingle & Scrutton (1974); when major regressions and accompanying hiatuses were postulated for the late Maastrichtian-mid Palaeocene, late Eocene-early Miocene, late Miocene-early Pliocene, and late Pliocene. Since this work, almost 200 additional dates have become available from the Tertiary rocks on the continental shelf (Siesser 1977a) and onshore areas (Siesser 1977b, and Siesser & Miles 1979), and several boreholes on the continental margin have been made public (du Toit 1976, McLachlan & McMillan 1979). New interpretations of these data have been published by Siesser & Dingle (1981), whose sea level curves have been amended to include Hendey's (1981a) data and are shown in Figure 186.

Definite lowermost Palaeocene (Lower Danian) has not been found either onshore or on the continental shelf around southern Africa, but 'middle' Palaeocene (Upper Danian/Montian) occurs in Mozambique (Grudja and Mangulane formations), at Richards Bay (Upper Danian), and off Cape St Lucia on the continental shelf. This pattern suggests that the regression initiated in late Cretaceous times continued into the Palaeocene (Lower Danian).

Eocene rocks are known onshore from Mozambique (Lower and Middle), Birbury (Lower), Bogenfels (Upper) and from several places on the continental shelf, with a short hiatus between Lower and Middle Eocene in the boreholes on the Agulhas Bank. This suggests that the Palaeogene transgression continued into the Lower Eocene, at the end of which there was a brief regression. The seas moved shoreward again in the Middle Eocene and reached a second Palaeogene high (e.g. Bogenfels) during the Upper Eocene. An assessment of the extent of these two Lower Tertiary transgressions has to be based on scanty data, but points to the earlier one being highest: 360 m (Needs Camp), 204 m (Birbury) and 163 m (Buntfeldschuh); versus 70 m (Bogenfels) and 28 m (Zululand-reworked into Miocene) for the Upper Eocene outcrops (Fig.186).

A major mid-Tertiary regression spans all or most of Oligocene time. Locally, some marine Oligocene rocks are reported in Mozambique (Inharrime Formation), but none onshore in South Africa or Namibia. They are similarly rare on the continental shelf, being found only in the J(c)-1 borehole, south of Plettenberg Bay (a single Upper Oligocene sample) and in Kudu 9A-1 (Lower Oligocene). Evidence from the J(c)-1 borehole where marine limestones overlie deltaic deposits led du Toit & Leith (1974) to suggest an early Oligocene transgression, but data from all other areas militate against this possibility. An alternate explanation for limestone over deltaic deposits is simple cessation of delta growth and the inevitable 'transgressive' covering by the sea, following delta subsidence and wave erosion. Siesser & Dingle (1981) report that erosion levels correlated with the mid-Tertiary regression can be recognised on seismic records down to depths of between ca.530 and 425 m in the sediment basins on the continental margin, and at ca.375 m on the flanks of the Agulhas Arch.

Miocene sea-level movements are a subject of some debate in southern Africa. In Mozambique, Flores (1973) and Frankel (1972) favour a middle-late Miocene transgression, followed by a late Miocene-Pliocene regression. Conversely, Förster (1975a) believes transgression occurred in early Miocene times, followed by regression, with short, local transgressions in middle Miocene-Pliocene times. A similar controversy exists in Zululand, where King (1953 et seq.) favours an early Miocene transgression, followed by erosion, then a Pliocene transgression. Frankel (1968) sees a middle or late Miocene to early Pliocene transgression in the same rocks. On the basis of onshore fossil dating by Stapleton (1977c), Maud & Orr (1975) and Siesser & Miles (1979), it appears that the seas only reached Zululand in late Miocene times at the earliest.

Other deposits, on the south and SW coasts (e.g. Alexandria-Bathurst, and Varswater formations) are partly Miocene, but their precise age is difficult to establish. Corvinus & Hendey (1978) and Hendey (1978) mention finding serpulid polychaete worm (cf. *Mercierella*) tubes in the Middle Miocene deposits at Arris Drift, on the Orange River about 30 km from the river mouth. This is a typically estuarine worm, and the implication is that the coastline was much closer to Arris Drift than it is today. However, estuaries can extend up rivers for a considerable distance, and it is difficult to estimate the actual height of sea-level based on these worm tubes. Nevertheless, the presence of the estuary, together with the height of the river terraces (some 40-50 m above present sea level), suggest a Middle Miocene sea level higher than today. Strong support for a middle-late Miocene age predominates on the shelf (rocks of early and late Miocene age occur, but are comparatively few in number).

In summary, we believe that the seas began to move landward in middle Miocene times, perhaps reaching the present coast during the middle Miocene, and certainly by late Miocene. This overall transgression continued into early Pliocene, although broken by a brief regressive pulse close to the Miocene-Pliocene boundary.

The Pliocene transgression is well-represented in the Mozambique, whilst some of the Alexandria Formation from the south coast is also Pliocene. Only Lower Pliocene has been definitely recognized among the Pliocene rocks on the continental shelf. In the Varswater/Saldanha area on the west coast, the transgressive Elandsfontyn Formation of ?early to middle Miocene age is separated from the transgressive early Pliocene (ca.5 my) Quartzose Sand Member of the Varswater Formation by the Gravel Member, which Hendey (1981a) considers a regressive lithofacies of late Miocene age (its vertebrate fauna indicates a maximum age of ca.12.5 my). Siesser & Rogers (1976) find evidence for a brief westward shift in the Benguela Current system in early Pliocene times which they suggest was caused by a short regression. Further regression with strong erosion is inferred for late Pliocene times, based on widespread regressive aeolian deposits onshore (e.g. Alexandria Formation aeolianties) and the destruction and re-working of offshore phosphorite beds.

We do not imply that the cycles of Tertiary sea-level changes outlined above are solely eustatically controlled, but consider that some element of the fluctuations may have been the result of periodic local isostatic imbalances. Thus the sea-level curve in Figure 186 should be considered as 'generalized', showing the gross movements of sea level around Southern Africa during the Tertiary. Local uplift or subsidence may have caused one area of the coast to submerge or emerge earlier than another area. The convergence of the Mio-Pliocene beach terraces along the north Namaqualand-south Namibia coast illustrates the degree of differential crustal movement that can be detected in well-documented regions (Fig.180). Nevertheless, it is interesting to compare our sea-level curve with that proposed by Quilty (1977) for Western Australia. He finds the first Tertiary transgression beginning in 'middle' Palaeocene times (P4) and continuing into earliest early Eocene. Sedimentation ceased in late Lower and early Middle Eocene. Late Eocene was a time of major transgression in Western Australia, and was followed by major regression during the early Oligocene. Seas transgressed again from late Oligocene to middle Miocene times. Middle Miocene was the second great Tertiary transgression in Australia. A late Middle or early Upper Miocene hiatus is followed by further transgression in the late Miocene which more or less continued into the Quaternary (the Pliocene is apparently not as well documented as earlier epochs).

The similarities between South Africa and Western Australia with respect to pre-Pliocene sea-level movements are striking. In both regions, the first major Tertiary transgression began in middle Palaeocene times and appears to have peaked around the Palaeocene-Eocene boundary. Little Middle Eocene sedimentation occurred in either region, but both experienced a major transgression during the late Eocene. Again, both regions experienced a major regression during the Oligocene. Here the only significant difference between the two areas arises: we believe this regression continued into the early Miocene around southern Africa whereas Quilty reports renewed transgression in middle Miocene times. The Western Australian sequence is broken by a brief hiatus in late Middle or early Upper Miocene, whereas we have suggested a regressive pulse closer to the Miocene-Pliocene boundary in southern Africa.

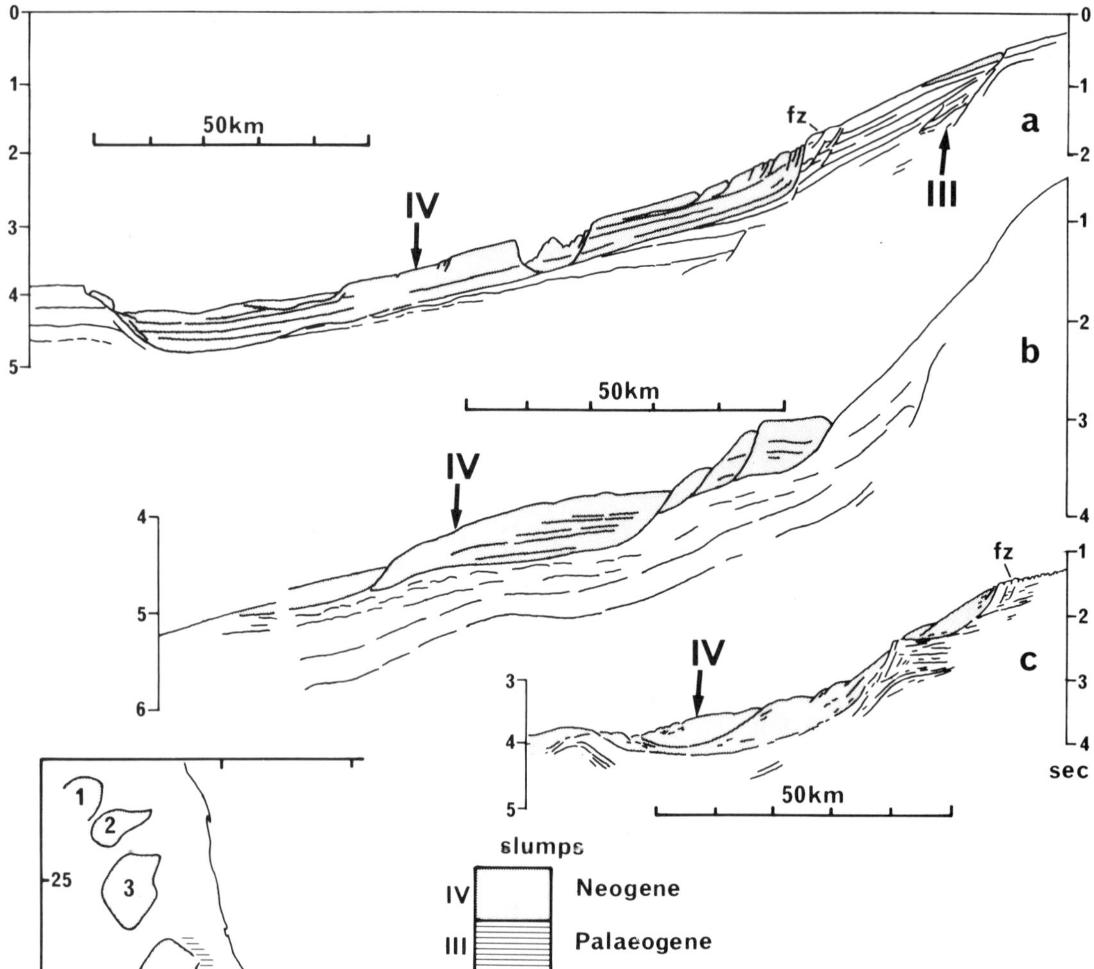


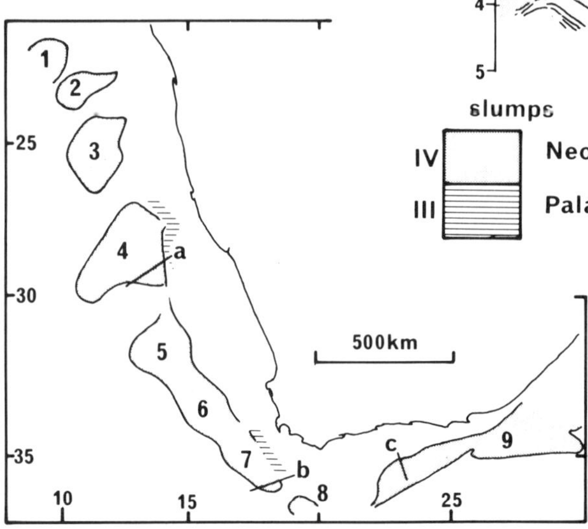


Figure 187. Cross-sections and distribution of Tertiary slumps on the continental margin of S. and S.W. Africa. Insert map shows the outline of Neogene features (cycle IV) and areas where Palaeogene features (cycle III) have been recognized. Table 66 sets out the stratigraphic relationships of the west coast slumps (Upper Cretaceous cycles I and II, to Tertiary cycles III and IV). Cross-section (b) is SW of Cape Town and shows the slumped mass that was penetrated by DSDP borehole 360. Data from Dingle (1977, 1980b). Slump locations: 1 = unnamed; 2 = Walvis Bay; 3 = Conception Bay; 4 = Chamais; 5 = Childs Bank; 6 = unnamed; 7 = Cape Town; 8 = unnamed; 9 = Agulhas. fz = fissured zone.

### 6.3.3 *Continental margin slumps and other allochthonous sedimentary features*

Tertiary sediments on the outer continental margins of the west and south coasts have been involved in allochthonism (slumps, slides, debris flows) on a massive scale (Fig.187). The full extent of these structures has not yet been determined and they may also occur on the east coast slope, but several examples are known in enough detail for their significance as mechanisms capable of modifying the morphology and sedimentary succession on the outer continental margin to be assessed (e.g. Dingle 1980b, Tab.66).

The best known, and probably largest of these structures lies on the outer part of the Agulhas Bank: the Agulhas Slump (Dingle 1977) (Fig.187c). It is at least 700 km in length along its head region, and in cross section displays all the classic morphology and internal structure of landslips: a fissured head region with a large tensional depression, a lens-shaped main body underlain by the concave zone of décollement (the glide plane), and a toe region which shows a variety of compressional or distensional features. The Agulhas Slump is divisible into two parts: a western

Table 66. Quantitative data on west and south coast continental margin Neogene Slumps (from Dingle 1980b).

| A. Structure (Fig.187) | Cross sectional area ($km^2$) | Average thickness (km) | Length of head (km) | Volume ($km^3$) | Area ($km^2$) |
|---|---|---|---|---|---|
| 1. Unnamed* | ? | ? | ? | ? | 18,000 |
| 2. Walvis Bay* | ? | ? | 50 | ? | 22,000 |
| 3. Conception Bay* | ? | ? | 75 | ? | 51,000 |
| 4. Chamais | 103.5 | 0.254 | 160 | 17,433 | 68,688 |
| 5. Childs Bank | 28.5 | 0.135 | 150 | 3,849 | 28,512 |
| 6. Unnamed | 28.0 | 0.135 | 180 | 4,549 | 33,696 |
| 7. Cape Town | 29.3 | 0.206 | 300 | 8,790 | 39,613 |
| 8. Unnamed | 11.3 | 0.206 | 30 | 339 | 4,170 |
| | | | | | 265,679+ |
| 9. Agulhas | 133.5 | 0.256 | 700 | 20,331 | 79,488 |

B. Slump stratigraphy (used on Figs. 149, 187)

| Age | Slumped sequence |
|---|---|
| present sea floor | |
| | IV |
| Pliocene | |
| | III |
| Reflector Davie (Palaeocene) | |
| | II |
| | I |
| Reflector AII (L.Albian) | |

* – original data from Summerhayes *et al.* 1979
\+ – total area along west coast

sector which was dammed behind the Agulhas Fracture Ridge and an eastern sector which spilled down the continental slope and into the deepwater Transkei Basin. The western sector has a strongly curved (in plan) and intensely fissured head region, but down slope movement of the thick sediment mass (maximum of about 540 m), which retained its original bedding and internal cohesion, has been relatively slight, probably less than 12 km. Compressional features can be seen in the toe region where it impinged against the marginal fracture ridge. In contrast, the eastern sector is thinner (maximum about 220 m) and the rock sequence has been jumbled and distended, particularly in the toe region which probably moved as a semi-radial debris flow across the floor of the NW Transkei Basin for at least 180 km, where it forms a slight but distinctive feature on the sea floor.

No mechanism for its generation has been established. Although Dingle (1977) tentatively postulated a seismic shock, the evidence for this is weak and based only on a known seismic history for the southern Cape graben fault lines onshore where a total of five shocks greater than magnitude 5.0 on the Richter Scale have been recorded since 1963 (see Theron 1974 for a summary). Its age is also not precisely known, but must be Pliocene or younger because limestones of this age have been affected in the fissured head region.

At least eight zones of slumping, affecting an area of over 265 000 $km^2$ of the continental slope and rise, have been identified on the west coast (Dingle 1980b, areas 1-8 on Fig.187). The longest sector (ca.750 km) is a relatively narrow zone (100-250 km wide) between 30 and 36½°S that comprises at least three individual features, the largest of which is the Cape Town Slump. The latter was first noted by Emery et al (1975) on four seismic profiles, and subsequently DSDP site 360 drilled through it (inadvertently it seems, as the significance of the structure was apparently not recognized by Bolli et al 1978). In many respects it is similar to the Agulhas Slump, having a rugged head region and retaining its internal structure remarkably clearly, as seen on seismic records. The glide plane at the base of the allochthonous mass shows up as a strong reflector, and in the DSDP cores is represented by a zone about 66 m thick in which soft sediment deformation structures are common (Fig.166): intensely folded layers, subrounded clasts set in less indurated matrix; flow structures with microfaulting, and large scale folding with reverse faulting in the sediment laminae. Attention is drawn to similar structures at DSDP site 397 on Leg 47 (von Rad et al 1979), which also drilled through a slump on the continental rise off NW Africa.

A further, very large allochthonous complex occurs west of the area between Lüderitz and the Orange River (Chamais Slump, 68 688 $km^2$). It has a maximum thickness of about 720 m and in its proximal region contains numerous large rotated blocks, some up to 20 km across. One particularly impressive feature is a detached unit (80 x 20 km ) which apparently moved independently over the surface of the main slump and is associated with huge fissures (up to 105 m deep) on the underlying surfaces. Structures as large as this clearly play an important role in the construction and periodic modification of the continental slope and rise, and the present-day macro- and mesomorphology of large parts of the outer margin along the west coast can be directly related to the presence of allochthonous sediment bodies of this nature.

All the examples mentioned so far have, in their thickest main body, retained coherent bedding, which although cut by occasional subsidiary rotational faults is by and large not distorted. In contrast, two much thinner (ca.50 m) allochthonous bodies in which no internal bedding has been preserved, and which seem to have moved downslope as debris flows, have been described from the continental slope west of Walvis Bay (Summerhayes et al 1979) (features 1 and 2 on Fig.187, Tab.66). These have been recognized on the basis of small topographic discontinuities and other microrelief features on high resolution seismic records. Although they involve moderately large areas of the sea floor (a total of 40 000 km$^2$) they are not of the same order of importance in mass downslope transport of sediment as the features farther south.

Lower Pliocene sediments occur at the top of the DSDP 360 sequence (Fig.166), and from seismic evidence (Bolli et al 1978, figs 6 & 7) it appears that these sediments were involved in the slumping (i.e. are not a subsequent deposit), suggesting, as for the Agulhas Slump, a Pliocene or younger age for the movement. Radiometric dating on piston cores suggest late Pleistocene ages (50 000 and 20 000 years BP) for the slumps north of Lüderitz (Summerhayes et al 1979). What little evidence we do have, therefore, does not preclude a Pleistocene age for all the Neogene slumps shown in Figure 187.

The question of trigger mechanisms for the west coast Neogene slumps is no easier to resolve than in the case of the Agulhas Slump, although several possibilities present themsevles. As a general statement, formulated for instance by such workers as Reading (1978, pp.373-374), Almagor (1979), and Summerhayes et al (1979), large allochthonous bodies will move downslope when the shear stress acting within a body of sediment exceeds the shear strength along any particular horizon, and that the following mechanisms capable of causing such disequilibrium situations to develop (i.e. trigger mechanisms) have been recognized: (i) slope steepening and support removal by downslope undercutting; (ii) thickening of the sediment pile by deposition or slumping farther upslope; (iii) increases in pore fluid pressure; (iv) thixotropy and fluidization of sediment by shock-induced strain (e.g. storm waves, earthquakes). Summerhayes et al (1979) and Dingle (1980b) discounted possibility (ii) as the main cause of slumping along the west coast because of the relatively low potential for high sediment input from most of the southern African hinterland, even during low sealevel stands. Possibility (iii) was considered by Summerhayes et al (1979), who cited pagoda structures within the Conception Bay slump as possible evidence of methane generation by degradation of high organic carbon concentrations in the sediments below the Benguela Current. At the temperatures and pressures found at these depths, however, the gas would normally remain in a hydrate or clathrate state (a solid, ice-like water/methane substance), until a lowering of the superincumbent pressure, or increase in bottom water temperature allowed a transformation of the clathrate to methane. The resultant rise in pore fluid pressure might have led to slope failure, and Summerhayes et al (1979) suggest that such conditions could have occurred during Pleistocene low sea-level stands. Evidence for pagoda structures has not been found in the areas farther south, and in any case Summerhayes et al were not convinced of the validity of the mechanism for the Namibian structures because of the water depth (> 2 000 m) at which the pagodas lie.

Evidence for possibility (iv), by seismic shock is more substantive and was discussed by Dingle (1980b). Since 1911 at least 18 seismic shocks with an intensity of at least magnitude 5 (Richter scale) have been recorded within 250 km of the present west coast, as well as numerous smaller events. 'Young' (presumably post-Jurassic) faults documented by Korn & Martin (1951) testify to mild but continuous Mesozoic and Cenozoic tectonic activity (including at least one 'Pleistocene' fault) along the west-coast margin. This evidence suggests that seismic activity was a possible mechanism for slope failure for all west-coast slumps, as well as the Agulhas Slump. A possible contributory mechanism that can be invoked in the case of the Cape Town (and perhaps the Chamais) slumps is slope steepening and support removal by downslope undercutting caused by the erosion of large quantities of sediment during vigorous bottom-current scouring (possibility (i) above). Evidence for such activity on a regional scale is found in the vicinity of DSDP site 361 where Eocene sediments lie at, or very close to, the sea floor and shear strength tests indicate that at least 500 m of overburden have been removed at this site, presumably by erosion (Bolli et al 1978, p.54). It is possible that the removal of such a large thickness of material from the continental rise west of Cape Town could have led to slope failure higher up: at least it would have made the slope material vulnerable to a large seismic shock. Although we have no dates from the sea floor west of the Chamais structure, seismic evidence suggests that horizons of Lower Tertiary age may crop out here also, possibly indicating severe Upper Tertiary sea-floor erosion similar to that off Cape Town. In summary, we can say that for the Neogene slumps along the west coast, seismic shocks with possible contributory factors such as methane-induced high pore-fluid pressure (north of Lüderitz) and downslope undercutting (Chamais and Cape Town areas) seem the most likely trigger mechanisms for slope failure.

Whatever the ultimate causes for the Neogene slumps it is significant that they originated partly in zones where massive slope failure had occurred during Upper Cretaceous and 'mid' Tertiary (post reflector Davie-pre-Neogene) times, and it is possible that some of the older glide planes could have been re-activated. Dingle (1980b) also suggests that there may be a correlation between periods of slumping and low sea level stands.

### 6.3.4 *Tertiary palaeoclimates*

World-wide, Tertiary climates show a progressive, if occasionally erratic decline in temperature, so that by mid-Oligocene times, the equable Cretaceous conditions had been replaced by a marked latitudinal thermal gradient which ranged, at least in the southern hemisphere, from tropical equatorial to glacial polar. During the later Tertiary, this trend intensified, with notable short reversals, so that by the end of the Pliocene, the whole of Antarctica was glaciated and north polar glaciation was underway. Much of this picture has been assembled from land-based studies, but significant advances and global syntheses have been generated by DSDP data (e.g. see Hallam 1981 for a summary). Furthermore, eustatic sea-level changes can be linked to this thermal history, so that a relatively well-integrated Tertiary palaeoenvironmental model can be produced that can be used as a background against which to interpret local phenomena (Fig.188).

In the case of southern Africa, there is a certain amount of marine relative water temperature data (DSDP borehole and onshore sections), as well as more qualitative terrestrial climatic information (from micro- and macrofloras, and, particularly, mammal population studies). Together with other lithofacies and geomorphological observations these allow a relatively satisfactory statement of Tertiary palaeoclimates to be made.

#### 6.3.4.1 *Palaeogene*

As a setting for early Tertiary palaeoclimates, it is necessary to briefly summarize Cretaceous conditions. Hallam (1981) has reviewed previous work, and from the evidence of both terrestrial floras (e.g. Vakhrameev 1964) and marine micro- and macrofossils (e.g. Stevens 1971, Douglas & Savin 1975), there emerges a picture of a mid-Cretaceous optimum with cooler, late Cretaceous temperatures. Cretaceous mean annual air temperatures were 10-15° higher than at present (Frakes 1979), with tropical-subtropical conditions extending to at least 45° either side of the Equator, and warm to cool conditions in higher latitudes (Hallam 1981). Surface-water temperatures in the NW Pacific reached a maximum of ca.28° in the Albian, and had fallen to 19° in Campanian-Maastrichtian times (Douglas & Savin 1975).

Palaeogene southern hemisphere ocean temperatures at a palaeolatitude of 60°S, which have been investigated by Shackleton & Kennett (1975) using oxygen and carbon isotope determinations on DSDP boreholes from south of Australia, show a progressive decline in surface temperature from a high of about 19° in the late Palaeocene/early Eocene, to about 11° at the end of the Eocene (Fig.188). At this point in time (38-39 my), there was a dramatic drop of about 5° over a period of perhaps 2-3 my, so that throughout the Oligocene, temperatures fluctuated around 6-7°C. This sharp temperature deterioration was the precursor to the onset of local glaciation in Antarctica in the late Oligocene (DSDP Scientific Staff 1973). Bottom temperatures were 1-2° lower than surface-water values during the Palaeogene.

Table 67. Southern Hemisphere Palaeogene vegetation sites. At these sites, Coetzee (1978a,b) has typified the vegetation as consisting of: Podocarpaceae, *Nothofagus*, Proteaceae, and ferns.

| | | |
|---|---|---|
| Late Oligocene | Ross Sea, Antarctica | Kemp & Barrett 1975 |
| Late Oligocene | McMurdo Sound, Antarctica | Cranwell 1969 |
| Late Eocene | Black Island, Antarctica | Kemp & Barrett 1975 |
| Palaeocene-Oligocene | 90° East Ridge, Indian Ocean | Kemp & Harris 1975 |
| Middle Eocene | Central Australia | Kemp 1976 |
| Late Cretaceous | Antarctica | Cranwell 1969 |

Frakes & Kemp (1973) have commented upon the maintenance of global warm Palaeocene-Eocene climates in terms of water mass circulation patterns. They deduced that because the Pacific Equatorial currents were deflected to high latitudes and circulation was complete through a Caribbean-Tethys waterway, equatorial temperatures were only slightly higher than at present, but that high latitude areas received a much higher proportion of the heat budget than they do today. A globally, equable warm climate persisted with a correspondingly-low latitude temperature gradient, and lower surface wind strengths. Progressive alterations in relative land mass geometries, and particularly the separation of Antarctica from Australia, and the establishment of circum-south polar circulation led to a reorganization of the heat budget, and deteriorating high latitude temperatures. Hallam (1981), quoting research on terrestrial Alaskan vegetation, refers to a drop in the mean annual temperature of 12-13° as recorded in Eocene to Oligocene floras.

Data on Palaeogene terrestrial vegetation in the southern hemisphere relates mostly to Australia and Antarctica. Kemp (1978) shows that in Eocene times, Australia was covered with rain forest that was a mixture of tropical and cool-temperate elements. She points to the high fern density and presence of epiphytic fungi as indicative of high humidity. This flora persisted into the Miocene. Coetzee (1978a, 1978b) has classified the Eocene to Oligocene (and probably the

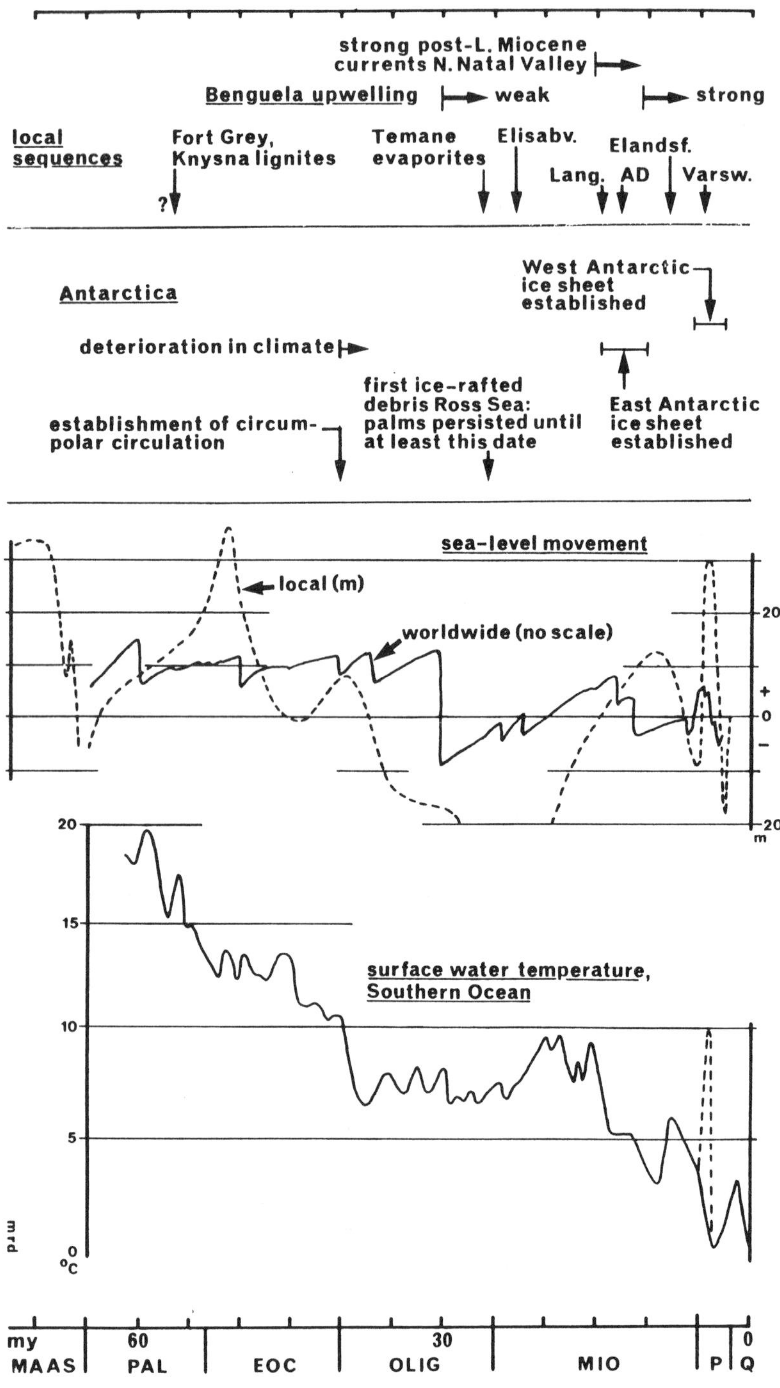


Figure 188. Southern Ocean surface temperature and Antarctic climatic variations during Tertiary time. Important southern African deposits, sea level movements and ocean current phenomena are plotted to the same scale for comparison. The establishment of circum-south polar circulation relates to final separation of Australia and Antarctica at about 40 my, and an early shallow-water connection between the S.Atlantic and Pacific Oceans through the Drake Passage (38-35 my) (Haq 1981). Total thermal isolation of Antarctica was not completed until further subsidence occurred south of Tasmania (33-30 my) and in the Drake Passage (16-15 my) (Haq 1981). Data from Ciesielski & Weaver (1974); Shackleton & Kennett (1975); Vail *et al.* (1977); Siesser & Dingle (1981); as well as numerous authors on southern African fossils sites quoted in the text. Abbreviations: Elisabv = Elisabethveld deposits; Lang = Langental beds; AD = Arris Drift beds; Elandsf = Elandsfontyn Formation; Varsw = Varswater Formation.

| | WEST COAST: water temperatures | WEST COAST: upwelling | WEST COAST: vegetation | WEST COAST: rainfall/temperature | WEST COAST: rainfall | SOUTH COAST: water temps. | EAST COAST | SOUTHERN OCEAN AREAS | SOUTHERN OCEAN AREAS: events |
|---|---|---|---|---|---|---|---|---|---|
| Pleistocene | | max. upwelling | fluctuating grassland & fynbos | max. aridity | ? Winter rainfall | | | | |
| Pliocene | Cape Basin: cool-temperate<br>S. Walvis Ridge: cold<br>N. Walvis Ridge: warm | increased upwelling | grassland replaced by fynbos & related shrub-land vegetation. Black rhino replaces white rhino<br>forests along rivers, grasslands between rivers | increasing aridity<br>marked cooling & dryness | ? Winter rainfall<br>Summer rainfall | ? | very warm | 1 - 2° | establishment of W.Antarctic ice cap |
| Miocene | Cape Basin: cool temperate<br>S. Walvis Ridge: cool - temperate<br>N. Walvis Ridge: tropical | large-scale up-welling<br>Spasmodic upwelling | open woodland, savanna<br>yellow woods & palms<br>1 at Arris Drift woodland; no medium or large grazers<br>2 at Lüderitz woodland; no medium or large grazers | increase in dryness<br>warm & wet | Summer rainfall<br>? monsoonal | warm-temp.<br>warm | warm - temp.<br>? tropical | 4 - 6°<br>6°<br>9° | establishment of E.Antarctic ice cap<br>last forests in Antarctica |
| Oligocene | Cape Basin: cool<br>S. Walvis Ridge: tropical-subtropical<br>N. Walvis Ridge: tropical | Spasmodic upwelling | yellow woods & palms | warm, slightly drier but still wet | | warm (podocarps) | warm v.dry in Mozambique evaporites | 6°<br>6° | First glaciation on Antarctica |
| Eocene-Palaeocene | Cape Basin: cool-temperate<br>S. Walvis Ridge: tropical-subtropical<br>N. Walvis Ridge: tropical | | yellow woods & palms | humid & wet | | warm (podocarps)<br>cool-temp. | warm<br>cold | 10-11°<br>? 19° | rainforest in Australia |

Figure 189. Southern African Tertiary palaeoclimates and vegetation. Most data are available from the west coast region, with mainly water temperatures from the S and E coasts. Southern Ocean temperatures and some important palaeoclimatic events are noted from Antarctica and Australia. See Figure 188 for further details. Numerous data sources cited in the text, with additional information from Deacon (personal communication 1980).

late Cretaceous and Palaeocene) southern continental floras as temperate-subtropical evergreen forests composed of Podocarpaceae, *Nothofagus,* Proteaceae, and ferns (Tab.67). Kemp & Barrett (1975) indicate that this vegetation type persisted on Antarctica until at least late Oligocene times, and that although Miocene strata do contain the same taxa, they are present in low concentrations, and there is evidence of reworking. Also, the Antarctic Eocene-Oligocene floras are indicative of cooler conditions than contemporary floras from New Zealand. Clearly, this early Tertiary flora persisted in Antarctica contemporaneously with local glaciation.

In southern Africa, evidence for warm, humid Palaeogene climates is fragmentary and somewhat inconsistent. Deacon (1983) mentions three floral assemblages from around South Africa, which he suggests indicate relatively warm climatic conditions: the Arnot pipe in Namaqualand; Fort Grey (eastern Cape Province), and the lignites at Knysna. The Arnot (Banke) structure is a kimberlite pipe which contains an upper sediment fill from which plant, vertebrate, and invertebrate fossils have been recovered during exploratory excavations for diamonds. Adamson (1931), Rennie (1931), and Kirchheimer (1934) recorded a leaf and pollen assemblage including *Podocarpus, Gunnera, Myrica, Ficus,* a proteaceous taxon like *Brabeium,* and *Protea.* Estes (1978) dated the deposits as late Eocene-Oligocene on the basis of the frog *Eoxenopoides reuningi.* Tankard & Rogers (1978) comment upon this assemblage, and suggest that it indicates a relatively dry climate with summer rainfall, and represents a transitional flora between temperate rainforest and sclerophyllous vegetation. Deacon (1983) is not convinced by this, and suggests that more data are needed to firmly establish evidence of dryness.

At Fort Grey, near East London, Adamson (1934) reported macroscopic remains of *Podocarpus, ?Widdringtonia,* and fruits of *Curtisia,* as well as other dicotyledons and monocotyledons. Du Toit (1954) has suggested a late Cretaceous-Eocene age for these fossils, and Deacon (1983) also believes that they represent a Palaeogene flora.

The Knysna lignites were first reported by Rogers (1909), and the flora of the lignite-bearing sands has been discussed by numerous authors, including Phillips

(1927), Levyns (1964) and Thiergart et al (1966). *Widdringtonia, Podocarpus, Gonioma* and *Curtisia* have all been reported, and significant points are the high percentage of Restionaceae and ferns, which suggest an estuarine site, and the absence of *Erica,* which is typical in modern wet sites. Deacon (1983) also comments on the lack of Compositae and Gramineae, on the basis of which he assigns a pre-Miocene age to the deposits. The suggestion of a dry climate by Thiergart et al (1966) on the basis of an abundance of carbonaceous fragments (charcoal), is contested by Deacon (1983), who proposes that this material is allochthonous, and derived from the adjacent mountainous hinterland.

Deacon (1983) sums up the floral evidence from these three proposed Palaeogene sites as indicating a vegetation in the Cape that was dominated by gymnosperms with Proteaceae, which was analogous to the Palaeogene Australian flora of dominant *Nothofagus* with Proteaceae. Neither the Cape nor Australia show any evidence for the presence of the Ericaceae or its sister family Eparidaceae, nor for dry summers, or extensive shrubland/heathland vegetation. On Figure 189, a flora of yellowwood with palms is shown for the southern Cape, and is based on Deacon (personal communication 1980) and Coetzee's (1978a) identification of palm pollens in her oldest samples from the SW Cape.

Water temperature information for the Palaeogene around southern Africa indicates generally cool-temperate conditions, although the temperature of the adjacent seas does not necessarily reflect the temperature of the adjacent land mass. This is well-illustrated by the present-day cold water Benguela Current (locally $<10°C$), and the adjacent hot Namib Desert, although in this case the sea temperatures do have a direct effect on precipitation. The Cape Basin DSDP sites (360 and 361) (Proto-Decima et al 1978), indicate cool to temperate nannofossil floras during Palaeocene-Eocene times, with very abundant *Isthmolithus recurvus* and *Neococcolithus dubius,* whilst a high *Chiasmolithus/Discoaster* ratio indicates cool water in the early to middle Oligocene, with a possible temperature minimum in the early Oligocene. Farther north, on the Walvis Ridge (site 362), the water temperature for the whole of the Tertiary was higher than that in the Cape Basin, an assessment based on the presence of relatively few *Chiasmolithus,* and more abundant sphenoliths and discoasters (Proto-Decima et al 1978, Siesser 1978b). Moreover, the Middle Eocene to Lower/Middle Miocene sediments contain tropical-subtropical foraminifera and warm water calcareous nannofossils. 150 km farther north still, on the northern side of the Walvis Ridge (site 363), tropical faunas and floras were deposited throughout early and middle Tertiary times.

Although large-scale upwelling of the Benguela Current along the west coast did not take place until the early Upper Miocene, Siesser (1978c) records a low and fluctuating phosphorus content in *Coccolithus pelagicus* skeletons between early Oligocene and Middle Miocene times. Because the onset of large-scale upwelling coincided with a small but significant increase in phosphorus uptake, Siesser (1978c) equates the earlier change in phosphorus content with the onset of weak and spasmodic welling up of cold Benguela water. This suggests that whilst significant aridification of the SW African coastal hinterland did not begin until the Upper Miocene, a small, but noticeable deterioration in terms of lower rainfall can be anticipated after early Oligocene times. We might expect this to be reflected in a change from tropical floras in the Palaeocene-Eocene, to subtropical-temperate floras in the Oligocene.

The only other substantive evidence available for west coast water temperatures is from the Eocene outcrops at Bogenfels. Here, Siesser & Salmon (1979) record mostly temperate-water coccoliths, but mention that one cold-water species (*Isthmolithus recurvus*) also occurs.

A period of humid tropical or subtropical climate has been postulated by Glass (1977) in pre-Middle Miocene times (possibly late Eocene or Oligocene) to account for the saprolite profiles that occur extensively in the SW Cape over granite (up to 60 m thick) and Malmesbury Group (up to 80 m thick). These form the basis of the kaolin mining operations in the Noordhoek/Kommetjie region of the Cape Peninsula, with further potentially exploitable deposits in the vicinity of Stellenbosch.

The rather sparse micropalaeontological data from the south coast (Agulhas Bank outcrops) indicate the following water palaeotemperatures (Siesser, personal data): cool-temperate during Palaeocene-early Eocene; warmer than present during late Eocene, late Oligocene, and early Miocene.

On the east coast, nannofossils and planktonic foraminifera indicate cool-temperate water temperatures in the Palaeocene (Stapleton 1975 remarked on a distinctly cold water assemblage in the Danian of Richards Bay, compared to the underlying Maastrichtian). In Mozambique, the Eocene rocks contain ooids, calcareous algae, and certain foraminifera which suggest warm oceanic waters. The Oligocene Temane Formation contains evaporites, which are clear indicators of warm temperatures.

In summary, the Palaeogene oceanic temperatures were generally cool to temperate around the whole subcontinent in the Palaeocene-early Eocene, and whilst they became warm throughout the rest of the period along the south and east coasts, they remained cool to temperate south of the Walvis Ridge along the west coast. On land, we have insufficient data to discriminate climatic changes because the localities are so poorly-dated, although the evidence suggests that at least between Namibia and the Eastern Cape, and pro-

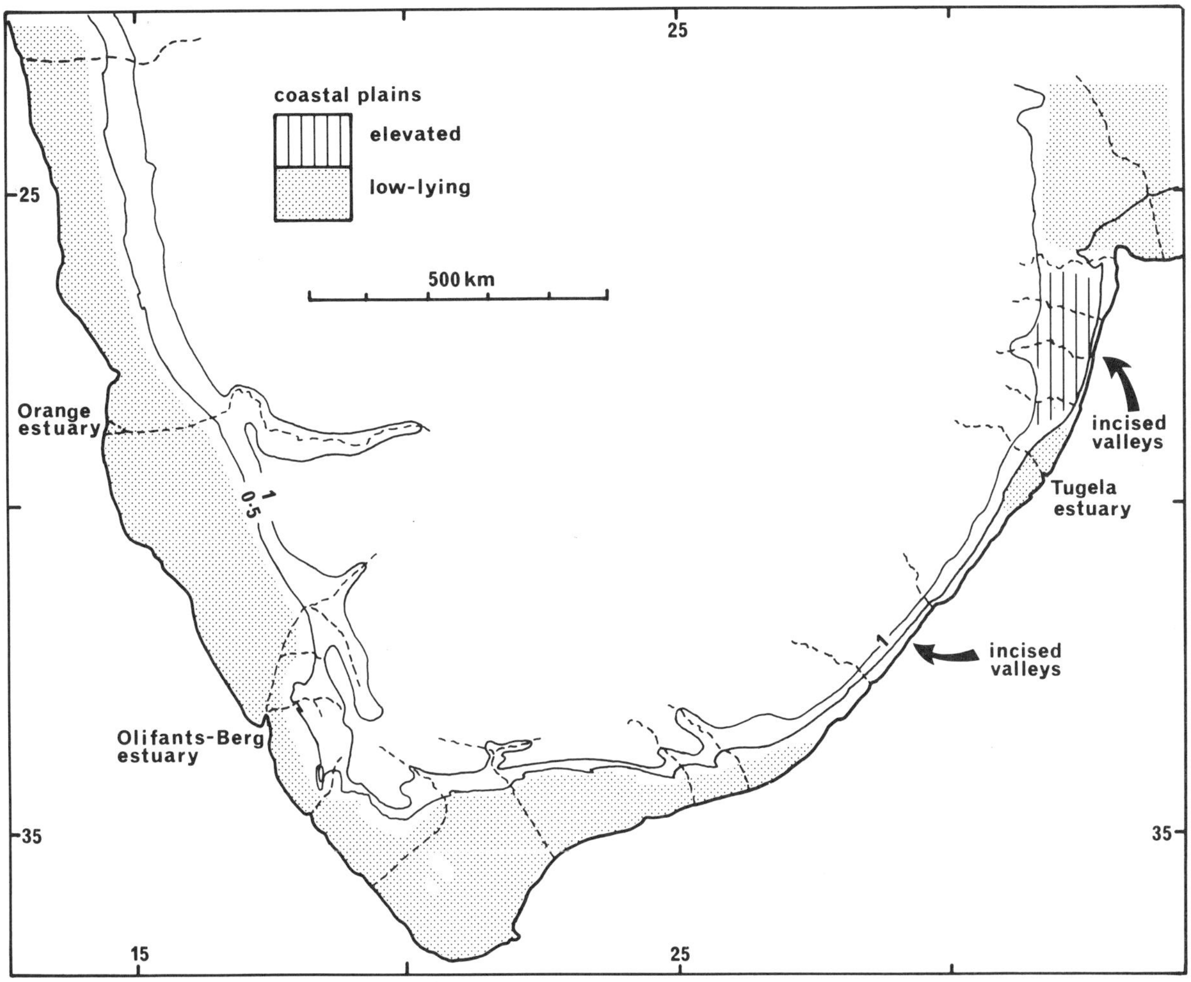


Figure 190. Speculative Oligocene palaeogeography for coastal southern Africa. Based on a nominal sea level lowering of 500 m. Postulated river courses shown by dashed lines. Such a reconstruction can obviously give only a crude outline of Oligocene coastal topography and configuration because post-Oligocene sediments have not been stripped off. The present coastline becomes the + 500 m topographic contour, and the 1 km line has also been added (i.e. the present day + 500 m contour). The map shows the contrasts that would have existed during this major regression between the very wide flat coastal plains of the south and west coasts, and the narrow, incised coastal fringes along much of the eastern margin. Similar effects would have been evident if smaller sea level falls were assumed (e.g. 300 m). See the text for a fuller discussion.

bably around to Mozambique, the flora was evergreen and palm forest, with summer rainfall.

On theoretical grounds (based on the assumption of weak Benguela upwelling) it might be expected that on the west coast, the tropical humid climate was progressively replaced from early Oligocene times by slightly drier, if not cooler weather, that persisted until the Miocene. This change coincided with an extended low sea level stand during early Oligocene to mid Miocene times, an event which would have created extensive coastal plains along the west (up to 200 km wide) and south (up to 250 km) margins, with a much more restricted seaward shift of the coastline along the east margin (Fig.190). Although wide rugged, rocky zones lay immediately adjacent to the inner edge of these coastal plains (in the vicinity of the present-day coastline) between Cape Columbine and Cape Agulhas, and between Plettenberg Bay and East London, the largest part of the newly-exposed areas would have been very flat, and crossed by slow, meandering rivers. The courses of these rivers as shown in Figure 190 are largely speculative, although, as previously discussed, the original cutting of the Cape Canyon probably dates from this period, so that a convergence of drainage patterns for the Olifants and Berg river systems is envisaged along the lines shown. Given the relatively warm and wet climate (compared to that of the present-day west coast), these very wide coastal plains must have been well-vegetated. The large gradient change in the vicinity of the present coast line (i.e. between about +380 and 450 m above the sea level shown in Figure 190) probably forced the west and south coast rivers to dump the bulk of their coarse sediment load along the inner margins of the mid-Tertiary coastal plains and (except in times of flood) carry only fine suspensate (muds) across the plains to the coast. On the eastern side of the sub-continent, coarse material was more likely to reach the coast because of the relatively narrow, steeper gradient coastal plains. Adjacent to the Lebombo Mountain line an elevated coastal plain would have been flanked on its seaward edge by a relatively steep gradient zone, and in this area, as well as that between about Port Alfred and just south of Durban, rivers probably incised their seaward courses, giving rise to the numerous upper slope and outer shelf canyons that occur along the east coast continental margin today.

6.3.4.2 *Neogene*

On a worldwide scale, Neogene times saw the onset of extensive ice cover in Antarctica, and the commencement of Arctic glaciation. Shackleton & Kennett (1975) indicate a 2-3° increase in southern ocean surface water temperatures during the early Miocene (relative to Oligocene values), followed by a sharp drop in the Middle-earliest Upper Miocene (14-10 my) of about 5°C (i.e. falling to a value of about 4°C). This resulted in the establishment of the East Antarctic ice sheet (in contrast to earlier local centres of glaciation), and the elimination of the remnants of the Palaeogene flora from the Ross Sea area (and probably the rest of Antarctica as well) (Kemp & Barrett 1975). Ciesielski & Weaver (1974), on the basis of silicoflagellate studies, postulate a temporary rise in high latitude Southern Ocean surface water temperatures in the late Miocene, a trend that is also indicated by the planktonic foraminiferal studies of Shackleton & Kennett (1975), who noted a rise to about 6°C. This short-lived amelioration was followed by a further deterioration in the latest Miocene-early Pliocene, when the West Antarctic ice sheet was established. The south polar ice sheet reached its present extent by about 3.8 my (Ciesielski & Weaver 1974) (Fig.188). Shackleton & Opdyke (1977) postulate that the initiation of northern hemisphere glaciation dates from this period (ca.3.2 my).

Local data on Neogene palaeoclimates come from several sources: terrestrial floras; terrestrial and marine vertebrates; marine macroinvertebrates; marine microfossils; and marine sedimentology and geochemistry. Looking first at the sea water temperature indicators, Proto-Decima et al (1978) postulate a continuation, throughout the Neogene, of the cool-temperate conditions in the Cape Basin (DSDP site 360) that they recorded for Palaeogene time. On the southern Walvis Ridge (DSDP site 362) they also record a continuity of the Palaeogene tropical-subtropical conditions (indicated by planktonic foraminifera and 'warm' water calcareous nannofossils) until middle to late Miocene times when there was a marked cooling, indicated by the dominance of cool-water microfossils. This trend continued into the Pliocene, by which time the assemblages were dominated by cold-water types. In contrast, DSDP site 363 on the north side of the Walvis Ridge shows tropical faunas and floras persisting into the Miocene.

Siesser (1978c) confirms these trends in terms of micropalaeontological and geochemical indicators that show the onset of intensified upwelling of cold nutrient-rich Atlantic Central Water in the Benguela Current system commencing at about 10 my (early Upper Miocene). The Walvis Ridge barrier prevented this phenomenon affecting the southernmost part of the Angola Basin. He relates the onset of west coast aridification to the event, and questions the postulate of van Zinderen Bakker (1975) that the Namib Desert is as old as the Oligocene, preferring a late Miocene initiation that has intensified into the Pleistocene.

Similar results have been achieved by Diester-Haas & Schrader (1979) who made a particular study of the Neogene diatom floras in the DSDP site 362 on the Walvis Ridge. Their evidence for initiation of cold water upwelling includes:

i) abundance and good preservation of opaline skeletons (diatoms and radiolaria);

ii) lowering of the planktonic/benthic foraminifera ratio;

iii) an increase in fish debris (otoliths, scales, bones and teeth;

iv) presence of phosphate grains.

Diatom studies show a steady increase in the number of frustules from Upper Miocene times (DSDP site 362, core 27), with a temporary reversal in abundance and preservation across the Mio-Pliocene boundary, which they equate with the 'Messinian event'. Both tropical and cool water types are present, suggesting mixed water assemblages. Clearly, they were dealing with populations from the warm Angola Current and cold Benguela current (a situation that has been shown to exist at present in the radiolarian assemblages from the Walvis Ridge abutment area by Robson 1983). Influxes of fresh-water diatoms and phytolithes in the Lower Pleistocene (core 1), Middle-Upper Pliocene (core 7), and Messinian (core 17) suggest increased land wind velocities at these times. Diester-Haas & Schrader (1979) suggest that their method of investigation of upwelling activity using diatom content cannot detect upwelling rates of $<2$ m/day, which at least gives a minimum value to the strength of the early Upper Miocene rates, and implies that the weak Oligocene upwelling noted by Siesser (1978c) was below this value. They also note that maximum opal values occur in the Pleistocene, indicating that the most intense cold water upwelling, possibly coupled with higher wind velocities occurred during this period. Embley & Morley (1980) have also shown that Pleistocene surface water temperatures in the Cape Basin were lower than at present: they reached a minimum low of

12°C (winter temperature) during the last glacial maximum (18 000 y BP), compared to a climatic optimum within the Holocene at 7 000 y BP of 18.4°C (winter) and 23.6°C (summer).

Micropalaeontological studies of samples from the south and east coast continental shelves (i.e. under the Agulhas Current) indicate water temperatures significantly higher than those for the Neogene west coast (Benguela Current):

i) Agulhas Bank: warm water in the late Oligocene and Miocene (foraminifera and nannofossils);

ii) East coast: warm water in the Miocene; very warm in the Pliocene (nannofossils).

To appreciate the significance of data from nearshore Neogene molluscan faunas, it is necessary to compare them with those of present day provincialism, and to note that in the Saldanha area of the SW Cape coast, the present faunas lie within the cold temperate province (Brown & Jarman 1978). It is, however, close to the southern boundary of this province, with, for instance, False Bay lying within the warm temperate province. At present, therefore, the coast between Saldanha and False Bay is a transition zone. For the Saldanha area, Hendey (1981a) quotes a present day mean annual sea temperature of 17°C, with an annual range of 8°C.

In the Miocene Gravel Member at Varswater, Tankard (1974b), and Hendey (1981a) consider that although the molluscan fauna consists of taxa presently extant around the South African coast, several of them indicate significantly warmer water (3-5°C higher) than presently found on the SW Cape coast: *Cellana capensis* (subtropical province north of Port Alfred); *Turbo sarmaticus, Barbatia obliquata, Ostrea atherstoni,* and *Striostrea margaritacea* (all these presently occur in False Bay, but not farther north up the SW Cape coast). Hendey (1981a) notes that sharks' teeth from the Gravel Member also indicate warm water types: *Carcharhinus melanopterus, C.limbatus* and *Squatina africana.* These assemblages are in marked contrast to the fauna from the overlying Quartzose Sand and Pelletal Phosphorite members of the Varswater Formation. In the former, the majority of the molluscan fauna lives today on the west coast, with the most abundant taxon (*Tricolia capensis*) indicative of cold temperature conditions. Only one species (*Pyrene albuginosa*) is indicative of warmer water (present range is False Bay to Natal). Hendey (1981a) concludes that the water temperature during the deposition of the Quartzose Sand member was probably similar to that of today in the Saldanha area. Similarly, although there are no identifiable molluscs in the Pelletal Phosphorite Member, the presence of the seal *Homiphoca capensis* (which has certain cold water morphological adaptations) also indicates a similar temperature range to that experienced in the area today. Hendey (1981b) suggests that the environmental conditions obtaining during the deposition of the Anyskop Marine Member were 'becoming warmer', but gives no additional information.

We do not have well-dated shallow-water molluscan assemblages from the south coast, but along the east coast, Stapleton (1977c) reported temperate-water foraminifera in the Upper Miocene rocks of Zululand, whilst Siesser & Miles (1979) consider the late Miocene Pecten Bed at Uloa to contain a warm-temperate planktonic foraminiferal assemblage. This is in line with King's (1953) recognition of warm-tropical molluscs in the same beds: *Aequipecten uloa, Pecten sapolwanaensis* and *Amusium umfolozianum.*

Data from Neogene terrestrial faunas and floras are relatively abundant, being most complete for the west coast. The Arris Drift and other coastal Namibian Miocene occurrences, are probably the oldest for which we have reliable data (early Miocene). In the former, Hendey (1978a) suggests a forested river setting, with dense undergrowth and warm ambient temperatures. He reached this conclusion on the presence of two proboscideans (elephants) (*Prodeinotherium*, and a gomphothere), a browser (low-crowned toothed rhinoceras), and a crocodile. The Namibian sites (between Langental and Lüderitz) indicate similar conditions, suggesting that the coastal zone of Namibia at least, was warm, well-watered, and well-vegetated in the early Miocene. Deacon (personal communication, 1980) suggests a woodland vegetation of trees and grassland.

Somewhat younger Miocene (Middle-Upper) is represented by the Elandsfontyn Formation and the overlying Gravel Member in the Saldanha area, and their mammalian faunas suggest to Hendey (1981a) a monsoon type climate (i.e. a mean annual temperature of about 18°C with heavy rains in summer and a drier winter). He also notes the presence of the horse *Hipparion* cf. *primigenium* in the Gravel Member, which indicates open woodland and savanna. Peats from the Elandsfontyn Formation in the Varswater Quarry yield *Curtisia dentata,* and as Tankard & Rogers (1978) point out, this genus is a rain forest tree that cannot tolerate aridity. It survives today in southernmost Africa only in moist mountain valleys of the SW Cape. Coetzee's (1978a) work on the Miocene floras from Noordhoek and Varswater shows that palm pollen is present in all levels, indicating warm damp conditions throughout the period, but that at certain levels conifers (podocarps) are also present in large quantities. Deacon (1983) comments that in these data he sees no evidence for summer dry climates, nor for the establishment of shrubland or heathland vegetation prior to the end of the Miocene, and suggests a warm, wet forested environment with non-seasonal rainfall (personal communication 1980). Evidence from both the mammals and the floras indicate a climatic deterioration across the Mio-Pliocene boundary, which in this area probably lies within or immediately above the Gravel Member (Hendey 1981b). There is still a predominance of browsers in the Quartzose Sand and Pelletal Phosphorite members (indicating persistence of forest or woodlands), but the presence of grazers indicates that grassland had become established locally. At some levels (for example in the upper channel of the Pelletal Phosphorite Member), *Damalacra neanica* and *D.acalla* (alcelaphine, hartebeest-like grazers) are more common, and some of the browsers show signs of ill-health. Hendey (1981b) considers the latter phenomenon to have been due to nutritional deficiencies, as manifest by hypoplasia in some of the giraffe population (etching of the tooth enamel), as well as irregular and excessive teeth wear in kobs. He surmises that these conditions may have been the result of the browsers having to eat grass (which is obviously not their natural food) because the climate was too dry to support an extensive enough tree vegetation to feed the whole browsing population. This possibility is reinforced by Klein's (1981) studies on mortality of the giraffes in floods, as a result of them having congregated in the river valley because of the scarcity of trees away from it, and Hendey's (1981b) observation that fires were common, for which he argues a long dry winter followed by fires in the first thunder storms of the summer. Finally, crocodiles have not been recorded in the Pliocene sediments, presumably because the climate was too cold.

From the palynological studies on peats from the Varswater Formation in the Saldanha embayment, and at Noordhoek, Coetzee (1978a) shows that in early Pliocene times there was evidence that the Miocene woodlands were being replaced by fynbos and related shrubland vegetation (typified by Proteaceae, Restionaceae, Ericacea, and Compositae, i.e. the development of modern vegetation), which argues for a cooler, considerably drier climate than in the Miocene. She relates this to the establishment and subsequent intensification of upwelling off the west coast. Hendey (1981b), commenting on this, in the light of the work of Taylor (1978) notes that dry summers in southern Africa are of recent (possibly Pleistocene) origin, and that although the fynbos is now centred on a winter rainfall area, it probably developed under summer rainfall conditions. The date for the establishment of the modern winter rainfall conditions is, therefore, uncertain.

In summary, this evidence points to the disappearance of extensive woodlands in the SW Cape and Namaqualand in early Pliocene times, and their replacement with grassland, and ultimately fynbos/shrubland in Upper Pliocene and Pleistocene times, this being in response to cooler and progressively more arid conditions that obtained as Benguela upwelling became established on a large scale. Maximum aridity was attained in the Pleistocene, so that at present the Lüderitz-Bogenfels area, which supported a varied large mammal population in the Miocene, now lies in the heart of the Namib Desert, an area of extreme aridity. The present annual mean rainfall in the Saldanha area is 253 mm (at Langebaan), with a mean annual temperature of 14.6°C (at Cape Columbine) (Boucher & Jarman 1977), which warrants the classification of semi-arid.

Whilst we do not have comparable data on terrestrial faunas and floras from the south and eastern coastal regions, the sea temperature data already mentioned probably indicates warmer and damper conditions in the Mio-Pliocene than on the contemporary west coast.

Data on the climate of the continental interior are more difficult to interpret than the coastal regions because of the lack of well-dated deposits. Certainly, aridification of the Namibia-Botswana-northern Cape Province area seems to have taken place in the Neogene. The Kalahari sands, calcretes and silcretes all attest to this, although their precise ages can only be inferred. The disappearance of large lakes in southern Africa also reflects progressive desiccation (e.g. lakes Mababe, Makgadikgadi and Ngami in Botswana), although drying up may have taken place only in the Pleistocene. Alluvial deposits, such as the High and Low-level Gravels, may provide some climatic evidence, but must be considered with caution. Too often, thick accumulations of coarse gravel have been equated directly with increased runoff related to high rainfall. Partridge (1969) points out that massive erosion and heavy alluviation are semi-arid phenomena, and that alluvial terraces are best developed in areas of sparse vegetation cover. Finally, the Plio-Pleistocene cave deposits of the Transvaal contain open plains-grassland antelope remains. Statistical treatment of them by Vrba (1974, 1975) shows that bush cover was more extensive in the late Pliocene than in the Pleistocene. She attributes this to either a decrease in rainfall, or a drop in temperature. Conversely, Brain (1958) has suggested an increase in rainfall for the same period, based on quartz grain angularity and chert/quartz ratios in breccias from the same cave deposits.

# 7

# Summary of Mesozoic and Tertiary igneous activity

In his 1972 paper on the Karoo Volcanic Cycle, Cox (p.330) concluded, inter alia, that 'all the Mesozoic igneous rocks of southern Africa . . . belong to a single long cycle of igneous activity'. He might have added Tertiary igneous rocks to this scheme as well because they are clearly associated with the kimberlite volcanism that forms the last gasp of the cycle. This statement may be justified in terms of the overall geochemical evolution of the magmas Cox envisaged, but to avoid oversimplification in a complex area further cognizance must be taken of the ages of the main centres of igneous activity and their relationship to local phenomena associated with the breakup of Gondwanaland. Sufficient data, both in terms of age determinations and regional setting are now available to make such a regional survey possible, and to discern tectonic/igneous relationships, even if it is beyond the scope of this book to propose geochemical models to link the various phenomena.

Figures 191 to 194 show the distribution in time and space of the various types of igneous rocks in the Mesozoic and Tertiary succession of southern Africa, as well as information on relevant igneous sequences from adjacent areas of Gondwanaland (South America and Antarctica). Although there is some local temporal overlap, which may be accounted for by regional differences in tectonic setting, Figure 192 clearly indicates a succession of igneous phases that can be distinguished in terms of both rock type (i.e. magma type) and style of emplacement (e.g. lavas, sills, dykes, diatremes) in relation to regional tectonic setting. We have numbered these phases I-VII, and Figure 194 summarizes their characteristics and relationships to plate tectonic events with particular reference to southern Africa, although the examples from South America and Antarctica are included for completeness. Two of these phases (IV and VI) are in fact periods of relative quiescence separating three periods of igneous activity that can be directly linked to major episodes in the plate tectonic development of southern Africa: phases I-III which include the main Karoo and Lebombo volcanism are synchronous with the Cape Orogeny and inter-continental rifting; phase V primarily relates to early drift and the opening of the South Atlantic; and phase VII includes post-drift alkaline and kimberlitic activity.

## 7.1 IGNEOUS PHASES I-III (Cape Orogeny and continental rifting)

As described in Chapter 2, Permian and early Triassic sediments in the main Karoo Basin contain abundant fine grained volcanic debris, presumably brought into the depositional environment by winds (e.g. Fuller 1970, Lock & Wilson 1975). Similar occurrences have been reported from the Permo-Triassic of Queen Maud Land and the Transantarctic Mountains (e.g. Elliot & Watts 1974), and here again no direct source has been identified for the ejecta. It is obvious, however, that a large province, with explosive volcanic centres existed in southern West Gondwana-West Antarctica from Lower Permian (?Sakmarian, uppermost Dwyka in South Africa, ?Artinskian, middle Buckley Formation in Antarctica) until at least Middle Triassic times. Elliot & Watts (1974) speculate that this volcanic source may have lain over West Antarctica, and de Wit (1977) has described evidence of a cordilleran volcanic arc along the Pacific-facing southern South America/Antarctic Peninsula sector of Gondwanaland, driven by subduction of the Phoenix plate (a suggestion made by Elliot 1975b). In southern Africa, the ?middle and late Triassic post-Cape Orogeny sediments (Elliot and Clarens formations) also contain abundant fine volcanic debris, so a source existed after the uplift of the Cape Trough (in the early-middle Triassic), perhaps implying that volcanism and the earlier (Permian) subsidence of the Cape Trough were not necessarily related. Cox (1978), on the other hand, suspects a direct link, not only between the earlier volcanism and the postulated west Gondwanaland subduction, but also between the latter, and the Karoo and Lebombo igneous activity.

Our phase I covers the relatively long period of early Permian to late Triassic time for which no known local centres of igneous activity have been identified in southern Africa, but during which large quantities of volcanic debris were supplied to the sedimentary environment. This phase of local igneous quiescence was

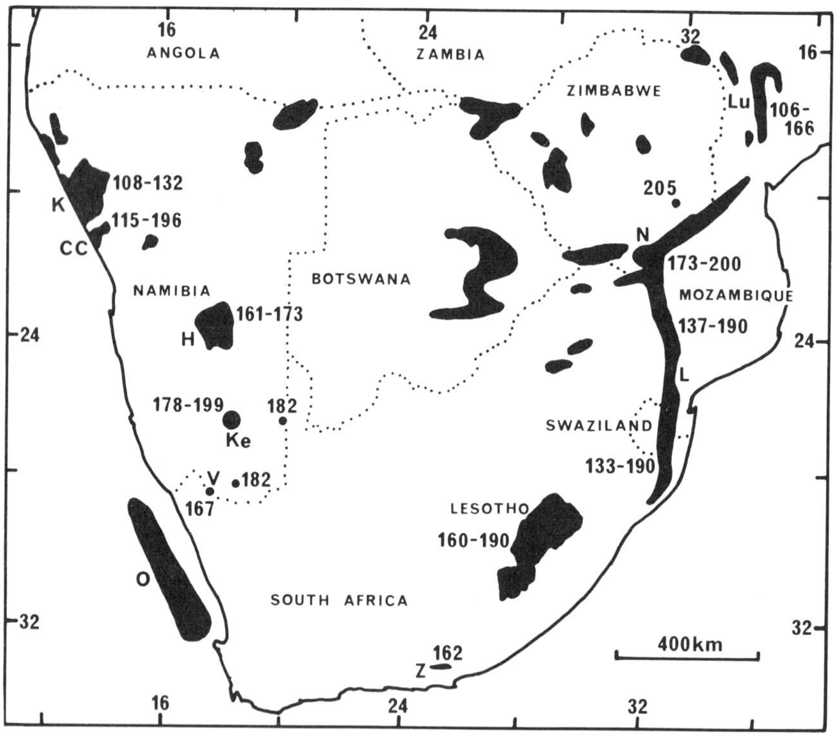


Figure 191. Distribution of Mesozoic basic igneous rocks in southern Africa. Ages and ranges for various sites quoted in my BP. Based mainly on Bristow (1980), with additions from Siedner & Miller (1968), Gidskehaug *et al.* (1975), and Siedner & Mitchell (1976). Other minor additions cited in the text. Abbreviations: K = Kaokoveld; CC = Cape Cross; H = Hoachanas; Ke = Keetmanshoop; V = Vioolsdrif; O = Orange Basin; Lu = Lupata; N = Nuanetsi; L = Lebombo; Z = Zuurberg.

abruptly brought to an end by widespread, mainly intrusive igneous activity (phase II) which marks the beginning of magmatism that is clearly related to the development of tensional zones across this part of Gondwanaland (Figs 192 & 193). Using de Wit's (1977) and Elliot's (1975a) reconstructions, rocks of our igneous phases II to V plot as shown in Figure 195. On the west coast (Namibia), phase II is represented by the Keetmanshoop dolerite sills (178-199 my) (Gidskehaug et al 1975, Richardson unpublished, in Bristow 1980), and the oldest dolerite dykes and alkaline intrusions (196 my) along the Cape Cross line, north of Walvis Bay (Siedner & Miller 1968). The latter were a precursor of more vigorous activity in phase V, but show an early establishment of geochemical characters that persisted from this rift setting through to an early post-drift situation (i.e. over a period of about 80 my). Phase II activity along what was to become the SE Atlantic continental margin was not, therefore, particularly extensive.

In contrast, widespread, and from field evidence, spectacular volcanic activity marked phase II in the Stormberg region of NE Cape. We have no published radiometric dates for the Stormberg vents, diatremes, and early lavas, but stratigraphic relationships show that they date between Norian (Elliot Formation, ca. 210 my) and the lowermost main Karoo lavas (ca. 190 my). Although minor rock types such as gabbros and andesites have been reported from this area, dolerites and low Mg tholeiitic lavas are volumetrically dominant, whilst geochemical analyses by Pemberton (1978), and Eales & Marsh (1979) have shown them to be in no way significantly different from the overlying main Karoo lavas and intrusions (Tab.68). Clearly, the explosive Stormberg episode was a precursor to the later, quieter, more widespread activity, and that the same magma sources were involved (Eales & Marsh 1979). Farther NE, the earliest manifestations of igneous activity along the eastern edge of the Kaapvaal Craton occur as carbonatitic intrusive complexes in the Nuanetsi region of SE Zimbabwe (e.g. Shawa, ca. 205 my, Nicolaysen et al 1962).

With progressive regional tensional stress, the dominantly intrusive phase II passes into the dominantly extrusive phase III which represents the classic Karoo volcanic sequence. As shown diagrammatically in Figure 194, two distinctive suites can be distinguished: linear eruptions along the edges of the cratonic blocks and major lines of crustal weakness; and more widespread and less heterogeneous intracratonic eruptions. In southern Africa, these are represented by the Lebombo and main Karoo (NE Cape-Lesotho) sequences, respectively. The history of the former has recently been reviewed by Bristow (1980), who shows that after the alkaline volcanism (our phase II), the centres of igneous activity migrated rapidly south and SE through the Nuanetsi, Tuli and north Lebombo region to Swaziland and the southern Lebombo. The earliest rocks are mafic (basalt, Sabie River Formation), which in Nuanetsi and the north Lebombo date from ca.

200 my, and continued to erupt until at least ca.173 my in the former, and ca.184 my in the latter (Jacobsen et al 1975). Farther south, in Swaziland, and in the southern Lebombo, the basal olivine-bearing rocks have not been directly dated, but Bristow (1980) concluded that they commenced somewhat later (ca.190 my), on the basis of the basalts overlying sediments (Elliot and Clarens formations) which contain clasts of the 197 my Dokolwayo Kimberlite in Swaziland. There is a small hiatus at the top of the Sabie River Basalt Formation, so that in the southern Lebombo, the contact with the overlying Jozini Rhyolites is a marked unconformity and farther north a smaller disconformity. The base of the rhyolites has been dated at ca.192 my in Nuanetsi (Manton, unpublished, quoted in Bristow 1980), and ca.177 my in the Lebombo, and more or less continuous eruptions appear to have taken place until about 160 my. Coeval with this activity, rhyolites dated at about 166 my were extruded in the Zambezi Graben (Flores 1964), and rhyolites and tholeiites have been reported from coastal Mozambique by Jaritz et al (1977) (157-177 my).

A significant geochemical characteristic of the mafic Lebombo rocks is their relatively high initial 87Sr/86Sr ratio (0.70968, average of Bristow's 1980 determinations). This contrasts with a value of 0.70570 for the main Karoo lavas (quoted by Ford & Kistler 1981), which were extruded more or less contemporaneously with the Lebombo Sabie River and Jozini formations, but to the west in an intra-cratonic setting. The main Karoo lavas date between ca.188 and 160 my, whilst their feeder dykes were emplaced between ca.190 and 137 my (though Bristow 1980 suspects argon loss and suggests that 155 my may be a more likely upper limit). No obvious tectonic control can be discerned in the main Karoo lava fields and sills. Contemporaneous and petrologically similar rocks occur in Namibia where the earliest lavas are those at Hoachanas (161-173 my). In terms of massive lava extrusion (which has been used to define phase III), there is a definite westward younging of activity across southern Africa: ca.200 my Lebombo; ca.188 my Karoo; 173 my Hoachanas. An unknown factor in this distribution is the precise age of the possible mid-Jurassic acid lavas reported from the Orange Basin by Gerrard & Smith (1980) (Fig.118). We relate these to the Argentinian acid lavas (see below), but stress here that the Orange Basin volcanics are tectonically analogous to the Lebombo suites; viz, they lie on the unstable side of major rift lines, and flank the basic main Karoo province on the craton (Fig.195).

The Zuurberg lavas of the Algoa Basin, which Marsh et al (1979) found geochemically indistinguishable from the tholeiites of the main Karoo lavas, have been dated at 162 my (Hill 1972). They lie adjacent to the faulted edges of the northern Algoa Basin, and although Marsh et al (1979) state they find no direct field evidence to link the igneous rocks to faulting, they consider that both faulting and magmatism were caused by tensional stress during continental rifting.

Extrusive igneous rocks (basic and acidic) occur extensively in areas of Gondwanaland adjacent to southern Africa. These have been plotted on a pre-drift reconstruction based on Elliot (1975a), as modified by de Wit (1977) (Fig.195) and several features on it should be noted. In Antarctica, Mesozoic tholeiites and basalts occur discontinuously in a 4 000 km zone along the western edge of East Antarctica, covering an estimated 100 000 $km^2$ and representing a volume of about 500 000 $km^3$ (Kyle et al 1981). These are the Ferrar Supergroup and they consist of basaltic lavas (Kirkpatrick Basalt Group), dolerite sills and dykes (Ferrar Dolerite Group) with a massive gabbroic intrusive complex (Dufek Massif). Kyle et al (1981), and Ford & Kistler (1980) recognize an important geochemical boundary at the northern edge of these outcrops which separate the northern low ratio 87Sr/86Sr (0.7044-0.7057) basalts and tholeiites in Queen Maud Land (the Weddell Sea Province of Ford & Kistler 1980) from the high ratio 87Sr/86Sr (0.7092-0.7142) tholeiites and dolerites of the Transantarctic Province which stretches from the Theron Mountains to Victoria Land. Kyle et al (1981) point out that the latter province lies along a remarkably straight zone that coincides with the Beardmore and Ross orogenic belts (of Pre-Cambrian and early Palaeozoic age, respectively) and is close to the continental side of the Antarctic extension of the Cape Fold Belt, variously referred to as the Weddell Orogenic Belt (Ford & Kistler 1980) or the Gondwanian Orogenic Belt (Elliot 1975a). Du Toit (1937) recognized these lineaments, which he referred to as part of his Samfrau Geosyncline. Plotting the igneous provinces onto Figure 195 shows that they lie adjacent to an extensive sequence of trans-Antarctic rift structures recently discovered by geophysical studies underlying the Weddell Sea (the Filchner Rift Zone, at least 15 km deep), and the Ross Sea (2-3 km deep) (Masolov et al 1981). In the reconstruction on Figure 195, the volcanic/rift lineaments of the Weddell Sea are continuations of the Lebombo-Zambezi regions of SE Africa, with a strong suggestion that the low ratio 87Sr/86Sr volcanics of Queen Maud Land are analogues to the main Karoo basalts, whilst the high ratio rocks of the Lebombo and Transantarctic Mountains lie on opposite sides of a complex inter-continental rift system that originally included the present-day Natal Valley, Mozambique Ridge, Zambezi Graben and Mozambique Channel, and the Filchner-Ross rifts. Radiometric age determinations (quoted in Ford & Kistler 1980, Kyle et al 1981) show that the igneous activity in East Antarctica has a spread of ca.153-181 my, which probably peaks around 179 my and possibly 165 my, suggesting an initiation somewhat later than that in the Lebombo-Karoo provinces, but with an upper limit of around the same age.

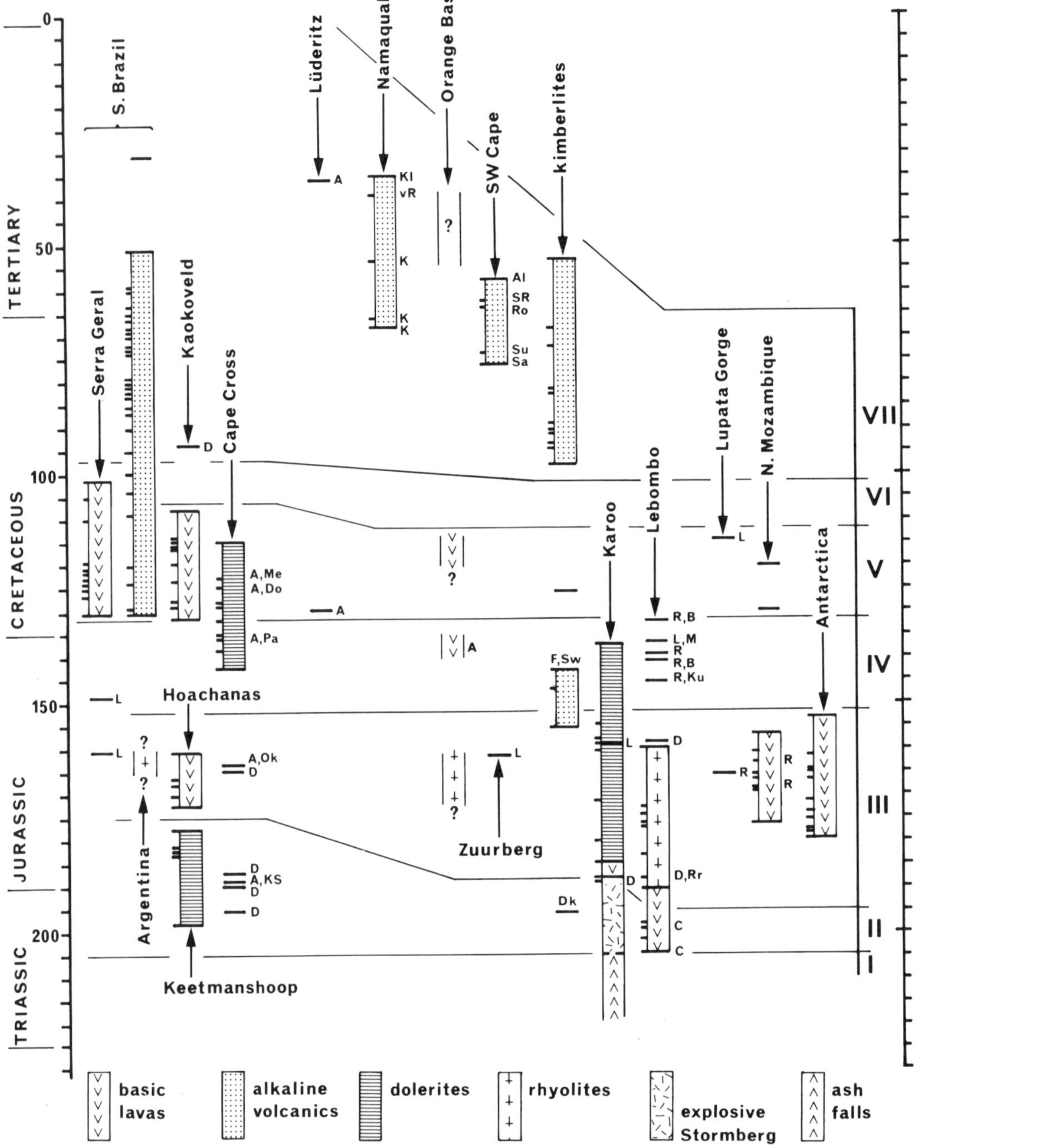


Figure 192. Temporal and spatial distribution of Mesozoic and Tertiary volcanism in southern Africa and adjacent Gondwanide localities. Age determinations shown as ticks on left side of columns, or as single bars. Igneous phases I-VII shown in right column. See text for discussion. Data from: Nicolaysen *et al.* (1962); McDougall (1963); Flores (1964, 1973); Siedner & Miller (1968); Fitch & Miller (1971); Dingle & Gentle (1972); Rex (1972); Hill (1972); Campos *et al.* (1974); Elliot (1975a); Gidskehaug *et al.* (1975); Marsh (1976); Urien *et al.* (1976); Siedner & Mitchell (1976); Pacca & Hiodo (1976); Jaritz *et al.* (1977); Botha & Hodgson (1977); Hawthorne *et al.* (1979); Moore (1979); Bristow (1980); Hargraves & Onstott (1980); Gerrard & Smith (1980). Ford & Kistler (1980); Kyle *et al.* (1981).

Abbreviations

Rock types: D = dolerite; A = alkaline rocks; K = kimberlite; L = basic lava.

Localities: Me = Messum; Do = Doros; Pa = Paresis; Ok = Okonjeje; KS = Klein Spitzkop; Kl = Klinghardt; vR = Vanrhynsdorp; Al = Alphard Banks; SR = Sout River; Ro = Robertson; Su = Sutherland; Sa = Saltpetrekop; Dk = Dokolwayo; F = Finsch; Sw = Swartruggens; B = Bumbeni; M = Movene; Ku = Kuleni; Rr = Rooirand.

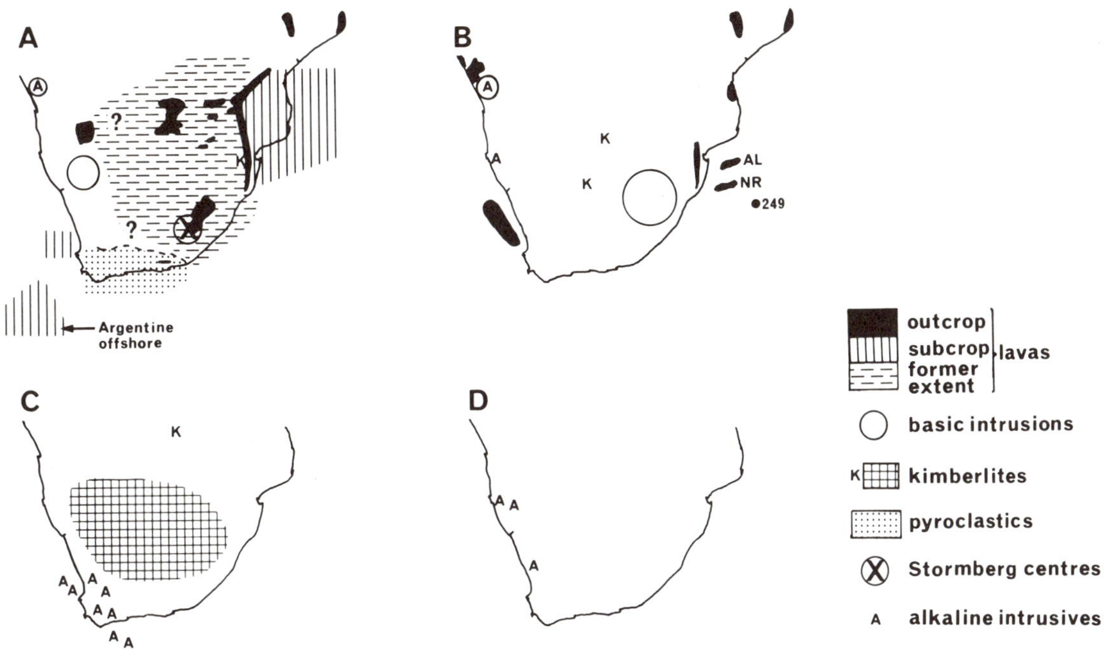


Figure 193. Distribution of igneous activity in southern Africa during Triassic to Tertiary time. The four time intervals selected are correlated with igneous phases I-VII. The numerous data sources are cited in the text and Figure 192.

A. Late Triassic to mid-Jurassic (200-160 my), phases I-III. The former extent of the Karoo lavas and their possible continuation eastwards is largely speculative.

B. Upper Jurassic to Lower Cretaceous (160-100 my), phases IV-VI. Extent of lavas in Orange Basin is speculative. Also, the ages of the Almirante Leite and Naude Ridge volcanics is not known, but have been placed here because of the large amounts of volcanic debris in Neocomian sediments on the Mozambique Ridge in DSDP 249.

C. Upper Cretaceous to early Palaeogene (100-50 my), early phase VII. Consists of the majority of kimberlites, and alkaline activity in the S.W.Cape.

D. Late Palaeogene (50-30 my), late phase VII. Alkaline activity in S.Namibia and Namaqualand.

Very extensive developments of mid-Jurassic acidic lavas and intrusions are known from the offshore areas of eastern Argentina (e.g. Urien et al 1976), but few petrological, geochemical, and chronological details of them have been published (see 4.5.3). They apparently have equivalents in the Orange Basin, where Gerrard & Smith (1980) report acid lavas below seismic reflector T as 'pre-graben' fill, and it seems probable that their evolution was related to the formation of extensive taphrogenic structures across what later became the Argentine and SW African continental margins. Onshore age equivalents adjacent to the Orange Basin are the Hoachanas lavas in central Namibia.

## 7.2 IGNEOUS PHASES IV-V (cessation of traphrogenesis and the initiation of continental drift)

Phase IV appears to have been a relatively short period (15 my) of quiescence, although locally considerable igneous activity occurred. In the southern Lebombo, the Kuleni rhyolitic lavas and dykes (ca.146 my), and the Bumbeni Complex acidic lavas, dykes and trachybasalts and trachyandesites (ca.133 my) (Bristow 1980) overlie the Jozini rhyolites with a strong unconformity, signifying a considerable period between their respective eruptions. This late phase of igneous activity along the Lebombo Line also produced the Movene basalts (ca.137 my) which are poorly exposed because of onlapping Cretaceous sediments under the Mozambique coastal plain. Waning dyke intrusion is also reflected by 137 my dates for some Karoo dolerites. A coeval precursor of later, more extensive activity are the 142-155 my dates for kimberlite plugs in the Finsch and Swartruggen clusters in the central part of the craton (Hargraves & Onstott 1980). Dolerite dyke swarms in coastal Namibia (Cape Cross area), with an age range of 136-143, fall within this phase, and belong to the group included in their 134.2 ± 0.9 my isochron by Siedner & Mitchell (1976). Amygdaloidal basic lavas below the ?Valanginian seismic reflector R may represent Upper Jurassic graben fill in the Orange Basin (Gerrard & Smith 1980) (Fig.118).

Phase V heralded a shift in the igneous centre of gravity, with extensive extrusion and intrusion of basic and alkaline magmas along the new, adjacent continental margins of SW Africa and SE South America. In Namibia, the Kaokoveld basaltic lavas of the Etendeka

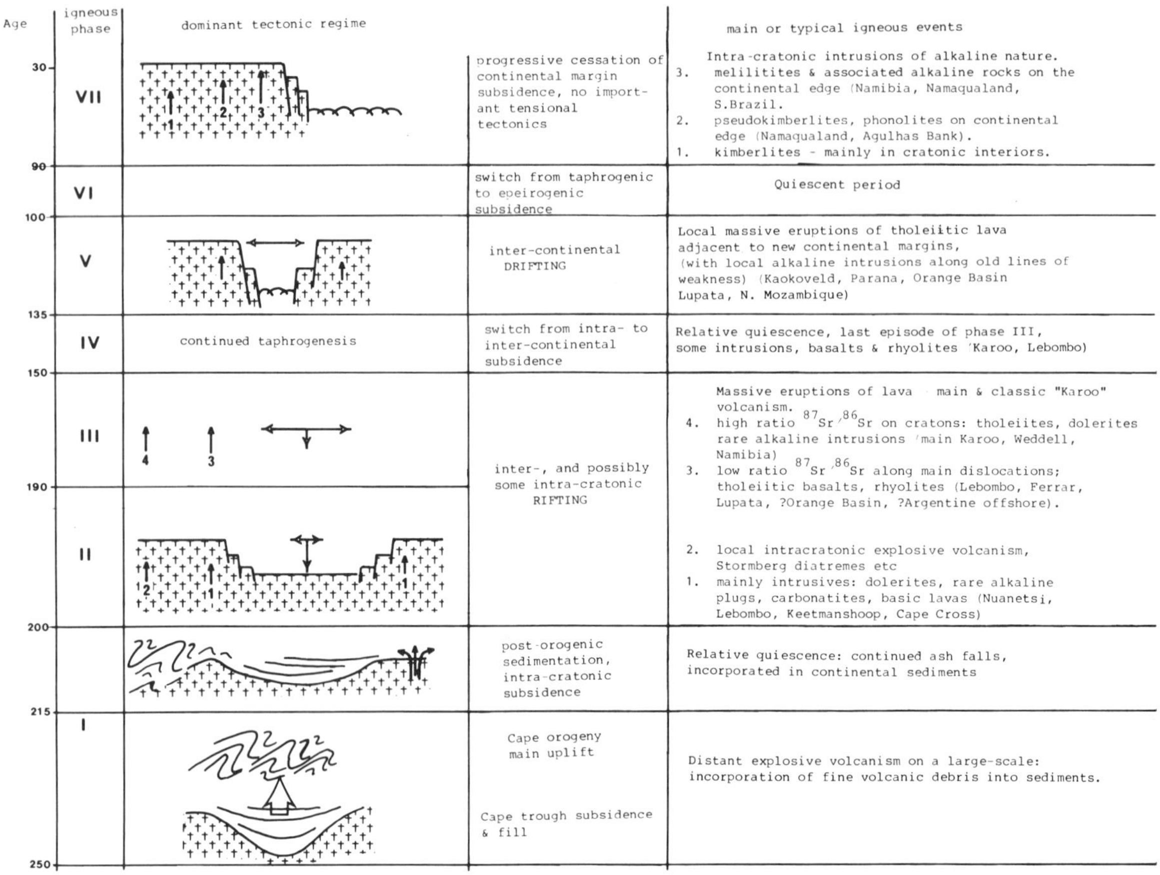


Figure 194. Mesozoic and Tertiary igneous activity in southern Africa related to the dominant tectonic regimes. Left columns show approximate ages in my and igneous phases I-VII (see Figure 192 for expansion of this theme). Column headed 'dominant tectonic regime' shows igneous centres (small arrows) relative to sites of continental crustal subsidence and tension. Relative rates of vertical and horizontal movement are shown by lengths of large arrows. Top two models (phases V and VII) show the generation of oceanic crust (hummocky topography). Numbers 1, 2, etc relate to specific igneous activity in time and space mentioned in right column text.

Plateau area were extruded over the period 108-132 my along a marginal fringe 300 km in length. Siedner & Miller (1968), and Gidskehaug et al (1975) claim a volcanic maximum at around 125 my, and this compares with an estimated age of drift intiation of ca. 127 my in the South Atlantic (see Chapter 1). Coeval with these extrusions, which culminated in rhyolites, are intrusions of dolerite dykes, and alkaline plugs and lavas along the Cape Cross Line which strikes inland, approximately normal to the Namibian margin for over 400 km parallel to the Damaraland Pre-Cambrian metamorphic belt. Whilst there can be no doubt that these activities were directly related to continental separation, a cautionary note needs to be sounded. Scrutton (1973a) made a review of the apparent relationship between continental separation and incidence of igneous activity, and came to the general conclusion that the climax of the latter frequently preceded the former by as much as 25 my, and appeared to corroborate the generally-held view that tensional rifting is 'normally' accompanied by extensive volcanism. In fact, this is not the case along the South Atlantic margins south of the Walvis Ridge because, with the exception of Barremian basic lavas reported by McLachlan & McMillan (1979) in the Orange Basin, no other extensive Lower Cretaceous basic lavas occur on the African or South American sides of the rift. The Kaokoveld (and their South American counterparts, the Serra Geral) lavas do not disprove this observation because although they are related to continental separation, they are centred on a curious tectonic anomaly, the Walvis Ridge/Rio Grande Rise-Torres Arch structure. The submarine expression of this anomalous effusive centre was formed by excessive volcanic activity as sea floor spreading proceeded, and remained a subaerial barrier straddling the South Atlantic until about Cenomanian times, and a shallow-water archipeligo until possibly as late as Maastrichtian times.

Table 68. Representative igneous rock analyses.

| % | 1D | 2D | 3D | 4D | 5Z | 6D | 7D | 8L | 9L | 10L | 11Ka | 12N | 13N | 14Ki | 15A | 16A |
|---|---|---|---|---|---|---|---|---|---|---|---|---|---|---|---|---|
| $SiO_2$ | 52.00 | 51.60 | 51.71 | 52.50 | 52.56 | 54.00 | 56.90 | 48.48 | 53.00 | 70.10 | 66.65 | 34.41 | 40.56 | 30.95 | 53.75 | 48.07 |
| $TiO_2$ | 01.00 | 00.95 | 00.98 | 01.00 | 01.07 | 01.02 | 00.93 | 02.93 | 01.60 | 00.57 | 01.11 | 06.00 | 03.32 | 02.34 | 00.70 | 02.06 |
| $Al_2O_3$ | 15.56 | 15.71 | 15.31 | 15.40 | 15.09 | 15.55 | 15.00 | 07.59 | 14.20 | 12.77 | 11.69 | 05.84 | 05.61 | 04.17 | 14.38 | 13.03 |
| $Fe_2O_3$ | 00.20 | 00.16 | 00.18 | 00.20 | 00.23 | 00.19 | 00.15 | 00.14 | 00.20 | 00.08 | 00.10 | 00.23 | 00.19 | tr | 00.18 | 00.19 |
| MgO | 07.20 | 07.11 | 06.83 | 07.10 | 05.85 | 06.15 | 04.10 | 16.30 | 05.70 | 00.32 | 00.96 | 16.28 | 15.32 | 32.30 | 06.64 | 08.59 |
| CaO | 10.30 | 10.60 | 10.84 | 10.30 | 07.58 | 08.97 | 08.20 | 06.49 | 09.10 | 01.70 | 02.40 | 13.08 | 09.06 | 08.92 | 10.60 | 09.70 |
| $Na_2O$ | 02.10 | 02.16 | 02.28 | 02.10 | 02.82 | 02.74 | 01.96 | 01.26 | 02.60 | 03.15 | 03.54 | 01.53 | 00.94 | 00.35 | 01.83 | 02.44 |
| $K_2O$ | 00.80 | 00.73 | 00.51 | 00.80 | 00.73 | 01.40 | 00.53 | 02.30 | 00.74 | 04.61 | 02.81 | 01.63 | 00.61 | 01.61 | 00.81 | 00.51 |
| $P_2O_5$ | 00.20 | 00.16 | 00.17 | 00.10 | 00.24 | 00.21 | 00.22 | 00.47 | 00.25 | 00.11 | 00.24 | 01.33 | 00.68 | – | 00.18 | 00.22 |
| $H_2O$ | * | * | * | * | 00.98 | 02.30 | 03.27 | 03.03 | * | 01.85 | 02.47 | * | 02.40 | 07.36 | 00.64 | 01.87 |
| LOI | * | * | * | * | 01.62 | * | * | * | * | * | * | * | 08.12 | 02.54 | * | 00.19 |
| Traces | * | | | * | | * | * | | | * | * | | | * | * | * |
| Sr | | 183 | 212 | | 385 | | | 992 | 297 | | | 1196 | 1029 | | | |
| Rb | | 13 | 12 | | 19 | | | 67 | 18 | | | 64 | 20 | | | |
| Zr | | 91 | 97 | | 108 | | | 403 | 142 | | | 574 | 315 | | | |
| Nb | | 4 | 7 | | 5 | | | 22 | 6 | | | 158 | 120 | | | |
| Y | | 24 | 25 | | 30 | | | 26 | 31 | | | 28 | 26 | | | |
| Zn | | * | * | | 101 | | | 106 | 100 | | | 123 | 97 | | | |
| Cu | | * | * | | 90 | | | * | 142 | | | 91 | 87 | | | |
| Ni | | 89 | 88 | | 52 | | | 898 | 79 | | | 547 | 595 | | | |
| Co | | * | * | | 42 | | | 78 | 48 | | | 99 | 56 | | | |
| Cr | | 271 | 264 | | 185 | | | 792 | 160 | | | 555 | 218 | | | |
| V | | 234 | 235 | | 263 | | | 188 | 267 | | | 305 | 250 | | | |
| La | | * | * | | 13 | | | * | * | | | * | * | | | |
| Ce | | * | * | | 25 | | | * | * | | | * | * | | | |
| Nd | | * | * | | 18 | | | * | * | | | * | * | | | |
| Ba | | * | * | | * | | | 959 | 286 | | | 836 | 843 | | | |
| Sc | | * | * | | * | | | 19 | 30 | | | * | * | | | |

All Fe expressed as $Fe_2O_3$; * – no analysis; traces in ppm rounded to nearest whole number

Key:

1. Karoo basalts, main Karoo Basin. Rhodes & Krohn (1972), N119
2. Lesotho Fm. lavas. Eales & Marsh (1979), N32
3. E.Cape Drakensberg dolerites. Eales & Marsh (1979), N17
4. Average Karoo dolerite, Walker & Poldervaart (1949), N43
5. Zuurberg basalts. Marsh *et al.* (1979), N4
6. Donnybrook basalt, Moshesh's Ford Fm. Lock *et al.* (1974), N1
7. Kraai River lava. Lock *et al.* (1974), N1
8. Picrite basalt, Lebombo Line. Bristow (1980), N9
9. Average S.Lebombo/Swaziland basalt. Bristow (1980), N26
10. Lebombo rhyolites. Wachendorf (1973), N32
11. Kaokoveld 'orthoclase porphyry'. Du Toit's No.47 (1954), N4
12. Average Namaqualand olivine melilitite and olivine nephelinite. Moore, Tab.3 (1979), N60
13. Average Namaqualand kimberlite. Moore, Tab.1c (1979), N7
14. Main Karoo kimberlite. Du Toit's No.58 (1954), ?N1
15. Transantarctic Mountain tholeiites. In Ford & Kistler (1980), N5
16. Queen Maud Land basalts. In Ford & Kistler (1980), N17

Abbreviations:
D = Drakensberg Group
Z = Zuurberg volcanics
L = Lebombo volcanics
Ka = Kaokoveld volcanics
N = Namaqualand volcanics
Ki = Kimberlites
A = Antarctic volcanics

The Walvis Ridge abutment lies adjacent to the Etendeka Plateau, and has a conjugate feature on the South American side in southern Brazil. Here the Rio Grande Rise abutment lies adjacent to the Torres Arch, and the basaltic Serra Geral lavas approach their closest to the coast. Inland, the lavas extend over a large part of the Parana Basin that occupies much of southern Brazil, and parts of Uruguay and Paraguay. Radiometric dates from the Serra Geral lavas show a close correspondence with the Kaokoveld lavas (101-132 my Pacca & Hiodo 1976, Campos et al 1974). There is no analogue to the Cape Cross Line on the Brazilian side, but a broad lineament of volcanic ring-complexes consisting of alkaline basalts and carbonatites (with kimberlites in the extreme east), that runs NE from the Walvis Ridge abutment zone into southern Angola (Beetz 1934), may be mirrored by a large number of alkaline intrusives to the north and NE of the Parana Basin in Sao Paulo State. Numerous dates have been derived for the latter, and they have a wide spread, but several are coe-

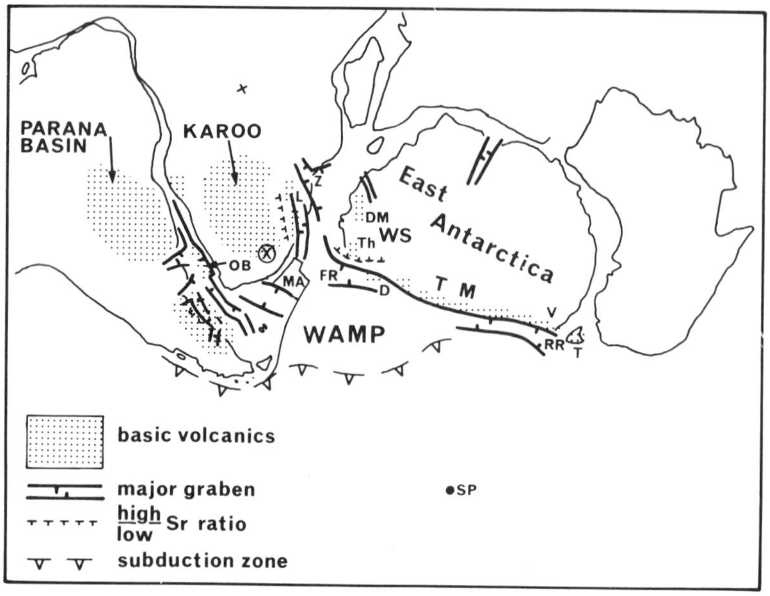


Figure 195. Major Mesozoic igneous and tectonic features in Gondwanaland. Refit is that of Figure 2 and is for late Triassic time. See text for discussion. Data mainly from Elliot (1975a); de Wit (1977); Ford & Kistler (1980); Bristow (1980); Masolov *et al.* (1981); Kyle *et al.* (1981). Abbreviations: DM = Dronning Maud Land theoleiites; WS = Weddell Sea Province; TM = Transantarctic Mountains Province; D = Dufek gabbro massif; Th = Theron Mountains; V = Victoria Land; T = Tasmania; FR = Filchner Rift; RR = Ross Rift; L = Lebombo Line; WAMP = West Antarctic microplates; MA = Maurice Ewing Bank/Agulhas Plateau Pre-Cambrian massif; OB = Orange Basin; Z = Zambezi Rift; SP = South Pole; ⊕ = Stormberg explosive volcanic province.

val with the Serra Geral lavas (100-131 my, Campos et al 1974). Only one published date is known from the Angolan structures: 112 ± 8 my (Lapido-Loureiro 1968).

## 7.3 IGNEOUS PHASES VI-VII (post-drift)

Large-scale lava extrusion ceased on the present continental areas in the South Atlantic between ca.100-108 my. This date was followed by a further period of quiescence (ca.10 my in length) that we have designated phase VI. A similar tailing off of activity in S Brazil is suggested by a relative lack of radiometric dates over the period 87-100 my (see Fig.192).

Igneous phase VII, which commenced in southern Africa at about 95 my, and continued sporadically until about 35 my (possibly 16 my), consisted almost exclusively of regional-scale intrusion of plug-like bodies of kimberlitic and alkaline affinities, with only minor alkaline extrusions. Kimberlite pipes have been dated as far back as 1 170 my in the case of the Premier pipe (Allsopp et al quoted in Hargraves & Onstott 1980), and in the Mesozoic occasional intrusions occurred at 197, 162, ca.150 and ca.125 my in the Finsch and Swartruggens pipes (Hargraves & Onstott 1980). The majority of the 150 or so plotted by Moore (1979) are Upper Cretaceous in age, and they occur over the whole of the central part of the subcontinent with a relatively small number of younger (Palaeogene) plugs in the NW Cape and Namaqualand. Although there are too few dates to confirm the trend, the impression certainly is that the younger intrusions lie progressively closer to the west coast (see Fig.168). Moore (1979) identifies three peaks in the distribution of the late Mesozoic-Palaeogene kimberlite dates: 90, 65 and 35 my with the earliest showing the larger cluster. As kimberlite emplacement tailed off in the late Cretaceous and early Tertiary, other small, local centres of alkaline volcanism, often consisting of a few plugs only affected the whole of the west coast from the Agulhas Bank and southern Cape (melilitites, phonolites, trachybasalts, etc.; Campanian-Palaeocene), Namaqualand (melilitites, pseudokimberlites; late Eocene-early Oligocene); and southern Namibia (various alkaline intrusives, and the Klinghardt phonolites in the vicinity of Lüderitz). Offshore, intrusions of Lower Tertiary age occur in the southern Orange Basin along an old taphrogenic (phase III) lineament. Their geochemical affinities are not known. Moore (1976, 1979) has pointed out the restriction of phonolite and olivine melilitites to the continental edges, and the preponderance of kimberlites in the interior, and considers continental edge flexing as a likely cause for this geochemical differentiation, whilst he links the times of emplacement to periods of sea level regression. Cox (1972) regards the kimberlites as originating from pockets of residual magma trapped at depth after the termination of the main Karoo volcanic activity, and that advanced crystallization of eclogite from picritic magmas was accompanied by a build up of volatiles that ultimately became the driving force for the creation of diatremes. He further postulated that only under the most stable tectonic conditions under cratonic areas would the pockets of residual magma remain sufficiently sealed to allow the necessary build up of pressure. Moore's (1976) ideas can be reconciled with this model by assuming that crustal flexing (which he attributes in some degree to sea level changes) allows these volatile-rich magma chambers to be tapped, and that progressively shallower magma sources are involved as the west coast is approached. Whilst these two suggestions have merit, they do not acount for the apparent chronological zonation of the intrusions. For example, why are the kimberlites close to the west coast younger than those in the interior?

Marsh (1973, 1975) has related the west coast alkaline igneous provinces at Cape Cross and Lüderitz to marginal offsets in the original rift patterns operating at the time of continental separation, and re-activated at later times. In the Cape Cross Line, intrusions are of two ages (Jurassic and early-middle Cretaceous). Whilst only the latter can be directly related to the SE Atlantic opening event, the earlier activity, which was of the same age as rift-related volcanicity in the eastern parts of the subcontinent (i.e. phases II and III), was along a line of weakness that may have been later exploited to form a marginal offset. In the much smaller Lüderitz province, only the nepheline syenites from the Granitberg foyaite complex (ca.130 my, Marsh 1975) have been dated as coeval with the younger Cape Cross intrusions, although on geomorphological grounds he considers the Drachenberg syenite complex and possibly the Garubberg carbonatites to be the same age. These intrusions lie in the vicinity of the southern margin of the Lüderitz Arch and may have been related to this older line of crustal weakness. In the youngest episode of volcanic activity along the Namaqualand-Namibia coast, Marsh (1973) suggests that reactivation of this lineament, which he postulates formed an important marginal offset-transform fault, accounts for the location of the phonolitic lavas in the Klinghardt Mountains (ca.37 and 36 my, Kröner 1973) and also probably for the melilitite and nephelinite plugs in the Pomona and Bogenfels areas which he considers the same age (Marsh 1975). Marsh's (1973) ideas are a special case of Moore's (1976) general model for tapping residual magmas marooned at the base of the crust, i.e. old lines of weakness that ran more or less normal to the line of continental separation, whilst the bulk of the kimberlitic/melilititic intrusions can be related only to approximately concentric zones around the continental edge.

Marsh (1976) concluded that in the Lüderitz area two (possibly three) distinct magma batches must have been available: one that gave rise to silica-poor, low pressure differentiates (Granitberg feldspathoidal rocks); and a critically oversaturated magma that differentiated towards silicic compositions (Drachenberg and Pomona syenites). This implies that during the initial opening of the South Atlantic, both magma types were independently tapped in close proximity, whilst during the later reactivation in the Tertiary, only the oversaturated magma erupted. Such relationships indicate that a parent magma suffered local contamination and/or crystal fractionation, or that distinct magnas were involved. The latter possibility implies mantle inhomogeneity.

Erlank et al (1980) have discussed the latter possibility in terms of southern African magmatism by drawing on previously published data, and their own results obtained as part of the International Geodynamics Project. They used two main approaches, namely Sr-isotope, and inter-element ratios, and concentrated their attention on the main Karoo area (including the Stormberg centres), and the Lebombo/Nuanetsi lines. Some of the regions (e.g. the southern Lebombo) show a large range in 87Sr/86Sr initial ratios, whilst others have a narrow spread of values, but all show different peaks, except for the main Karoo and Nuanetsi (Fig. 196). Consdering data such as these in terms of magma sources, Erlank et al (1980) suggest various alternatives. For example: the low initial Sr-isotope ratios of the Lebombo rhyolites (0.703-0.704) compared to the higher values of the bulk of the southern Lebombo mafic rocks (main peak at 0.707-0.708) suggests that the former (together with the mafics at the secondary peak 0.702-0.703) could have been derived directly from the mantle, whereas the latter could be ascribed to crustal contamination. As they point out, the problem with this hypothesis is why only the mafic rocks suffered contamination, as both they and the rhyolites were erupted along the same narrow zone of crustal weakness. Alternatively, the rhyolites may have been derived from the (?lower) crust, and that the mafics were contaminated with a further, as yet unknown, crustal source. However, there is no evidence for crustal contamination of the Lebombo mafics (or the main Karoo basalts) as shown by their lack of positive correlation between initial Sr-isotope ratio, and elements such as $SiO_2$, $K_2O$, and Rb, which are typically enriched in the continental crust relative to the mantle. Lateral mantle heterogeneity is suspected.

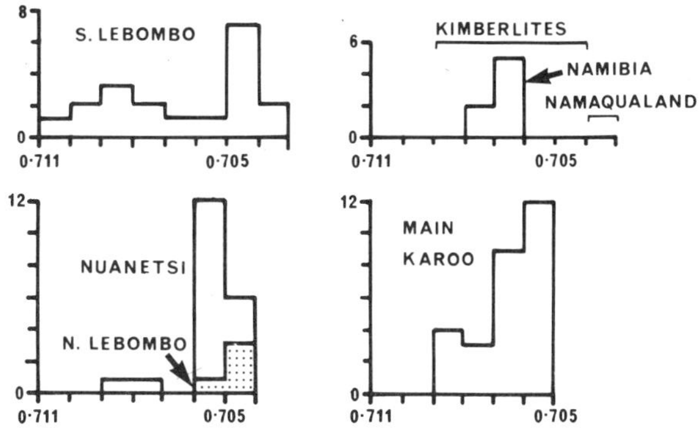


Figure 196. Histograms of initial 87Sr/86Sr ratios for Mesozoic and Tertiary mafic rocks (<54 % $SiO_2$) in southern Africa. Compiled from data in Erlank *et al.* (1980). Reference to Namaqualand is for a range of values for olivine melilitites. The number of determinations for kimberlites and olivine melilitites is not known.

The case for heterogeneity is strengthened by data that Erlank et al (1980) assemble and synthesize on elemental variations. Data originally reported by Lock et al (1974), Robey (1976) and Pemberton (1978) for the Stormberg mafic rocks show that certain ratios (e.g. $TiO_2$/Zr and Zr/Nb) show significant differences for each of the three main sequences of mafic rocks in the area: Drumbo Basalts of the Moshesh's Ford Formation; Kraai River Formation; and the Lesotho Formation. Erlank et al (1980) conclude that such variations also strongly suggest heterogeneous, and not originally homogeneous mantle sources that were modified by either fractionation or contamination. Further, rare earth element data from the Lebombo Line shows marked regional trends: in the southern Lebombo, the basalts have rare earth contents similar to those from the Moshesh's Ford Formation, and the main Karoo basalts (which are comparable with other continental tholeiites), whereas north of Swaziland, the mafic rocks of the Lebombo show conspicuous enrichment in these elements.

Erlank et al (1980) concluded that geochemical data such as summarized above, indicate that the overall geochemical compositional variability of the basalts in the Lebombo and Karoo regions are primarily due to lateral mantle heterogeneity (possibly vertically arranged) and not due to variable contamination or fractionation of a common magma. Whether this phenomenon developed progressively before the Triassic-Tertiary eruptions that concern us here, or whether one event of dehomogenization of the mantle can be identified, is not clear. However, they indicate that heterogeneity is evident by 2 000 my, and there are indications, of which the Mesozoic eruptions form an important part of the evidence, that variability in initial Sr-isotope ratios has developed progressively, implying that local geochemical evolution of the mantle is an ongoing process.

APPENDIX 1

# Upper Permian to Lower Triassic tetrapods

DICYNODON ZONE
Order Cotylosauria
Suborder Diadectomorpha
*Paeiasaurus serridens* Owen 1876
*Anthodon* sp.
*Spondylolestes rubidgei* Broom 1937
*Owenetta rubidgei* Broom 1939
Order Millerosauria
*Milleretta rubidgei* (Broom 1938)
*Millerosaurus nuffieldi* Watson 1957
*M.pricei* (Watson 1957)
*M.ornatus* Broom 1948
Order Therapsida
Suborder Anomodontia
Infraorder Dicynodontia
*Dicynodon lacerticeps* (Owen 1876)
*D.ingens* Broom 1907
*D.kitchingi* Broom 1937
*D.leontops* Broom 1913
*D.ictidops* Broom 1913
*D.namus* Broom 1913
*D.microdon* Broom 1936
*D.osborni* Broom 1921
*D.benjamini* Broom 1948
*D.duvenhagei* Broom 1948
*D.cadlei* Broom 1940
*D.weatherbyi* Broom 1941
*D.sidneyi* Broom 1940
*D.macrorhynchus* Broom 1921
*D.venteri* Broom 1925
*D.watsoni* Broom 1921
*D.woodwardi* Broom 1921
*D.leontocephalus* Broom 1950
*D.grahami* Broom 1940
*D.halli* Watson 1914
*D.lissops* Broom 1913
*Oudenodon baini* Owen 1860
*Dicynodontoides parringtoni* Broom 1940
*Dinanomodon rubidgei* Broom 1938
*Aulacephalodon baini* (Owen 1844)
*Pelanomodon moschops* (Broom 1913)
*P.halli* Broom 1950
*Proaulacocephalodon miltoni* Toerien 1955
*Emydorhynchus palustris* Broom 1913
*Newtonella platyceps* Broom 1937
Suborder Theriodontia
Infraorder Gorgonopsia
*Arctops watsoni* Brink & Kitching 1953
*A.minor* Broom 1948
*Cyonosaurus rubidgei* (Broom 1937)
*C.longiceps* Olson 1937
*C.kitchingi* (Broom 1948)
*Lycaenops minor* (Brink & Kitching 1953)
*L.kingwilli* (Broom 1948)
*L.angusticeps* Broom 1942
*L.ornatus* Broom 1925
*Nanogorgon gracilis* Broom & Robinson 1948
*Rubidgea* sp.
*R.major* Brink & Kitching 1953
*R.platyrhinus* Brink & Kitching 1953
*Leontocephalus rubidgei* (Broom 1940)
*L.cadlei* (Broom 1940)
*Pseudohipposaurus kitchingi* (Broom 1948)
*Prorubidgea alticeps* (Brink & Kitching 1953)
*P.brodei* (Broom 1941)
*P.robusta* Brink & Kitching 1953
*P.maccabei* Broom 1940
*Arctognathus curvimda* (Owen 1876)
*Cerdognathus greyi* Broom 1915
*Dinogorgon pricei* (Broom & George 1950)
*D.quinquemolaris* van Huene 1950
*Aloposaurus tenuis* (Brink & Kitching 1953)
*Aelurosaurus wilmanae* (Broom 1940)
*Scylacognathus parvus* Broom 1913
*Broomicephalus laticeps* Brink & Kitching 1953
*Rubidgeina angusticeps* Broom 1942
*Clelandina scheepersi* (Brink & Kitching 1953)
*Sycosaurus vanderhorsti* (Broom & George 1950)
*Paragalerhinus rubidgei* (Broom 1936)
*Ictidorhinus martinsi* Broom 1913
Infraorder Scaloposauria
*Ictidosuchops intermedius* (Broom 1938)
*I.baurioides* Broom 1940
*I.rubidgei* Broom 1938
*Tetracynodon* sp.
*T.tenuis* Broom & Robinson 1948
*Homodontosaurus kitchingi* Broom 1949
*Scaloporhinus angulorugatus* Boonstra 1953
*Nanictidops kitchingi* Broom 1940
*Lycideops longiceps* Broom 1931
*Pelictosuchus paucidens* Broom 1940
Infracorder Therocephalia
*Whaitsia paltyceps* Haughton 1918
*Moschorhinus kitchingi* Broom 1920
*Cerdosuchus aulodon* Broom 1936
*Notaelurodon kitchingi* Broom 1936
*Cerdosuchoides brevidens* Broom 1936
*Akidnognathus parvus* Haughton 1918
*Promoschorhynchus platyrhinus* Brink 1954
Infraorder Cynodontia
*Cyrbasiodon boycei* Broom 1931
*Procynosuchus delaharpeae* Broom 1937
*Cynosaurus suppostus* (Owen 1876)
Order Eosuchia
Suborder Younginiformes
*Youngina* sp.
*Y.capensis* Broom 1914
*Palaeagama vielhaueri* Broom 1926
*Saurosteronon baini* Huxley 1868

LYSTROSAURUS ZONE
Order Cotylosauria
Suborder Diadectomorpha
*Procolophon* sp.
*P.trigoniceps* Owen 1876
*Owenetta* sp.
Order Therapsida
Suborder Anomodontia
Infraorder Dicynodontia
*Lystrosaurus declivis* (Owen 1860)
*L.murrayi* (Huxley 1859)
*L.oviceps* Haughton 1915
*L.maccaigi* (Seeley 1898)
*L.platyceps* (Seeley 1898)
*L.curvatus* (Owen 1876)
*L.rubidgei* Broom 1940
*Myosaurus gracilis* Haughton 1917
Suborder Theriodontia
Infraorder Scaloposauria
*Scaloposaurus constrictus* Owen 1876
*Zorillodontops gracilis* Cluver 1969
*Ericiolacerata parva* Watson 1931
*Olivieria parringtoni* Brink 1968
*Tetracynodon darti* Sigogrean 1963
*Regisaurus jacobi* Mendrez 1971
Infraorder Therocephalia
*Moschorhinus kitchingi* Broom 1920
Infraorder Cynodontia
*Cyrbasiodon boycei* Broom 1931
*Thrinaxodon liorhinus* Seeley 1894
*Galesaurus planiceps* Owen 1859
*Diademodon tetragonus* Seeley 1894
Order Thecodontia
Suborder Proterosuchia
*Proterosuchus vanhoepeni* (Haughton 1924)
*P.fergusi* Broom 1903
Order Eosuchus
Suborder Younginiformes
*Heleosuchus griesbachi* (Owen 1876)
*Paliguana whitei* Broom 1903
*Noteosuchus colletti* (Watson 1911)
Order Prolacertilia
Suborder Parathecodontia
*Prolacerta broomi* Parrington 1935
*Prolacerata* sp.
Order Araeoscelidia
*Aenigmasaurus grallator* Parrington 1953
Amphibia: Labyrinthodonta
*Limnoiketes paludinatans* (Parrington 1948)
*Lydekkerina huxleyi* (Lydekker 1889)
*L.putterilli* Broom 1930
*Uranocentrodon senekalensis* (van Hoepen 1911)
*Laccocephalus inspecratus* Watson 1919

CYNOGNATHUS ZONE
Order Cotylosauria
Suborder Diadectomorpha
*Microtheledon parvus* (Broom 1905)
*Myocephalus crassidens* Broom 1936
*Thelegnathus browni* Broom 1905
Order Therapsida
Suborder Anomodontia
Infraorder Dicynodontia
*Kannemeyeria simocephalus* (Weithofer 1888)
*Kombuisia frerensis* Hotton 1974
Suborder Theriodontia
Infraorder Cynodontia
*Cynognathus creteronotus* Seeley 1895
*Diademodon tetragonus* Seeley 1894
*Trirachodon berryi* Seeley 1894
*T.kannemeyeri* Seeley 1895
*Tribolodon frerensis* Seeley 1895
Infraorder Bauriamorpha
*Aelurosuchus browni* Broom 1906
*Melinodon simus* Broom 1905
*Microgomphodon oligocynus* Seeley 1895
*Sesamodon browni* Broom 1905
*Bauria cynops* Broom 1909
*Watsoniella breviceps* Broili & Schröder 1935
Order Thecodontia
Suborder Proterosuchia
*Erythrosuchus africanus* Broom 1905
*Euparkeria capensis* Broom 1913
Order Rhynchocephalea
Suborder Sphenodontia
*Palacrodon browni* Broom 1906
Suborder Rhychosauria
*Howesia browni* Broom 1905
*Mesosuchus browni* Watson 1912
Amphibia; Labyrinthodonta
*Micropholis stowi* Owen 1876
*Microposaurus rasei* Haughton 1925
*Trematosuchus sobeyi* (Haughton 1915)
*Laidleria gracilis* Kitching 1957
*Rhytidosteus capensis* Owen 1884
*Batrachosuchus browni* Broom 1903
*B.watsoni* Haughton 1925
*Kestrosaurus dreyeri* Haughton 1925
*Parotosaurus haughtoni* (Boili & Schröder 1935)

APPENDIX 2

# Fossils from the New (E) Varswater Quarry, Langebaan Road

| | GM | QSM | PPM1 | PPM2 | Inferred habitat preference |
|---|---|---|---|---|---|
| PHYLUM PROTOZOA | | | | | |
| ORDER FORAMINIFERA | | | | | |
| *Ammonia beccarii* | – | – | – | • | marine (littoral) and estuarine |
| *Elphidium advenum* | – | – | – | • | |
| *Cibicides lobatulus* | – | – | – | • | marine (littoral; attached to vegetation, etc.) |
| *Planorbulina mediterranensis* | – | – | – | • | |
| *Rosalina* cf. *bradyi* | – | – | – | • | |
| PHYLUM BRACHIOPODA (lamp shells) | | | | | |
| *Kraussina rubra* | • | – | – | – | marine (rocky shore) |
| PHYLUM ECHINODERMATA | | | | | |
| CLASS ECHINOIDEA (sea urchins) | | | | | |
| *Parechinus angulosus* | • | – | – | – | marine (rocky shore) |
| PHYLUM ARTHROPODA | | | | | |
| CLASS CRUSTACEA | | | | | |
| ORDER CIRRIPEDIA (acorn barnacles) | | | | | |
| Gen. & sp. indet. | • | – | – | – | marine (rocky shore) |
| ? ORDER DECAPODA (lobsters, etc.) | | | | | |
| Gen. & sp. indet. | • | – | – | – | marine, estuarine and freshwater |
| SUBCLASS OSTRACODA | | | | | |
| Family Cytheridae | | | | | |
| *Gomphocythere expansa* | – | • | – | – | freshwater |
| Family Cyprididae | | | | | |
| *Zonocypris cordata* | – | • | – | – | |
| Family not det. | | | | | |
| Gen. & at least 3 spp. not det. | – | • | – | – | |
| PHYLUM MOLLUSCA | | | | | |
| CLASS GASTROPODA | | | | | |
| ORDER ARCHAEOGASTROPODA | | | | | |
| Family Patellidae (limpets) | | | | | |
| *Cellana capensis* | • | – | – | – | marine (rocky shore, warm water) |
| ? *Cellana* sp. | • | – | – | – | marine (rocky shore) |
| *Patella granularis* | • | – | – | – | |
| Family Fissurellidae (keyhole limpets) | | | | | |
| *Diodora parviforata* | • | – | – | – | marine (rocky shore) |
| Family Haliotidae (perlemoens or abalones) | | | | | |
| *Haliotis saldanhae* | • | – | – | – | marine (rocky shore) |
| *Haliotis* sp. | • | – | – | – | |
| Family Trochidae (top shells) | | | | | |

| | GM | QSM | PPM1 | PPM2 | Inferred habitat preference |
|---|---|---|---|---|---|
| *Oxystele tigrina* | ● | – | – | – | |
| *Oxystele variegata* | – | ● | – | – | marine (rocky shore) |
| *Gibbula benzi* | – | ● | – | – | |
| Family Turbinidae (turban shells) | | | | | |
| *Turbo sarmaticus* | ● | – | – | – | marine (rocky shore, warm water) |
| Family Phasianellidae (pheasant shells) | | | | | |
| *Tricolia neritina* | ● | ● | – | – | marine (rocky shore) |
| *Tricolia capensis* | – | ● | – | – | marine (rocky shore, cold water) |
| ORDER MESOGASTROPODA | | | | | |
| Family Littorinidae (periwinkles) | | | | | |
| *Littorina* cf. *knysnaensis* | – | ● | – | – | marine (rocky shore) |
| ? *Littorina* sp. | ● | – | – | – | |
| Family Hydrobiidae (snails) | | | | | |
| *Tomichia ventricosa* | – | ● | – | – | fresh or brack water |
| Family Assimineidae (snails) | | | | | |
| *Assiminea* sp. | – | ● | – | – | mainly estuarine |
| ORDER NEOGASTROPODA | | | | | |
| Family Muricidae (rock shells or whelks) | | | | | |
| *Ocenebra scrobiculata* | ● | – | – | – | marine (rocky shore) |
| Family Thaisidae (rock shells or whelks) | | | | | |
| *Thais dubia* | ● | – | – | – | marine (rocky shore) |
| Family Columbellidae (dove shells) | | | | | |
| *Pyrene albuginosa* | – | ● | – | – | marine (rocky shore, warm water) |
| Family Nassariidae (plough shells, dog whelks) | | | | | |
| *Bullia* sp. | ● | – | – | – | marine and/or estuarine |
| *Bullia digitalis* | – | ● | – | – | marine (sandy shore) |
| *Bullia laevissima* | – | ● | – | – | marine (sandy shore and estuarine) |
| *Bullia* sp. nov. | – | ● | – | – | marine and/or estuarine |
| *Nassarius* cf. *analogicus* | – | ● | – | – | marine (sandy shore and estuarine) |
| *Nassarius* sp. B | – | ● | – | – | marine and/or estuarine |
| *Nassarius* sp. C | | | | | marine and/or estuarine |
| Family Marginellidae | | | | | |
| *Marginella* sp. | – | ● | – | – | marine and/or estuarine |
| Family Turridae (screw shells) | | | | | |
| *'Crassispira'* sp. | ● | – | – | – | |
| *'Clavatula'* sp. | ● | – | – | – | marine |
| *'Turris'* sp. | ● | – | – | – | |
| ORDER ENTOMOTAENIATA | | | | | |
| Family Pyramidellidae | | | | | |
| *Turbonilla kraussi* | ● | – | – | – | marine |
| ? *Pyramidella* sp. | ● | – | – | – | |
| ORDER BASOMMATOPHORA | | | | | |
| Family Siphonariidae (false limpets) | | | | | |
| *Siphonaria* sp. | ● | – | – | – | marine (rocky shore) |
| Family Ferrissiidae (snails) | | | | | |
| *Burnupia capensis* | – | ● | – | – | on rushes in fresh-water or estuaries |
| Family Planorbidae (snails) | | | | | |
| *Ceratophallus natalensis* | – | ● | – | – | freshwater (subtropical, tropical) |

| | GM | QSM | PPM1 | PPM2 | Inferred habitat preference |
|---|---|---|---|---|---|
| *Bulinus 'tropicus'* | – | ● | – | – | freshwater |
| ORDER STYLOMMATOPHORA | | | | | |
| Family Succineidae (snails) | | | | | |
| *Succinea* sp. | – | ● | – | – | terrestrial or semi-aquatic (freshwater) |
| Family Endodontidae (snails) | | | | | |
| *Trachycystis* cf. *capensis* | – | ● | – | – | terrestrial |
| CLASS AMPHINEURA | | | | | |
| Family Chitonidae (chitons) | | | | | |
| *Chiton nigrovirescens* | – | ● | – | – | marine (rocky shore, cold water) |
| CLASS PELECYPODA | | | | | |
| Family Donacidae (sand mussels) | | | | | |
| *Donax serra* | cf. | ● | – | – | marine (sandy shore) |
| *Donax* sp(p). | ● | – | – | – | |
| Family Carditidae (false cockles) | | | | | |
| *Cuna aquaedulcensis* | – | ● | – | – | marine |
| Family Arcidae (ark shells) | | | | | |
| *Barbatia obliquata* | ● | – | – | – | marine (rocky shore, warm water) |
| Family Mytilidae (mussels) | | | | | |
| ? *Perna* sp. | ● | – | – | – | marine (rocky shore) |
| Family Veneridae (venus shells) | | | | | |
| *Pitar* sp. | ● | – | – | – | marine |
| MOLLUSCA INDET. | – | – | – | ● | |
| PHYLUM CHORDATA | | | | | |
| CLASS CHONDRICHTHYES | | | | | |
| ORDER SELACHII (sharks) | | | | | |
| Family Hexanchidae | | | | | |
| *Notidanus serratissimus* | ● | – | – | – | marine |
| Family Carcharhinidae | | | | | |
| *Carcharhinus melanopterus* | ● | – | – | – | marine (warm water) |
| *Carcharhinus limbatus* | ● | – | – | – | |
| *Galaeorhinus* sp. | ● | – | – | – | marine |
| *Prionace glauca* | ● | – | – | – | |
| *Negaprion* or *Hypoprion* sp. | ● | – | – | – | |
| Family Odontaspidae | | | | | |
| *Odontaspis accutissima* | ● | – | – | – | |
| *Odontaspis* sp. B | ● | – | – | – | |
| *Odontaspis* sp. C | ● | – | – | – | |
| Family Otodontidae | | | | | |
| *Megaselachus megalodon* | ● | – | – | – | |
| Family Carcharodontidae | | | | | |
| *Carcharodon* sp. | ● | – | – | – | |
| Family Isuridae | | | | | |
| *Isurus* sp. | ● | – | – | – | |
| Family Squalidae | | | | | |
| *Squalus* sp. | ● | – | – | – | |
| Family Squatinidae | | | | | |
| *Squatina africana* | ● | – | – | – | marine (warm water) |
| *Squatina* sp. B | ● | – | – | – | marine |
| SELACHII not det. | ● | ● | ● | – | |

| | GM | QSM | PPM1 | PPM2 | Inferred habitat preference |
|---|---|---|---|---|---|
| ORDER BATOIDEA (skates, rays) | | | | | |
| Family Rajidae (skates) | | | | | |
| *Raja* sp. | • | – | – | – | marine |
| Family Trygonidae (sting rays) | | | | | |
| Gen. & sp. not det. | • | – | – | – | marine |
| Family Myliobatidae (eagle rays) | | | | | |
| *Myliobatis* sp. | • | – | – | – | marine |
| CLASS OSTEICHTHYES | | | | | |
| ORDER PERCIFORMES | | | | | |
| Family Sparidae (mussel-crackers) | | | | | |
| Gen. & sp. not det. | • | – | – | – | marine |
| ORDER SILURIFORMES | | | | | |
| Family Tachysuridae (catfish) | | | | | |
| ? *Tachysurus* sp. | – | • | • | – | marine and estuarine |
| OSTEICHTHYES not det. | • | • | • | • | |
| CLASS AMPHIBIA | | | | | |
| ORDER ANURA (frogs, toads) | | | | | |
| Gen. & sp(p). not det. | – | • | • | – | terrestrial and freshwater |
| CLASS REPTILIA | | | | | |
| ORDER CHELONIA (tortoises, turtles) | | | | | |
| *Chersina* sp. | • | • | • | – | terrestrial |
| Gen. & sp. not det. | – | • | – | – | marine or freshwater |
| Gen. & sp. not det. | – | – | • | – | marine or freshwater |
| ORDER SQUAMATA | | | | | |
| SUBORDER LACERTILIA | | | | | |
| Family Chameleontidae (chameleons) | | | | | |
| Gen. & sp(p). not det. | – | • | • | – | bushes and trees |
| Family Gekkonidae (geckos) | | | | | |
| Gen. & sp(p). not det. | – | • | • | – | terrestrial |
| Family Varanidae (leguans, monitor lizards) | | | | | |
| *Varanus* sp. | – | • | – | – | terrestrial or freshwater |
| SUBORDER OPHIDIA (snakes) | | | | | |
| Gen. & sp(p). not det. | – | • | • | – | terrestrial |
| SQUAMATA not det. | – | • | • | – | |
| CLASS AVES | | | | | |
| ORDER STRUTHIONIFORMES | | | | | |
| Family Struthionidae (ostriches) | | | | | |
| *Struthio* sp. | – | – | • | – | terrestrial (open country) |
| ORDER SPHENISCIFORMES | | | | | |
| Family Spheniscidae (penguins) | | | | | |
| *Inguza predemersus* | – | • | cf. | – | marine (mainly coastal, cold water) |
| *Dege hendeyi* | cf. | • | • | – | marine (mainly coastal, cold water) |
| ORDER PODICIPEDIFORMES | | | | | |
| Family Podicipedidae (grebes) | | | | | |
| Gen. & sp. not det. | – | • | • | – | ?freshwater |
| ORDER PROCELLARIIFORMES (petrels etc.) | | | | | |
| Gen. & at least 3 spp. not det. | – | • | • | – | marine (pelagic) |
| ORDER PELECANIFORMES (pelicans etc.) | | | | | |
| Fam., gen. & sp. not det. | – | • | – | – | |
| Family Phalacrocoracidae (cormorants) | | | | | |
| Gen. & at least 2 spp. not det. | – | • | • | – | marine and/or freshwater |

| | GM | QSM | PPM1 | PPM2 | Inferred habitat preference |
|---|---|---|---|---|---|
| ? Family Sulidae (gannets) | | | | | |
| Gen. & sp. not det. | – | ● | – | – | marine |
| ORDER CICONIIFORMES | | | | | |
| Family Ciconidae (storks) | | | | | |
| Gen. & sp. not det. | – | ● | ● | – | terrestrial (open country and/or marsh) |
| Family Threskiornithidae (ibises, spoonbills) | | | | | |
| Gen. & sp. not det. | – | ● | – | – | terrestrial (open country and fresh-water) |
| ORDER ANSERIFORMES | | | | | |
| Family Anatidae (ducks, geese) | | | | | |
| Gen. & at least 4 spp. not det. | – | ● | ● | – | freshwater and estuarine |
| ORDER FALCONIFORMES | | | | | |
| Family Falconidae (falcons) | | | | | |
| Gen. & sp. not det. | – | ● | – | – | aerial, aboreal and/or terrestrial |
| Family Accipitridae (hawks, eagles etc.) | | | | | |
| Gen. & at least 4 spp. not det. | – | ● | | – | aerial, arboreal and/or terrestrial |
| ?Gypaetinae (vultures) gen. & sp. not det. | – | ● | – | – | terrestrial (open country) |
| ORDER GALLIFORMES | | | | | |
| Family Phasianidae (game birds) | | | | | |
| Gen. & at least 2 spp. (francolin & quail) not det. | – | ● | ● | – | terrestrial (open country with low cover) |
| ORDER GRUIFORMES | | | | | |
| Family Gruidae (cranes) | | | | | |
| Gen. & sp. not det. | – | – | ● | – | terrestrial (open country, marsh) |
| Family Rallidae (rails) | | | | | |
| Gen. & sp. not det. | – | ● | ● | – | terrestrial (dense cover) and freshwater |
| Family Otidae (bustards) | | | | | |
| Gen. & sp. not det. | – | ● | – | – | terrestrial (open, mainly dry country) |
| ORDER CHARADRIIFORMES (shorebirds) | | | | | |
| Fam., gen. & at least 10 spp. not det. | – | ● | ● | – | freshwater, estuarine and/or marine |
| Family Pteroclidae (sandgrouse) | | | | | |
| Gen. & sp. not det. | – | ● | – | – | terrestrial (near water) |
| ORDER COLUMBIFORMES | | | | | |
| Family Columbidae (pigeons, doves) | | | | | |
| Gen. & at least 2 spp. not det. | – | ● | ● | – | terrestrial and arboreal |
| ORDER PSITTACIFORMES (parrots) | | | | | |
| Fam., gen. & at least 2 spp. not det. | – | ● | ? | – | arboreal (forests, woodlands) |
| ORDER STRIGIFORMES | | | | | |
| Family Strigidae (owls) | | | | | |
| Gen. & at least 2 spp. not det. | – | ● | – | – | arboreal and terrestrial |
| ORDER COLIIFORMES | | | | | |
| Family Coliidae (colies) | | | | | |
| Gen. & sp. not det. | – | ● | – | – | arboreal (bushland and woodland) |

| | GM | QSM | PPM1 | PPM2 | Inferred habitat preference |
|---|---|---|---|---|---|
| ORDER CORACIIFORMES (rollers etc.) | | | | | |
| Fam., gen. & sp. not det. | – | • | – | – | arboreal and terrestrial |
| ? Family Alcedinidae (kingfishers) | | | | | |
| Gen. & at least 2 spp. not det. | – | • | – | – | freshwater |
| ORDER PICIFORMES (woodpeckers etc.) | | | | | |
| Fam., gen. & at least 2 spp. not det. | – | • | – | – | arboreal and terrestrial |
| ORDER APODIFORMES | | | | | |
| Family Apodidae (swifts) | | | | | |
| Gen. & sp. not det. | – | – | • | – | aerial |
| ORDER PASSERIFORMES (songbirds) | | | | | |
| Fam., gen. & at least 9 spp. not det. | – | • | • | – | varied |
| AVES not det. | – | • | • | – | |
| CLASS MAMMALIA | | | | | |
| ORDER INSECTIVORA | | | | | |
| Family Chrysochloridae (golden moles) | | | | | |
| *Chrysochloris* sp. | – | • | • | – | fossorial |
| Family Soricidae (shrews) | | | | | |
| *Myosorex* sp. | – | • | – | – | terrestrial |
| *Suncus* sp. | – | • | – | – | |
| Soricidae gen. & sp(p). not det. | – | • | • | – | |
| Family Macroscelididae (elephant shrews) | | | | | |
| *Elephantulus* sp. | – | • | • | – | |
| ORDER CHIROPTERA (bats) | | | | | |
| Family Vespertilionidae | | | | | |
| *Eptesicus* sp. | – | • | – | – | aerial |
| ORDER PRIMATES | | | | | |
| Family Cercopithecidae (monkeys etc.) | | | | | |
| Gen. & sp. indet. | – | • | – | – | arboreal and terrestrial |
| ORDER PHOLIDOTA (pangolins) | | | | | |
| *Phataginus* sp. | – | • | – | – | terrestrial |
| ORDER TUBULIDENTATA (aardvarks) | | | | | |
| Gen. & sp. not det. | – | • | • | – | terrestrial |
| ORDER CARNIVORA | | | | | |
| Family Canidae (foxes, jackals etc.) | | | | | |
| Gen. & sp. not det. (? aff. *'Canis brevirostris'*) | – | – | • | – | terrestrial |
| *Vulpes* sp. (fox) | – | – | • | – | |
| Family Ursidae (bears) | | | | | |
| *Agriotherium africanum* | – | – | • | – | terrestrial (?woodlands) |
| Family Mustelidae (weasels, martens etc.) | | | | | |
| *Plesiogulo monspessulanus* (wolverine) | – | • | ? | – | terrestrial (?woodlands) |
| *Mellivora benfieldi* (honey badger) | – | – | • | – | terrestrial |
| *Enhydriodon africanus* (otter) | – | – | • | – | freshwater |
| Family Phocidae (seals) | | | | | |
| *Homiphoca capensis* | cf. | • | • | – | marine and ?estuarine |
| Family Viverridae (mongooses etc.) | | | | | |
| *'Viverra' leakeyi* (?aff. *Civettictis*) (civet) | – | • | • | – | terrestrial |
| Viverrinae gen. & sp. not det. (?aff. *Pseudocivetta*) (civet) | – | • | • | – | |
| *Genetta* sp. (genet) | – | • | – | – | |
| *Herpestes* spp. A, B (mongooses) | – | • | • | – | |
| Herpestinae spp. C, D, E (mongooses) | – | • | – | – | |
| Herpestinae not det. | – | – | • | – | |

| | GM | QSM | PPM1 | PPM2 | Inferred habitat preference |
|---|---|---|---|---|---|
| Family Hyaenidae (hyaenas) | | | | | |
| *'Adcrocuta' australis* | – | ● | ? | – | |
| *Ictitherium preforfex* | – | – | ● | – | |
| *Hyaena abronia* | – | ● | ● | – | |
| *Hyaenictitherium namaquense* | – | ● | – | – | terrestrial |
| *Euryboas* sp. | – | ● | ● | – | |
| Hyaenidae sp. E | – | – | ● | – | |
| Hyaenidae not det. | – | – | ● | – | |
| Family Felidae (cats) | | | | | |
| *'Machairodus'* sp. (sabre-tooth) | – | ● | – | – | |
| *Homotherium* sp. (sabre-tooth) | – | ● | ● | – | |
| *Felis* sp. (wildcat-like) | – | ● | – | – | |
| *Felis* aff. *issiodorensis* (lynx-like) | – | ● | ● | – | terrestrial |
| *Felis obscura* (lynx-like) | – | – | ● | – | |
| *Dinofelis diastemata* (false sabre-tooth) | – | ● | ● | – | |
| Felidae not det. | – | – | ● | – | |
| CARNIVORA not det. | | | | | |
| Gen. & sp. not det. (Canidae or Viverridae) | – | ● | – | – | |
| Gen. & sp. not det. (?Procyonidae) | – | ● | – | – | |
| Gen. & sp. not det. (?Lutrinae) | – | ● | – | – | |
| ORDER PROBOSCIDEA (elephants & kin) | | | | | |
| Family Gomphotheriidae | | | | | |
| *Anancus* sp. | – | ● | ● | – | |
| Family Elephantidae | | | | | terrestrial (woodlands) |
| *Mammuthus subplanifrons* | – | ● | ● | – | |
| ORDER HYRACOIDEA (hyraxes or dassies) | | | | | |
| Family Procaviidae | | | | | |
| *Procavia* cf. *antiqua* | – | ● | ? | – | terrestrial or arboreal |
| ORDER PERISSODACTYLA | | | | | |
| Family Equidae (horses) | | | | | |
| *Hipparion* cf. *primigenium* | ● | – | – | – | |
| *Hipparion* cf. *baardi* | – | ● | ● | – | |
| *Hipparion* cf. *namaquense* | – | – | – | ? | terrestrial (grasslands) |
| Family Rhinocerotidae (rhinos) | | | | | |
| *Ceratotherium praecox* | – | ● | ● | – | |
| ORDER ARTIODACTYLA | | | | | |
| Family Tayassuidae (peccaries) | | | | | |
| *Pecarichoerus*? (or *Barberahyus africanus*) | – | – | ● | – | terrestrial |
| Family Suidae (pigs) | | | | | |
| *Nyanzachoerus* cf. *pattersoni* (or *kanamensis*) | – | ● | – | – | terrestrial (?wood- |
| *Nyanzachoerus* cf. *jaegeri* | – | – | ● | – | lands) |
| Family Hippopotamidae (hippos) | | | | | |
| Gen. & sp. not det. | – | – | ● | – | freshwater and terrestrial |
| Family Giraffidae (giraffes) | | | | | |
| *Sivatherium hendeyi* | – | ● | ● | – | |
| *Palaeotragus* cf. *germaini* | – | – | ● | – | terrestrial (woodlands) |
| *Giraffa* sp. | – | ● | ● | – | |
| Family Bovidae (buffaloes, antelopes etc.) | | | | | |
| *Tragelaphus* sp. A (nyala-like) | – | ● | ● | – | |
| *Tragelaphus* sp. B (nyala-like) | – | – | ● | – | terrestrial (woodlands) |
| *Mesembriportax* (or *Miotragocerus*) *acrae* (kudu-like relative of nilgai) | – | ● | ● | – | |
| *Simatherium demissum* (buffalo) | – | ● | ● | – | terrestrial (?grasslands) |

| | GM | QSM | PPM1 | PPM2 | Inferred habitat preference |
|---|---|---|---|---|---|
| *Kobus subdolus* (kob-like) | – | – | ● | – | terrestrial (woodlands near freshwater) |
| *Kobus* sp. B (kob-like) | – | – | ● | – | terrestrial (woodlands near freshwater) |
| *Damalacra neanica* (hartebeest-like) | – | – | ● | – | terrestrial (grasslands) |
| *Damalacra acalla* (hartebeest-like) | – | ● | ● | – | terrestrial (grasslands) |
| *Raphicerus paralius* (steenbok) | – | ● | ● | – | terrestrial (?grasslands) |
| *Gazella* sp. (gazelle) | – | ● | ● | – | terrestrial (?grasslands) |
| Ovibovini gen. & at least 2 spp. not det. | – | ● | ● | – | terrestrial |
| ORDER LAGOMORPHA | | | | | |
| Family Laporidae (hares, rabbits) | | | | | |
| *Pronolagus* sp. | – | ● | ● | – | terrestrial |
| ORDER RODENTIA | | | | | |
| Family Bathyergidae (rodent moles) | | | | | |
| *Bathyergus* sp. | – | ● | ● | – | fossorial |
| *Cryptomys* sp. | – | ● | ? | – | fossorial |
| Family Hystricidae (porcupines) | | | | | |
| Gen. & sp. not det. A | – | ● | – | – | terrestrial |
| Gen. & sp. not det. B | – | – | ● | – | terrestrial |
| Family Cricetidae (rats, mice, gerbils etc.) | | | | | |
| *Mystromys* sp. A | – | ● | – | – | terrestrial |
| *Mystromys* cf. *darti* | – | ● | – | – | terrestrial |
| *Mystromys* cf. *hausleitneri* | – | ● | – | – | terrestrial |
| *Gerbillus* or *Desmodillus* sp. | – | ● | – | – | terrestrial |
| *Dendromus* sp. | – | ● | – | – | terrestrial (?woodlands) |
| *Steatomys* or *Malacothrix* sp. | – | ● | – | – | terrestrial |
| Family Muridae (rats, mice) | | | | | |
| *Aethomys* spp. A, B | – | ● | – | – | terrestrial |
| *Mus* spp. A, B | – | ● | – | – | terrestrial |
| *Rhabdomys* sp. | – | ● | – | – | terrestrial |
| *Euryotomys pelomyoides* | – | ● | – | – | terrestrial (?near freshwater) |
| Family Muscardinidae (dormice) | | | | | |
| *Graphiurus* sp. | – | ● | – | – | terrestrial (?woodlands) |
| RODENTIA not det. | – | ● | ● | – | |
| ORDER CETACEA (whales, dolphins) | | | | | |
| Gen. & spp. not det. | ● | ● | ● | – | marine |

Key:
GM = Gravel Member
QM = Quartzose Sand Member
PPM1 = Pelletal Phosphorite Member – river channel fill
PPM2 = Pelletal Phosphorite Member – undifferentiated shallow marine sands
(After Hendey 1981a)

APPENDIX 3

# Offshore commercial boreholes with partial data release

| No. | S.latitude | E.longitude |
|---|---|---|
| PB-A1 | 34.1611 | 20.3367 |
| MB-A1 | 34.2314 | 22.1984 |
| G(a)-A1 | 34.5523 | 23.7207 |
| G(a)-A2 | 34.5741 | 23.7604 |
| G(a)-A3 | 34.5632 | 23.7717 |
| G(a)-B1 | 34.3841 | 23.7961 |
| G(a)-C1 | 34.7174 | 23.0491 |
| F-1 | 35.0012 | 22.2428 |
| G(b)-Gemsbok 1 | 34.4842 | 24.2433 |
| J(c)-1 | 29.4615 | 31.5944 |
| E-1 | 34.6628 | 21.2517 |
| D-A1 | 35.0919 | 20.9189 |
| F-B1 | 34.4061 | 22.8112 |
| G(b)-Springbok 1 | 34.6184 | 24.2857 |
| F-D1 | 34.8567 | 22.9440 |
| F-E1 | 34.9251 | 22.3676 |
| E-B1 | 34.6625 | 21.2500 |
| H(b)-Hartebeest 1 | 34.2774 | 25.9219 |
| Kudu9-A1 | 28.5481 | 14.5791 |
| F-F2 | 35.3276 | 22.2975 |
| B(a)-A2 | 31.9083 | 17.6870 |
| B(a)-A1 | 31.8667 | 17.6103 |
| A-A1 | 31.2192 | 16.9212 |
| E-C1 | 34.8880 | 21.4252 |
| D-B1 | 34.6416 | 20.9193 |
| Gb-C1 | 34.4104 | 24.4002 |
| Ga-D1 | 34.3473 | 23.4314 |
| Ga-G1 | 34.5776 | 23.4502 |
| Ga-E1 | 34.7009 | 23.8324 |
| K-B1 | 30.7107 | 15.4481 |
| K-A1 | 30.8076 | 16.0166 |
| K-A2 | 30.8342 | 16.0091 |
| K-D1 | 31.7314 | 16.3372 |
| A-C1 | 32.5000 | 16.8917 |
| A-C2 | 32.3319 | 16.8236 |

(Du Toit & Leith 1974, du Toit 1976, McLachlan & McMillan 1979, Gerrard & Smith 1980)

# References


Adamson, R.S. 1931. Notes on some petrified wood from Banke, Namaqualand. *Trans.R.Soc.S.Afr.,* 19:255-258.

Adamson, R.S. 1934. Fossil plants from Fort Grey, near East London. *Ann.S.Afr.Mus.,* 31:67-96.

Allsopp, H.L. & P.Kolbe 1965. Isotopic age determinations on the Cape Granite and intruded Malmesbury sediments, Cape Peninsula, South Africa. *Geochim.Cosmochim.Acta.,* 29:1115-1130.

Almagor, G. 1979. A review: geotechnical studies at continental margins. *Geol.Surv.Israel Rept.,* MG/79/3:1-100.

Ames, L.L. 1959. The genesis of carbonate apatite. *Econ.Geol.,* 54:829-841.

Anderson, H.M. 1974. A brief review of the flora of the Molteno 'Formation' (Triassic), South Africa. *Palaeont. afr.,* 17:1-10.

Anderson, H.M. & J.M.Anderson 1970. A preliminary review of the biostratigraphy of the uppermost Permian, Triassic and lowermost Jurassic of Gondwanaland. *Palaeont.afr.,* 13 (supplement): 22pp.

Anderson J.M. 1981. World Permo-Triassic correlations: their biostratigraphic basis. In: M.M.Cresswell & P.Vella (eds), *Gondwana Five.* Proc. 5th Int.Gondwana Symp.Wellington, 1980, Rotterdam, Balkema: 3-17.

Anderson, J.M. & R.U.Schwyzer 1977. The biostratigraphy of the Permian and Triassic Part 4 – Palaeomagnetic evidence for large scale intra-Gondwana plate movement during the Carboniferous to Jurassic. *Trans.geol.soc.S.Afr.,* 80:211-234.

Anderson, W. 1907. The Upper Cretaceous rocks of Natal and Zululand. *Rep.geol.Surv.Natal Zululand,* 3:47-64.

Andrews, C.W. 1910. Description of a new Plesiosaur (*Plesiosaurus capensis* sp.nov.) from the Uitenhage beds of Cape Colony. *Ann.S.Afr.Mus.,* 7:309-323.

Atherstone, W.G. 1857. Geology of Uitenhage. *The Eastern Province Monthly Mag.,* Grahamstown, 1(11):580-595.

Baily, W.H. 1855. Description of some Cretaceous fossils, South Africa. *Q.Jl.geol.Soc.Lond.,* 11:454-465.

Bain, A.G. 1845. On the discovery of the fossil remains of bidental and other reptiles in South Africa. *Proc.geol.Soc. Lond.,* 4:499-500.

Bain, A.G. 1856. On the geology of southern Africa. *Trans. geol.Soc.Lond.,* Ser.2,7:175-192.

Balme, B.E. 1970. Palynology of the Permian and Triassic strata in the Salt Range and Surghar Range, West Pakistan. *Univ.Kans.Dept.Geol.Sp.Publ.,* 4:305-453.

Bang, N.D. & A.F.Pearce 1976. Large-scale circulation of surface water of the south Indian Ocean. In: A.E.F.Heydorn (ed.), *Ecology of the Agulhas Current Region – an assessment of biological responses to environmental parameters in the south West Indian Ocean.* Proc.Marine & Freshwater Conf., Pretoria:4-10.

Banks, M.R. 1978. Correlation chart for the Triassic System of Australia. *Bull. Bur.Min.Resources, Geol.Geophys.,* 156C: 39pp.

Barker, P.F. 1977. Correlations between sites on the eastern Falkland Plateau by means of seismic reflection profiles, Leg 36, DSDP. In: P.Barker et al, *Initial Rep.Deep Sea Drilling Project,* 36, Washington, US Government Printing Office:971-990.

Barker, P.F. 1979. The history of ridge-crest off-set at the Falkland-Agulhas Fracture Zone from a small-circle geophysical profile. *Geophys.J.R.astr.Soc.,* 59:131-145.

Barker, P.F., I.W.D.Dalziel, M.G.Dinkelman, D.H.Elliot, A.M. Gombos, A.Lonardi, G.Plafker, J.Tarney, R.W.Thompson, R.C.Tjalsma, C.C.von der Borch & S.W.Wise 1977a. *Initial Rep.Deep Sea Drilling Project,* 36, Washington, US Government Printing Office.

Barker, P.F., I.W.D.Dalziel, M.G.Dinkelman, D.H.Elliot, A.M. Gombos, A.Lonardi, G.Plafker, J.Tarney, R.W.Thompson, R.C.Tjalsma, C.C.von der Borch, S.W.Wise, W.Harris & W.V.Sliter 1977b. Evolution of the southwestern Atlantic Ocean Basin: Results of Leg 36, Deep Sea Drilling Project. In: P.Barker et al, *Initial Rep.Deep Sea Drilling Project,* 36, Washington, US Government Printing Office:993-1014.

Barker, P.F. & R.A.Jahn 1980. A marine geophysical reconnaissance of the Weddell Sea. *Geophys.J.R.astr.Soc.,* 63: 271-283.

Barrett, D.M. 1977. Agulhas Plateau off southern Africa: a geophysical study. *Bull.geol.Soc.Am.,* 88:749-763.

Barron, E.J., C.G.A.Harrison & W.W.Hay 1978. A revised reconstruction of the southern continents. *EOS,* 59:436-449.

Baturin, G.N. 1969. Authigenic phosphate concentration in Recent sediments of the southwest African shelf. *Dokl. Earth Sci.Sect.English Transl.,* 189:227-230.

Baturin, G.N. 1970. Recent authigenic phosphorite formation in the south-west African shelf. *Rep.Inst.geol.Sci.,* 70(13): 87-98.

Baturin, G.N. & V.T.Dubinchuk 1974. Microstructures of Agulhas Bank phosphorites. *Mar.Geol.,* 16:M63-70.

Beaumont, C. & J.F.Sweeney 1978. Graben generation of major sedimentary basins. *Tectonophysics,* 50:T19-23.

Beck, R.H. & P.Lehner 1974. Ocean, new frontier in exploration. *Bull.Am.Assoc.Petrol.Geol.,* 58:376-395.

Beckinsale, R.D., J.Tarney, D.P.F.Darbyshire & M.J.Humm 1977. Rb-Sr and K-Ar age determinations on samples of the Falkland Plateau basement at site 330, DSDP. In: P. Barker et al, *Initial Rep.Deep Sea Drilling Project,* 36, Washington, US Government Printing Office:923-927.

Beetz, P.R.W. 1934. Geology of south west Angola between Cunene and the Lunda Axis. *Trans.geol.Soc.S.Afr.,* 36: 137-176.

Beetz, W. 1926. Die Tertiärablagerungen der Küstennamib. In: E.Kaiser (ed.), *Die Diamantenwüste Südwestafrikas,* 2:1-54, Berlin, Dietrich Reimer (Ernst Vohsen).

Belderson, R.H. 1961. *The size distribution characteristics of the Recent shallow marine sediments off Durban, South Africa.* Unpublished MSc thesis, Univ. of Natal, Durban.

Beltan, L. & A.Tintori 1981. The genus *Saurichthys* (Pisces, Actinopterigii) during the Gondwana period. In: M.M. Cresswell & P.Vella (eds), *Gondwana Five,* Proc.5th Int.

Gondwana Symp.Wellington, 1980, Rotterdam, Balkema: 53-59.
Bendix-Almgreen, S.E. 1976. Palaeovertebrate faunas of Greenland. In: A.Escher & W.S.Watt (eds), *Geology of Greenland,* Copenhagen, Geological Survey of Greenland: 534-573.
Benton, M.J. 1979. Ecological succession among late Palaeozoic and Mesozoic tetrapods. *Palaeogeog.Palaeoclimat. Palaeoecol.,* 26:127-150.
Bergh, H.W. 1977. Mesozoic sea floor off Dronning Maud Land, Antarctica. *Nature,* 269:686-687.
Bergh, H.W. & D.M.Barrett 1980. Agulhas Basin magnetic bight. *Nature,* 287:591-595.
Beukes, N.J. 1970. Stratigraphy and sedimentology of the Cape Sandstone Stage Karroo System. In: *Second Gondwana Symposium: Proceedings and Papers,* Pretoria, Council for Scientific and Industrial Research:321-341.
Bharadwaj, D.C. 1970. Palynological subdivision of the Gondwana sequence of India: In: *Second Gondwana Symposium: Proceedings and Papers,* Pretoria, Council for Scientific and Industrial Research:531-536.
Bigarella, J.J. 1970. Continental drift and palaeocurrent analysis. In: *Second Gondwana Symposium: Proceedings and Papers,* Pretoria, Council for Scientific and Industrial Research:73-97.
Birch, G.F. 1975. *Sediments on the continental margin off the west coast of South Africa.* Unpublished PhD thesis, Univ.Cape Town, 210pp.
Birch, G.F. 1977a. Phosphorites from the Saldanha Bay region. *Trans.R.Soc.S.Afr.,* 42:223-240.
Birch, G.F. 1977b. Surficial sediments on the continental margin off the west coast of South Africa. *Mar.Geol.,* 23: 305-337.
Birch, G.F. 1978. Penecontemporaneous phosphatization by replacement and precipitation mechanisms on the western margin of southern Africa. *Tech.Rept.Joint Geol.Surv. Univ.Cape Town Mar.Geol.Prog.,* 10:112-127.
Birch, G.F. 1979a. Phosphatic rocks on the western margin of South Africa. *J.Sed.Petrol.,* 49:93-110.
Birch, G.F. 1979b. Phosphorite pellets and rock from the western continental margin and adjacent coastal terrace of South Africa. *Mar.Geol.,* 33:91-116.
Birch, G.F. 1980. A model of penecontemporaneous phosphatization by diagenetic and authigenic mechanisms from the western margin of southern Africa. In: Y.K.Bentor (ed.), *Marine Phosphorites. Soc.Econ.Pal.Min.Spec.Pub.,* 29:79-100.
Birch, G.F., J.Thomson, J.M.McArthur & W.C.Burnett 1983. Pleistocene phosphorites of the west coast of South Africa. *Nature,* 302:601-603.
Birklund, T. & K.Perch-Nielsen 1976. Late Palaeozoic-Mesozoic evolution of central East Greenland. In: A.Escher & W.S.Watt (eds), *Geology of Greenland,* Copenhagen, Geological Survey of Greenland:302-339.
Bishop, W.W. 1967. The later Tertiary in East Africa – volcanics, sediments and faunal inventory. In: W.W.Bishop & J.C.Clark (eds), *Background to evolution in Africa,* Chicago, University Press:31-56.
Böhm, J. 1926. Über Tertiäire Versteinerungen von der Bogenfelser Diamantfeldern. In: E.Kaiser (ed.), *Die Diamantwenwüste Südwestafrikas,* 2, Berlin, Dietrich Reimer (Ernst Vohsen):55-87.
Böhm, J. & W.Weissermel 1913. Über Tertiäire Versteimerungen von der Bogenfelser Diamantfeldern. *Beitr.geol.Erforsch. dt.Schutzgeb.,* 5:1-58.
Bolli, H.M., W.B.F.Ryan, J.B.Foresman, W.E.Hottman, H. Kagami, J.F.Longoria, B.K.McKnight, M.Melguen, J.Natland, F.Proto-Decima & W.G.Siesser 1978. *Initial Rep. Deep Sea Drilling Project,* 40, Washington, US Government Printing Office.
Boocock, C. & O.J.van Straten 1962. Notes on the geology and hydrogeology of the Central Kalahari region, Bechuanaland Protectorate. *Trans.geol.Soc.S.Afr.,* 65:125-171.
Boonstra, L.D. 1969. The fauna of the Tapinocephalus Zone (Beaufort Beds of the Karroo). *Ann.S.Afr.Mus.,* 56:1-75.
Botha, B.J.V. 1968. The stratigraphy of the Red Beds Stage, Karroo System, at Elliot, C.P. *Trans.geol.Soc.S.Afr.,* 71: 101-118.
Botha, B.J.V. 1971. Kussinglawa en pypamandels in die Stormberg basalt naby Barkly-Oos. *Tydskrif.Natuurwet.,* 11:68-77.
Botha, B.J.V. & F.D.I.Hodgson 1977. Karoo dolerites in north-western Damaraland. *Trans.geol.Soc.S.Afr.,* 79:186-190.
Botha, B.J.V. & W.Linström 1978. A note on the stratigraphy of the Beaufort Group in north-western Natal. *Trans.geol. Soc.S.Afr.,* 81:35-40.
Botha, B.J.V. & J.C.Theron 1966. New evidence for the early commencement of Stormberg volcanism. *Tydskrif.Natuurwet.,* 6:469-473.
Boucher, C. & M.L.Jarman 1977. The vegetation of the Langebaan area, South Africa. *Trans.R.Soc.S.Afr.,* 42:241-272.
Bourdon, M. & P.Magnier 1969. Notes on the Tertiary fossils at Birbury, Cape Province. *Trans.geol.Soc.S.Afr.,* 72:123-125.
Brain, C.K. 1958. The Transvaal ape-man-bearing deposits. *Mem.Transvaal Mus.,* 11:1-131.
Brain, C.K. 1976. A re-interpretation of the Swartkrans site and its remains. *S.Afr.J.Sci.,* 72:141-146.
Bremner, J.M. 1975. Faecal pellets, glauconite, phosphorite and bedrock from the Kunene-Walvis continental margin. *Tech.Rept.Joint Geol.Surv. Univ.Cape Town Mar.Geol. Prog.,* 7:57-68.
Bremner, J.M. 1978a. *Sediments on the continental margin off South West Africa between latitudes 17° and 25°S.* Unpub. PhD thesis, Univ.Cape Town, 300pp.
Bremner, J.M. 1978b. Contemporary phosphorite formation off South West Africa. *Tech.Rept.Joint Geol.Surv. Univ. Cape Town Mar.Geol.Prog.,* 10:75-84.
Bremner, J.M. 1980a. Physical parameters of the diatomaceous mud belt off South West Africa. *Mar.Geol.,* 34:M67-76.
Bremner, J.M. 1980b. Concretionary phosphorite from SW Africa. *J.geol.Soc.Lond.,* 137:773-786.
Brenner, P.W. & H.J.Oertli 1976. Lower Cretaceous ostracodes (Valanginian to Hauterivian) from the Sundays River Formation, Algoa Basin, South Africa. *Bull.Centre Recherche Pau, Societè Nationale des Pètroles d'Aquitaine,* 10:471-533.
Bristow, J.W. 1980. *The geochronology and geochemistry of Karoo volcanics in the Lebombo and adjacent areas.* Unpub. PhD thesis, Univ.Cape Town, 257pp.
Brock, A., P.L.McFadden & T.C.P.Partridge 1977. Preliminary palaeomagnetic results from Makapansgat and Swartkrans. *Nature,* 266:249-250.
Broom, R. 1904. On the occurrence of an opisthocoelian dinosaur (*Algoasaurus bauri*) in the Cretaceous beds of South Africa. *Geol.Mag.,* 51:445-447.
Broom, R. 1906. The Permian and Triassic faunas of South Africa. *Geol.Mag.,* 5:29-30.
Broom, R. 1909a. An attempt to determine the horizons of the fossil vertebrates of the Karroo. *Ann.S.Afr.Mus.,* 7: 285-289.
Broom, R. 1909b. The fossil fishes of the Upper Karroo beds of South Africa. *Ann.S.Afr.Mus.,* 7:251-268.
Broom, R. 1910. Observations on some specimens of South African fossil reptiles preserved in the British Museum. *Trans.R.Soc.S.Afr.,* 2:19-25.
Brown, A.C. & N.Jarman 1978. Coastal marine habitats. In: M.J.A.Werger (ed.), *Biogeography and ecology of southern Africa,* The Hague, W.Junk:1239-1277.

Brown, D.A., K.S.W.Campbell & K.A.W.Crook 1968. *The geological evolution of Australia and New Zealand,* Oxford, Pergamon Press.

Brown, J.T. & C.E.Gow 1976. Plant fossils as indicators of the rate of deposition of the Kirkwood Formation in the Algoa Basin. *S.Afr.J.Sci.,* 72:278-279.

Bryan, G.M. & E.S.W.Simpson 1971. Seismic refraction measurements on the continental shelf between Orange River and Cape Town. *Inst.geol.Sci.Rep.,* 70/16:186-198.

Bullard, E., J.E.Everett & A.G.Smith 1965. The fit of the continents around the Atlantic. *Phil.Trans.R.Soc.Lond.,* A258:41-51.

Bullen-Newton, R. 1913. On some Kainozoic shells from South Africa. *Rec.Albany Mus.,* 2:315-352.

Burger, A.J. & F.J.Coertze 1973. Radiometric age measurements on rocks from southern Africa to the end of 1971. *Bull.Geol.Surv.S.Afr.,* 58:46pp.

Butzer, K.W. 1973. On the geology of a late Pliocene *Mammuthus* site, Virginia, Orange Free State. *Navors.nas.Mus.,* Bloemfontein, 2:386-393.

Butzer, K.W. 1974. Palaeo-ecology of South African australopithecines: Taung revisited. *Current Anthropology,* 15: 367-426.

Butzer, K.W. 1976. Lithostratigraphy of the Swartkrans Formation. *S.Afr.J.Sci.,* 72:136-141.

Butzer, K.W., D.M.Helgren, G.J.Fock & R.Stuckenrath 1973. Alluvial terraces of the lower Vaal River, South Africa: a reappraisal and reinvestigation. *J.Geol.,* 81:341-362.

Cahen, L. & J.Lepersonne 1952. Equivalence entre le systeme du Kalahari du Congo Belge et les Kalahari Beds d'Afrique Australe. *Mem.Soc.Belg.Geol.,* 8:1-64.

Calvert, S.E. & N.B.Price 1971a. Upwelling and nutrient regeneration in the Benguela Current, October 1968. *Deep Sea Res.,* 18:505-523.

Calvert, S.E. & N.B.Price 1971b. Recent sediments on the south west African shelf. *Inst.geol.Sci.Rep.,* 70(16):171-185.

Campos, C.W.M., F.C.Ponte & K.Miura 1974. Geology of the Brazilian continental margin. In: C.A.Burk & C.L.Drake (eds), *The Geology of Continental Margins,* New York, Springer-Verlag:447-461.

Carrington, A.J. & B.F.Kensley 1969. Pleistocene molluscs from the Namaqualand coast. *Ann.S.Afr.Mus.,* 52:189-223.

Carter, A.N. 1966. Age of the Aliwal Shoal, South Africa. *Nature,* 211:507-508.

Cayeaux, L. 1934. The phosphatic nodules of the Agulhas Bank. *Ann.S.Afr.Mus.,* 31:105-136.

Chapman, F. 1904. Foraminifera and ostracoda from the Cretaceous of East Pondoland, South Africa. *Ann.S.Afr.Mus.,* 4:221-237.

Chapman, F. 1916. Foraminifera and ostracoda from the Upper Cretaceous of Needs Camp, Buffalo River, Cape Province. *Ann.S.Afr.Mus.,* 12:107-118.

Chapman, F. 1923. On some foraminifera and ostracoda from the Cretaceous of the Umzamba River, Pondoland. *Trans. geol.Soc.S.Afr.,* 26:107-118.

Chapman, F. 1930. On a foraminiferal limestone of Upper Eocene age from the Alexandria Formation, South Africa. *Ann.S.Afr.Mus.,* 28:291-296.

Chester, R. & L.R.Johnson 1971. Atmospheric dusts collected off the West African coast. *Nature,* 229:105-107.

Chetty, P. & R.W.E.Green 1977. Seismic refraction observations in the Transkei Basin and adjacent areas. *Mar.Geophys. Res.,* 3:197-208.

Chowdhury, M.K.R., M.V.A.Sastry, S.C.Shah, G.Singh & S.C. Ghosh 1975. Triassic floral succession in the Gondwana of Peninsular India. In: K.S.W.Campbell (ed.), *Gondwana Geology,* Canberra, Australian National University Press: 149-158.

Ciesielski, P.F. & F.M.Weaver 1974. Early Pliocene temperature changes in the Antarctic Seas. *Geology,* 2:511-515.

Cleverly, R.W. 1979. The volcanic geology of the Lebombo Monocline in Swaziland. *Trans.geol.Soc.S.Afr.,* 82:343-348.

Cleverly, R.W. & J.W.Bristow 1979. Revised volcanic stratigraphy of the Lebombo monocline. *Trans.geol.Soc.S.Afr.,* 82:227-230.

Cluver, M.A. 1977. Aspects of dicynodont classification. In: *Guide to 1977 Karoo Symposium,* Johannesburg, Bernard Price Inst.Pal.Res., Univ.Witwatersrand.

Cluver, M.A. 1978. *Fossil Reptiles of the South African Karoo,* Cape Town, South African Museum, 54pp.

Coetzee, D.S. 1979. *Styl en intensiteit van vervorming in die Kaap- en Karoo-opeenvolging vanaf Meiringspoort tot by Beaufort Wes langs 22½° oosterlengte.* Unpub. MSc thesis, Univ.Stellenbosch.

Coetzee, D.S. 1983. The deformation style between Meiringspoort and Beaufort West. In: Geodynamics of the Cape Fold Belt. *Geol.Soc.S.Afr.Spec.Pub.*

Coetzee, J.A. 1978a. Climatic and biological changes in southwestern Africa during the late Cainozoic. In: E.M.van Zinderen Bakker & J.A.Coetzee (eds), *Palaeoecology of Africa* 10/11, Rotterdam, A.A.Balkema:13-29.

Coetzee, J.A. 1978b. Late Cainozoic palaeo-environments of southern Africa. In: E.M.van Zinderen Bakker (ed.), *Antarctic glacial history and world palaeoenvironments,* Rotterdam, A.A.Balkema:115-127.

Coetzee, J.A. 1981. A palynological record of very primitive angiosperms in Tertiary deposits of the south-western Cape Province, South Africa. *S.Afr.J.Sci.,* 77:341-343.

Colbert, E.H. 1970. Antarctic Gondwana tetrapods. In: *Second Gondwana Symposium: Proceedings and Papers,* Pretoria, Council for Scientific and Industrial Research:659-664.

Colbert, E.H. 1981. The distribution of tetrapods and the breakup of Gondwana. In: M.M.Cresswell & P.Vella (eds), *Gondwana Five,* Proc. 5th Int.Gondwana Symp., Wellington, 1980, Rotterdam, A.A.Balkema:277-282.

Coleman, J.M. & L.E.Garrison 1977. Geological aspects of marine slope stability, northwestern Gulf of Mexico. *Mar. Geotechnol.,* 2:9-44.

Collet, L.W. 1905. Les contrétions phosphatées de l'Agulhas Bank. *Proc.R.Soc.Edinb.,* 25:862-893.

Collignon, M. 1966. *Atlas des fossiles caractéristiques de Madagascar (Ammonites). XIV* (Santonien), Tananarive, Service Geologique.

Cooke, H.B.S. 1947. The development of the Vaal River and its deposits. *Trans.geol.Soc.S.Afr.,* 46:243-260.

Cooper, M.R. 1973. *A revision of the Lower Cretaceous (Upper Valanginian) Cephalopoda from the Uitenhage Formation of South Africa.* Unpub. MSc thesis, Univ.Natal, 221pp.

Cooper, M.R. 1974a. In discussion of: Van Eeden, O.R. The correlation of the subdivisions of the Karroo system. *Trans. geol.Soc.S.Afr.,* 77:377-380.

Cooper, M.R. 1974b. The Cretaceous stratigraphy of South-Central Africa. *Ann.S.Afr.Mus.,* 66:81-107.

Cooper, M.R. 1979. Cretaceous Trigoniidae (Mollusca, Bivalves) from the Brenton Formation, Knysna. *Ann.S.Afr.Mus.,* 78: 49-67.

Cooper, M.R. 1981. Revision of the Late Valanginian cephalopoda from the Sundays River Formation of South Africa, with special reference to the genus *Olcostephanus. Ann.S. Afr.Mus.,* 83:147-366.

Corvinus, G. & Q.B.Hendey 1978. A new Miocene vertebrate locality at Arrisdrift in South West Africa (Namibia). *N.Jb. Geol.Paläont.Mh.,* 1978(4):193-205.

Cosgriff, J.W. 1965. A new genus of Temnospondyli from the Triassic of Western Australia. *Trans.R.Soc.West.Australia,* 48:65-90.

Cox, C.B. 1967. Changes in terrestrial vertebrate faunas during the Mesozoic. In: W.B.Harland et al (eds), *The Fossil Record,* London, Geological Society of London:77-89.

Cox, C.B. 1973. Triassic tetrapods. In: A.Hallam (ed.), *Atlas of*

*Palaeobiogeography,* Amsterdam, Elsevier:213-223.
Cox, K.G. 1972. The Karroo volcanic cycle. *Jl.geol.Soc.Lond.,* 128:311-336.
Cox, K.G. 1978. Flood basalts, subduction and the breakup of Gondwanaland. *Nature,* 274:47-49.
Cradock, C. 1974. Tectonic evolution of the Pacific margin of Gondwanaland. In: K.S.W.Campbell (ed.), *Gondwana Geology,* Canberra, Australian National University Press:609-618.
Cranwell, L.M. 1969. Antarctic and circum-Antarctic palynological contributions. *Antarctic J. U.S.,* 4:197-198.
Crick, G.C. 1907a. The cephaloda from the tributaries of the Manuan Creek, Zululand. *Rep.geol.Surv.Natal Zululand,* 3: 163-234.
Crick, G.C. 1907b. Note on a Cretaceous ammonite from the mouth of the Umpenyati River, Natal. *Rep.geol.Surv.Natal Zululand,* 3:250.
Crockett, R.N. & M.T.Jones 1975. Some aspects of the geology of the Waterberg System in eastern Botswana. *Trans.geol. Soc.S.Afr.,* 78:1-10.
Cromer, J.B. & R.Littlejohn 1977. Content, composition, and thermal history of organic matter in Mesozoic sediments, Falkland Plateau. In: P.Barker et al, *Initial Rep. Deep Sea Drilling Project,* 36, Washington, Government Printing Office:941-944.
Currie, J.B., H.W.Patnode & R.P.Trump 1962. Development of folds in sedimentary strata. *Bull.geol.Soc.Am.,* 73:655-674.
Dalla Salda, L. 1980. Some relationships between the cratonic areas of the Rio de la Plata, South America and the western Kalahari, Africa. *Ann.Rept.Precamb.Res. Unit, Univ.Cape Town,* 17:88-111.
Daves, O. 1972. Pleistocene shorelines in the southern and south-eastern Cape Province (2). *Ann.Natal. Mus.,* 21:225-279.
Davis, G.L. 1977. The ages and uranium contents of zircons from kimberlites and associated rocks. *Extended Abstracts, Second International Kimberlite Conference, Santa Fe, New Mexico.*
Dawson, J.B. 1971. Advances in kimberlite geology. *Earth Sci. Rev.,* 7:187-214.
Deacon, H.J. 1983. The comparative evolution of Mediterranean-type ecosystems: a southern perspective. *MEDCON Review Pap.2* (in press).
De Beer, J.H. 1978. Deep electrical resistivity in southern Africa – Part 2. *Trans.geol.Soc.S.Afr.,* 81:143-154.
De Beer, J.H. 1983. Geophysical studies in the southern Cape Province, and models of the lithosphere in the Cape Fold Belt. In: Geodynamics of the Cape Fold Belt. *Geol.Soc.S. Afr., Spec.Pub.*
De Beer, J.H., J.S.V.van Zijl & F.K.Bahnemann 1974. Plate tectonic origin for the Cape Fold Belt? *Nature,* 252:675-676.
De la Cruz, M.A. 1978. Marine geophysical and geological investigations in Saldanha Bay. *Joint Geol.Surv. Univ.Cape Town Mar.Geol.Prog.Bull.,* 9:115pp.
De Sitter, L.U. 1956. *Structural Geology.* New York, McGraw-Hill, 552pp.
De Swart, A.M.J. 1974. Geomorphological dating of Cave openings in South Africa. *Nature,* 250:683pp.
De Swart, A.M.J. & G.Bennet 1974. Structural and physiographic development of Natal since the late Jurassic. *Trans. geol.Soc.S.Afr.,* 77:309-322.
De Swart, A.M.J., O.Fletcher & P.Toshek 1974. Note on orogenic style in the Cape Fold Belt. *Trans.geol.Soc.S.Afr.,* 77:53-58.
De Swart, A.M.J. & D.M.Rowsell 1974. Note on the relationship between diagenesis and deformation in the Cape Fold Belt. *Trans.geol.Soc.S.Afr.,* 77:239-246.
De Villiers, J. 1944. A review of the Cape orogeny. *Ann. Univ. Stellenbosch* 22, Sec.A:183-208.
De Villiers, J. 1956. Die Drie Sintaksisse in die Suidwestelike Kaapprovinsie. *Tegnikon,* 9:75-87.
De Wit, M.J. 1977. The evolution of the Scotia Arc as a key to the reconstruction of southwestern Gondwanaland. *Tectonophysics,* 37:53-81.
Diester-Haas, L. & H.-J.Schrader 1979. Neogene coastal upwelling history off northwest and southwest Africa. *Mar. Geol.,* 29:39-53.
Dietz, R.S. & J.C.Holden 1970. Reconstruction of Pangaea: breakup and dispersion of continents, Permian to present. *J.Geophys.Res.,* 75:4939-4956.
Dingle, R.V. 1969a. Marine Neocomian ostracoda from South Africa. *Trans.R.Soc.S.Afr.,* 38:139-163.
Dingle, R.V. 1969b. Upper Senonian ostracods from the coast of Pondoland, South Africa. *Trans.R.Soc.S.Afr.,* 38:347-385.
Dingle, R.V. 1971a. Tertiary sedimentary history of the continental shelf off southern Cape Province, South Africa. *Trans. geol.Soc.S.Afr.,* 74:173-186.
Dingle, R.V. 1971b. Some Cretaceous ostracodal assemblages from the Agulhas Bank (South African continental margin). *Trans.R.Soc.S.Afr.,* 39:393-418.
Dingle, R.V. 1973a. Post-Palaeozoic stratigraphy of the eastern Agulhas Bank, South African continental margin. *Mar.Geol.,* 15:1-23.
Dingle, R.V. 1973b. Mesozoic palaeogeography of the southern Cape, South Africa. *Palaeogeog.Palaeoclimat.Palaeoecol.,* 13:203-213.
Dingle, R.V. 1973c. The geology of the continental shelf between Lüderitz and Cape Town (southwest Africa), with special reference to Tertiary strata. *Jl.geol.Soc.Lond.,* 129: 337-363.
Dingle, R.V. 1974. Agulhas Bank phosphorites: a review of 100 years of investigation. *Trans.geol.Soc.S.Afr.,* 77:261-264.
Dingle, R.V. 1976. Palaeogene ostracods from the continental shelf off Natal, South Africa. *Trans.R.Soc.S.Afr.,* 42:35-79.
Dingle, R.V. 1977. The anatomy of a large submarine slump on a sheared continental margin (SE Africa). *Jl.geol.Soc. Lond.,* 134:293-310.
Dingle, R.V. 1978. South Africa. In: M.Moullade & A.E.M. Nairn (eds), *The Phanerozoic Geology of the World II, The Mesozoic,* A, Amsterdam, Elsevier:401-434.
Dingle, R.V. 1979. Sedimentary basins and basement structures on the continental margin of southern Africa. *Bull.Geol. Surv.S.Afr.,* 63:29-45.
Dingle, R.V. 1980a. Sedimentary basins on the continental margins of southern Africa and an assessment of their hydrocarbon potential. *Erdöl und Kohle Erdgas Petrochemie,* 33:457-463.
Dingle, R.V. 1980b. Large allochthonous sediment masses and their role in the construction of the continental slope and rise off southwestern Africa. *Mar.Geol.,* 37:333-354.
Dingle, R.V. 1980c. Marine Santonian and Campanian ostracods from a borehole at Richards Bay, Zululand. *Ann.S.Afr. Mus.,* 82:1-70.
Dingle, R.V. 1981. The Campanian and Maastrichtian ostracoda of South East Africa. *Ann.S.Afr.Mus.,* 85:1-181.
Dingle, R.V. 1982a. Some aspects of Cretaceous ostracod biostratigraphy of South Africa and relationships with other Gondwanide localities. *Cret.Res.,* 3:367-389.
Dingle, R.V. 1982b. Continental margin subsidence: a comparison between the east and west coasts of Africa. In: R.A. Scrutton (ed.), *Dynamics of Passive Margins,* American Geophysical Union, Geodynamics ser., 6:59-71.
Dingle, R.V. & F.Camden-Smith 1979. Acoustic stratigraphy and current-generated bedforms in deep ocean basins off southeastern Africa. *Mar.Geol.,* 33:239-260.

in the Knysna Outlier, Cape Province. *Trans.R.Soc.S.Afr.*, 40:279-298.

Dingle, R.V. & R.I.Gentle 1972. Early Tertiary volcanic rocks on the Agulhas Bank, South African continental shelf. *Geol. Mag.*, 109:127-136.

Dingle, R.V., S.W.Goodlad & A.K.Martin 1978. Bathymetry and stratigraphy of the northern Natal Valley (SW Indian Ocean): a preliminary account. *Mar.Geol.*, 28:89-106.

Dingle, R.V. & H.C.Klinger 1972. The stratigraphy and ostracod fauna of the Upper Jurassic sediments from Brenton,

Dingle, R.V., A.R.Lord & Q.B.Hendey 1979. New sections in the Varswater Formation (Neogene) of Langebaan Road, south-western Cape, South Africa. *Ann.S.Afr.Mus.*, 78:81-92.

Dingle, R.V. & R.A.Scrutton 1974. Continental breakup and the development of post-Paleozoic sedimentary basins around southern Africa. *Bull.geol.Soc.Am.*, 85:1467-1474.

Dingle, R.V. & W.G.Siesser 1977. Geology of the continental margin between Walvis Bay and Ponta do Ouro. *Geol.Surv. S.Afr.Marine Geoscience series 2* (map and explanatory notes).

Dingle, R.V. & E.S.W.Simpson 1976. The Walvis Ridge: a review. In: C.L.Drake (ed.), *Geodynamics: Progress and Prospects,* Washington, American Geophysical Union:160-176.

Douglas, R.G. & Savin, S.M. 1975. Oxygen and carbon isotope analyses of Cretaceous and Tertiary microfossils from Shasky Rise and other sites in the North Pacific. In: R.L. Larson et al, *Init.Rep.Deep Sea Drilling Project,* 32:509-520.

DSDP Scientific Staff 1973. Leg 28. *Geotimes,* 18(6):19-24.

Dubey, A.K. & P.R.Cobbold 1977. Non-cylindrical flexural slip folds in nature and experiment. *Tectonophysics,* 38: 223-239.

Duncan, R.A., R.B.Hargraves & G.P.Brey 1978. Age, palaeomagnetism and chemistry of melilitite basalts in the southern Cape, South Africa. *Geol.Mag.*, 115:317-327.

Dunn, E.J. 1878. *Geological Report of the Stormberg Coalfield,* Cape Town, Parliamentary report G4.

Du Plessis, A. 1977. Sea floor spreading south of the Agulhas Fracture Zone. *Nature,* 270:719-721.

Du Plessis, A. 1979. The evolution of the south-eastern Atlantic Ocean: a review. *Bull.Geol.Surv.S.Afr.*, 63:1-27.

Du Plessis, A. & E.S.W.Simpson 1974. Magnetic anomalies associated with the southeastern continental margin of South Africa. *Mar.Geophys.Res.*, 2:99-110.

Du Preez, J.W. 1944. Lithology, structure and mode of deposition of the Cretaceous deposits in the Oudtshoorn area. *Ann.Univ.Stellenbosch* 22(A):209-237.

Du Toit, A.L. 1904. Geological Survey of Elliot and Xalanga, Tembuland. *Ann.Rep.Geol.Comm.Cape of Good Hope* for 1903, 8:169-205.

Du Toit, A.L. 1905. The geological survey of the divisions of Aliwal North, Herschel, Barkly East and part of Wodehouse. *Ann.Rep.Geol.Comm.Cape of Good Hope* for 1904, 9:71-181.

Du Toit, A.L. 1911. Report on the goelogical survey of MacLear and portions of Engcobo, Mount Fletcher, Qumbu and Mount Frere. *Ann.Rep.Geol.Comm.Cape of Good Hope* for 1910, 15:69-110.

Du Toit, A.L. 1912a. Report on the geological survey of part of the Stormbergen. *Ann.Rep.Geol.Comm.Cape of Good Hope* for 1911, 16:112-136.

Du Toit, A.L. 1912b. Geological survey of part of the Transkei. *Ann.Rep.Geol.Comm. Cape of Good Hope* for 1911, 87-109.

Du Toit, A.L. 1913. The geology of Pondoland. *Ann.Rep.Geol. Surv.S.Afr.* (for 1912), 7:153-179.

Du Toit, A.L. 1917. Report on the phosphate of Saldanha Bay. *Mem.Geol.Surv.S.Afr.*, 10:38pp.

Du Toit, A.L. 1918. The zones of the Karroo System and their distribution. *Proc.geol.Soc.S.Afr.*, 21:17-37.

Du Toit, A.L. 1920a. *The geology of Pondoland and portions of Alfred and Lower Umzimkulu counties, Natal. An Explanation of Cape Sheet 28 (Pondoland). Geol.Surv.S.Africa.* 44pp.

Du Toit, A.L. 1920b. The Karroo dolerites of South Africa. *Trans.geol.Soc.S.Afr.*, 23:1-42.

Du Toit, A.L. 1926. *Geology of South Africa,* Edinburgh, Oliver & Boyd:463pp.

Du Toit, A.L. 1927. The fossil flora of the Upper Karroo beds. *Ann.S.Afr.Mus.*, 22:289-420.

Du Toit, A.L. 1929. The volcanic belt of the Lebombo – a region of tension. *Trans.R.Soc.S.Afr.*, 18:189-217.

Du Toit, A.L. 1933. Crustal movements as a factor in the geographical evolution of South Africa. *S.Afr.geogr.J.*, 16:3-20.

Du Toit, A.L. 1937. *Our Wandering Continents,* Edinburgh, Oliver & Boyd:366pp.

Du Toit, A.L. 1954. *The Geology of South Africa,* Edinburgh, Oliver & Boyd:611pp.

Du Toit, S.R. 1976. *Mesozoic Geology of the Agulhas Bank, South Africa.* Unpub. PhD thesis, Univ.Cape Town:160pp.

Du Toit, S.R. & M.J.Leith 1974. The J(c)-1 borehole on the continental shelf near Stanger, Natal. *Trans.geol.Soc.S.Afr.*, 77:247-252.

Du Toit, S.R., H.de la R.Winter, H.C.Klinger & I.R.McLachlan 1975. *Preliminary stratigraphic framework for the Jurassic and Cretaceous rocks in the Republic of South Africa presented by the Jurassic-Cretaceous Working Group.* Unpub. report to South African Committee for Stratigraphy by the Jurassic-Cretaceous Working Group.

Eales, H.V. & P.W.K.Booth 1974. The Birds River Gabbro Complex, Dordrecht District. *Trans.geol.Soc.S.Afr.*, 77:1-16.

Eales, H.V. & J.S.Marsh 1979. High-Mg tholeiitic rocks and their significance in the Karroo central province. *S.Afr.J. Sci.*, 75:400-404.

Eisenhart, W. 1974. *The fossil cercopithecoids of Makapansgat and Sterkfontein.* Unpub. BA thesis, Harvard University.

Ellenberger, P. 1970. Les niveaux paleontologiques de prémière apparition des mammifères primordiaux en Afrique du Sud et leur ichnologie. Establissement de zones stratigraphique detaillées dans le Stormberg du Lesotho (Afrique du Sud) (Trias superieur à Jurassique). In: *Second Gondwana Symposium Proceedings and Papers,* Pretoria, Council for Scientific and Industrial Research:343-370.

Elliot, D.H. 1975a. Gondwana basins of Antarctica. In: K.S.W. Campbell (ed.), *Gondwana Geology,* Canberra, Australian National University Press:493-536.

Elliot, D.H. 1975b. Tectonics of Antarctica: a review. *Am.J.Sci.*, 275A:45-106.

Elliot, D.H. & M.Johnson 1972. The Gondwanide orogeny: new data from South Africa and the problem of the Falkland Islands. *Geol.Soc.Am.Abs.*, 1972 Ann.Mtgs., 498.

Elliot, D.H. & D.R.Watts 1974. The nature and origin of volcaniclastic material in some Karroo and Beacon rocks. *Trans. geol.Soc.S.Afr.*, 77:109-111.

Embley, R.W. & J.J.Morley 1980. Quaternary sedimentation and palaeoenvironmental studies off Namibia (South-West Africa). *Mar.Geol.*, 36:183-204.

Emery, K.O., E.Uchupi, C.O.Bowin, J.Phillips & E.S.W.Simpson 1975. Continental margin off western Africa: Cape St Francis (South Africa) to Walvis Ridge (south west Africa). *Bull.Am.Assoc.Petrol.Geol.*, 59:3-59.

Engelbrecht, L.N.J., F.J.Coertze & A.A.Snyman 1962. *Die geologie van die gebied tussen Port Elizabeth en Alexandria, Kaapprovinsie,* Explanation to sheet 3325D (Port Elizabeth), 3326 (Alexandria) and 3425B. *Geol.Surv.S.Afr.*, 54pp.

Eriksson, P.G. 1979. Mesozoic sheetflow and playa sediments of the Clarens Formation in the Kamberg area of the Natal Drakensberg. *Trans.geol.Soc.S.Afr.*, 82:257-258.

Erlank, A.J., H.L.Allsopp, A.R.Duncan & J.W.Bristow 1980. Mantle heterogeneity beneath southern Africa: evidence from the volcanic record. *Phil.Trans.R.Soc.Lond.*, A297: 295-307.

Estes, R. 1978. Relationships of the South African fossil frog *Eoxenopoides reuingi* (Anura, Pipidae). *Ann.S.Afr.Mus.*, 73:49-80.

Etheridge, R. 1907. Cretaceous fossils of Natal.2. The Umzinene River deposit, Zululand. *Rep.geol.Surv.Natal Zululand*, 3:67-90.

Falcon, R.M.S. 1975. Application of palynology in sub-dividing the coal-bearing formations of the Karoo sequence in southern Africa. *S.Afr.J.Sci.*, 71:336-344.

Falvey, D.A. 1974. The development of continental margins in plate tectonic theory. *Austral.Petrol.Expl.Assoc.J.*, 14: 95-106.

Falvey, D.A. & M.F.Middleton 1981. Passive continental margins: evidence for a prebreakup deep crustal metamorphic subsidence mechanism. *Oceanol.Acta*, 4(suppl.):103-114.

Fitch, F.J. 1972. In discussion of: K.G.Cox, The Karroo volcanic cycle. *Jl.geol.Soc.Lond.*, 128:333.

Fitch, F.J. & J.A.Miller 1971. Potassium-argon radio ages of Karroo volcanic rocks from Lesotho. *Bull.volcan.*, 35:64-84.

Flemming, B.W. 1978. Underwater sand dunes along the southeast African continental margin – observations and implications. *Mar.Geol.*, 26:177-198.

Flemming, B.W. 1980a. *Guide to the Cainozoic deposits of the Saldanha Bay/Langebaan Lagoon area.* Council for Scientific and Industrial Research Rept. SEA/8002, Stellenbosch, 69pp.

Flemming, B.W. 1980b. Sand transport and bedform patterns on the continental shelf between Durban and Port Elizabeth (southeast African continental margin). *Sediment.Geol.*, 26:179-205.

Flemming, B.W. 1981. Factors controlling shelf sediment dispersal along the southeast African continental margin. *Mar. Geol.*, 42:259-277.

Flint, J.C.E. & R.E.Gould 1975. A note on the fossil megafloras of the Nymboidea and Red Cliff Coal Measures, southern Clarence-Moreton Basin, N.S.W. *J.Proc.R.Soc. N.S.W.*, 108:70-74.

Flores, G. 1964. On the age of the Lupata rocks, lower Zambezi River, Mozambique. *Trans.geol.Soc.S.Afr.*, 68:110-118.

Flores, G. 1973. The Cretaceous and Tertiary sedimentary basins of Mozambique and Zululand. In: G.Blant (ed.), *Sedimentary Basins of the African Coasts, Part 2 south and east coast,* Paris, Association of African Geological Surveys:81-111.

Ford, A.B. & R.W.Kistler 1980. K-Ar age, composition and origin of Mesozoic mafic rocks related to Ferrar Group, Pensacola Mountains, Antarctica. *N.Z.J.Geol.Geophy.*, 23:371-390.

Förster, R. 1975a. The geological history of the sedimentary basin of southern Mozambique, and some aspects of the origin of the Mozambique Channel. *Palaeogeog.Palaeoclimat.Palaeoecol.*, 17:267-287.

Förster, R. 1975b. Die geologische ent wicklung von Süd-Mozambique seit der Unterkreide und die ammonitenfauna von Unterkreide und Cenoman. *Geol.Jahrbuch*, B12:3-324.

Frakes, L.A. 1979. *Climates Throughout Geologic Time.* Amsterdam, Elsevier.

Frakes, L.A. & E.M.Kemp 1973. Palaeogene continental positions and evolution of climate. In: D.H.Tarling & S.K.Runcorn (eds), *Implications of continental drift to the earth sciences.* London, Academic Press:539-558.

Francheteau, J. & X.Le Pichon 1972. Marginal fracture zones as structural framework of the continental margins in the South Atlantic Ocean. *Bull.Am.Assoc.Petrol.Geol.*, 56:991-1007.

Francisconi, O. & R.O.Kowsmann 1976. Preliminary structural study of the south Brazilian continental margin. *Ann.Brazil. Acad.Sci.*, 48(suppl.):89-100.

Frankel, J.J. 1952. Silcrete near Albertinia, Cape Province. *S.Afr.J.Sci.*, 173-182.

Frankel, J.J. 1960. The geology along the Umfolozi River, south of Mtubatuba, Zululand. *Trans.geol.Soc.S.Afr.*, 63: 231-252.

Frankel, J.J. 1966. The basal rocks of the Tertiary at Uloa, Zululand, South Africa. *Geol.Mag.*, 103:214-230.

Frankel, J.J. 1967. In discussion of: McCarthy, M.J. Stratigraphical and sedimentological evidence from the Durban region of major sea-level movements since the late Tertiary. *Trans.geol.Soc.S.Afr.*, 70:191-194.

Frankel, J.J. 1968. Tertiary sediments in the lower Umfolozi River Valley, Zululand. *Trans.geol.Soc.S.Afr.*, 71:135-146.

Frankel, J.J. 1972. Distribution of Tertiary sediments in Zululand and southern Mozambique, southeast Africa. *Bull.Am. Ass.Petrol.Geol.*, 56:2415-2425.

Frankel, J.J. & L.E.Kent 1937. Grahamstown surface quartzites (silcretes). *Trans.geol.Soc.S.Afr.*, 40:1-42.

Fuller, A.O. 1970. The occurrence of laumonite in strata of the Karroo System, South Africa. In: *Second Gondwana Symposium: Proceedings and Papers,* Pretoria, Council for Scientific and Industrial Research:455-456.

Fuller, A.O. 1979. Phosphate occurrences on the western and southern coastal areas and continental shelves of southern Africa. *Econ.Geol.*, 74:221-231.

Galton, P.M. & M.A.Cluver 1976. *Anchisaurus capensis* (Broom) and a revision of the Anchisauridae (Reptilia, Saurischia). *Ann.S.Afr.Mus.*, 69:121-159.

Gardiner, B.G. & R.A.Jubb 1975. A preliminary catalogue of identifiable fossil fish material from South Africa. *Ann.S. Afr.Mus.*, 67(11):381-440.

Geological Survey of South Africa 1970. Geological map of the Republic of South Africa 1:1 000 000 scale. Pretoria, Government Printer.

Geological Survey of South Africa 1977. Umtata Sheet 3128. 1:250 000 scale. Pretoria, Government Printer.

Gerrard, I. 1958. *The olivine-melilitites from the Western Cape,* Unpubl. MSc thesis, Univ.Cape Town.

Gerrard, I. & G.C.Smith 1980. *The Post Palaeozoic succession and structure of the south-western African continental margin.* Johannesburg, Southern Oil Exploration Corp., 28pp.

Gevers, T.W. 1929. The volcanic vents of the western Stormberg. *Trans.geol.Soc.S.Afr.*, 31(for 1928):43-62.

Gevers, T.W. 1936. The Etjo Beds of northern Hereroland, South West Africa. *Trans.geol.Soc.S.Afr.*, 39:317-329.

Gevers, T.W. & J.de V.Little 1946. Upper Cretaceous beds between Intongazi and Umkandandhlobu Rivers, Alfred County, Natal. *Trans.geol.Soc.S.Afr.*, 48:27-29.

Ghosh, P.K. & N.D.Mitra 1970. A review of recent progress in the studies of the Gondwanas of India. In: *Second Gondwana Symposium: Proceedings and Papers,* Pretoria, Council for Scientific and Industrial Research:29-47.

Gidskehaug, A., K.M.Creer & J.G.Mitchell 1975. Palaeomagnetism and K-Ar ages of the south-west African basalts and their bearing on the time of initial rifting of the South Atlantic Ocean. *Geophys.J.R.astr.Soc.*, 42:1-20.

Glass, J. 1977. Deep weathering of the southwestern Cape granite and Malmesbury Group: palaeoclimatic implications. *Tech.Rept.Joint Geol.Surv.Univ.Cape Town Mar.Geol.Prog.*, 9:118-135.

Goslin, J., J.Mascle, J.C.Sibuet & H.Hoskins 1974. Geophysical study of the easternmost Walvis Ridge, South Atlantic: morphology and shallow structure. *Bull.geol.Soc.Am.*, 85: 619-632.

Gough, D.I. 1973. Possible linear plume under southernmost Africa. *Nature Phys.Sci.*, 245:93-94.

Grant, R.E. & G.A.Cooper 1973. Brachiopods and Permian correlations. *Mem. Can. Petrol. Geol. Soc.*, 2:572-593.

Gray, D.R. 1977a. Differentiation associated with discrete crenulation cleavage. *Lithos*, 10:89-101.

Gray, D.R. 1977b. Morphological classification of crenulation cleavage. *J. Geol.*, 85:229-235.

Gray, D.R. 1978. Cleavages in deformed psammitic rocks from south-eastern Australia, their nature and origin. *Bull. geol. Soc. Amer.*, 89:577-590.

Green, A.H. 1883. *Report on the coals of the Cape Colony.* Cape Town, Parliamentary report.

Green, D. 1969. Interrelationship of the stratigraphy of the Karroo System in the Republic of Botswana and South West Africa. *Trans. geol. Soc. S. Afr.*, 72:1-7.

Green, D., R.N.Crockett & M.T.Jones 1980. Tectonic control of Karoo sedimentation in mid-eastern Botswana. *Trans. geol. Soc. S. Afr.*, 83:213-219.

Greenman, L. 1966. *The geology of Area 2615C, Lüderitz, South West Africa.* Unpub. BSc thesis, Univ. Cape Town.

Greenway, M.E. 1972. The Geology of the Falkland Islands. *Brit. Ant. Surv. Sci. Rept.*, 76, NERC, London.

Gresse, P.G. 1976. *The structure and metamorphism of the pre-Cape rocks in the coastal strip south of George.* Unpub. MSc thesis, Univ. Stellenbosch.

Gresse, P.G. 1983. Lithostratigraphy and structure of the Kaaimans Group. In: Geodynamics of the Cape Fold Belt. *Geol. Soc. S. Afr. Spec. Pub.*

Griesbach, C.L. 1871. On the geology of Natal in South Africa. *Q. Jl. geol. Soc. Lond.*, 27:53-72.

Griffiths, J. 1978. A fragmentary specimen of *Saurichthys* sp. from the Upper Beaufort series of South Africa. *Ann. S. Afr. Mus.*, 76(8):299-307.

Grine, F.E. & Q.B.Hendey 1981. Earliest primate remains from South Africa. *S. Afr. J. Sci.*, 77:374-376.

Grove, A.T. 1969. Landforms and climatic change in the Kalahari and Ngamiland. *Geogr. J.*, 135:191-212.

Gulbrandsen, R.A. 1966. Chemical composition of phosphorites of the Phosphoria Formation. *Geochim. et Cosmochim. Acta*, 30:769-778.

Hälbich, I.W. 1977. Fold profiles and tectonic shortening in the Cape Fold Belt. *Trans. geol. Soc. S. Afr.*, 80:273-266.

Hälbich, I.W. 1978. Fold profiles and tectonic shortening in the Cape Fold Belt. Discussion and reply. *Trans. geol. Soc. S. Afr.*, 81:403-408.

Hälbich, I.W. 1983a. Stress and strain in various structural zones of the Cape Fold Belt. In: Geodynamics of the Cape Fold Belt. *Geol. Soc. S. Afr. Spec. Pub.*

Hälbich, I.W. 1983b. A tectogenesis of the Cape Fold Belt. In: Geodynamics of the Cape Fold Belt. *Geol. Soc. S. Afr. Spec. Pub.*

Hälbich, I.W. 1983c. Disharmonic folding, detachment and thrusting in the Cape Fold Belt. In: Geodynamics of the Cape Fold Belt. *Geol. Soc. S. Afr. Spec. Pub.*

Hälbich, I.W. 1983d. Intraformational folding in the Cape Fold Belt. In: Geodynamics of the Cape Fold Belt. *Geol. Soc. S. Afr. Spec. Pub.*

Hälbich, I.W. 1983e. Geodynamics of the Cape Fold Belt in the Republic of South Africa – a summary. I.G.P. Working Group 9. *Am. Geophys. Un. Spec. Pub.*

Hälbich, I.W. & D.H.Cornell 1983. Metamorphic zoning in the Cape Fold Belt. In: Geodynamics of the Cape Fold Belt. *Geol. Soc. S. Afr. Spec. Pub.*

Hälbich, I.W., F.J.Fitch & J.A.Miller 1983. Dating the Cape Orogeny. In: Geodynamics of the Cape Fold Belt. *Geol. Soc. S. Afr. Spec. Pub.*

Hales, A.L. & D.I.Gough 1960. Isostatic anomalies and crustal structure in the southern Cape. *Geophys. J. R. astr. Soc.*, 3: 225-236.

Hales, A.L. & J.B.Nation 1973. A seismic refraction study in the southern Indian Ocean. *Bull. Seismolog. Soc. Am.*, 63: 1951-1966.

Hallam, A. 1981. *Facies interpretation and the stratigraphic record.* San Francisco, Freeman, 291pp.

Hallam, C.D. 1964. The geology of the coastal diamond deposits of southern Africa (1959). In: S.H.Haughton (ed.), *The Geology of some ore deposits in southern Africa.* Johannesburg, Geol. Soc. S. Afr. Vol.2:671-728.

Halle, T.G. 1912. On the geological structure and history of the Falkland Islands. *Bull. Geol. Inst. Univ. Uppsala*, 11:115-229.

Hamilton, W.R. & J.A.van Couvering 1977. Lower Miocene mammals from South West Africa. *Namib Bull. (Suppl. 2, Transv. Mus. Bull.)*:9-11.

Haq, B.U. 1981. Paleogene paleoceanography: early Cenozoic oceans revisited. *Oceanol. Acta*, 4(suppl.):71-82.

Hargraves, R.B. & T.C.Onstott 1980. Paleomagnetic results from some southern African kimberlites, and their tectonic significance. *J. Geophys. Res.*, 85:3587-3596.

Harris, W.K. 1977. Palynology of cores from Deep Sea Drilling sites 327, 328 and 330, South Atlantic Ocean. In: P.Barker et al, *Initial Rep. Deep Sea Drilling Project*, 36, Washington, US Government Printing Office:761-767.

Harrison, J.V. & N.L.Falcon 1934. Collapse structures. *Geol. Mag.*, 71:529-539.

Hart, G.F. 1970. The stratigraphic subdivision and equivalents of the Karroo sequence as suggested by palynology. *IUGS Symposium* Buenos Aires 1967, Paris, UNESCO:23-32.

Hart, T.J. & R.J.Currie 1960. The Benguela Current. *Discovery Rept.*, 31:123-298.

Hartnady, C.J. 1966. *Structure of the pre-Cape beds near Worcester.* Unpub. Hons. thesis, Univ. Cape Town.

Hartnady, C.J. 1969. Structural analysis of some pre-Cape formations in the Western Province. *Precamb. Res. Unit Univ. of Cape Town Bull.*, 6:70pp.

Hartnady, C.J., A.R. Newton & J.N.Theron 1974. Stratigraphy and structure of the Malmesbury Group in the southwestern Cape. In: A.Kröner (ed.), *Contributions to the Precambrian geology of southern Africa. Precamb. Res. Unit., Univ. of Cape Town Bull.*, 15:193-213.

Haughton, S.H. 1915. On some dinosaur remains from Bushmanland. *Trans. R. Soc. S. Afr.*, 5:259-264.

Haughton, S.H. 1919. A review of the reptilian fauna of the Karroo System of South Africa. *Trans. geol. Soc. S. Afr.*, 22: 1-25.

Haughton, S.H. 1924. The fauna and stratigraphy of the Stormberg series. *Ann. S. Afr. Mus.*, 12:323-495.

Haughton, S.H. 1925. The Tertiary deposits of the south-eastern districts of the Cape Province. *Trans. geol. Soc. S. Afr.*, 28:27-32.

Haughton, S.H. 1928. *The geology of the country between Grahamstown and Port Elizabeth. An explanation of the Cape Sheet No.9 (Port Elizabeth).* Geol. Surv. S. Afr., 45pp.

Haughton, S.H. 1930a. Note on the occurrence of Upper Cretaceous marine beds in South West Africa. *Trans. geol. Soc. S. Afr.*, 33:61-63.

Haughton, S.H. 1930b. On the occurrence of Upper Cretaceous marine fossils near Bogenfels, S.W.Africa. *Trans. R. Soc. S. Afr.*, 18:361-365.

Haughton, S.H. 1931a. On a collection of fossil frogs from the clays at Banke. *Trans. R. Soc. S. Afr.*, 19:233-249.

Haughton, S.H. 1931b. The late Tertiary and Recent deposits of the west coast of South Africa. *Trans. Geol. Soc. S. Afr.*, 34:19-58.

Haughton, S.H. 1932. On the phosphate deposits near Langebaanweg, Cape Province. *Trans. geol. Soc. S. Afr.*, 35:119-124.

Haughton, S.H. 1956. Phosphatic-glauconitic deposits off the west coast of South Africa. *Ann. S. Afr. Mus.*, 42:329-334.

Haughton, S.H. 1963. *Stratigraphic history of Africa south of*

*the Sahara.* Edinburgh, Oliver & Boyd, 365pp.
Haughton, S.H. 1969. *Geological History of Southern Africa.* Johannesburg, Geological Society of South Africa:535pp.
Haughton, S.H., J.J.Blignault, P.J.Roussow, J.J.Spies & S.Zagt 1953. Results of an investigation into the possible presence of oil in the Karroo rocks in parts of the Union of South Africa. *Mem.Geol.Surv.S.Afr.,* 45:130pp.
Haughton, S.H., H.F.Frommurze & D.J.L.Visser 1937. *The geology of portion of the coastal belt near the Gamtoos Valley Cape Province. An explanation of Sheet Nos.151 North and 151 South (Gamtoos River).* Geol.Surv.S.Afr., 64pp.
Haughton, S.H. & A.W.Rogers 1924. The volcanic rocks south of Zuurberg. *Trans.R.Soc.S.Afr.,* 11:235-249.
Hausmann, J.F.L. 1837. Beiträg zur Kunde der geognostischen constitution von Süd-Afrika. *Göttingishe gelehrte Anzeigen,* 1449-1462.
Hawthorne, J.B., A.J.Carrington, C.R.Clement & E.W.Skinner 1979. Geology of the Dokolwayo Kimberlite and associated paleoalluvial diamond deposits. In: F.R.Boyd & H.O.A.Meyer (eds), *Kimberlites, Diatremes and Diamonds.* Proc.2nd Int. *Kimb.Conf.* 1: Washington Am.Geophys.Un., 59-70.
Heath, D.C. 1966. *Die geologie van die Karoosisteem in die gebied Mariental-Asab, Suidwes-Afrika.* Unpub. MSc thesis, Univ.OFS.
Hendey, Q.B. 1970. A review of the geology and palaeontology of the Plio/Pleistocene deposits at Langebaanweg, Cape Province. *Ann.S.Afr.Mus.,* 56:75-117.
Hendey, Q.B. 1973. Fossil occurrences at Langebaanweg, Cape Province. *Nature,* 244:13-14.
Hendey, Q.B. 1974a. The late Cenozoic Carnivora of the southwestern Cape Province. *Ann.S.Afr.Mus.,* 63:1-369.
Hendey, Q.B. 1974b. Faunal dating of the late Cenozoic of southern Africa. *Quat.Res.,* 4:149-161.
Hendey, Q.B. 1976. The Pliocene fossil occurrences in 'E' Quarry, Langebaanweg, South Africa. *Ann.S.Afr.Mus.,* 69: 215-247.
Hendey, Q.B. 1978a. Preliminary report on the Miocene vertebrates from Arrisdrift, South West Africa. *Ann.S.Afr.Mus.,* 76:1-41.
Hendey, Q.B. 1978b. The age of fossils from Baard's Quarry, Langebaanweg, South Africa. *Ann.S.Afr.Mus.,* 75:1-24.
Hendey, Q.B. 1981a. Palaeoecology of the late Tertiary fossil occurrences in 'E' Quarry, Langebaanweg, South Africa, and a reinterpretation of their geological context. *Ann.S. Afr.Mus.,* 84:1-104.
Hendey, Q.B. 1981b. Geological succession at Langebaanweg, Cape Province, and global events of the late Tertiary. *S.Afr. J.Sci.,* 77:33-38.
Henriques da Silva, G. 1967. Fósseis do Miocénico marinho das plantações Santiago (Magude). *Rev.Estudos Gerais Univ. Moçambique,* Sér.6, 4:1-21.
Hill, R.S. 1972. *The Geology of the northern Algoa Basin, Port Elizabeth.* Unpub. MSc thesis, Univ. Stellenbosch:68pp.
Hiller, N. & N.Stavrakis 1980. Distal alluvial fan deposits in the Beaufort Group of the eastern Cape Province. *Trans.geol. Soc.S.Afr.,* 83:353-360.
Hodgson, F.D.I. 1970. The geology of the Karroo System in the southern Kaokoveld, South West Africa. In: *Second Gondwana Symposium: Proceedings and Papers.* Pretoria, Council for Scientific and Industrial Research:233-240.
Hofmeyer, P.K. 1971. *The abundances and distribution of some trace elements in some selected South African shales.* Unpub. PhD thesis, Univ.Cape Town (2 vols.).
Hopwood, A.J. 1929. New and little known mammals from the Miocene of Africa. *Am.Mus.Novit.,* 344:1-9.
Hoschek, G. 1969. The stability of staurolite and chloritoid and their significance in metamorphism in pelitic rocks. *Contr.Min.Petrol.,* 22:208-233.
Hoskins, H., C.U.Rogers & A.O. Woo 1974. Data report on oblique reflection-refraction radio-sonobuoy profiles on the African Atlantic continental margin (R/V Atlantis II cruises 67 & 75). *Woods Hole Oceanographic Inst.Tech. Rept.,* 73-74.
Howell, F.C. 1978. Hominidae. In: V.J.Maglio & H.B.S.Cooke (eds), *Evolution of African Mammals,* Cambridge Mass: Harvard Univ.Press:154-248.
Hoyt, J.H., B.L.Oostdam & D.D.Smith 1969. Offshore sediments and valleys of the Orange River (South and South West Africa). *Mar.Geol.,* 7:69-84.
Hughes, A.R. & P.V.Tobias 1977. A fossil skull probably of the genus *Homo* from Sterkfontein, Transvaal. *Nature,* 265:310-312.
Hutchinson, G.E. 1950. Survey of existing knowledge of biogeochemistry. 3. The biogeochemistry of vertebrate excretion. *Bull.Am.Mus.Nat.Hist.,* 96:1-554.
Jacobsen, J.B.E. & T.S.McCarthy 1975. Possible late Karroo carbonatite and basalt intrusions at Messina, Transvaal. *Trans.geol.Soc.S.Afr.,* 78:153-159.
Jacobsen, J.B.E., D.C.Rex & W.J.Sevenster 1975. K-Ar ages of some mafic dykes from the Messina district, Transvaal. *Trans.geol.Soc.S.Afr.,* 78:359-360.
Jansen, H. 1975. The Soutpansberg Trough (northern Transvaal) – an aulacogen. *Trans.geol.Soc.S.Afr.,* 78:129-136.
Jaritz, W., H.Krenzer, P.Muller & W.Harre 1977. Die vulkanitserien in Kuskengebiete von Nordmoçambique. *Geol.Jb.,* 76:147-165.
Jenkeyns, H.C. 1980. Cretaceous anoxic events: from continents to oceans. *J.geol.Soc.Lond.,* 137:171-188.
Johnson, M.R. 1966. *The stratigraphy of the Cape and Karoo Systems in the Eastern Cape Province.* Unpub. MSc thesis, Grahamstown, Rhodes Univ.
Johnson, M.R. 1976. *Stratigraphy and Sedimentology of the Cape and Karoo Sequences in the Eastern Cape Province.* Unpub. PhD thesis, Grahamstown, Rhodes Univ.: 336pp.
Johnson, M.R., B.J.V.Botha, P.J.Hugo, A.W.Keyser, B.R.Turner & H.de la R.Winter 1976. *Preliminary Report on stratigraphic nomenclature in the Karoo sequence.* Unpub. report to South African Committee for Stratigraphy by the Karoo working group.
Jones, D.L. & G.Plafker 1977. Mesozoic megafossils from DSDP Hole 327A and site 330 on the eastern Falkland Plateau. In: P.Barker et al, *Initial Rep. Deep Sea Drilling Project,* 36, Washington, US Government Printing Office: 845-855.
Jones, T.R. 1867. Some secondary fossils from South Africa. *Q.Jl.geol.Soc.Lond.,* 23:139-175.
Jones, T.R. 1901. On the Enon conglomerate of the Cape of Good Hope, and its fossil Estheriae. *Geol.Mag.,* 48:350-354.
Jubb, R.A. 1973. Brief synthesis of present information on the geographical and stratigraphical distribution of fossil fish within the Stormberg stage, South Africa. *Palaeont. afr.,* 16:17-23.
Kaiser, E. 1926. *Die Diamantenwüste Südwestafrikas, I.* Berlin, Dietrich Reimer, 321pp.
Kemp, E.M. 1976. Early Tertiary pollen from Napperby, Central Australia. *Bur.Min.Res.J.Aust.Geol.Geophys.,* 1:109-114.
Kemp, E.M. 1978. Tertiary climatic evolution and vegetation history in the southeast Indian Ocean region. *Palaeogeog. Palaeoclimat.Palaeoecol.,* 24:169-208.
Kemp, E.M. & P.J.Barrett 1975. Antarctic glaciation and early Tertiary vegetation. *Nature,* 258:507-508.
Kemp, E.M. & W.K.Harris 1975. The vegetation of Tertiary islands on the Ninetyeast Ridge. *Nature,* 258:303-307.
Kennedy, W.J. & H.C.Klinger 1971. A major intra-Cretaceous unconformity in eastern south Africa. *Jl.geol.Soc.Lond.,* 127:183-186.

Kennedy, W.J. & H.C.Klinger 1975. Cretaceous faunas from Zululand and Natal, South Africa. Introduction, Stratigraphy. *Bull.Brit.Mus.Nat.Hist.* Geology, 25:265-315.

Kennedy, W.J. & H.C.Klinger (in press). Cretaceous stratigraphy of Natal and Zululand. *Bull.Geol.Surv.S.Afr.*

Kensley, B.F. 1972. Pliocene marine invertebrates from Langebaanweg, Cape Province. *Ann.S.Afr.Mus.*, 60:173-190.

Kensley, B.F. 1975. A late Pleistocene raised beach from Milnerton, Cape. Unpub. contribution to the Conf.S.Afr.Soc. Quat.Res.Cape Town, 1975. Quoted in Rogers 1979.

Kensley, B.F. 1977. A second assemblage of Pliocene invertebrates from Langebaanweg, Cape. *Ann.S.Afr.Mus.*, 72:189-210.

Kent, P.E. 1974. Continental margin of East Africa – a region of vertical tectonics. In: C.A.Burk & C.L.Drake (eds), *Geology of Continental Margins.* Berlin, Springer:313-320.

Kent, P.E. 1977. The Mesozoic development of aseismic continental margins. *Jl.Geol.Soc.Lond.*, 134:1-18.

Keyser, A.W. 1973. A new Triassic vertebrate fauna from South West Africa. *Palaeont.afr.*, 16:1-15.

Keyser, A.W. 1979. A review of the biostratigraphy of the Beaufort Group in the Karoo Basin of South Africa. *Geocongress 79 Abstr.*, Part 2:13-29. Geol.Soc.S.Africa.

Keyser, A.W. 1981. The stratigraphic distribution of the Dicynodontia of Africa reviewed. In: M.M.Cresswell & P. Vella (eds), *Gondwana Five,* Proc.5th Int.Gondwana Symp. Wellington, 1980, Rotterdam, Balkema:61-63.

Keyser, A.W. & A.R.I.Cruikshank 1979. The origins and classification of Triassic dicynodonts. *Trans.geol.Soc.S.Afr.*, 82: 81-108.

Keyser, A.W. & R.M.H.Smith 1979. Vertebrate biozonation of the Beaufort Group with special reference to the western Karoo Basin. *Ann.Geol.Surv.S.Afr.*, 12(for 1977-78):1-35.

Keyser, U. 1972. The occurrence of diamonds along the coast between the Orange River and the Port Nolloth Reserve. *Bull.Geol.Surv.S.Afr.*, 64:23pp.

Kilmurray, J.O. 1975. Las Sierras Australes de la Provincia de Buenos Aires. Las Fases de Deformation y Interpretatin Estratigrafica. *Rev.Asoc.Geol.Argentina,* 30(4):331-348.

King, L.C. 1953. A Miocene fauna from Zululand. *Trans.geol. Soc.S.Afr.*, 56:59-91.

King, L.C. 1966. An extensive Pliocene Formation in Natal and Mozambique – its geomorphological implications. *Trans.geol.Soc.S.Afr.*, 69:201-210.

King, L.C. 1967. *Scenary of South Africa.* Edinburgh, Oliver & Boyd:308pp.

King, L.C. 1970. Uloa revisited: a review of the essential data from this classic locality and a statement of Tertiary history of the Zululand coastal area. *Trans.geol.Soc.S.Afr.*, 73:151-157.

King, L.C. 1972. Geomorphic significance of the Late Cretaceous limestones at Needs Camp, near East London. *Trans. geol.Soc.S.Afr.*, 75:1-3.

King, L.C. 1973. Pliocene marine fossils from the Alexandria Formation in the Paterson District, Eastern Cape Province, and their geomorphic significance. *Trans.geol.Soc.S.Afr.*, 75:159-160.

King, L.C. & R.R.Maud 1965, The geology of Durban and environs. *Geol.Surv.S.Afr.Bull.*, 42.

Kingsley, C.S. 1964. *Die geologie van die Serie Stormberg in die Sisteem Karroo, suid van Bethlehem, O.V.S.* Unpub. MSc thesis, Univ.OFS:84pp.

Kirchheimer, F. 1934. On pollen from the Upper Cretaceous dysodil of Banque, Namaqualand (South Africa). *Trans.R. Soc.S.Afr.*, 21:41-50.

Kitchin, F.L. 1908. The invertebrate fauna and palaeontological relationships of the Uitenhage series. *Ann.S.Afr.Mus.*, 7:21-250.

Kitching, J.W. 1970. A short review of the Beaufort zoning in South Africa. In: *Second Gondwana Symposium: Proceedings and Papers.* Pretoria, Council for Scientific and Industrial Research:309-312.

Kitching, J.W. 1977. The distribution of the Karoo vertebrate fauna. *Mem. Bernard Price Inst.Palaeont.Res.*, 1:131pp.

Klein, R.G. 1981. Ungulate mortality and sedimentary facies in the late Tertiary Varswater Formation, Langebaanweg, South Africa. *Ann.S.Afr.Mus.*, 84:233-254.

Klinger, H.C. 1977. Cretaceous deposits near Bogenfels, South West Africa. *Ann.S.Afr.Mus.*, 73:81-92.

Klinger, H.C., E.G.Kauffman & W.J.Kennedy 1980. Upper Cretaceous ammonites and inoceramids from the off-shore Alphard Group of South Africa. *Ann.S.Afr.Mus.*, 82:293-320.

Klinger, H.C. & W.J.Kennedy 1977. Upper Cretaceous ammonites from a borehole near Richards Bay, South Africa. *Ann. S.Afr.Mus.*, 72:69-107.

Klinger, H.C. & W.J.Kennedy 1979. Cretaceous faunas from Southern Africa Lower Cretaceous ammonites, including a new bochianitid genus, from Umgazana, Transkei. *Ann.S. Afr.Mus.*, 78:11-19.

Klinger, H.C. & W.J.Kennedy 1980. The Umzamba Formation at its type section Umzamba Estuary (Pondoland, Transkei), the ammonite content and palaeogeographical distribution. *Ann.S.Afr.Mus.*, 81:207-222.

Klinger, H.C., W.J.Kennedy & R.V.Dingle 1972. A Jurassic ammonite from South Africa. *N.Jb.Geol.Paläont.Mh.*, 11: 653-659.

Klinger, H.C., W.J.Kennedy & W.G.Siesser 1976. *Yabeiceras* (Coniacian ammonite) from the Alphard Group off the southern Cape coast. *Ann.S.Afr.Mus.*, 69:161-168.

Klinger, H.C. & B.E.Lock 1978. Upper Cretaceous sediments from the Igoda River mouth, East London, South Africa. *Ann.S.Afr.Mus.*, 77:71-83.

Knox, G.J. 1977. Caliche profile formation, Saldanha Bay (South Africa). *Sedimentology,* 24:657-674.

Kolodny, Y. 1969. Are marine phosphorites forming today? *Nature,* 224:1017-1019.

Kolodny, Y. & I.R.Kaplan 1970. Uranium isotopes in sea-floor phosphorites. *Geochim.Cosmochim.Acta,* 34:3-24.

Korn, H. & H.Martin 1951. The seismicity of southwest Africa. *Trans.geol.Soc.S.Afr.*, 54:85-88.

Krige, A.V. 1927. An examination of the Tertiary and Quaternary changes of sea-level in South Africa, with special stress on the evidence in favour of a recent world-wide sinking of ocean-level. *Ann. Univ.Stellenbosch,* 5, section A(1):1-81.

Kröner, A. 1973. Comments on 'Is the African plate stationary?' *Nature,* 243:29-30.

Krynau, J.R. 1983. Granite intrusion and regional metamorphism in the Kaaimans Group. In: Geodynamics of the Cape Fold Belt. *Geol.Soc.S.Afr.Spec.Pub.*

Krynau, J.R. & P.G.Gresse 1980. The Kaaimans Group in the George area, Cape Province, a model for the origin of deformation and metamorphism in the southern Cape Fold Belt. *Trans.geol.Soc.S.Afr.*, 83:23-28.

Kyle, P.R., D.H.Elliot & J.F.Sutter 1981. Jurassic Ferrar Supergroup tholeiites from the Transantarctic Mountains, Antarctica, and their relationship to the initial fragmentation of Gondwana. In: M.M.Cresswel & P.Vella (eds), *Gondwana Five,* Proc.5th Int.Gondwana Symp., Wellington 1980, Rotterdam, Balkema:283-287.

Lambert, G. & V.Scheibnerovà 1974. Albian foraminifera of Zululand (South Africa) and Great Artesian Basin (Australia). *Micropalaeontology,* 20:76-96.

Lamont, G.T. 1947. *The Geology of part of the Vanrhynsdorp Division, Cape Province.* Unpub. PhD thesis, Univ.Cape Town.

Lang, W.D. 1908. Polyzoa and anthozoa from the Upper Cretaceous limestones of Needs Cape, Buffalo River. *Ann.S.Afr. Mus.*, 7:1-11.

Lapido-Loureiro, F.E.de V. 1968. Sub-volcanic carbonatite structures of Angola. *Rept.23rd Int.Geol.Congr.Prague,* 2: 147-161.

Larson, R.L. & T.W.C.Hilde 1975. A revised time scale of magnetic reversals for the Early Cretaceous, Late Jurassic, *J. Geophys.Res.*, 80:2586-2594.

Larson, R.L. & J.W.Ladd 1973. Evidence for the opening of the south Atlantic in the early Cretaceous. *Nature,* 246: 209-212.

Larson, R.L. & W.C.Pitman 1972. World-wide correlation of Mesozoic magnetic anomalies and its implications. *Bull. geol.Soc.Am.,* 83:3645-3662.

Lawson, A.C. 1927. Folded mountains and isostasy. *Bull.geol. Soc.Am.,* 38:253-265.

Lehner, P. & P.A.C.de Ruiter 1977. Structural history of Atlantic margin of Africa. *Bull.Am.Assoc.Petrol.Geol.,* 61:961-981.

Leith, M.J. & D.M.Rowsell 1979. Burial history and temperature-depth conditions for hydrocarbon generation and migration on the Agulhas Bank, South Africa. *Geol.Soc.S. Afr.Spec.Publ.,* 6:205-217.

Le Pichon, X. & D.E.Hayes 1971. Marginal off-sets, fracture zones and the early opening of the South Atlantic. *J.Geophys.Res.,* 76: 6283-6293.

Le Roux, D.M. 1977. *Struktuurondersoek van die Kaapse Plooigordel in die omgewing van Barrydale.* Unpub. MSc thesis, Univ.Stellenbosch.

Le Roux, D.M. 1983. Geological structure in the Little Karoo about 21°E longitude. In: Geodynamics of the Cape Fold Belt. *Geol.Soc.S.Afr.Spec.Pub.*

Le Roux, J.P. 1977. *The stratigraphy, sedimentology and structure of the Kango Group north of Oudtshoorn, Cape Province.* Unpub. MSc thesis, Univ.Stellenbosch.

Le Roux, J.P. 1983. Structural evolution of the Kango Group. In: Geodynamics of the Cape Fold Belt. *Geol.Soc.S.Afr. Spec.Publ.*

Le Roux, J.S. 1974. *Paleogeologiese en paleogeografiese aspekte van die Etage Rooilaevan die sisteem Karroo.* Unpub. DSc thesis, Bloemfontein, Univ.OFS:295pp.

Levyns, M.R. 1964. Migrations and origin of the Cape Flora. *Trans.R.Soc.S.Afr.,* 37:85-107.

Leyden, R., M.Ewing & E.S.W.Simpson 1971. Geophysical reconnaissance on the African Shelf: Cape Town to East London. *Bull.Am.Ass.Petrol.Geol.,* 55:651-657.

Lock, B.E. 1973. Tertiary limestones at Needs Camp, near East London. *Trans.geol.Soc.S.Afr.,* 76:1-5.

Lock. B.E. 1978. The Cape Fold Belt of South Africa; tectonic control of sedimentation. *Proc.Geol.Ass.,* 89:263-281.

Lock, B.E. 1980. Flat-plate subduction and the Cape Fold Belt of South Africa. *Geology,* 8:35-39.

Lock, B.E., A.L.Paverd & T.J.Broderick 1974. Stratigraphy of the Karroo volcanic rocks of the Barkly East district. *Trans. geol.Soc.S.Afr.,* 77:117-129.

Lock, B.E., R.Shone, A.T.Coates & C.J.Hatton 1975. Mesozoic Newark-type sedimentary basins within the Cape Fold Belt of southern Africa. *9th Int.Sediment.Congress,* 4:217-226, Nice.

Lock, B.E. & J.D.Wilson 1975. Discussion of 'The nature and origin of volcaniclastic material in some Karroo and Beacon rocks' by Elliot, D.H. & D.R.Watts, *Trans.geol.Soc. S.Afr.,* 78:171.

Ludbrook, N.H. 1978, Australia. In: Moullade, M. & A.E.M. Nairn (eds), *The Phanerozoic Geology of the World II, The Mesozoic, A.* Amsterdam, Elsevier:209-249.

Ludwig, W.J., J.E.Nafe, E.S.W.Simpson & S.Sacks 1968a. Seismic-refraction measurements on the southeast African continental margin. *J.Geophys.Res.,* 73:3707-3719.

Ludwig, W.J., J.I.Ewing & M.Ewing 1968b. Structure of the Argentine continental margin. *Bull.Am.Assoc.Petrol.Geol.,* 52:2337-2368.

Lutjeharms, J.R.E. & H.R.Valentine 1981. Ocean circulation studies in the vicinity of southern Africa: preliminary results using FGGE drifters and remote sensing. *Adv.Space Res.,* 1:211-223.

Mabbutt, J.A. 1955. Erosion surfaces in Little Namaqualand and the ages of surface deposits in the south western Kalahari. *Trans.geol.Soc.S.Afr.,* 58:13-29.

Mabbutt, J.A. 1956. The physiography and surface geology of the Hopefield fossil site. *Trans.R.Soc.S.Afr.,* 35:21-58.

Makrides, M. 1979. *Foraminifera of the Upper Cretaceous Mzamba Formation, Transkei, southern Africa.* Unpub. MSc thesis, Univ.Witwatersrand, Johannesburg:86pp.

Mallick, D.I.J., F.Habgood & A.C.Skinner 1981. A geological interpretation of Landsat imagery and air photography of Botswana. *Overseas Geol.Miner.Resour.,* London, HMSO: 56:35pp.

Manton, W.I. 1968. The origin of associated basic and acid rocks in the Lebombo-Nuanetsi Igneous Province, southern Africa, as implied by strontium isotopes. *J.Petrol.,* 9:23-39.

Marais, J.A.H. & M.R.Johnson 1965. Thicknesses of the Cape and Karroo Systems in the Eastern Cape. *Ann.Geol.Surv. S.Afr.,* 76:145-152.

Marsh, J.S. 1973. Relationships between transform directions and alkaline igneous rock lineaments in Africa and South America. *Earth Planet.Sci.Lett.,* 18:317-323.

Marsh, J.S. 1975. The Lüderitz Alkaline Province, South West Africa I: Descriptive petrology of the Granitberg Foyaite complex. *Trans.geol.Soc.S.Afr.,* 78:215-224.

Marsh, J.S. 1976. The Lüderitz Alkaline Province, South West Africa, III: The Pomona and Drachenberg Syenite complexes. *Trans.geol.Soc.S.Afr.,* 79:168-176.

Marsh, J.S., B.E.Lock & W.H.Fuchter 1979. New chemical analyses of the Suurberg volcanic rocks and their significance in relation to Mesozoic volcanism in Southern Africa. *S.Afr.J.Sci.,* 75:227-229.

Martin, A.K., S.W.Goodlad, C.J.H.Hartnady & A.du Plessis 1982a. Early Cretaceous palaeopositions of the Falkland Plateau relative to southern Africa using Mesozoic sea-floor spreading anomalies. *Geophys.J.R.astr.Soc.,* 71:567-579.

Martin, A.K., S.W.Goodlad & D.A.Salmon 1982b. Sedimentary basin in-fill in the northernmost Natal Valley, hiatus development and Agulhas Current palaeo-oceanography. *J.geol.Soc. Lond.,* 139:183-201.

Martin, A.K., C.J.H.Hartnady & S.W.Goodlad 1981. A revised fit of South America and South Central Africa. *Earth Planet. Sci.Lett.,* 54:293-305.

Martin, H. 1970. In discussion of: F.D.I.Hodgson. The geology of the Karroo System in the southern Kaokoveld, South West Africa. In: *Second Gondwana Symposium: Proceedings and Papers,* Pretoria: Council for Scientific and Industrial Research, p.235.

Martin, H. 1973a. The Atlantic margin of southern Africa between latitude 17° south and the Cape of Good Hope. In: A.E.M.Nairn & F.G.Stehli (eds), *The Ocean Basins and Margins Vol.1. The South Atlantic.* New York, Plenum Press:277-300.

Martin, H. 1973b. Palaeozoic, Mesozoic and Cenozoic deposits on the coast of South West Africa. In: G.Blant (ed.), *Sedimentary Basins of the African Coasts, Part 2 south and east coast,* Paris, Association of African Geological Surveys:7-15.

Martin, H. 1976. A geodynamic model for the evolution of the continental margin of southwestern Africa. *Ann.Braz.Acad. Sci.,* 48(suppl.):169-177.

Martini, E. 1971. Standard Tertiary and Quaternary calcareous nannoplankton zonation. In: A.Farinacci (ed.), *Proc.II Planktonic Conf.Roma 1970,* Rome, Edizioni Technoscienza: 739-785.

Martini, J.E.J. 1974. The presence of ash beds and volcanic fragments in the greywackes of the Karroo System in the south-

ern Cape Province. *Trans.geol.Soc.S.Afr.,* 77:113-116.
Masolov, V.N., R.G.Kurinin & G.E.G.Sevmorgeo 1981. Crustal structures and tectonic significance of Antarctic rift zones (from geophysical evidence). In: M.M.Cresswell & P.Vella (eds), *Gondwana Five,* Proc.5th Int.Gondwana Symp., Wellington, 1980, Rotterdam, Balkema:303-309.
Maud, R.R. 1961. A preliminary review of the structure of coastal Natal. *Trans.geol.Soc.S.Afr.,* 64:247-256.
Maud, R.R. 1968. Quaternary geology and soil formation in coastal Natal. *Z.Geomorph.,* Suppl.7:155-199.
Maud, R.R. & W.N.Orr 1975. Aspects of post-Karroo geology in the Richards Bay area. *Trans.geol.Soc.S.Afr.,* 78:101-109.
McCarthy, M.J. 1967. Stratigraphical and sedimentological evidence from the Durban region of major sea-level movements since the late Tertiary. *Trans.geol.Soc.S.Afr.,* 70:135-165.
McCarthy, M.J. 1970. An occurrence of pillow lava in a basal flow of the Drakensberg volcanic stage, Ndedema Valley, Natal. In: *Second Gondwana Symposium: Proceedings and Papers,* Pretoria, Council for Scientific and Industrial Research:433-439.
McConnell, R.B. 1980. A resurgent taphrogenic lineament of Precambrian origin in eastern Africa. *J.geol.Soc.Lond.,* 137: 483-489.
McDougall, I. 1963. Potassium-argon age measurements on dolerites from Antarctica and South Africa. *J.Geophys. Res.,* 68:1535-1545.
McElhinny, M.W. & P.J.Burek 1971. Mesozoic palaeomagnetic stratigraphy. *Nature,* 232:98-102.
McGowran, B. & A.C.Moore 1971. A reptilian tooth and Upper Cretaceous microfossils from the Lower Quarry at Needs Camp, South Africa. *Trans.geol.Soc.S.Afr.,* 74:103-105.
McLachlan, I.R., P.W.Brenner & I.K.McMillan 1976a. The stratigraphy and micropalaeontology of the Cretaceous Brenton Formation and the PB-A/1 well, near Knysna, Cape Province. *Trans.geol.Soc.S.Afr.,* 79:341-370.
McLachlan, I.R., I.K.McMillan & P.W.Brenner 1976b. Micropalaeontological study of the Cretaceous beds at Mbotyi and Mngazana, Transkei, South Africa. *Trans.geol.Soc.S. Afr.,* 79:321-340.
McLachlan, I.R. & I.K.McMillan 1976. Review and stratigraphic significance of southern Cape Mesozoic palaeontology. *Trans. geol.Soc.S.Afr.,* 79:197-212.
McLachlan, I.R. & I.K.McMillan 1979. Microfaunal biostratigraphy, chronostratigraphy and history of Mesozoic and Cenozoic deposits on the coastal margin of South Africa. *Geol.Soc.S.Afr.Spec.Publ.,* 6:161-181.
McQuillan, R. & D.A.Ardus 1977. *Exploring the Geology of Shelf Seas,* London, Graham & Trotman Ltd:234pp.
Merensky, H. 1910. The diamond deposits of Lüderitz-land, German South West Africa. *Trans.geol.Soc.S.Afr.,* 12:13-23.
Miall, A.J. 1977. A review of the braided river depositional environment. *Earth Sci.Rev.,* 13:1-62.
Midgley, D.C. & M.V.Pitman 1969. Surface water resources of South Africa. *Hydrol.Res.Unit, Dept.Civil Eng.,* Rept.2/69, 128pp. Johannesburg, Univ.Witwatersrand.
Miller, A.K. & W.M.Furnish 1956. Tertiary nautiloids dredged near Cape of Good Hope. *Ann.S.Afr.Mus.,* 42:327-328.
Miller, H. 1981. Pre-Andean orogenies of southern South America in the context of Gondwana. In: M.M.Cresswell & P.Vella (eds), *Gondwana Five,* Proc.5th Int.Gondwana Symp., Wellington 1980, Rotterdam, Balkema:237-242.
Miller, R.McG. 1963. *Geology of the Knysna District.* Unpub. BSc thesis, Univ.Cape Town:35pp.
Miller, R.McG. 1975. Notes on the revised map of the geology around Knysna. *Trans.geol.Soc.S.Afr.,* 78:367.
Moore, A.E. 1976. Controls of post-Gondwanaland alkaline volcanism in southern Africa. *Earth Planet.Sci.Lett.,* 31: 291-296.
Moore, A.E. 1979. *The Geochemistry of the Olivine Melilitites and related rocks of Namaqualand-Bushmanland, South Africa.* Unpub. PhD thesis, Univ. Cape Town, 2 vols.
Mountain, E.D. 1939. Pebble beds in the Lower Beaufort Series of the East London division. *S.Afr.J.Sci.,* 36:164-169.
Mountain, E.D. 1946. *The geology of an area east of Grahamstown. An explanation of Sheet 136 (Grahamstown).* Geol. Surv.S.Afr., 56pp.
Mountain, E.D. 1962. *The geology of the country around Port Alfred. An explanation of sheets 3326D (Port Alfred) and 3327C.* Geol.Surv.S.Afr., 28pp.
Mountain, E.D. 1974. *The geology of the area around East London, Cape Province. An explanation of sheet map 3227D (East London), 3228 (Kei Mouth).* Geol.Surv.S.Afr.
Murray, J. & A.F.Renard 1891. *Deep-sea Deposits.* Rep.Sci. Res.HMS Challenger 1873-1876, London, HMSO, 525pp.
Murray, L.G. 1970. Exploration and sampling methods employed in the offshore diamond industry. *9th Commonwealth Min. Metall.Congr.London,* 2, London:71-94.
Murray, L.G., R.H.Joynt, D.O'C.O'Shea, R.W.Foster & L.Kleinjan 1970. The geological environment of some diamond deposits off the coast of South West Africa. *Inst.Geol.Sci. London,* Rept. 70/12:119-142.
Musacchio, E.A. 1979. Datos paleobiogeographico de algunas asociaciones de foraminiferos, ostracodos y carofitas del Jurasico medio y el Cretacico inferior de Argentina. *Ameghiniana,* 16:247-271.
Netterburg, F. 1969. *The geology and engineering properties of South African calcretes.* Unpub. PhD thesis, Univ.Witwatersrand:1070pp.
Netterburg, F. 1971. Calcrete in road construction, *CSIR Research Rep.286, Nat.Inst.Road Res.Bull.,* 10:73pp. Pretoria, Council for Scientific and Industrial Research.
Netterburg, F. 1975. Pedogenic materials. In: *Sixth Regional Conf. for Afr.on Soil Mechanics and Foundation Engineering.* Durban, 1:291-295.
Netterberg, F. 1980. Geology of southern African calcretes: 1. Terminology, description, macrofeatures, and classification. *Trans.geol.Soc.S.Afr.,* 83:255-283.
Neumayr, M. 1885. *Phylloceras semistriatum* Orb. von Mossambique. *Denkschr.Akad.Wiss.Wien,* 50:139.
Newton, A.R. 1966. Some photogeological principles and their application in the Cape Province. *S.Afr.J.Photogramm.,* 2: 272-295.
Newton, A.R. 1973. A gravity-folding model for the Cape Fold Belt. *Trans.geol.Soc.S.Afr.,* 76:145-152.
Newton, A.R. 1975. *Structural applications of Remote Sensing in the Cape Province, and a new model for the evolution of the Cape Fold Belt.* Unpub. PhD thesis, Univ.Cape Town.
Newton, A.R. 1980. Intraplate Nature of the Cape Fold Belt. *S.Afr.J.Sci.,* 76:272-273.
Newton, R.B. 1896. On the occurrence of *Alectryonia ungulata* in SE Africa, with a notice of previous researches on the Cretaceous conchology of Southern Africa. *J.Conchology,* 8:136-151.
Newton, R.B. 1913. On some Cainozoic shells from South Africa. *Rec.Albany Mus.,* 2:315-352.
Newton, R.B. 1924. A contribution to the palaeontology of Portuguese East Africa. *Trans.geol.Soc.S.Afr.,* 26:141-159.
Nicolaysen, L.O., A.J.Burger & R.L.Johnson 1962. The age of Shawa carbonatite complex. *Trans.geol.Soc.S.Afr.,* 65:293-294.
Norton, I.O. & J.G.Sclater 1979. A model for the evolution of the Indian Ocean and the breakup of Gondwanaland. *J. Geophys.Res.,* 84:6803-6830.
Oberholzer, W.F. 1963. Noticia explicativa das folhas SUL-G-36, C' D e J, Carte geológica de Moçambique. *Moçambique Servicos Geol.Minas Bol.,* 31:32pp.

Ochev, V.G. 1966. Stratigraphic outline of continental Triassic deposits on the Russian platform and the Cisuralian region. *Dokl.Akad.Nauk.S.S.S.R.,* 171:84-86.

Olson, E.C. 1957. Catalogue of localities of Permian and Triassic terrestrial vertebrates of the territories of the USSR. *J.Geol.,* 65:196-226.

Orr, W.N. & J.R. Chapman 1974. Danian (Palaeocene) marine rocks at Richards Bay, South Africa. *S.Afr.J.Sci.,* 70:247-249.

O'Shea, D.O'C. 1971. *An outline of the inshore submarine geology of South West Africa and Namaqualand.* Unpub. MSc thesis, Univ.Cape Town.

Pacca, I.G. & F.Y.Hiodo 1976. Paleomagnetic analysis of Mesozoic Serra Geral basaltic lava flows in southern Brazil. *An. Acad.Braz.Sci.,* 48(suppl.):207-214.

Parker, R.J. 1971. The petrography and major element geochemistry of phosphorite nodule deposits on the Agulhas Bank, South Africa. *SANCOR Mar.Geol.Prog.Bull.,* 2, Cape Town, Dept.Geology, Univ.Cape Town. 94pp.

Parker, R.J. 1975. The petrology and origin of some glauconitic and glauco-conglomeratic phosphorites from the South African continental margin. *J.Sed.Petrol.,* 45:230-242.

Parker, R.J. & W.G.Siesser 1972. Petrology and origin of some phosphorites from the South African continental margin. *J.Sed.Petrol.,* 42:434-440.

Parker, R.J. & E.S.W.Simpson 1972. South African Agulhas Bank phosphorites. *Phosphorus & Potassium,* 58:18 & 27.

Parr, W.J. 1958. The foraminifera of the Bluff Beds, Durban, and some other South African calcareous coastal rocks. *Trans.geol.Soc.S.Afr.,* 61:103-136.

Parsons, B. & J.G.Sclater 1977. An analysis of the variations of ocean floor bathymetry and heat flow with age. *J.Geophys. Res.,* 82:803-827.

Parson, L.M., D.G.Roberts & P.R.Miles 1981. Magnetic anomalies in the Somali Basin, NW Indian Ocean. *Geophys. J. R.astr.Soc.,* 65:260.

Partridge, T.C. 1969. Fluvial features and climatic change during the Quaternary in South Africa. *S.Afr.Archaeol.Bull.,* 24:106-116.

Partridge, T.C. 1973. Geomorphological dating of cave openings at Makapansgat, Sterkfontein, Swartkrans and Taung. *Nature,* 246:75-78.

Partridge, T.C. 1975. Stratigraphic, geomorphological and palaeo-environmental studies of the Makapansgat Limeworks and Sterkfontein hominid sites: a progress report of work carried out between 1965 and 1975. *S.Afr.Soc. Quat.Res.Conf.Proc.* (unpublished).

Partridge, T.C. 1978. Re-appraisal of lithostratigraphy of Sterkfontein hominid site. *Nature,* 275:282-287.

Partridge, T.C. & A.B.A.Brink 1967. Gravels and terraces of the lower Vaal Basin. *S.Afr.geogr.J.,* 49:21-38.

Passarge, S. 1904. *Die Kalahari.* Berlin, Reimer, 822pp.

Pemberton, J. 1978. *The geochemistry and petrology of Karoo basalts of the Barkly East area, north-eastern Cape.* MSc thesis, Rhodes Univ., Grahamstown.

Phillips, J.F.W. 1927. Fossil *Widdringtonia* in lignite of the Knysna series, with a note on fossil leaves of several other species. *S.Afr.J.Sci.,* 24:188-197.

Plows, W.J. 1921. The Cretaceous rocks of Pondoland. *Ann. Durban Mus.,* 3:58-66.

Plumstead, E.P. 1970. A review of contributions to the knowledge of Gondwana megaplant fossils and floras of Africa published since 1950. *IUGS Symposium,* Buenos Aires 1967, Paris, UNESCO:139-148.

Pomeral, C. 1975. *Stratigraphie et Paléogéographie Ere Mésozoique,* Paris, Doin éditeurs. 383pp.

Postuma, J.A. 1971. *Manual of planktonic foraminifera,* Amsterdam, Elsevier, 420pp.

Potgieter, T.C. 1950. The structure and petrology of the George granite plutons and the invaded pre-Cape sedimentary rocks. *Ann. Univ. Stellenbosch,* 26:323-412.

Pretorius, D.A. 1979. The crustal architecture of southern Africa. *Trans.geol.Soc.S.Afr.,* Annex.to vol.76 (1973):1-60.

Proto Decima, F., F.Medizza & L.Todesco 1978. Southeastern Atlantic Leg 40 calcareous nannofossils. In: H.M.Bolli et al, *Initial Rep. Deep Sea Drilling Project,* 40, Washington, US Government Printing Office:571-634.

Quilty, P.G. 1977. Cenozoic sedimentation cycles in western Australia. *Geology,* 5:336-340.

Rabinowitz, P.D. 1976. A geophysical study of the continental margin of southern Africa. *Bull.Geol.Soc.Am.,* 87:1643-1653.

Rabinowitz, P.D., M.F.Coffin & D.Falvey 1981. The Madagascar Africa separation. *Eos,* 62:297.

Rabinowitz, P.D. & J.LaBrecque 1979. The Mesozoic South Atlantic Ocean and evolution of its continental margin. *J. Geophys.Res.,* 84:5973-6002.

Ramsay, J.G. 1967. *Folding and Fracturing of rocks.* New York, McGraw-Hill, 568pp.

Rayner, D.H. 1971. Data on the environment and preservation of late Palaeozoic tetrapods. *Proc. Yorks.geol.Soc.,* 38:437-495.

Reading, H.G. (ed.) 1978. *Sedimentary Environments and Facies,* Oxford, Blackwell, 557pp.

Reeves, C.V. 1972. Rifting in the Kalahari? *Nature,* 237:95-96.

Reeves, C.V. 1978. A failed Gondwana spreading axis in southern Africa. *Nature,* 273:222.

Reineck, H.E. & I.B.Singh 1973. *Depositional Sedimentary Environments.* Berlin, Springer-Verlag. 439pp.

Rennie, J.V.L. 1931. Note on fossil leaves from the Banke clays. *Trans.R.Soc.S.Afr.,* 19:251-253.

Reuning, E. 1923. Geologische Uebersichtskarte des Mittleren Teiles von Südwestafrika der intrusionsverband der Granite des Mittleren Hererolandes und des Angrenzenden küsten gebietes in Südwestafrika. *Geol. Rund.,* 14(3):232-239.

Reuning, E. & F.von Huene 1925. Fossilführende Karrooschichten in nördlichen Südwestafrika. *Neues Jb.Min.Geol.Paläontol.,* 52B:94-122.

Rex, D.C. 1972. Potassium-argon age determinations on volcanic and associated rocks from the Antarctic Peninsula and Dronning Maud Land. In: R.J.Adie (ed.), *Antarctic geology and geophysics,* Oslo, Universitetsforlaget:133-136.

Rhodes, R.C. 1974. A gravity model for the Cape Fold Belt: Discussion. *Trans.geol.Soc.S.Afr.,* 77:207-208.

Rhodes, R.C. & D.H.Krohn 1972. Tectonic control over regional geochemical variation in the Karroo basaltic province of South Africa. *Trans.geol.Soc.S.Afr.,* 75:11-22.

Riek, E.F. 1974. Upper Triassic insects from the Molteno 'Formation', South Africa. *Palaeont.afr.,* 17:19-31.

Rigassi, D.A. & G.Dixon 1972. Cretaceous of the Cape Province, Republic of South Africa. *Proc.Ibadan Univ.Conf. on African Geology,* 1970:513-527.

Robey, J.van A. 1976. *Aspects of the geochemistry of the Karoo dolerites and basalts of the North Eastern Cape Province, South Africa.* MSc thesis, Rhodes Univ., Grahamstown.

Robson, S. 1983. The distribution of Recent radiolaria in surficial sediments of the continental margin off northern Namibia. *J.Brit.Micropalaeont.Soc.* (in press).

Rogers, A.W. 1906. Geological survey of parts of the divisions of Uitenhage and Alexandria, with appendix on occurrence of the wood beds on Loerie and Gamtoos Rivers. *Ann.Rep. Geol.Comm.Cape of Good Hope,* for 1905:11-46.

Rogers, A.W. 1909. Notes on a journey to Knysna. *Ann.Rep. Geol.Comm.Cape of Good Hope,* for 1908:130-134.

Rogers, A.W. 1910. The Swartkops bore-hole. *Ann.Rep.Geol. Comm.Cape of Good Hope,* for 1909:111-116.

Rogers, A.W. 1915. The occurrence of dinosaurs in Bushmanland. *Trans.R.Soc.S.Afr.,* 5:265-272.

Rogers, A.W. & A.L.du Toit 1904a. Geological surveys of parts of the division of Ceres, Sutherland and Calvinia. *Ann.Rep. Geol.Comm. Cape of Good Hope,* for 1903:1-70.

Rogers, A.W. & A.L.du Toit 1904b. The Sutherland volcanic pipes and their relationships to other vents in South Africa. *Trans.Phil.Soc.S.Afr.,* 15:61-83.

Rogers, A.W. & A.L.du Toit 1909. *An Introduction to the Geology of the Cape Colony,* London, Longmans:491pp.

Rogers, A.W. & E.H.L.Schwarz 1900. Report on the southern districts between Breede River and George. *Ann.Rep.Geol. Comm. Cape of Good Hope,* for 1898:57-64.

Rogers, A.W. & E.H.L.Schwarz 1901. Report on the surveys of parts of the Uitenhage and Port Elizabeth divisions. *Ann. Rep.Geol.Comm. Cape of Good Hope,* for 1900:1-18.

Rogers, A.W. & E.H.L.Schwarz 1902a. Report on a journey from Swellendam to Mossel Bay. *Ann.Rep.Geol.Comm. Cape of Good Hope,* for 1901:8-21.

Rogers, A.W. & E.H.L.Schwarz 1902b. General survey of the rocks in the southern parts of the Transkei and Pondoland including a description of the Cretaceous rocks of Eastern Pondoland. *Ann.Rep.Geol.Comm. Cape of Good Hope,* for 1901:25-46.

Rogers, J. 1977. Sedimentation on the continental margin off the Orange River and the Namib Desert. *Joint Geol.Surv. Univ.Cape Town Mar.Geol.Prog.Bull.,* 7:212pp. Cape Town, Dept.Geology.

Rogers, J. 1979. The sedimentary succession at the Koeberg nuclear power station, Melkbosstrand. Extended Abstract for Geokongres 1979, Port Elizabeth. Geological Society of South Africa. 9pp.

Rogers, J. 1980. First report on Cenozoic sediments between Cape Town and Elands Bay. *Geol.Surv.Rept.* (Open file no.165), Pretoria, 64pp.

Romer, A.S. 1947. Review of the Labyrinthodontia. *Bull. Mus.Comp.Zool.Harv.,* 99(1):1-368.

Romer, A.S. 1970. Tetrapod vertebrates and Gondwanaland. In: *Second Gondwana Symposium: Proceedings and Papers,* Pretoria, Council for Scientific and Industrial Research:111-124.

Romer, A.S. 1975. Intercontinental correlation of Triassic Gondwana vertebrate faunas. In: K.S.W.Campbell (ed), *Gondwana Geology,* Canberra, Australian National University Press:469-473.

Rooseboom, A. 1978. Sedimentafvoer in suider-Afrikaanse riviere. *Water S.Afr.,* 4:14-17.

Ruddock, A. 1973. The Tertiary limestones of the southern coastal regions of Cape Province, South Africa. In: G. Blant (ed), *Sedimentary Basins of the African coasts, Part 2 south and east coast,* Paris, Association of African Geological Surveys:49-62.

Rust, I.C. 1973. The evolution of the Palaeozoic Cape Basin, southern margin of Africa. In: A.E.M.Nairn & F.G.Stehli (eds), *The Ocean Basins and Margins Volume 1. The South Atlantic.* New York, Plenum Press:247-276.

Rust, I.C. 1975. Tectonic and sedimentary framework of Gondwana Basins in Southern Africa. In: K.S.W.Campbell (ed.), *Gondwana Geology,* Canberra, Australian National University Press:537-564.

Ryan, P.J. 1967. A basin analysis of the Waterberg coal-field. Unpub. report of the *Geol.Surv.S.Afr.*

Ryan, P.J. 1968. *Stratigraphy of the Ecca Series and the lowermost Beaufort Beds (Permian) in the Great Karoo basin of South Africa.* Unpub. PhD thesis, Univ. Witwatersrand.

SACS (South African Committee for Stratigraphy) 1980. Stratigraphy of South Africa. Part 1 (compiled by L.E.Kent). Lithostratigraphy of the Republic of South Africa, South West Africa/Namibia and the Republics of Bophuthatswana, Transkei and Venda. *Handbook Geol.Surv.S.Afr.,* 8:690pp.

Saito, T., L.H.Burckle & J.D.Hayes 1974. Implications of some pre-Quaternary sediment cores and dredgings. In: W.W.Hay (ed.), *Studies in paleo-oceanography.* Soc.Econ. Paleont.Minteral.Spec.Pub.20:6-36.

Salmon, D. 1979. Some Tertiary foraminifers in deep sea piston cores from the Central Terrace and Agulhas Passage, SW Indian Ocean. *Tech.Rept.Joint Geol.Surv. Univ.Cape Town Mar.Geol.Prog.,* 11:80-88.

Sauvan, P., J.Esquevin & G.Chennaux 1975. Transformations induites pardes intrusions doleritiques dans une serie Argileuse: l'ecca de Bergville (Afrique du Sud). *Bull.Centre Rech.Pau-SNPA,* 9:261-351.

Schlanger, S.O. & H.C.Jenkyns 1976. Cretaceous oceanic anoxic events: causes and consequences. *Geol.Mijnb.,* 55: 179-184.

Schoch, A.E. & A.J.Burger 1976. U-Pb zircon age of the Saldanha quartz prophyry, western Cape Province. *Trans. geol.Soc.S.Afr.,* 79:239-241.

Scholtz, D.L. 1946. On the younger pre-Cambrian granite plutons of the Cape Province. *Proc.geol.Soc.S.Afr.,* 49: 35-82.

Schwarz, E.H.L. 1900. Knysna, between the Gouwkamma (Homtini) and the Blue Krantz Rivers. *Ann.Rep.Geol. Comm. Cape of Good Hope,* for 1899:51-64.

Schwarz, E.H.L. 1903. Geological Survey of parts of the Mataliele Division, Griqualand East. *Ann.Rep.Geol.Comm. Cape of Good Hope,* for 1902:11-61.

Schwarz, E.H.L. 1904. Geological Survey of parts of Prince Albert, Willowmore, and Uniondale. *Ann.Rep.Geol.Comm. Cape of Good Hope,* for 1903:71-137.

Schwarz, E.H.L. 1905. Geological Survey of the Divisions of Tulbagh, Ceres, and Worcester. *Ann.Rep.Geol.Comm. Cape of Good Hope,* for 1904:261-290.

Schwarz, E.H.L. 1906. Geological Survey of the coastal plateau in the divisions of George, Knysna, Uniondale and Humansdorp. *Ann.Rep.Geol.Comm. Cape of Good Hope,* for 1905:49-93.

Schwarz, E.H.L. 1908. The Alexandria Formation (Uppermost Cretaceous) on the south coast of Africa. *Trans. geol.Soc.S.Afr.,* 11:107-115.

Schwarz, E.H.L. 1915. New Cretaceous fossils from Brenton, Knysna. *Rec.Albany Mus.,* 3:120-126.

Scogings, J.L. & C.J.Lenz 1961. The Karroo System of sediments of eastern Swaziland. *Bull.Geol.Surv.Mines Dept. Swaziland,* 1:84-91.

Scott, L. 1976. Palynology of Lower Cretaceous deposits from the Algoa Basin (Republic of South Africa). *Pollen et Spores,* 18(4):563-609.

Scrutton, R.A. 1973a. The age relationship of igneous activity and continental breakup. *Geol.Mag.,* 110:227-234.

Scrutton, R.A. 1973b. Structure and evolution of the seafloor south of South Africa. *Earth Planet.Sci.Lett.,* 19:250-256.

Scrutton, R.A. 1976a. Continental breakup and deep crustal structure at the margins of southern Africa. *Ann.Braz. Acad.Sci.,* 48(suppl.):275-286.

Scrutton, R.A. 1976b. Crustal structure at the continental margin south of South Africa. *Geophys.J.R.astr.Soc.,* 44:601-623.

Scrutton, R.A. & R.V.Dingle 1974. Basement control over sedimentation on the continental margin west of southern Africa. *Trans.geol.Soc.S.Afr.,* 77:253-260.

Scrutton, R. A. & A.du Plessis 1973. Possible marginal fracture ridge south of South Africa. *Nature Phys.Sci.,* 242: 180-182.

Seeley, H.G. 1892. Researches on the structure, organization and classification of the fossil Reptilia. *Phil.Trans.R.Soc.,* 182:311-370.

Seeley, H.G. 1895. Researches on the structure, organization and classification of the fossil Reptilia. *Phil. Trans. R. Soc.,* 186:1-148.

Ségoufin, J. 1978. Anomalies magnétiques Mésozoiques dans le bassin de Mozambique. *C.R.Acad.Sc.Paris,* 287D:109-112.

Ségoufin, J. & P.Patriat 1980. Existence d'anomalies Mésozoiques dans le bassin de Somalie. Implications pour les relations Afrique-Antarctique-Madagascar. *C.R.Acad.Sc. Paris,* 291B:85-88.

Seward, A.C. 1903. Fossil floras of Cape Colony. *Ann.S.Afr. Mus.,* 4:1-122.

Seward, A.C. 1907. Notes on fossil plants from South Africa. *Geol.Mag.,* 54:481-487.

Shackleton, J.N. & J.P.Kennett 1975. Paleotemperature history of the Cenozoic and the initiation of Antarctic glaciation: oxygen and carbon isotope analyses in DSDP sites 277, 279 and 281. In: J.P.Kennett et al, *Initial Rep.Deep Sea Drilling Project,* 29, Washington, US Government Printing Office:743-755.

Shackleton, J.N. & N.D.Opdyke 1977. Oxygen isotope and palaeomagnetic evidence for early Northern hemisphere glaciation. *Nature,* 270:216-219.

Shannon, L.V. 1966. Hydrology of the south and west coast of South Africa. *Investl.Rep.Div.Sea Fish.S.Afr.,* 5:1-22.

Shone, R.W. 1976. *The sedimentology of the Mesozoic Algoa Basin.* Unpub. MSc thesis, Univ.Port Elizabeth,·48pp.

Shone, R.W. 1978. A case for lateral gradation between the Kirkwood and Sundays River Formations, Algoa Basin. *Trans.geol.Soc.S.Afr.,* 81:319-326.

Siedner, G. & J.A.Miller 1968. K-Ar age determinations on basaltic rocks from South West Africa and their bearing on continental drift. *Earth Planet.Sci.Lett.,* 4:451-458.

Siedner, G. & J.G.Mitchell 1976. Episodic Mesozoic volcanism in Namibia and Brazil: a K-Ar isochron study bearing on the opening of the South Atlantic. *Earth Planet.Sci.Lett.,* 30:292-302.

Siesser, W.G. 1971. Petrology of some South African coastal and offshore carbonate rocks and sediments. *SANCOR Mar.Geol.Prog.Bull.,* 3, Cape Town, Dept.Geology, Univ. Cape Town:232pp.

Siesser, W.G. 1972a. Petrology of the Cainozoic Coastal Limestones of the Cape Province, South Africa. *Trans.geol.Soc. S.Afr.,* 75:177-185.

Siesser, W.G. 1972b. Environmental discrimination of coastal limestones by conventional and electron microscope methods. *Trans.geol.Soc.S.Afr.,* 75:295-298.

Siesser, W.G. 1972c. Petrology of the South African coastal limestones. *Trans.geol.Soc.S.Afr.,* 75:177-185.

Siesser, W.G. 1977a. Biostratigraphy and micropalaeontology of continental margin samples. *Tech.Rept.Joint Geol.Surv. Univ.Cape Town Mar.Geol.Prog.,* 9:108-117.

Siesser, W.G. 1977b. Upper Eocene age of marine sediments on Bogenfels, South West Africa based on calcareous nannofossils. In: *Papers on Biostratigraphic Research.* Bull.Geol. Surv.S.Afr., 60:72-74.

Siesser, W.G. 1978a. Age of phosphorite on the South African continental margin. *Mar.Geol.,* 26:M17-28.

Siesser, W.G. 1978b. Leg 40 results in relation to continental shelf and onshore geology. In: H.M.Bolli et al, *Initial Reports of the Deep Sea Drilling Project,* 40, Washington, US Government Printing Office:965-979.

Siesser, W.G. 1978c. Aridification of the Namib Desert: evidence from oceanic cores. In: E.M.van Zinderen Bakker (ed.), *Antarctic glacial history and world palaeoenvironments,* Rotterdam, Balkema:105-113.

Siesser, W.G. 1982. Cretaceous calcareous nannoplankton in South Africa. *J.Paleont.,* 56:335-350.

Siesser, W.G. & R.V.Dingle 1981. Tertiary sea-level movements around southern Africa. *J.Geol.,* 89:83-96.

Siesser, W.G. & G.A.Miles 1979. Calcareous nannofossils and planktic foraminifers in Tertiary limestones, Natal and Eastern Cape, South Africa. *Ann.S.Afr.Mus.,* 79:139-158.

Siesser, W.G. & J.Rogers 1976. Authigenic pyrite and gypsum in South West African continental slope sediments. *Sedimentol.,* 23:567-577.

Siesser, W.G. & D.Salmon 1979. Eocene marine sediments in the Sperrgebiet, South West Africa. *Ann.S.Afr.Mus.,* 79:9-34.

Simpson, E.S.W. 1974. *Chart 125A,* Stellenbosch, National Res.Inst.Oceanology.

Simpson, E.S.W. 1977. Evolution of the South Atlantic. *Trans. geol.Soc.S.Afr.,* 80(annexure):1-15.

Simpson, E.S.W. & R.V.Dingle 1973. Offshore sedimentary basins on the southeastern continental margin of South Africa. In: G.Blant (ed.), *Sedimentary Basins of the African coasts, Part 2 south and east coast,* Paris, Association of African Geological Surveys:63-67.

Simpson, E.S.W. & A.du Plessis 1968. Bathymetric, magnetic and gravity data from the continental margin of southwestern Africa. *Can.J.Earth.Sci.,* 5:1119-1123.

Simpson, E.S.W., R.Schlich, J.Gieskes, W.A.Girdley, L.Leclaire, B.Vaughn, C.Moore, C.Müller, J.Sigal, T.L.Vallier, S.M.White & B.Zobel 1974. *Initial Rep.Deep Sea Drilling Project,* 25, Washington, US Government Printing Office.

Simpson, E.S.W., G.J.Sclater, B.Parsons, I.Norton & L.Meinke 1979. Mesozoic magnetic lineations in the Mozambique Basin. *Earth Planet.Sci.Lett.,* 43:260-264.

Simpson, G.G. 1973. Tertiary penguins (Sphenisciformes, Spheniscidae) from Ysterplaats, Cape Town, South Africa. *S.Afr.J.Sci.,* 69:342-344.

Simpson, G.G. 1979. Tertiary penguins from the Duinefontein site Cape Province, South Africa. *Ann.S.Afr.Mus.,* 79: 1-7.

Singer, R. & A.O.Fuller 1962. The geology and description of a fossiliferous deposit near Zwartklip in False Bay. *Trans. R.Soc.S.Afr.,* 36:205-211.

Sliter, W.V. 1977. Cretaceous foraminifers from the southwestern Atlantic Ocean, Leg 36, Deep Sea Drilling Project. In: P.Barker et al, *Initial Rep.Deep Sea Drilling Project,* 36,Washington, US Government Printing Office:519-573.

Smit, P.J. 1977. *Die geohidrologie in die opnanggebied van die Moloporivier in die noordelik Kalahari.* Unpub. PhD thesis, Bloemfontein Univ.OFS.

Smith, A.G. & A.Hallam 1970. The fit of the southern continents. *Nature,* 225:139-144.

Smith, R.M.H. 1980. The lithology, sedimentology and taphonomy of flood-plain deposits of the Lower Beaufort (Adelaide Subgroup) strata near Beaufort West. *Trans.geol.Soc. S.Afr.,* 83:399-413.

Smitter, Y.H. 1955. Foraminifera from the Upper Cretaceous beds occurring near the Itongazi River, Natal. *Palaeont.afr.,* 3:103-107.

Soares, A.F. & G.Henriques da Silva 1970. Contribuição para o estudo da geologia do Maputo. Estratigrafia e paleontologia da região de Madubula e suas relações com aréas vizinhas. *Revta Ciênc.geol.Lourenço Marques* (A),3:1-85.

Soares, P.C. & P.M.B.Landim 1976. Comparison between the tectonic evolution of the intracratonic and marginal basins in south Brazil. *Ann.Brazil Acad.Sci.,* 48(suppl.):313-324.

SOEKOR, 1976. *The structure of the Mesozoic succession of the Agulhas Bank.* Johannesburg, Technical Department, Southern Oil Exploration Corporation, 13pp.

Söhnge, P.G. 1934. The Worcester Fault. *Trans.geol.Soc.S.Afr.,* 37:253-277.

Söhnge, P.G., D.J.L.Visser & C.van Riet Lowe 1937. The geology and archaeology of the Vaal River Basin. *Geol. Surv. S.Afr. Mem.*, 35:184pp.

Spath, L.F. 1921. On Cretaceous Cephalopoda from Zululand. *Ann. S.Afr. Mus.*, 12:217-321.

Spath, L.F. 1925. On Upper Albian Ammonoidea from Portuguese East Africa. With an appendix on Upper Cretaceous ammonites from Maputoland. *Ann. Transv. Mus.*, 11:179-200.

Spath, L.F. 1930. On the Cephalopoda of the Uitenhage beds. *Ann. S.Afr. Mus.*, 28:131-157.

Spath, L.F. 1934. *Catalogue of the Fossil Cephalopoda in the British Museum (Natural History). Part IV The Ammonoidea of the Trias.* London, British Museum (Nat.Hist.), 512pp.

Spath, L.F. 1954. The Upper Cretaceous Cephalopod fauna of Grahamland. *Scient. Rep. Falkld. Isl. Dep. Surv.*, 3:1-60.

Spence, D.L. 1970. *A seismic refraction study of sedimentary structure on the Agulhas Bank, south of Cape Infanta.* Unpub. MSc thesis, Univ. Witwatersrand.

Spies, J.J., L.N.J. Engelbrecht, S.J. Malherbe & J.J. Viljoen 1963. *Die geologie van die gebied tussen Bredasdorp en Gansbaai. Explanation of sheets 3419C and 3419D (Gansbaai) and 3420C (Bredasdorp).* Geol.Surv.S.Afr., 39pp.

Stapleton, R.P. 1975. Planktonic foraminifera and calcareous nannofossils at the Cretaceous-Tertiary contact in Zululand. *Palaeont. afr.*, 18:53-69.

Stapleton, R.P. 1977a. Carboniferous unconformity in southern Africa. *Nature*, 268:222-223.

Stapleton, R.P. 1977b. Hot spores. *Nuclear Active*, 17:8-9.

Stapleton, R.P. 1977c. Planktonic foraminifera and the age of the Uloa *Pecten* bed. In: *Papers on Biostratigraphic Research. Bull. Geol. Surv. S.Afr.*, 60:11-17.

Stapleton, R.P. 1978a. Microflora from a possible Permo-Triassic transition in South Africa. *Rev. Palaeobotany Palynology*, 25:253-258.

Stapleton, R.P. 1978b. Organic metamorphism and uranium occurrences in the Beaufort Group of South Africa. *Econ. Geol.*, 73:283-285.

Stapleton, R.P. & E.M. Beer 1977. Micropalaeontological age determinations for the Brenton Beds. In: *Papers on Biostratigraphic Research.* Bull.Geol.Surv.S.Afr., 60:1-10.

Stavrakis, N. 1980a. Opaque heavy minerals of the Katberg sandstone, South Africa. *Trans. geol. Soc. S.Afr.*, 83:17-21.

Stavrakis, N. 1980b. Sedimentation of the Katberg sandstone and adjacent formations in the south-eastern Karoo Basin. *Trans. geol. Soc. S.Afr.*, 83:361-374.

Steckler, M.S. & A.B. Watts 1978. Subsidence of the Atlantic-type continental margin off New York. *Earth Planet. Sci. Lett.*, 41:1-13.

Steinmetz, J.C. 1981. The origin of the late Oligocene *Braarudosphaera* chalks in the South Atlantic Ocean. *Eos*, 26(17): 297.

Stevens, G.R. 1971. The relationship of isotopic temperatures and faunal realms to Jurassic-Cretaceous paleogeography, particularly in the SW Pacific. *J. R. Soc. New Zealand*, 1: 145-158.

Stewart, A.G. 1973. *Petrology and petrography of samples from two Algoa Basin cores.* Unpub. MSc thesis, Univ. Cape Town, 118pp.

Stocken, C.G. 1962. The diamond deposits of the Sperrgebiet, South West Africa. *Field Excursion Guide, 5th Ann. Congr. Geol. Soc. S.Afr.*

Stow, G.W. 1871. On some points in South African geology, Part 1, No.1: the Jurassic formations. *Q. Jl. geol. Soc. Lond.*, 27:497-514.

Stratten, T. 1970. Late Karroo igneous rocks of the southern Lebombo Mountain range. In: *Second Gondwana Symposium: Proceedings and Papers*, Pretoria, Council for Scientific and Industrial Research:441-444.

Stromer, E. 1922. Erste mitteilung über Tertiäre Wirbeltier-Reste aus Deutsch-Südwestafrika. *Sber. bayer. Akad. Wiss.*, 1921:331-340.

Stromer, E. 1926. Reste land- und süsswasser-bewohnender Wirbeltiere aus dem Diamantenfelden Deutsch-Südwestafrika. In: E. Kaiser (ed), *Die Diamantenwüste Südwestafrikas*, 2, Berlin, Dietrich Reimer:107-153.

Stromer, E. 1931a. Reste süsswasser- und land bewohnender Wirbeltiere aus dem Diamantenfelden Klein-Namaqualandes (Südwest-Afrika). *Sber. bayer, Akad. Wiss.*, 1931:17-47.

Stromer, E. 1931b. *Palaeothentoides africanus*, nov.gen., nov. spec., ein erstes Beuteltier aus Afrika. *Sber. bayer. Akad. Wiss.*, 1931:177-190.

Summerhayes, C.P. 1970. *Phosphate deposits on the northwest African continental shelf and slope.* Unpub. PhD thesis, Univ. London. 282pp.

Summerhayes, C.P. 1972. Aspects of the mineralogy and geochemistry of Agulhas Bank sediments. Parts I and II. *SANCOR Mar. Geol. Prog. Tech. Rept.*, 4:83-95, Univ. Cape Town.

Summerhayes, C.P. 1973a. Distribution and origin of phosphate in sediments from the Agulhas Bank, South Africa. *Tech. Rept. Joint Geol. Surv. Univ. Cape Town Mar. Geol. Prog.*, 5:94-114.

Summerhayes, C.P. 1973b. Distribution, origin and economic potential of phosphatic sediments from the Agulhas Bank, South Africa. *Trans. geol. Soc. S.Afr.*, 76:271-277.

Summerhayes, C.P., G.F. Birch, J. Rogers & R.V. Dingle 1973. Phosphate in the sediments off southwestern Africa. *Nature*, 243:509-511.

Summerhayes, C.P., B.D. Bornhold & P.W. Embley 1979. Surficial slides and slumps on the continental slope and rise of southwest Africa: a reconnaissance study. *Mar. Geol.*, 31:265-277.

Surange, K.R. 1975. Indian Lower Gondwana Floras. In: K.S.W. Campbell (ed.), *Gondwana Geology*, Canberra, Australian National University Press:135-147.

Szalay, F.S. & E. Delson 1979. *Evolutionary History of the Primates*, London, Academic Press.

Taljaard, M.S. 1936. South African melilite basalts and their relations. *Trans. geol. Soc. S.Afr.*, 39:281-316.

Tankard, A.J.T. 1974a. Petrology and origin of the phosphorite and aluminium phosphate rock of the Langebaanweg-Saldanha area, south-western Cape Province. *Ann. S.Afr. Mus.*, 65:217-249.

Tankard, A.J.T. 1974b. Varswater Formation of the Langebaanweg-Saldanha area, Cape Province. *Trans. geol. Soc. S.Afr.*, 77:263-283.

Tankard, A.J.T. 1975a. *The late Cenozoic History and Palaeoenvironments of the coastal margin of the south-western Cape Province, South Africa.* Unpub. PhD thesis, Grahamstown, Rhodes Univ.

Tankard, A.J.T. 1975b. The marine Neogene Saldanha Formation. *Trans. geol. Soc. S.Afr.*, 78:257-264.

Tankard, A.J.T. & J. Rogers 1978. Late Cenozoic palaeoenvironments on the west coast of southern Africa. *J. Biogeog.*, 5:319-337.

Tasch, P., M.V.A. Sastry, S.C. Shah, B.R.J. Rao, C.N. Rao & S.C. Ghosh 1975. Estheriids of the Indian Gondwanas: significance for continental fit. In: K.S.W. Campbell (ed.), *Gondwana Geology*, Canberra, Australian National Univ. Press: 443-452.

Tate, R. 1867. On some secondary fossils from South Africa. *Q. Jl. geol. Soc. Lond.*, 23:139-175.

Taylor, H.C. 1978. Capensis. In: M.J.A. Werger (ed), *Biogeography and ecology of southern Africa.* The Hague, Junk: 171-229.

Theron, J.C. 1970. *Some geological aspects of the Beaufort Series in the Orange Free State.* Unpub. DSc thesis, Univ. OFS.

Theron, J.C. 1975. Sedimentological evidence for the extension of the African continent southwards during the late Permian-early Triassic times. In: K.S.W.Campbell (ed.), *Gondwana Geology,* Canberra, Australian National Univ. Press:61-71.
Theron, J.N. 1962. On the nature of the Cape folding in the district of Willowmore, CP. *Ann. Univ. Stellenbosch,* 37: 347-424.
Theron, J.N. 1969. The Baviaanskloof range – a South African nappe. *Trans.geol.Soc.S.Afr.,* 72:29-30.
Theron, J.N. 1974. Die seismiese geskiedenis van die suidweslike Kaap Provinsie. *Geol.Surv.S.Afr.Seismol.Ser.,* 4:9-16.
Thiergart, F., R.Frantz & K.Raukopf 1966. Summary of the pollen flora of Tertiary deposits near Knysna (Cape Province). In: E.M.van Zinderen Bakker (ed.), *Palaeoecology of Africa* Vol.1, Rotterdam, Balkema:100-102.
Thompson, R.W. 1977. Mesozoic sedimentation on the eastern Falkland Plateau. In: P.Barker et al, *Initial Rep.Deep Sea Drilling Project,* 36, Washington, US Government Printing Office:877-891.
Timofeev, P.P. & L.I.Bogolyubova 1981. Organic matter in Cretaceous 'black shales' from the Atlantic and Pacific oceans and their relation to the problems of oil and gas formation. *Bull. Centres Rech.Explor.-Prod.Elf-Aquitaine,* 5:575-600.
Tobias, P.V. 1973. Implications of the new age estimates of the early South African hominids. *Nature,* 246:79-83.
Torquato, J.R. 1976. Geotectonic correlation between SE Brazil and SW Africa. *Ann.Braz.Acad.Sci.,* 48(suppl.): 351-363.
Truswell, J.F. 1970. *An introduction to the Historical Geology of South Africa,* Cape Town, Purnell:167pp.
Truswell, J.F. 1975. Cape diastrophism, *Nature,* 257:255.
Truswell, J.F. 1977. *The geological evolution of South Africa.* Cape Town, Purnell.
Tucholke, B.E., R.E.Houtz & D.M.Barrett 1981. Continental crust beneath the Agulhas Plateau, southwest Indian Ocean. *J.Geophys.Res.,* 85(B5):3791-3806.
Turner, B.R. 1969. *The stratigraphy and sedimentary history of the Molteno stage in part of the north-east Cape Province.* Unpub. MSc thesis, Univ.Witwatersrand:195pp.
Turner, B.R. 1970. Facies analysis of the Molteno sedimentary cycle. In: *Second Gondwana Symposium: Proceedings and Papers,* Pretoria, Council for Scientific and Industrial Research:313-319.
Turner, B.R. 1971. The geology and coal resources of the north-eastern Cape Province. *Bull. Geol.Surv.S.Afr.,* 52: 74pp.
Turner, B.R. 1972a. Revision of the stratigraphic position of cynodonts from the upper part of the Karroo (Gondwana) System of Lesotho. *Geol.Mag.,* 109:349-360.
Turner, B.R. 1972b. Silica diagenesis in the Molteno sandstone. *Trans.geol.Soc.S.Afr.,* 75:55-66.
Turner, B.R. 1975a. *The stratigraphy and sedimentary history of the Molteno Formation in the main Karroo Basin of South Africa and Lesotho.* Unpub. PhD thesis, Univ.Witwatersrand, 314pp.
Turner, B.R. 1975b. Statistical appraisal of Molteno (Triassic) sedimentary cycles from the upper part of the Karroo (Gondwana) System in South Africa. *J.Sed.Petrol.,* 45:95-104.
Turner, B.R. 1977. Fluviatile cross-bedding patterns in the upper Triassic Molteno Formation of the Karroo (Gondwana) Supergroup in South Africa and Lesotho. *Trans. geol.Soc.S.Afr.,* 80:241-252.
Turner, B.R. 1978a. Palaeohydraulics of clast transport during deposition of the Upper Triassic Molteno Formation in the main Karoo Basin of South Africa. *S.Afr.J.Sci.,* 74: 171-173.
Turner, B.R. 1978b. Sedimentary patterns of uranium mineralization in the Beaufort Group of the southern Karoo (Gondwana) Basin, South Africa. In: A.D.Miall (ed.), *Fluvial sedimentology. Mem. Can.Soc.Petrol.geol.,* 5:831-848.
Turner, B.R. (ed.) 1979. *Geology of the uraniferous Beaufort Group near Beaufort West.* Guidebook to Geokongress 79, Johannesburg, Geol.Soc.S.Afr., 71pp.
Uchupi, E. & J.A.Austin 1981. The continent-ocean transition off south-western Africa. *Eos,* 62(17):407.
Urien,C.M., L.R.Martins & J.J.Zambrano 1976. The geology and tectonic framework of southern Brazil, Uruguay and northern Argentina continental margin: their behaviour during the southern Atlantic opening. *Ann.Brazil Acad. Sci.,* 48(suppl.):365-376.
Urien, C.M. & J.J.Zambrano 1973. The geology of the basins of the Argentine continental margin and Malvinas Plateau. In: A.E.M.Nairn & F.S.Stehli (eds), *The Ocean Basins and Margins. Volume 1 The South Atlantic,* New York, Plenum Press:135-169.
Vail, J.F. 1964. Mesozoic igneous activity in Central Africa. *Proc.22 Int.geol.Congr.,* 16:212-225, New Delhi.
Vail, P.R., R.M.Mitchum & S.Thompson 1977. Seismic stratigraphy and global changes in sea level. Part 4: Global cycles of relative changes of sea level. *Mem.Am.Ass.Petrol. Geol.,* 26:83-97.
Vakhrameev, V.A. 1964. Jurassic and early Cretaceous floras of Eurasia and the palaeofloristic provinces of this period. *Trans.geol.Inst.Moscow,* 102:1-263 (in Russian).
Van Andel, T.H. & S.E.Calvert 1971. Evolution of sediment wedge, Walvis Shelf, South West Africa. *J.Geol.,* 79:585-602.
Van Dijk, D.E., D.K.Hobday & A.J.Tankard 1978. Permo-Triassic lacustrine deposits in the eastern Karoo Basin, Natal, South Africa. *Spec.Publs.int.Ass.sediment,* 2:225-239.
Van Eeden, O.R. 1973. A correlation of the subdivisions of the Karoo System. *Trans.geol.Soc.S.Afr.,* 76:201-206.
Van Eeden, O.R., H.N.Visser, J.S.van Zyl, F.J.Coertze & J.T. Wessels 1955. The geology of the eastern Soutpansberg and the lowveld to the north. *Geol.Surv.S.Afr.* sheet expl. 42:126pp.
Van Heeden, J. 1979. The morphology and taxonomy of *Euskelosaurus* (Reptilia: Saurischia; Late Triassic) from South Africa. *Navor.Nasion.Mus.Bloemfontein,* 4(2):21-84.
Van Hinte, J.E. 1976a. A Cretaceous time scale. *Bull.Am. Ass.Petrol.Geol.,* 60:269-287.
Van Hinte, J.E. 1976b. A Jurassic time scale. *Bull.Am.Ass. Petrol.Geol.,* 60:489-497.
Van Hoepen, E.C.N. 1926. Oor die Krytafsettinge van Soeloeland. *S.Afr.J.Sci.,* 23:216-222.
Van Hoepen, E.C.N. 1965. New and little known Zululand and Pondoland ammonites. *Ann.Geol.Surv.S.Afr.,* 4:158-172.
Van Riet Lowe, C. 1952. The Vaal River chronology – an up-to-date summary. *S.Afr.Archaeol. Bull.,* 7:135-149.
Van Zinderen Bakker, E.M. 1975. The origin and palaeoenvironment of the Namib Desert biome. *J.Biogeog.,* 2:65-73.
Veeh, H.H., S.E.Calvert & N.B.Price 1974. Accumulation of uranium in sediments and phosphorites on the South West African shelf. *Mar.Chem.,* 2:189-202.
Visser, H.N. & A.E.Schoch 1973. The geology and mineral resources of the Saldanha Bay area. *Mem.Geol.Surv.S.Afr.,* 63:150pp.
Visser, H.N. & S.W.van der Merwe 1959. The northeastern Springbok Flats Coal-Field. Records of boreholes 1 to 27 drilled for the Department of Mines. *Bull.Geol.Surv.S.Afr.,* 31:97pp.
Visser, J.N.J. & B.J.V.Botha 1980. Meander channel, point bar, crevasse splay and aeolian deposits from the Elliot For-

mation, in Barkly Pass, North-eastern Cape. *Trans.geol. Soc.S.Afr.,* 83:55-62.

Von Rad, U., W.B.F.Ryan, M.A.Arthur, P.Cepek, M.B.Cita, C.Cornfold, L.Garifal, N.Hamilton, B.Lopatin, G.F.Lutze, F.W.McCoy, G.Mountain, M.Sarnthein, O.E.Weser, J.K.Whelan & F.H.Wind 1979. *Initial Rep.Deep Sea Drilling Project,* 47(1), Washington, US Government Printing Office:835pp.

Vrba, E.S. 1974. Chronological and ecological implications of the fossil Bovidae at the Sterkfontein Australopithecine site. *Nature,* 250:19-23.

Vrba, E.S. 1975. Some evidence of chronology and palaeoecology of Sterkfontein, Swartkrans and Kromdraai from the fossil Bovidae. *Nature,* 254:301-304.

Wachendorf, H. 1967. Zur Unterkreide-stratigraphie von Süd-Moçambique. *N.Jb.Geol.Paläont.Abh.,* 129:272-303.

Wachendorf, H. 1973. The rhyolitic lava flows of the Lebombos (SE Africa). *Bull. Volcanologique,* 37:515-529.

Walker, F. & A.Poldervaart 1949. Karroo dolerites of the Union of South Africa. *Bull.geol.Soc.Am.,* 60:591-706.

Waterhouse, J.B. 1976. World correlations for Permian marine faunas. *Pap.Dept.Geol.Univ.Queensland,* 7(2):1-232.

Watson, D.M.S. 1914. The zones of the Beaufort Beds of the Karoo System in South Africa. *Geol.Mag.,* 1:203-208.

Watters, B.R. 1967. *The basal beds of the Stormberg volcanics at Majuba Nek.* Unpub. BSc thesis, Univ.Cape Town. 89pp.

Watts, D.R. & A.M.Bramall 1981. Palaeomagnetic evidence for a displaced terrain in western Antarctica. *Nature,* 293:638-641.

Wegener, A. 1924. *The Origins of Continents and Oceans,* London, Dover, 212pp.

Winter, H.de la R. 1973. Geology of the Algoa Basin, South Africa. In: G.Blant (ed.), *Sedimentary Basins of the African coasts, Part 2 south and east coast,* Paris, Association of African Geological Surveys:17-48.

Winter, H.de la R. 1979. An application of basic principles of stratigraphy to the Jurassic-Cretaceous interval in southern Africa. *Geol.Soc.S.Afr.Spec.Publ.,* 6:183-196.

Winter, H.de la R. & J.J.Venter 1970. Lithostratigraphic correlation of recent deep boreholes in the Karroo-Cape sequence. In: *Second Gondwana Symposium: Proceedings and Papers,* Pretoria, Council for Scientific and Industrial Research:395-408.

Woods, H. 1908. Echinoidea, Brachiopoda and Lamellibranchia from the Upper Cretaceous limestones of Needs Camp, Buffalo River. *Ann.S.Afr.Mus.,* 7:13-19.

Wright, E.P. 1978. Geological studies in the northern Kalahari. *Geogr.J.,* 144:235-249.

Wybergh, W. 1919. The coastal limestones of the Cape Province. *Trans.geol.Soc.S.Afr.,* 22:46-67.

Wybergh, W. 1920. The limestone resources of the Union. *Mem.Geol.Surv.S.Afr.,* 2:149pp.

Wyley, A. 1859. *Report of the Geological Surveyor upon a journey made by him during the year 1858 in two directions across the Colony and its results.* Parliamentary Rep. G54, Cape of Good Hope, Cape Town.

Zambrano, J.J. & C.M.Urien 1974. Pre-Cretaceous basins in the Argentine continental shelf. In: C.A.Burk & C.L.Drake (eds), *The Geology of Continental Margins,* Berlin, Springer-Verlag:463-469.

# Author index

# Subject index

*Italics denote page numbers for figures*